MODERN TOPICS IN METRICAL FIXED POINT THEORY

MODERN TOPICS IN METRICAL FIXED POINT THEORY

Mihai Turinici
A I Cuza University of Iaşi, Romania

World Scientific

NEW JERSEY • LONDON • SINGAPORE • BEIJING • SHANGHAI • HONG KONG • TAIPEI • CHENNAI

Published by

World Scientific Publishing Co. Pte. Ltd.
5 Toh Tuck Link, Singapore 596224
USA office: 27 Warren Street, Suite 401-402, Hackensack, NJ 07601
UK office: 57 Shelton Street, Covent Garden, London WC2H 9HE

Library of Congress Control Number: 2024061783

British Library Cataloguing-in-Publication Data
A catalogue record for this book is available from the British Library.

MODERN TOPICS IN METRICAL FIXED POINT THEORY

ISBN 978-981-98-0726-0 (hardcover)
ISBN 978-981-98-0727-7 (ebook for institutions)
ISBN 978-981-98-0728-4 (ebook for individuals)

For any available supplementary material, please visit
https://www.worldscientific.com/worldscibooks/10.1142/14160#t=suppl

To the memory of my parents

IOSIF TURINICI (1922–1991)

ŞTEFANIA TURINICI (1927–2019)

With infinite Love and Respect

We all carry inside us

People who came before us

LIAM CALLANAN

This book is a reprint of the original

Mihai Turinici

MODERN DIRECTIONS IN

METRICAL FIXED POINT THEORY

Revised edition

PIM, Iaşi (Romania), 2023

Preface

Let (X, d) be a (complete) metric space; and $T : X \to X$ be a selfmap of X. We say that $z \in X$ is *fixed* under T, provided $z = Tz$. Roughly speaking, the *Metrical Fixed Point Theory* consists in a collection of metrical techniques leading us to a fixed point of T by starting from the iterates of T at certain points of X. Essentially, there are two directions in this theory:

D-1) The *Banach contractive* one, having as model conditions like

$d(Tx, Ty) \leq \alpha d(x, y)$, $x, y \in X$ (for some constant $\alpha \in [0, 1[$)

introduced by Banach [Fund. Math., 3 (1922), 133–181]

D-2) The *Caristi-Kirk contractive* one, having as model conditions like

$d(x, Tx) \leq \varphi(x) - \varphi(Tx)$, $x \in X$ (for some lsc function $\varphi : X \to R_+$)

due to Caristi and Kirk [L. Notes Math., vol. 490, pp. 74–83, Springer, Berlin, 1975].

Now, these fixed point results found a multitude of applications in *Operator Equations Theory*; so, they were the subject of many extensions. A systematic study of all these is absolutely impossible; since – according to some realistic evaluations – there must be more than 15,000 of such extensions. So, if we are desperately looking for a "solid" reason to study these objects, the only hint to be provided is the following: all contractive conditions of this type belong to the same Creation as any other immaterial product of our (past, present or future) Civilization ...

Having these precise, it is our aim in the following to discuss a number of interesting (according to our criteria) themes from this multitude. The exposition has four chapters. A description of these may be given as below.

CHAPTER 1: PRELIMINARIES

In *Section 1.1*, after introducing the Zermelo-Fraenkel set theory and naive Cantor set theory, the outlines for the construction of natural numbers in both systems are being sketched. Then, some questions involving the Bernays-Tarski Dependent Choice principle (DC) were underlined.

Further, in *Section 1.2*, the concepts of convergence and Cauchy structure are introduced, under the lines in Kasahara [Publ. Res. Inst. Math. Sci. Kyoto Univ., 12 (1976), 427–437] and Turinici [Bul. Inst. Polit. Iaşi (Sect.: Mat., Mec. Teor., Fiz.), 53(57) (2007), 393–411], respectively.

The objective of *Section 1.3* is to define two basic families of (real valued) functions: the Meir-Keeler and Matkowski admissible ones.

In *Section 1.4*, some classes of natural numbers (involving progressions and almost progressions) are being considered.

Then, in *Section 1.5*, the concepts of geometric and asymptotic Meir-Keeler as well as Matkowski admissible relations are introduced.

Further, in *Section 1.6*, some preliminary facts involving the topological structures are discussed, with a particular focusing on the class of metric and gauge spaces.

Finally, in *Section 1.7*, some direct extensions of Banach contraction principle are given, as well as its Bessaga type converse.

CHAPTER 2: CLASSES OF CONTRACTIVE MAPS

In *Section 2.1*, a lot of implicit methods is proposed for a class of anticipative Rhoades contractions on bi-relational metric spaces, with a special emphasis on some related statements obtained by Dung and Hang [Vietnam J. Math., 43 (2015), 743–755], Chandok et al. [J. Egyptian Math. Soc., 23 (2015), 95–101], and Wardowski [Proc. Amer. Math. Soc., 146 (2018), 1585–1598].

Further, in *Section 2.2*, a fixed point result is given for a class of functional contractions over local Branciari ordered metric spaces.

The objective of *Section 2.3* is to get a lot of fixed point results concerning Matkowski regular operators acting on quasi-ordered metric spaces.

Finally, in *Section 2.4*, a functional extension is given for the diagonal fixed point result in Cirić and Presić [Acta Math. Comenianae, 76 (2007), 143–147] involving multistep type iterations.

CHAPTER 3: MAXIMAL TECHNIQUES

In *Section 3.1*, it is shown that the fixed point result over Banach lattices due to Agarwal and Khamsi [Nonlin. Anal., 74 (2011), 141–145] is reducible to certain ordering principles introduced by Turinici [Demonstr. Math., 22 (1989), 213–228].

Next, in *Section 3.2*, a simplification of the multivalued Tarski fixed point statement in Zhu and Li [JOTA, 132 (2007), 493–507] is provided, over a (partial order) separable framework.

Finally, in *Section 3.3*, a Matkowski type functional version is given for the 1967 contraction mapping principle in Gheorghiu [Stud. Cerc. Mat., 19 (1967), 119–122]; and then, as a by-product of this, an appropriate functional extension is proposed for the homotopical fixed point result in gauge spaces due to Frigon [L. Notes Nonlin. Anal., 16 (2017), 9–91].

CHAPTER 4: VARIA

In *Section 4.1*, some fixed point theorems are given for a class of functional contractions over quasi-ordered quasi-metric spaces, that may be viewed as a (geometric type) "full" quasi-metric counterpart of some recent contributions in the area due to Jleli and Samet [Fixed Point Th. Appl., 2012, 2012:210].

The objective of *Section 4.2* is to establish that the common fixed point result in Cirić et al. [Fixed Point Th. Appl., Volume 2008, Article ID 131294] involving Jungck contractions in ordered metric spaces is nothing but a particular case of the fixed point one due to Agarwal et al. [Appl. Anal., 87 (2008), 109–116].

Further, in *Section 4.3*, some fixed point results are given for a class of Meir-Keeler contractive maps with scattered spectrum acting on relational metric spaces.

In *Section 4.4*, a metric version is proposed for the local PPF (past-present-future) dependent fixed point result in Bernfeld et al. [Applicable Anal., 6 (1977), 271–280], as well as for its global variants.

The exposition ends with *Section 4.5*, where some fixed point results are given for a class of Meir-Keeler contractive maps acting on order pseudometric spaces; these extend – in a partial way – some related statements in the area due to Roldán and Shahzad [Fixed Point Th. Appl., 2014, 2014:218].

Finally, we sincerely hope that the present work will be of some avail to certain people working in *Metrical Fixed Point Theory*. For all of them – as well as for the others – we do wish all the best.

Mihai Turinici
December, 2024
Iaşi, ROMANIA

Contents

Preface vii

1. PRELIMINARIES 1

1.1 Dependent Choice Principles 1
1.1.1 Zermelo-Fraenkel set theory 1
1.1.2 Cantor set theory 13
1.1.3 Dependent Choice equivalents 26
1.2 Conv-Cauchy Structures 31
1.2.1 General conv-Cauchy structures 31
1.2.2 Pseudometric conv-Cauchy structures 32
1.2.3 Non-Cauchy asymptotic sequences 36
1.2.4 Equivalence criteria 46
1.2.5 Convergent real sequences 49
1.3 Admissible Functions 57
1.3.1 Matkowski and Meir-Keeler functions 57
1.3.2 Sequential counterparts 59
1.3.3 Strongly Matkowski admissible criteria 67
1.3.4 Complete functions 74
1.4 Classes of Natural Progressions 85
1.4.1 Sub-aprogressions 85
1.4.2 Admissible sets 87
1.4.3 Scattered sets 89
1.4.4 Uniformly admissible relations 93
1.4.5 Cauchy criteria over rss 94
1.5 Geometric-Asymptotic Relations 103
1.5.1 Geometric relations 103

1.5.2 Asymptotic versions 106
1.5.3 Duality principle 110
1.5.4 Boyd-Wong type relations 112
1.5.5 Rhoades type relations 115
1.5.6 Plane Rhoades relations 124
1.5.7 Complete relations 138
1.6 Topological Structures 147
1.6.1 Topology and cotopology 147
1.6.2 Topological constructions 148
1.6.3 Comparison criteria 152
1.6.4 Basic concepts 156
1.6.5 Metrical structures 157
1.6.6 Gauge spaces 163
1.7 Banach Contractions and Relatives 167
1.7.1 Meir-Keeler fixed point principle 167
1.7.2 Main result 170
1.7.3 Functional counterparts 175
1.7.4 Banach contraction principle 183

2. CLASSES OF CONTRACTIVE MAPS 187

2.1 Anticipative Rhoades Contractions in bi-Relational Metric Spaces 187
2.1.1 Dependent Choice principles 187
2.1.2 Conv-Cauchy structures 187
2.1.3 Admissible functions 187
2.1.4 Statement of the problem 187
2.1.5 Main result 196
2.1.6 Analytic methods in KSRDC-theory 200
2.1.7 Particular aspects 216
2.2 Functional Contractions in Local Branciari Ordered Metric Spaces 225
2.2.1 Dependent Choice principles 225
2.2.2 Conv-Cauchy structures 225
2.2.3 Banach contractions and relatives 225
2.2.4 Statement of the problem 225
2.2.5 Local Branciari metric spaces 231
2.2.6 Main result 236
2.2.7 Particular cases 242
2.3 Matkowski Regular Maps in Ordered Metric Spaces 245

2.3.1 Dependent Choice principles 245
2.3.2 Conv-Cauchy structures 245
2.3.3 Banach contractions and relatives 245
2.3.4 Statement of the problem 245
2.3.5 Main result 249
2.3.6 Altering rational contractions 251
2.4 Fixed Points via Multistep Type Iterations 259
2.4.1 Banach contractions and relatives 259
2.4.2 Cirić-Presić result 259
2.4.3 Previous Tasković approach 263
2.4.4 Main result 267
2.4.5 Related Abbas-Ilić-Nazir statement 272

3. MAXIMAL TECHNIQUES 277

3.1 Caristi-Kirk Results in Banach Lattices 277
3.1.1 Dependent Choice principles 277
3.1.2 Pseudometric maximal principles 277
3.1.3 Vectorial Caristi-Kirk principles 290
3.1.4 Application to normed lattices 302
3.2 Tarski Fixed Point Results over Separable Structures . . . 307
3.2.1 Dependent Choice principles 307
3.2.2 Statement of the problem 307
3.2.3 Separable structures 310
3.2.4 Fang spaces 322
3.2.5 Main result 329
3.2.6 Cardinal extensions 332
3.3 Homotopic Metric Interval Maps in Gauge Spaces 337
3.3.1 Dependent Choice principles 337
3.3.2 Conv-Cauchy structures 337
3.3.3 Admissible functions 337
3.3.4 Topological structures 337
3.3.5 Banach contractions and relatives 337
3.3.6 Gheorghiu functional contractions 337
3.3.7 Homotopic fixed points 346
3.3.8 Kang-Park principles 353

4. VARIA 363

4.1 Geometric Functional Contractions in Ordered Quasi-Metric Spaces . . . 363
4.1.1 Dependent Choice principles . . . 363
4.1.2 Conv-Cauchy structures . . . 363
4.1.3 Geometric-asymptotic relations . . . 363
4.1.4 Banach contractions and relatives . . . 363
4.1.5 Statement of the problem . . . 364
4.1.6 Main result . . . 376
4.1.7 Particular cases . . . 382
4.2 Factorial Jungck Contractions in Ordered Metric Spaces . . . 391
4.2.1 Dependent Choice principles . . . 391
4.2.2 Conv-Cauchy structures . . . 391
4.2.3 Admissible functions . . . 391
4.2.4 Banach contractions and relatives . . . 391
4.2.5 Leader contractions . . . 391
4.2.6 Jungck-Matkowski maps . . . 399
4.2.7 Particular cases . . . 410
4.2.8 Related approach . . . 413
4.3 Kannan Spectral Contractive Maps in Relational Metric Spaces . . . 417
4.3.1 Dependent Choice principles . . . 417
4.3.2 Conv-Cauchy structures . . . 417
4.3.3 Classes of natural progressions . . . 417
4.3.4 Geometric-asymptotic relations . . . 417
4.3.5 Banach contractions and relatives . . . 417
4.3.6 Statement of the problem . . . 417
4.3.7 Main result . . . 429
4.3.8 Particular cases . . . 436
4.4 PPF Dependent Fixed Points and the Unexpected Way to Nirvana . . . 445
4.4.1 Dependent Choice principles . . . 445
4.4.2 Banach contractions and relatives . . . 445
4.4.3 Meir-Keeler contractions . . . 445
4.4.4 Metrical fixed point principles . . . 451
4.4.5 Razumikhin chains . . . 457
4.4.6 Conditional BLR theorem . . . 460
4.4.7 Standard systems . . . 465
4.4.8 Isometric systems . . . 472

4.4.9 Linear case 480

4.5 Meir-Keeler Maps in Order Pseudometric Spaces 491

4.5.1 Dependent Choice principles 491

4.5.2 Conv-Cauchy structures 491

4.5.3 Geometric-asymptotic relations 491

4.5.4 Banach contractions and relatives 491

4.5.5 Order pseudometric spaces 491

4.5.6 Meir-Keeler contractions 496

4.5.7 Main result 504

4.5.8 Further aspects 507

Bibliography 509

Index 523

POLONIUS:

What do you read, my lord?

HAMLET:

Words, Words, Words ...

W. SHAKESPEARE: ***HAMLET*****, Act II, Scene 2**

Chapter 1

PRELIMINARIES

1.1 Dependent Choice Principles

Abstract. Some preliminary facts are discussed about Zermelo-Fraenkel set theory, classes of relations, construction of natural numbers, Cantor set theory, and Dependent Choice principles.

Subject-Class 2020. 47H10 (Primary), 54H25 (Secondary).

Keywords. Zermelo-Fraenkel set theory, Axiom of Choice, natural number, Cantor set theory, Peano axioms, Dependent Choice principles.

1.1.1 *Zermelo-Fraenkel set theory*

The ultimate goal of any *Set Theory* is to give

(ug-1) an appropriate description of (general) sets
(ug-2) a simple and useful theory of natural numbers
(ug-3) a good introduction of ordinals and cardinals.

For the needs of metrical fixed point theory, the first and second of these objectives will suffice. As a rule, these are to be attained within the setting of axiomatic set theory or naive set theory.

In the following, the former of these frameworks will be sketched. Precisely, a semi-formalized description is given for the (complete) Zermelo-Fraenkel axiomatic system (abbreviated: (ZF-com)), under the lines in Cohen [63, Ch. II].

(A) The formal system of (ZF-com) consists of the logical symbols

(ls-1) $\neg$ (non), & (and), $\vee$ (or),
(ls-2) $\Longrightarrow$ (implies), $\Longleftrightarrow$ [if and only if (in short: iff)],
(ls-3) $\forall$ (for all), $\exists$ (exists),

(ls-4) (,) (parentheses), $x, y, ..., X, Y, ...$ (variable symbols),
(ls-5) $\in$ (belongs to) [the specific primitive symbol of our system],
(ls-6) $=$ (equal) [the specific derived symbol of our system].

All these are subjected to the rules of forming valid sentences. Concerning this aspect, let Φ and Ψ be two sentences. Then, by definition

$\neg\Phi$ is true iff Φ is false
$\Phi \vee \Psi$ is true iff (Φ=true) or (Ψ=true)
$\Phi \& \Psi$ is true iff (Φ=true) and (Ψ=true)
$\Phi \Longrightarrow \Psi$ is identical with $(\neg\Phi \vee \Psi)$
(hence, $\Phi \Longrightarrow \Psi$ is true whenever Φ=false)
$\Phi \Longleftrightarrow \Psi$ is identical with $(\Phi \Longrightarrow \Psi)\&(\Psi \Longrightarrow \Phi)$.

All objects in (ZF-com) are sets. However, in the sentence $x \in y$, we say that the set x is an *element* of the set y.

Finally, we accept (as a logical principle)

(LE) *(Leibniz equality rule)*:
$\forall x \forall y(x = y \Longrightarrow \forall z(z \in x \Longleftrightarrow z \in y))$.

Having these precise, we may now describe the axioms of (ZF-com).

(A-I) *Axiom of Extensionality* (in short: AExt):

$$\forall x \forall y(\forall z(z \in x \Longleftrightarrow z \in y) \Longrightarrow x = y).$$

Note that, by (LE), we may write (AExt) as

(AExt-var) $\forall x \forall y(\forall z(z \in x \Longleftrightarrow z \in y) \Longleftrightarrow x = y)$;

which tells us that a set is determined by its elements. An equivalent way of expressing this axiom is the following. Define the sentence:

$x \subseteq y$ iff (by definition) $\forall z(z \in x \Longrightarrow z \in y)$;

then, (AExt-var) may be written in the double inclusion way

(AExt-2in) $\forall x \forall y((x \subseteq y \& y \subseteq x) \Longleftrightarrow x = y)$.

(A-II) *Axiom of Empty Set* (in short: AEset).
Define the sentence

$x \notin y$ iff $\neg(x \in y)$ (meaning: $x \in y$ is false).

The axiom in question writes

$\exists x \forall y(y \notin x)$ (there exists a set without elements).

This uniquely determined set x is referred to as the *empty set*; we denote it as $\emptyset$.

(A-III) *Axiom of Pairs* (in short: APairs):

$$\forall x \forall y \exists z \forall w (w \in z \iff (w = x \vee w = y)).$$

As before, the set z defined by this axiom is uniquely determined, and written as $\{x, y\}$. In particular, when $x = y$, we denote $\{x, y\}$ as $\{x\}$. Concerning this aspect, define the *almost singleton* (in short: *asingleton*) and *singleton* properties as:

x is *asingleton* iff $\forall u \forall v \ (u \in x \& v \in x \Longrightarrow u = v)$
x is *singleton* iff (x=asingleton & $x \neq \emptyset$).

Then, define

$(x, y) = \{\{x\}, \{x, y\}\}$ (the *couple* (or, *ordered pair*) of x and y).

The following property is immediate:

$$\forall x \forall y \forall u \forall v ((x, y) = (u, v) \iff x = u \& y = v).$$

By a *relation*, we mean any set $\mathcal{R}$ of ordered pairs. Then, by a *function*, we mean any relation f with

$(x, y) \in f$ and $(x, z) \in f$ imply $y = z$.

(A-IV) *Axiom of Union* (in short: AUni):

$$\forall x \exists y \forall z (z \in y \iff \exists t (z \in t \& t \in x)).$$

As before, the set y defined by this axiom is uniquely determined, and denoted as $\cup x$. Note that $\cup \emptyset = \emptyset$; because

$$z \in \cup \emptyset \iff \exists t (z \in t \& t \in \emptyset)$$

and the right member of this equivalence is false. When $x = \{a, b\}$, we put $a \cup b := \cup \{a, b\}$.

(A-V) *Axiom of Power* (in short: APow):

$$\forall x \exists y \forall z (z \in y \iff z \subseteq x)).$$

As before, the set y defined by this axiom is uniquely determined, and denoted as $\exp[x]$ (the class of all subsets in x).

(A-VI) *Axiom of Separation* (in short: ASep).

Let us enumerate the countably many formulae in our system with at least one independent variable; any object of this type will be written as

$Q(x) := A_n(x; t_1, ..., t_k)$. [Note that this enumeration is only a structural indication concerning these formulae; because we have not yet defined the natural numbers.]

The axiom in question writes

$$(Q(x)\text{=formula}) \Longrightarrow \forall x \exists y \forall z (z \in y \Longleftrightarrow z \in x \& Q(z));$$

and says that, for every set x there exists a set y consisting of all $z \in x$ with the property $Q(z)$; in short: $y = \{z \in x; Q(z)\}$.

This axiom allows us defining, for each set x,

$\exp(x) = \{z \in \exp[x]; z \neq \emptyset\}$
(the class of all *nonempty subsets* in x).

Further, by (ASep), we may introduce for each nonempty set x,

$\cap x = \{z \in \cup x; z \in t, \forall t \in x\}$
(the *intersection* of x).

To explain it, we start from the intuitively clear observation: for each nonempty set x, an element of $\cap x$ must belong to $\cup x$. Unfortunately, the case of $x = \emptyset$ does not enter in this construction, because

the formula $(z \in t, \forall t \in \emptyset)$ (identical with $\forall t (t \in \emptyset \Longrightarrow z \in t)$)
is true for each z; so, it cannot separate sets.

A way for solving it is *local* in nature. Precisely, we formally define

(for each nonempty set a and each $x \in \exp[a]$):
$\cap x$=the above convention, when $x \neq \emptyset$; $\cap x = a$, when $x = \emptyset$.

Moreover, by the same separation axiom, we may define the *domain* and *codomain* of any relation $\mathcal{R}$ as

Dom($\mathcal{R}$)=the class of all $x \in \cup \cup \mathcal{R}$
for which there exists $y \in \cup \cup \mathcal{R}$ with $(x, y) \in \mathcal{R}$
Codom($\mathcal{R}$)=the class of all $y \in \cup \cup \mathcal{R}$
for which there exists $x \in \cup \cup \mathcal{R}$ with $(x, y) \in \mathcal{R}$;

hence, in particular, the *domain* and *codomain* of any function f, denoted as Dom(f) and Codom(f), respectively. If

$\text{Dom}(f) = x$ and $\text{Codom}(f) \subseteq y$,

we then say that f is a *function from x to y*.

Finally, (ASep) allows us introducing the *Cartesian product* of two sets

$$a \times b = \{z \in \exp[\exp[a \cup b]]; z = (x, y), \text{ for some } x \in a, y \in b\};$$

as well as the class of all functions between two sets

$$\mathcal{F}(a, b) = \{f \in \exp[a \times b]; f = \text{function}, \mathrm{Dom}(f) = a\};$$

as usual, we use for $f \in \mathcal{F}(a, b)$ the writing $f : a \to b$.

(A-VI-a) *Axiom of Replacement* (in short: ARep).

A stronger variant of the (ASep) is stated as follows. Let us enumerate the countably many formulae in our system with at least two independent variables; any object of this type is written as

$$P(x, y) := A_n(x, y; t_1, ..., t_k).$$

[This enumeration is just a structural indication concerning these formulae; because we have not yet defined the natural numbers.] For any such formula, define the property

$$P(x, y)\text{=functional, iff}$$
$$(\forall x \exists y P(x, y))\&(\forall r \forall s \forall t(P(r, s)\&P(r, t) \Longrightarrow s = t));$$

in this case, for each x, we uniquely define y as a function of x; written: $y = f(x)$. Note that, the functional formula $P(x, y)$ is highly non-constructive; because, for each set x, a verification of $y = f(x)$ may require an investigation of all sets.

The axiom in question writes

$$(P(x, y)\text{=functional}) \Longrightarrow \forall u \exists v \forall r(r \in v \Longleftrightarrow \exists s(s \in u\&P(s, r));$$

it says that, whenever the formula $P(x, y)$ is functional and f denotes its attached function, one gets: for each set u, the range of f on u is a set. Note that Axiom of Replacement is to be used in higher constructive stages of our axiomatic system, involving ordinal and cardinal theory.

(A-VII) *Axiom of Russell* (in short: ARuss):

$\forall x(x \notin x)$ (there is no set belonging to itself).

The Axiom of Russell (ARuss) assures us that most contradictory sets are being eliminated from this theory.

A basic consequence of this axiom is contained in

Proposition 1.1. *We have*

Axiom of Replacement implies Axiom of Separation.

The reciprocal inclusion is not in general true.

Proof. i): Given the formula $Q(.)$ and the set x, define the associated formula (of two independent variables)

$$P(u, v) \text{ iff } (Q(u)\&v = u) \vee (\neg Q(u)\&v = x).$$

Clearly, P is functional; and its attached function f is expressed as

$$f(u) = u, \text{ if } Q(u) \text{ is true; } f(u) = x, \text{ if } Q(u) \text{ is false.}$$

Let s denote the range of f on x. By (ARep), s is a set; moreover, by (A-Russ), $x \notin s \cap x$. Given $v \in s \cap x$, we have $[v = f(y)$, for some $y \in x]$. As $v \neq x$, it results that $Q(y)$=true and $(v =)f(y) = y$; so that, $v \in x$ and $Q(v)$=true. Conversely, let $v \in x$ be such that $Q(v)$=true. This yields $v = f(v)$; hence, $v \in s$; in which case, $v \in s \cap x$. Putting these together, gives $s \cap x = \{y \in x; Q(y)\}$.

ii): See the (already quoted) Cohen's book for details. □

Another application of this axiom is as follows. For each set x, denote

$$\text{suc}(x) := x \cup \{x\} \text{ (referred to as: the } \textit{successor} \text{ of } x).$$

The motivation of this convention is as follows. Define the *strict inclusion*

$$x \subset y \text{ iff } x \subseteq y \text{ and } x \neq y.$$

Proposition 1.2. *The following properties are valid:*

(12-1) $x \subset \text{suc}(x)$, *for each* x

(12-2) $x \subset y \subset \text{suc}(x)$ *is false, for each* x *and* y.

Proof. i): Suppose by absurd that $x = \text{suc}(x) = x \cup \{x\}$. This yields $\{x\} \subseteq x$; whence, $x \in x$; in contradiction with (ARuss).

ii): Suppose that $x \subset y \subset \text{suc}(x) = x \cup \{x\}$. If $x \in y$, then $\text{suc}(x) \subseteq y$; contradiction. If $x \notin y$, then $y \subset x$; again a contradiction. □

(A-VII-a) A stronger variant of Axiom of Russell (ARuss) is *Axiom of Regularity* (in short: AReg):

$$\forall x(x \neq \emptyset) \Longrightarrow \exists y(y \in x\&\forall z(z \in x \Longrightarrow z \notin y).$$

This axiom says that each nonempty set x admits an element y which is minimal with respect to the relation $(\in)$.

Proposition 1.3. *Under the above setting, we have*

Axiom of Regularity implies Axiom of Russell.

Proof. Given the set x, let us apply the Axiom of Regularity to $u = \{x\}$:

there exists $y \in u$ (hence, $y = x$), such that,
for all $z \in u$ (hence, $z = x$), we have $z \notin y$; hence $x \notin x$.

And, from this, we are done. □

As before, this axiom is to be used in higher constructive stages of our axiomatic system, involving ordinal and cardinal theory.

(A-VIII) *Axiom of Infinity* (in short: AInfty):

$$\exists x((\emptyset \in x)\&\forall y(y \in x \Longrightarrow \text{suc}(y) \in x)).$$

Any set given by this axiom will be referred to as *inductive*.

(A-IX) *Axiom of Choice* (in short: AC).

Given the nonempty set a, let us say that $f \in \mathcal{F}(a, \cup a)$ is *selective*, if

(sel) $f(x) \in x$, for each $x \in a$ with $x \neq \emptyset$.

In this case, the axiom in question writes

$$\forall a(a \neq \emptyset \Longrightarrow \exists f \in \mathcal{F}(a, \cup a)(f = \text{selective})).$$

In other words: for each nonempty set a, there exists a selective function $f \in \mathcal{F}(a, \cup a)$.

This axiom allows us constructing such functions by an infinite amount of choosing procedures, even if the predicative mechanism of these choices is not given.

(B) Having these described, put for simplicity

(ZF-com) = the complete Zermelo-Fraenkel system, consisting of axioms (A-I)-(A-V), (A-VI-a)+(A-VII-a), and (A-VIII)+(A-IX)
(ZF) (the Zermelo-Fraenkel system) = the complete Zermelo-Fraenkel system (ZF-com) where (A-VI-a)+(A-VII-a) are substituted by the couple (A-VI)+(A-VII)
(ZF-AC) (the strongly reduced Zermelo-Fraenkel system) = the Zermelo-Fraenkel system (ZF) without Axiom of Choice (AC).

As we will see, most of the results in metrical fixed point theory are holding over the strongly reduced system (ZF-AC), completed with a "local" theory of relations.

Let X be a nonempty set. By a *relation* over X, we mean any part $\mathcal{R} \subseteq X \times X$; in this case, $(X, \mathcal{R})$ is called a *relational structure*. For simplicity, we sometimes write $(x, y) \in \mathcal{R}$ as $x\mathcal{R}y$. Note that $\mathcal{R}$ may be regarded as a mapping between X and $\exp[X]$ (= the class of all subsets in X). In fact, denote

$$X(x, \mathcal{R}) = \{y \in X; x\mathcal{R}y\} \text{ (the \textit{section} of } \mathcal{R} \text{ through } x), \ x \in X;$$

then, the desired mapping representation is $[\mathcal{R}(x) = X(x, \mathcal{R}),\ x \in X]$. Further, call $\mathcal{R}$, *proper* when $\mathcal{R}(x)$ is nonempty, for each $x \in X$. In this case, $\mathcal{R}$ is a mapping from X to $\exp(X)$ (= the class of all nonempty subsets in X); and $(X, \mathcal{R})$ is referred to as a *proper relational structure*; a basic example of such object is

$$\mathcal{I} = \{(x, x); x \in X\} \text{ [the \textit{identical relation} over } X].$$

Given the relations $\mathcal{R}$, $\mathcal{S}$ over X, define their *product* $\mathcal{R} \circ \mathcal{S}$ as

$$(x, z) \in \mathcal{R} \circ \mathcal{S}, \text{ if there exists } y \in X \text{ with } (x, y) \in \mathcal{R},\ (y, z) \in \mathcal{S}.$$

Also, for each relation $\mathcal{R}$ in X, denote

$$\mathcal{R}^{-1} = \{(x, y) \in X \times X; (y, x) \in \mathcal{R}\} \text{ (the \textit{inverse} of } \mathcal{R}).$$

Finally, given the relations $\mathcal{R}$ and $\mathcal{S}$ on X, let us say that $\mathcal{R}$ is *coarser* than $\mathcal{S}$ (or, equivalently: $\mathcal{S}$ is *finer* than $\mathcal{R}$), provided

$$\mathcal{R} \subseteq \mathcal{S}; \text{ that is: } x\mathcal{R}y \text{ implies } x\mathcal{S}y.$$

Given a relation $\mathcal{R}$ on X, the following properties are to be discussed:

(P1) $\mathcal{R}$ is *reflexive*: $\mathcal{I} \subseteq \mathcal{R}$
(P2) $\mathcal{R}$ is *irreflexive*: $\mathcal{R} \cap \mathcal{I} = \emptyset$
(P3) $\mathcal{R}$ is *transitive*: $\mathcal{R} \circ \mathcal{R} \subseteq \mathcal{R}$
(P4) $\mathcal{R}$ is *symmetric*: $\mathcal{R}^{-1} = \mathcal{R}$
(P5) $\mathcal{R}$ is *antisymmetric*: $\mathcal{R}^{-1} \cap \mathcal{R} \subseteq \mathcal{I}$.

This yields the relations to be used; the following ones are important:

(C0) $\mathcal{R}$ is *amorphous* (i.e.: it has no specific properties)
(C1) $\mathcal{R}$ is a *quasi-order* (reflexive and transitive)
(C2) $\mathcal{R}$ is a *strict order* (irreflexive and transitive)
(C3) $\mathcal{R}$ is an *equivalence* (reflexive, transitive, symmetric)
(C4) $\mathcal{R}$ is a *(partial) order* (reflexive, transitive, antisymmetric)
(C5) $\mathcal{R}$ is the *trivial* relation (in the sense: $\mathcal{R} = X \times X$).

Given the ordered structure $(A, \leq)$, call it *totally ordered*, provided

for each $a, b \in A$, we have either $a \leq b$ or $b \leq a$;
or, equivalently:
for each $a, b \in A$ with $a \neq b$, we have either $a < b$ or $b < a$.

Also, given $B \in \exp(A)$, let us say that $b \in B$ is a *first element* of it, when

$b \leq c$, $\forall c \in B$; or, equivalently:
$b < c$, $\forall c \in B$, $b \neq c$.

In this case, we say that $(A, \leq)$ is *well ordered*, provided

each nonempty subset B of A has a first element.

(C) With these preliminaries, we may now give the outlines for the construction of natural numbers in our axiomatic system. Some preliminaries are needed. For each property $P(x)$ (depending on the variable x), define the concept

$P(x)$ is inductive: $P(\emptyset) \& \forall y(P(y) \Longrightarrow P(\mathrm{suc}(y)))$.

Some concrete examples may be given as below. Call the (arbitrary) set y

(is) *in-subs*, if $x \in y$ implies $(x \subset)\mathrm{suc}(x) \subseteq y$
(or, equivalently (by definition): $x \in y$ iff $\mathrm{suc}(x) \subseteq y$)
(sis) *subs-in-subs*, provided ($x =$ in-subs and $x \subset y$) imply $x \in y$.

Proposition 1.4. *The following are valid:*

(14-1) the in-subs property is inductive
(14-2) $x =$ in-subs and $x \subset \mathrm{suc}(y)$ imply $x \subseteq y$
(14-3) $x =$ in-subs, $y =$ subs-in-subs, and $x \subset y$ imply $\mathrm{suc}(x) \subseteq y$
(14-4) the subs-in-subs property is inductive.

Proof. i): Clearly, $\emptyset =$ in-subs (in a vacuous way). Suppose that $y =$ in-subs; we claim that $\mathrm{suc}(y) =$ in-subs. Let the set x be such that

$x \in \mathrm{suc}(y) = y \cup \{y\}$; hence, either $x \in y$ or $x = y$.

The second alternative gives $\mathrm{suc}(x) = \mathrm{suc}(y)$; while, from the first one, we get (by the accepted upon y hypothesis)

$\mathrm{suc}(x) \subseteq y \subset \mathrm{suc}(y)$; hence, $\mathrm{suc}(x) \subset \mathrm{suc}(y)$.

ii): Suppose that x = in-subs, and $x \subset \text{suc}(y) = y \cup \{y\}$. If (by absurd) $x \cap \{y\} \neq \emptyset$, then $y \in x$. As x = in-subs, we get $\text{suc}(y) \subseteq x$; so that (by the imposed hypothesis) $\text{suc}(y) \subset \text{suc}(y)$; contradiction. So, $x \cap \{y\} = \emptyset$; wherefrom (again by the imposed hypothesis) $x = x \cap y$; hence, $x \subseteq y$.

iii): Suppose that x = in-subs, y = subs-in-subs, and $x \subset y$. By the admitted upon y hypothesis, $x \in y$; whence, $\{x\} \subseteq y$. Combining with the posed hypothesis, yields $\text{suc}(x) \subseteq y$; and the assertion follows.

iv): Clearly, $\emptyset$ = subs-in-subs (in a vacuous way). Suppose that y = subs-in-subs; we claim that $\text{suc}(y)$ = subs-in-subs. Let x be such that x = in-subs and $x \subset \text{suc}(y) = y \cup \{y\}$. By the preceding stage, $x \subseteq y$. The case $x = y$ gives $x \in \text{suc}(x) = \text{suc}(y)$. Suppose now that $x \subset y$. By the imposed (upon y) hypothesis, $x \in y$; and this, via $y \subset \text{suc}(y)$, gives $x \in \text{suc}(y)$. □

We are now in position to give a basic answer to the posed question.

Theorem 1.1. *There exists a (nonempty) set $\mathcal{N}$, with*

(11-a) $\mathcal{N}$ is inductive: $\emptyset \in \mathcal{N}$ and ($X \in \mathcal{N} \Longrightarrow \text{suc}(X) \in \mathcal{N}$)
(11-b) $\mathcal{N}$ has the induction property:
$\mathcal{M} \subseteq \mathcal{N}$ and $\mathcal{M}$ = inductive imply $\mathcal{M} = \mathcal{N}$
(11-c) $\mathcal{N} \subseteq \mathcal{B}$, whenever $\mathcal{B}$ is inductive;
so that, $\mathcal{N}$ is uniquely determined by this inductive property
(11-d) each $Y \in \mathcal{N}$ is in-subs (so: $X \in \mathcal{N}$, $X \in Y$ imply $\text{suc}(X) \subseteq Y$)
(11-e) each $Y \in \mathcal{N}$ is subs-in-subs; whence, in particular
(by the preceding stage): $X \in \mathcal{N}$ and $X \subset Y$ imply $X \in Y$
(11-f) the successor suc(.) is a bijection from $\mathcal{N}$ to $\mathcal{N} \setminus \{\emptyset\}$; so that,
$\text{pred} := \text{suc}^{-1}$ (the predecessor) is a bijection from $\mathcal{N} \setminus \{\emptyset\}$ to $\mathcal{N}$
(11-g) $(\forall X, Y \in \mathcal{N})$: $X \in Y \Longleftrightarrow X \subset Y \Longleftrightarrow \text{suc}(X) \subseteq Y$
(11-h) for each $X, Y \in \mathcal{N}$, we have
$X \subset \text{suc}(Y) \Longleftrightarrow X \subseteq Y$; hence, $X \subset Y \Longleftrightarrow X \subseteq \text{pred}(Y)$.

Proof. From Axiom of Infinity, there exists some inductive set $\mathcal{A}$

$$\emptyset \in \mathcal{A} \text{ and } (X \in \mathcal{A} \text{ implies } \text{suc}(X) \in \mathcal{A}).$$

(Here, as precise, $\text{suc}(X) := X \cup \{X\}$ is the *successor* of X). Denote

$$\Gamma = \text{the family of all inductive subsets in } \mathcal{A}.$$

Clearly, Γ is nonempty, because $\mathcal{A} \in \Gamma$. In this case, $\mathcal{N} := \cap\Gamma$ is well defined; we claim that it fulfills all conditions above.

i): For the moment, $\mathcal{N}$ is inductive as intersection of inductive subsets.

ii): As $\mathcal{M} \subseteq \mathcal{N} \subseteq \mathcal{A}$, we must have $\mathcal{M} \subseteq \mathcal{A}$; hence, $\mathcal{M} \in \exp[\mathcal{A}]$. This, along with $\mathcal{M} =$ inductive, yields $\mathcal{M} \supseteq \mathcal{N}$; whence, $\mathcal{M} = \mathcal{N}$.

iii): Let the inductive set $\mathcal{B}$ be given. Clearly, $\mathcal{M} := \mathcal{N} \cap \mathcal{B}$ is an inductive subset of $\mathcal{N}$; and this yields $\mathcal{M} = \mathcal{N}$; hence, $\mathcal{N} \subseteq \mathcal{B}$.

iv): Let $\mathcal{M}$ be the subset of all $Y \in \mathcal{N}$ with the in-subs property. Clearly, $\emptyset \in \mathcal{M}$ (in a vacuous way). On the other hand, the in-subs property is transferrable from Y to $\text{suc}(Y)$. Hence, by the induction property, $\mathcal{M} = \mathcal{N}$.

v): Let $\mathcal{P}$ stand for the subset of all $Y \in \mathcal{N}$ with the subs-in-subs property. Clearly, $\emptyset \in \mathcal{P}$ (in a vacuous way). On the other hand, by a previous result, the subs-in-subs property is transferable from Y to $\text{suc}(Y)$. Hence, by the induction property, $\mathcal{P} = \mathcal{N}$.

vi): There are two properties to be discussed.

vi-1): First, we establish that suc(.) is injective. In fact, let $X, Y \in \mathcal{N}$ be such that $\text{suc}(X) = \text{suc}(Y)$. Combining with $X \in \text{suc}(X)$ we get

$$X \in \text{suc}(Y) = Y \cup \{Y\}; \text{ hence, } X \in Y \text{ or } X = Y.$$

The first alternative gives (by the in-subs property)

$$\text{suc}(X) \subseteq Y; \text{ or, equivalently (by hypothesis) } \text{suc}(Y) \subseteq Y;$$

absurd, via $Y \subset \text{suc}(Y)$. Hence, the second alternative must hold; and this gives us the desired fact.

vi-2): Denote $\mathcal{K} := \text{suc}(\mathcal{N}) \cup \{\emptyset\}$. Clearly, $\emptyset \in \mathcal{K}$, and

$$\begin{aligned} &\text{suc}(\mathcal{K}) = \text{suc}(\text{suc}(\mathcal{N})) \cup \{\text{suc}(\emptyset)\} \subseteq \\ &\text{suc}(\mathcal{N}) \cup \{\text{suc}(\emptyset)\} = \text{suc}(\mathcal{N}) \subseteq \mathcal{K}; \end{aligned}$$

whence (putting these together) $\mathcal{K}$ is inductive. But then, $\mathcal{K} = \mathcal{N}$; that is: $\text{suc}(\mathcal{N}) = \mathcal{N} \setminus \{\emptyset\}$; wherefrom, all is clear.

vii): Clearly, $X \in Y \iff X \subset Y$ follows from in-subs and subs-in-subs properties. On the other hand, $X \in Y \iff \text{suc}(X) \subseteq Y$ follows from the in-subs property.

viii): Two cases occur.

viii-1): The right to left inclusion is clear. For the left to right inclusion, suppose that $X \subset \text{suc}(Y) = Y \cup \{Y\}$. If $Y \in X$, then (see above) $Y \subset X \subset \text{suc}(Y)$; in contradiction with the successor definition. Hence, $X \subseteq Y$; and the claim follows.

viii-2): Denoting $V = \text{pred}(Y)$, the underlying relation becomes

$$X \subset \text{suc}(V) \iff X \subseteq V.$$

This however is clear, by the preceding stage. □

For the moment, $(\mathcal{N}, \subseteq)$ is (partially) ordered; or, equivalently: $(\mathcal{N}, \subset)$ is strictly ordered. Further properties of it are collected in

Theorem 1.2. *Under the obtained facts we have, in addition,*

(12-a) the structure $(\mathcal{N}, \subset)$ is well ordered:
each nonempty subset Δ of $\mathcal{N}$ admits a first element A (in the sense: $A \subset X$ (or, equivalently, $A \in X$), for each $X \in \Delta$, $X \neq A$)
(12-b) the structure $(\mathcal{N}, \subseteq)$ is totally ordered:
for each $X, Y \in \mathcal{N}$, we have either $X \subseteq Y$ or $Y \subseteq X$
(12-c) for each $Y \in \mathcal{N}$, the structure $(Y, \supset)$ (or, equivalently, $(Y, \ni)$) is well ordered: each nonempty subset Γ of Y has a last element B (in the sense: $X \subset B$ (or, equivalently, $X \in B$) for each $X \in \Gamma$, $X \neq B$).

Proof. i): Let Δ be a nonempty subset of $\mathcal{N}$. If $\emptyset \in \Delta$, the conclusion is clear. Assume now that $\emptyset \notin \Delta$; hence, $\Delta = \Delta \setminus \{\emptyset\}$. As a consequence,

$\mathcal{M} = \{X \in \mathcal{N}; X \subset Y, \forall Y \in \Delta\}$ is nonempty (as $\emptyset \in \mathcal{M}$).

Clearly, a property like

(for each $X \in \mathcal{N}$): $X \in \mathcal{M}$ implies $\mathrm{suc}(X) \in \mathcal{M}$

yields $\mathcal{M} = \mathcal{N}$; impossible, under $\Delta = \Delta \setminus \{\emptyset\}$= nonempty. Hence, there must be some $G \in \mathcal{N}$ such that (under the notation $H := \mathrm{suc}(G)$)

(p-1) $G \in \mathcal{M}$; whence (see above) $H \subseteq Y$, for all $Y \in \Delta$
(p-2) $H \notin \mathcal{M}$; that is: $H \subset Z$ is false, for some $Z \in \Delta$.

From these relations, $H = Z \in \Delta$; whence, $H \in \Delta$. Combining with (p-1), it follows that H is the first element of Δ; and the conclusion follows.

ii): The assertion is obtainable from the previous one, by simply applying it to all subsets with two elements. Another proof to be proposed is the following. Let us say that the element $Y \in \mathcal{N}$ is *admissible*, provided

for each $X \in \mathcal{N}$, we have either $X \subseteq Y$ or $Y \subseteq X$.

It is not hard to see that the class $\mathcal{L}$ of all such subsets is inductive

$\emptyset \in \mathcal{L}$ and [$Y \in \mathcal{L}$ implies $\mathrm{suc}(Y) \in \mathcal{L}$].

As $\mathcal{N}$ has the induction property, $\mathcal{L} = \mathcal{N}$; and the conclusion follows.

iii): Let Γ be a nonempty subset of Y. As a consequence of this,

($\forall G \in \Gamma$): $G \in Y$; hence (see above) $G \subset Y$; wherefrom
$\mathcal{P} = \{X \in \mathcal{N}; G \subset X, \forall G \in \Gamma\}$ is nonempty (as $Y \in \mathcal{P}$).

By the well ordered property, $\mathcal{P}$ admits a first element, P (in the sense: $P \subseteq Y, \forall Y \in \mathcal{P}$). Denote $Q = \text{pred}(P)$. From the very definition of $\mathcal{P}$

$$G \subset P \text{ (hence, } G \subseteq Q\text{), for all } G \in \Gamma.$$

On the other hand,

$$G = Q \text{ (hence, } Q \in \Gamma\text{), for some } G \in \Gamma;$$

for, otherwise, $Q \in \mathcal{P}$; contradiction. Hence, Q is the first element of Γ in $(Y, \supset)$; and the assertion follows. □

As precise, $\mathcal{N}$ is uniquely determined as the minimal element within the class of inductive sets. Usually, we denote $\emptyset$ as 0 (= the null element) and $\mathcal{N}$ as N. Each element of this set is referred to as a *natural number.*

1.1.2 *Cantor set theory*

The axiomatic Zermelo-Fraenkel set theory we just sketched has, as primary objective, the elimination of some paradoxes from naive Cantor set theory; especially, the ones involving Russell formula ($X \in X$) and ordinal/cardinal numbers. But, as long as these issues are not being encountered, the naive Cantor set theory (completed with Peano axiomatic system) serves as an excellent substitute for the Zermelo-Fraenkel axiomatic system in question. It is our aim in the following to verify this, in a synthetic manner. As usual, we accept (as a logical principle)

(LE-CST) *(Leibniz equality rule)*:
$\forall x \forall y (x = y \Longrightarrow \forall z (z \in x \longleftrightarrow z \in y)).$

(A) Formally, the naive Cantor set theory is not based upon general axioms. However, after an inspection of its basic lines, we conclude that it is founded upon three axioms, expressed as below.

(A-I-CST) *Axiom of Extensionality* (in short: AExt-CST):

$$\forall x \forall y (\forall z (z \in x \Longleftrightarrow z \in y) \Longrightarrow x = y).$$

Note that, by (LE-CST), we may write (AExt-CST) as

$$\text{(AExt-CST-var)} \quad \forall x \forall y (\forall z (z \in x \Longleftrightarrow z \in y) \Longleftrightarrow x = y);$$

which tells us that a set is determined by its elements. An equivalent way of expressing this axiom is the following. Define the sentence:

$$x \subseteq y \text{ iff (by definition) } \forall z (z \in x \Longrightarrow z \in y);$$

then, (AExt-CST-var) may be written in the double inclusion way

(AExt-CST-2in) $\forall x \forall y((x \subseteq y \& y \subseteq x) \Longleftrightarrow x = y)$.

(A-II-CST) *Axiom of Separation* (in short: ASep-CST):

$(Q(x) = \text{formula}) \Longrightarrow \exists y \forall z(z \in y \Longleftrightarrow Q(z))$.

This axiom says that, there exists a set y consisting of all z with the property $Q(z)$; in short: $y = \{z; Q(z)\}$. Note that, unlike its corresponding axiom in (ZF), no initial restriction about the sets z above is provided. A dramatic consequence of this is that, for an appropriate construction of such sets, we must avoid at least two classes of such formulae $Q(x)$.

(Avd-1) Firstly, we must avoid the sentences $Q(x)$ with

$Q(x)$ is *non-separable*: $(\forall x)Q(x)$.

Because, in this case, $\{x; Q(x)\}$ coincides with the class of all sets; so that, it has no relevance for us. For example, the formula $Q(x)$ defined as $(x = x)$ is non-separable; but there are many other examples of such objects.

(Avd-2) Secondly, we must avoid the sentences $Q(x)$ with

$Q(x)$ is *set-contradictory*: $\{x; Q(x)\}$ is contradictory.

For example, the Russell formula $R(x)$ defined as $(x \notin x)$ is set-contradictory; because the associated Russell Set $\mathcal{R} = \{x; x \notin x\}$ is contradictory, in the sense: $[\mathcal{R} \in \mathcal{R}$ iff $\mathcal{R} \notin \mathcal{R}]$. Note that, a way of avoiding constructions generated by the formula $R(x)$ is that described by the Axiom of Russell; according to which, $R(x)$ is taken as non-separable. This yields us a general way of eliminating such cases, by accepting the principle:

(SC-NS) each set-contradictory sentence is non-separable.

Starting from (ASep-CST), we may now construct all sets encountered in the system (ZF). A list of these is given below.

(i) *Construction of Empty Set*:

$\emptyset = \{x; x \neq x\}$.

The set defined by this axiom is uniquely determined; in addition, we have $\forall x(x \notin \emptyset)$.

(ii) *Construction of Pairs*:

$(\forall x \forall y)\ \{x, y\} := \{z; z = x \vee z = y\}$.

The set defined by this construction is uniquely determined. In particular, when $x = y$, we denote $\{x, y\}$ as $\{x\}$. Then, define

$(\forall x \forall y)\ (x, y) := \{\{x\}, \{x, y\}\}$
(the *couple* (or, *ordered pair*) of x and y).

The following property is immediate:

$$\forall x \forall y \forall u \forall v((x, y) = (u, v) \iff x = u \& y = v)).$$

By a *relation*, we mean any set $\mathcal{R}$ of ordered pairs. Then, by a *function*, we mean any relation f with

$(x, y) \in f$ and $(x, z) \in f$ imply $y = z$.

(iii) *Construction of Union*:

$$(\forall x)\ \cup x := \{z; \exists t(z \in t \& t \in x)\}.$$

As before, this set is uniquely determined. Note that $\cup\emptyset = \emptyset$; because $z \in \cup\emptyset \iff \exists t(z \in t \& t \in \emptyset)$ and the right member of this equivalence is false. In particular, when $x = \{a, b\}$, we put $a \cup b := \cup\{a, b\}$.

(iv) *Construction of Cartesian Product*:

$$(\forall x \forall y)\ x \times y := \{z; \exists u \exists v(u \in x \& v \in y \& z = (u, v))\}.$$

(v) *Construction of Power Set*:

$$\forall x : \exp[x] = \{z; z \subseteq x\}.$$

As before, the set defined by this axiom is uniquely determined. Put also $\exp(x) = \{y \in \exp[x]; y \neq \emptyset\}$ (the class of all nonempty subsets in x). By definition, a *relation* between x and y is any $\mathcal{R} \in \exp[x \times y]$. As usual, $(u, v) \in \mathcal{R}$ will be denoted as $u\mathcal{R}v$. This allows us defining the *domain* and *codomain* of any such relation $\mathcal{R}$ as

$\text{Dom}(\mathcal{R}) = \{u \in x; \exists v \in y, (u, v) \in \mathcal{R}\}$
$\text{Codom}(\mathcal{R}) = \{v \in y; \exists u \in x, (u, v) \in \mathcal{R}\}.$

Further, let us say that $f \in \exp[x \times y]$ is a *function*, if

$(u, v) \in f$ and $(u, w) \in f$ imply $v = w$.

If in addition $\text{Dom}(f) = x$, we say that f is a *function from x to y*. Denote

$\mathcal{F}(x, y) =$ the set of all functions from x to y;

as usual, we write $f \in \mathcal{F}(x, y)$ as $f : x \to y$.

(vi) *Construction of Intersection*:

$$(\forall x) \cap x := \{z; z \in t, \forall t \in x\}.$$

Note that, whenever $x \neq \emptyset$, the introduced set is uniquely determined. In particular, when $x = \{a, b\}$, we put $a \cap b := \cap\{a, b\}$.

However, in case of $x = \emptyset$, this construction does not work, because

the formula $(z \in t, \forall t \in \emptyset)$ (identical with $\forall t(t \in \emptyset \Longrightarrow z \in t)$)
is true, for all z; hence, non-separable;

and (see above) any construction involving non-separable formulae is not relevant here; so that $\cap\emptyset$ is, formally, undefined. A way of completing this construction (for empty set case) is, as precise, *local* in nature; namely,

(A-def) for the given nonempty set A, we formally put $\cap\emptyset = A$.

This convention is exclusively working at the level of $\exp[A]$, for an appropriate handling of unions and intersections of subsets in A.

(A-III-CST) *Axiom of Choice* (in short: AC-CST).

Given the nonempty set a, let us say that $f \in \mathcal{F}(a, \cup a)$ is a *selection*, if

(sel) $f(x) \in x$, for each $x \in a$ with $x \neq \emptyset$.

In this case, the axiom in question writes

$$\forall a(a \neq \emptyset \Longrightarrow \exists f(f \in \mathcal{F}(a, \cup a)\&(f = \text{selection}));$$

and says that: for each nonempty set a, there exists at least one selection $f \in \mathcal{F}(a, \cup a)$. It allows us constructing the function f by an infinite amount of choosing procedures, even if the predicative mechanism of these choices is not given.

(B) Having these precise, we may now pass to the (axiomatic) construction of natural numbers in (CST). The exposition follows, in part, the lines in Becheanu et al. [24, Ch. III] and Vaisman [260, Ch. IV]; see also Halmos [106, Ch. 12].

Let N be a (nonempty) set, $0 \in N$ be a fixed element, and $s : N \to N$ be a function; we say that the triple $(N, 0, s)$ is a *Peano system*, provided

(pea-1) $0 \notin s(N)$ (meaning: $s(n) \neq 0$, $\forall n \in N$)
(pea-2) $s(.)$ is injective: $(s(n) = s(m)$ implies $n = m)$
(pea-3) N has the *induction* property:
if $M \subseteq N$ is *inductive* $(0 \in M$ and $s(M) \subseteq M)$, then $M = N$.

The conditions (pea-1)–(pea-3) will be then referred to as the *Peano axioms* of our Cantor set theory. An immediate consequence of these is

Proposition 1.5. *Let $(N, 0, s)$ be a Peano system. Then,*

$s(.)$ is a bijection between N and $N \setminus \{0\}$.

Proof. Denote $P = s(N) \cup \{0\}$. Clearly, P is inductive, because $0 \in P$ and $s(P) \subseteq s(N) \subseteq P$. By the induction property, we then have $P = N$; so, $s : N \to N \setminus \{0\}$ is surjective. Combining with the injective property gives the desired fact. □

As a consequence of this, the inverse function pre $:= s^{-1}$ is well defined as a bijection from $N \setminus \{0\}$ to N. In this case, for each $n \in N \setminus \{0\}$, pre(n) is referred to as the *predecessor* of n.

Note that a Peano system is not uniquely determined. So, it is natural to ask of how different two such objects may be. The following result (referred to as: Peano Systems Isomorphism theorem; in short: (PS-Iso)) is an adequate answer to this.

Theorem 1.3. *Let $(N, 0, s)$ be a Peano system; and (Γ, θ, σ) be a triple, where Γ is a nonempty set, $\theta \in \Gamma$ is an element, and $\sigma : \Gamma \to \Gamma$ is a function. Then,*

(21-a) *There exists a unique $G : N \to \Gamma$, with the compatible property*

(comp) $G(0) = \theta$ *and* $G(s(n)) = \sigma(G(n))$, $\forall n \in N$.

(21-b) *If, in addition, (Γ, θ, σ) is a Peano system, then G is a bijection. Hence: any two Peano systems $(N, 0, s)$ and (Γ, θ, σ) are identical up to a bijection $G : N \to \Gamma$, fulfilling the compatible relations (comp) above.*

Proof. There are several steps to be passed.

i): Let $\mathcal{G}$ be the class of all subsets $E \subseteq N \times \Gamma$, with

(st-2-ind) E is *starting* ($(0, \theta) \in E$) and
double inductive ($(n, \xi) \in E$ implies $(s(n), \sigma(\xi)) \in E$).

Clearly, $\mathcal{G}$ is nonempty, because $N \times \Gamma$ is an element of it. Denote

$G = \cap\mathcal{G}$ (the minimal subset of $N \times \Gamma$ fulfilling (st-2-ind)).

We claim that G has the properties:

(pro-1) G is *starting* ($(0, \theta) \in G$) and
double inductive ($(n, \xi) \in G$ implies $(s(n), \sigma(\xi)) \in G$)

(pro-2) $\mathrm{Dom}(G) = N$ ($\forall n \in N$, $\exists \xi \in \Gamma$: $(n, \xi) \in G$)
(pro-3) $G(n) =$ singleton, $\forall n \in N$; hence, $G \in \mathcal{F}(N, \Gamma)$.

The first assertion is evident. The second assertion follows at once from

$P = \{n \in N; G(n) \neq \emptyset\}$ is inductive; hence, $P = N$.

The third assertion is a consequence of the fact that

$Q = \{n \in N; G(n) = \text{singleton}\}$ is inductive; hence, $Q = N$.

Two cases must be considered.

i-1) Clearly, $G(0) =$ singleton is equivalent with

$G(0) = \{\theta\}$; that is: $\zeta \neq \theta$ implies $\zeta \notin G(0)$.

For, let $\zeta \neq \theta$ be such that $(0, \zeta) \in G$. Putting $G^* := G \setminus \{(0, \zeta)\}$, we have

i-1-1) G^* is starting, in view of $(0, \theta) \in G^*$

i-1-2) G^* is double inductive; because, given $(n, \xi) \in G^*$,

$(s(n), \sigma(\xi)) = (0, \zeta)$ implies $s(n) = 0$, contradiction.

But then, $G^* \supseteq G$; impossible, via $G^* \subset G$.

i-2) Assume that $G(m) =$ singleton; we have to establish that $G(s(m)) =$ singleton. Denote $G(m) = \eta$; hence, $(m, \eta) \in G$; so that $(s(m), \sigma(\eta)) \in G$. We claim that $G(s(m)) = \{\sigma(\eta)\}$. Suppose not:

$G(s(m)) \neq \{\sigma(\eta)\}$; hence, there exists $\zeta \neq \sigma(\eta)$ with $(s(m), \zeta) \in G$.

Putting $G^* = G \setminus \{(s(m), \zeta)\}$, we have

i-2-1) G^* is starting, because

$(0, \theta) = (s(m), \zeta)$ implies $0 = s(m)$; contradiction

i-2-2) G^* is double inductive; for, given $(n, \xi) \in G^*$,

$(s(n), \sigma(\xi)) = (s(m), \zeta) \Longrightarrow s(n) = s(m)$ (hence, $n = m$), $\sigma(\xi) = \zeta$;
so (as $(m, \xi) \in G^*$):
$\xi = G(m) = \eta$ and (as a consequence) $\sigma(\eta) = \sigma(\xi) = \zeta$; contradiction.

But then, $G^* \supseteq G$; impossible, in view of $G^* \subset G$.

ii): By the above step, we have that G is an element of $\mathcal{F}(N, \Gamma)$. Suppose that there exists another function $H : N \to \Gamma$, with

$H(0) = \theta$ and $[H(s(n)) = \sigma(H(n)), \forall n \in N]$.

Then $P = \{n \in N; G(n) = H(n)\}$ is inductive, because $G(0) = H(0)$ and

$$G(n) = H(n) \Longrightarrow G(s(n)) = \sigma(G(n)) = \sigma(H(n)) = H(s(n));$$

and, from this, $P = N$; which tells us that $G = H$.

iii): Suppose now that (Γ, θ, σ) is a Peano system too; we have to establish that G is bijective (meaning: surjective and injective).

Part-1): Let us prove that G is surjective: $G(N) = \Gamma$. It will suffice verifying that $G(N)$ is an inductive part of Γ with respect to the Peano system (Γ, θ, σ). Clearly, $\theta = G(0) \in G(N)$. Suppose now that $\xi \in G(N)$; hence, $\xi = G(n)$, for some $n \in N$; then,

$$\sigma(\xi) = \sigma(G(n)) = G(s(n)) \in G(N).$$

Putting these together, yields the inductive property; as desired.

Part-2): Further, let us prove that G is injective. To do this, it will suffice proving that $\Delta = \{\xi \in \Gamma; G^{-1}(\xi) = \text{singleton}\}$ is an inductive part of Γ with respect to the Peano system (Γ, θ, σ). Firstly, we claim that $\theta \in \Delta$. Suppose not: $G(n) = \theta$, for some $n \in N \setminus \{0\}$. Clearly, $n = s(m)$, for some $m \in N$; so that $G(n) = G(s(m)) = \sigma(G(m)) \neq \theta$; contradiction; hence the claim. Secondly, assume that $\xi \in \Delta$; we have to establish that $\sigma(\xi) \in \Delta$. By hypothesis, $G^{-1}(\xi) = \{n\}$, for some $n \in N$; and this yields $G(s(n)) = \sigma(G(n)) = \sigma(\xi)$. Suppose that $G(m) = \sigma(\xi)$, for some $m \in N$. As $\sigma(\xi) \neq \theta$, it results (by the preceding stage) that $m \neq 0$; whence, $m = s(p)$, for some $p \in N$. But then,

$\sigma(\xi) = G(m) = G(s(p)) = \sigma(G(p))$;
wherefrom (as σ is injective) $\xi = G(p)$; that is, $p \in G^{-1}(\xi)$;
so that (by the inductive hypothesis) $p = n$.

This, by the above, gives $m = s(p) = s(n)$; wherefrom, $\sigma(\xi) \in \Delta$. Putting these together yields the inductive property; wherefrom, $\Delta = \Gamma$, as desired. The proof is thereby complete. □

Remark 1.1.

An alternate proof of this last part may be given as below. By the preceding step, we have promised a function $\Lambda : \Gamma \to N$, with the properties

$$\Lambda(\theta) = 0 \text{ and } [\Lambda(\sigma(\xi)) = s(\Lambda(\xi)), \forall \xi \in \Gamma].$$

We now claim that

$$(\Lambda(G(n)) = n, \forall n \in N) \text{ and } (G(\Lambda(\xi)) = \xi, \forall \xi \in \Gamma);$$

wherefrom, both G and H are bijective. Clearly, it will suffice that the former of these relations holds; and this happens whenever

$$Q = \{n \in N; \Lambda(G(n)) = n\} \text{ is inductive.}$$

For the moment, $0 \in Q$ (cf. the very definition of G and Λ). Suppose now that $n \in Q$; that is, $\Lambda(G(n)) = n$. Then,

$$\Lambda(G(s(n))) = \Lambda(\sigma(G(n))) = s(\Lambda(G(n))) = s(n); \text{ that is, } s(n) \in Q;$$

and our assertion is proved.

Remark 1.2.

In the Zermelo-Fraenkel axiomatic system there exists at least one Peano system with some additional properties. However, in the Cantor set theory, the existence of such objects is assured in this axiomatic way only.

(C) Let $(N, 0, s)$ be a Peano system. As usual, we shall denote ($s(0) = 1$, $s(1) = 2$, and so on). It is our aim in the following to introduce an algebraic and order structure over the ambient set N. By definition, any function $\varphi : N \times N \to N$ will be referred to as a *composition rule* over N. Usually, for each $x, y \in N$, $\varphi(x, y)$ is denoted as $x + y$ (the *additive* notation) or xy (the *multiplicative* notation).

Theorem 1.4. *There exists a unique composition rule* $\varphi : N \times N \to N$ *such that (under the additive notation)*

$$(\forall m \in N)\text{: } m + 0 = m,\ m + s(n) = s(m + n),\ \forall n \in N.$$

Proof. Fix $m \in N$ and the triple (Γ, θ, σ), where ($\Gamma = N$, $\theta = m$, $\sigma = s$).

i): By the Peano Systems Isomorphism theorem (PS-Iso), there exists a unique function $f_m : N \to N$, such that

$$f_m(0) = m,\ f_m(s(n)) = s(f_m(n)),\ \forall n \in N.$$

Then, if we put

$$m + n = f_m(n),\ n \in N,$$

the relations in our statement are true, under the composition rule

$$\varphi(m, n) = f_m(n),\ m, n \in N.$$

ii): Let $\psi : N \times N \to N$ be another compositions rule fulfilling

$$(\forall m \in N)\text{: } \psi(m, 0) = m,\ \psi(m, s(n)) = s(\psi(m, n)),\ \forall n \in N.$$

Fix $m \in N$ and denote $(g_m(n) = \psi(m,n),\ n \in N)$. By the relations above

$$g_m(0) = m,\ g_m(s(n)) = s(g_m(n)),\ \forall n \in N.$$

This, along with the uniqueness of f_m (see above) gives $g_m = f_m$. □

By definition, $(m,n) \mapsto m+n$ is referred to as the *addition* (or, *sum*) between m and n. Some basic properties of this operation are contained in

Proposition 1.6. *Under the above conventions,*

(22-1) $0 + n = n,\ \forall n \in N$
(22-2) $s(m) + n = s(m+n),\ \forall m, n \in N$
(22-3) $(m+n) + p = m + (n+p),\ \forall m, n, p \in N$
(22-4) $m + n = n + m,\ \forall m, n \in N.$

Proof. i): The subset $Q = \{n \in N; 0+n = n\}$ is inductive; so that, $Q = N$.
ii): Fix some $m \in N$. The subset $Q = \{n \in N; s(m) + n = s(m+n)\}$ is inductive; so that, $Q = N$.
iii): Fix $m, n \in N$. The subset $Q = \{p \in N; (m+n)+p = m+(n+p)\}$ is inductive; so that, $Q = N$.
iv): Fix some $m \in N$. The subset $Q = \{n \in N; m+n = n+m\}$ is inductive; so that, $Q = N$. □

Theorem 1.5. *There exists a unique composition rule* $\varphi : N \times N \to N$ *such that (under the multiplicative notation)*

$$(\forall m \in N):\ m0 = 0,\ ms(n) = mn + m,\ \forall n \in N.$$

Proof. Fix some $m \in N$, and consider the function

$$(\alpha_m : N \to N):\ \alpha_m(n) = n + m (= m+n),\ n \in N.$$

Then, let us consider the triple (Γ, θ, σ), where $(\Gamma = N,\ \theta = 0,\ \sigma = \alpha_m)$.
i): By the Peano Systems Isomorphism theorem (PS-Iso), there exists a unique function $f_m : N \to N$, such that

$$f_m(0) = 0,\ f_m(s(n)) = \sigma(f_m(n)) = f_m(n) + m,\ \forall n \in N.$$

Then, if we put

$$mn = f_m(n),\ n \in N,$$

relations in the statement are true, under the composition rule

$$\varphi(m,n) = f_m(n),\ m, n \in N.$$

ii): Let $\psi : N \times N \to N$ be a composition rule fulfilling

$$(\forall m \in N)\text{: } \psi(m,0) = 0,\ \psi(m,s(n)) = \psi(m,n) + m,\ \forall n \in N.$$

Fix $m \in N$ and denote $(g_m(n) = \psi(m,n),\ n \in N)$. By the relations above

$$g_m(0) = 0,\ g_m(s(n)) = g_m(n) + m,\ \forall n \in N.$$

This, along with the uniqueness of f_m (see above), gives $g_m = f_m$. □

By definition, $(m,n) \mapsto mn$ is referred to as the *multiplication* (or, *product*) between m and n. Some of its basic properties are contained in

Proposition 1.7. *Under the above conventions,*

(23-1) $0n = 0,\ \forall n \in N$
(23-2) $s(m)n = mn + n,\ \forall m,n \in N$
(23-3) $m(n+p) = mn + mp,\ \forall m,n,p \in N$
(23-4) $(mn)p = m(np),\ \forall m,n,p \in N$
(23-5) $mn = nm,\ \forall m,n \in N$
(23-6) *(*$1n = n1$ *and)* $1n = n,\ \forall n \in N$*, where* $1 = s(0)$.

Proof. i): The subset $Q = \{n \in N; 0n = 0\}$ is inductive; so that, $Q = N$.

ii): Fix some $m \in N$. The subset $Q = \{n \in N; s(m)n = mn + n\}$ is inductive; so that, $Q = N$.

iii): Fix $m, n \in N$. The subset $Q = \{p \in N; m(n+p) = mn + mp\}$ is inductive; so that, $Q = N$.

iv): Fix $m, n \in N$. The subset $Q = \{p \in N; (mn)p = m(np)\}$ is inductive; so that, $Q = N$.

v): Fix $m \in N$. The subset $Q = \{n \in N; mn = nm\}$ is inductive; so, $Q = N$.

vi): The first half is clear, by the above. For the second half, we have

$$1n = s(0)n = 0n + n = 0 + n = n, \text{ for each } n \in N;$$

and the assertion is proved. □

The following supplementary properties of these operations are valid.

Proposition 1.8. *Under the preceding facts, we also have*

(24-1) $m + n = 0$ *iff (*$m = 0$ *and* $n = 0$*)*
(24-2) $m + n = m + p$ *iff* $n = p$
(24-3) $mn = 0$ *iff (*$m = 0$ *or* $n = 0$*)*
(24-4) *(*$\forall m \neq 0$*):* $mn = mp$ *iff* $n = p$

(24-5) $mn = 1$ *iff* ($m = 1$ *and* $n = 1$).

Proof. i): Suppose that $n \neq 0$; hence, $n = s(v)$, for some $v \in N$. Then,

$$0 = m + n = m + s(v) = s(m + v);\ \text{impossible, via } s(N) = N \setminus \{0\}.$$

ii): The subset $Q = \{m \in N; m + n = m + p \Longrightarrow n = p\}$ is inductive; so that, necessarily, $Q = N$.

iii): Suppose that $m \neq 0$ and $n \neq 0$. By the quoted property of $s(.)$, we must have $n = s(v)$, for some $v \in N$. Then,

$$0 = mn = ms(v) = mv + m;\ \text{wherefrom } mv = m = 0;\ \text{contradiction.}$$

iv): Fix some $m \neq 0$. The subset $Q = \{n \in N; mn = mp \Longrightarrow n = p\}$ is inductive; so that, $Q = N$.

v): As $mn = 1 = s(0) \neq 0$, we must have $m \neq 0$ and $n \neq 0$; whence, $m = s(u)$, $n = s(v)$, for some $u, v \in N$. In this case,

$$s(0) = 1 = mn = s(u)n = un + n = un + s(v) = s(un + v);$$

wherefrom (as $s(.)$=injective) $0 = un + v$. This, by a preceding fact, yields $un = 0$ (hence, $u = 0$, in view of $n \neq 0$) and $v = 0$; whence, $m = n = 1$. □

(D) Passing to the order structure of N, define the relations over N

(s-ineq) $n < m$ iff $n + p = m$, for some $p \in N$, $p \neq 0$;
(ineq) $n \leq m$ iff either $n < m$ or $n = m$.

Note that, by this very definition,

$$n \leq m \text{ iff } n + p = m, \text{ for some } p \in N.$$

Some immediate properties of this couple of relations are contained in the statement below. Let us remember that the inverse function $\text{pre} := s^{-1}$ is well defined as a bijection from $N \setminus \{0\}$ to N. In this case, for each $n \in N \setminus \{0\}$, $\text{pre}(n)$ is referred to as the *predecessor* of n. Note that, as

$$n = s(\text{pre}(n)) = s(\text{pre}(n) + 0) = \text{pre}(n) + 1, \text{ where } 1 = s(0),$$

it is natural to denote ($\text{pre}(n) = n - 1$, $n \in N \setminus \{0\}$).

Proposition 1.9. *Under these conventions,*

(25-1) ($<$) *is irreflexive and transitive; hence, a strict order*
(25-2) ($\leq$) *is reflexive, transitive, antisymmetric; hence, an order*
(25-3) ($\forall n, \forall m$): $n < m$ *implies* $s(n) \leq m$ *and* $n \leq \text{pre}(m)$

(25-4) $(\forall n \in N)$*:* $n < s(n)$*, and there is no* $m \in N$ *with* $n < m < s(n)$
(25-5) $m < n$ $(m \leq n) \Longrightarrow m + p < n + p$ $(m + p \leq n + p)$*,* $\forall p \in N$
(25-6) $m < n$ $(m \leq n) \Longrightarrow mp < np$ $(mp \leq np)$*,* $\forall p \in N \setminus \{0\}$*.*

Proof. i): The irreflexive part is evident, by definition. For the transitive part, let $m, n, p \in N$ be such that $m < n$, $n < p$; hence

$$m + u = n,\ n + v = p, \text{ for some } u, v \in N \setminus \{0\}.$$

This yields $m + u + v = p$, with $u + v \neq 0$; whence, $m < p$.

ii): Evident, by the preceding step.

iii): If $n < m$, then $n + p = m$, for some $p \in N \setminus \{0\}$. From this choice, $p = s(q)$, for some $q \in N$; and then

$s(n) + q = s(n + q) = n + s(q) = n + p = m$; wherefrom $s(n) \leq m$
$s(n + q) = n + s(q) = n + p = m$; wherefrom $n + q = \text{pre}(m)$;
so that (by definition), $n \leq \text{pre}(m)$.

iv): The first part is evident, in view of

$$s(n) = s(n + 0) = n + s(0); \text{ and } 1 = s(0) \neq 0.$$

Passing to the second part, assume that there exists $m \in N$ with $n < m < s(n)$. The first half yields (by the above) $s(n) \leq m$; and this, combined with the second half, yields $s(n) \leq m < s(n)$; hence, $s(n) < s(n)$; contradiction.

v): Let $u \in N \setminus \{0\}$ be such that $m + u = n$. Then,

$$(m + p) + u = n + p; \text{ hence, } m + p < n + p.$$

The second half of this is immediate.

vi): Let $u \in N \setminus \{0\}$ be such that $m + u = n$. Then, for each $p \in N \setminus \{0\}$,

$$mp + up = np, \text{ and } up \neq 0 \text{ (as } u \neq 0\text{); whence, } mp < np.$$

The second half of this is immediate. □

Let us say that the subset A of N is *bounded (above)* if

$a \leq u$, $\forall a \in A$, for some $u \in N$; or, equivalently
(passing to $v := s(u)$): $a < v$, $\forall a \in A$, for some $v \in N$.

Theorem 1.6. *The following properties of the structure* $(N, \leq)$ *are valid:*

(24-a) $(N, \leq)$ *is well ordered: each (nonempty) subset* $A \subseteq N$ *has a first element* $a \in A$ *(in the sense:* $a \leq n$*, for each* $n \in A$*)*

(24-b) $(N, \leq)$ *is totally ordered:* $\forall m, n \in A$*, either* $m \leq n$ *or* $m > n$

(24-c) *Each (nonempty) bounded above subset $A \subseteq N$ has a last element $b \in A$ (in the sense: $n \leq b$, for each $n \in A$).*

Proof. i): Let the (nonempty) subset $A \subseteq N$ be given. If $0 \in A$, we are done; because $0 \leq n$, $\forall n \in A$. Suppose now that $0 \notin A$; hence, $0 < a$, $\forall a \in A$. The subset $P = \{n \in N; n < a, \forall a \in A\}$ is therefore nonempty, via $0 \in P$. If the inductive property holds

$(\forall n)$ $n \in P$ implies $s(n) \in P$

then, $P = N$; in contradiction with $A =$ nonempty. Hence, there must be some $m \in P$ with $s(m) \notin P$. Note that, by definition

$m < a$ (hence, $s(m) \leq a$), $\forall a \in A$.

This yields $s(m) = b$, for some $b \in A$; for, otherwise, $s(m) \in P$; contradiction. Summing up, $b \in A$ is the first element of A.

ii): The assertion is immediately obtainable from the previous one, by simply applying it to all subsets with two elements. However, for completeness reasons, we provide a proof of it. Define the subset of N

$P = \{n \in N; \text{for each } m \in N, \text{ we have } n \leq m \text{ or } n > m\}.$

Clearly, $0 \in P$. Suppose now that $n \in P$; we claim that $s(n) \in P$. Let $m \in N$ be arbitrary fixed. If $n < m$ then (see above) $s(n) \leq m$. Further, if $n = m$, then $s(n) = s(m) > m$. Finally, if $n > m$, then $s(n) > n > m$; whence, $s(n) > m$. Putting these together, yields $s(n) \in P$; whence (by the inductive property) $P = N$.

iii): Let $A \subseteq N$ be (nonempty) bounded above. The subset $P = \{n \in N; n > a, \forall a \in A\}$ is, by this choice, nonempty. So, by a preceding stage, there exists a first element, $p \in P$. Clearly, $p \in N \setminus \{0\}$, in view of $A \neq \emptyset$; hence, $q := \text{pred}(p)$ exists. In addition (by a previous fact)

$p > a$, $\forall a \in A$ implies $(p >)q \geq a$, $\forall a \in A$.

If ($q > a$, $\forall a \in A$), then $q \in P$; in contradiction with $p =$ first element of P. So, $q = a$, for some $a \in A$; whence, q is the last element of A. □

(E) Denote, for each $a, b \in N$, $a \leq b$

$N[a,b] = \{n \in N; a \leq n \leq b\}$, $N]a,b[= \{n \in N; a < n < b\}$,
$N[a,b[= \{n \in N; a \leq n < b\}$, $N]a,b] = \{n \in N; a < n \leq b\}$.

For each $n \in N(1, \leq)$, let $N(n, >) := N[0, n-1]$ stand for the *initial interval* (in N) induced by n. Any set U with $U \sim N$ (in the sense: there exists a bijection from U to N) will be called *effectively denumerable.* In addition, given some natural number $n \in N(1, \leq)$, any set V with $V \sim N(n, >)$ will be said to be n*-finite*; when n is generic here, we say that V is *finite.* Finally, we say that the (nonempty) set W is (at most) *denumerable* iff it is either effectively denumerable or finite.

Let X be a nonempty set. By a *sequence* in X we mean any mapping $x : N \to X$. For simplicity reasons, it will be useful to denote it as $(x(n); n \geq 0)$, or $(x_n; n \geq 0)$; moreover, when no confusion can arise, we further simplify this notation as $(x(n))$ or (x_n), respectively. Any sequence $(y_n := x_{i(n)}; n \geq 0)$ with $(i(n))$ being *divergent* ($i(n) \to \infty$ as $n \to \infty$) will be referred to as a *subsequence* of (x_n); and any sequence $(y_n := x_{i(n)}; n \geq 0)$ with $(i(n))$ = strictly ascending (hence, divergent) will be referred to as a *strict subsequence* of (x_n). Note that, under such a convention, the relations "subsequence of" and "strict subsequence of" are transitive; that is:

> (z_n) = (strict) subsequence of (y_n) and (y_n) = (strict) subsequence of (x_n) imply (z_n) = (strict) subsequence of (x_n).

Finally, let $\mathcal{R} \subseteq X \times X$ be a relation over X. The above developments allow us to introduce the powers of this relation $\mathcal{R}$ as

$$\mathcal{R}^0 = \mathcal{I},\ \mathcal{R}^{n+1} = \mathcal{R}^n \circ \mathcal{R},\ n \in N.$$

The following properties will be useful in the sequel:

$$\mathcal{R}^{m+n} = \mathcal{R}^m \circ \mathcal{R}^n,\ (\mathcal{R}^m)^n = \mathcal{R}^{mn},\ \forall m, n \in N.$$

1.1.3 *Dependent Choice equivalents*

Remember that, according to our previous conventions,

> (ZF-com) = the complete Zermelo-Fraenkel system, consisting of axioms (A-I)-(A-V), (A-VI-a)+(A-VII-a), and (A-VIII)+(A-IX)
> (ZF) (the Zermelo-Fraenkel system) = the system (ZF-com), with (A-VI-a)+(A-VII-a) substituted by (A-VI)+(A-VII)
> (ZF-AC) (the strongly reduced Zermelo-Fraenkel system) = the Zermelo-Fraenkel system (ZF) without Axiom of Choice (AC).

For the metrical fixed point theory, (ZF) is our working axiomatic system. An outstanding part of (ZF) is the *Axiom of Choice* (abbreviated: (AC)); which, in a convenient manner, may be written as

(AC) For each couple (J, X) of nonempty sets and each function $F : J \to \exp(X)$, there exists a (selective) function $f : J \to X$ (in the sense: $f(\nu) \in F(\nu)$, for each $\nu \in J$).

There are many logical equivalents of (AC); see, for instance, Moore [171, Appendix 2], or Rubin and Rubin [201, Part I]. Sometimes, when the set X is endowed with denumerable type structures, the case of $J = N$ will suffice for all choice reasonings; and the existence of such a selective function may be determined by using a weaker form of (AC), called: *Dependent Choice* principle (in short: (DC)). Some preliminaries are needed.

Let X be a nonempty set and $\mathcal{R}$ be some relation over X; then, $(X, \mathcal{R})$ will be referred to as a *relational structure.* Note that $\mathcal{R}$ may be regarded as a mapping between X and $\exp[X]$ (= the class of all subsets in X). In fact, let us simplify the string $(x, y) \in \mathcal{R}$ as $x\mathcal{R}y$; and put

$$X(x, \mathcal{R}) = \{y \in X; x\mathcal{R}y\} \text{ (the } \textit{section} \text{ of } \mathcal{R} \text{ through } x\text{)}, \ x \in X;$$

then, the desired mapping representation is $(\mathcal{R}(x) = X(x, \mathcal{R}); x \in X)$. Call the relation $\mathcal{R}$ over X, *proper* when

$$(X(x, \mathcal{R}) =)\mathcal{R}(x) \text{ is nonempty, for each } x \in X.$$

Then, $\mathcal{R}$ is to be viewed as a mapping between X and $\exp(X)$; and the couple $(X, \mathcal{R})$ will be referred to as a *proper relational structure.*

For each natural number $k \geq 1$, call the map $F : N(k, >) \to X$, a *k-sequence*; if $k \geq 1$ is generic, we talk about a *finite* sequence. The following result, referred to as the *Finite Dependent Choice property* (in short: (DC-fin)) is available in the *strongly reduced* Zermelo-Fraenkel system (ZF-AC). Given $a \in X$, let us say that the k-sequence $F : N(k, >) \to X$ (where $k \geq 2$) is $(a, \mathcal{R})$*-iterative*, provided

$$F(0) = a \text{ and } F(i)\mathcal{R}F(i + 1), \text{ for all } i \in N(k - 1, >).$$

Proposition 1.10. *Let the relational structure $(X, \mathcal{R})$ be proper. Then, for each $k \geq 2$, the following property holds:*

(P(k)) for each $a \in X$, there exists an $(a, \mathcal{R})$-iterative k-sequence.

Proof. Clearly, $(P(2))$ is true; just take $b \in \mathcal{R}(a)$ and define the 2-sequence $F : N(2, >) \to X$ as: $F(0) = a$, $F(1) = b$. Assume that $(P(k))$ is true, for some $k \geq 2$; we claim that $(P(k+1))$ is true as well. In fact, let $F : N(k,>) \to X$ be an $(a,\mathcal{R})$-iterative k-sequence, assured by hypothesis. As $\mathcal{R}$ is proper, $\mathcal{R}(F(k-1))$ is nonempty; let u be some element of it. The map $G : N(k+1,>) \to X$ introduced as

$$G(i) = F(i), i \in N(k,>);\ G(k) = u$$

is an $(a,\mathcal{R})$-iterative $(k+1)$-sequence; and then, we are done. □

Now, it is natural to see what happens when k "tends to infinity". At a first glance, the following *Dependent Choice* principle (in short: (DC)) is obtainable in (ZF-AC) by means of this "limit" process. Given $a \in X$, let us say that the sequence $(x_n; n \geq 0)$ in X is *$(a;\mathcal{R})$-iterative*, provided

$$x_0 = a \text{ and } (x_n \mathcal{R} x_{n+1} \text{ (meaning: } x_{n+1} \in \mathcal{R}(x_n)),\ \forall n).$$

Proposition 1.11. *Let the relational structure $(X,\mathcal{R})$ be proper. Then, for each $a \in X$ there is at least an $(a,\mathcal{R})$-iterative sequence in X.*

Formally, the "argument" involved here consists in the possibility of constructing an (infinite) sequence via finite sequences. This, ultimately, cannot be done within (ZF-AC); so, the limit process in question does not work here; whence, (DC) is not obtainable from the axioms of (ZF-AC). On the other hand, this principle – proposed, independently, by Bernays [25] and Tarski [221] – is deductible from (AC), but not conversely; cf. Wolk [267]. Denote, for simplicity

> (ZF-AC+DC) (the *reduced* Zermelo-Fraenkel system) = the strongly reduced Zermelo-Fraenkel system (ZF-AC) completed with the Dependent Choice principle (DC).

According to the developments in Moskhovakis [173, Ch. 8], and Schechter [211, Ch. 6], the reduced system (ZF-AC+DC) it comprehensive enough so as to cover the "usual" mathematics; see also Moore [171, Appendix 2, Table 4]. This reduced system will be largely used along our exposition devoted to metrical fixed point theory. So, determining various equivalents of (DC) may be not without profit.

Let $(\mathcal{R}_n; n \geq 0)$ be a sequence of relations on X. Given $a \in X$, let us say that the sequence $(x_n; n \geq 0)$ in X is *$(a;(\mathcal{R}_n; n \geq 0))$-iterative*, provided

$$x_0 = a \text{ and } [x_n \mathcal{R}_n x_{n+1} \text{ (that is: } x_{n+1} \in \mathcal{R}_n(x_n)),\ \forall n].$$

The following *Diagonal Dependent Choice* principle (in short: (DDC)) is now available.

Proposition 1.12. *Let $(\mathcal{R}_n)$ be a sequence of proper relations on X. Then, for each $a \in X$ there exists an $(a; (\mathcal{R}_n))$-iterative sequence in X.*

Clearly, (DDC) includes (DC); to which it reduces when $(\mathcal{R}_n; n \geq 0)$ is constant. The reciprocal of this is also true. In fact, letting the premises of (DDC) hold, put $P = N \times X$; and let $\mathcal{S}$ be the relation over P

$$\mathcal{S}(i, x) = \{i+1\} \times \mathcal{R}_i(x), \quad (i, x) \in P.$$

It will suffice applying (DC) to $(P, \mathcal{S})$ and $b := (0, a) \in P$ to get the conclusion in the statement; we do not give details.

Summing up, (DDC) is provable in (ZF-AC+DC). This is valid as well for its variant, referred to as: the *Selected Dependent Choice* principle (in short: (SDC)).

Proposition 1.13. *Let the map $F : N \to \exp(X)$ and the relation $\mathcal{R}$ over X fulfill*

$$(\forall n \in N)\text{: } \mathcal{R}(x) \cap F(n+1) \neq \emptyset, \quad \forall x \in F(n).$$

Then, for each $a \in F(0)$ there exists a sequence $(x(n); n \geq 0)$ in X with

$$x(0) = a, \ \ x(n) \in F(n), \ x(n+1) \in \mathcal{R}(x(n)), \ \forall n.$$

As before, (SDC) $\Longrightarrow$ (DC) ($\Longleftrightarrow$ (DDC)); just take $(F(n) = X;\, n \in N)$. The reciprocal is also true: (DDC) $\Longrightarrow$ (SDC). This follows from

Proof. (Proposition 1.13) Let the premises of (SDC) be true. Define a sequence of relations $(\mathcal{R}_n; n \geq 0)$ over X as: for each $n \geq 0$,

$$\mathcal{R}_n(x) = \mathcal{R}(x) \cap F(n+1), \ \text{ if } x \in F(n),$$
$$\mathcal{R}_n(x) = \{x\}, \text{ otherwise } (x \in X \setminus F(n)).$$

Clearly, $\mathcal{R}_n$ is proper for all n. So, by (DDC), we have that, for the starting $a \in F(0)$, there exists an $(a, (R_n; n \geq 0))$-iterative sequence $(x(n); n \geq 0)$ in X. Combining with the definition above gives the conclusion. □

In particular, when $\mathcal{R} = X \times X$, the regularity condition imposed in (SDC) holds. The corresponding variant of the underlying statement is just (AC(N)) (= the *Denumerable Axiom of Choice*). Precisely, we have

Proposition 1.14. *Let $F : N \to \exp(X)$ be a function. Then, for each $a \in F(0)$ there exists a function $f : N \to X$ with $f(0) = a$ and $(f(n) \in F(n), \forall n)$.*

As a consequence of the above facts, (DC) $\Longrightarrow$ (AC(N)) in (ZF-AC). A direct verification of this is obtainable by an application of (DC) to the relational structure $(Q, \mathcal{T})$, where

$$Q = N \times X,\ \mathcal{T}(n, x) = \{n+1\} \times F(n+1),\ \ n \in N,\ x \in X;$$

we do not give details. The reciprocal inclusion is not true; see, for instance, Moskhovakis [173, Ch. 8, Sec. 8.25].

1.2 Conv-Cauchy Structures

Abstract. Some preliminary facts are discussed about convergence and Cauchy structures over abstract and pseudometric spaces.

Subject-Class 2020. 47H10 (Primary), 54H25 (Secondary).

Keywords. Convergence and Cauchy structure, asymptotic sequence, superior and inferior limit.

1.2.1 *General conv-Cauchy structures*

Let X be a nonempty set. Call the subset Y of X, *almost-singleton* (in short: *asingleton*) provided $y_1, y_2 \in Y$ implies $y_1 = y_2$; and *singleton* if, in addition, Y is nonempty; note that in this case $Y = \{y\}$, for some $y \in X$.

Let $\mathcal{S}(X)$ be the class of all sequences (x_n) in X. By a *(sequential) convergence structure* on X we mean any part $\mathcal{C}$ of $\mathcal{S}(X) \times X$, with (cf. Kasahara [141]):

(conv-1) $\mathcal{C}$ is *hereditary*:
$((x_n); x) \in \mathcal{C} \Longrightarrow ((y_n); x) \in \mathcal{C}$, for each subsequence (y_n) of (x_n)
(conv-2) $\mathcal{C}$ is *reflexive*: for each $u \in X$,
the constant sequence $(x_n = u; n \geq 0)$ fulfills $((x_n); u) \in \mathcal{C}$.

When the reflexive property is not present, then $\mathcal{C}$ will be called a *(sequential) almost convergence structure* on X. But, in the following, this concept will be not in general considered.

Given $(x_n) \in \mathcal{S}(X)$, $x \in X$, write $((x_n); x) \in \mathcal{C}$ as $x_n \xrightarrow{\mathcal{C}} x$; this reads:

(x_n), *$\mathcal{C}$-converges* to x (also referred to as: x is the *$\mathcal{C}$-limit* of (x_n)).

The set of all such x is denoted $\mathcal{C} - \lim_n(x_n)$; clearly, two alternatives hold

(conv) (x_n) is *$\mathcal{C}$-convergent*, when $\mathcal{C} - \lim_n(x_n)$ is nonempty
(div) (x_n) is *$\mathcal{C}$-divergent*, when $\mathcal{C} - \lim_n(x_n)$ is empty.

The following condition is to be optionally considered here:

(conv-3) $\mathcal{C}$ is *separated*:
$\mathcal{C} - \lim_n(x_n)$ is an asingleton, for each sequence (x_n);

when it holds, $\mathcal{C} - \lim_n(x_n) = \{z\}$ will be also written as $\mathcal{C} - \lim_n(x_n) = z$.

Further, let us say that the subset $\mathcal{H} \subseteq \mathcal{S}(X)$ is a (sequential) *Cauchy structure* on X, provided (cf. Turinici [244])

(Cauchy-1) $\mathcal{H}$ is *hereditary*:
$(x_n) \in \mathcal{H}$ implies $(y_n) \in \mathcal{H}$, for each subsequence (y_n) of (x_n)
(Cauchy-2) $\mathcal{H}$ is *reflexive*: for each $u \in X$,
the constant sequence $(x_n = u; n \geq 0)$ fulfills $(x_n) \in \mathcal{H}$.

Each element of $\mathcal{H}$ will be referred to as a $\mathcal{H}$-*Cauchy* sequence in X.

As before, when the reflexive condition is not holding, then $\mathcal{H}$ is referred to as a *(sequential) almost Cauchy structure*. But, in the following, this concept will be not in general considered.

Finally, call the couple $(\mathcal{C}, \mathcal{H})$, a *(sequential) conv-Cauchy structure* on X. The optional conditions about this couple to be considered here are

(CC-1) $(\mathcal{C}, \mathcal{H})$ is *regular*: each $\mathcal{C}$-convergent sequence is $\mathcal{H}$-Cauchy
(CC-2) $(\mathcal{C}, \mathcal{H})$ is *complete*: each $\mathcal{H}$-Cauchy sequence is $\mathcal{C}$-convergent.

1.2.2 *Pseudometric conv-Cauchy structures*

A standard way of introducing such structures is the *(pseudo) metrical* one. By a *pseudometric* over X we shall mean any map $d : X \times X \to R_+$. Given such an object, the following properties are to be optionally used:

(ref) d is *reflexive*: $x = y$ implies $d(x, y) = 0$
(tri) d is *triangular*: $d(x, z) \leq d(x, y) + d(y, z)$, $\forall x, y, z \in X$
(suf) d is *sufficient*: $d(x, y) = 0$ implies $x = y$
(sym) d is *symmetric*: $d(x, y) = d(y, x)$, $\forall x, y \in X$.

This yields a lot of pseudometrics to be used; the following are important:

(p1) d is *r-pseudometric* (reflexive)
(p2) d is *t-pseudometric* (triangular)
(p3) d is *rs-pseudometric* (reflexive, sufficient)
(p4) d is *almost semimetric* (reflexive, triangular)
(p5) d is *semimetric* (reflexive, triangular, symmetric)
(p6) d is *quasi-metric* (reflexive, triangular, sufficient)
(p7) d is *metric* (reflexive, triangular, symmetric, sufficient).

Let $d(.,.)$ be a reflexive pseudometric (in short: *r-pseudometric*) on X; in this case, (X, d) is called a *r-pseudometric space*. Given the sequence (x_n) in X and the point $x \in X$, we say that (x_n), *d-converges* to x (written as: $x_n \xrightarrow{d} x$) provided

$d(x_n, x) \to 0$ as $n \to \infty$; that is:
$\forall \varepsilon > 0, \exists i = i(\varepsilon)$: $i \leq n \Longrightarrow d(x_n, x) < \varepsilon$.

By this very definition, we have the hereditary and reflexive properties:

(d-conv-1) $(\xrightarrow{d})$ is *hereditary*:
$x_n \xrightarrow{d} x$ implies $y_n \xrightarrow{d} x$, for each subsequence (y_n) of (x_n)
(d-conv-2) $(\xrightarrow{d})$ is *reflexive*: for each $u \in X$,
the constant sequence $(x_n = u; n \geq 0)$ fulfills $x_n \xrightarrow{d} u$.

As a consequence, $(\xrightarrow{d})$ is a sequential convergence on X. The set of all such limit points of (x_n) will be denoted $d - \lim_n(x_n)$; or $\lim_n(x_n)$, when d is understood. Concerning this concept, there are two possible alternatives

(d-conv) (x_n) is *d-convergent*, when $\lim_n(x_n)$ is nonempty
(d-div) (x_n) is *d-divergent*, when $\lim_n(x_n)$ is empty.

The following condition is to be optionally considered here:

(d-conv-3) $(\xrightarrow{d})$ is *separated* (referred to as: d is *separated*):
$\lim_n(x_n)$ is an asingleton, for each sequence (x_n);

when it holds, $\lim_n(x_n) = \{z\}$ is written as $\lim_n(x_n) = z$. Note that, from our setting, d is not separated. However, this is retainable if (in addition)

d is triangular, sufficient, symmetric [hence, a *metric* on X].

A basic question to be posed here is to indicate conditions under which

(1-cont) d is *1-continuous*: $x_n \xrightarrow{d} x$ implies $d(x_n, y) \to d(x, y)$, $\forall y \in X$
(2-cont) d is *2-continuous*: $y_n \xrightarrow{d} y$ implies $d(x, y_n) \to d(x, y)$, $\forall x \in X$
(cont) d is *continuous*: $x_n \xrightarrow{d} x$, $y_n \xrightarrow{d} y$ imply $d(x_n, y_n) \to d(x, y)$.

Note that, when $d =$ pseudometric, these are independent conditions. However, we do have (under the r-pseudometric property of d)

continuous implies 1-continuous and 2-continuous.

The following property is immediate; so, we do not give details.

Proposition 1.15. *Suppose that $d(.,.)$ is semimetric. Then,*
(21-1) *The mapping $(x, y) \mapsto d(x, y)$ is d-Lipschitz, in the sense*

$$|d(x, y) - d(u, v)| \leq d(x, u) + d(y, v),\ \forall (x, y), (u, v) \in X \times X$$

(21-2) *$d(.,.)$ is continuous; hence, 1-continuous and 2-continuous.*

Further, call the sequence (x_n), *d-Cauchy* provided

$d(x_m, x_n) \to 0$ as $m, n \to \infty$, $m < n$; that is,
$\forall \varepsilon > 0$, $\exists j = j(\varepsilon)$: $j \leq m < n \Longrightarrow d(x_m, x_n) < \varepsilon$;

the class of all these will be denoted as $Cauchy(d)$. As before, we have the hereditary and reflexive properties

(d-Cauchy-1) $Cauchy(d)$ is *hereditary*: (x_n) is d-Cauchy implies (y_n) is d-Cauchy, for each subsequence (y_n) of (x_n)
(d-Cauchy-2) $Cauchy(d)$ is *reflexive*: for each $u \in X$, the constant sequence $(x_n = u; n \geq 0)$ is d-Cauchy;

hence, $Cauchy)(d)$ is a Cauchy structure on X. Finally, as already precise, the couple $((\xrightarrow{d}), Cauchy(d))$ is referred to as a *conv-Cauchy structure* on X (attached to d). Note that, by the imposed (upon d) conditions, the underlying object is not regular. But, when $d(.,.)$ is a semimetric, the considered property holds.

Given the sequence (x_n) in X, it follows by definition that

(equi) $x_n \xrightarrow{d} x$ iff $[y_n \xrightarrow{d} x$, for each subsequence (y_n) of $(x_n)]$.

This tells us that if the right member of (equi) is false, the initial sequence (x_n) is not d-convergent. A natural question is to establish under which conditions upon (x_n) is its d-convergence assured under a partial validity of the right member in (equi). The following answer to this is to be noted.

Proposition 1.16. *Letting (X, d) be a semimetric space, suppose that the sequence $(x_n; n \geq 0)$ in X and the point $v \in X$ are such that*

(22-1) $(x_n; n \geq 0)$ is d-Cauchy
(22-2) $y_n \xrightarrow{d} v$, for some subsequence $(y_n = x_{k(n)}; n \geq 0)$ of (x_n).

Then, necessarily, $x_n \xrightarrow{d} v$ as $n \to \infty$.

Proof. Let $\varepsilon > 0$ be arbitrary fixed. From the d-Cauchy property, there exists some rank $m(\varepsilon) \geq 0$, such that

$m(\varepsilon) \leq i \leq j \Longrightarrow d(x_i, x_j) < \varepsilon/2.$

On the other hand, by the sub-sequential convergence property, we have that, for the same $\varepsilon > 0$, there exists an index $n(\varepsilon) \geq m(\varepsilon)$, such that

$$n(\varepsilon) \leq n \Longrightarrow d(y_n, v) = d(x_{k(n)}, v) < \varepsilon/2.$$

Finally, as $(k(n); n \geq 0)$ is divergent, there exists some $p(\varepsilon) \geq n(\varepsilon)$, with

$$k(n) \geq n(\varepsilon), \text{ for all } n \geq p(\varepsilon).$$

Combining these we have, for each $n \geq p(\varepsilon)(\geq n(\varepsilon) \geq m(\varepsilon))$

$$d(x_n, v) \leq d(x_n, x_{k(n)}) + d(x_{k(n)}, v) < \varepsilon/2 + \varepsilon/2 = \varepsilon;$$

and the conclusion follows. □

Let (X, d) be a r-pseudometric space. In the following, some classes of sequences (related to the d-Cauchy ones) are introduced. Call the sequence (x_n) in X,

(d-asy) *d-asymptotic*, if $\lim_n d(x_n, x_{n+1}) = 0$,
(d-to-asy) *d-total-asymptotic*, if $\lim_n d(x_n, x_{n+i}) = 0$, $\forall i \in N(1, \leq)$.

Clearly, for each sequence (x_n) in X

$$d\text{-Cauchy} \Longrightarrow d\text{-total-asymptotic} \Longrightarrow d\text{-asymptotic};$$

but, the converse inclusions are not in general true.

Define, for each $\gamma > 0$, $H \in \exp(N)$,

$$\mathcal{Z}(\gamma, H) := \{i \in N; n \geq i, h \in H \Longrightarrow d(x_n, x_{n+h}) < \gamma\}$$
$$Z(\gamma, H) := \min \mathcal{Z}(\gamma, H) \text{ (where } \min(\emptyset) = \infty);$$

we say that $Z(\gamma, H)$ is the *asymptotic rank* attached to γ and H. Clearly,

$$(\gamma_1 \geq \gamma_2, H_1 \subseteq H_2) \longrightarrow$$
$$\mathcal{Z}(\gamma_1, H_1) \supseteq \mathcal{Z}(\gamma_2, H_2),\ Z(\gamma_1, H_1) \leq Z(\gamma_2, H_2).$$

Finally, we say that the sequence (x_n) is *d-telescopic Cauchy*, when

$$\sum_n d(x_n, x_{n+1})(= d(x_0, x_1) + d(x_1, x_2) + \ldots) < \infty.$$

An appropriate setting to discuss the relationships between such concepts is the triangular one. In this direction, we have

Proposition 1.17. *Supposed that, in addition, d is triangular on X. Then, for each sequence (x_n) in X,*

(23-1) (x_n) is d-asymptotic iff (x_n) is d-total-asymptotic
(23-2) (x_n) is d-telescopic Cauchy implies (x_n) is d-Cauchy.

Proof. i): Let $i \in N(1, \leq)$ be arbitrary fixed. By the triangular inequality,

$(d(x_n, x_{n+i}) \leq \rho_n + ... + \rho_{n+i-1}, \forall n)$,
where $(\rho_n := d(x_n, x_{n+1}); n \geq 0)$.

By the imposed hypothesis, the right member of this relation tends to zero as $n \to \infty$; wherefrom, all is clear.

ii): By hypothesis, we have under the notation $(\rho_n := d(x_n, x_{n+1}))$,

$\sigma_n := \sum_{k \geq n} \rho_k \to 0$ as $n \to \infty$.

In this case, for each $n, m \geq 0$ with $n < m$,

$$d(x_n, x_m) \leq \rho_n + \rho_{n+1} + ... + \rho_{m-1} \leq \sigma_n;$$

and this, along with the property of (σ_n), gives the needed conclusion. □

Finally, note that some extensions of these facts are possible under the lines discussed in the 2001 PhD Thesis by Hitzler [113, Ch. 1, Sec. 1.2]; since, however, these are beyond our scope, further details are not necessary.

1.2.3 *Non-Cauchy asymptotic sequences*

Let $(X, d, \leq)$ be a quasi-ordered semimetric space. Given the sequence (x_n) in X, call it *ascending* when $[i \leq j$ implies $x_i \leq x_j]$. In the following, a characterization is established for the ascending d-asymptotic sequences in X that are not endowed with the d-Cauchy property.

(A) Remember that $(x_n; n \geq 0)$ is *d-asymptotic*, provided

$r_n := d(x_n, x_{n+1}) \to 0$ as $n \to \infty$.

Clearly, each d-Cauchy sequence is d-asymptotic too; the reciprocal is not in general true. So, the d-Cauchy criteria we are looking for are to be sought within the class of d-asymptotic sequences. Precisely, suppose that

(for each ascending d-asymptotic sequence (x_n)):
(x_n) is not d-Cauchy implies (x_n) has the property (π).

Then, the d-Cauchy criterion in question writes

(for each ascending d-asymptotic sequence (x_n)):
(x_n) is not endowed with the property (π) implies (x_n) is d-Cauchy.

To get examples of such properties, we need a simplified form of a previous convention. Given the d-asymptotic sequence $(x_n; n \geq 0)$ and $\varepsilon > 0$, define

$i \in N$ is *ε-regular*, provided $[i \leq n$ implies $d(x_n, x_{n+1}) < \varepsilon]$.

The class $\mathcal{Z}(\varepsilon)$ of all these numbers is nonempty; so that

$(\forall \varepsilon > 0)$: $Z(\varepsilon) = \min \mathcal{Z}(\varepsilon)$ is well defined, as an element of N;
with, in addition: $d(x_n, x_{n+1}) < \varepsilon$, for all $n \geq Z(\varepsilon)$.

Define the subsets of $N^2 := N \times N$

$$(\leq; N) = \{(m,n) \in N^2; m \leq n\},\ (<; N) = \{(m,n) \in N^2; m < n\};$$

these are the graph over N of the relations $(\leq)$ and $(<)$, respectively. Further, let us say that the subset Θ of R^0_+ is $(>)$-*cofinal* in R^0_+, when

for each $\varepsilon > 0$, there exists $\theta \in \Theta$ with $\varepsilon > \theta$;

note that, in this case, the following hereditary property holds

$\beta \in \Theta$ implies $\Theta(\beta, >)$ is $(>)$-cofinal in R^0_+.

Given this subset, define (as before)

$$(\geq; \Theta) = \{(\beta, \gamma) \in \Theta \times \Theta; \beta \geq \gamma\},\ (>; \Theta) = \{(\beta, \gamma) \in \Theta \times \Theta; \beta > \gamma\};$$

these are just the graph over Θ of the relations $(\geq)$ and $(>)$, respectively. In addition, for each $(\beta, \gamma) \in (>; \Theta)$, denote

$$B(\beta, \gamma) = \{n \in N(2, \leq); 3^{-n} < \min\{\beta - \gamma, \gamma\};$$

clearly, $B(\beta, \gamma)$ is nonempty and hereditary:

$i \in B(\beta, \gamma)$ and $i \leq j$ imply $j \in B(\beta, \gamma)$.

In this case, it makes sense denoting

$$b(\beta, \gamma) = \min B(\beta, \gamma),\ (\beta, \gamma) \in (>; \Theta).$$

If no confusion arise, we write $b(\beta, \gamma)$ as b, for simplicity; hence, by definition

($b \geq 2$ and) for each $k \in N$, we have $3^{-b-k} < \min\{\beta - \gamma, \gamma\}$.

From the perspective of our initial problem, the points $\beta \in \Theta$ appearing here must be chosen according to some extra conditions. Let (x_n) be a sequence in X. Define (with Θ as before)

(d-1) $\beta \in \Theta$ is (x_n)-*admissible*, when
$E(j; \beta) := \{(m,n) \in (<; N); j \leq m < n, d(x_m, x_n) > \beta\} \neq \emptyset$, $\forall j$
(d-2) $\text{adm}(\Theta; (x_n))$= the class of all these points $\beta \in \Theta$.

Proposition 1.18. *Let (x_n) be a sequence in X that is not d-Cauchy; and Θ be $(>)$-cofinal in R^0_+. Then, the following are valid*

(31-1) $\text{adm}(\Theta; (x_n))$ *is nonempty*
(31-2) $\beta \in \text{adm}(\Theta; (x_n))$ *implies* $\Theta(\beta, \geq) \subseteq \text{adm}(\Theta; (x_n))$*;*
so that, $\text{adm}(\Theta; (x_n))$ *is* $(>)$*-cofinal in* R_+^0
(31-3) for each $\beta \in \text{adm}(\Theta; (x_n))$ *and each* $\gamma \in \Theta$*, we have*
$(\beta, \gamma) \in (>; \Theta)$ *iff* $(\beta, \gamma) \in (>; \text{adm}(\Theta; (x_n)))$.

Proof. i): By definition, the d-Cauchy property of our sequence writes:

$$\forall \varepsilon \in R_+^0, \exists a \in N, \forall (m, n) \in (<; N): a \leq m < n \Longrightarrow d(x_m, x_n) \leq \varepsilon.$$

As Θ is a $(>)$-cofinal part in R_+^0, this property may be also written as

$$\forall \theta \in \Theta, \exists \alpha \in N, \forall (m, n) \in (<; N): \alpha \leq m < n \Longrightarrow d(x_m, x_n) \leq \theta.$$

The negation of our property means: there exists $\beta \in \Theta$ such that

$$E(j; \beta) := \{(m, n) \in (<; N); j \leq m < n, d(x_m, x_n) > \beta\} \neq \emptyset, \forall j.$$

But then, $\beta \in \text{adm}(\Theta; (x_n))$; and the first conclusion follows.

ii): Further, as

$$(\forall j): \beta_1 \geq \beta_2 \text{ implies } E(j; \beta_1) \subseteq E(j; \beta_2),$$

it is clear that the first half of our second conclusion holds; wherefrom (by a previous observation) the second part of underlying conclusion holds too.

iii): Evident, by the preceding stage. □

Finally, the following conventions are needed. For each sequence of ranks $(\lambda(k); k \geq 0)$ in $N(1, \leq)$, define the property

$(\lambda(k); k \geq 0)$ is *strictly ascending*: $i < j$ implies $\lambda(i) < \lambda(j)$;
hence, $(\lambda(k))$ is divergent ($\lim_k \lambda(k) = \infty$).

Then, for each sequence (r_n) in R and each point $r \in R$, let us write

$r_n \to r+$ (or: $\lim_n(r_n) = r+$) if $r_n \to r$ and ($\forall n$, $r_n > r$)
$r_n \to r++$ (or: $\lim_n(r_n) = r++$) if $r_n \to r$ and ($\forall\forall n$, $r_n > r$).

Here, given a property $\pi(k)$ depending on $k \in N$, let us say that it holds for *nearly all* k [written: ($\forall\forall k$, $\pi(k)$)], provided

the set $\{k \in N; \pi(k)\}$ includes a final interval of N:
there exists $c = c(\pi) \in N$ such that ($\pi(k)$ is true, for all $k \geq c$).

The following result, referred to as *Boyd-Wong non-Cauchy Criterion* (in short: (BW-n-CC)) is now available.

Theorem 1.7. *Let the ascending sequence $(x_n; n \geq 0)$ in X be such that*

(31-i) (x_n) *is d-asymptotic* $(r_n := d(x_n, x_{n+1}) \to 0$ *as* $n \to \infty)$
(31-ii) $(x_n; n \geq 0)$ *is not d-Cauchy.*

Further, let the $(>)$*-cofinal subset* Θ *of* R^0_+*, the* (x_n)*-admissible couple* (β, γ) *in* $(>; \mathrm{adm}(\Theta; (x_n)))$ *(with the associated rank* $b = b(\beta, \gamma)$*), and the strictly ascending rank sequence* $(\lambda(k); k \geq 0)$ *in* $N(1, \leq)$ *be (arbitrary) fixed. There exists then a rank sequence* $(J(k); k \geq 0)$ *in* $N(1, \leq)$ *and a couple of rank sequences* $(m(k); k \geq 0)$ *and* $(n(k); k \geq 0)$ *in* $N(1, \leq)$*, with*

(31-a) $k + 1 \leq J(k) \leq m(k) < m(k) + 3\lambda(k) < n(k)$*, and*
$x_{m(k)} \leq x_{n(k)}$*, for all* k
(31-b) $J(k) \leq m(k) < m(k) + 2\lambda(k) < n(k) - 1 < n(k)$*, and*
$d(x_{m(k)}, x_{n(k)}) > \gamma$*,* $d(x_{m(k)}, x_{n(k)-1}) \leq 3^{-b-k} + \gamma$*,* $\forall k$
(31-c) $U_k := d(x_{m(k)}, x_{n(k)}) \to \gamma+$ *as* $k \to \infty$
(31-d) for each couple of rank sequences $(\mu(k); k \geq 0)$ *and* $(\nu(k); k \geq 0)$ *in* N *with* $(\mu(k), \nu(k) \leq 3\lambda(k), \forall k)$*, we have that*
$V_k := d(x_{m(k)+\mu(k)}, x_{n(k)+\nu(k)}) \to \gamma+$ *as* $k \to \infty$
(31-e) for each couple of ranks (i, j) *in* $N \times N$ *with* $(i, j \leq 3\lambda(0))$*,*
$S_k := d(x_{m(k)+i}, x_{n(k)+j}) \to \gamma+$ *as* $k \to \infty$
(31-f) for each couple of ranks (i, j) *in* $N \times N$*, one has*
$T_k := d(x_{m(k)+i}, x_{n(k)+j}) \to \gamma++$ *as* $k \to \infty$*.*

Proof. Let the $(>)$-cofinal (in R^0_+) subset Θ of R^0_+, and the strictly ascending rank sequence $(\lambda(k); k \geq 0)$ in $N(1, \leq)$ be given. As (x_n) is not d-Cauchy, the relation $(>; \mathrm{adm}(\Theta; (x_n)))$ is nonempty (cf. a previous observation). Let (β, γ) be an element of it (with the associated rank $b = b(\beta, \gamma) \geq 2$); note that, as a consequence of $\beta \in \mathrm{adm}(\Theta; (x_n))$, we have

$$E(j) := \{(m, n) \in (<; N); j \leq m < n, d(x_m, x_n) > \beta\} \neq \emptyset, \forall j.$$

With these data, define the rank sequence $(J(k); k \geq 0)$ in $N(1, \leq)$, as

$$(J(k) = 1 + k + Z(3^{-2b-k}/\lambda(k)); k \geq 0).$$

Then, define the sequence of nonempty relations over N

$(A(k) = E(J(k)); k \geq 0)$; hence, by definition,
$A(k) := \{(m, n) \in (<; N); J(k) \leq m < n, d(x_m, x_n) > \beta\}$, $k \geq 0$.

By the triangle inequality (and the choice of b)

$(\forall k)$: $(m,n) \in A(k)$ implies $(m,n) \in (<;N)$, $J(k) \leq m < n$, and
$d(x_{m+s}, x_{n+t}) \geq d(x_m, x_n) - d(x_m, x_{m+s}) - d(x_n, x_{n+t}) >$
$\beta - 3^{-2b-k+1} - 3^{-2b-k+1} > \beta - 3^{-b-k} > \gamma$, $\forall s,t \in N[0, 3\lambda(k)]$;

which tells us that

$(\forall k)$: $B(k) := \{(m,n) \in (<;N); J(k) \leq m < n, d(x_{m+s}, x_{n+t}) > \gamma,$
$\forall s,t \in N[0, 3\lambda(k)]\}$ is a nonempty relation over N.

Having this precise, denote for each $k \geq 0$

$$m(k) = \min \mathrm{Dom}(B(k)),\ n(k) = \min B(k)(m(k)).$$

By this very definition, we get

(pro-1) $(\forall k)$: $k+1 \leq J(k) \leq m(k) < n(k)$; hence, $x_{m(k)} \leq x_{n(k)}$,
(pro-2) $(\forall k)$: $d(x_{m(k)+s}, x_{n(k)+t}) > \gamma$, $\forall s,t \in N[0, 3\lambda(k)]$.

We claim that the rank sequence $(J(k))$ and the couple of rank-sequences $[(m(k)), (n(k))]$ fulfill all conclusions in the statement.

i): By (pro-1), we get the first, second, and third part of (31-a).

ii): Concerning the fourth part, suppose by contradiction that

$$(m(k) <)n(k) \leq m(k) + 3\lambda(k), \text{ for some } k \geq 0.$$

By $m(k) \geq J(k) \geq Z(3^{-2b-k}/\lambda(k))$, and the triangle inequality,

$$d(x_{m(k)}, x_{n(k)}) \leq 3^{-2b-k+1} < 3^{-b-k} < \gamma; \text{ in contradiction with (pro-2)};$$

whence, the underlying part in (31-a) holds too.

iii): The first and second part of (31-b) are directly obtainable from the preceding stage and (pro-2), respectively. Concerning its third part, let $k \geq 0$ be arbitrary fixed. By definition, $n(k)$ is the minimum of all ranks $p \in N$ with

$(m(k), p) \in B(k)$; that is:
$J(k) \leq m(k) < p$ and $d(x_{m(k)+s}, x_{p+t}) > \gamma$, $\forall s,t \in N[0, 3\lambda(k)]$.

As $m(k) < m(k) + 2\lambda(k) < n(k) - 1$, we must have

(pro-3) $d(x_{m(k)+u}, x_{n(k)-1+v}) \leq \gamma$, for some $u,v \in N[0, 3\lambda(k)]$.

By $m(k) \geq J(k) \geq Z(3^{-2b-k}/\lambda(k))$, and the triangle inequality,

$$\begin{aligned}&d(x_{m(k)}, x_{n(k)-1}) \leq \\ &d(x_{m(k)}, x_{m(k)+u}) + d(x_{n(k)-1}, x_{n(k)-1+v}) + \\ &d(x_{m(k)+u}, x_{n(k)-1+v}) \leq 3^{-2b-k+1} + 3^{-2b-k+1} + \gamma \leq 3^{-b-k} + \gamma;\end{aligned}$$

proving that the third part of (31-b) follows.

iv): From these facts and triangular inequality,

$$\gamma < d(x_{m(k)}, x_{n(k)}) \le d(x_{m(k)}, x_{n(k)-1}) + r_{n(k)-1} \le \\ 3^{-b-k} + \gamma + r_{n(k)-1}, \text{ for all } k.$$

Passing to limit as $k \to \infty$ in this double inequality gives (31-c).

v): By the definition of $[(m(k)), (n(k))]$, $[(\mu(k)), (\nu(k))]$, and (pro-2),

$V_k > \gamma$, for all $k \ge 0$; so, the first half of (31-d) holds.

Moreover, from a metrical property of d, and the choice of b

$$|d(x_{m(k)}, x_{n(k)}) - d(x_{m(k)+\mu(k)}, x_{n(k)+\nu(k)})| \le \\ d(x_{m(k)}, x_{m(k)+\mu(k)}) + d(x_{n(k)}, x_{n(k)+\nu(k)}) \le \\ 3^{-2b-k+1} + 3^{-2b-k+1} < 3^{-2b-k+2} \le 3^{-b-k}, \text{ for all } k \ge 0.$$

Passing to limit as $k \to \infty$ in the relation between the extremal members and taking the preceding stage into account, gives the second half of (31-d).

vi): By the strict ascending property of $(\lambda(k); k \ge 0)$, the sequences $(\mu(k) = i; k \ge 0)$ and $(\nu(k) = j; k \ge 0)$, fulfill $(\mu(k), \nu(k) \le 3\lambda(k), \forall k)$; and this, along with the preceding stage, yields (31-e).

vii): Let $(i, j) \in N \times N$ be given. By the strict ascending property of the rank sequence $(\lambda(k); k \ge 0)$, there exists an index $L = L(i, j) \in N$ with: $i, j \le 3\lambda(k)$, for all $k \ge L$. Then, define the couple of sequences $(\mu(k); k \ge 0)$ and $(\nu(k); k \ge 0)$ as

$$(\mu(k) = \nu(k) = 0;\ k \le L), \text{ and } (\mu(k) = i,\ \nu(k) = j;\ k > L).$$

Clearly, $(\mu(k), \nu(k) \le 3\lambda(k), \forall k)$; and this, by the preceding stage, yields

$$T_k^* := d(x_{m(k)+\mu(k)}, x_{n(k)+\nu(k)}) \to \gamma+ \text{ as } k \to \infty.$$

Combining with $[T_k = T_k^*$, for all $k > L]$ yields (31-f). □

(B) Given the $(>)$-cofinal (in R^0_+) subset Θ of R^0_+ and the strictly ascending rank sequence $(\lambda(k); k \ge 0)$ in $N(1, \le)$, let us say that $[\gamma; (J(k)); (m(k)); (n(k))]$ is a *Boyd-Wong* $[\Theta; (\lambda(k))]$*-system* attached to the ascending sequence (x_n), provided $\gamma \in \Theta$ and

(BW-1) $(\forall k)$: $k + 1 \le J(k) \le m(k) < m(k) + 3\lambda(k) < n(k)$, and $x_{m(k)} \le x_{n(k)}$

(BW-2) for each couple of rank sequences $(\mu(k))$ and $(\nu(k))$ in N with $(\mu(k), \nu(k) \leq 3\lambda(k), \forall k)$, we have that
$V_k := d(x_{m(k)+\mu(k)}, x_{n(k)+\nu(k)}) \to \gamma+$ as $k \to \infty$
(BW-3) for each couple of ranks (i, j) in $N \times N$ with $(i, j \leq 3\lambda(0))$,
$S_k := d(x_{m(k)+i}, x_{n(k)+j}) \to \gamma+$ as $k \to \infty$
(BW-4) for each couple of ranks (i, j) in $N \times N$, one has
$T_k := d(x_{m(k)+i}, x_{n(k)+j}) \to \gamma++$ as $k \to \infty$.

From the result above, one then gets the following characterization of the d-Cauchy property for ascending d-asymptotic sequences.

Theorem 1.8. *Let the ascending sequence $(x_n; n \geq 0)$ in X be d-asymptotic. Then, the following conditions/properties are equivalent*

(32-a) $(x_n; n \geq 0)$ is not d-Cauchy
(32-b) for each $(>)$-cofinal subset Θ of R^0_+ and each strictly ascending rank sequence $(\lambda(k); k \geq 0)$ in $N(1, \leq)$, there is at least one Boyd-Wong $[\Theta, (\lambda(k))]$-system attached to (x_n).

Proof. There are two steps to be passed.

Step 1. Suppose that the first condition in this statement holds. As a consequence of it, we have (by a previous auxiliary fact) that $(>; \text{adm}(\Theta; (x_n)))$ is nonempty; let (β, γ) be an element of it. Given the data $[\Theta; (\beta, \gamma); (\lambda(k))]$, we can apply the Boyd-Wong non-Cauchy Criterion to them. From its conclusions, it is now clear that $[\gamma; (J(k)); (m(k)); (n(k))]$ is a Boyd-Wong $[\Theta; (\lambda(k))]$-system attached to (x_n); as desired.

Step 2. Suppose that the second condition in this statement holds. For the given $(>)$-cofinal (in R^0_+) subset Θ of R^0_+ and the strictly ascending rank sequence $(\lambda(k); k \geq 0)$ in $N(1, \leq)$, there must be one Boyd-Wong $[\Theta, (\lambda(k))]$-system $[\gamma; (J(k)); (m(k)); (n(k))]$ attached to (x_n), in the sense of (BW-1)–(BW-4); hence,

$$d(x_{m(k)}, x_{n(k)}) \to \gamma+, \text{ as } k \to \infty \text{ [in view of (BW-3)]}.$$

This necessarily gives us that $(x_n; n \geq 0)$ is not d-Cauchy; for otherwise–under a d-Cauchy condition upon our starting sequence–it follows that

$$d(x_{m(k)}, x_{n(k)}) \to 0, \text{ as } k \to \infty;$$

in contradiction with the limit property above. □

In particular, when $\Theta = R^0_+$ and $(\lambda(k) = 1 + k; k \geq 0)$, the obtained statement covers the 1969 one in Boyd and Wong [35]; so, it is natural that this result be referred to in the proposed way. Further aspects may be found in Reich [194]; see also Khan et al. [144].

(C) Returning to the general setting, we stress that Theorem 1.8 is a *characterization* of d-Cauchy property for an ascending d-asymptotic sequence; so – technically – it is difficult to be directly applicable in practice. As a consequence, it would be useful for us to formulate a lot of sufficient versions of this (characterization) result with a simpler handling. Some conventions are needed. Letting $(m(k))$ and $(n(k))$ be a couple of ranks in N, define the property

$(m(k))$ and $(n(k))$ are divergent and separated:
$\lim_k(m(k)) = \lim_k(n(k)) = \infty$ and $(m(k) < n(k), \forall k)$;

Then, given a property $\pi(k)$ depending on $k \in N$, we say that it holds for *many* k [written: $(\exists\exists k, \pi(k))$], provided

the set $\{k \in N; \pi(k)\}$ is unbounded: $\sup\{k \in N; \pi(k)\} = \infty$.

Denote, for simplicity reasons,

(den-1) Lim :: either of the operators lim sup or lim inf
(den-2) ∇ :: either of the operators $\forall\forall$ or $\exists\exists$.

The first result in this series [referred to as: *Boyd-Wong limiting Cauchy Criterion*; in short: (BW-limi-CC)] is to be formulated as below.

Theorem 1.9. *Let the ascending d-asymptotic sequence* $(x_n; n \geq 0)$ *in* X *be such that the limiting Boyd-Wong condition is holding*

(BW-limi) for each $(>)$*-cofinal subset* Θ *of* R^0_+*, each* $\gamma \in \Theta$*, and each index* $h \in N(1, \leq)$*, there exists a rank sequence* $(\mu(k); k \geq 0)$*, such that: for each couple of divergent separated rank sequences* $(m(k))$ *and* $(n(k))$*, fulfilling* $(\mathrm{Lim}_k d(x_{m(k)+p}, x_{n(k)+q}) \leq \gamma$*, for some* $p, q \in N[0, h])$*, one has* $(\nabla k,\ d(x_{m(k)+\mu(k)}, x_{n(k)+\mu(k)}) \leq \gamma)$.

Then, necessarily, the sequence $(x_n; n \geq 0)$ *is d-Cauchy.*

As a matter of fact, we have four results here, according to the possible choices of Lim and ∇. However, only two are important: the ones given by the couples $(\limsup, \forall\forall)$ and $(\liminf, \exists\exists)$.

The second result in this series [referred to as: *Boyd-Wong local Cauchy Criterion*; in short: (BW-loc-CC)] may be stated as below.

Theorem 1.10. *Let the ascending d-asymptotic sequence $(x_n; n \geq 0)$ in X be such that the local Boyd-Wong condition is being satisfied*

> *(BW-loc) for each $(>)$-cofinal subset Θ of R^0_+, each $\gamma \in \Theta$, and each index $h \in N(1, \leq)$, there exists a rank sequence $(\mu(k); k \geq 0)$ such that: for each couple of divergent separated rank sequences $(m(k))$ and $(n(k))$ with $(\lim_n d(x_{m(k)+p}, x_{n(k)+q}) = \gamma+$, for some $p, q \in N[0, h])$, we must have $(\nabla k,\ d(x_{m(k)+\mu(k)}, x_{n(k)+\mu(k)}) \leq \gamma)$.*

Then, necessarily, the sequence $(x_n; n \geq 0)$ is d-Cauchy.

Note that we have two results here, according to the possible choices of ∇. But, the $(\exists\exists)$ variant of it is weaker than the $(\forall\forall)$ variant of the same.

Technically speaking, these results are nothing else than particular cases of the Boyd-Wong non-Cauchy Criterion (BW-n-CC), as results from

Proposition 1.19. *Under the precise conventions, we have for each ascending d-asymptotic sequence (x_n) in X:*

> *(32-1) Boyd-Wong non-Cauchy Criterion (BW-n-CC) includes Boyd-Wong limiting Cauchy Criterion (BW-limi-CC)*
> *(32-2) Boyd-Wong limiting Cauchy Criterion (BW-limi-CC) includes Boyd-Wong local Cauchy Criterion (BW-loc-CC).*

Proof. There are two steps to be passed.

Step 1. [(BW-limi-CC) $\Longrightarrow$ (BW-loc CC)]

Let $[\Theta; \gamma; h]$ be given; and $(\mu(k))$ be the rank sequence indicated by (BW-limi-CC). Then, let the couple of rank sequences $[(m(k)); (n(k))]$ be as in the premise of (BW-loc-CC). By the trivial relation

> (for each pair $p, q \in N[0, h]$):
> $\lim_k d(x_{m(k)+p}, x_{n(k)+q}) = \gamma+$ implies $\mathrm{Lim}_k d(x_{m(k)+p}, x_{n(k)+q}) \leq \gamma$,

the premise of (BW-limi-CC) is acceptable; wherefrom, by its conclusion, $(\nabla k,\ d(x_{m(k)+\mu(k)}, x_{n(k)+\mu(k)}) \leq \gamma)$; as desired.

Step 2. [(BW-n-CC) $\Longrightarrow$ (BW-limi CC)]

Let the ascending d-asymptotic sequence (x_n) in X be such that (BW-limi) is holding; but [contrary to the conclusion] (x_n) is not d-Cauchy.

(I) Fix the $(>)$-cofinal subset Θ of R^0_+. As already precise, the relation $(>; \mathrm{adm}(\Theta; (x_n)))$ is not empty; let (β, γ) be an element of it. Then, take some index $h \in N(1, \leq)$. Starting from $[\Theta; \gamma; h]$, we have, by the admitted hypothesis, that there exists a rank sequence $(\mu(k); k \geq 0)$ such that the implication of (BW-limi) is holding.

(II) Further, starting from the rank sequence $(\mu(k); k \geq 0)$, define

$(\nu(k) = \max\{\mu(0), ..., \mu(k)\}; k \geq 0)$, $(\lambda(k) = h + k + \nu(k); k \geq 0)$;
hence, $(\lambda(k); k \geq 0)$ is strictly ascending in $N(h, \leq) \subseteq N(1, \leq)$.

(III) Given the data $[\Theta; (\beta, \gamma); (\lambda(k))]$, let us apply them the Boyd-Wong non-Cauchy Criterion (BW-n-CC). It then follows that there exist a rank sequence $(J(k); k \geq 0)$ in $N(1, \leq)$ and a couple of rank sequences $(m(k))$ and $(n(k))$ in $N(1, \leq)$, so that

(BW-a) $k + 1 \leq J(k) \leq m(k) < m(k) + 3\lambda(k) < n(k)$, $\forall k$;
hence, $(m(k))$ and $(n(k))$ are divergent, separated
(BW-b) $\lim_k d(x_{m(k)+p}, x_{n(k)+q}) = \gamma+$, $\forall p, q \in N[0, 3h]$
(in view of $p, q \leq 3h \leq 3\lambda(0)$); hence, in particular,
$(\mathrm{Lim}_k d(x_{m(k)+p}, x_{n(k)+q}) \leq \gamma$, $\forall p, q \in N[0, h])$
(BW-c) $\lim_k d(x_{m(k)+\mu(k)}, x_{n(k)+\mu(k)}) = \gamma+$
(in view of $(\mu(k) \leq \nu(k) \leq \lambda(k) \leq 3\lambda(k)$, $\forall k)$); hence, in particular,
$(d(x_{m(k)+\mu(k)}, x_{n(k)+\mu(k)}) > \gamma$, $\forall k)$.

(IV) Finally, by the very choice of the rank sequence $(\mu(k); k \geq 0)$, we have in particular [by the limiting Boyd-Wong condition (BW-limi)]

(BW-lim) for each couple of rank sequences $(m(k))$ and $(n(k))$
taken as in (BW-b) (the last part) we have
$(\nabla k$, $d(x_{m(k)+\mu(k)}, x_{n(k)+\mu(k)}) \leq \gamma)$.

This, however, is impossible in view of (BW-c) (the last part). Hence, our working assumption is not acceptable; and the conclusion follows. □

Remark 1.3.
An open problem to be posed is that of

(for each ascending d-asymptotic sequence): Boyd-Wong local Cauchy Criterion implies Boyd-Wong non-Cauchy Criterion.

Note that, if this were true, one must have (by the inclusions above)

(for each ascending d-asymptotic sequence):
(BW-n-CC) $\Longrightarrow$ (BW-limi-CC) $\Longrightarrow$ (BW-loc-CC) $\Longrightarrow$ (BW-n-CC);
hence, these results would be mutually equivalent.

We conjecture that the answer is negative. A motivation of this is related to the Boyd-Wong non-Cauchy Criterion (BW-n-CC) being a characterization result for the underlying class of sequences.

An interesting particular case of Boyd-Wong limiting Cauchy criterion (BW-limi-CC) was obtained by Abtahi [4]; and referred to as: *Abtahi Cauchy Criterion*; in short: (A-CC)]:

Theorem 1.11. *Let the ascending d-asymptotic sequence (x_n) in X fulfill*

(Abt-cond) for each $\varepsilon > 0$, there exists a sequence $(\nu(k))$ of ranks in N, such that, for each couple of rank sequences $(m(k))$ and $(n(k))$: $(\limsup_k d(x_{m(k)}, x_{n(k)}) \leq \varepsilon) \Longrightarrow (\forall\forall k,\ d(x_{m(k)+\nu(k)}, x_{n(k)+\nu(k)}) \leq \varepsilon)$.

Then, necessarily, $(x_n; n \geq 0)$ is d-Cauchy.

Clearly, (A-CC) is deductible from (BW-limi-CC), under

$\Theta = R_+^0$, $h = 1$, $(\mu(k) = \nu(k); k \geq 0)$, $p = q = 0$; as well as (in addition) (Lim, ∇) is taken as identical with $(\limsup, \forall\forall)$.

Moreover, according to author's developments in the quoted paper and the ones in Abtahi et al. [5], (A-CC) is a powerful instrument in deriving various results in metrical fixed point theory; hence, the same conclusion is valid for (BW-limi-CC). Putting these together, it seems that Boyd-Wong local Cauchy Criterion (BW-loc-CC) is somehow useless with respect to the (stronger) Boyd-Wong limiting Cauchy criterion (BW-limi-CC) – especially by the particular variant of it described via (A-CC). However, the situation is opposite: in many situations, (BW-limi-CC) [or (A-CC)] does not work while (BW-loc-CC) does; we will clarify these aspects elsewhere.

1.2.4 *Equivalence criteria*

Let $\chi : R_+ \to R_+$ be a function; we call it *normal*, provided

(nor-1) χ is *reflexive sufficient* ($\chi(t) = 0$ iff $t = 0$)
(nor-2) χ is *semi-regressive* ($\chi(t) \leq t$, for each $t \in R_+$)
(nor-3) χ is *increasing* and *subadditive*
($\chi(t_1 + t_2) \leq \chi(t_1) + \chi(t_2)$, $\forall t_1, t_2 \in R_+$).

Some compatible type properties of such functions are concentrated in

Proposition 1.20. *Suppose that $\chi : R_+ \to R_+$ is normal. Then, the following are true, in the reduced system (ZF-AC+DC)*

(41-1) χ is zero-compatible:
for each $\varepsilon > 0$, there exists $\delta > 0$, such that $\chi(t) < \delta$ implies $t < \varepsilon$
(41-2) χ is sequentially zero-compatible: $\chi(t_n) \to 0$ iff $t_n \to 0$.

Proof. i): Suppose that this assertion is not true: there exists $\varepsilon > 0$, with

for each $\delta > 0$, $H(\delta) := \{t \geq \varepsilon; \chi(t) < \delta\}$ is nonempty.

Let $(\delta_n; n \geq 0)$ be a strictly descending sequence in R_+^0 with $\delta_n \to 0$. (For example, we may take $(\delta_n = 2^{-n}; n \geq 0)$; clearly, this is not obtained by a sequential choice). From the Denumerable Axiom of Choice (deductible in (ZF-AC+DC)), there exists a sequence (t_n) in R_+ with

$(\forall n)$: $t_n \in H(\delta_n)$; that is: $t_n \geq \varepsilon$, and $\chi(t_n) < \delta_n$.

By the increasing property of χ, we get

$0 < \chi(\varepsilon) < \delta_n$, $\forall n$; whence, $0 < \chi(\varepsilon) \leq 0$;

a contradiction. Therefore, our working assumption is not acceptable; and the conclusion follows.

ii): Clearly, $t_n \to 0$ implies $\chi(t_n) \to 0$, by the semi-regressive property. So, it remains to establish that $\chi(t_n) \to 0$ implies $t_n \to 0$. Let $\varepsilon > 0$ be arbitrary fixed; and $\delta > 0$ be given by the zero-compatible property of χ. As $\chi(t_n) \to 0$, there exists an index $n(\delta)$ such that (by the quoted property)

$n \geq n(\delta)$ implies $\chi(t_n) < \delta$; hence, $t_n < \varepsilon$.

As $\varepsilon > 0$ was arbitrarily chosen, this yields $t_n \to 0$; as claimed. □

As an application of these facts we may get equivalence properties between the conv-Cauchy structures attached to certain pseudometrics.

Let X be a nonempty set. Remember that the pseudometric $d : X \times X \to R_+$ is a *semimetric* on X, provided [d is reflexive, triangular, symmetric]; in this case, (X, d) will be referred to as a *semimetric space*. Fix such an object; and let $\chi : R_+ \to R_+$ be a function. For the moment,

(chi-d) $e(x, y) = \chi(d(x, y))$, $x, y \in X$

defines a pseudometric over X. Further properties of this object are available under the normality properties we just discussed. Precisely, we have

Proposition 1.21. *Let $\chi : R_+ \to R_+$ be normal. Then, in (ZF-AC+DC),*

(42-1) *the associated pseudometric $e(.,.)$ is a semimetric on X*

(42-2) *the conv-Cauchy structures attached to d and e are equivalent, in the sense: for each sequence (x_n) in X and each $x \in X$,*

(conv-eq) $x_n \xrightarrow{d} x$ *iff* $x_n \xrightarrow{e} x$

(C-eq) (x_n) *is d-Cauchy iff* (x_n) *is e-Cauchy.*

Proof. i): Evident.

ii): The equivalence of convergence structures is clear, from the sequential zero-compatible property of χ, by simply noting that

$$x_n \stackrel{d}{\longrightarrow} x \text{ means } d(x_n, x) \to 0 \text{ and } x_n \stackrel{e}{\longrightarrow} x \text{ means } \chi(d(x_n, x)) \to 0;$$

so, it remains to establish the equivalence of Cauchy structures. The left to right implication is trivial, via $e \leq d$. To establish the right to left implication, suppose that (x_n) is e-Cauchy. Let $\varepsilon > 0$ be arbitrary fixed; and $\delta > 0$ be given by the zero-compatible property of χ. From the e-Cauchy property, there must be some index $n(\delta)$ such that

$n(\delta) \leq n \leq m$ implies $e(x_n, x_m) = \chi(d(x_n, x_m)) < \delta$;
whence (by the underlying property) $d(x_n, x_m) < \varepsilon$;

and from this, we are done. □

Now, a basic example of normal function is

$(\chi_1 : R_+ \to R_+)$: $\chi_1(t) = \alpha t$, $t \in R_+$, for some $\alpha \in]0, 1]$.

This is a linear function. For a nonlinear example of this type, define

$(\chi_2 : R_+ \to R_+)$: $\chi_2(t) = t/(1 + t)$, $t \in R_+$.

It is not hard to see that $\chi_2(.)$ is a normal function. Some other examples of this type are available by starting from the observation

if $\xi : R_+ \to R_+$ and $\eta : R_+ \to R_+$ are normal, then $\xi \circ \eta$ is normal too.

In particular, if χ_1 and χ_2 are the normal functions above, then

$\chi_1^n \circ \chi_2^m$ is normal, for each $n, m \in N$.

Concerning the introduced concept, the following problem is sometimes useful in applications. Denote, for simplicity

$\mathcal{F}_0(in, re)(R_+) =$ the class of all $\omega \in \mathcal{F}(R_+)$ with
$\omega(0) = 0$, $\omega =$ increasing and regressive ($\omega(t) < t$, $\forall t \in R_+^0$).

Fix some function $\omega \in \mathcal{F}_0(in, re)(R_+)$. Usually, the normal functions χ to be considered are of the form

$\chi(t) = t - \omega(t)$, $t \in R_+$ (the *complement* of ω);

we then say that ω is *complementary normal.* In this case, we may ask of which properties has the function ω. An appropriate answer is given by

Proposition 1.22. *Suppose that* $\omega \in \mathcal{F}_0(in, re)(R_+)$ *is complementary normal [in the sense:* $(\chi(t) = t - \omega(t); t \geq 0)$ *is normal]. Then,*

ω is Matkowski admissible: $\lim_n \omega^n(t) = 0$, $\forall t \in R_+$.

Proof. Given $t_0 \in R_+$, consider the iterative sequence $(t_{n+1} = \omega(t_n); n \geq 0)$; clearly, (t_n) is descending. If $t_k = 0$ for some k, then $(t_n = 0, \forall n \geq k)$; and all is clear. Suppose now that $(t_n > 0, \forall n)$; then, (t_n) is a strictly descending sequence in R_+^0. From the iterative relation,

$(\forall n)$: $t_{n+1} = t_n - \chi(t_n)$; so that, $\chi(t_n) = t_n - t_{n+1}$.

This yields

$\sum_n \chi(t_n) < \infty$; whence, $\chi(t_n) \to 0$ as $n \to \infty$.

Combining with a previous result, gives $\lim_n t_n = 0$; as desired. □

Remark 1.4.

The reciprocal of this inclusion is not in general true; that is

if $\omega \in \mathcal{F}_0(in, re)(R_+)$ is Matkowski admissible, then
ω is not complementary normal.

A motivation of this bad property is due to the subadditive property of the complementary function $(\chi(t) = t - \omega(t); t \geq 0)$ being not in general assured under the described setting.

Finally, note that all examples above are illustrations of these results. Further aspects will be delineated elsewhere.

1.2.5 *Convergent real sequences*

In the following, the convergence of real sequences is discussed.

Let R stand for the real numbers set. Remember that one may introduce an addition $[(x, y) \mapsto x + y]$, multiplication $[(x, y) \mapsto xy]$, and (total) ordering structure $[x \leq y$ iff $y - x \geq 0]$ on R by successive extensions through Z (the integer numbers set), and Q (the rational numbers set). This, among others, allow us to define the *modulus* function $|.|$ and the associated map $d(.,.)$ over R, as

$(|x| = \max\{x, -x\}; x \in R)$; $(d(x, y) = |x - y|; x, y \in R)$.

By the well-known modulus properties, $d(.,.)$ is reflexive sufficient, triangular and symmetric; hence, a *metric* (referred to as: the *usual metric* of R). In terms of this metric, the convergence of a sequence (r_n) in R towards some point $r \in R$ means

$$\forall \varepsilon > 0,\ \exists n(\varepsilon) \in N\text{: } n \geq n(\varepsilon) \text{ implies } r - \varepsilon < r_n < r + \varepsilon.$$

In addition to this, we have the possibility of defining an *unbounded convergence* as follows: given the sequence (r_n), let us say that

$$r_n \to -\infty \text{ when, } \forall \alpha \in R,\ \exists n(\alpha) \in N\text{: } n \geq n(\alpha) \text{ implies } r_n < \alpha$$
$$r_n \to \infty \text{ when, } \forall \beta \in R,\ \exists n(\beta) \in N\text{: } n \geq n(\beta) \text{ implies } r_n > \beta.$$

Then, define an *extended convergence* property for a real sequence (r_n) as:

(r_n) is *extended convergent* (in short: *ext-convergent*), when
$r_n \to r$ for some $r \in R$ or $r_n \to s$, for some $s \in \{-\infty, \infty\}$.

The auxiliary fact below is immediate; so, we do not give details.

Proposition 1.23. *Let (r_n) be a real sequence. Then,*

(51-1) $r_n \to \sup\{r_n; n \geq 0\}$ *if (r_n) is ascending ($i \leq j \Longrightarrow r_i \leq r_j$)*
(51-2) $r_n \to \inf\{r_n; n \geq 0\}$ *if (r_n) is descending ($i \leq j \Longrightarrow r_i \geq r_j$).*

Having these precise, define for a sequence (r_n) in R

$$\liminf_n(r_n) = \sup_m \Phi_*(m), \text{ where } (\Phi_*(m) = \inf\{r_i; i \geq m\}; m \geq 0),$$
$$\limsup_n(r_n) = \inf_m \Phi^*(m), \text{ where } (\Phi^*(m) = \sup\{r_j; j \geq m\}; m \geq 0).$$

Clearly, $(\Phi_*(m); m \geq 0)$ = ascending, $(\Phi^*(m); m \geq 0)$ = descending, and $[\Phi_*(m) \leq \Phi^*(m), \forall m]$; wherefrom $\Phi_*(p) \leq \Phi^*(q)$, $\forall (p, q)$; so that,

(pro-1) $\liminf_n(r_n) \leq \limsup_n(r_n)$, for each sequence (r_n).

A natural question to be posed is what happens when equality occurs. Given the property $\pi(n)$ depending on $n \in N$, we say that it holds for *almost all* $n \in N$ [written: $(\forall\forall n, \pi(n))$], if

$$\exists h = h(\pi) \in N, \text{ such that: } \pi(n) \text{ holds, for all } n \geq h.$$

The appropriate answer to our question is contained in

Theorem 1.12. *Let (r_n) be a sequence in R. Then,*

(51-a) *if* $r_n \to r$*, for some* $r \in R \cup \{-\infty, \infty\}$*, then* $\liminf_n(r_n) = \limsup_n(r_n) = r$.

(51-b) *conversely, if* $\liminf_n(r_n) = \limsup_n(r_n) = r$*, for some* $r \in R \cup \{-\infty, \infty\}$*, then* $r_n \to r$.

Proof. There are two parts to be passed.

Part 1. Suppose that $r_n \to r$, for some $r \in R \cup \{-\infty, \infty\}$; we want to establish that $\liminf_n(r_n) = \limsup_n(r_n) = r$. The case of $r \in \{-\infty, \infty\}$ is clear; so, we may assume that $r \in R$. Let $\alpha, \beta \in R$ be such that $\beta < r < \alpha$. By definition, $(\forall\forall n, \beta < r_n < \alpha)$; wherefrom (by a direct reasoning)

$$\beta \leq \liminf_n(r_n) \leq \limsup_n(r_n) \leq \alpha.$$

This, by the arbitrariness of (α, β) with the prescribed condition, gives $\liminf_n(r_n) = \limsup_n(r_n) = r$.

Part 2. Suppose that $\liminf_n(r_n) = \limsup_n(r_n) = r$ for some $r \in R \cup \{-\infty, \infty\}$; we claim that $r_n \to r$. As before, the case of $r \in \{-\infty, \infty\}$ is clear; so, assume that $r \in R$. Let $\alpha, \beta \in R$ be such that

$$\beta < r < \alpha; \text{ that is: } \beta < \liminf_n(r_n) = \limsup_n(r_n) < \alpha.$$

By the infimum and supremum definition, $(\forall\forall n, \beta < r_n < \alpha)$. This, by the arbitrariness of (α, β) with the prescribed condition, gives $r_n \to r$; and the claim follows. □

Returning to the general setting, note that for each sequence (r_n),

(pro-2) $\liminf_n r_n = -\limsup_n(-r_n)$, $\limsup_n r_n = -\liminf_n(-r_n)$.

The following additive type properties are useful in the sequel.

Theorem 1.13. *Let the above conventions be in use. Then,*

(52-a) *For each couple of sequences* (r_n) *and* (s_n) *in* R*,*

(52-a-1) $\limsup_n(r_n + s_n) \leq \limsup_n r_n + \limsup_n s_n$,
whenever the right member exists
(52-a-2) $\limsup_n(r_n + s_n) = \limsup_n r_n + \lim_n s_n$,
whenever (s_n) *is convergent (hence, the right member exists)*

(52-b) *For each couple of sequences* (r_n) *and* (s_n) *in* R_+^0*,*

(52-b-1) $\limsup_n(r_n s_n) \leq \limsup_n r_n \cdot \limsup_n s_n$,
whenever the right member exists

(52-b-2) $\limsup_n (r_n s_n) = \limsup_n r_n \cdot \lim_n s_n$,
whenever (s_n) *is convergent in* R^0_+ *(hence, the right member exists).*

Proof. i-1): Define the sequences in $R \cup \{\infty\}$

$$U_n = \sup_{j \geq n}(r_j + s_j),\ V_n = \sup_{j \geq n} r_j,\ W_n = \sup_{j \geq n} s_j,\ n \geq 0.$$

Clearly, (U_n), (V_n), (W_n) are descending, and $[U_n \leq V_n + W_n, \forall n]$. Passing to limit (= infimum) as $n \to \infty$, we are done.

i-2): Denote $\lambda := \lim_n s_n$; and assume that $\lambda \in R$. Let $\alpha, \beta \in R$ be such that $\alpha < \lambda < \beta$. By definition,

$$(\forall\forall n,\ \alpha \leq s_n \leq \beta);\ \text{so},\ (\forall\forall n,\ r_n + \alpha \leq r_n + s_n \leq r_n + \beta).$$

Passing to supremum beyond these ranks, gives $(\forall\forall n,\ V_n + \alpha \leq U_n \leq V_n + \beta)$; wherefrom, passing to limit as $n \to \infty$ and taking into account the conditional arbitrariness of (α, β), all is clear.

ii-1): Define the sequences in $R^0_+ \cup \{\infty\}$

$$U_n = \sup_{j \geq n}(r_j s_j),\ V_n = \sup_{j \geq n} r_j,\ W_n = \sup_{j \geq n} s_j,\ n \geq 0.$$

Clearly, (U_n), (V_n), (W_n) are descending, and $[U_n \leq V_n W_n, \forall n]$. Passing to limit (= infimum) as $n \to \infty$, we are done.

ii-2): Denote $\lambda := \lim_n s_n$; and assume that $\lambda \in R^0_+$. Let $\alpha, \beta \in R^0_+$ be such that $\alpha\lambda < \lambda < \beta\lambda$. By definition,

$$(\forall\forall n,\ \alpha\lambda \leq s_n \leq \beta\lambda);\ \text{so},\ (\forall\forall n,\ \alpha\lambda r_n \leq r_n s_n \leq \beta\lambda r_n).$$

Passing to supremum beyond these ranks, gives $(\forall\forall n,\ \alpha\lambda V_n \leq U_n \leq \beta\lambda V_n)$; wherefrom, passing to limit as $n \to \infty$ and taking into account the conditional arbitrariness of (α, β), all is clear. □

The following sub-sequential properties of these limits are holding.

Theorem 1.14. *Given the real sequence* (r_n), *let* $u := \liminf_n r_n$ *and* $v := \limsup_n r_n$ *stand for its extreme limits. Then,*

(53-a) there exists a subsequence $(s_n := r_{i(n)}; n \geq 0)$ *of* (r_n),
with $s_n \to u$ *as* $n \to \infty$
(53-b) there exists a subsequence $(t_n := r_{j(n)}; n \geq 0)$ *of* (r_n),
with $t_n \to v$ *as* $n \to \infty$.

Proof. As before, it will suffice verifying one of these parts, the second one say. Three alternatives occur.

Alter 1. Suppose that $v = \infty$; whence

$$\sup\{r_m; m > n\} = \sup\{r_m; m \geq n\} = \infty, \forall n.$$

Let $(\alpha_n; n \geq 0)$ be a strictly ascending sequence in R^0_+ with $\lim_n \alpha_n = \infty$; for example, $(\alpha_n = 2^n; n \geq 0)$. For each n, there exists $m > n$ with $r_m > \alpha_n$. The selfmap $P(.)$ of N, defined as

$$P(n) = \min\{m \in N(n, <); r_m > \alpha_n\}, \ n \geq 0$$

is therefore well constructed, with

$$n < P(n), \ \alpha_n < r_{P(n)}, \text{ for all } n.$$

Denoting $(i(n) = P^n(0); n \geq 0)$, we therefore have

$$i(n) < i(n+1), \ \alpha_{i(n)} < r_{i(n+1)}, \ \forall n.$$

The subsequence $(t_n := r_{i(n)}; n \geq 0)$ is well defined and $t_n \to \infty$ as $n \to \infty$.

Alter 2. Suppose that $v = -\infty$; we then have (by our notations)

$$-\infty \leq u \leq v = -\infty; \text{ so that } r_n \to -\infty \text{ as } n \to \infty.$$

The subsequence of (r_n) to be determined is just (r_n).

Alter 3. Suppose that $-\infty < v < \infty$; whence

$$-\infty < \sup\{r_m; m > n\} \leq \sup\{r_m; m \geq n\} < \infty, \forall n.$$

Let $(\beta_n; n \geq 0)$ be a strictly descending sequence in R^0_+ with $\lim_n \beta_n = 0$. For each n, there exists $p > n$ with

$$\sup\{r_k; k \geq p\} < v + \beta_n; \text{ as well as (by definition)}$$
$$\sup\{r_k; k \geq p\} \geq v > v - \beta_n.$$

As a consequence of these relations, we have for each n

there exists $q \geq p$ (hence, $q > n$) with $v - \beta_n < r_q < v + \beta_n$.

The selfmap $Q(.)$ of N, defined as

$$Q(n) = \min\{m \in N(n, <); v - \beta_n < r_m < v + \beta_n\}, \ n \geq 0$$

is therefore well constructed, with

$$n < Q(n), \ v - \beta_n < r_{Q(n)} < v + \beta_n, \text{ for all } n.$$

Denoting $(i(n) = Q^n(0); n \geq 0)$, we therefore have

$$i(n) < i(n+1), \ v - \beta_{i(n)} < r_{i(n+1)} < v + \beta_{i(n)}, \ \forall n.$$

This, along with

$\lim_n(v - \beta_{i(n)}) = \lim_n(v + \beta_{i(n)}) = v$,

tells us that the subsequence $(t_n := r_{i(n)}; n \geq 0)$ is well defined and $t_n \to v$ as $n \to \infty$. The proof is complete. □

The following relative type statement (referred to as: Cesaro-Stolz theorem; in short: (CS-th)) is useful in practice.

Theorem 1.15. *Let the real sequences (r_n) and (s_n) be such that*

(s_n) is strictly positive, strictly ascending, and $s_n \to \infty$.

Then, the following conclusions hold

(54-a) $\limsup_n(r_n/s_n) \leq \limsup_n(r_{n+1} - r_n)/(s_{n+1} - s_n)$,
$\liminf_n(r_n/s_n) \geq \liminf_n(r_{n+1} - r_n)/(s_{n+1} - s_n)$
(54-b) $\lim_n(r_n/s_n) = \lim_n(r_{n+1} - r_n)/(s_{n+1} - s_n)$,
whenever the right member limit exists.

Proof. It will suffice proving the first part; and this, in turn, is reduced to its first half (because its second half is obtainable by passing to the sequence $(-r_n)$). Denote $L := \limsup_n(r_{n+1} - r_n)/(s_{n+1} - s_n)$. The case of $L = \infty$ is clear; so, without loss, one may assume that $L < \infty$. Let $\lambda > L$ be arbitrary fixed. By definition, there must be some $i = i(\lambda) \in N$, such that

$r_{k+1} - r_k < \lambda(s_{k+1} - s_k)$, $\forall k \geq i$.

Adding these inequalities, we derive

$r_{i+m} - r_i < \lambda(s_{i+m} - s_i)$, for all $m \geq 1$;
so that
$r_{i+m}/s_{i+m} \leq \lambda + (r_i - \lambda s_i)/s_{i+m}$, for all $m \geq 1$.

Passing to superior limit as $m \to \infty$, gives

$(\limsup_n(r_n/s_n) =)\ \limsup_m r_{i+m}/s_{i+m} \leq \lambda$;

and this, along with the arbitrariness of $\lambda > L$, gives the desired fact. □

Remember that, for each sequence (r_n) in R, and each element $r \in R$,

$r_n \to r+$ $(r_n \to r-)$, when $r_n \to r$ and $[r_n > r$ $(r_n < r)$, $\forall n]$.

Further, by a *strict subsequence* of (r_n), we mean any subsequence $(s_n = r_{i(n)}; n \geq 0)$ of (r_n) with $(i(n))$ being strictly ascending (hence, divergent).

Proposition 1.24. *Let the sequence (r_n) in R and the number $\varepsilon \in R$ be such that $r_n \to \varepsilon+$. Then, there exists a strict subsequence $(r_n^* := r_{i(n)}; n \geq 0)$ of (r_n), with the properties*

$(r_n^*; n \geq 0)$ *is strictly descending and* $r_n^* \to \varepsilon+$.

Proof. Put $i(0) = 0$. As $\varepsilon < r_{i(0)}$ and $r_n \to \varepsilon+$, we have that

$A(i(0)) := \{n > i(0); r_n < r_{i(0)}\}$ is not empty;
hence, $i(1) := \min(A(i(0)))$ is an element of it, and $r_{i(1)} < r_{i(0)}$.

Likewise, as $\varepsilon < r_{i(1)}$ and $r_n \to \varepsilon+$, we have that

$A(i(1)) := \{n > i(1); r_n < r_{i(1)}\}$ is not empty;
hence, $i(2) := \min(A(i(1)))$ is an element of it, and $r_{i(2)} < r_{i(1)}$.

This procedure may continue indefinitely; and yields (without any choice technique) a strictly ascending rank sequence $(i(n); n \geq 0)$ (hence, $i(n) \to \infty$ as $n \to \infty$) for which the attached strict subsequence $(r_n^* := r_{i(n)}; n \geq 0)$ of $(r_n; n \geq 0)$ fulfills

$r_{n+1}^* < r_n^*$, for all n; hence, (r_n^*) is (strictly) descending.

On the other hand, by this very subsequence property,

$(r_n^* > \varepsilon, \forall n)$, and $\lim_n r_n^* = \lim_n r_n = \varepsilon$.

Putting these together, we get the desired conclusion. □

A bi-dimensional counterpart of these facts may be given along the lines below. Let $\pi(t, s)$ (where $t, s \in R$) be a logical property involving pairs or real numbers. Given the couple of real sequences $(t_n; n \geq 0)$ and $(s_n; n \geq 0)$, call the strict subsequences $(t_n^*; n \geq 0)$ of (t_n) and $(s_n^*; n \geq 0)$ of (s_n), *compatible* when

$(t_n^* = t_{i(n)} n \geq 0)$, and $(s_n^* = s_{i(n)}; n \geq 0)$,
for the same strictly ascending rank sequence $(i(n); n \geq 0)$.

Proposition 1.25. *Let the couple of real sequences $(t_n; n \geq 0)$, $(s_n; n \geq 0)$, and the couple of real numbers (a, b) be such that*

$(t_n \to a+,\ s_n \to b+$ as $n \to \infty)$ and $(\pi(t_n, s_n)$ is true, $\forall n)$.

There exists then a couple of strict subsequences $(t_n^*; n \geq 0)$ and $(s_n^*; n \geq 0)$ of $(t_n; n \geq 0)$ and $(s_n; n \geq 0)$ respectively, with

(53-1) $(t_n^*; n \geq 0)$ *and* $(s_n^*; n \geq 0)$ *are strictly descending, compatible*
(53-2) $(t_n^* \to a+,\ s_n^* \to b+,$ *as* $n \to \infty)$, *and* $(\pi(t_n^*, s_n^*)$ *holds,* $\forall n)$.

Proof. By the preceding statement, $(t_n; n \geq 0)$ admits a strict subsequence $(T_n := t_{i(n)}; n \geq 0)$, with

(T_n) is strictly descending, and $T_n \to a+$.

Denote $(S_n := s_{i(n)}; n \geq 0)$; clearly,

(S_n) is a strict subsequence of $(s_n; n \geq 0)$ with $S_n \to b+$ as $n \to \infty$.

Moreover, by this very construction $[\pi(T_n, S_n)$ holds, for all $n]$. Again by that statement, there exists a strict subsequence $(s_n^* := S_{j(n)} = s_{i(j(n))}; n \geq 0)$ of $(S_n; n \geq 0)$ (hence, of $(s_n; n \geq 0)$ as well), with

(s_n^*) is strictly descending, and $s_n^* \to b+$.

Denote further $(t_n^* := T_{j(n)} = t_{i(j(n))}; n \geq 0)$; this is a strict subsequence of $(T_n; n \geq 0)$ (hence, of $(t_n; n \geq 0)$ as well), with

$(t_n^*; n \geq 0)$ is strictly descending, and $(t_n^* \to a+$, as $n \to \infty)$;

Finally, by this construction (and a previous relation) $[\pi(t_n^*, s_n^*)$ holds, for all $n]$. Summing up, the couple of subsequences $(t_n^*; n \geq 0)$ and $(s_n^*; n \geq 0)$ has all needed properties; and the conclusion follows. □

1.3 Admissible Functions

Abstract. Some preliminary facts are discussed about Matkowski and Meir-Keeler admissible real functions.

Subject-Class 2020. 47H10 (Primary), 54H25 (Secondary).

Keywords. Regressive function, Matkowski and Meir-Keeler admissible property, Boyd-Wong and Geraghty conditions, locally normal/subnormal functions.

1.3.1 *Matkowski and Meir-Keeler functions*

Remember that, for each sequence (r_n) in R and each $r \in R$, we denoted

$r_n \to r+$ ($\lim_n r_n = r+$) if $r_n \to r$ and $r_n > r$, for all n
$r_n \to r++$ ($\lim_n r_n = r++$) if $r_n \to r$ and $r_n > r$, for almost all n.

Here, given the property $\pi(n)$ depending on $n \in N$, we say that it holds for *almost all* $n \in N$ [written: $(\forall\forall n, \pi(n))$], if

$\exists h = h(\pi) \in N$, such that: $\pi(n)$ holds, for all $n \geq h$.

Denote by $\mathcal{F}(re)(R^0_+, R)$ the family of all $\varphi \in \mathcal{F}(R^0_+, R)$ that are *regressive* [$\varphi(t) < t$, $\forall t \in R^0_+$]. For each $\varphi \in \mathcal{F}(re)(R^0_+, R)$, define the properties

(n-d-a) φ is *non-diagonally admissible*:
there are no strictly descending sequences $(t_n; n \geq 0)$ in R^0_+ and no elements ε in R^0_+ with $t_n \to \varepsilon+$, $\varphi(t_n) \to \varepsilon+$
(M-a) φ is *Matkowski admissible*: for each $(t_n; n \geq 0)$ in R^0_+ with $(t_{n+1} \leq \varphi(t_n), \forall n)$ we have $\lim_n t_n = 0$
(str-M-a) φ is *strongly Matkowski admissible*: for each $(t_n; n \geq 0)$ in R^0_+ with $(t_{n+1} \leq \varphi(t_n), \forall n)$ we have $\sum_n t_n < \infty$.

(The conventions (M-a) and (str-M-a) are taken from Matkowski [164], [165], and Turinici [248], respectively). Clearly, for each $\varphi \in \mathcal{F}(re)(R^0_+, R)$,

strongly Matkowski admissible implies Matkowski admissible;

but the converse is not in general true.

To get circumstances under which such properties hold, we need some conventions. Given $\varphi \in \mathcal{F}(re)(R^0_+, R)$, define the global properties

(MK-a) φ is *Meir-Keeler admissible*:
$\forall \varepsilon > 0, \exists \delta > 0$, such that $(\varepsilon < s < \varepsilon + \delta)$ implies $\varphi(s) \leq \varepsilon$

(R-a) φ is *Rhoades admissible*: for each $\varepsilon > 0$, there exists $\delta > 0$, such that $(t, s > 0,\ t \le \varphi(s)$ and $\varepsilon < s < \varepsilon + \delta)$ imply $t \le \varepsilon$.

(Note that the former property is taken from the developments in Meir and Keeler [168]; but the latter one seems to be new.)

Theorem 1.16. *We have, in (ZF-AC+DC), the inclusions over the class* $\mathcal{F}(re)(R^0_+, R)$

$$(MK\text{-}a) \Longrightarrow (R\text{-}a) \Longrightarrow (M\text{-}a) \Longrightarrow (n\text{-}d\text{-}a) \Longrightarrow (MK\text{-}a).$$

Hence, (MK-a), (R-a), (M-a), (n-d-a) are equivalent over $\mathcal{F}(re)(R^0_+, R)$.

Proof. i): (MK-a) $\Longrightarrow$ (R-a). Let $\varepsilon > 0$ be given; and $\delta > 0$ be assured by the Meir-Keeler admissible property. Take the numbers $t, s > 0$ with $t \le \varphi(s)$ and $\varepsilon < s < \varepsilon + \delta$. As φ is Meir-Keeler admissible, this yields $\varphi(s) \le \varepsilon$; wherefrom (as $t \le \varphi(s)$), we must have $t \le \varepsilon$; hence the claim.

ii): (R-a) $\Longrightarrow$ (M-a). Let (s_n) be a sequence in R^0_+ with $(s_{n+1} \le \varphi(s_n), \forall n)$. Clearly, (s_n) is strictly descending in R^0_+; hence, $\sigma := \lim_n s_n$ exists in R_+. Suppose by contradiction that $\sigma > 0$; and let $\rho > 0$ be given by the Rhoades admissible property of φ. By the above convergence relations, there exists some rank $n(\sigma)$, such that $[n \ge n(\sigma)$ implies $\sigma < s_n < \sigma + \rho]$. But then, putting $(t_n := \varphi(s_n); n \ge 0)$, we get (for the same ranks)

$\sigma < s_{n+1} \le t_n < s_n < \sigma + \rho$; hence, in particular, $t_n > \sigma$;
in contradiction with the admitted hypothesis.

This proves that $\sigma = 0$, as claimed.

iii): (M-a) $\Longrightarrow$ (n-d-a). Suppose that φ is not endowed with the non-diagonally admissible property: there must be a strictly descending sequence (t_n) in R^0_+ and an $\varepsilon > 0$, such that $[t_n \to \varepsilon+$ and $\varphi(t_n) \to \varepsilon+$, as $n \to \infty]$. Put $i(0) = 0$. As $\varepsilon < \varphi(t_{i(0)})$ and $t_n \to \varepsilon+$, we have that

$A(i(0)) := \{n > i(0); t_n < \varphi(t_{i(0)})\}$ is not empty;
hence, $i(1) := \min(A(i(0)))$ is an element of it, and $t_{i(1)} < \varphi(t_{i(0)})$.

Likewise, as $\varepsilon < \varphi(t_{i(1)})$ and $t_n \to \varepsilon+$, we have that

$A(i(1)) := \{n > i(1); t_n < \varphi(t_{i(1)})\}$ is not empty;
hence, $i(2) := \min(A(i(1)))$ is an element of it, and $t_{i(2)} < \varphi(t_{i(1)})$.

This procedure may continue indefinitely; and yields (without any choice technique) a strictly ascending rank sequence $(i(n); n \ge 0)$ for which the attached subsequence $(s_n := t_{i(n)}; n \ge 0)$ of (t_n) fulfills $[s_{n+1} < \varphi(s_n)(<$

s_n), for all n]. On the other hand, by this very subsequence property, [($s_n > \varepsilon$, $\forall n$) and $\lim_n s_n = \lim_n t_n = \varepsilon$]. The obtained relations contradict our hypothesis; and we are done.

iv): (n-d-a) $\Longrightarrow$ (MK-a). Suppose that φ is not Meir-Keeler admissible: there exists $\varepsilon > 0$, such that

$$H(\delta) := \{t \in R^0_+; \varepsilon < t < \varepsilon + \delta, \varphi(t) > \varepsilon\} \text{ is not empty, for each } \delta > 0.$$

Taking a strictly descending sequence $(\delta_n; n \geq 0)$ in R^0_+ with $\delta_n \to 0$, we get by the Denumerable Axiom of Choice (AC(N)) [deductible, as precise, in (ZF-AC+DC)], a sequence $(t_n; n \geq 0)$ in R^0_+, so as

$$(t_n \in H(\delta_n), \forall n); \text{ or, equivalently (by definition, and } \varphi = \text{regressive)}$$
$$(\varepsilon < \varphi(t_n) < t_n < \varepsilon + \delta_n, \forall n); \text{ hence, } \varphi(t_n) \to \varepsilon+ \text{ and } t_n \to \varepsilon+.$$

By a previous result, there exists a subsequence $(r_n := t_{i(n)})$ of (t_n), with

$$(r_n) \text{ is strictly descending and (by the above) } r_n \to \varepsilon+, \ \varphi(r_n) \to \varepsilon+.$$

Since this contradicts the posed hypothesis, the assertion follows. □

1.3.2 *Sequential counterparts*

In the following, some sequential versions of the introduced admissible properties are provided.

(A) For any $\varphi \in \mathcal{F}(re)(R^0_+, R)$ and any $s \in R^0_+$, put

$$\Lambda^+\varphi(s) = \inf_{0<\varepsilon<s} \Phi(s+)(\varepsilon); \text{ where } \Phi(s+)(\varepsilon) = \sup \varphi(]s, s+\varepsilon[)$$
$$\Lambda^\pm\varphi(s) = \inf_{0<\varepsilon<s} \Phi(s\pm)(\varepsilon); \text{ where } \Phi(s\pm)(\varepsilon) = \sup \varphi(]s-\varepsilon, s+\varepsilon[).$$

From the regressive property of φ, these limit quantities fulfill

$$(-\infty \leq) \ \Lambda^+\varphi(s) \leq \Lambda^\pm\varphi(s) \leq s, \ \ \forall s \in R^0_+;$$

but the case of extremal values being attained cannot be avoided.

The following properties of introduced concepts will be useful.

Proposition 1.26. *Given $\varphi \in \mathcal{F}(re)(R^0_+, R)$ and $s \in R^0_+$, we have in the reduced system (ZF-AC+DC),*

(21-1) $\limsup_n(\varphi(t_n)) \leq \Lambda^+\varphi(s)$,
for each sequence (t_n) in R^0_+ with $t_n \to s+$
(21-2) there exists a strictly descending sequence (r_n) in R^0_+ with $r_n \to s+$ and $\varphi(r_n) \to \Lambda^+\varphi(s)$

(21-3) $\limsup_n(\varphi(t_n)) \leq \Lambda^{\pm}\varphi(s)$,
for each sequence (t_n) *in* R_+^0 *with* $t_n \to s$
(21-4) there exists a sequence (r_n) *in* R_+^0 *with*
$r_n \to s$ *and* $\varphi(r_n) \to \Lambda^{\pm}\varphi(s)$.

Proof. i): Two steps must be passed.

i-1): Given ε in $]0, s[$, there exists a rank $p(\varepsilon) \geq 0$ such that $s < t_n < s + \varepsilon$, for all $n \geq p(\varepsilon)$; hence

$$\limsup_n(\varphi(t_n)) \leq \sup\{\varphi(t_n); n \geq p(\varepsilon)\} \leq \Phi(s+)(\varepsilon).$$

Passing to infimum over ε, yields the claim.

i-2): Denote, for simplicity,

$$\alpha = \Lambda^+\varphi(s); \text{ hence, } \alpha = \inf_{0<\varepsilon<s} \Phi(s+)(\varepsilon) \text{ and } -\infty \leq \alpha \leq s.$$

Then, define $(\beta_n := \Phi(s+)(2^{-n-1}s); n \geq 0)$; it is a sequence in R, because

$$(\forall n): -\infty < \varphi(t) < t < s + 2^{-n-1}s, \text{ for each } t \in]s, s + 2^{-n-1}s[.$$

Moreover, we have by definition that

$$(\beta_n) \text{ is descending, } (\beta_n \geq \alpha, \ \forall n), \ \inf_n \beta_n = \alpha; \text{ hence } \lim_n \beta_n = \alpha.$$

Further, denote

$$(\gamma_n = \beta_n - 3^{-n}; n \geq 0); \text{ hence, } (\gamma_n < \beta_n, \forall n), \ \lim_n \gamma_n = \lim_n \beta_n = \alpha.$$

From the supremum definition,

$$H_n := \{t \in]s, s + 2^{-n-1}s[; \varphi(t) > \gamma_n\} \neq \emptyset, \text{ for all } n \geq 0.$$

This, along with Denumerable Axiom of Choice (deductible in (ZF-AC+DC)), yields a sequence (t_n) with $(t_n \in H_n, \forall n)$; that is:

$$(\forall n): t_n \in]s, s + 2^{-n-1}s[, \text{ and } \gamma_n < \varphi(t_n) \leq \beta_n;$$

so, putting these together, $[t_n \to s+$ and $\varphi(t_n) \to \alpha]$. By a previous result, there exists a subsequence $(r_n := t_{i(n)})$ of (t_n), with

$$(r_n) \text{ is strictly descending and } r_n \to \varepsilon+, \ \varphi(r_n) \to \alpha.$$

Clearly, the obtained sequence $(r_n; n \geq 0)$ has all desired properties.

ii): As before, two steps must be passed.

ii-1): Given $\varepsilon \in]0, s[$, there exists a rank $p(\varepsilon) \geq 0$ such that $s - \varepsilon < t_n < s + \varepsilon$, for all $n \geq p(\varepsilon)$; hence

$\limsup_n(\varphi(t_n)) \leq \sup\{\varphi(t_n); n \geq p(\varepsilon)\} \leq \Phi(s\pm)(\varepsilon)$.

Taking the infimum over ε in this relation, we get the desired conclusion.

ii-2) Denote, for simplicity,

$\alpha = \Lambda^{\pm}\varphi(s)$; hence, $\alpha = \inf_{0<\varepsilon<s} \Phi(s\pm)(\varepsilon)$ and $-\infty \leq \alpha \leq s$.

Then, define $(\beta_n := \Phi(s\pm)(2^{-n-1}s); n \geq 0)$; this is a sequence in R, via

$(\forall n)$: $-\infty < \varphi(t) < t < s + 2^{-n-1}s$, $\forall t \in]s - 2^{-n-1}s, s + 2^{-n-1}s[$.

In addition, (β_n) is descending and

$(\beta_n \geq \alpha,\ \forall n)$, $\inf_n \beta_n = \alpha$; hence $\lim_n \beta_n = \alpha$.

Further, denote

$(\gamma_n = \beta_n - 3^{-n}; n \geq 0)$; hence, $(\gamma_n < \beta_n, \forall n)$, $\lim_n \gamma_n = \lim_n \beta_n = \alpha$.

By the same reasoning as before, one may construct a sequence (r_n) with

$(\forall n)$: $r_n \in]s - 2^{-n-1}s, s + 2^{-n-1}s[$, $\varphi(r_n) > \gamma_n$, and $\varphi(r_n) \leq \beta_n$;

so, putting these together, $[r_n \to s$ and $\varphi(t_n) \to \alpha]$; as claimed. □

Call the function $\varphi \in \mathcal{F}(re)(R^0_+, R)$,

(BW-a) *Boyd-Wong admissible* [35], if $\Lambda^+\varphi(s) < s$, for all $s > 0$
(bi-BW-a) *strongly (or: bilateral) Boyd-Wong admissible*, provided $\Lambda^{\pm}\varphi(s) < s$, for all $s \in R^0_+$.

Clearly, for any such functions we have

strongly Boyd-Wong admissible implies Boyd-Wong admissible;

but the reciprocal is not in general true. In particular, $\varphi \in \mathcal{F}(re)(R^0_+, R)$ is Boyd-Wong admissible, provided

φ is *upper semicontinuous at the right* on R^0_+: $\Lambda^+\varphi(s) \leq \varphi(s)$, $\forall s \in R^0_+$.

This, e.g., is fulfilled when φ is *continuous at the right* on R^0_+; for, in such a case, $[\Lambda^+\varphi(s) = \varphi(s)$, for each $s \in R^0_+]$. Similar conclusions are to be formulated with respect to the strong Boyd-Wong property.

Having these precise, let us introduce the sequential properties below. Call the function $\varphi \in \mathcal{F}(re)(R^0_+, R)$,

(s-BW-a) *sequentially Boyd-Wong admissible*, provided
for each strictly descending sequence $(t_n; n \geq 0)$ in R_+^0
and each $\varepsilon > 0$ with $t_n \to \varepsilon+$, we have $\limsup_n(\varphi(t_n)) < \varepsilon$
(s-R-a) *sequentially Rhoades admissible*, provided
for each strictly descending sequence $(t_n; n \geq 0)$ in R_+^0
and each $\varepsilon > 0$ with $t_n \to \varepsilon+$, we have $\liminf_n(\varphi(t_n)) < \varepsilon$.

Theorem 1.17. *The following inclusions are valid, in (ZF-AC+DC), over the functional class $\mathcal{F}(re)(R_+^0, R)$*

(21-a) (BW-a) is equivalent with (s-BW-a)
(21-b) (s-BW-a) is equivalent with (s-R-a)
(21-c) (s-R-a) implies (n-d-a), (R-a), and (MK-a)
(21-d) (s-BW-a) implies (n-d-a), (R-a), and (MK-a).

Proof. i): The left to right inclusion is immediate, by the preceding statement. Concerning the right to left inclusion, suppose that (BW-a) is not true; hence (by a previous evaluation) $\Lambda^+\varphi(\varepsilon) = \varepsilon$, for some $\varepsilon > 0$. By an auxiliary fact, there exists a strictly descending sequence (r_n) in R_+^0, with

$$r_n \to \varepsilon+ \text{ and } (\limsup_n \varphi(r_n) =) \lim_n \varphi(r_n) = \Lambda^+\varphi(\varepsilon) = \varepsilon.$$

But then, φ does not satisfy (s-BW-a); contradiction; hence the assertion.

ii): The left to right inclusion is immediate, by definition. For the right to left inclusion, one may proceed as below. Suppose by absurd that $\varphi \in \mathcal{F}(re)(R_+^0, R)$ is not sequentially Boyd-Wong admissible; hence,

$\limsup_n \varphi(t_n) = \varepsilon$, for some strictly descending sequence $(t_n; n \geq 0)$ in R_+^0 and some $\varepsilon > 0$ with $t_n \to \varepsilon+$.

Combining with a preceding fact, yields

$$\varepsilon = \limsup_n \varphi(t_n) \leq \Lambda^+\varphi(\varepsilon) \leq \varepsilon; \text{ hence, } \Lambda^+\varphi(\varepsilon) = \varepsilon.$$

This, by a previous auxiliary fact, yields

$(\liminf_n \varphi(r_n) =) \lim_n \varphi(r_n) = \varepsilon$,
for some strictly descending sequence $(r_n; n \geq 0)$ in R_+^0 with $r \to \varepsilon+$.

But then, the sequential Rhoades admissible property of φ is contradicted. Hence, our working assumption is not true; and the assertion follows.

iii): Let us establish the first relation in this series. Suppose by absurd that (n-d-a) fails: there exists a strictly descending sequence (t_n) in R_+^0

and a number $\varepsilon > 0$, such that $[t_n \to \varepsilon+,\ \varphi(t_n) \to \varepsilon+,$ as $n \to \infty]$. This, along with $[\liminf_n \varphi(t_n) = \limsup_n \varphi(t_n) = \varepsilon]$, contradicts the property (s-R-a). Hence, our working assumption is not true; and the conclusion is clear. The remaining relations follow at once in view of

(n-d-a) $\Longleftrightarrow$ (R-a) $\Longleftrightarrow$ (MK-a) (see above);

however, for completeness, we provide a proof of the second one. Suppose that $\varphi \in \mathcal{F}(re)(R^0_+, R)$ is not Rhoades admissible; that is, for some $\varepsilon > 0$,

$$H(\delta) := \{(t,s) \in R^0_+ \times R^0_+; t \le \varphi(s), \varepsilon < s < \varepsilon + \delta, t > \varepsilon\} \neq \emptyset,\ \forall \delta > 0.$$

Taking a strictly descending sequence $(\delta_n; n \ge 0)$ in R^0_+ with $\delta_n \to 0$, we get by the Denumerable Axiom of Choice (AC(N)) [deductible, as precise, in (ZF-AC+DC)], a sequence $((t_n, s_n); n \ge 0)$ in $R^0_+ \times R^0_+$, so as

$(\forall n)$: $((t_n, s_n) \in H(\delta_n)$; that is, $\varepsilon < t_n \le \varphi(s_n) < s_n < \varepsilon + \delta_n$.

As a consequence, $[t_n \to \varepsilon+,\ s_n \to \varepsilon+,$ as $n \to \infty]$. On the other hand,

$(\varepsilon < \varphi(s_n) < \varepsilon + \delta_n, \forall n)$ implies $\lim_n \varphi(s_n) = \varepsilon$.

By a previous result, there exists a subsequence $(s^*_n := s_{i(n)})$ of (s_n), with

(s^*_n) is strictly descending and $s^*_n \to \varepsilon+$;
hence, necessarily, $(\liminf_n \varphi(s^*_n) =) \lim_n \varphi(s^*_n) = \varepsilon$;

in contradiction with the sequential Rhoades admissible property of φ. The remaining inclusions are obtainable in a similar way.

iv): Evident, by the above stages. □

Remark 1.5.

From the preceding facts,

(BW-MK) each Boyd-Wong admissible function in $\mathcal{F}(re)(R^0_+, R)$ is Meir-Keeler admissible.

A shorter verification of this is as follows. (cf. Boyd and Wong [35]):

Suppose that $\varphi \in \mathcal{F}(re)(R^0_+, R)$ is Boyd-Wong admissible, and fix $\gamma > 0$; hence $\Lambda^+\varphi(\gamma) < \gamma$. By definition, there exists $\beta = \beta(\gamma)$ in $]0, \gamma[$ with $[\gamma < t < \gamma + \beta$ implies $\varphi(t) < \gamma]$; proving that φ is Meir-Keeler admissible.

(B) A functional property that is stronger than the Boyd-Wong admissible one may be constructed as follows. Let $\varphi \in \mathcal{F}(re)(R^0_+, R)$ be a function; we call it *Geraghty admissible* [96], provided

$(t_n; n \geq 0)$= sequence in R^0_+ and $\varphi(t_n)/t_n \to 1$ imply $t_n \to 0$.

Proposition 1.27. *Let $\varphi \in \mathcal{F}(re)(R^0_+, R)$ be Geraghty admissible. Then, φ is necessarily Boyd-Wong admissible.*

Proof. Suppose that $\varphi \in \mathcal{F}(re)(R^0_+, R)$ is not Boyd-Wong admissible: there exists $s \in R^0_+$ with $\Lambda^+\varphi(s) = s$. Combining with a preceding fact, there must be a strictly descending sequence $(r_n; n \geq 0)$ in R^0_+, with

$$r_n \to s+ \text{ and } \varphi(r_n) \to s;\text{ whence } \varphi(r_n)/r_n \to 1;$$

in contradiction with φ = Geraghty admissible. □

Remark 1.6.

The reverse inclusion is not in general true. In fact, the function $\varphi \in \mathcal{F}(re)(R^0_+, R)$ given as $[\varphi(t) = t(1-e^{-t}); t > 0]$ is continuous; hence, Boyd-Wong admissible. On the other hand, taking $(t_n = n+1; n \geq 0)$ in R^0_+, we have $[\varphi(t_n)/t_n \to 1$ and $t_n \to \infty]$; so that, φ is not Geraghty admissible.

(C) In the following, a particular version of these facts is formulated, in terms of increasing functions from the class $\mathcal{F}(R^0_+, R)$.

Let $\mathcal{F}(re, in)(R^0_+, R)$ stand for the class of all $\varphi \in \mathcal{F}(re)(R^0_+, R)$, with the increasing property. The following characterization of our previous Matkowski properties imposed to φ is now available. Given $t > 0$, let

$$\varphi^0(t) = t,\ (\varphi^{n+1}(t) = \varphi(\varphi^n(t));\ n \geq 0),$$

stand for the iterates of φ at this point. Note that such a construction may be non-effective; because, $\varphi^2(t) = \varphi(\varphi(t))$ is undefined whenever $\varphi(t) \leq 0$.

Proposition 1.28. *For each $\varphi \in \mathcal{F}(re, in)(R^0_+, R)$, we have*

(23-1) φ is Matkowski admissible, iff
$(\forall t > 0)$: $\lim_n \varphi^n(t) = 0$, whenever $(\varphi^n(t); n \geq 0)$ exists
(23-2) φ is strongly Matkowski admissible, iff
$(\forall t > 0)$: $\sum_n \varphi^n(t) < \infty$, whenever $(\varphi^n(t); n \geq 0)$ exists.

The proof is immediate, by the increasing property of φ and

$$(\forall t > 0)\text{: } (\varphi^n(t); n \geq 0) \text{ exists iff } \{\varphi^n(t); n \geq 0\} \subseteq R^0_+.$$

As before, we need sufficient conditions over $\mathcal{F}(re, in)(R^0_+, R)$ under which this property holds. For each $\varphi \in \mathcal{F}(re, in)(R^0_+, R)$, denote

$\varphi(s+0) := \lim_{t\to s+}\varphi(t)$ (the *right limit* of φ at s), $s \in R^0_+$;
clearly, $[\varphi(s) \le \varphi(s+0) \le s$, for all $s > 0]$.

Further, denote (over the same class of functions)

$M(\varphi)$ = the subset of all $s > 0$ with $\varphi(s+0) = s$.

We stress that the extremal case of $M(\varphi) = \emptyset$ cannot be avoided; just take the function $\varphi \in \mathcal{F}(re, in)(R^0_+, R)$ as

φ is right continuous on R^0_+: $\varphi(s+0) = \varphi(s)(< s)$, for each $s \in R^0_+$.

Concerning the opposite extremal case $M(\varphi) = R^0_+$, the following simple (negative) answer is available (see also Jachymski [122]):

Proposition 1.29. *For each function $\varphi \in \mathcal{F}(re, in)(R^0_+, R)$, we have*

$M(\varphi)$ is at most denumerable in R^0_+; hence, $M(\varphi) = R^0_+$ is false.

Proof. By the increasing property of our function φ, the subset

$\Gamma = \{s \in R^0_+; \varphi$ is right discontinuous at $s\}$

appears as (at most) countable; cf. Natanson [177, Ch. 8, Sec. 1]; so, necessarily, $\Theta := R^0_+ \setminus \Gamma$ is nonempty in R^0_+, and

$\varphi(s+0) = \varphi(s)(< s)$, for each $s \in \Theta$; whence $M(\varphi) \cap \Theta = \emptyset$.

This tells us that, necessarily, $M(\varphi) \subseteq \Gamma$; and conclusion follows. □

The following completion of a previous result is now available.

Proposition 1.30. *Suppose that the function $\varphi \in \mathcal{F}(re, in)(R^0_+, R)$ fulfills*

φ is strongly regressive: $\varphi(s+0) < s$, for each $s > 0$.

Then, φ is Matkowski admissible.

Proof. The conclusion follows at once in view of

$\Lambda^+\varphi(s) = \varphi(s+0)$, $\forall s \in R^0_+$, whenever φ = increasing,

and a previous fact involving Boyd-Wong admissible functions; however, for completeness, we shall provide an argument. Given $s_0 > 0$, assume that the iterative sequence $(s_n = \varphi^n(s_0); n \ge 0)$ exists (in R^0_+). By the regressive property of φ, (s_n) is strictly descending; hence, $s := \lim_n s_n$ exists, with $(s_n > s, \forall n)$. Suppose by contradiction that $s > 0$. Combining with $[\varphi(s+0) = \lim_n \varphi(s_n) = \lim_n s_{n+1}]$, yields $\varphi(s+0) = s$; contradiction. Hence, $s = 0$; and we are done. □

Remark 1.7.

The reverse inclusion is not true. In fact, let us consider the function $\varphi \in \mathcal{F}(re, in)(R^0_+, R)$, according to (for some $r > 0$):

$$(\varphi(t) = t/2, \text{ if } t \leq r), \ (\varphi(t) = r, \text{ if } t > r).$$

Clearly, φ is Matkowski admissible. On the other hand, [$\varphi(r + 0) = r$; whence, φ is not strongly regressive]; and this proves our claim. For an extended example, see Turinici [228] and the references therein.

Now, it is natural to establish the connection between the introduced class and the Meir-Keeler one. An appropriate answer to this is as follows.

Proposition 1.31. *We have, for each $\varphi \in \mathcal{F}(re, in)(R^0_+, R)$:*

Matkowski admissible is equivalent with Meir-Keeler admissible.

Proof. The conclusion follows from the one relative to $\mathcal{F}(re)(R^0_+, R)$: but, for completeness reasons, we will give a (different) reasoning for it.

i) (cf. Jachymski [120]). Assume that $\varphi \in \mathcal{F}(re, in)(R^0_+, R)$ is Matkowski admissible; we have to establish that it is Meir-Keeler admissible. If the underlying property fails, then (for some $\gamma > 0$):

$$\forall \beta > 0, \ \exists t \in]\gamma, \gamma + \beta[, \text{ such that } \varphi(t) > \gamma.$$

As φ is increasing, this yields (by the arbitrariness of β)

$$(\varphi(t) > \gamma, \ \forall t > \gamma); \text{ whence, by induction: } (\varphi^n(t) > \gamma, \ \forall n, \ \forall t > \gamma).$$

Taking some $t > \gamma$ and passing to limit, one gets $0 \geq \gamma$; contradiction.

ii) Assume that $\varphi \in \mathcal{F}(re, in)(R^0_+, R)$ is Meir-Keeler admissible; we have to establish that it is Matkowski admissible. Given $s_0 > 0$, suppose that the iterative sequence $(s_n = \varphi^n(s_0); n \geq 0)$ exists (in R^0_+). As φ = regressive, (s_n) is strictly descending; hence, $s := \lim_n s_n$ exists, with (in addition) $s_n > s$, for all n. Suppose by contradiction that $s > 0$; and let $r > 0$ be assured by the Meir-Keeler admissible property of φ. By definition, there exists an index $n(r) \geq 0$, such that [$n \geq n(r)$ implies $s < s_n < s + r$]. This, by the Meir-Keeler property, gives (for the same ranks)

$$s < s_{n+1} = \varphi(s_n) \leq s; \text{ contradiction.}$$

Hence, $s = 0$; wherefrom φ is Matkowski admissible. □

Note that these examples do not exhaust the properties of such functions. Further aspects may be found in Du [81].

1.3.3 *Strongly Matkowski admissible criteria*

In the following, a lot of criteria for the strongly Matkowski admissible property will be given.

(I) Call $h \in \mathcal{F}(R^0_+)$, *γ-locally int-normal* (where $\gamma > 0$), provided

(lo-in-1) $h(.)$ is decreasing on $]0, \gamma[$
(lo-in-2) $H(t) := \int_0^t h(\xi)d\xi < \infty$, for each $t \in]0, \gamma[$.

Note that, by the condition (lo-in-1),

$$\int_0^t h(\xi)d\xi := \lim_{s \to 0+} \int_s^t h(\xi)d\xi \text{ exists in } R_+ \cup \{\infty\}, \text{ for each } t \in]0, \gamma[;$$

so, condition (lo-in-2) is meaningful. Moreover, by these definitions,

(p1) $H(.)$ is strictly increasing on $]0, \gamma[$ $(t_1 < t_2 \Longrightarrow H(t_1) < H(t_2))$
(p2) $H(.)$ is continuous on $]0, \gamma[$, and $H(0 + 0) := \lim_{t \to 0+} H(t) = 0$.

When $\gamma > 0$ is generic here, we then say that $h \in \mathcal{F}(R^0_+)$ is *locally int-normal.* Clearly, the invariance property holds

$$h \text{ is locally int-normal} \Longrightarrow ah + b \text{ is locally int-normal}, \forall a, b > 0.$$

Given $\varphi \in \mathcal{F}(re)(R^0_+, R)$, let us associate it the function $g \in \mathcal{F}(R^0_+)$, as

$$(g(t) = t/(t - \varphi(t)); t > 0); \text{ in short: } g = \text{id}/(\text{id} - \varphi).$$

Further, call $g \in \mathcal{F}(R^0_+)$, *γ-locally int-subnormal* (where $\gamma > 0$), provided

$$(g(t) \leq h(t),\ t \in R^0_+), \text{ where } h \in \mathcal{F}(R^0_+) \text{ is } \gamma\text{-locally int-normal.}$$

When $\gamma > 0$ is generic, we say that $g \in \mathcal{F}(R^0_+)$ is *locally int-subnormal.*

The following locally int-subnormal type strongly Matkowski criterion (in short: (str-M-lis)) is available.

Theorem 1.18. *Let the function $\varphi \in \mathcal{F}(re)(R^0_+, R)$ be such that its associated function $g = \text{id}/(\text{id} - \varphi)$ is locally int-subnormal. In addition, suppose that one of the conditions below is holding*

(31-i) φ is Matkowski admissible
(equivalently: Meir-Keeler admissible)
(31-ii) φ is Boyd-Wong admissible.

Then, φ is strongly Matkowski admissible (see above).

Proof. There are two steps to be passed.

i): By the imposed condition, there exists $\gamma > 0$ and $h \in \mathcal{F}(R^0_+)$ with

$$(g(t) \le h(t),\ t \in R^0_+), \text{ and } h \text{ is } \gamma\text{-locally int-normal.}$$

Moreover, denoting by $H(.)$ the associated to $h(.)$ function, we have

$$H \text{ is strictly increasing, continuous on }]0,\gamma[\text{ and } H(0+0) = 0.$$

Let the sequence (t_n) in R^0_+ be such that $[t_{n+1} \le \varphi(t_n),\ \forall n]$; clearly, (t_n) is strictly descending. By the Matkowski admissible condition, we have that $t_n \to 0$ as $n \to \infty$. Note that, as a consequence of this,

$$\text{there exists } m = m(\gamma) \in N, \text{ such that } t_n < \gamma,\ \forall n \ge m.$$

Having these precise, let $i \ge m$ be arbitrary fixed. By the above choice,

$$t_i - \varphi(t_i) \le t_i - t_{i+1}; \text{ whence, } 1 \le (t_i - t_{i+1})/(t_i - \varphi(t_i)).$$

Combining with $h =$ decreasing yields (by the definition of $H(.)$)

$$t_i \le (t_i - t_{i+1})g(t_i) \le (t_i - t_{i+1})h(t_i) \le H(t_i) - H(t_{i+1}).$$

Taking the relation between extremal members gives (by the preceding fact)

$$\sum_{n \ge m} t_n \le H(t_m) - H(0+0) < \infty; \text{ proving that } \sum_n t_n \text{ converges.}$$

ii): Evident, in view of every Boyd-Wong admissible function from the class $\mathcal{F}(re)(R^0_+, R)$ being Matkowski admissible. □

(II) In the following, a basic particular case of this result is discussed. Given $\varphi \in \mathcal{F}(re)(R^0_+, R)$, let us introduce the property

(z-con) φ is *zero-contractive*: $\limsup_{t \to 0+}[\varphi(t)/t] < 1$.

The following zero-contractive type strongly Matkowski admissible criterion (in short: (str-M-zc)) is to be stated.

Theorem 1.19. *Let the function $\varphi \in \mathcal{F}(re)(R^0_+, R)$ be zero contractive. In addition, suppose that one of the conditions below is holding*

(32-i) φ is Matkowski admissible
(equivalently: Meir-Keeler admissible)
(32-ii) φ is Boyd-Wong admissible.

Then, φ is strongly Matkowski admissible.

Proof. For technical reasons, two proofs of this result will be given.

(Standard) i): By the zero-contractive property, there must be some $\lambda \in]0,1[$ and some $\gamma > 0$, such that $\varphi(t) \leq \lambda t$, for all $t \in]0,\gamma[$. Let the sequence $(t_n; n \geq 0)$ in R_+^0 be such that $[t_{n+1} \leq \varphi(t_n), \forall n]$; hence, (t_n) is strictly descending. From the Matkowski admissible condition, $t_n \to 0$ as $n \to \infty$; so, there must be some index $m = m(\gamma)$, such that

$(\forall n \geq m)$: $0 < t_n < \gamma$; whence, $\varphi(t_n) \leq \lambda t_n$.

Combining with the inductive property of our sequence, one gets

$t_{n+1} \leq \lambda t_n$, for all $n \geq m$; wherefrom $\sum_n t_n < \infty$.

ii): Evident, in view of every Boyd-Wong admissible function from the class $\mathcal{F}(re)(R_+^0, R)$ being Matkowski admissible.

(Relative) Let the function $\varphi \in \mathcal{F}(re)(R_+^0, R)$ be zero-contractive. By definition, there exist $\lambda \in]0,1[$ and $\gamma > 0$, such that $\varphi(t) \leq \lambda t$, for all $t \in]0,\gamma[$. This yields, for its associated function $g := \text{id}/(\text{id} - \varphi)$,

$g(t) := t/(t-\varphi(t)) \leq 1/(1-\lambda)$, $t \in]0,\gamma[$; so, g is γ-locally int-subnormal.

Combining with the locally int-subnormal type strongly Matkowski admissible criterion (str-M-lis) gives our conclusion. □

A limiting version of this statement is to be obtained under the lines below. Call the function $h \in \mathcal{F}(R_+^0)$, *lim-right-normal* provided

the right limit $h(s+0) := \lim_{t \to s+} h(t)$ exists in R_+, $\forall s \in R_+$.

Note at this moment that the invariance property holds

h is lim-right-normal implies $ah + b$ is lim-right-normal, $\forall a, b > 0$.

Given $\varphi \in \mathcal{F}(re)(R_+^0, R)$, let us associate it the function $g \in \mathcal{F}(R_+^0)$, as $g = \text{id}/(\text{id} - \varphi)$. Further, call $g \in \mathcal{F}(R_+^0)$, *lim-right-subnormal* provided

$g(t) \leq h(t)$, $t \in R_+^0$, where $h \in \mathcal{F}(R_+^0)$ is lim-right-normal.

Proposition 1.32. *Let the function $\varphi \in \mathcal{F}(re)(R_+^0, R)$ be such that the associated function $g = \text{id}/(\text{id} - \varphi)$ is lim-right-subnormal. Then, φ is strongly Matkowski admissible.*

Proof. By the imposed condition,

$g(t) \leq h(t) \leq k(t) := h(t) + 2$, $t \in R_+^0$,

for some function $h \in \mathcal{F}(R^0_+)$ with h = lim-right-normal (hence, so is $k := h + 2$). From this relation, we derive

$$\varphi(t) \leq t(k(t) - 1)/k(t) \text{ [hence, } \varphi(t)/t \leq (k(t) - 1)/k(t)], \ t \in R^0_+.$$

Taking $s > 0$ and passing to superior limit as $t \to s+$ in the first half of this, yields (via $1 \leq k(s + 0) - 1 < k(s + 0) < \infty$)

$$\limsup_{t \to s+} \varphi(t) \leq s(k(s + 0) - 1)/k(s + 0) < s, \ \forall s > 0;$$

which (by the arbitrariness of $s > 0$) tells us that φ is Boyd-Wong admissible; hence, Matkowski admissible. Further, passing to lim sup as $t \to 0+$ in the second half of this, yields (via $1 \leq k(0 + 0) - 1 < k(0 + 0) < \infty$)

$$\limsup_{t \to 0+} [\varphi(t)/t] \leq (k(0 + 0) - 1)/k(0 + 0) < 1;$$
proving that φ is zero-contractive.

Summing up, (str-M-zc) is applicable here; and we are done. □

Remark 1.8.

The following monotone variant of this result is to be noted. Given $h \in \mathcal{F}(R^0_+)$, call it *boundedly monotone*, when

h is either (increasing) or (decreasing, bounded).

As before, the invariance property holds

h is boundedly monotone $\Longrightarrow ah + b$ is boundedly monotone, $\forall a, b > 0$.

It is now clear that

$(\forall h \in \mathcal{F}(R^0_+))$: h is boundedly monotone implies h is lim-right-normal;

so, the preceding statement holds when (in particular) *lim-right-normal* is replaced with *boundedly monotone.*

Remark 1.9.

The converse inclusion is not in general true. For, take $h \in \mathcal{F}(R^0_+)$ as

for each $n \geq 0$: $h(t) = t - n$, if $n < t \leq n + 1$.

Clearly, h is lim-right-normal. On the other hand, h is bounded on R^0_+. Finally, h is not monotone (i.e.: increasing or decreasing); because

$(\forall n \geq 1)$: $h(n) = 1 > 0 = h(n + 0)$; whence, h is not increasing;
$(\forall n \geq 0)$: h is strictly increasing on $]n, n + 1]$; so, h is not decreasing;

and this proves our claim.

(III) A global version of the integral type result above may be stated as follows. Call the function $h \in \mathcal{F}(R^0_+)$, *int-normal* provided

(i-n-1) $h(.)$ is decreasing on R^0_+
(i-n-2) $H(t) := \int_0^t h(\xi)d\xi < \infty$, for each $t > 0$.

Note that, by the former condition (i-n-1),

$\int_0^t h(\xi)d\xi := \lim_{s\to 0+} \int_s^t h(\xi)d\xi$ exists in $R_+ \cup \{\infty\}$, for each $t > 0$;

so, the latter condition (i-n-2) is meaningful. Moreover, by these definitions,

(p–1) $H(.)$ is strictly increasing on R^0_+ $(t_1 < t_2 \Longrightarrow H(t_1) < H(t_2))$
(p–2) $H(.)$ is continuous on R^0_+ and $H(0+0) := \lim_{t\to 0+} H(t) = 0$.

In addition, we have that the invariance property holds

h is int-normal implies $ah + b$ is int-normal, for each $a, b > 0$.

Given $\varphi \in \mathcal{F}(re)(R^0_+, R)$, let us associate it the function $g \in \mathcal{F}(R^0_+)$, as $g = \text{id}/(\text{id} - \varphi)$. Further, call $g \in \mathcal{F}(R^0_+)$, *int-subnormal* provided

$g(t) \leq h(t)$, $t \in R^0_+$, where $h \in \mathcal{F}(R^0_+)$ is int-normal.

The following int-subnormal type strongly Matkowski criterion (in short: (str-M-is)) is available.

Theorem 1.20. *Let the function $\varphi \in \mathcal{F}(re)(R^0_+, R)$ be such that the associated function $g = \text{id}/(\text{id} - \varphi)$ is int-subnormal. Then, φ is strongly Matkowski admissible (see above).*

Proof. For technical reasons, we provide two proofs of this statement.
(Standard) By the imposed condition,

$g(t) \leq h(t)$, $t \in R^0_+$, for some int-normal function $h \in \mathcal{F}(R^0_+)$.

Let the sequence (t_n) in R^0_+ fulfill $[t_{n+1} \leq \varphi(t_n), \forall n]$; so, (t_n) is strictly descending. Further, let $i \geq 0$ be arbitrary fixed. By the above choice,

$t_i - \varphi(t_i) \leq t_i - t_{i+1}$; whence, $1 \leq (t_i - t_{i+1})/(t_i - \varphi(t_i))$.

Combining with $h =$ decreasing yields (by the definition of $H(.)$)

$$t_i \leq (t_i - t_{i+1})g(t_i) \leq (t_i - t_{i+1})h(t_i) \leq H(t_i) - H(t_{i+1}).$$

Passing to limit in the relation between extremal members gives

$\lim_i t_i = 0$; whence, φ is Matkowski admissible.

On the other hand, from the same relation,

$\sum_{i \le n} t_i \le H(t_0) - H(t_{n+1})$, $\forall n$;
wherefrom, by a limit process, $\sum_n t_n \le H(t_0) - H(0+0) < \infty$;

which tells us that the series $\sum_n t_n$ converges.

(Relative) By the imposed condition, we have for $g = \mathrm{id}/(\mathrm{id} - \varphi)$,

$g(t) \le h(t) \le k(t) := h(t) + 2$, $t \in R_+^0$, for some $h \in \mathcal{F}(R_+^0)$
with h=int-normal; hence, so is $k := h + 2$.

This yields in a direct mode

$\varphi(t) \le t(k(t) - 1)/k(t)$ (hence, $\varphi(t)/t \le (k(t) - 1)/k(t)$), $t \in R_+^0$.

Passing to superior limit as $t \to s+$ (where $s > 0$ is arbitrary fixed) in the first half of this relation, gives

$\limsup_{t \to s+} \varphi(t) \le s(k(s+0) - 1)/k(s+0) < s$, $\forall s > 0$;

which tells us that φ is Boyd-Wong admissible; hence, Matkowski admissible. On the other hand, evidently,

$(\forall g \in \mathcal{F}(R_+^0))$: g is int-subnormal implies g is locally int-subnormal.

Summing up, the locally int-subnormal type strongly Matkowski admissible criterion (str-M-lis) is applicable here; and, from this, we are done. □

In particular, if g is int-normal, the obtained criterion is nothing else than the 1975 one due to Altman [14].

For the moment, (str-M-lis) includes both (str-M-zc) and (str-M-is) criteria. So, we may ask of which relationships exist between these stronger criteria. A partial answer to this is to be given along the lines below.

Example 1.1.

Let the function $h \in \mathcal{F}(R_+^0)$ be taken according to

h is int-normal, $(h(t) \ge 2, \forall t \in R_+^0)$, and $h(0+0) = \infty$;

note that there are many functions of this type, like

$h(t) = 2 + t^{-\nu}$, $t \in R_+^0$, for some $\nu \in]0, 1[$.

Then, let the function $\varphi \in \mathcal{F}(re)(R_+^0, R)$ be introduced as

$\varphi(t) = t(h(t) - 1)/h(t) = t[1 - 1/h(t)]$, $t \in R^0_+$.

By this very convention,

(pro-1) $g := \text{id}/(\text{id} - \varphi) = h$; so that, g is int-subnormal.

Consequently, the int-subnormal type strongly Matkowski criterion (str-M-is) is applicable to this context; wherefrom φ is strongly Matkowski admissible. On the other hand, φ is Boyd-Wong admissible, in view of

(pro-2) $\Lambda^+\varphi(s) = \varphi(s+0) = s[h(s+0) - 1]/h(s+0) < s$, $\forall s \in R^0_+$.

But, as (cf. the assumptions upon h)

(pro-3) $\lim_{t\to 0+}[\varphi(t)/t] = \lim_{t\to 0+}[1 - 1/h(t)] = 1$,

the zero-contractive type strongly Matkowski admissible criterion (str-M-zc) is not applicable here. This tells us that the (str-M-is) criterion is more general than the (str-M-zc) one.

(IV) Finally, an extended result in this area is available. Let $\Gamma : R^0_+ \to R$ be a function with the properties

Γ is continuous and strictly decreasing;

note that, by a classical statement,

(q-1) $J := \Gamma(R^0_+)$ is an interval of R; namely, $J =]\Gamma(\infty), \Gamma(0+)[$
(q-2) $\Gamma^{-1} : J \to R^0_+$ exists and is continuous strictly decreasing.

Theorem 1.21. *Let the function $\varphi \in \mathcal{F}(re)(R^0_+)$ be such that*

(34-i) φ *is Matkowski admissible*
(34-ii) $\liminf_{t\to 0+}[\Gamma(\varphi(t)) - \Gamma(t)] > 0$
(34-iii) $\int_{\Gamma(b)}^{\Gamma(0+)} \Gamma^{-1}(s)ds < \infty$, *for some/all* $b > 0$.

Then, φ *is strongly Matkowski admissible.*

Proof. Let the sequence $(t_n; n \geq 0)$ in R^0_+ be such that

$(t_{n+1} \leq \varphi(t_n), \forall n)$; hence, $(t_n; n \geq 0)$ is strictly descending and $\lim_n(t_n) = 0$ (as φ = Matkowski admissible).

By the Cesaro-Stolz theorem (CS-th) (see above)

$\liminf_n[\Gamma(t_n)/n] \geq \liminf_n[\Gamma(t_{n+1}) - \Gamma(t_n)] \geq$
$\liminf_n[\Gamma(\varphi(t_n)) - \Gamma(t_n)] \geq \liminf_{t\to 0+}[\Gamma(\varphi(t)) - \Gamma(t)] > 0$.

Let $\lambda \in]0,1[$ be in the open interval of extremal terms. By definition, there must be some index $m = m(\lambda)$, such that

$\Gamma(t_n) > \lambda n$, for all $n \geq m$; hence, $\lim_n \Gamma(t_n) = \lim_n \lambda n = \infty$;

note that, as a consequence of this, $\Gamma(0+) = \infty$; whence, $J =]\Gamma(\infty), \infty[$. Fix in the following some $b > 0$; then, take $a \in]0, b[$ with

$(\Gamma(a) >)\ \lambda\Gamma(a) > \Gamma(b)$; whence, $\lambda\Gamma(a) \in J$; possible, in view of
$\Gamma(0+) = \infty > \Gamma(a) > \lambda\Gamma(a) > \Gamma(b) > \Gamma(\infty)$.

Fix $p \geq m$ with $(p >)\lambda p > \Gamma(a)$; note that

$(\forall n \geq p)$: $\Gamma(t_n) > \lambda n > \Gamma(a)$; whence, $t_n < \Gamma^{-1}(\lambda n) < a$.

Finally, combining with the posed hypothesis,

$\int_{\Gamma(a)}^{\infty} \Gamma^{-1}(\lambda s)ds = (1/\lambda) \int_{\lambda\Gamma(a)}^{\infty} \Gamma^{-1}(r)dr < (1/\lambda) \int_{\Gamma(b)}^{\infty} \Gamma^{-1}(r)dr < \infty$;

so (via integral Cauchy convergence criterion and $\lambda n \geq \lambda p > \Gamma(a)$, $\forall n \geq p$)

$\sum_{n>p} t_n \leq \sum_{n>p} \Gamma^{-1}(\lambda n) < \infty$; whence, $\sum_n t_n$ converges.

The proof is thereby complete. □

In particular, if we take $\Gamma(.)$ as

$\Gamma(t) = \int_1^t f(s)ds$, $t \in R_+^0$,
where $f \in \mathcal{F}(R_+^0)$ is locally Lebesgue integrable,

one gets the related statement in Timofte [226]. Further aspects may be found in Brauer [37]; see also Drewniak and Drobot [79].

1.3.4 *Complete functions*

In the following, some extensions of the developments above to the class of complete functions will be considered.

I) Define the functional class

$\mathcal{F}_0(R_+) =$ the family of all $\widehat{\chi} \in \mathcal{F}(R_+)$ with $\widehat{\chi}(0) = 0$;

any member of it will be referred to as a *complete function*. Then, for any $\widehat{\chi} \in \mathcal{F}_0(R_+)$, let $\chi \in \mathcal{F}(R_+^0, R)$ stand for its restriction $[\chi(t) = \widehat{\chi}(t);\ t > 0]$. Returning to the initial class, define the property

$\widehat{\chi}$ is *regressive*: $\widehat{\chi}(t) < t$, $\forall t \in R_+^0$;

let $\mathcal{F}_0(re)(R_+)$ denote the class of all regressive functions in $\mathcal{F}_0(R_+)$. Clearly, for each $\widehat{\chi} \in \mathcal{F}_0(re)(R_+)$, its associated function χ is regressive too; hence, an element of $\mathcal{F}(re)(R_+^0, R)$.

As a rule, any property defined upon the complete class $\mathcal{F}_0(re)(R_+)$ generates a corresponding one over the restricted class $\mathcal{F}(re)(R_+^0, R)$. For example, let us introduce the properties over $\mathcal{F}_0(re)(R_+)$

(M-a-0) $\widehat{\chi}$ is *Matkowski admissible*: each sequence $(t_n; n \geq 0)$ in R_+^0 with $(t_{n+1} \leq \widehat{\chi}(t_n); n \geq 0)$ fulfills $t_n \to 0$
(s-M-a-0) $\widehat{\chi}$ is *strongly Matkowski admissible*: each sequence $(t_n; n \geq 0)$ in R_+^0 with $(t_{n+1} \leq \widehat{\chi}(t_n); n \geq 0)$ fulfills $\sum_n t_n < \infty$.

As both these concepts involve the points in R_+^0 only, we have

(p1) $\widehat{\chi}$ is Matkowski admissible iff so is χ
(p2) $\widehat{\chi}$ is strongly Matkowski admissible iff so is χ.

Likewise, define the related properties over $\mathcal{F}_0(re)(R_+)$

(n-d-a-0) $\widehat{\chi}$ is *non-diagonally admissible*:
there are no sequences $(t_n; n \geq 0)$ in R_+^0
and no elements $\varepsilon \in R_+^0$ with $t_n \to \varepsilon+$, $\widehat{\chi}(t_n) \to \varepsilon+$
(MK-a-0) $\widehat{\chi}$ is *Meir-Keeler admissible*:
$\forall \varepsilon > 0$, $\exists \delta > 0$, such that $\varepsilon < s < \varepsilon + \delta \Longrightarrow \widehat{\chi}(s) \leq \varepsilon$.

As before, both these concepts involve the points in R_+^0 only; so that,

(p3) $\widehat{\chi}$ is non-diagonally admissible iff so is χ
(p4) $\widehat{\chi}$ is Meir-Keeler admissible iff so is χ.

The relationships between these corresponding properties over $\mathcal{F}(R_+^0, R)$ were already clarified. Taking them into account, one gets a corresponding statement involving the introduced properties over $\mathcal{F}_0(re)(R_+)$.

Theorem 1.22. *We have, in (ZF-AC+DC), the inclusions over* $\mathcal{F}_0(re)(R_+)$

$$\text{(M-a-0)} \Longrightarrow \text{(n-d-a-0)} \Longrightarrow \text{(MK-a-0)} \Longrightarrow \text{(M-a-0)}.$$

Hence, (M-a-0), (n-d-a-0), and (MK-a-0) are equivalent on $\mathcal{F}_0(re)(R_+)$.

II) In the following, some particular cases of these properties are given. For any $\widehat{\varphi} \in \mathcal{F}_0(re)(R_+)$ and any $s \in R_+^0$, put

$\Lambda^+\widehat{\varphi}(s) = \inf_{0<\varepsilon<s} \widehat{\Phi}(s+)(\varepsilon)$;
where $\widehat{\Phi}(s+)(\varepsilon) = \sup \widehat{\varphi}(]s, s+\varepsilon[)$, $0 < \varepsilon < s$
$\Lambda^{\pm}\widehat{\varphi}(s) = \inf_{0<\varepsilon<s} \widehat{\Phi}(s\pm)(\varepsilon)$;
where $\widehat{\Phi}(s\pm)(\varepsilon) = \sup \widehat{\varphi}(]s-\varepsilon, s+\varepsilon[)$, $0 < \varepsilon < s$.

Since these concepts involve the points in R^0_+ only, we necessarily have

$$\Lambda^+\widehat{\varphi}(s) = \Lambda^+\varphi(s),\ \Lambda^{\pm}\widehat{\varphi}(s) = \Lambda^{\pm}\varphi(s), \text{ for all } s > 0;$$

where φ is the associated to $\widehat{\varphi}$ function. As a consequence, the properties of these objects over $\mathcal{F}_0(re)(R_+)$ are, practically, identical with the ones of corresponding objects over $\mathcal{F}(re)(R^0_+, R)$. To describe them, note that, from the regressive property of $\widehat{\varphi}$, one has an evaluation like

$$(0 \leq)\ \Lambda^+\widehat{\varphi}(s) \leq \Lambda^{\pm}\widehat{\varphi}(s) \leq s,\ \ \forall s \in R^0_+;$$

but the case of extremal values being attained cannot be avoided.

The following consequence of this remark is our starting point.

Proposition 1.33. *Let $\widehat{\varphi} \in \mathcal{F}_0(re)(R_+)$, $s \in R^0_+$ be given. Then,*

(41-1) $\limsup_n(\widehat{\varphi}(t_n)) \leq \Lambda^+\widehat{\varphi}(s)$,
for each sequence (t_n) in R^0_+ with $t_n \to s+$
(41-2) there exists a strictly descending sequence (r_n) in R^0_+ with $r_n \to s+$ and $\widehat{\varphi}(r_n) \to \Lambda^+\widehat{\varphi}(s)$
(41-3) $\limsup_n(\widehat{\varphi}(t_n)) \leq \Lambda^{\pm}\widehat{\varphi}(s)$,
for each sequence (t_n) in R^0_+ with $t_n \to s$
(41-4) there exists a sequence (r_n) in R^0_+ with $r_n \to s$ and $\widehat{\varphi}(r_n) \to \Lambda^{\pm}\widehat{\varphi}(s)$.

Now, define two basic properties for a function $\widehat{\varphi} \in \mathcal{F}_0(re)(R_+)$:

(BW-a-0) $\widehat{\varphi}$ is *Boyd-Wong admissible* [35]: $\Lambda^+\widehat{\varphi}(s) < s$, $\forall s > 0$
(bi-BW-a-0) $\widehat{\varphi}$ is *strongly (or: bilateral) Boyd-Wong admissible*: $\Lambda^{\pm}\widehat{\varphi}(s) < s$, for all $s \in R^0_+$.

Clearly, for any such functions we have

strongly Boyd-Wong admissible implies Boyd-Wong admissible;

but the reciprocal inclusion is not in general true. By the preceding discussion involving the associated functions

$\widehat{\varphi} \in \mathcal{F}_0(re)(R_+)$ is Boyd-Wong admissible iff
its associated function $\varphi \in \mathcal{F}(re)(R^0_+, R)$ is Boyd-Wong admissible.

In particular, $\widehat{\varphi} \in \mathcal{F}_0(re)(R_+)$ is Boyd-Wong admissible, provided

$\widehat{\varphi}$ is *upper semicontinuous at the right* on R^0_+: $\Lambda^+\widehat{\varphi}(s) \leq \widehat{\varphi}(s)$, $\forall s \in R^0_+$.

This, e.g., is fulfilled when $\widehat{\varphi}$ is *continuous at the right* on R^0_+; for, in such a case, $[\Lambda^+\widehat{\varphi}(s) = \widehat{\varphi}(s),\ \forall s \in R^0_+]$. Similar conclusions are to be formulated with respect to the strong Boyd-Wong property; we do not give details.

A sequential version of this concept may be introduced as below. Call the function $\widehat{\varphi} \in \mathcal{F}_0(re)(R_+)$, *sequentially Boyd-Wong admissible*, provided

(s-BW-a-0) for each strictly descending sequence $(t_n; n \geq 0)$ in R^0_+ and each $\varepsilon > 0$ with $t_n \to \varepsilon+$, we have $\limsup_n(\widehat{\varphi}(t_n)) < \varepsilon$.

By the same discussion involving the associated function,

$\widehat{\varphi} \in \mathcal{F}_0(re)(R_+)$ is sequentially Boyd-Wong admissible iff its associated function $\varphi \in \mathcal{F}(re)(R^0_+, R)$ is sequentially Boyd-Wong admissible.

As a consequence, the comparison of these properties over $\mathcal{F}_0(re)(R_+)$ is equivalent with the comparison of corresponding properties over $\mathcal{F}(re)(R^0_+, R)$. This, combined with a preceding result, yields

Theorem 1.23. *The following inclusions are valid, in (ZF-AC+DC), for a generic function* $\widehat{\varphi} \in \mathcal{F}_0(re)(R_+)$

(42-a) the Boyd-Wong admissible property (BW-a-0) is equivalent with the sequential Boyd-Wong admissible property (s-BW-a-0)
(42-b) the Boyd-Wong admissible property (BW-a-0) (or, equivalently, the sequential Boyd-Wong admissible property (s-BW-a-0)) implies the Meir-Keeler admissible property (MK-a-0) (or, equivalently, the Matkowski admissible property (M-a-0)).

Finally, denote, for simplicity

$\mathcal{F}_0(in)(R_+)$ = the class of all increasing $\omega \in \mathcal{F}_0(R_+)$
$\mathcal{F}_0(re, in)(R_+) = \mathcal{F}_0(re)(R_+) \cap \mathcal{F}_0(in)(R_+)$.

Clearly, the introduced properties remain with the same content. The only formal difference refers to the Matkowski admissible properties. Precisely, for a function $\widehat{\varphi} \in \mathcal{F}_0(re, in)(R_+)$,

$\widehat{\varphi}$ is Matkowski admissible iff $\widehat{\varphi}^n(t) \to 0$ as $n \to \infty$, for all $t \geq 0$
$\widehat{\varphi}$ is strongly Matkowski admissible iff $\sum_n \widehat{\varphi}^n(t) < \infty$, for all $t \geq 0$.

Here, for each $n \geq 0$, $\widehat{\varphi}^n$ stands for the n-th iterate of $\widehat{\varphi}$.

III) A basic application involving these concepts may be described along the lines below. Fix some $\omega \in \mathcal{F}_0(re, in)(R_+)$, and let

$$\omega^* = \text{id} - \omega \text{ (that is: } (\omega^*(t) = t - \omega(t); t \in R_+))$$

denote its *complement.* We say that ω is *complementary coercive*, in case

ω^* is coercive: $\lim_{t\to\infty} \omega^*(t) = \infty$; or, equivalently:
$[\omega^* \leq a] := \{t \in R_+; \omega^*(t) \leq a\}$ is bounded, for each $a \geq 0$.

Note that, under such a condition, the function $\gamma \in \mathcal{F}(R_+)$ given as

$$\gamma(s) = \sup A(s) \text{ where } A(s) = \{t \in R_+; t \leq \omega(s+t)\},\ s \in R_+$$

[referred to as the *right complementary inverse* (in short: *rc-inverse*) associated to ω] is well-defined. In fact, it will suffice noting that,

$(\forall s \in R_+)$: $A(s) = \{t \in R_+; \omega^*(s+t) \leq s\}$; so, $A(s)$ is bounded in R_+.

Clearly, $\gamma(0) = 0$; hence, $\gamma \in \mathcal{F}_0(R_+)$. In addition, by the properties of ω,

(p-1) (Increasing): $s_1 \leq s_2$ implies $A(s_1) \subseteq A(s_2)$; so, γ is increasing
(p-2) (Representation formula): $\gamma(s) \leq \omega(s + \gamma(s))$, for all $s \in R^0_+$.

To verify this, let $t \in A(s)$ be arbitrary fixed; hence, $t \leq \omega(s+t)$. By the definition of supremum (and $\omega =$ increasing), $(t \leq \omega(s+\gamma(s)),\ \forall t \in A(s))$; wherefrom (again by the underlying definition) the conclusion follows.

Further properties of $\gamma(.)$ are available under extra conditions upon $\omega(.)$.

Proposition 1.34. *Let the function $\omega \in \mathcal{F}_0(re, in)(R_+)$ fulfill, in addition*

ω is strongly regressive: $\omega(t+0) < t$, $\forall t > 0$.

Then, the following extra property of $\gamma(.)$ is available

$(\lim_{t\to 0+} \gamma(t) =)\gamma(0+0) = 0$; *that is:* $s_n \to 0+$ *implies* $\gamma(s_n) \to 0$.

Proof. As γ is increasing, $\delta := \gamma(0+0)$ exists. Assume by contradiction that $\delta > 0$. Let (s_n) be a sequence in R^0_+ with $s_n \to 0+$; whence, $(\gamma(s_n) \geq \delta,\ \forall n)$, and $\gamma(s_n) \to \delta$. From the preceding representation formula

$$\gamma(s_n) \leq \omega(s_n + \gamma(s_n)), \text{ for all } n.$$

As $n \to \infty$, we have $s_n + \gamma(s_n) \to \delta+$; so that, passing to limit yields $\delta \leq \omega(\delta + 0) < \delta$; contradiction. Hence, $\gamma(0+0) = 0$, as claimed. □

Sometimes, we may ask for the associated rc-inverse $\gamma(.)$ a property like

γ is strongly regressive: $\gamma(t+0) < t$, for all $t \in R_+^0$;

the initial function ω will be then referred to as *rc-regressive*. A useful answer in this direction is obtained by imposing regularity conditions upon

$(\eta(t) = \omega(2t); t \geq 0)$ [the *double function* attached to ω].

Note that, by the increasing property of ω (and η)

$(\forall t > 0)$: $\eta(t+0) = \inf_{s>0} \eta(t+s) = \inf_{s>0} \omega(2t+2s) = \omega(2t+0)$.

Theorem 1.24. *Let the function $\omega \in \mathcal{F}_0(re, in)(R_+)$ be such that*

(43-i) the double function η is strongly regressive ($\eta(t+0) < t$, $\forall t > 0$); expressed as: ω is double strongly regressive
(43-ii) the double function η is complementary coercive ($\eta^* := id - \eta$ is coercive); expressed as: ω is double complementary coercive.

Then,
(43-a) *its associated complement $\omega^* := id - \omega$ is coercive; so, the right complementary inverse*

$\gamma(s) = \sup A(s)$ *where* $A(s) = \{t \in R_+;\ t \leq \omega(s+t)\}$, $s \in R_+$

is well defined as an element of $\mathcal{F}_0(R_+)$
(43-b) *The right complementary inverse $\gamma(.)$ fulfills*

(p-1) γ is increasing and $\gamma(s) \leq \omega(s+\gamma(s))$, for all $s \in R_+$
(p-2) γ is zero-continuous: $\gamma(t) \to 0 = \gamma(0+0)$ as $t \to 0$

(43-c) *Finally, we have the properties*

(p-3) γ is strongly regressive; hence, in particular, regressive
(p-4) γ is complementary coercive: $\gamma^* := id - \gamma$ is coercive.

Proof. By a preceding relation,

$(\forall t > 0)$: $(\omega(t) \leq)\omega(t+0) < t/2$; whence, $t - \omega(t) > t - t/2 = t/2$; proving that ω is complementary coercive: $\omega^* := id - \omega$ is coercive.

i): Clearly, the right complementary inverse $\gamma : R_+ \to R_+$ is well defined.
ii): The properties (p-1) and (p-2) of γ are directly obtainable by the preceding facts, in view of

ω is double strongly regressive implies ω is strongly regressive.

iii): Fix $s > 0$; and let (s_n) be a sequence with $s_n \to s+$ as $n \to \infty$. As γ is increasing, we have (denoting $r := \gamma(s+0)$)

$$(\gamma(s_n) \geq r, \forall n) \text{ and } \gamma(s_n) \to r \text{ as } n \to \infty.$$

From the above representation formula,

$$\text{(rep-seq)} \quad \gamma(s_n) \leq \omega(s_n + \gamma(s_n)), \forall n.$$

Denote for simplicity $u = s + r$. By the posed hypotheses,

$$s_n + \gamma(s_n) \to u+; \text{ whence } \omega(s_n + \gamma(s_n)) \to \omega(u+0).$$

Passing to limit in (rep-seq) gives (as ω is double strongly regressive)

$$r \leq \omega(u+0) < (1/2)u = (1/2)(s+r); \text{ wherefrom, } r < s;$$

and the conclusion follows.

iv): Let $a \geq 0$ be arbitrary fixed. We have to establish that $A := \{t \in R_+; t \leq \gamma(t) + a\}$ is bounded. But (from the properties of ω and γ)

$$A \subseteq \{t \in R_+; t \leq \omega(t + \gamma(t)) + a\} \subseteq \{t \in R_+; t \leq \omega(2t) + a\} =$$
$$\{t \in R_+; t \leq \eta(t) + a\} = \{t \in R_+; \eta^*(t) \leq a\};$$

so that, as the last subset is bounded, we are done. □

IV) The obtained properties of the right complementary inverse $\gamma \in \mathcal{F}_0(R_+)$ associated to the initial function $\omega \in \mathcal{F}_0(re, in)(R_+)$ are basic tools for getting a lot of relative type statements involving real sequences, to be used further. The first result in this series involves the conservation of boundedness and zero-convergence.

Theorem 1.25. *Let the sequences $(a_n; n \geq 0)$ and $(b_n; n \geq 0)$ in R_+ and the function $\omega \in \mathcal{F}_0(re, in)(R_+)$ be such that*

(44-i) $a_n \leq b_n + \omega(a_n)$, *for all* n
(44-ii) ω *is strongly regressive* ($\omega(t+0) < t$, $\forall t > 0$)
(44-iii) ω *is complementary coercive* ($\omega^* := id - \omega$ *is coercive*).

Then, necessarily,

(44-a) (b_n) *is bounded implies* (a_n) *is bounded*
(44-b) (b_n) *is zero-convergent implies* (a_n) *is zero-convergent.*

Proof. i): By definition, $\beta := \sup\{b_n; n \geq 0\}$ exists in R_+. This, by the imposed hypothesis, yields

$(\forall n)$: $a_n \leq \beta + \omega(a_n)$; that is $\omega^*(a_n) \leq \beta$.

Combining with the coerciveness of ω^*, yields the desired fact.

ii): Suppose that (b_n) is zero-convergent. In particular, (b_n) is bounded; hence, by the preceding stage, (a_n) is bounded too. From a classical principle involving such objects, (a_n) has convergent subsequences. Let $(a_{i(n)}; n \geq 0)$ be one of these; hence, $a_{i(n)} \to a$ as $n \to \infty$, for some $a \in R_+$. Suppose by contradiction that $a > 0$. Denoting $(A_{i(n)} = \max\{a_{i(n)}, a + 2^{-n}\}; n \geq 0)$, we have

$$(a_{i(n)} \leq b_{i(n)} + \omega(A_{i(n)}),\ n \geq 0) \text{ and } (A_{i(n)} \to a+ \text{ as } n \to \infty).$$

This, by a limit process, yields (via strongly regressive property of ω)

$$a \leq \omega(a + 0) < a; \text{ absurd; whence, } a = 0.$$

In other words, all convergent subsequences $(a_{i(n)}; n \geq 0)$ of (a_n) fulfill $\lim_n a_{i(n)} = 0$. But then, $\lim_n a_n = 0$; and the conclusion follows. □

A bi-dimensional version of this result is stated as follows. By a *pseudometric* over N, we mean any mapping $e : N \times N \to R_+$. If, in addition,

$e(.,.)$ is reflexive: $e(n,n) = 0$, $\forall n$

we say that $e(.,.)$ is a *r-pseudometric.* Given the r-pseudometric $e(.,.)$, call it *Cauchy,* in case of $[e(n,m) \to 0$ as $n,m \to \infty$, $n \leq m]$; and *asymptotic,* provided $[e(n,n+1) \to 0$ as $n \to \infty]$. Clearly,

$e(.,.)$ is Cauchy implies $e(.,.)$ is asymptotic;

the reciprocal is not in general true.

Theorem 1.26. *Let the r-pseudometrics $a(.,.)$ and $b(.,.)$ over N and the function $\omega \in \mathcal{F}_0(re,in)(R_+)$ be such that*

(45-i) $a(n,m) \leq b(n,m) + \omega(a(n,m))$, for all $n,m \in N$, $n \leq m$
(45-ii) ω is strongly regressive ($\omega(t+0) < t$, $\forall t > 0$)
(45-iii) ω is complementary coercive ($\omega^* := id - \omega$ is coercive).

Then, necessarily,

(45-a) $b(.,.)$ is asymptotic implies $a(.,.)$ is asymptotic
(45-b) $b(.,.)$ is Cauchy implies $a(.,.)$ is Cauchy.

Proof. There are two steps to be passed.

i): Suppose that

$$b(.,.) \text{ is asymptotic } [(\beta_n := b(n, n+1); n \geq 0) \text{ fulfills } \beta_n \to 0].$$

We have to establish that

$$a(.,.) \text{ is asymptotic } [(\alpha_n := a(n, n+1); n \geq 0) \text{ fulfills } \alpha_n \to 0].$$

This, however, by the sequential condition

$$\text{(seq)} \quad \alpha_n \leq \beta_n + \omega(\alpha_n), \text{ for all } n \in N,$$

is deductible from the preceding statement; so that, our claim follows.

ii): Now, let us show that

$$b(.,.) \text{ is Cauchy implies } a(.,.) \text{ is Cauchy.}$$

The last property means (by definition)

$$\forall \varepsilon > 0, \text{ there exists } j(\varepsilon), \text{ such that } j(\varepsilon) \leq n \leq m \text{ implies } a(n, m) \leq \varepsilon;$$

or, equivalently [passing to the successor $J(\varepsilon) := j(\varepsilon) + 1$ and remembering that $a(.,.)$ is r-pseudometric]

$$\forall \varepsilon > 0, \text{ there exists } j(\varepsilon), \text{ such that } j(\varepsilon) < n < m \text{ implies } a(n, m) \leq \varepsilon.$$

The negation of this means: there exists $\varepsilon > 0$, such that

$$C(j) = \{(n, m) \in N \times N; j < n < m, a(n, m) > \varepsilon\} \neq \emptyset, \forall j.$$

Denote, for each j

$$n(j) = \min \text{Dom}(C(j)), \; m(j) = \min C(j)(n(j)).$$

Fix $j(0) \in N$. By this construction, there exists a couple of ranks $(j(1) = n(j(0)), j(2) = m(j(0)))$, with $j(0) < j(1) < j(2)$, $a(j(1), j(2)) > \varepsilon$. Further, given $j(2)$, there exists a couple of ranks $(j(3) = n(j(2)), j(4) = m(j(2)))$, with $j(2) < j(3) < j(4)$, $a(j(3), j(4)) > \varepsilon$. The procedure may continue indefinitely (without any choice techniques); and yields a strictly ascending sequence of ranks $(j(n); n \geq 0)$ [hence, $\lim_n j(n) = \infty$], with

$$\alpha_n := a(j(2n+1), j(2n+2)) > \varepsilon, \forall n.$$

On the other hand, by the Cauchy property

$$\beta_n := b(j(2n+1), j(2n+2)) \to 0, \text{ as } n \to \infty.$$

This, in combination with the sequential condition

$$\text{(seqq)} \quad \alpha_n \leq \beta_n + \omega(\alpha_n), \text{ for all } n \in N$$

gives a contradiction with respect to the preceding statement. Hence, necessarily, $a(.,.)$ is Cauchy; as claimed. □

An interesting question to be posed is that of such conclusions being retainable when some of the basic hypotheses upon ω would be dropped. We conjecture that the answer is negative.

V) A good illustration of the statement above may be described as below. Let (X, d) and (J, g) be a couple of semimetric spaces. Denote

$\mathcal{F}(0, cont)(R_+, R_+ \cup \{\infty\})$ = the class of all $\psi \in \mathcal{F}(R_+, R_+ \cup \{\infty\})$ with $(\lim_{t\to 0+}\psi(t) =)\psi(0+0) = 0 = \psi(0)$.

The following relative type statement is in effect for us.

Theorem 1.27. *Let the sequences (x_n) in X, (t_n) in J, and the functions $\psi \in \mathcal{F}(0, cont)(R_+, R_+ \cup \{\infty\})$, $\omega \in \mathcal{F}_0(re, in)(R_+)$ be such that*

(46-i) $d(x_n, x_m) \leq \psi(g(t_n, t_m)) + \omega(d(x_n, x_m))$, whenever $n \leq m$
(46-ii) ω is strongly regressive ($\omega(t+0) < t$, $\forall t > 0$)
(46-iii) ω is complementary coercive ($\omega^* := id - \omega$ is coercive).

Then, necessarily,

(46-a) (t_n) is g-asymptotic implies (x_n) is d-asymptotic
(46-b) (t_n) is g-Cauchy implies (x_n) is d-Cauchy.

Proof. There are two steps to be passed.

i): Denote, for simplicity,

$$(\alpha_n := d(x_n, x_{n+1}); n \geq 0),\ (\delta_n := g(t_n, t_{n+1}); n \geq 0);$$

note that, by the imposed condition,

$$\alpha_n \leq \psi(\delta_n) + \omega(\alpha_n),\ \forall n.$$

We have to establish that $[\delta_n \to 0$ implies $\alpha_n \to 0]$. By the choice of ψ, the sequence (in $R_+ \cup \{\infty\}$) $(\beta_n := \psi(\delta_n); n \geq 0)$ fulfills $\beta_n \to 0$ as $n \to \infty$; hence, there exists $i \in N$ such that

$$B_n := \beta_{n+i} < \infty, \text{ for all } n.$$

Denote also $(A_n = \alpha_{n+i}; n \geq 0)$. By these remarks, (A_n) and (B_n) are sequences over R_+ with (by the above)

$$[A_n \leq B_n + \omega(A_n),\ \forall n], \text{ and } [B_n \to 0 \text{ as } n \to \infty].$$

By the above sequential type relative result we get

$$A_n \to 0; \text{ whence, } \alpha_n \to 0;$$

and the first conclusion follows.

ii): Suppose that

$$(t_n) \text{ is } g\text{-Cauchy: } g(t_n, t_m) \to 0 \text{ as } n, m \to \infty,\ n < m.$$

We have to establish that

$$(x_n) \text{ is } d\text{-Cauchy: } d(x_n, x_m) \to 0 \text{ as } n, m \to \infty,\ n < m.$$

A direct way of doing this is the recursion to the double sequential relative statement we just established. However, for completeness reasons, we will use a different reasoning. Suppose by contradiction that this last property is not true. As (x_n) is d-asymptotic, there exists, by a previous auxiliary fact, a number $\delta > 0$, a rank sequence $(J(k); k \geq 0)$, and a couple of rank-sequences $(m(k))$, $(n(k))$, with

$$\text{(p-1)}\quad k+1 \leq J(k) \leq m(k) < n(k),\ d(x_{m(k)}, x_{n(k)}) > \delta,\ \forall k \geq 0$$
$$\text{(p-2)}\quad u_k := d(x_{m(k)}, x_{n(k)}) \to \delta+ \text{ as } k \to \infty.$$

Denote also $(v_k := g(t_{m(k)}, t_{n(k)}); k \geq 0)$; clearly,

$$v_k \to 0 \text{ as } k \to \infty \text{ (as } (t_n) \text{ is } g\text{-Cauchy).}$$

By the imposed hypothesis,

$$u_k \leq \psi(v_k) + \omega(u_k),\ \forall k.$$

This, along with the preceding stage, yields

$$u_k \to 0 \text{ as } k \to \infty; \text{ contradiction.}$$

Hence, our working hypothesis cannot be accepted; wherefrom the assertion follows. The proof is thereby complete. □

An interesting question is that of this conclusion being retained when the hypothesis about ω is to be substituted by the weaker one

$$\omega \text{ is Matkowski admissible (or, equivalently: Meir-Keeler admissible).}$$

The answer to this is negative, in general; we do not give details.

1.4 Classes of Natural Progressions

Abstract. Some preliminary facts are discussed about sub-aprogressions, admissible and scattered sets, as well as uniformly admissible relations over natural numbers. Then, a characterization is given for the non-Cauchy property of asymptotic sequences in relational metric spaces.

Subject-Class 2020. 47H10 (Primary), 54H25 (Secondary).

Keywords. Chain, sub-aprogression, admissible and scattered set, regularity, relational metric space, Cauchy criteria.

1.4.1 *Sub-aprogressions*

Let N be the set of natural numbers. Remember that the map $d : N \times N \to N$ given as $(d(n,m) = |n-m|,\ n,m \in N)$ is a *(standard) metric* over N. We say that the part $U \in \exp(N)$ is an *unbounded chain*, provided

$$\sup(U) = \infty; \text{ meaning: } U(n,<) := \{u \in U; n < u\} \neq \emptyset,\ \forall n \in N;$$

the class of all these is denoted as $\text{unb}(N)$. Note that, any $U \in \text{unb}(N)$ may be represented as

$$U = \{u_n; n \geq 0\}, \text{ where } u_0 = \min(U) \text{ and}$$
$$(u_n; n \geq 0) \text{ is strictly ascending } (i < j \text{ implies } u_i < u_j).$$

Given $U \in \text{unb}(N)$, let $\Sigma_U \in \mathcal{F}(U)$ stand for the *(immediate) successor* function attached to U, as $[\Sigma_U(p) = \min U(p,<),\ p \in U]$. Given $h \in N(1,\leq)$, let us say that U is a *h-arithmetic-progression* (in short: *h-aprogression*), when $[d(p,\Sigma_U(p)) = h,\ \forall p \in U]$. In particular, under the precise representation $U = \{u_n; n \geq 0\}$, this definition writes:

$$(u_{n+1} - u_n = h,\ \forall n); \text{ referred to as: } (u_n; n \geq 0) \text{ is } h\text{-}aprogression.$$

When $h \in N(1,\leq)$ is generic, we say that $U \in \text{unb}(N)$ is an *aprogression*.

An extension of this concept is as follows. Given $k \in N(1,\leq)$, let us say that $U \in \text{unb}(N)$ is a *k-sub-aprogression*, if $[d(p,\Sigma_U(p)) \leq k,\ \forall p \in U]$. In particular, under the representation $U = \{u_n; n \geq 0\}$, this writes:

$$u_{n+1} - u_n \leq k,\ \forall n; \text{ referred to as: } (u_n; n \geq 0) \text{ is } k\text{-}sub\text{-}aprogression.$$

The class of all such $k \in N(1,\leq)$ will be denoted as $\text{sapr}(U)$; and referred to as the *sub-aprogression domain* of U. Clearly,

$$\text{sapr}(U) \text{ is hereditary: } k \in \text{sapr}(U),\ q \geq k \Longrightarrow q \in \text{sapr}(U).$$

When $\text{sapr}(U)$ is nonempty, we say that U is *sub-aprogression*.

A natural question is to determine whether any unbounded chain U of N is sub-aprogression. The answer is negative, as results from

Example 1.2.

Let $n \mapsto \varphi(n)$ be a sequence in R_+ with $\lim_n \varphi(n) = \infty$; and U be an unbounded chain of N; hence, $U = \{u_n; n \geq 0\}$, where $u_0 = \min(U)$ and $(u_n; n \geq 0)$ is strictly ascending. Suppose that

$$u_{n+1} - u_n \geq \varphi(n), \text{ for almost all } n \in N.$$

[Here, given the property $\pi(n)$ depending on $n \in N$, we say that π holds for *almost all* $n \in N$ [written: $(\forall\forall n, \pi(n))$], if there exists $h = h(\pi) \in N$ such that $\pi(n)$ holds, $\forall n \geq h$.] We claim that $\text{sapr}(U) = \emptyset$; which means: U is not sub-aprogression. Suppose not: there exists $k \in N(1, \leq)$, such that U is k-sub-aprogression: $[u_{n+1} - u_n \leq k$, for all $n \in N]$. This, along with the choice of our sequence, gives $[k \geq \varphi(n)$, for almost all $n \in N]$. Passing to limit as $n \to \infty$ in this relation gives $k = \infty$; contradiction.

A relative type version of these developments is to be introduced as follows. Given $U \in \text{unb}(N)$, let us define the concepts

U is *relatively aprogression*:
$U \supseteq V$, for some aprogression $V \in \text{unb}(N)$
U is *relatively sub-aprogression*:
$U \supseteq V$, for some sub-aprogression $V \in \text{unb}(N)$.

Clearly, for each unbounded chain U of N,

relatively aprogression implies relatively sub-aprogression;

but the reciprocal is not in general true.

By these definitions, it trivially follows that any sub-aprogression is relatively sub-aprogression. So, it is natural to ask whether the converse inclusion is valid too. The positive answer to this is contained in

Proposition 1.35. *For each* $U \in \text{unb}(N)$, *we have*

(11-1) *relatively sub-aprogression* $\Longrightarrow$ *sub-aprogression*
(11-2) *relatively sub-aprogression* $\Longleftrightarrow$ *sub-aprogression.*

Proof. i): Suppose that $U \in \text{unb}(N)$ is relatively sub-aprogression:

$U \supseteq V$, where $V \in \text{unb}(N)$ is sub-aprogression.

Without loss, assume that $\min(V) = \min(U)$; for, otherwise, when $\min(V) > \min(U)$, one has that $W := V \cup \{\min(U)\}$ is a subset of U with the sub-aprogression property. Fix some $k \in \text{sapr}(V)$; hence, $[d(v, \Sigma_V(v)) \leq k$, for all $v \in V]$. Let $u \in U$ be arbitrary fixed. By the minimum convention above, one has

$$v \leq u < \Sigma_V(v), \text{ for some (uniquely determined) } v \in V.$$

As $\Sigma_V(v) \in U$, we must get (by the definition of immediate successor)

$$v \leq u < \Sigma_U(u) \leq \Sigma_V(v); \text{ so, } d(u, \Sigma_U(u)) \leq d(v, \Sigma_V(v)) \leq k.$$

Hence, $k \in \text{sapr}(U)$; and we are done.

ii): Evident, by the above observation. □

Concerning these facts, remember that for each U in $\text{unb}(N)$

U is relatively aprogression implies U is relatively sub-aprogression.

This, along with our preceding developments tells us that

$$(\forall U \in \text{unb}(N)): U\text{=relatively aprogression} \Longrightarrow U\text{=sub-aprogression}.$$

So, we may ask whether the converse inclusion holds

$$(\forall U \in \text{unb}(N)): U \text{ is sub-aprogression} \Longrightarrow U \text{ is relatively aprogression}.$$

This is a difficult question; some partial aspects of it were discussed in Wagstaff Jr [262].

1.4.2 *Admissible sets*

In what follows, the concept of *admissible* set is introduced; and a characterization of it in terms of (relative) sub-aprogressions is being proposed.

Remember that the (nonempty) part U of N is an *unbounded chain* of N, provided $\sup(U) = \infty$; the class of all these is denoted as $\text{unb}(N)$. Note that any $U \in \text{unb}(N)$ is to be written as $U = \{u_n; n \geq 0\}$, where $u_0 = \min(U)$ and the sequence (u_n) in N is strictly ascending.

(A) Given $h \in N(1, \leq)$, let us say that $U \in \text{unb}(N)$ is *h-admissible*, if

(h-adm) for each $n \in N$ there exists $u \in U$ with $n < u \leq n + h$.

The class of all these $h \in N(1, \leq)$ will be denoted as $\text{adm}(U)$; referred to as the *admissible domain* of U. Clearly,

$$\text{adm}(U) \text{ is hereditary: } h \in \text{adm}(U),\ p \geq h \Longrightarrow p \in \text{adm}(U).$$

When adm(U) is nonempty, we say that U is *admissible.*

A stronger version of the concept above may be given as follows. Given $h \in N(1, \leq)$, let us say that $U \in \text{unb}(N)$ is *h-super-admissible*, if

(h-sup-adm) $\forall n \in N(1, \leq)$, $\exists u \in U$, with $n < u \leq n + h$.

The class of all such $h \in N(1, \leq)$ will be denoted as supadm(U); referred to as the *super-admissible domain* of U. Clearly,

supadm(U) is hereditary: $h \in \text{supadm}(U)$, $p \geq h \Longrightarrow p \in \text{supadm}(U)$.

When supadm(U) is nonempty, we say that U is *super-admissible.*

The connections between these two concepts is described in

Proposition 1.36. *Let the unbounded chain U of N be given. Then,*

(21-1) *($\forall h \geq 1$): U is h-admissible implies U is h-super-admissible*

(21-2) *($\forall h \geq 1$): U is h-super-admissible implies U is $(h+1)$-admissible*

(21-3) *U is admissible iff U is super-admissible.*

Proof. i): Evident, by definition.

ii): Suppose that U is h-super-admissible; and let $n \in N$ be arbitrary fixed. If $n > 0$ (that is: $n \geq 1$) then, by definition, there exists $u \in U$ with $n < u \leq n + h < n + h + 1$. Suppose now that $n = 0$. Again by definition, there exists $v \in U$ with $1 < v \leq 1 + h$; whence, $0 < v \leq 1 + h$. This shows that U is $(h + 1)$-admissible; and the assertion follows.

iii): Evident, by definition. □

(B) As a consequence, the admissible and super-admissible concepts are equivalent. However, for technical reasons, it will be more convenient working with the admissible concept. A good characterization of it in terms of (relative) sub-aprogressions may be given as follows.

Theorem 1.28. *Let the unbounded chain U of N be given. Then,*

(21-a) *U is admissible iff U is sub-aprogression*

(21-b) *U is admissible iff U is relatively sub-aprogression.*

(21-c) *U is admissible when U is relatively aprogression.*

Proof. i): Two inclusions must be verified.

i-1): Suppose that U is admissible; that is: [U is h-admissible, for some $h \in N(1, \leq)$]; we have to establish that U is sub-aprogression. By definition, $U = \{u_n; n \geq 0\}$, where $u_0 = \min(U)$ and the sequence (u_n) in N is strictly ascending. We claim that (u_n) is h-sub-aprogression; and, from this, all is done. Let $i \in N$ be arbitrary fixed. From the h-admissible property, there exists, for $u_i \in N$, some $r = u_j$ in U, with $u_i < r = u_j \leq u_i + h$; hence, in particular, $i < j$. From the last inequality, one gets $i + 1 \leq j$; so that, $u_i < u_{i+1} \leq u_j \leq u_i + h$. This yields $u_{i+1} - u_i \leq h$; and proves our claim.

i-2): Suppose that U is sub-aprogression; we have to derive that U is admissible. By definition, the strictly ascending sequence $(u_n; n \geq 0)$ appearing in the representation $U = \{u_n; n \geq 0\}$ is sub-aprogression; so, there exists $h \in N(1, \leq)$ such that $(u_n; n \geq 0)$ is h-sub-aprogression. Denote $k = \max\{u_0, h\}$; we claim that U is k-admissible. Let $n \in N$ be arbitrary fixed. The property is clear for $n < u_0$; for, in such a case, $[n < u_0 \leq k \leq n + k]$; so, without loss, one may assume that $n \geq u_0$. As (u_n) is strictly ascending, there exists a uniquely determined rank $i \in N$ with $(u_{i+1} - u_i \leq h$ and) $u_i \leq n < u_{i+1}$; and this, along with $[u_{i+1} \leq u_i + h \leq n + k]$ proves the desired fact, with $r = u_{i+1}$.

ii): As precise, the concepts of sub-aprogression and relatively sub-aprogression are equivalent over the class $\text{unb}(N)$; so, we are done.

iii): Evident, in view of each relatively aprogression in N being endowed with the sub-aprogression property. □

Concerning the obtained facts, a natural question is to determine whether any unbounded chain of N is admissible. A negative answer to this is available by the characterization of admissible sets we just derived and a previous example involving sub-aprogressions.

1.4.3 *Scattered sets*

In what follows, another characterization of the above concepts in terms of scattered sets of natural numbers is proposed. Remember that the standard metric over N is $(d(n, m) = |n - m|, n, m \in N)$.

For each $a \in N$ and $B \in \exp(N)$, denote

$$\text{dist}(a, B) - \inf\{d(a, b); b \in B\} = \min\{d(a, b); b \in B\}.$$

(The min-notation is due to the fact that the infimum in the right part is effectively attained.) Note that, as a consequence of this,

$$\text{(for each } a \in N,\ B \in \exp(N)\text{): } \text{dist}(a, B) = 0 \text{ iff } a \in B.$$

Having these precise, denote

$$(A, B \in \exp(N))\text{: } \Gamma(A, B) = \sup\{\text{dist}(a, B); a \in A\}.$$

We therefore constructed a map $\Gamma : \exp(N) \times \exp(N) \to N \cup \{\infty\}$; the basic properties of it are concentrated in

Proposition 1.37. *Under these conventions, we have*

(31-1) $(\forall A, B \in \exp(N))$: $\Gamma(A, B) = 0$ *iff* $A \subseteq B$,
(31-2) $(\forall A, B, C \in \exp(N))$: $\Gamma(A, C) \leq \Gamma(A, B) + \Gamma(B, C)$.

Proof. i): The right to left inclusion is clear; so, it remains to establish the left to right inclusion. Let $A, B \in \exp(N)$ be such that $\Gamma(A, B) = 0$; i.e.,

$\text{dist}(a, B) = 0$, for each $a \in A$.

By a previous observation involving our point to set distance,

$a \in B$, for each $a \in A$; whence, $A \subseteq B$.

ii): Let the parts $A, B, C \in \exp(N)$ be arbitrary fixed. For each $a \in A$, $b \in B$, and $c \in C$, we have (by the triangular inequality)

$d(a, c) \leq d(a, b) + d(b, c)$; whence,
$\text{dist}(a, C) \leq d(a, b) + \text{dist}(b, C) \leq d(a, b) + \Gamma(B, C)$,

if we take the infimum over $c \in C$ and remember the definition of $\Gamma(B, C)$. Passing to infimum over $b \in B$ in this last relation, gives

$\text{dist}(a, C) \leq \text{dist}(a, B) + \Gamma(B, C)$, $\forall a \in A$;

wherefrom, taking the supremum over $a \in A$, we are done. □

As a consequence of this, the application

$\Delta(A, B) = \max\{\Gamma(A, B), \Gamma(B, A)\}$, $A, B \in \exp(N)$

is a *generalized metric* over $\exp(N)$, in the sense described by Luxemburg [163] and Jung [130]. In fact, it is nothing else than the well-known Hausdorff-Pompeiu (generalized) metric over $\exp(N)$; cf. Kuratowski [155, Ch. 2, Sec. 15.7].

We may now introduce a basic notion. Let U be an unbounded chain of N. Given $k \in N(1, \leq)$, let us say that U is k-*scattered*, provided

$\Gamma(N, U) \leq k$ (or, equivalently: $\Delta(N, U) \leq k$):
$\forall n \in N$, $\exists u \in U$, such that $d(n, u) \leq k$;

the class of all these writes $\text{sca}(U)$ (the *scattering domain* of U). Clearly,

$\text{sca}(U)$ is *hereditary*: $k \in \text{sca}(U)$, $q \geq k \implies q \in \text{sca}(U)$;

if sca(U) is nonempty, then U is called *scattered* (in N).

In the following, we provide a useful characterization of scattered unbounded chains of N. Take some $U \in \text{unb}(N)$; hence, $U = \{u_n; n \geq 0\}$, where $u_0 = \min(U)$ and the sequence $(u_n; n \geq 0)$ in N is strictly ascending. Given $k \in N(1, \leq)$, remember that U is *k-sub-aprogression*, provided

its associated sequence (u_n) is k-sub-aprogression ($u_{n+1} - u_n \leq k$, $\forall n$).

The class sapr(U) of all such $k \in N(1, \leq)$ will be referred to as the *sub-aprogression domain* of U. Clearly,

sapr(U) is hereditary: $k \in \text{sapr}(U)$, $q \geq k \Longrightarrow q \in \text{sapr}(U)$;

if sapr(U) is nonempty, we say that U is *sub-aprogression.*

Theorem 1.29. *Let $U = \{u_n; n \geq 0\}$ (where $u_0 = \min(U)$ and $(u_n; n \geq 0)$ is strictly ascending) be an unbounded chain of N. Then, for each $k \geq 1$,*

(31-a) U is k-scattered $\Longrightarrow$ U is $(2k+1)$-sub-aprogression
(31-b) U is k-sub-aprogression implies U is $\max\{u_0, k\}$-scattered.

Hence, putting these together, the generic equivalence holds:

(31-c) U is scattered iff U is sub-aprogression.

Proof. i) Suppose that U is k-scattered; and let the rank $i \geq 0$ be arbitrary fixed. Given the natural number $n = u_i + k + 1$, we have

(for all $p \leq i$): $d(n, u_p) \geq d(n, u_i) = k + 1 > k$.

By a previous observation,

$\text{dist}(n, U) = d(n, u_j)$, for some $j \geq 0$.

This, along with $\text{dist}(n, U) \leq k$, yields (by the preceding evaluation)

$(|n - u_j| =) d(n, u_j) \leq k$, for some $j > i$.

If $u_j < n$, then (as $u_i < u_{i+1} \leq u_j$), one gets (by the choice of n)

$$u_{i+1} - u_i < n - u_i = k + 1 < 2k + 1;$$

and, if $u_j \geq n$, the distance relation involving (n, u_j) yields

$(u_{i+1} \leq)\ u_j \leq n + k = u_i + 2k + 1$; whence: $u_{i+1} - u_i \leq 2k + 1$.

This, by the arbitrariness of i, tells us that U is $(2k + 1)$-sub-aprogression.

ii) Let U be k-sub-aprogression; and put $h := \max\{u_0, k\}$. Clearly,

$$\text{(for all } n \leq u_0\text{): } \operatorname{dist}(n, U) \leq d(n, u_0) = u_0 - n \leq u_0 \leq h.$$

Further, let $n > u_0$ be arbitrary fixed. Since (u_n) is unbounded, there exists some (unique) rank $p = p(n) \geq 0$ such that $u_p \leq n < u_{p+1}$. This yields

$$d(n, u_p) = n - u_p < u_{p+1} - u_p \leq k; \text{ hence, } \operatorname{dist}(n, U) \leq k \leq h.$$

Putting these together, it follows that U is h-scattered.

iii) Evident, by definition (and the above steps). □

Now, a natural question is to determine the connections between admissible and scattered sets. The answer to this is contained in

Proposition 1.38. *We have, over the class* unb(N), *that*

(32-1) the concepts of admissible, sub-aprogression, and scattered are equivalent to each other
(32-2) the concepts of admissible, relatively sub-aprogression, sub-aprogression and scattered are equivalent to each other.

Proof. i): By our previous facts, we have for each $U \in$ unb(N)

U is admissible iff U is sub-aprogression
U is scattered iff U is sub-aprogression;

and this proves the first conclusion.

ii): By the same previous facts, we also have for each $U \in$ unb(N)

U is relatively sub-aprogression iff U is sub-aprogression.

Combining with the first stage, we get the second conclusion. □

Technically speaking, the core of our second conclusion consists in the concepts of relatively sub-aprogression and sub-aprogression being equivalent. A first proof of this was already carried out in a preceding place. Here, we propose another proof of the same based on scattering methods.

Proposition 1.39. *Let* $U \in$ unb(N) *be arbitrary fixed. Then,*

(33-1) U is relatively sub-aprogression implies U is scattered; hence, U is sub-aprogression
(33-2) U is relatively sub-aprogression iff U is sub-aprogression
(33-3) the concepts of admissible, relatively sub-aprogression, sub-aprogression and scattered are all equivalent for U.

Proof. i): Let $U \in$ unb(N) be relatively sub-aprogression; hence,

$U \supseteq V$, where $V \in \mathrm{unb}(N)$, $V =$ sub-aprogression.

By definition, V is k-sub-aprogression, for some $k \in \mathrm{sapr}(V)$ (fixed in the sequel). This tells us that $V = \{v_n; n \geq 0\}$, where

(p-1) $v_0 = \min(V)$ and (v_n) is strictly ascending (hence, unbounded)
(p-2) $v_{i+1} - v_i \leq k$, for all i.

Combining with a preceding statement, we have

V is h-scattered $[\Delta(N, V) \leq h]$, where $h = \max\{v_0, k\}$.

But then, from the triangular property of our generalized metric Δ,

$\Delta(N, U) \leq \Delta(N, V) + \Delta(V, U) = \Delta(N, V) \leq h$;

whence, U is h-scattered. This, again by the preceding statement, tells us that U is sub-aprogression; and conclusion follows.

ii): Evident, by the above discussed facts. □

1.4.4 *Uniformly admissible relations*

Let $\mathcal{S} \subseteq N \times N$ be a (nonempty) relation over N. Given $h \in N(1, \leq)$, let us say that $\mathcal{S}$ is *h-uniformly-admissible*, (in short: *h-uadmissible*), provided

$\forall (m, n) \in (<; N)$, $\exists p \in N$: $m\mathcal{S}p$, $n < p \leq n + h$.

Here, for simplicity reasons, we denoted

$(<; N) = \{(m, n) \in N \times N; m < n\}$; (the *graph* of $(<)$ over N).

The class of all such elements h will be denoted as $\mathrm{uad}(\mathcal{S})$ (the *uadmissible region* of $\mathcal{S}$). Clearly,

$\mathrm{uad}(\mathcal{S})$ is *hereditary*: $h \in \mathrm{uad}(\mathcal{S})$, $q \geq h \Longrightarrow q \in \mathrm{uad}(\mathcal{S})$;

when $\mathrm{uad}(\mathcal{S})$ is nonempty, we say that $\mathcal{S}$ is *uadmissible*. A functional way of expressing the previous convention is the following. Let $\mathcal{S}$ be a h-uadmissible relation over N (where $h \geq 1$); this writes:

$\forall (m, n) \in (<; N)$, $\mathcal{L}(m, n) := \{q \in N[1, h]; m\mathcal{S}(n + q)\}$ is nonempty;

where $N[a, b] = \{n \in N; a \leq n \leq b\}$, $a, b \in N$, $a \leq b$. We therefore associated a multivalued function $\mathcal{L} : (<; N) \to \exp(N[1, h])$ to the relation $\mathcal{S}$. Let $L : (<; N) \to N[1, h]$ stand for its minimal selection

$L(m, n) = \min \mathcal{L}(m, n)$, $(m, n) \in (<; N)$;

we then say that $L(.)$ is the associated (univalent) function attached to the h-uadmissible relation $\mathcal{S}$ over N. Clearly, no choice techniques were used here; hence, this construction is valid over the strongly reduced system (ZF-AC). By the very definition above, we have

$$m\mathcal{S}(n + L(m,n)), \text{ for each } (m,n) \in (<; N).$$

Concerning some concrete classes of uadmissible relations, a particular setting to solve this question is that characterized as follows. Denote

$$\mathcal{S}(0) := \{n \in N; 0\mathcal{S}n\} \text{ (the } \textit{zero-section} \text{ of } \mathcal{S}).$$

Proposition 1.40. *Suppose that*

$\mathcal{S}$ is translation invariant: $(m,n) \in \mathcal{S} \Longrightarrow (m+j, n+j) \in \mathcal{S}$, $\forall j \in N$.

Then,

(41-1) $(\forall h \geq 1)$: $\mathcal{S}$ is h-uadmissible iff $\mathcal{S}(0)$ is h-super-admissible
(41-2) $(\forall h \geq 1)$: $\mathcal{S}$ is h-uadmissible iff $\mathcal{S}(0)$ is $(h+1)$-admissible
(41-3) $\mathcal{S}$ is uadmissible iff $\mathcal{S}(0)$ is admissible.

Proof. i): Assume that $\mathcal{S}$ is h-uadmissible. Taking $m = 0$, gives

$$\forall n \geq 1, \exists p \in \mathcal{S}(0)\text{: } n < p \leq n + h;$$

whence, $\mathcal{S}(0)$ is h-super-admissible. Conversely, assume that $\mathcal{S}(0)$ is h-super-admissible; and let $m, n \in N$ be such that $m < n$; hence, $n - m > 0$ (or, equivalently: $n - m \geq 1$). By hypothesis, there exists $r \in \mathcal{S}(0)$ with $n - m < r \leq n - m + h$. Denote $p = m + r$. By the preceding relation, $n < p \leq n + h$. In addition, as $\mathcal{S}$ is translation invariant and $0\mathcal{S}r$, we get $m\mathcal{S}(m + r)$; that is: $m\mathcal{S}p$. Summing up, $\mathcal{S}$ is h-admissible; as claimed.

ii): Evident, from a preceding result.

iii): Evident, by definition. □

Some concrete examples of such relations will be given a bit further.

1.4.5 *Cauchy criteria over rss*

Let X be a nonempty set. Take a *semimetric* (that is: reflexive, triangular, symmetric map) $d : X \times X \to R_+$; and let $\mathcal{G}$ be a relation over X; the triplet $(X, d, \mathcal{G})$ will be termed a *relational semimetric space.*

In the following, a basic d-Cauchy criterion will be stated over the structure $(X, d, \mathcal{G})$. Some preliminaries are needed.

Denote, for each sequence $(x_n; n \geq 0)$ in X, and each $p, q \in N$, $p \leq q$,

$$x[p,q] = \{x_i; i \in N[p,q]\},\ x]p,q[= \{x_i; i \in N]p,q[\},$$
$$x]p,q] = \{x_i; i \in N]p,q]\},\ x[p,q[= \{x_i; i \in N[p,q[\}.$$

Remember that the d-Cauchy property of (x_n) writes:

(d-C) $\forall \varepsilon \in R^0_+$, $\exists i = i(\varepsilon) \in N$: $i \leq m < n \Longrightarrow d(x_m, x_n) < \varepsilon$.

Also, let us say that (x_n) is *d-asymptotic*, provided

(d-asy) $r_n := d(x_n, x_{n+1}) \to 0$ as $n \to \infty$.

Clearly, each d-Cauchy sequence is d-asymptotic too; the reciprocal is not in general true. This tells us that the d-Cauchy criteria we are looking for are to be sought in the class of d-asymptotic sequences. The general scheme of handling these is as below. Let (π) be a property attached to the sequences in X. Suppose that the following inclusion is valid

(for each d-asymptotic sequence (x_n)):
(x_n) is not d-Cauchy implies (x_n) has the property (π).

Then, the underlined d-Cauchy criterion writes

(for each d-asymptotic sequence (x_n)):
(x_n) is not endowed with the property (π) implies (x_n) is d-Cauchy.

To get a property of this type, we need some conventions. Given the d-asymptotic sequence $(x_n; n \geq 0)$ and $\varepsilon > 0$, let us say that $i \in N(1, \leq)$ is *ε-regular*, provided

$i \leq n$ implies $d(x_n, x_{n+1}) < \varepsilon$.

The class $\mathcal{Z}(\varepsilon)$ of all these numbers is nonempty; so that

$(\forall \varepsilon > 0)$: $Z(\varepsilon) = \min \mathcal{Z}(\varepsilon)$ is well defined, as an element of $N(1, \leq)$;
with, in addition: $d(x_n, x_{n+1}) < \varepsilon$, for all $n \geq Z(\varepsilon)$.

Further, note that by this very definition,

$(\forall j \geq 1)$: $\Delta_n(j) := \text{diam}(x[n, n+j]) < \varepsilon$,
for all $n \geq Z(\varepsilon/j)$; whence, $\Delta_n(j) \to 0$ as $n \to \infty$.

Let $(m(k); k \geq 0)$ and $(n(k); k \geq 0)$ be a couple of rank sequences; we call it *separated*, provided

$(m(k) < n(k)$, for all $k)$, and $\lim_k m(k) = \lim_k n(k) = \infty$.

Technically speaking, the d-Cauchy criterion (for d-asymptotic sequences) we are looking for is essentially related to the existence of certain couples of rank sequences $(m(k))$ and $(n(k))$, with the supplementary properties

$(m(k))$ and $(n(k))$ are separated, and $x_{m(k)}\mathcal{G}x_{n(k)}$, for all k.

To this end, a special requirement upon our sequence is in order. Denote

$$(\le; N) = \{(m,n) \in N^2; m \le n\},\ (<; N) = \{(m,n) \in N^2; m < n\};$$

these are just the graph over N of the relations $(\le)$ and $(<)$, respectively. Given $h \in N(1, \le)$, let us say that (x_n) is $(\mathcal{G}, h)$-*uniformly-admissible* (in short: $(\mathcal{G}, h)$-*uadmissible*), provided

for each $(m,n) \in (<; N)$, we have $\mathcal{G}(x_m) \cap x]n, n+h] \neq \emptyset$; that is: there exists $p \in N$, such that $x_m\mathcal{G}x_p$ and $n < p \le n+h$.

The class of all such elements h will be denoted as $\mathrm{uad}(x_n)$ (the *uadmissible region* of (x_n)). Clearly,

$\mathrm{uad}(x_n)$ is *hereditary*: $h \in \mathrm{uad}(x_n)$, $q \ge h$ imply $q \in \mathrm{uad}(x_n)$;

when $\mathrm{uad}(x_n)$ is nonempty, we say that (x_n) is $\mathcal{G}$-*uadmissible*.

A local characterization of this property may be given along the lines below. Denote by $\widehat{\mathcal{G}}$ the (x_n)-*trace* of $\mathcal{G}$ over $(<; N)$

$(m, n \in (<; N))$: $m\widehat{\mathcal{G}}n$ iff $x_m\mathcal{G}x_n$.

Then, by this very definition,

(x_n) is $(\mathcal{G}, h)$-uadmissible iff $\widehat{\mathcal{G}}$ is h-*uadmissible*:
$\forall (m,n) \in (<; N)$, we have $\widehat{\mathcal{G}}(m) \cap N]n, n+h] \neq \emptyset$; that is:
$\exists p \in N$ such that $m\widehat{\mathcal{G}}p$ and $n < p \le n+h$.

This, in particular, holds under the requirements

(adm-1) $\widehat{\mathcal{G}}$ is *translation invariant*:
$m\widehat{\mathcal{G}}n$ implies $(m+j)\widehat{\mathcal{G}}(n+j)$, $\forall j \in N$
(adm-2) $\widehat{\mathcal{G}}(0) = \{n \in N; 0\widehat{\mathcal{G}}n\}$ is $(h+1)$-*admissible*:
for each $n \in N$ there exists $r \in \widehat{\mathcal{G}}(0)$ with $n < r \le n+h+1$.

To express these conditions in terms of our initial sequence, define

$\mathrm{spec}[(x_n)] = \{n \in N; x_0\mathcal{G}x_n\}$ (the *spectrum* of (x_n)).

Then, the underlying conditions are

(adm-1-seq) (x_n) is *$\mathcal{G}$-translated*:
$x_m \mathcal{R} x_n$ implies $x_{m+j} \mathcal{G} x_{n+j}$, for each $j \in N$
(adm-2-seq) (x_n) is *spectrally $(h+1)$-admissible*:
spec$[(x_n)]$ is $(h+1)$-admissible.

These, however, are just an illustration of such requirements upon (x_n) being meaningful; and not as mandatory conditions upon our data.

The following auxiliary fact represents our basic step towards an appropriate answer to the posed question.

Theorem 1.30. *Let the sequence $(x_n; n \geq 0)$ in X and the natural number $h \geq 1$ be such that*

(51-i) (x_n) is d-asymptotic
(51-ii) (x_n) is $(\mathcal{G}, h)$-uadmissible.

Further, let $\widehat{\mathcal{G}}$ stand for the (x_n)-trace of $\mathcal{G}$ over $(<; N)$ (see above). Then, (x_n) is d-Cauchy, if and only if it fulfills the relational condition

(d-C-rela) $\forall \varepsilon \in R^0_+$, $\exists j = j(\varepsilon) \in N$, such that
$(j \leq m < n,\ m\widehat{\mathcal{G}}n)$ implies $d(x_m, x_n) \leq \varepsilon$.

Proof. By a simple handling, the d-Cauchy property of (x_n) writes:

(d-C) $\forall \varepsilon \in R^0_+$, $\exists i = i(\varepsilon) \in N$: $i \leq m < n \Longrightarrow d(x_m, x_n) \leq \varepsilon$.

We claim that, by the imposed conditions, this is equivalent with its relational form (d-C-rela). In fact, the inclusion (d-C) $\Longrightarrow$ (d-C-rela) is clear; so, we have to establish the reverse inclusion: (d-C-rela) $\Longrightarrow$ (d-C). To this end, let $\varepsilon > 0$ be arbitrary fixed. By (d-C-rela), there exists some index $j_1(\varepsilon) \in N$, such that

(d-C-rela-1) $(j_1(\varepsilon) \leq m < n,\ m\widehat{\mathcal{G}}n)$ imply $d(x_m, x_n) \leq \varepsilon/2$.

On the other hand, by the d-asymptotic property of (x_n), one may construct the index $j_2(\varepsilon) = Z(\varepsilon/2h)$; which, by definition, fulfills

(d-asy) $q \geq j_2(\varepsilon)$ implies $\Delta_q(h) := \text{diam}(x[q, q+h]) < \varepsilon/2$.

Denote $j(\varepsilon) = \max\{j_1(\varepsilon), j_2(\varepsilon)\}$; and let the couple of ranks (m, n) be such that $j(\varepsilon) \leq m < n$. If $n \leq m + h$ then, by $j_2(\varepsilon) \leq m < n$ and (d-asy),

$$d(x_m, x_n) \leq \Delta_m(h) = \text{diam}(x[m, m+h]) < \varepsilon/2 < \varepsilon;$$

and we are done. Suppose now that $n > m + h$. By the h-uadmissible property of $\widehat{\mathcal{G}}$, we must have $\widehat{\mathcal{G}}(m) \cap N]n-h,n] \neq \emptyset$. Letting p be some point in this intersection, we have (from the above choice)

$$(m\widehat{\mathcal{G}}p \text{ and}) \ n-h < p \leq n; \text{ whence, } m < n-h < p \leq n < p+h.$$

Then, by $j(\varepsilon) \leq m < p$ and (d-C-rela-1),

$$d(x_m, x_n) \leq d(x_m, x_p) + d(x_p, x_n) \leq \varepsilon/2 + \Delta_p(h) < \varepsilon/2 + \varepsilon/2 = \varepsilon;$$

and the assertion follows. □

We are now in position to state the desired answer. Let us say that the subset Θ of R^0_+ is $(>)$-*cofinal* in R^0_+, when:

for each $\varepsilon \in R^0_+$, there exists $\theta \in \Theta$ with $\varepsilon > \theta$.

Further, given the sequence (r_n) in R and the point $r \in R$, let us define

$r_n \to r+$ (also written: $\lim_n r_n = r+$), if $r_n \to r$ and $(r_n > r, \forall n)$.

The following statement (referred to as Boyd-Wong relational non-Cauchy Criterion; in short: (BW-r-n-CC)) is now available.

Theorem 1.31. *Let the sequence $(x_n; n \geq 0)$ in X and the natural number $h \geq 1$ be such that*

(52-i) (x_n) is d-asymptotic
(52-ii) (x_n) is $(\mathcal{G}, h)$-uadmissible
(52-iii) (x_n) is not d-Cauchy.

Further, let $\widehat{\mathcal{G}}$ stand for the (x_n)-trace of $\mathcal{G}$ over $(<; N)$; and take a subset Θ of R^0_+ with the $(>)$-cofinal property. There exist then a number $\gamma \in \Theta$, a rank $J := J(\gamma, h) \geq 1$, and a couple $(m(k); k \geq 0)$, $(n(k); k \geq 0)$ of rank-sequences, with the properties

(52-a) $(\forall k)$: $J + k \leq m(k) < m(k) + 3h < n(k)$
(52-b) the rank sequences $(m(k))$ and $(n(k))$ are separated and $x_{m(k)}\mathcal{G}x_{n(k)}$, $d(x_{m(k)}, x_{n(k)}) > \gamma$, for all k
(52-c) $(\forall k)$: $m(k) < p(k)=$ the predecessor of $n(k)$ in $\widehat{\mathcal{G}}(m(k))$, and $d(x_{m(k)}, x_{p(k)}) \leq \gamma + \Delta_{m(k)}(3h) + \Delta_{n(k)-2h}(5h)$
(52-d) $U_k := d(x_{m(k)}, x_{n(k)}) \to \gamma+$ as $k \to \infty$
(52-e) for each $s, t \in N[0, 3h]$, we have $V_k(s,t) := d(x_{m(k)+s}, x_{n(k)+t}) \to \gamma+$, as $k \to \infty$

(52-f) for each $p, q \in N$, *we have*
$V_k(p,q) := d(x_{m(k)+p}, x_{n(k)+q}) \to \gamma$, *as* $k \to \infty$.

Proof. There are several steps to be passed.

Step 1. By a previous auxiliary fact, the d-Cauchy property of (x_n) writes under the relational form

(d-C-rela) $\forall \varepsilon \in R_+^0$, $\exists j(\varepsilon) \in N$, such that:
$(j(\varepsilon) \le m < n,\ m\widehat{\mathcal{G}}n) \Longrightarrow d(x_m, x_n) \le \varepsilon$.

On the other hand, as Θ is a $(>)$-cofinal part in R_+^0, the relational d-Cauchy property (d-C-rela) may be also written as

(d-C-rela-Th) $\forall \theta \in \Theta$, $\exists j(\theta) \in N$, such that:
$(j(\theta) \le m < n,\ m\widehat{\mathcal{G}}n)$ imply $d(x_m, x_n) \le \theta$.

Step 2. By the posed hypotheses (and our preceding observation)

(x_n) is not d-Cauchy; hence, the negation of (d-C-rela-Th) holds.

By definition, there exists $\beta \in \Theta$ such that

(rela-1) $E(j) := \{(m,n) \in (<; N); j \le m < n, m\widehat{\mathcal{G}}n, d(x_m, x_n) > \beta\}$
is a nonempty relation over N, for all $j \ge 0$.

As (x_n) is d-asymptotic, we may construct the map $Z : R_+^0 \to N(1, \le)$ with the precise properties. Let the number $\gamma \in \Theta$ be given according to

$\beta > 3\gamma$ (possible, since Θ is $(>)$-cofinal in R_+^0).

Further; let us take

(rela-2) $J(\gamma, h) = Z(\gamma/3h)$; hence, in particular
$\Delta_q(3h) := \mathrm{diam}(x[q, q+3h]) < \gamma$, $\forall q \ge J(\gamma, h)$.

Denote for simplicity $[J := J(\gamma, h)]$; and put, for each $k \ge 0$,

$A(k) = E(J + k)$; that is, $A(k) := \{(m,n) \in (<; N);$
$J + k \le m < n, m\widehat{\mathcal{G}}n, d(x_m, x_n) > \beta\}$.

Clearly, by the preceding nonemptiness property (rela-1),

$A(k)$ is a nonempty relation over N, for each $k \ge 0$.

From the triangle inequality, we have for all $k \ge 0$,

$$d(x_{m+s}, x_{n+t}) \geq d(x_m, x_n) - d(x_m, x_{m+s}) - d(x_n, x_{n+t}) >$$
$$\beta - \gamma - \gamma = \beta - 2\gamma > \gamma, \forall s, t \in N[0, 3h], \forall (m, n) \in A(k);$$

which tells us that

$B(k) := \{(m, n) \in (<; N);$
$J + k \leq m < n, m\widehat{\mathcal{G}}n, d(x_{m+s}, x_{n+t}) > \gamma, \forall s, t \in N[0, 3h]\}$
is a nonempty relation over N, for all $k \geq 0$.

Having this precise, denote for each $k \geq 0$

$$m(k) = \min \mathrm{Dom}(B(k)),\ n(k) = \min B(k)(m(k)).$$

By this very definition, we get

(pro-1) $(\forall k)$: $J+k \leq m(k) < n(k)$, $m(k)\widehat{\mathcal{G}}n(k)$ (so that: $x_{m(k)}\mathcal{R}x_{n(k)}$),
(pro-2) $(\forall k)$: $d(x_{m(k)+s}, x_{n(k)+t}) > \gamma$, $\forall s, t \in N[0, 3h]$.

Step 3. We claim that the couple (γ, J) and the couple of rank-sequences $(m(k); k \geq 0)$, $(n(k); k \geq 0)$ fulfill all written conclusions.

i): For the moment, the first and second part of (52-a) is fulfilled. Concerning the third part of the same, suppose by contradiction that

$$(m(k) <)n(k) \leq m(k) + 3h, \text{ for some } k \geq 0.$$

Then by (rela-2) (and $m(k) \geq J + k \geq J$)

$$d(x_{m(k)}, x_{n(k)}) \leq \Delta_{m(k)}(3h) < \gamma;$$

in contradiction with (pro-2); whence, the third part of (52-a) holds too.

ii): Further, by (pro-1) an (pro-2), it is clear that (52-b) follows.

iii): Let $k \geq 0$ be arbitrary fixed. By definition,

$$(n(k) \in \widehat{\mathcal{G}}(m(k)) \text{ and}) \ d(x_{m(k)+s}, x_{n(k)+t}) > \gamma, \forall s, t \in N[0, 3h].$$

On the other hand, by the preceding step,

$$m(k) < n(k) - 2h < n(k) - h < n(k);$$

and this yields (as $\widehat{\mathcal{G}}$ is h-uadmissible)

$\widehat{\mathcal{G}}(m(k)) \cap N]n(k) - 2h, n(k) - h] \neq \emptyset;$
whence, $\widehat{\mathcal{G}}(m(k)) \cap N]n(k) - 2h, n(k)[\neq \emptyset.$

In this case, $p(k) = \max \widehat{\mathcal{G}}(m(k)) \cap N]n(k) - 2h, n(k)[$ exists as the predecessor of $n(k) \in \widehat{\mathcal{G}}(m(k))$; precisely,

(pred-1) $p(k), n(k) \in \widehat{\mathcal{G}}(m(k))$ and $\widehat{\mathcal{G}}(m(k)) \cap N]p(k), n(k)[= \emptyset$

(pred-2) in addition, $m(k) < n(k) - 2h < p(k) < n(k)$;
hence, $n(k) - 2h < p(k) < p(k) + 3h < n(k) + 3h$.

This, along with $[J + k \leq m(k) < p(k)]$, tells us by the very definition of $n(k)$ [as minimal element in $B(k)(m(k))$], that

(pro-3) $d(x_{m(k)+u}, x_{p(k)+v}) \leq \gamma$, for some $u, v \in N[0, 3h]$.

Combining with the triangular inequality, gives

(pro-4) $d(x_{m(k)}, x_{p(k)}) \leq d(x_{m(k)}, x_{m(k)+u}) + d(x_{p(k)}, x_{p(k)+v}) + d(x_{m(k)+u}, x_{p(k)+v}) \leq \gamma + \Delta_{m(k)}(3h) + \Delta_{n(k)-2h}(5h)$, $\forall k \geq 0$,

and conclusion (52-c) is proved.

iv): By the definition of $(B(k); k \geq 0)$, we have $(U_k > \gamma, \forall k)$. Further, by the very definition of these ranks,

$$m(k) < n(k) - 2h < p(k) < n(k), \forall k \geq 0.$$

Combining with the triangular inequality and (pro-4), gives

$$\gamma < d(x_{m(k)}, x_{n(k)}) \leq d(x_{m(k)}, x_{p(k)}) + d(x_{p(k)}, x_{n(k)}) \leq \gamma + \Delta_{m(k)}(3h) + \Delta_{n(k)-2h}(5h) + \Delta_{n(k)-2h}(2h), \forall k \geq 0.$$

Passing to limit as $k \to \infty$ yields (52-d).

v): Let $s, t \in N[0, 3h]$ be arbitrary fixed. By the definition of $(B(k); k \geq 0)$, we have $(V_k(s, t) > \gamma, \forall k)$. Moreover, from a metrical property of d,

$$|d(x_{m(k)}, x_{n(k)}) - d(x_{m(k)+s}, x_{n(k)+t})| \leq d(x_{m(k)}, x_{m(k)+s}) + d(x_{n(k)}, x_{n(k)+t}) \leq \Delta_{m(k)}(3h) + \Delta_{n(k)}(3h), \text{ for all } k \geq 0.$$

Passing to limit in the relation between the extremal members of this relation gives (by the above) the conclusion (52-e).

vi): Let $p, q \in N$ be arbitrary fixed. By the same metrical property of d we just evoked, one may write

$$|d(x_{m(k)}, x_{n(k)}) - d(x_{m(k)+p}, x_{n(k)+q})| \leq d(x_{m(k)}, x_{m(k)+p}) + d(x_{n(k)}, x_{n(k)+q}) \leq \Delta_{m(k)}(p) + \Delta_{n(k)}(q), \text{ for all } k \geq 0.$$

Passing to limit in the relation between the extremal members of this relation gives (52-f); and completes the argument. □

By definition, the quadruple $[\gamma; J; (m(k); k \geq 0); (n(k); k \geq 0)]$ given by this result will be referred to as a *Boyd-Wong* (Θ, h)*-system* attached to (x_n). By the result in question, the following characterization is available.

Theorem 1.32. *Let the sequence* $(x_n; n \geq 0)$ *in* X *be d-asymptotic and* $\mathcal{G}$*-uadmissible. Then, the following conditions/properties are equivalent*

(53-a) $(x_n; n \geq 0)$ *is not d-Cauchy*
(53-b) for each $(>)$*-cofinal subset* Θ *of* R^0_+ *and each* $h \in \mathrm{uad}(x_n)$*, there is at least one Boyd-Wong* $[\Theta, h]$*-system attached to* (x_n)*.*

Proof. By the preceding result, the first condition includes the second one. Conversely, assuming that the second condition holds, fix some $(>)$-cofinal subset Θ of R^0_+ as well as some $h \in \mathrm{uad}(x_n)$. From the imposed hypothesis, there must be one Boyd-Wong $[\Theta, h]$-system $[\gamma; J; (m(k); k \geq 0); (n(k); k \geq 0)]$ attached to (x_n); hence (from the written conclusions)

$$(\gamma \in \Theta \text{ and}) \ d(x_{m(k)}, x_{n(k)}) \to \gamma+ \text{ as } k \to \infty.$$

This necessarily gives us that $(x_n; n \geq 0)$ is not d-Cauchy; for otherwise – by the very definition of d-Cauchy sequence – it follows that

$$d(x_{m(k)}, x_{n(k)}) \to 0 \text{ as } k \to \infty;$$

in contradiction with the limiting property above. □

In particular, when $\Theta = R^0_+$ and $\mathcal{G} = X \times X$ (the trivial relation over X), the obtained statement covers the 1969 one in Boyd and Wong [35]; so, it is natural that this result be referred to in the proposed way. Further aspects may be found in Reich [194]; see also Khan et al. [144].

1.5 Geometric-Asymptotic Relations

Abstract. Some preliminary facts are discussed about geometric and asymptotic Meir-Keeler relations over the (positive) real numbers.

Subject-Class 2020. 47H10 (Primary), 54H25 (Secondary).

Keywords. Relation, (strongly) Matkowski admissible property, geometric/asymptotic Meir-Keeler condition, simulation/manageable couple.

1.5.1 *Geometric relations*

Let $\Omega \in \exp(R^0_+ \times R^0_+)$ be a relation over R^0_+; as a rule, we write $(t, s) \in \Omega$ as $t\Omega s$. The following global property upon this object is our starting point

(u-diag) Ω is *upper diagonal*: $t\Omega s$ implies $t < s$.

Denote the class of all these relations as $\text{udiag}(R^0_+)$. We stress that all subsequent constructions are related to this setting, combined with

(1-decr) Ω is *first variable decreasing*:
$t_1, t_2, s \in R^0_+$, $t_1 \geq t_2$ and $t_1\Omega s$ imply $t_2\Omega s$
(2-incr) Ω is *second variable increasing*:
$t, s_1, s_2 \in R^0_+$, $s_1 \leq s_2$ and $t\Omega s_1$ imply $t\Omega s_2$
(addi) Ω in *additive*:
$t_1, s_1, t_2, s_2 \in R^0_+$, $t_1\Omega s_1$, $t_2\Omega s_2$ imply $(t_1 + t_2)\Omega(s_1 + s_2)$
(1-s-clo) Ω is *first variable sup-closed*:
$(\forall s \in R^0_+)$: $\sup \Omega^{-1}(s) \in \Omega^{-1}(s)$, whenever $\Omega^{-1}(s) \neq \emptyset$
(2-i-clo) Ω is *second variable inf-closed*:
$(\forall t \in R^0_+)$: $\inf \Omega(t) \in \Omega(t)$, whenever $\Omega(t) \neq \emptyset$.

The first group of these constructions involves two sequential properties upon $\text{udiag}(R^0_+)$, described as

(M-ad) Ω in *Matkowski admissible*:
$(t_n; n \geq 0)$ in R^0_+ and $(t_{n+1}\Omega t_n, \forall n)$ imply $\lim_n t_n = 0$
(s-M-ad) Ω is *strongly Matkowski admissible*:
$(t_n; n \geq 0)$ in R^0_+ and $(t_{n+1}\Omega t_n, \forall n)$ imply $\sum_n t_n < \infty$.

The former of these admissible properties may be viewed as an abstraction of some techniques due to Matkowski [164]; see also Rhoades [195]. And the latter of these admissible conditions – introduced (in a formal way) by Turinici [248] – appears as a relational version of certain developments in Browder [40] and Reich [194].

In a strong connection with this, the second group of such objects involves the following (geometric) properties for upper diagonal relations

(g-mk) Ω has the *geometric Meir-Keeler property*:
$\forall \varepsilon > 0$, $\exists \delta > 0$, $\forall (t, s)$: $t\Omega s$, $\varepsilon < s < \varepsilon + \delta \Longrightarrow t \leq \varepsilon$
(g-bila-s) Ω is *geometric bilateral separable*:
$\forall \beta > 0$, $\exists \gamma \in]0, \beta[$, $\forall (t, s)$: $t, s \in]\beta - \gamma, \beta + \gamma[\Longrightarrow (t, s) \notin \Omega$
(g-left-s) Ω is *geometric left separable*:
$\forall \beta > 0$, $\exists \gamma \in]0, \beta[$, $\forall t$: $t \in]\beta - \gamma, \beta[\Longrightarrow (t, \beta) \notin \Omega$.

The former of these properties is related to the developments in Meir and Keeler [168], subsequently refined by Matkowski [166] and Cirić [59]. Note that it may be also written in the (apparently extended) form

(g-mk-c) Ω has the *complete geometric Meir-Keeler property*:
$\forall \varepsilon > 0$, $\exists \delta > 0$, $\forall (t, s)$: $t\Omega s$, $s < \varepsilon + \delta \Longrightarrow t \leq \varepsilon$.

Technically speaking, the geometric Meir-Keeler property is strongly related with the Matkowski admissible property. Remember that, for the sequence $(r_n; n \geq 0)$ in R_+ and the point $r \in R_+$, we denoted

$r_n \to r+$ if $r_n \to r$ and $r_n > r$, for all n
$r_n \to r++$ if $r_n \to r$ and $r_n > r$, for almost all n.

Here, given the property $\pi(n)$ depending on the variable $n \in N$, we say that it holds for *almost all* $n \in N$ [written: $(\forall\forall n, \pi(n))$], if

$\exists h = h(\pi) \in N$, such that: $\pi(n)$ holds, for all $n \geq h$.

The following basic equivalence result is available.

Theorem 1.33. *The following hold, in the reduced system (ZF-AC+DC):*

(11-a) (for each $\Omega \in \text{udiag}(R^0_+)$*):*
Ω is geometric Meir-Keeler implies Ω is Matkowski admissible
(11-b) (for each first variable decreasing $\Omega \in \text{udiag}(R^0_+)$*):*
Ω is Matkowski admissible implies Ω is geometric Meir-Keeler.

Hence, summing up

(11-c) (for each first variable decreasing $\Omega \in \text{udiag}(R^0_+)$*):*
Ω is geometric Meir-Keeler iff Ω is Matkowski admissible.

Proof. Three basic stages must be passed.

i): Suppose that $\Omega \in \mathrm{udiag}(R_+^0)$ is geometric Meir-Keeler; we have to establish that Ω is Matkowski admissible. Let $(t_n; n \geq 0)$ be a sequence in R_+^0, fulfilling $(t_{n+1}\Omega t_n$, for all $n)$. By the upper diagonal property, we get $(t_{n+1} < t_n, \forall n)$; so, (t_n) is strictly descending. As a consequence, $\tau = \lim_n t_n$ exists in R_+, with $(t_n > \tau, \forall n)$. Assume by contradiction that $\tau > 0$; and let $\sigma > 0$ be the number assured by the geometric Meir-Keeler property. By definition, there exists an index $n(\sigma)$, with

$$(t_{n+1}\Omega t_n \text{ and}) \ \tau < t_n < \tau + \sigma, \text{ for all } n \geq n(\sigma).$$

This, by the quoted property, gives (for the same ranks) $\tau < t_{n+1} \leq \tau$; contradiction. Hence, necessarily, $\tau = 0$; and the conclusion follows.

ii): Suppose that the first variable decreasing $\Omega \in \mathrm{udiag}(R_+^0)$ is Matkowski admissible; we have to establish that Ω is geometric Meir-Keeler. Suppose by contradiction that this is not true; i.e. (for some $\varepsilon > 0$)

$$H(\delta) := \{(t,s) \in \Omega; \varepsilon < s < \varepsilon + \delta, t > \varepsilon\} \text{ is nonempty, for each } \delta > 0.$$

Taking a zero converging sequence $(\delta_n; n \geq 0)$ in R_+^0, we get by the Denumerable Axiom of Choice (AC(N)) [deductible, as precise, in (ZF-AC+DC)], a sequence $((t_n, s_n); n \geq 0)$ in $R_+^0 \times R_+^0$, so as

$(\forall n)$: $(t_n, s_n) \in H(\delta_n)$;
or, equivalently (by the upper diagonal property)
$(t_n \Omega s_n$ and$)$ $\varepsilon < t_n < s_n < \varepsilon + \delta_n$.

Note that, as a direct consequence,

$$(t_n \Omega s_n, \text{ for all } n), \text{ and } t_n \to \varepsilon+, \ s_n \to \varepsilon+, \text{ as } n \to \infty.$$

Put $i(0) = 0$. As $\varepsilon < t_{i(0)}$ and $s_n \to \varepsilon+$ as $n \to \infty$, we have that

$A(i(0)) := \{n > i(0); s_n < t_{i(0)}\}$ is not empty; hence,
$i(1) := \min(A(i(0)))$ is an element of it, and $s_{i(1)} < t_{i(0)}$; so that,
$s_{i(1)}\Omega s_{i(0)}$ (as $t_{i(0)}\Omega s_{i(0)}$ and Ω is first variable decreasing).

Likewise, as $\varepsilon < t_{i(1)}$ and $s_n \to \varepsilon+$ as $n \to \infty$, we have that

$A(i(1)) := \{n > i(1); s_n < t_{i(1)}\}$ is not empty; hence,
$i(2) := \min(A(i(1)))$ is an element of it, and $s_{i(2)} < t_{i(1)}$; so that,
$s_{i(2)}\Omega s_{i(1)}$ (as $t_{i(1)}\Omega s_{i(1)}$ and Ω is first variable decreasing).

This procedure may continue indefinitely; and yields (without any choice technique) a strictly ascending rank sequence $(i(n); n \geq 0)$ in N for which the attached subsequence $(r_n := s_{i(n)}; n \geq 0)$ of $(s_n; n \geq 0)$ fulfills

$r_{n+1}\Omega r_n$, for all n; whence $r_n \to 0$ (as Ω is Matkowski admissible).

On the other hand, by our subsequence property,

$(r_n > \varepsilon, \forall n)$ and $\lim_n r_n = \lim_n s_n = \varepsilon$; that is: $r_n \to \varepsilon+$.

The obtained relation is in contradiction with the previous one. Hence, the working condition cannot be true; and we are done.

iii): Evident, by the above. □

1.5.2 *Asymptotic versions*

In the following, sequential characterizations are provided for the properties appearing in our geometric concepts above. Given the upper diagonal relation Ω over R^0_+, let us introduce the (asymptotic type) conventions

(a-mk) Ω is *asymptotic Meir-Keeler*:
there are no strictly descending sequences (t_n) and (s_n) in R^0_+ and no elements ε in R^0_+, with $((t_n, s_n) \in \Omega, \forall n)$ and $(t_n \to \varepsilon+, s_n \to \varepsilon+)$
(a-bila-sep) Ω is *asymptotic bilateral separable*:
there are no sequences (t_n) and (s_n) in R^0_+ and no elements β in R^0_+, with $((t_n, s_n) \in \Omega, \forall n)$ and $(t_n \to \beta, s_n \to \beta)$
(a-left-sep) Ω is *asymptotic left separable*:
there are no strictly ascending sequences (t_n) in R^0_+ and no elements β in R^0_+, with $((t_n, \beta) \in \Omega, \forall n)$ and $(t_n \to \beta-)$.

Here, for each sequence (r_n) in R and each element $r \in R$, we denoted

$r_n \to r+$ $(r_n \to r-)$, when $r_n \to r$ and $[r_n > r$ $(r_n < r)$, $\forall n]$.

Remark 1.10.
Concerning the bilateral concept, let us define

(a-bila-sep-str) Ω is *strongly asymptotic bilateral separable*:
there are no sequences (t_n) and (s_n) in R^0_+ and no elements β in R^0_+, with $((t_n, s_n) \in \Omega, \forall n)$ and $\liminf_n t_n \geq \beta$, $\limsup_n s_n \leq \beta$.

We claim that

(Ω is strongly asymptotic bilateral separable) iff
(Ω is asymptotic bilateral separable).

In fact, the left to right inclusion is clear. Concerning the converse inclusion, suppose that Ω is not strongly asymptotic bilateral separable: there are sequences $(t_n; n \geq 0)$ and $(s_n; n \geq 0)$ in R^0_+ and elements β in R^0_+, with

$((t_n, s_n) \in \Omega$ (so, $t_n < s_n$), $\forall n)$ and $\liminf_n t_n \geq \beta$, $\limsup_n s_n \leq \beta$.

By these relations, we have

$\beta \leq \liminf_n t_n \leq \limsup_n t_n \leq \limsup_n s_n \leq \beta$,
$\beta \leq \liminf_n t_n \leq \liminf_n s_n \leq \limsup_n s_n \leq \beta$;
wherefrom $t_n \to \beta$, $s_n \to \beta$, as $n \to \infty$.

Combining these relations, yields a contradiction with respect to Ω being asymptotic bilateral separable; so that our assertion is proved.

Remark 1.11.

Denote, for simplicity,

$\mathcal{D}(R^0_+) := \{(t,t); t > 0\}$ (the *diagonal* of $R^0_+ \times R^0_+$).

Given the upper diagonal relation Ω over R^0_+, an equivalent way of expressing its asymptotic bilateral separable property reads:

(a-bila-sep-alm) Ω is *almost asymptotic bilateral separable*: for each sequence $((t_n, s_n); n \geq 0)$ in $R_+ \times R_+$ with $(\forall\forall n, (t_n, s_n) \in \Omega)$ and $((t_n, s_n); n \geq 0) =$ convergent, we cannot have $\lim_n (t_n, s_n) \in \mathcal{D}(R^0_+)$.

A graphical extension of this may be given as below. Letting $(X, \mathcal{R})$ be a relational structure and $\mathcal{Q} : X \times X \to R_+ \times R_+$ be a map, define

(a-bila-sep-gr) Ω is *asymptotic bilateral separable* modulo $(X, \mathcal{R}; Q)$: for each sequence $((x_n, y_n))$ in $\mathcal{R}$ with $(\forall\forall n, Q(x_n, y_n) \in \Omega)$ and $(\mathcal{Q}(x_n, y_n)) =$ convergent, we cannot have $\lim_n Q(x_n, y_n) \in \mathcal{D}(R^0_+)$.

The following generic relationship is valid (for an arbitrary upper diagonal relation $\Omega \subseteq R^0_+ \times R^0_+$), in the reduced system (ZF-AC+DC):

(a-bila-sep-alm) $\Longleftrightarrow$ (a-bila-sep-gr), for all admissible $(X, \mathcal{R}; Q)$.

In fact, the left to right inclusion yields by using the sequence $((t_n, s_n) = Q(x_n, y_n); n \geq 0)$; while the right to left inclusion follows by taking

$X = R_+$, $\mathcal{R} = R_+ \times R_+$, $Q =$ the identity map of $\mathcal{F}(R_+ \times R_+)$.

Remark 1.12.

Concerning the left concept above, let us consider the related condition

(a-left-sep-str) Ω is *strongly asymptotic left separable*:
there are no sequences (t_n) in R^0_+ and no elements β in R^0_+, with $((t_n, \beta) \in \Omega, \forall n)$ and $(t_n \to \beta)$.

We claim that, for each $\Omega \in \text{udiag}(R^0_+)$,

(Ω is strongly asymptotic left separable)
is equivalent with (Ω is asymptotic left separable).

In fact, the left to right inclusion is clear. Concerning the reciprocal one, suppose that $\Omega \in \text{udiag}(R^0_+)$ is not strongly asymptotic Meir-Keeler: there is a sequence (t_n) in R^0_+ and an element β in R^0_+, with $((t_n, \beta) \in \Omega, \forall n)$ and $t_n \to \beta$. By the upper diagonal property of Ω, $(t_n < \beta, \forall n)$; so that, $(t_n \to \beta-)$. By an auxiliary fact, there exists a strictly ascending subsequence $(t^*_n = t_{i(n)}; n \geq 0)$ of $(t_n; n \geq 0)$, with $((t^*_n, \beta) \in \Omega, \forall n)$ and $(t^*_n \to \beta-)$. This, however, contradicts the asymptotic left separated property of Ω; and therefore, our assertion is proved.

Returning to the general setting above, the relationships with the corresponding geometric type notions are described in the statement below.

Theorem 1.34. *The following generic relationships are valid (for a generic $\Omega \in \text{udiag}(R^0_+)$), in the reduced system (ZF-AC+DC):*

(21-a) geometric Meir-Keeler $\Longleftrightarrow$ asymptotic Meir-Keeler
(21-b) geometric bilateral separable is equivalent with asymptotic bilateral separable (hence, with strongly asymptotic bilateral separable)
(21-c) geometric left separable is equivalent with asymptotic left separable (hence, with strongly asymptotic left separable).

Proof. There are three steps to be passed.

i-1): Let $\Omega \in \text{udiag}(R^0_+)$ be a geometric Meir-Keeler relation; but – contrary to the conclusion – assume that Ω does not have the asymptotic Meir-Keeler property: there are two strictly descending sequences (t_n) and (s_n) in R^0_+ and an element ε in R^0_+, with $((t_n, s_n) \in \Omega, \forall n)$ and $(t_n \to \varepsilon+, s_n \to \varepsilon+)$. Let $\delta > 0$ be the number given by the geometric Meir-Keeler property of Ω. By definition, there exists a (common) rank $n(\delta)$, such that

$$n \geq n(\delta) \text{ implies } \varepsilon < t_n < \varepsilon + \delta,\ \varepsilon < s_n < \varepsilon + \delta.$$

From the second relation, we must have (by the hypothesis about Ω)

$$t_n \leq \varepsilon, \text{ for all } n \geq n(\delta); \text{ in contradiction with the first relation.}$$

Hence, Ω is asymptotic Meir-Keeler; as asserted.

i-2): Let $\Omega \in \text{udiag}(R^0_+)$ be asymptotic Meir-Keeler. If Ω is not endowed with the geometric Meir-Keeler property, we must have (for some $\varepsilon > 0$)

$$H(\delta) := \{(t,s) \in \Omega; \varepsilon < s < \varepsilon + \delta,\ t > \varepsilon\} \neq \emptyset, \text{ for each } \delta > 0.$$

Taking a zero converging sequence $(\delta_n; n \geq 0)$ in R_+^0, we get by the Denumerable Axiom of Choice (AC(N)) [deductible, as precise, in (ZF-AC+DC)], a sequence $((t_n, s_n); n \geq 0)$ in $R_+^0 \times R_+^0$, so as

$(\forall n)$: $(t_n, s_n) \in H(\delta_n)$; that is (by the upper diagonal property)
$((t_n, s_n) \in \Omega$ and) $\varepsilon < t_n < s_n < \varepsilon + \delta_n$.

Note that, as a direct consequence,

$(t_n \Omega s_n$, for all n), and $(t_n \to \varepsilon+,\ s_n \to \varepsilon+)$, as $n \to \infty$.

By a previous result, there exists a compatible couple of subsequences $(t_n^* := t_{i(n)}; n \geq 0)$ of $(t_n; n \geq 0)$ and $(s_n^* := s_{i(n)}; n \geq 0)$ of $(s_n; n \geq 0)$, with

$(t_n^* \Omega s_n^*, \forall n)$; (t_n^*), (s_n^*) are strictly descending; $t_n^* \to \varepsilon+$ and $s_n^* \to \varepsilon+$.

This, however, is in contradiction with respect to the posed hypothesis upon Ω; wherefrom, our assertion follows.

ii-1): Let $\Omega \in \text{udiag}(R_+^0)$ be geometric bilateral separable; but – contrary to the conclusion – assume that Ω is not asymptotic bilateral separable: there are two sequences $(t_n; n \geq 0)$ and $(s_n; n \geq 0)$ in R_+^0 and an element $\beta \in R_+^0$, with $((t_n, s_n) \in \Omega, \forall n)$ and $(t_n \to \beta,\ s_n \to \beta)$. Let $\gamma > 0$ be the number given by the geometric bilateral separable property of Ω. By definition, there exists a (common) rank $n(\gamma)$, such that $(n \geq n(\gamma)$ implies $t_n, s_n \in]\beta - \gamma, \beta + \gamma[)$. This, along with $[t_n \Omega s_n, \forall n \geq n(\gamma)]$ contradicts the geometric bilateral separable property of Ω.

ii-2): Let $\Omega \in \text{udiag}(R_+^0)$ be an asymptotic bilateral separable relation. If Ω is not geometric bilateral separable, we must have (for some $\beta > 0$)

$$B(\gamma) := \{(t,s) \in \Omega; t, s \in]\beta - \gamma, \beta + \gamma[\} \neq \emptyset, \text{ for each } \gamma \in]0, \beta[.$$

Taking a strictly descending sequence $(\gamma_n; n \geq 0)$ in $]0, \beta[$ with $\gamma_n \to 0+$, we get by the Denumerable Axiom of Choice (AC(N)) [deductible, as precise, in the system (ZF-AC+DC)], a sequence $((t_n, s_n); n \geq 0)$ in Ω, so as

$(\forall n)$: $(t_n, s_n) \in B(\gamma_n)$; or, equivalently (by the definition above)
$(t_n, s_n) \in \Omega$ and $t_n, s_n \in]\beta - \gamma_n, \beta + \gamma_n[$.

By the second half of this last relation, we must have $(t_n \to \beta,\ s_n \to \beta)$; and this, along with the first half of the same, contradicts the imposed hypothesis. Hence, necessarily, Ω is geometric bilateral separable.

iii-1): Let $\Omega \in \text{udiag}(R^0_+)$ be a geometric left separable relation; but – contrary to the conclusion – assume that Ω is not asymptotic left separable: there is a strictly ascending sequence $(t_n; n \geq 0)$ and an element $\beta \in R^0_+$, with $((t_n, \beta) \in \Omega, \forall n)$ and $(t_n \to \beta-)$. Let $\gamma \in]0, \beta[$ be the number given by the geometric left separable property of Ω. By definition, there exists a rank $n(\gamma)$, such that $[n \geq n(\gamma)$ implies $\beta - \gamma < t_n < \beta]$. This, along with $[t_n \Omega \beta, \forall n \geq n(\gamma)]$ contradicts the geometric left separable property of Ω.

iii-2): Let $\Omega \in \text{udiag}(R^0_+)$ be an asymptotic left separable relation. If Ω is not geometric left separable. we must have (for some $\beta > 0$)

$$L(\gamma) := \{t \in]\beta - \gamma, \beta[; (t, \beta) \in \Omega\} \neq \emptyset, \text{ for each } \gamma \in]0, \beta[.$$

Taking a strictly descending sequence $(\gamma_n; n \geq 0)$ in $]0, \beta[$ with $\gamma_n \to 0+$, we get by the Denumerable Axiom of Choice (AC(N)) [deductible, as precise, in the system (ZF-AC+DC)], a sequence $(t_n; n \geq 0)$ in R^0_+, so as

$$(\forall n): t_n \in L(\gamma_n); \text{ that is (by the definition above)}$$
$$\beta - \gamma_n < t_n < \beta, \text{ and } (t_n, \beta) \in \Omega.$$

By the first half of this last relation, we must have $t_n \to \beta-$; and this, along with an auxiliary fact, tells us that there exists a subsequence $(t^*_n := t_{i(n)}; n \geq 0)$ of $(t_n; n \geq 0)$, with $(t^*_n; n \geq 0)$ is strictly ascending, and $t^*_n \to \beta-$. On the other hand, by the second half of our underlying relation, $[(t^*_n, \beta) \in \Omega, \forall n]$. Putting these together yields a contradiction with the asymptotic left separable property of Ω. □

1.5.3 *Duality principle*

The sequential, geometric, and asymptotic properties above were attached to subsets of $R^0_+ \times R^0_+$. But, any such object, Ω say, is characterized by its *(complementary) characteristic function* $f_\Omega : R^0_+ \times R^0_+ \to R$, introduced as

$$f_\Omega(t, s) = 0, \text{ if } (t, s) \in \Omega; \; f_\Omega(t, s) = 1 \text{ if } (t, s) \notin \Omega.$$

From a larger perspective, the pair $(0, 1)$ may be replaced by any pair (a, b) of (distinct) points in $R(\pm\infty) := R \cup \{-\infty, \infty\}$ [the set of all *extended real numbers*]. The details of this construction may be described as follows.

For each relation Ω on R^0_+, we associate the function $\xi : R^0_+ \times R^0_+ \to R(\pm\infty)$, by the convention

$$\xi(t, s) = 0, \text{ if } (t, s) \in \Omega; \; \xi(t, s) = -\infty, \text{ if } (t, s) \notin \Omega.$$

It will be referred to as the *function* generated by Ω; clearly,

$(t,s) \in \Omega$ iff $\xi(t,s) \geq 0$.

Conversely, given a function $\xi : R^0_+ \times R^0_+ \to R(\pm\infty)$, we associate it a relation Ω over R^0_+, according to

$\Omega = \{(t,s) \in R^0_+ \times R^0_+; \xi(t,s) \geq 0\}$ (in short: $\Omega = [\xi \geq 0]$);
referred to as: the *the positive section* of ξ.

Note that the correspondence between ξ and $[\xi \geq 0]$ is not injective; because, for the function $\eta := g\xi$ (where $g > 0$), its associated relation $[\eta \geq 0]$ is identical with the relation $[\xi \geq 0]$ attached to ξ.

Now, call the function $\xi : R^0_+ \times R^0_+ \to R(\pm\infty)$, *upper diagonal* provided:

(u-diag) $\xi(t,s) \geq 0$ implies $t < s$.

All subsequent constructions are carried out within this setting. In particular, the following properties for upper diagonal functions ξ are considered:

(M-ad) ξ in *Matkowski admissible*:
$(t_n; n \geq 0)$ in R^0_+ and $(\xi(t_{n+1}, t_n) \geq 0, \forall n)$ imply $\lim_n t_n = 0$
(s-M-ad) ξ in *strongly Matkowski admissible*:
$(t_n; n \geq 0)$ in R^0_+ and $(\xi(t_{n+1}, t_n) \geq 0, \forall n)$ imply $\sum_n t_n < \infty$.

In a direct connection with this, the second group of such objects involves the following (geometric) properties for upper diagonal functions

(g-mk) ξ is *geometric Meir-Keeler*:
$\forall \varepsilon > 0, \exists \delta > 0, \forall (t,s)$: $\xi(t,s) \geq 0, \varepsilon < s < \varepsilon + \delta \Longrightarrow t \leq \varepsilon$
(g-bila-s) ξ is *geometric bilateral separable*:
$\forall \beta > 0, \exists \gamma \in]0, \beta[, \forall (t,s)$: $t, s \in]\beta - \gamma, \beta + \gamma[\Longrightarrow \xi(t,s) < 0$
(g-left-s) ξ is *geometric left separable*:
$\forall \beta > 0, \exists \gamma \in]0, \beta[, \forall t$: $t \in]\beta - \gamma, \beta[\Longrightarrow \xi(t, \beta) < 0$.

Finally, the asymptotic versions of these geometric properties are as follows (again over the class of upper diagonal functions)

(a-mk) ξ is *asymptotic Meir-Keeler*:
there are no strictly descending sequences (t_n) and (s_n) in R^0_+ and no elements ε in R^0_+, with $(\xi(t_n, s_n) \geq 0, \forall n)$ and $(t_n \to \varepsilon+,\ s_n \to \varepsilon+)$
(a-bila-s) ξ is *asymptotic bilateral separable*:
there are no sequences (t_n) and (s_n) in R^0_+ and no elements $\beta \in R^0_+$, with $(\xi(t_n, s_n) \geq 0, \forall n)$ and $(t_n \to \beta,\ s_n \to \beta)$

(a-left-s) ξ is *asymptotic left separable*:
there are no strictly ascending sequences (t_n) in R^0_+
and no elements β in R^0_+, with $(\xi(t_n, \beta) \geq 0, \forall n)$ and $(t_n \to \beta-)$.

By the above observations, one gets the following *Duality Principle*:

(DP-1) the relationships between geometric Meir-Keeler and Matkowski admissible properties of upper diagonal functions are simple translations of the ones involving upper diagonal relations, and vice versa

(DP-2) the relationships between the geometric concepts and their asymptotic counterparts of upper diagonal functions are simple translations of the ones involving upper diagonal relations, and vice versa.

Summing up, any concept (like the ones above) about (upper diagonal) elements in the class $\mathcal{F}(R^0_+ \times R^0_+, R(\pm\infty))$ is reducible to a corresponding concept about (upper diagonal) elements in the class $\exp(R^0_+ \times R^0_+)$. For the rest of our exposition, it will be convenient working with relations in $\exp(R^0_+ \times R^0_+)$, and not with functions in $\mathcal{F}(R^0_+ \times R^0_+, R(\pm\infty))$; this, however, is but a methodology question.

1.5.4 *Boyd-Wong type relations*

The first class of relations to be discussed is the Boyd-Wong one. Let $\chi \in \mathcal{F}(re)(R^0_+, R)$ be a function. Define the (attached) *Boyd-Wong* relation $\Omega := \Omega[\chi]$ over R^0_+, as

$(t, s \in R^0_+)$: $(t, s) \in \Omega$ iff $t \leq \chi(s)$; clearly, Ω is upper diagonal.

The proposed terminology is suggested by the developments in Boyd and Wong [35]; see also Matkowski [164].

Theorem 1.35. *The following conclusions involving $\Omega = \Omega[\chi]$ are valid:*

(41-a) Ω is first variable decreasing, geometric/asymptotic left separable, and first variable sup-closed
(41-b) Ω is second variable increasing, when χ is increasing
(41-c) Ω has the geometric Meir-Keeler property whenever
χ is Rhoades admissible (or, equivalently: Meir-Keeler admissible)
(41-d) Ω is Matkowski admissible when
χ is Matkowski admissible (or, equivalently: Meir-Keeler admissible)
(41-e) Ω is strongly Matkowski admissible whenever
χ is strongly Matkowski admissible
(41-f) Ω is geometric/asymptotic bilateral separable when
χ is strongly Boyd-Wong admissible

(41-g) Ω is additive, if χ is super-additive:
$\chi(t+s) \geq \chi(t) + \chi(s)$, $\forall t, s \in R_+^0$
(41-h) Ω is second variable inf-closed when χ is right normal on R_+^0: $\liminf_n \chi(r_n) \leq \chi(r)$, *for each $r > 0$ and each strictly descending sequence (r_n) with $r_n \to r+$.*

Proof. i): The first part is clear; and the remaining ones are a direct consequence of the immediate relation

$$\Omega^{-1}(s) =]0, \chi(s)] \text{ (and } \chi(s) < s\text{), for each } s \in R_+^0.$$

ii), iii), iv), v): Evident.

vi): Suppose that Ω is not asymptotic bilateral separable: there are sequences $(t_n; n \geq 0)$ and $(s_n; n \geq 0)$ in R_+^0 and elements $\beta \in R_+^0$, with $((t_n, s_n) \in \Omega, \forall n)$ and $(t_n \to \beta,\ s_n \to \beta)$. By the definition of our relation,

$$(t_n \leq \chi(s_n),\ \forall n), \text{ and } t_n \to \beta,\ s_n \to \beta.$$

Passing to lim sup as $n \to \infty$, yields $[\beta \leq \Lambda^{\pm}\chi(\beta) < \beta]$; a contradiction.

vii): Take the couples (t_1, s_1) and (t_2, s_2) with

$$(t_1, s_1), (t_2, s_2) \in \Omega; \text{ that is: } t_1 \leq \chi(s_1),\ t_2 \leq \chi(s_2).$$

Adding these inequalities yields (via χ = super-additive)

$$t_1 + t_2 \leq \chi(s_1) + \chi(s_2) \leq \chi(s_1 + s_2); \text{ so that: } (t_1 + t_2)\Omega(s_1 + s_2);$$

and the claim follows.

viii): Let the point $t \in R_+^0$ be such that $\Omega(t) := \{s \in R_+^0; t \leq \chi(s)\} \neq \emptyset$. Note that, by the upper diagonal property,

$$\Omega(t) \subseteq]t, \infty[; \text{ whence, } \gamma := \inf \Omega(t) \geq t.$$

Suppose by contradiction that $\gamma \notin \Omega(t)$. By definition, there exists a strictly descending sequence (s_n) in R_+^0 with

$$(s_n \in \Omega(t) \text{ (that is: } t \leq \chi(s_n)),\ \forall n), \text{ and } s_n \to \gamma+.$$

This yields (via χ = right normal)

$$t \leq \liminf_n \chi(s_n) \leq \chi(\gamma); \text{ that is, } t\Omega\gamma; \text{ whence, } \gamma \in \Omega(t);$$

a contradiction; wherefrom, the conclusion is clear. □

Remark 1.13.

A natural question to be posed is that of certain relations Ω over R^0_+ being represented as $\Omega[\chi]$, where the functions χ are like before. The following partial answer will be useful in this direction.

Let Ω be an upper diagonal relation over R^0_+. As usual, denote

$$\Omega(t) = \{s \in R^0_+; (t,s) \in \Omega\},\ t \in R^0_+,$$
$$\mathrm{Dom}(\Omega) = \{t \in R^0_+; \Omega(t) \neq \emptyset\},$$
$$\Omega^{-1}(s) = \{t \in R^0_+; (t,s) \in \Omega\},\ s \in R^0_+,$$
$$\mathrm{Codom}(\Omega) = \{s \in R^0_+; \Omega^{-1}(s) \neq \emptyset\}.$$

Let us associate to Ω the function $\chi = \chi[\Omega] : R^0_+ \to R$ as

$$\chi(s) = \sup \Omega^{-1}(s),\ s \in \mathrm{Codom}(\Omega);\ \chi(s) = 0,\ s \in R^0_+ \setminus \mathrm{Codom}(\Omega).$$

Note that, by the upper diagonal property of Ω we have

$$\chi(s) \leq s,\ \forall s \in \mathrm{Codom}(\Omega), \text{ and } \chi(s) < s,\ \forall s \in R^0_+ \setminus \mathrm{Codom}(\Omega);$$

hence, the obtained function χ is not in general regressive. To get such a property, further conditions upon Ω must be imposed. Precisely, suppose

(1-s-clo) Ω is first variable sup-closed:
$(\forall s \in R^0_+)$: $\sup \Omega^{-1}(s) \in \Omega^{-1}(s)$, whenever $\Omega^{-1}(s) \neq \emptyset$.

In this case, we have by definition

$$\chi(s) \in \Omega^{-1}(s) \text{ (hence, } \chi(s) < s\text{), for all } s \in \mathrm{Codom})\Omega);$$

wherefrom, combining with the above,

$$\chi(s) < s, \text{ for all } s \in R^0_+; \text{ so that: } \chi \text{ is regressive.}$$

Note that, as a direct consequence,

$$(t,s) \in \Omega \text{ implies } s \in \mathrm{Codom}(\Omega),\ t \leq \chi(s);$$

so that (under the precise property)

$$\Omega \subseteq \Omega[\chi]; \text{ where (as established) } \chi \in \mathcal{F}(R^0_+, R) \text{ is regressive.}$$

This tells us that, all reasonings upon Ω may be transferred to the Boyd-Wong relation $\Omega[\chi]$. However, for an efficient handling of this last object, further properties of the function χ must be admitted. Some related aspects may be found in Lim [162], and the references therein.

1.5.5 *Rhoades type relations*

The second class of relations to be considered is the Rhoades one. Let (ψ, φ) be a pair of functions in $\mathcal{F}(R_+^0, R)$, fulfilling

(norm) (ψ, φ) is *normal*:
ψ is increasing and φ is *strictly positive* ($\varphi(R_+^0) \subseteq R_+^0$).

Define the *Rhoades* associated relation $\Omega = \Omega[\psi, \varphi]$ on R_+^0, as

$(t, s) \in \Omega$ iff $\psi(t) \leq \psi(s) - \varphi(s)$.

Note that Ω is upper diagonal. In fact, let $t, s \in R_+^0$ be such that $(t, s) \in \Omega$; that is: $\psi(t) \leq \psi(s) - \varphi(s)$. By the strict positivity of φ, one gets $\psi(t) < \psi(s)$; and this, along with the increasing property of ψ, shows that $t < s$; whence, the assertion. The introduced convention is related to the developments in Rhoades [197], refined by Dutta and Choudhury [84].

Further properties of this relation are available under certain supplementary conditions about the normal couple (ψ, φ), like the ones below:

(as-pos) φ is *asymptotic positive*:
for each strictly descending sequence (t_n) in R_+^0 and each $\varepsilon > 0$ with $t_n \to \varepsilon+$, we must have $\limsup_n(\varphi(t_n)) > 0$
(lim-bd-osc) (ψ, φ) is *limit-bounded oscillating*:
for each sequence (t_n) in R_+^0 and each $\beta > 0$ with $t_n \to \beta$, we have $\limsup_n(\varphi(t_n)) > \psi(\beta + 0) - \psi(\beta - 0)$
(bd-le-osc) (ψ, φ) is *bounded left oscillating*:
for each $\beta > 0$, we have $\varphi(\beta) > \psi(\beta) - \psi(\beta - 0)$.

Note that concrete examples of the first property are available under

(φ = increasing or continuous) implies φ = asymptotic positive.

In fact, let the strictly descending sequence $(t_n; n \geq 0)$ in R_+^0 and the number $\varepsilon > 0$ be such that $t_n \to \varepsilon+$. When φ = increasing, we have

$\varphi(t_n) \geq \varphi(\varepsilon) > 0$, $\forall n$; whence $\limsup_n(\varphi(t_n)) > 0$.

On the other hand, when φ = continuous,

$\limsup_n(\varphi(t_n)) = \lim_n(\varphi(t_n)) = \varphi(\varepsilon) > 0$; and conclusion follows.

Proposition 1.41. *Let* (ψ, φ) *be a normal couple of functions in* $\mathcal{F}(R_+^0, R)$*; and* $\Omega := \Omega[\psi, \varphi]$ *be the associated upper diagonal relation. Then,*

(51-1) if φ is asymptotic positive, then
Ω is asymptotic (or, equivalently: geometric) Meir-Keeler (hence, Matkowski admissible)
(51-2) if (ψ, φ) is limit-bounded oscillating, then the associated relation Ω is asymptotic/geometric bilateral separable
(51-3) if (ψ, φ) is bounded left oscillating, then the associated relation Ω is asymptotic/geometric left separable
(51-4) Ω is additive, if $\psi - \varphi$ is super-additive (see above) and ψ is sub-additive: $\psi(t+s) \le \psi(t) + \psi(s)$, $\forall t, s \in R_+^0$
(51-5) Ω is first variable sup-closed when ψ is left continuous ($\psi(t-0) = \psi(t)$, $\forall t \in R_+^0$)
(51-6) Ω is second variable inf-closed when $\psi - \varphi$ is right normal on R_+^0: $\liminf_n(\psi - \varphi)(r_n) \le (\psi - \varphi)(r)$, for each $r > 0$ and each strictly descending sequence (r_n) with $r_n \to r+$.

Proof. i): Suppose that Ω is not asymptotic Meir-Keeler: there exist strictly descending sequences (t_n) and (s_n) in R_+^0 and elements ε in R_+^0 with $((t_n, s_n) \in \Omega, \forall n)$ and $(t_n \to \varepsilon+,\ s_n \to \varepsilon+)$. By the first half of this,

$$(0 <)\varphi(s_n) \le \psi(s_n) - \psi(t_n),\ \forall n.$$

Passing to limit and noting that $\lim_n \psi(s_n) = \lim_n \psi(t_n) = \psi(\varepsilon + 0)$, yields $\lim_n \varphi(s_n) = 0$; in contradiction with the asymptotic positivity of φ.

ii): Suppose that Ω is not asymptotic bilateral separable: there exist sequences (t_n) and (s_n) in R_+^0 and elements β in R_+^0, with $((t_n, s_n) \in \Omega, \forall n)$ and $(t_n \to \beta,\ s_n \to \beta)$. By the first half of this,

$$(0 <)\varphi(s_n) \le \psi(s_n) - \psi(t_n),\ \forall n.$$

Passing to lim sup as $n \to \infty$, yields

$$\limsup_n \varphi(s_n) \le \psi(\beta + 0) - \psi(\beta - 0);$$

in contradiction with (ψ, φ) being limit-bounded oscillating.

iii): Suppose by contradiction that Ω is not asymptotic left separable: there exist strictly ascending sequences (t_n) in R_+^0 and elements β in R_+^0, with $((t_n, \beta) \in \Omega, \forall n)$ and $(t_n \to \beta-)$. By the first half of this, we get

$$\psi(t_n) \le \psi(\beta) - \varphi(\beta),\ \forall n.$$

Passing to lim sup as $n \to \infty$, yields (as ψ = increasing)

$$\psi(\beta - 0) \le \psi(\beta) - \varphi(\beta);\ \text{that is: } \varphi(\beta) \le \psi(\beta) - \psi(\beta - 0);$$

in contradiction with (ψ, φ) being bounded left oscillating.

iv): Take the couples (t_1, s_1) and (t_2, s_2) with

$$(t_1, s_1), (t_2, s_2) \in \Omega; \text{ that is: } \psi(t_1) \leq (\psi - \varphi)(s_1),\ \psi(t_2) \leq (\psi - \varphi)(s_2).$$

This yields (via $\psi - \varphi =$ super-additive, $\psi =$ sub-additive)

$$\psi(t_1 + t_2) \leq \psi(t_1) + \psi(t_2) \leq (\psi - \varphi)(s_1) + (\psi - \varphi)(s_2) \leq \\ (\psi - \varphi)(s_1 + s_2); \text{ that is: } (t_1 + t_2)\Omega(s_1 + s_2);$$

and the claim follows.

v): Evident, in view of

$$\Omega^{-1}(s) = \{t \in R_+^0; \psi(t) \leq \psi(s) - \varphi(s)\},\ s \in R_+^0.$$

vi): Let the point $t \in R_+^0$ be such that

$$\Omega(t) = \{s \in R_+^0; \psi(t) \leq \psi(s) - \varphi(s)\} \neq \emptyset.$$

Note that, by the upper diagonal property,

$$\Omega(t) \subseteq]t, \infty[; \text{ whence, } \gamma := \inf \Omega(t) \geq t.$$

Suppose by contradiction that $\gamma \notin \Omega(t)$. By definition, there exists a strictly descending sequence (s_n) in R_+^0 with

$$\psi(t) \leq (\psi - \varphi)(s_n),\ \forall n, \text{ and } s_n \to \gamma+.$$

This yields (via $(\psi - \varphi) =$ right normal)

$$\psi(t) \leq \liminf_n (\psi - \varphi)(s_n) \leq (\psi - \varphi)(\gamma); \text{ that is, } t\Omega\gamma; \text{ contradiction;}$$

and the conclusion follows. □

Finally, some other properties of this associated relation are available under certain extra sequential conditions upon the normal couple (ψ, φ). Let us say that the couple (ψ, φ) is *finitely-asymptotic regular*, provided

(f-a-reg-1) φ is asymptotic positive and $\psi(0 + 0) > -\infty$
(f-a-reg-2) φ is *asymptotic expansive*: $\forall$ strictly descending sequence (t_n) in R_+^0 with $\lim_n t_n = 0$ and $\sum_n \varphi(t_n) < \infty$, we have $\sum_n t_n < \infty$.

Likewise, call (ψ, φ), *infinitely-asymptotic regular* provided

(i-a-reg-1) φ is increasing (hence, asymptotic positive) (i-a-reg-2) (ψ, φ) is *asymptotic μ-dominating* (for some $\mu > 1$): $\forall$ strictly descending sequence $(t_n; n \geq 0)$ in R_+^0 with $\lim_n t_n = 0$ and $\sum_n [\varphi(t_n)/(1 + \psi(t_0) - \psi(t_n))]^\mu < \infty$, we have $\sum_n t_n < \infty$.

Proposition 1.42. *Let (ψ, φ) be a normal couple; and $\Omega := \Omega[\psi, \varphi]$ stand for its associated relation. Then*

(52-1) Ω is first variable decreasing
(52-2) Ω is second variable increasing if $\psi - \varphi$ is increasing
(52-3) Ω is strongly Matkowski admissible if (ψ, φ) is finitely-asymptotic regular
(52-4) Ω is strongly Matkowski admissible if (ψ, φ) is infinitely-asymptotic regular.

Proof. i), ii): Evident.

iii): Let $(t_n; n \geq 0)$ be a sequence in R_+^0, with

$$t_{n+1} \Omega t_n \text{ (that is: } \psi(t_{n+1}) \leq \psi(t_n) - \varphi(t_n)\text{), for all } n.$$

Note that, by a re-arrangement in this relation, we get

$$(0 <)\varphi(t_n) \leq \psi(t_n) - \psi(t_{n+1}), \text{ for all } n.$$

From the upper diagonal property of Ω, $(t_n; n \geq 0)$ is strictly descending; hence, $\tau := \lim_n t_n$ exists. We claim that $\tau = 0$; that is: $\lim_n t_n = 0$. For, suppose by contradiction that $\tau > 0$. Passing to lim sup as $n \to \infty$ in the re-arranged relation, we get (as ψ = increasing)

$$0 \leq \limsup_n \varphi(t_n) \leq \psi(\tau + 0) - \psi(\tau + 0) = 0;$$

impossible, via φ = asymptotic positive. On the other hand,

$$\sum_n (\psi(t_n) - \psi(t_{n+1})) < \infty; \text{ because } \psi(0 + 0) > -\infty.$$

This, combined with the re-arranged inequality once again, gives

$$(t_n \to 0 \text{ and}) \sum_n \varphi(t_n) < \infty; \text{ whence, } \sum_n t_n < \infty,$$

if we remember that φ is asymptotic expansive.

iv): Let $(t_n; n \geq 0)$ be a sequence in R_+^0, with

$$t_{n+1} \Omega t_n \text{ (that is: } \psi(t_{n+1}) \leq \psi(t_n) - \varphi(t_n)\text{), for all } n.$$

Note that, by a simple re-arrangement in the last relation,

$$(0 <)\varphi(t_n) \leq \psi(t_n) - \psi(t_{n+1}), \text{ for all } n.$$

By the upper diagonal property of Ω, $(t_n; n \geq 0)$ is strictly descending; hence, $\tau := \lim_n t_n$ exists. We claim that $\tau = 0$; that is: $\lim_n t_n = 0$. For, suppose by contradiction that $\tau > 0$. Passing to $\limsup$ as $n \to \infty$ in the re-arranged relation, we get (as ψ = increasing)

$$0 \leq \limsup_n \varphi(t_n) \leq \psi(\tau + 0) - \psi(\tau + 0) = 0;$$

impossible, via φ = asymptotic positive; hence the claim. Further, adding the first $(n+1)$ re-arranged relations, we derive (via φ = increasing)

$$(n+1)\varphi(t_{n+1}) \leq (n+1)\varphi(t_n) \leq \varphi(t_0) + ... + \varphi(t_n) \leq$$
$$\psi(t_0) - \psi(t_{n+1}) \leq 1 + \psi(t_0) - \psi(t_{n+1}), \text{ for all } n \geq 0.$$

By an appropriate transformation of the (additively generated) inequality,

$$[\varphi(t_{n+1})/(1 + \psi(t_0) - \psi(t_{n+1}))]^\mu \leq 1/(n+1)^\mu, \text{ for all } n \geq 0.$$

As $\sum_n 1/(n+1)^\mu < \infty$, it follows that

$$\textstyle\sum_n [\varphi(t_{n+1})/(1 + \psi(t_0) - \psi(t_{n+1}))]^\mu < \infty; \text{ whence, } \sum_n t_n < \infty,$$

as (ψ, φ) is asymptotic μ-dominating. This proves our assertion. □

Remark 1.14.

Concerning the finitely-asymptotic regular property of (ψ, φ), the following are to be stressed. Pick the function $\psi : R_+^0 \to R$ according to the limit condition $\psi(0+0) > -\infty$.

Case 1. Let the function $\varphi : R_+^0 \to R_+^0$ be taken as constant

$$\varphi(t) = \alpha, \ t > 0; \text{ where } \alpha > 0 \text{ is arbitrary fixed.}$$

Clearly, φ is increasing; hence asymptotic positive. Moreover, the asymptotic expansive condition is fulfilled in a trivial way; because

there are no sequences $(t_n; n \geq 0)$ in R_+^0 with $\sum_n \varphi(t_n) < \infty$;

so that, the Rhoades relation $\Omega := \Omega[\psi, \varphi]$ is strongly Matkowski admissible in a trivial way. A confirmation of this is also due to the fact that

there are no sequences $(t_n; n \geq 0)$ in R_+^0, with $(t_{n+1}\Omega t_n$, for all $n)$.

For, if $(t_n; n \geq 0)$ is such a sequence, then

$$\alpha = \varphi(t_n) \leq \psi(t_n) - \psi(t_{n+1}), \text{ for all } n;$$

wherefrom, passing to limit as $n \to \infty$, one gets $\alpha = 0$; contradiction. Consequently, the relational type constructions like in Wardowski [264], [265], are not working here.

Case 2. As a consequence of this discussion, the case of $\varphi =$ constant is non-effective for us. To get examples with an effective character, let us start from the class of functions $\varphi : R^0_+ \to R^0_+$ given as

φ is *strongly asymptotic expansive*:
($\varphi(t) \geq \alpha t$, if $\varphi(t) \leq \beta$), where $\alpha, \beta > 0$ are constants.

Note that φ is asymptotic expansive. Indeed, let the strictly descending sequence (t_n) in R^0_+ be such that $\lim_n t_n = 0$ and $\sum_n \varphi(t_n) < \infty$. In particular, $\lim_n \varphi(t_n) = 0$; so, there must be some rank $n(\beta)$ such that

$\varphi(t_n) \leq \beta$, for all $n \geq n(\beta)$.

This, along with the strong asymptotic expansive property, gives

$\varphi(t_n) \geq \alpha t_n$, for all $n \geq n(\beta)$; wherefrom, $\sum_n t_n < \infty$.

On the other hand, φ is asymptotic positive. In fact, let the strictly descending sequence (t_n) in R^0_+ and the number $\varepsilon > 0$ be such that $t_n \to \varepsilon+$; but – contrary to the desired conclusion – we have

$\limsup_n(\varphi(t_n)) = 0$; hence, $\lim_n \varphi(t_n) = 0$ (as φ is strictly positive).

By definition, there exists some rank $n(\beta)$ such that [$\varphi(t_n) \leq \beta$, for all $n \geq n(\beta)$]. This, via strong asymptotic expansive property, gives

$\varphi(t_n) \geq \alpha t_n > \alpha\varepsilon$, for all $n \geq n(\beta)$;
wherefrom, $\limsup_n \varphi(t_n) \geq \alpha\varepsilon > 0$;

in contradiction with the working hypothesis. In particular, the function

$\varphi = \Gamma_\nu$, (where $\Gamma_\nu(t) = t^\nu$, $t > 0$), for some $\nu \in]0, 1]$

is (strictly positive, increasing, and) strongly asymptotic expansive (just take $\alpha = \beta = 1$ in the corresponding definition); so that, this argument is applicable here.

Remark 1.15.

Passing to the discussion of infinitely-asymptotic regular properties, let $\varphi \in \mathcal{F}(R^0_+)$ be again taken as constant

$\varphi(t) = \alpha$, $t > 0$; where $\alpha > 0$ is arbitrary fixed.

Clearly, φ is increasing. Moreover, the domination property becomes

ψ is *asymptotic μ-dominating* (for some $\mu > 1$):
$\forall$ strictly descending sequence $(t_n; n \geq 0)$ in R_+^0 with
$\lim_n t_n = 0$ and $\sum_n [1 + \psi(t_0) - \psi(t_n)]^{-\mu} < \infty$, we have $\sum_n t_n < \infty$.

A basic particular case when this property holds is

$\exists \lambda \in]0, 1[$, such that ψ is *λ-polynomial*: $\lim_{t \to 0+} \psi(t) t^\lambda = 0$.

In fact, put $\mu := 1/\lambda$ (hence, $\mu > 1$); and let the sequence $(t_n; n \geq 0)$ in R_+^0 be as in the premise above. From the posed assumption, we have

$\lim_n [1 + \psi(t_0) - \psi(t_n)] t_n^\lambda = 0$; or, equivalently,
$\lim_n [1 + \psi(t_0) - \psi(t_n)]^\mu t_n = 0$.

The sequence in the left-hand side of this last relation is therefore bounded; so that, there must be some $C > 0$, with

$[1 + \psi(t_0) - \psi(t_n)]^\mu t_n \leq C$, for all n; or, equivalently:
$t_n / C \leq [1 + \psi(t_0) - \psi(t_n)]^{-\mu}$, for all n.

But, from this, the desired assertion is clear; hence the conclusion.

In the following, some basic (and useful) particular choices for the couple (ψ, φ) above are to be discussed.

Part-Case I) The construction of Boyd-Wong relation (involving some $\chi \in \mathcal{F}(re)(R_+^0, R))$ is a particular case of this one, corresponding to

$\psi(t) = t$, $\varphi(t) = t - \chi(t)$, $t \in R_+^0$.

Part-Case II) Let $\psi \in \mathcal{F}(R_+^0, R)$ be an increasing function and $C > 0$ be a constant. Taking $\varphi \in \mathcal{F}(R_+^0, R)$ as $(\varphi(t) = C;\ t \in R_+^0)$, the couple (ψ, φ) of functions over $\mathcal{F}(R_+^0, R)$ appears as a normal one, with $\varphi =$ asymptotic positive. As a consequence, the relation $\Omega := \Omega[\psi, \varphi]$ fulfills

Ω is first variable decreasing, second variable increasing,
and asymptotic Meir-Keeler (hence, Matkowski admissible).

Moreover, under the extra condition

ψ is *asymptotic μ-dominating* (for some $\mu > 1$):
$\forall$ strictly descending sequence (t_n) in R_+^0 with $\lim_n t_n = 0$
and $\sum_n [1 + \psi(t_0) - \psi(t_n)]^{-\mu} < \infty$, we have $\sum_n t_n < \infty$

one derives that $\Omega := \Omega[\psi, \varphi]$ is strongly Matkowski admissible.

Part-Case III) Let $\lambda : R^0_+ \to]1, \infty[$ and $\mu : R^0_+ \to]0, 1[$ be a couple with $\lambda(.)$ = increasing. Define a relation $\Omega := \Omega[[\lambda, \mu]]$ over R^0_+ as

$$t\Omega s \text{ iff } \lambda(t) \leq [\lambda(s)]^{\mu(s)}.$$

This will be referred to as the *Jleli-Samet relation* attached to (λ, μ). (The proposed conventions comes from the developments in Jleli and Samet [129], corresponding to $\mu(.)$ = constant.) By a direct calculation, it is evident that

$$t\Omega s \text{ iff } t\Omega[\psi, \varphi]s; \text{ where } \psi(t) = \log[\log(\lambda(t))],\ \varphi(t) = -\log(\mu(t)),\ t > 0.$$

Hence, this construction is entirely reducible to the standard one in this series. Further aspects may be found in Suzuki and Vetro [220].

Part-Case IV) Let the couple (ψ, α) over $\mathcal{F}(R^0_+, R)$ be admissible; i.e.:

(admi-1) $\psi(.)$ is increasing, right continuous, strictly positive

(admi-2) $-\alpha(.)$ is right lsc on R^0_+, and $\gamma := \psi - \alpha$ is strictly positive.

Proposition 1.43. *Let the functions (ψ, α) be as before. Then,*

(53-1) *The couple (ψ, γ) (where $\gamma := \psi - \alpha$) is normal over $\mathcal{F}(R^0_+, R)$, with γ = asymptotic positive*

(53-2) *The associated to (ψ, γ) relation*

$$t\Omega s \text{ iff } \psi(t) \leq \psi(s) - \gamma(s) \text{ (that is: } \psi(t) \leq \alpha(s))$$

is upper diagonal and asymptotic (hence, geometric) Meir-Keeler.

Proof. i): By definition, ψ is increasing and γ is strictly positive.

ii): Suppose by contradiction that $\gamma(.)$ is not asymptotic positive: there exist $\varepsilon > 0$ and a strictly descending sequence (t_n) in R^0_+, with

$$t_n \to \varepsilon+ \text{ and } \limsup_n(\gamma(t_n)) = 0; \text{ whence, } \lim_n(\gamma(t_n)) = 0.$$

The last relation gives

$$\lim_n(-\alpha(t_n)) = -\psi(\varepsilon) \text{ (as } \psi \text{ is right continuous).}$$

Combining with $-\alpha(.)$ being right lsc on R^0_+, yields (by this limit process)

$$-\alpha(\varepsilon) \leq -\psi(\varepsilon); \text{ or, equivalently: } \gamma(\varepsilon) \leq 0;$$

in contradiction with the strict positivity of γ. Hence, our working assumption is not acceptable; and the claim follows.

iii): Evident, by our previous facts. □

Part-Case V) Let $\psi \in \mathcal{F}(R^0_+, R)$ and $\Delta \in \mathcal{F}(R)$ be a couple of functions. The following condition upon these objects will be considered here

(BV-c) (ψ, Δ) is a *Bari-Vetro couple*:
ψ is increasing and Δ is regressive ($\Delta(r) < r$, for all $r \in R$).

In this case, by definition,

$\varphi(t) := \psi(t) - \Delta(\psi(t)) > 0$, for all $t > 0$;

so that, (ψ, φ) is a normal couple of functions in $\mathcal{F}(R^0_+, R)$. Let $\Omega := \Omega[\psi, \Delta]$ be the (associated) *Bari-Vetro relation* over R^0_+, introduced as

$t\Omega s$ iff $\psi(t) \le \Delta(\psi(s))$; or, equivalently: $\psi(t) \le \psi(s) - \varphi(s)$.

(This convention is related to the developments in Di Bari and Vetro [75].) From (BV-c), Ω is an upper diagonal relation over R^0_+. It is natural then to ask: under which extra assumptions about our data we have that Ω is an asymptotic Meir-Keeler relation. The simplest one may be written as

(a-reg) Δ is *asymptotic regressive*:
for each descending sequence (r_n) in R and each $\alpha \in R$ with $r_n \to \alpha$, we have that $\liminf_n \Delta(r_n) < \alpha$.

Note that, by the non-strict character of the descending property above,

Δ is asymptotic regressive implies Δ is regressive.

Proposition 1.44. *Let the functions $(\psi \in \mathcal{F}(R^0_+, R), \Delta \in \mathcal{F}(R))$ be such that the conditions below hold*

(ψ, Δ) is an asymptotic Bari-Vetro couple; that is:
ψ is increasing and Δ is asymptotic regressive.

Then,

(54-1) *the couple (ψ, φ) is normal and φ is asymptotic positive*

(54-2) *the associated relation Ω is upper diagonal, and asymptotic Meir-Keeler (hence, geometric Meir-Keeler).*

Proof. There are two steps to be passed.

i): By a previous observation, (ψ, φ) is a normal couple; so, it remains only to establish that φ is asymptotic positive. Let the strictly descending sequence $(t_n; n \ge 0)$ in R^0_+ and the number $\varepsilon > 0$ be such that $t_n \to \varepsilon+$; we must derive that $\limsup_n(\varphi(t_n)) > 0$. Denote, for simplicity

$(r_n = \psi(t_n),\ n \geq 0);\ \alpha = \psi(\varepsilon + 0).$

By our conditions (and ψ = increasing), (r_n) is descending and $r_n \to \alpha$. This, combined with an additive property of lim sup operator, yields

$$\limsup_n \varphi(t_n) = \limsup_n [r_n - \Delta(r_n)] =$$
$$\lim_n r_n + \limsup_n(-\Delta(r_n)) = \alpha - \liminf_n \Delta(r_n) > 0;$$

hence the claim.

ii): The assertion follows from (ψ, φ) being a normal couple with (φ = asymptotic positive), and Ω being identical with $\Omega[\psi, \varphi]$. However, for completeness reasons, we provide an argument for this. Assume, by contradiction, that there exist a couple of strictly descending sequences (t_n) and (s_n) in R^0_+, and a number $\varepsilon > 0$, with the properties

$$t_n \to \varepsilon+,\ s_n \to \varepsilon+, \text{ and } t_n \Omega s_n \text{ [meaning: } \psi(t_n) \leq \Delta(\psi(s_n))], \forall n.$$

From the increasing property of ψ, one has (under $\alpha := \psi(\varepsilon + 0)$)

$(u_n := \psi(t_n))$ and $(v_n := \psi(s_n))$ are descending sequences in R, with $(u_n \to \alpha,\ v_n \to \alpha$, as $n \to \infty)$ and $(u_n \leq \Delta(v_n),\ \forall n)$.

Passing to lim inf in this relation yields (via Δ = asymptotic regressive)

$$\alpha = \liminf_n u_n \leq \liminf_n \Delta(v_n) < \alpha; \text{ contradiction.}$$

Hence, our working assumption is not acceptable; and we are done. □

In particular, when ψ and Δ are continuous, our statement reduces to the one in Jachymski [126]; see also Suzuki [219].

Finally, note that many relations of this type – including the ones in Argoubi et al. [18], Jachymski [126], Roldán et al. [198], to quote only a few – are ultimately reducible to the relations $\Omega[\chi]$ and $\Omega[\psi, \varphi]$ for appropriate functions χ and (ψ, φ). However, none of these geometric techniques, are appropriate tools for handling contractive conditions like in Khan et al. [144], or Turinici [227]; further aspects will be delineated elsewhere.

1.5.6 *Plane Rhoades relations*

In what follows, a bi-dimensional extension of our concepts is considered. Let $(\psi : R^0_+ \to R,\ \Phi : R^0_+ \times R^0_+ \to R)$ be a pair of functions, with

(p-n-couple) (ψ, Φ) is a *plane normal couple*:
ψ is increasing, and Φ is *strictly positive* $(\Phi(t, s) > 0,\ \forall t, s > 0)$.

Define the *plane Rhoades* associated relation $\Omega = \Omega[\psi, \Phi]$ over R^0_+, as

$(t,s) \in \Omega$ iff $\psi(t) \leq \psi(s) - \Phi(t,s)$.

Note that, by the plane normal couple condition, Ω is upper diagonal. In fact, let $t, s \in R^0_+$ be such that

$(t,s) \in \Omega$; that is: $\psi(t) \leq \psi(s) - \Phi(t,s)$.

By the strict positivity of Φ, one gets $\psi(t) < \psi(s)$; and this, along with the increasing property of ψ shows that $t < s$; whence, the assertion. The introduced convention is related to the developments in Khojasteh et al. [146] (see below).

Now, further properties of this relation are available under some (sequential) extra conditions upon (ψ, Φ). The first group of these writes

(d-a-pos) Φ is *diagonal asymptotic positive*:
for each strictly descending sequence (t_n) in R^0_+
and each $\varepsilon > 0$ with $t_n \to \varepsilon+$, we must have $\limsup_n \Phi(t_{n+1}, t_n) > 0$
(a-pos) Φ is *asymptotic positive*: for each couple
of strictly descending sequences (t_n) and (s_n) in R^0_+ and each $\varepsilon > 0$
with $(t_n \to \varepsilon+,\ s_n \to \varepsilon+)$, we have $\limsup_n \Phi(t_n, s_n) > 0$.

Clearly (under (ψ, Φ) = plane normal couple)

Φ is increasing in each variable $\Longrightarrow$
Φ is asymptotic positive $\Longrightarrow$ Φ is diagonal asymptotic positive.

Indeed, for each couple of strictly descending sequences (t_n) and (s_n) in R^0_+ and each $\varepsilon > 0$ with $(t_n \to \varepsilon+,\ s_n \to \varepsilon+)$, we have

$\Phi(t_n, s_n) \geq \Phi(\varepsilon, \varepsilon) > 0,\ \forall n$; whence, $\limsup_n \Phi(t_n, s_n) \geq \Phi(\varepsilon, \varepsilon) > 0$.

In the same context, it will be useful to introduce the combined notions

(d-a-p-nor) (ψ, Φ) is *diagonal asymptotic plane normal couple*:
(ψ, Φ) is a plane normal couple with Φ = diagonal asymptotic positive
(a-p-nor) (ψ, Φ) is *asymptotic plane normal couple*:
(ψ, Φ) is a plane normal couple with Φ = asymptotic positive.

Note that, under such a setting, the second inclusion above means

(ψ, Φ) is asymptotic plane normal couple implies
(ψ, Φ) is diagonal asymptotic plane normal couple.

The reciprocal inclusion is not in general true, as shown in

Example 1.3.

Fix $\alpha > 0$; and let the function $g : R^0_+ \to R^0_+$ be defined as

$g(t) = 1$, if $0 < t \leq \alpha$; $g(t) = t - \alpha$, if $t > \alpha$.

Then, let $\Phi : R^0_+ \times R^0_+ \to R^0_+$ be introduced according to

$\Phi(t, s) = \max\{t, s\}$, if $t, s > 0$, $t \neq s$; $\Phi(r, r) = g(r)$, $r > 0$.

Clearly, Φ is *diagonal asymptotic positive*; for, given the strictly descending sequence $(t_n; n \geq 0)$ in R^0_+ and the number $\varepsilon > 0$ with $t_n \to \varepsilon+$, we have

$\Phi(t_{n+1}, t_n) = \max\{t_{n+1}, t_n\} = t_n > \varepsilon$, for all n;
whence, $\limsup_n \Phi(t_{n+1}, t_n) \geq \varepsilon > 0$.

On the other hand, Φ is not asymptotic positive. In fact, take the couple of sequences (t_n) and (s_n) in R^0_+ as

$(t_n = r_n; n \geq 0)$, $(s_n = r_n; n \geq 0)$; where (r_n) is a sequence in R^0_+ with:
(r_n) is strictly descending and $r_n \to \alpha+$ as $n \to \infty$.

Then (again by definition)

$\Phi(t_n, s_n) = g(r_n) = r_n - \alpha \to 0$, as $n \to \infty$;
whence $\limsup_n \Phi(t_n, s_n) = \lim_n \Phi(t_n, s_n) = 0$;

and our assertion follows. This, according to our conventions, means

(ψ, Φ) is a diagonal asymptotic plane normal couple,
but not an asymptotic plane normal couple;

and concludes the argument.

The second group of conditions is of asymptotic regular type. Namely, call (ψ, Φ), *finitely-asymptotic regular* provided

(f-a-reg-1) Φ is diagonal asymptotic positive and $\psi(0+0) > -\infty$
(f-a-reg-2) Φ is *diagonal asymptotic expansive*:
for each strictly descending sequence $(t_n; n \geq 0)$ in R^0_+ with $\lim_n t_n = 0$ and $\sum_n \Phi(t_{n+1}, t_n) < \infty$, we have $\sum_n t_n < \infty$.

Also, let us say that (ψ, Φ) is *infinitely-asymptotic regular*, provided

(i-a-reg-1) Φ is increasing in each variable

(i-a-reg-2) (ψ, Φ) is *asymptotic μ-dominating* (for some $\mu > 1$): $\forall$ strictly descending sequence $(t_n; n \geq 0)$ in R^0_+ with $\lim_n t_n = 0$ and $\sum_n [\Phi(t_{n+1}, t_n)/(1 + \psi(t_0) - \psi(t_{n+1}))]^\mu < \infty$, we have $\sum_n t_n < \infty$.

Proposition 1.45. *Let (ψ, Φ) be a plane normal couple; and $\Omega := \Omega[\psi, \Phi]$ stand for its associated plane Rhoades relation. Then*

(61-1) Ω is first variable decreasing if $t \mapsto \psi(t) + \Phi(t, s)$ is increasing, for each $s \in R^0_+$
(61-2) Ω is second variable increasing if $s \mapsto \psi(s) - \Phi(t, s)$ is increasing, for each $t \in R^0_+$
(61-3) Ω is asymptotic Meir-Keeler (hence, geometric Meir-Keeler; hence, Matkowski admissible), if Φ is asymptotic positive
(61-4) Ω is strongly Matkowski admissible if (ψ, Φ) is finitely-asymptotic regular
(61-5) Ω is strongly Matkowski admissible if (ψ, Φ) is infinitely-asymptotic regular.

Proof. i), ii): Evident.

iii): Suppose that Ω does note have the asymptotic Meir-Keeler property: there exist strictly descending sequences (t_n) and (s_n) in R^0_+ and elements ε in R^0_+ with $(t_n \Omega s_n, \forall n)$ and $(t_n \to \varepsilon+,\ s_n \to \varepsilon+)$. By the first half of this, we get

$$(0 <)\Phi(t_n, s_n) \leq \psi(s_n) - \psi(t_n), \forall n.$$

Passing to the superior limit as $n \to \infty$, and noting that

$$\lim_n \psi(s_n) = \lim_n \psi(t_n) = \psi(\varepsilon + 0),$$

one gets $\limsup_n \Phi(t_n, s_n) = 0$; in contradiction with the asymptotic positive property of Φ. So, Ω has the asymptotic Meir-Keeler property.

iv): Let $(t_n; n \geq 0)$ be a sequence in R^0_+, with

$$t_{n+1} \Omega t_n \text{ (that is: } \psi(t_{n+1}) \leq \psi(t_n) - \Phi(t_{n+1}, t_n)), \text{ for all } n.$$

By a re-arrangement in this relation, one gets

$$(0 <)\Phi(t_{n+1}, t_n) \leq \psi(t_n) - \psi(t_{n+1}), \text{ for all } n.$$

From the upper diagonal property of Ω, $(t_n; n \geq 0)$ is strictly descending; hence, $\tau := \lim_n t_n$ exists, and $(t_n > \tau, \forall n)$. We claim that $\tau = 0$; that is: $\lim_n t_n = 0$. For, suppose that $\tau > 0$. Passing to $\limsup$ as $n \to \infty$ in the re-arranged relation, yields (via ψ = increasing)

$$(0 \leq) \limsup_n \Phi(t_{n+1}, t_n) \leq \psi(\tau + 0) - \psi(\tau + 0) = 0;$$

in contradiction with (Φ = diagonal asymptotic positive); hence the claim. On the other hand,

$$\sum_n (\psi(t_n) - \psi(t_{n+1}) < \infty; \text{ because } \psi(0 + 0) > -\infty.$$

This, combined with the re-arranged inequality once again, gives

$$(t_n \to 0 \text{ and}) \sum_n \Phi(t_{n+1}, t_n) < \infty; \text{ whence, } \sum_n t_n < \infty,$$

if we remember that Φ is diagonal asymptotic expansive.

v): Let $(t_n; n \geq 0)$ be a sequence in R^0_+, with

$$t_{n+1} \Omega t_n \text{ (that is: } \psi(t_{n+1}) \leq \psi(t_n) - \Phi(t_{n+1}, t_n)), \text{ for all } n.$$

By a re-arrangement in this relation, one gets

$$(0 <) \Phi(t_{n+1}, t_n) \leq \psi(t_n) - \psi(t_{n+1}), \text{ for all } n.$$

From the upper diagonal property of Ω, (t_n) is strictly descending; hence, $\tau := \lim_n t_n$ exists, and $(t_n > \tau, \forall n)$. We claim that $\tau = 0$; that is: $\lim_n t_n = 0$. For, suppose that $\tau > 0$. Passing to lim sup as $n \to \infty$ in the re-arranged relation, yields (via ψ = increasing)

$$(0 \leq) \limsup_n \Phi(t_{n+1}, t_n) \leq \psi(\tau + 0) - \psi(\tau + 0) = 0;$$

in contradiction with (Φ = diagonal asymptotic positive); hence the claim. Adding the first $(n+1)$ relations gives (via Φ = increasing in both variables)

$$(n + 1)\Phi(t_{n+1}, t_n) \leq \psi(t_0) - \psi(t_{n+1}) \leq 1 + \psi(t_0) - \psi(t_{n+1}), \forall n.$$

By a small transformation of the (additively generated) inequality,

$$[\Phi(t_{n+1}, t_n)/(1 + \psi(t_0) - \psi(t_{n+1}))]^\mu \leq 1/(n + 1)^\mu, \text{ for all } n \geq 0.$$

As $\sum_n 1/(n + 1)^\mu < \infty$, it follows that

$$\sum_n [\Phi(t_{n+1}, t_n)/(1 + \psi(t_0) - \psi(t_{n+1}))]^\mu < \infty; \text{ whence, } \sum_n t_n < \infty,$$

if we remember that (ψ, Φ) is asymptotic μ-dominating. □

Remark 1.16.

Concerning the finitely-asymptotic regular property of (ψ, Φ), the following are to be discussed. Let $\psi : R^0_+ \to R$ be such that $\psi(0 + 0) > -\infty$.

Case 1. Take the function $\Phi : R^0_+ \times R^0_+ \to R$ as constant

$$\Phi(t, s) = \alpha, \ t, s > 0; \text{ where } \alpha > 0 \text{ is arbitrary fixed.}$$

Clearly, Φ is asymptotic positive. Moreover, Φ is diagonal asymptotic expansive; but, in a trivial way, because

there are no sequences $(t_n; n \geq 0)$ in R_+^0 with $\sum_n \Phi(t_{n+1}, t_n) < \infty$.

As a consequence, the plane Rhoades relation Ω is strongly Matkowski admissible in a trivial way. This is confirmed by the (negative) assertion

(non-M) there are no strictly descending sequences $(t_n; n \geq 0)$ in R_+^0, with $(t_{n+1}\Omega t_n$, for all $n)$.

For, if $(t_n; n \geq 0)$ is such a sequence, then

$\alpha = \Phi(t_{n+1}, t_n) \leq \psi(t_n) - \psi(t_{n+1})$, for all n;

wherefrom, passing to limit as $n \to \infty$, one gets $\alpha = 0$; contradiction.

Case 2. As a consequence of this discussion, the case of $\Phi =$ constant is non-effective for us. To get examples with an effective character, let us start from the class of functions $\Phi : R_+^0 \times R_+^0 \to R$, taken as

Φ is strongly asymptotic expansive:
$(\Phi(t, s) \geq \alpha(t + s)$, if $\Phi(t, s) \leq \beta)$, where $\alpha, \beta > 0$ are constants.

Note that any such function is diagonal asymptotic expansive. In fact, let the strictly descending sequence $(t_n; n \geq 0)$ in R_+^0 be such that

$\lim_n t_n = 0$ and $\sum_n \Phi(t_{n+1}, t_n) < \infty$.

By the convergence property, we necessarily have $\lim_n \Phi(t_{n+1}, t_n) = 0$; so, there must be some rank $n(\beta)$ such that

$\Phi(t_{n+1}, t_n) \leq \beta$, for all $n \geq n(\beta)$.

This, along with the strong asymptotic expansive property, gives

$\Phi(t_{n+1}, t_n) \geq \alpha(t_{n+1} + t_n)$, for all $n \geq n(\beta)$;

wherefrom, $\sum_n t_n < \infty$; and the claim follows.

In particular, the function $\Phi_{\lambda,\mu} : R_+^0 \times R_+^0 \to R$ introduced as (for some couple $\lambda, \mu \in]0, 1]$)

$\Phi_{\lambda,\mu}(t, s) = \Gamma_\lambda(t) + \Gamma_\mu(s)$, $t, s > 0$ (where $\Gamma_\nu(r) = r^\nu$, $r > 0$, $\nu > 0$)

is strictly positive and increasing in each variable (hence, asymptotic positive). Moreover, $\Phi_{\lambda,\mu}$ is strongly asymptotic expansive (just take $\alpha = \beta = 1$ in the corresponding definition); so that, this argument is applicable here.

Remark 1.17.

Passing to the infinitely-asymptotic regular property, let $\Phi : R^0_+ \times R^0_+ \to R$ be again taken as constant

$\Phi(t, s) = \alpha$, $t, s > 0$; where $\alpha > 0$ is arbitrary fixed.

Clearly, Φ is strictly positive and asymptotic positive. Moreover, the domination property becomes

ψ is *asymptotic μ-dominating* (for some $\mu > 1$):
$\forall$ strictly descending sequence $(t_n; n \geq 0)$ in R^0_+ with
$\lim_n t_n = 0$ and $\sum_n [1 + \psi(t_0) - \psi(t_n)]^{-\mu} < \infty$, we have $\sum_n t_n < \infty$.

A basic particular case when this property holds is

there exists $\lambda \in]0, 1[$, such that
ψ is *λ-polynomial*: $\lim_{t \to 0+} \psi(t) t^\lambda = 0$.

In fact, put $\mu := 1/\lambda$ (hence, $\mu > 1$); and let the sequence (t_n) in R^0_+ be as in the premise above. From the posed assumption, we have

$\lim_n [1 + \psi(t_0) - \psi(t_n)] t_n^\lambda = 0$; or, equivalently:
$\lim_n [1 + \psi(t_0) - \psi(t_n)]^\mu t_n = 0$.

The sequence in the left-hand side of this relation is therefore bounded; so that, there must be some $C > 0$, with

$[1 + \psi(t_0) - \psi(t_n)]^\mu t_n \leq C$, for all n; or, equivalently:
$t_n / C \leq [1 + \psi(t_0) - \psi(t_n)]^{-\mu}$, for all n.

But, from this, the desired assertion is clear; hence the conclusion.

In the following, two particular cases of such constructions are discussed.

Part-Case I) Given the functions $\psi : R^0_+ \to R$ and $\xi : R^0_+ \times R^0_+ \to R$, we say that (ψ, ξ) is a *simulation couple*, if

(diag) ψ is increasing and ξ is *ψ-diagonal*:
$\xi(t, s) < \psi(s) - \psi(t)$, for each $t, s > 0$.

Having this couple, define an associated relation $\Omega(\psi, \xi)$ over R^0_+, as

$(t, s) \in \Omega(\psi, \xi)$ iff $\xi(t, s) \geq 0$.

This concept may be viewed as an extension of the related one defined (over pseudometric structures) in the survey paper by Turinici [257].

Some useful properties of this associated relation are available under (sequential) extra conditions upon (ψ, ξ). The first group of these writes

(d-a-neg) ξ is *diagonal asymptotic negative*:
for each strictly descending sequence (t_n) in R_+^0 and each $\varepsilon > 0$ with $t_n \to \varepsilon+$, we have $\liminf_n \xi(t_{n+1}, t_n) < 0$
(a-neg) ξ is *asymptotic negative*: for each couple of strictly descending sequences (t_n) and (s_n) in R_+^0 and each $\varepsilon > 0$ with $t_n \to \varepsilon+$, $s_n \to \varepsilon+$, we have $\liminf_n \xi(t_n, s_n) < 0$.

In this context, it will be useful to introduce the combined notions

(d-a-sim) (ψ, ξ) is *diagonal asymptotic simulation couple*:
(ψ, ξ) is a simulation couple with ξ = diagonal asymptotic negative
(a-sim) (ψ, ξ) is *asymptotic simulation couple*:
(ψ, ξ) is a simulation couple with ξ = asymptotic negative.

To get the relationships between these, the following *duality* principle will be useful. Given the couple (ψ, ξ) where $\psi : R_+^0 \to R$ and $\xi : R_+^0 \times R_+^0 \to R$, let us construct another couple (ψ, Φ), where $\Phi : R_+^0 \times R_+^0 \to R$ is taken as

$$\Phi(t, s) = \psi(s) - \psi(t) - \xi(t, s),\ t, s > 0;$$

we say that (ψ, Φ) is the *dual* of (ψ, ξ). Then, given such a couple, we denote by $\Omega[\psi, \Phi]$ the associated relation over R_+^0

$$(t, s) \in \Omega[\psi, \Phi] \text{ iff } \psi(t) \leq \psi(s) - \Phi(t, s).$$

Conversely, given the couple (ψ, Φ) where $\psi : R_+^0 \to R$ and $\Phi : R_+^0 \times R_+^0 \to R$, let us construct another couple (ψ, ξ), where $\xi : R_+^0 \times R_+^0 \to R$, as

$$\xi(t, s) = \psi(s) - \psi(t) - \Phi(t, s),\ t, s > 0;$$

we say that (ψ, ξ) is the *dual* of (ψ, Φ).

Proposition 1.46. *Let these conventions be in use. Then,*

(62-1) *If (ψ, ξ) is a (diagonal) asymptotic simulation couple that is: a simulation couple where ξ is (diagonal) asymptotic negative), then the dual couple (ψ, Φ) is a (diagonal) asymptotic plane normal one (i.e.: a plane normal couple with Φ = (diagonal) asymptotic positive); moreover, the associated relations $\Omega(\psi, \xi)$ and $\Omega[\psi, \Phi]$ are identical:*

$$(\forall t, s > 0)\text{: } (t, s) \in \Omega(\psi, \xi) \textit{ iff } (t, s) \in \Omega[\psi, \Phi].$$

(62-2) *If (ψ, Φ) is a (diagonal) asymptotic plane normal couple (that is: a plane normal couple where Φ is (diagonal) asymptotic positive), then the dual couple (ψ, ξ) is a (diagonal) asymptotic simulation one (i.e.: a simulation couple with ξ = (diagonal) asymptotic negative); moreover, the associated relations $\Omega[\psi, \Phi]$ and $\Omega(\psi, \xi)$ are identical:*

$$(\forall t, s > 0)\text{: } (t, s) \in \Omega[\psi, \Phi] \text{ iff } (t, s) \in \Omega(\psi, \xi).$$

Proof. There are two stages to be completed.

i) By the very definition of simulation couple

$\Phi(t, s) = \psi(s) - \psi(t) - \xi(t, s) > 0$, for all $t, s > 0$;
whence, Φ is strictly positive;

and this, along with ψ = increasing, tells us that (ψ, Φ) is a plane normal couple. Moreover, by these conventions, one has for each $t, s > 0$

$(t, s) \in \Omega(\psi, \xi)$ iff $\xi(t, s) \geq 0$
$(t, s) \in \Omega[\psi, \Phi]$ iff $\psi(t) \leq \psi(s) - \Phi(t, s)$.

This, along with

$\xi(t, s) \geq 0$ iff $\psi(t) \leq \psi(s) - \Phi(t, s)$

gives us the identity between these relations.

To complete the argument, we have two alternatives to discuss.

Alter i-1. Given the strictly descending sequence (t_n) in R^0_+ and the number $\varepsilon > 0$ with $t_n \to \varepsilon+$, we have (as ξ = diagonal asymptotic negative)

$$\limsup_n \Phi(t_{n+1}, t_n) = \limsup_n [\psi(t_n) - \psi(t_{n+1}) - \xi(t_{n+1}, t_n)] = \psi(\varepsilon + 0) - \psi(\varepsilon + 0) - \liminf_n \xi(t_{n+1}, t_n) > 0;$$

whence, Φ is diagonal asymptotic positive; so that, (ψ, Φ) is a diagonal asymptotic plane normal couple.

Alter i-2. For each couple of strictly descending sequences (t_n) and (s_n) in R^0_+ and each number $\varepsilon > 0$ with $t_n \to \varepsilon+$, $s_n \to \varepsilon+$, we have (as ξ = asymptotic negative)

$$\limsup_n \Phi(t_n, s_n) = \limsup_n [\psi(s_n) - \psi(t_n) - \xi(t_n, s_n)] = \psi(\varepsilon + 0) - \psi(\varepsilon + 0) - \liminf_n \xi(t_n, s_n) > 0.$$

This, by definition, tells us that Φ = asymptotic positive; so that, (ψ, Φ) = asymptotic plane normal couple.

ii) By the very definition of a plane normal couple

$$\xi(t,s) = \psi(s) - \psi(t) - \Phi(t,s) < \psi(s) - \psi(t), \text{ for all } t,s > 0;$$

and this, along with ψ = increasing, tells us (ψ, ξ) is a simulation couple. Moreover, by these conventions, one has for each $t, s > 0$

$$(t,s) \in \Omega[\psi, \Phi] \text{ iff } \psi(t) \leq \psi(s) - \Phi(t,s),$$
$$(t,s) \in \Omega(\psi, \xi) \text{ iff } \xi(t,s) \geq 0.$$

This, along with

$$\psi(t) \leq \psi(s) - \Phi(t,s) \text{ iff } \xi(t,s) \geq 0$$

gives us the identity between these relations.

Finally, as before, we have two alternatives to discuss

Alter ii-1. Given the strictly descending sequence (t_n) in R_+^0 and the number $\varepsilon > 0$ with $t_n \to \varepsilon+$, we have (as Φ = diagonal asymptotic positive)

$$\liminf_n \xi(t_{n+1}, t_n) = \liminf_n [\psi(t_n) - \psi(t_{n+1}) - \Phi(t_{n+1}, t_n)] =$$
$$\psi(\varepsilon + 0) - \psi(\varepsilon + 0) - \limsup_n \Phi(t_{n+1}, t_n) < 0;$$

whence, ξ is diagonal asymptotic negative; so that, (ψ, ξ) is a diagonal asymptotic simulation couple.

Alter ii-2. For each couple of strictly descending sequences (t_n) and (s_n) in R_+^0 and each number $\varepsilon > 0$ with $t_n \to \varepsilon+$, $s_n \to \varepsilon+$, we have (as Φ = asymptotic positive)

$$\liminf_n \xi(t_n, s_n) = \liminf_n [\psi(s_n) - \psi(t_n) - \Phi(t_n, s_n)] =$$
$$\psi(\varepsilon + 0) - \psi(\varepsilon + 0) - \limsup_n \Phi(t_n, s_n) < 0.$$

This, by definition, tells us that ξ is asymptotic negative; so, (ψ, ξ) is an asymptotic simulation couple. □

Remark 1.18.

The obtained facts yield an appropriate answer to the comparison problem between these concepts of simulation couples. Precisely, given the simulation couple (ψ, ξ), where $\psi : R_+^0 \to R$, $\xi : R_+^0 \times R_+^0 \to R$, we have

(ψ, ξ) is an asymptotic simulation couple implies
(ψ, ξ) is a diagonal asymptotic simulation couple.

The converse inclusion is not in general true. For, let $\psi : R_+^0 \to R$ be increasing. Take some $\alpha > 0$; and let $g : R_+^0 \to R_+^0$ be defined as

$$g(t) = 1, \text{ if } 0 < t \leq \alpha; \; g(t) = t - \alpha, \text{ if } t > \alpha.$$

Then, let $\Phi = \Phi_g : R^0_+ \times R^0_+ \to R^0_+$ be introduced according to

$$\Phi(t,s) = \max\{t,s\}, \text{ if } t,s > 0,\ t \neq s;\ \Phi(r,r) = g(r),\ r > 0.$$

As precise in a previous place,

Φ is diagonal asymptotic positive; but not asymptotic positive.

Now, let (ψ, ξ) be the associated to (ψ, Φ) couple; that is,

$$(\psi = \text{as before and})\ \xi(t,s) = \psi(s) - \psi(t) - \Phi(t,s),\ t,s > 0.$$

By the previous facts, (ψ, ξ) is a simulation couple; with, in addition

ξ is diagonal asymptotic negative; but not asymptotic negative.

This, according to our conventions, means

(ψ, ξ) is a diagonal asymptotic simulation couple;
but not an asymptotic simulation couple;

and concludes our assertion.

Let $I(.)$ stand for the identity function of $\mathcal{F}(R^0_+)$:

$I(t) = t$, $t > 0$ (hence, $I(.)$ is increasing).

Further, let $\xi : R^0_+ \times R^0_+ \to R$ be a function. According to a general convention, we say that (I, ξ) is an *asymptotic simulation couple*, when

(asc-1) ξ is *I-diagonal*: $\xi(t,s) < s - t$, for each $t,s > 0$
(whence: (I, ξ) is a *simulation couple*)
(asc-2) ξ is *asymptotic negative*:
for each couple of strictly descending sequences (t_n) and (s_n) in R^0_+ and each ε in R^0_+ with $t_n \to \varepsilon+$, $s_n \to \varepsilon+$, we have $\liminf_n \xi(t_n, s_n) < 0$.

A variant of this construction is as follows. Let us say that (I, ξ) is a *strong asymptotic simulation couple*, when

(sasc-1) (I, ξ) is a simulation couple (see above)
(sasc-2) ξ is *strong asymptotic negative*:
for each couple of sequences (t_n) and (s_n) in R^0_+ and each ε in R^0_+ with $(t_n \to \varepsilon,\ s_n \to \varepsilon)$, we have $\limsup_n \xi(t_n, s_n) < 0$.

This concept (also written as: $\xi : R^0_+ \times R^0_+ \to R$ is a *simulation function*) has been introduced by Khojasteh et al. [146]. According to these authors, it may be viewed as a unifying tool for many metrical fixed point theorems; so, a discussion of its basic lines may be of interest. Clearly,

(I, ξ) = strong asymptotic simulation couple implies
(I, ξ) = asymptotic simulation couple.

Concerning the converse inclusion, the natural problem to be posed is

for each asymptotic simulation couple (ψ, ξ) we have – or not – that (I, ξ) is a strong asymptotic simulation couple.

The answer to this is negative; as shown in the example below.

Example 1.4.
Let $\psi : R^0_+ \to R$ be increasing discontinuous; hence,

$\alpha := \sup\{\psi(t+0) - \psi(t-0); t \in R^0_+\} > 0$
(clearly, the alternative $\alpha = \infty$ cannot be avoided).

Let $\beta \in]0, \alpha[$ be fixed; so, by the very definition above,

there exists $\theta > 0$ such that $\psi(\theta + 0) - \psi(\theta - 0) > \beta$.

Given the couple (ψ, β), define a function $\xi : R^0_+ \times R^0_+ \to R$ as

$\xi(t, s) = \psi(s) - \psi(t) - \beta$, $t, s > 0$.

i): We claim that (ψ, ξ) is an asymptotic simulation couple.
i-1): In fact, as $\beta > 0$,

ξ is ψ-diagonal; whence, (ψ, ξ) is a simulation couple.

i-2): Moreover, ξ is asymptotic negative. For, let the couple of strictly descending sequences (t_n) and (s_n) in R^0_+ and the number ε in R^0_+ be such that $t_n \to \varepsilon+$, $s_n \to \varepsilon+$. Then, by definition

$\lim_n \xi(t_n, s_n) = \psi(\varepsilon + 0) - \psi(\varepsilon + 0) - \beta = -\beta < 0;$

and the assertion follows.

ii): On the other hand, we assert that (I, ξ) is not a strong asymptotic simulation couple.

ii-1): First, (I, ξ) is not a simulation couple. For, assume that

(I, ξ) is a simulation couple; that is (by definition)
$\psi(s) - \psi(t) - \beta < s - t$, for each $t, s > 0$.

Passing to limit as $(s \to \theta+, t \to \theta-)$, one gets (by the choice of our data)

$(0 <)\psi(\theta + 0) - \psi(\theta - 0) - \beta \leq 0$; contradiction;

and the assertion follows.

ii-2): In addition, ξ is not strong asymptotic negative. Indeed, let the couple of sequences (t_n) and (s_n) in R^0_+ and the number $\theta > 0$ (described as before) be such that $(t_n \to \theta-,\ s_n \to \theta+)$. Then, by definition

$$\lim_n \xi(t_n, s_n) = \psi(\theta + 0) - \psi(\theta - 0) - \beta > 0;$$

and the claim follows. Putting these together, we are done.

Summing up, the class of asymptotic simulation couples includes, effectively, the class of strong asymptotic simulation couples we just defined.

Part-Case II) Given the functions $\psi : R^0_+ \to R$ and $\eta : R^0_+ \times R^0_+ \to R$, we say that (ψ, η) is an *asymptotic manageable couple*, if

(amc-1) (ψ, η) is a *simulation couple*; i.e. (see above): ψ is increasing and η is ψ-*diagonal* ($\eta(t, s) < \psi(s) - \psi(t)$, for each $t, s > 0$)
(amc-2) η is *asymptotic subunitary*: for each couple of strictly descending sequences (t_n) and (s_n) in R^0_+ and each ε in R^0_+ with $t_n \to \varepsilon+$, $s_n \to \varepsilon+$, we have $\liminf_n [t_n + \eta(t_n, s_n)]/s_n < 1$.

The relationships between this concept and the preceding one (of asymptotic simulation couple) are described in the statement below.

Proposition 1.47. *Let the functions $\psi : R^0_+ \to R$ and $\eta : R^0_+ \times R^0_+ \to R$ be such that (ψ, η) is an asymptotic manageable couple. Then,*

(63-1) *η is asymptotic negative; so that (by definition) (ψ, η) is an asymptotic simulation couple*

(63-2) *the associated relation $\Omega := \Omega[\psi, \eta]$ on R^0_+, introduced as*

$$(t, s \in R^0_+)\text{: } t\Omega s \text{ iff } \eta(t, s) \geq 0$$

is upper diagonal and asymptotic Meir-Keeler (hence, geometric Meir-Keeler; hence, Matkowski admissible).

Proof. There are two steps to be passed.

i): Let the couple of strictly descending sequences (t_n) and (s_n) in R^0_+ and the number $\varepsilon > 0$ be such that $t_n \to \varepsilon+$ and $s_n \to \varepsilon+$. By the imposed condition, we derive

$$\liminf_n [t_n + \eta(t_n, s_n)]/s_n < 1.$$

On the other hand, for each $n \geq 0$ one has

$$R_n := \eta(t_n, s_n)/s_n = (t_n + \eta(t_n, s_n))/s_n + (-t_n/s_n).$$

By an auxiliary fact involving extremal limits

$$\alpha := \liminf_n R_n = \liminf_n (t_n + \eta(t_n, s_n))/s_n - 1 < 0.$$

Fix some number $\lambda \in]\alpha, 0[$. Combining with another property of such objects, there exists a subsequence $(R_n^* := R_{i(n)}; n \geq 0)$ of (R_n), defined by a corresponding couple of subsequences $(t_n^* := t_{i(n)}; n \geq 0)$ and $(s_n^* := s_{i(n)}; n \geq 0)$ of (t_n) and (s_n) respectively, such that (as $n \to \infty$)

$$R_n^* := \eta(t_n^*, s_n^*)/s_n^* \to \alpha; \text{ hence, } \eta(t_n^*, s_n^*) \to \alpha\varepsilon (< \lambda\varepsilon < 0).$$

This, by definition, yields an index m with

$$\eta(t_n^*, s_n^*) < \lambda\varepsilon < 0, \forall n \geq m; \text{ wherefrom, } \liminf_n \eta(t_n, s_n) \leq \lambda\varepsilon < 0;$$

proving that η is asymptotic negative. Combining with (ψ, η) being a simulation couple, one derives that (ψ, η) is an asymptotic simulation couple.

ii): The assertion follows by a statement involving asymptotic simulation couples; however, for completeness reasons, we provide an argument for this.

ii-1): Let $t, s > 0$ be such that $t\Omega s$; that is: $\eta(t, s) \geq 0$. By the simulation couple property, we derive

$$\psi(t) - \psi(s) < -\eta(t, s) \leq 0; \text{ whence, } \psi(t) < \psi(s).$$

Combining with ψ = increasing, yields $t < s$; so that, Ω is upper diagonal.

ii-2): Suppose by contradiction that there exists a couple of strictly descending sequences (t_n) and (s_n) in R_+^0, and a number $\varepsilon > 0$, with

$$t_n \to \varepsilon+, \ s_n \to \varepsilon+, \text{ and } t_n \Omega s_n \text{ (that is: } \eta(t_n, s_n) \geq 0\text{), for each } n.$$

Then, clearly,

$$\liminf_n [t_n + \eta(t_n, s_n)]/s_n \geq \liminf_n (t_n/s_n) = 1;$$

in contradiction with η being asymptotic subunitary. Hence, Ω has the asymptotic Meir-Keeler property; and, from this, we are done. □

Let $I(.)$ stand for the identity function of $\mathcal{F}(R_+^0)$:

$$I(t) = t, \ t > 0 \text{ (hence, } I(.) \text{ is increasing).}$$

Further, let $\eta : R_+^0 \times R_+^0 \to R$ be a function. We say that (I, η) is a *strong asymptotic manageable couple*, when

(samc-1) η is *I-diagonal*: $\eta(t,s) < s - t$, for each $t, s > 0$;
(whence: (I, η) is a *simulation couple*)
(samc-2) η is *strong asymptotic subunitary*:
for each bounded sequence (t_n) in R^0_+, and each descending sequence (s_n) in R^0_+ we have $\limsup_n [t_n + \eta(t_n, s_n)]/s_n < 1$.

This concept (also written as: $\eta : R^0_+ \times R^0_+ \to R$ is a *manageable function*) was introduced by Du and Khojasteh [82]. We claim that it is a particular case of the preceding one; precisely, that the inclusion below holds

(I, η) is a strong asymptotic manageable couple implies
(I, η) is an asymptotic manageable couple.

In fact, let the couple of strictly descending sequences (t_n) and (s_n) in R^0_+ and the number $\varepsilon > 0$ be such that $t_n \to \varepsilon+$, $s_n \to \varepsilon+$. Clearly,

$$(t_n; n \geq 0) \text{ is bounded, } (s_n; n \geq 0) \text{ is descending;}$$

and then, by the imposed hypothesis

$$\liminf_n [t_n + \eta(t_n, s_n)]/s_n \leq \limsup_n [t_n + \eta(t_n, s_n)]/s_n < 1;$$

so that, η is asymptotic subunitary. This, added to (I, η) being a simulation couple, proves our claim.

Note that the simulation type contractions like in Alsulami et al. [13] or Khojasteh et al. [146] are all reducible to such objects. However, none of these techniques can handle the contractive conditions appearing in Khan et al. [144], or Turinici [227]; further aspects will be delineated elsewhere.

1.5.7 *Complete relations*

The relations encountered until now are (nonempty) subsets of $R^0_+ \times R^0_+$. But, in practice, we are sometimes led to relations appearing as (nonempty) parts of $R_+ \times R_+$; which may be viewed as *completions* of the preceding ones. So, it is natural to determine whether this completion process is technically relevant.

By a *complete relation*, we mean any (nonempty) part $\widehat{\Omega}$ of $R_+ \times R_+$. Given such an object, denote

$$\Omega = \widehat{\Omega} \cap (R^0_+ \times R^0_+) \text{ (the } \textit{restriction} \text{ of } \widehat{\Omega} \text{ to } R^0_+ \times R^0_+).$$

The relationships between $\widehat{\Omega}$ and Ω are to be discussed over special subclasses of $\exp(R_+ \times R_+)$. Namely, for any complete relation $\widehat{\Omega}$, define

(zfirst) $\widehat{\Omega}$ is *zero first*: $(0, s) \in \widehat{\Omega}$ for all $s \in R_+$
(zsecond) $\widehat{\Omega}$ is *zero second*: $(t, 0) \in \widehat{\Omega}$ iff $t = 0$.

The subclasses of all these will be denoted as zfirst(R_+) and zsecond(R_+), respectively; most of the next concepts will be subordinated to them.

I) For each complete relation $\widehat{\Omega}$, define the property

$\widehat{\Omega}$ is *upper diagonal*: $(t, s) \in \widehat{\Omega}$ and $s > 0$ imply $t < s$;

the class of all these relations will be denoted as udiag(R_+).

Proposition 1.48. *For each complete relation* $\widehat{\Omega}$, *we have*

$\widehat{\Omega}$ *is upper diagonal iff (its associated relation)* Ω *is upper diagonal.*

Proof. The left to right inclusion is clear; so, it remains to establish the right to left one. Let $(t, s) \in \widehat{\Omega}$ with $s > 0$ be given. If $t = 0$, then $t < s$; and we are done. Assume that $t > 0$. We therefore have $(t, s) \in \Omega$; and this yields $t < s$ (as Ω is upper diagonal). The proof is complete. □

II) The next global properties to be considered are

(1-decr) $\widehat{\Omega}$ is *first variable decreasing*:
$t_1, t_2, s \in R_+$, $t_1 \geq t_2$ and $(t_1, s) \in \widehat{\Omega}$ imply $(t_2, s) \in \widehat{\Omega}$
(2-incr) $\widehat{\Omega}$ is *second variable increasing*:
$t, s_1, s_2 \in R_+$, $s_1 \leq s_2$ and $(t, s_1) \in \widehat{\Omega}$ imply $(t, s_2) \in \widehat{\Omega}$.

Proposition 1.49. *We have, in the class* zfirst$(R_+) \cap$ zsecond(R_+),

(72-1) $\widehat{\Omega}$ *is first variable decreasing iff (its associated relation)* Ω *is first variable decreasing*
(72-2) $\widehat{\Omega}$ *is second variable increasing iff (its associated relation)* Ω *is second variable increasing.*

Proof. As before, it will suffice establishing the right to left part of our first and second inclusion.

i): Let $t_1, t_2, s \in R_+$ be such that $t_1 \geq t_2$ and $(t_1, s) \in \widehat{\Omega}$. If $s = 0$, then $t_1 = 0$ (as $\widehat{\Omega}$ is zero second); and this, via $t_1 \geq t_2$, gives $t_2 = 0$; hence, $(t_2, s) \in \widehat{\Omega}$. Suppose now that $s > 0$. The case of $t_2 = 0$ is clear, as $\widehat{\Omega}$ is zero first; so, assume $t_2 > 0$ (hence, $t_1 > 0$). In this case, the desired fact results from the posed hypothesis upon Ω.

ii): Let $t, s_1, s_2 \in R_+$ be such that $s_1 \leq s_2$ and $(t, s_1) \in \widehat{\Omega}$. If $s_1 = 0$, then $t = 0$ (as $\widehat{\Omega}$ is zero second); wherefrom, $(t, s_2) \in \widehat{\Omega}$ (as $\widehat{\Omega}$ is zero first). Suppose now that $s_1 > 0$ (hence, $s_2 > 0$). The case of $t = 0$ yields $(t, s_2) \in \widehat{\Omega}$, via $\widehat{\Omega}$ being zero first; so, without loss, assume that $t > 0$. Then, the desired fact results from the posed hypothesis upon Ω. □

III) Passing to the sequential properties, the following are of interest

(M-ad) $\widehat{\Omega}$ in *Matkowski admissible*:
$(t_n; n \geq 0)$ in R_+ and $((t_{n+1}, t_n) \in \widehat{\Omega}, \forall n)$ imply $\lim_n t_n = 0$
(s-M-ad) Ω is *strongly Matkowski admissible*:
$(t_n; n \geq 0)$ in R_+ and $((t_{n+1}, t_n) \in \widehat{\Omega}, \forall n)$ imply $\sum_n t_n < \infty$.

Proposition 1.50. *We have, for each upper diagonal complete relation $\widehat{\Omega}$ belonging to the class* zsecond(R_+),

(73-1) $\widehat{\Omega}$ is Matkowski admissible iff
(its associated relation) Ω is Matkowski admissible
(73-2) $\widehat{\Omega}$ is strongly Matkowski admissible iff
(its associated relation) Ω is strongly Matkowski admissible.

Proof. Again as before, it will suffice establishing that the right to left part of each inclusion is available.

i): Let $(t_n; n \geq 0)$ in R_+ be such that $((t_{n+1}, t_n) \in \widehat{\Omega}, \forall n)$. If $t_k = 0$ for some k, then (as $\widehat{\Omega}$ is zero second), $t_n = 0$, for all $n \geq k$; and we are done. Suppose now that $(t_n > 0, \forall n)$; meaning: (t_n) is a sequence in R^0_+. Then, the desired conclusion follows from the hypothesis upon Ω.

ii): Let $(t_n; n \geq 0)$ in R_+ be such that $((t_{n+1}, t_n) \in \widehat{\Omega}, \forall n)$. If $t_k = 0$ for some k, then (as $\widehat{\Omega}$ is zero second), $t_n = 0$, for all $n \geq k$; and we are done. Suppose now that $(t_n > 0, \forall n)$; meaning: (t_n) is a sequence in R^0_+. Then, the desired conclusion follows from the hypothesis upon Ω. □

IV) Further, the following geometric properties must be considered:

(g-mk) $\widehat{\Omega}$ has the *geometric Meir-Keeler property*:
$\forall \varepsilon > 0$, $\exists \delta > 0$, $\forall (t, s)$: $(t, s) \in \widehat{\Omega}$, $\varepsilon < s < \varepsilon + \delta \Longrightarrow t \leq \varepsilon$
(g-bila-s) $\widehat{\Omega}$ is *geometric bilateral separable*:
$\forall \beta > 0$, $\exists \gamma \in]0, \beta[$, $\forall (t, s)$: $t, s \in]\beta - \gamma, \beta + \gamma[\Longrightarrow (t, s) \notin \widehat{\Omega}$
(g-left-s) $\widehat{\Omega}$ is *geometric left separable*:
$\forall \beta > 0$, $\exists \gamma \in]0, \beta[$, $\forall t$: $t \in]\beta - \gamma, \beta[\Longrightarrow (t, \beta) \notin \widehat{\Omega}$.

Proposition 1.51. *We have, for each upper diagonal complete relation $\widehat{\Omega}$,*

(74-1) $\widehat{\Omega}$ *has the geometric Meir-Keeler property iff*
(its associated relation) Ω *has the geometric Meir-Keeler property*
(74-2) $\widehat{\Omega}$ *is geometric bilateral separable iff*
(its associated relation) Ω *is geometric bilateral separable*
(74-3) $\widehat{\Omega}$ *is geometric left separable iff*
(its associated relation) Ω *is geometric left separable.*

Proof. Like before, it will suffice establishing that the right to left part of each inclusion takes place.

i): Given $\varepsilon > 0$, let $\delta > 0$ be given by the geometric Meir-Keeler property for Ω. Then, let $(t, s) \in \widehat{\Omega}$ be such that $\varepsilon < s < \varepsilon + \delta$. If $t = 0$, then $t < \varepsilon$. If $t > 0$, then $(t, s) \in \Omega$; and conclusion $t \leq \varepsilon$ follows from the hypothesis imposed upon Ω.

ii): Given $\beta > 0$, let $\gamma \in]0, \beta[$ be given by the geometric bilateral separable property for Ω. Then, assume that

$t, s \in]\beta - \gamma, \beta + \gamma[$; hence, $t, s > 0$.

By the hypothesis about Ω, one derives $(t, s) \notin \Omega$. But then, $(t, s) \notin \widehat{\Omega}$; for, otherwise, $(t, s) \in \widehat{\Omega}$ and $t, s > 0$ imply $(t, s) \in \Omega$; contradiction.

iii): Given $\beta > 0$, let $\gamma \in]0, \beta[$ be given by the geometric left separable property for Ω. Then, assume that

$t \in]\beta - \gamma, \beta[$; hence, $t > 0$.

By the hypothesis about Ω, one derives $(t, \beta) \notin \Omega$. But then, $(t, \beta) \notin \widehat{\Omega}$; for, otherwise, $(t, \beta) \in \widehat{\Omega}$ and $t, \beta > 0$ imply $(t, \beta) \in \Omega$; contradiction. □

V) Now, combining these with a previous fact, one gets the following useful equivalence result involving complete relations.

Theorem 1.36. *Under these conditions, one has in (ZF-AC+DC), for each upper diagonal complete relation* $\widehat{\Omega}$*:*

(71-a) $\widehat{\Omega} \in \text{zsecond}(R_+)$ *and* $\widehat{\Omega}$ *is geometric Meir-Keeler imply* $\widehat{\Omega}$ *is Matkowski admissible*
(71-b) $\widehat{\Omega} \in \text{zfirst}(R_+) \cap \text{zsecond}(R_+)$, $\widehat{\Omega}$ *is first variable decreasing, and* $\widehat{\Omega}$ *is Matkowski admissible, imply* $\widehat{\Omega}$ *is geometric Meir-Keeler.*

Hence, summing up

(71-c) for each first variable decreasing $\widehat{\Omega} \in \text{zfirst}(R_+) \cap \text{zsecond}(R_+)$, *we have:* $\widehat{\Omega}$ *is geometric Meir-Keeler iff* $\widehat{\Omega}$ *is Matkowski admissible.*

VI) Now, as in the standard case, each geometric concept involving complete relations admits an asymptotic counterpart, as follows

(a-mk) $\widehat{\Omega}$ is *asymptotic Meir-Keeler*:
there are no strictly descending sequences (t_n) and (s_n) in R^0_+ and no elements ε in R^0_+, with $((t_n, s_n) \in \widehat{\Omega}, \forall n)$ and $(t_n \to \varepsilon+, \ s_n \to \varepsilon+)$
(a-bila-sep) $\widehat{\Omega}$ is *asymptotic bilateral separable*:
there are no sequences $(t_n; n \geq 0)$ and $(s_n; n \geq 0)$ in R^0_+ and no elements β in R^0_+, with $((t_n, s_n) \in \widehat{\Omega}, \forall n)$ and $(t_n \to \beta, \ s_n \to \beta)$
(a-left-sep) $\widehat{\Omega}$ is *asymptotic left separable*:
there are no strictly ascending sequences (t_n) in R^0_+
and no elements β in R^0_+, with $((t_n, \beta) \in \widehat{\Omega}, \forall n)$ and $(t_n \to \beta-)$.

Here, for each sequence (r_n) in R and each element $r \in R$, we denoted

$$r_n \to r+ \ (r_n \to r-), \text{ when } r_n \to r \text{ and } [r_n > r \ (r_n < r), \ \forall n].$$

Note that, in all these concepts, only sequences and elements in R^0_+ are present. Hence, these asymptotic type concepts for a complete relation $\widehat{\Omega}$ are identical with the corresponding ones relative to the associated relation Ω. This, along with a preceding equivalence result, yields

Theorem 1.37. *The following relationships are valid, for an arbitrary upper diagonal complete relation $\widehat{\Omega}$, in the reduced system (ZF-AC+DC):*

(72-a) geometric Meir-Keeler $\Longleftrightarrow$ asymptotic Meir-Keeler
(72-b) geometric bilateral separable is equivalent with asymptotic bilateral separable
(72-c) geometric left separable $\Longleftrightarrow$ asymptotic left separable.

VII) Finally, let us give some basic examples of such relations.
VII-1) Define the functional class

$$\mathcal{F}_0(R_+) = \text{the family of all } \widehat{\chi} \in \mathcal{F}(R_+) \text{ with } \widehat{\chi}(0) = 0.$$

Then, for any such function $\widehat{\chi}$, let $\chi \in \mathcal{F}(R^0_+, R_+)$ stand for its restriction

$$\chi(t) = \widehat{\chi}(t), \text{ for each } t > 0.$$

Returning to the initial class, define the property

$\widehat{\chi}$ is *regressive*: $\widehat{\chi}(t) < t, \ \forall t \in R^0_+$;

as usual, we let $\mathcal{F}_0(re)(R_+)$ denote the class of all regressive functions in $\mathcal{F}_0(R_+)$. Clearly, for any such function $\widehat{\chi}$, its associated function χ is regressive too; hence, an element of $\mathcal{F}(re)(R^0_+, R_+)$. Conversely, if the restriction $\chi \in \mathcal{F}(R^0_+, R_+)$ is regressive, then the original function $\widehat{\chi} \in \mathcal{F}_0(R_+)$ is regressive as well.

Now, as a rule, any property defined upon $\mathcal{F}_0(re)(R_+)$ generates a corresponding one over the class $\mathcal{F}(re)(R^0_+, R_+)$. For example, let us say that

(M-ad) $\widehat{\chi}$ is *Matkowski admissible*, provided each sequence $(t_n; n \geq 0)$ in R^0_+ with $(t_{n+1} \leq \widehat{\chi}(t_n); n \geq 0)$ fulfills $t_n \to 0$
(s-M-ad) $\widehat{\chi}$ is *strongly Matkowski admissible*, provided each sequence $(t_n; n \geq 0)$ in R^0_+ with $(t_{n+1} \leq \widehat{\chi}(t_n); n \geq 0)$ fulfills $\sum_n t_n < \infty$.

As both these concepts involve the points in R^0_+ only, we have

(p1) $\widehat{\chi}$ is Matkowski admissible iff (its associated function) χ is Matkowski admissible
(p2) $\widehat{\chi}$ is strongly Matkowski admissible iff (its associated function) χ is strongly Matkowski admissible.

Likewise, let us say that $\widehat{\chi} \in \mathcal{F}_0(re)(R_+)$ is *Meir-Keeler admissible* if

$$\forall \varepsilon > 0, \exists \delta > 0\colon \varepsilon < t < \varepsilon + \delta \Longrightarrow \widehat{\chi}(t) \leq \varepsilon.$$

As before, this concept involves the points in R^0_+ only; so, we have

$\widehat{\chi}$ is Meir-Keeler admissible iff (its associated function) χ is Meir-Keeler admissible.

Having these precise, fix some function $\widehat{\chi}$ in $\mathcal{F}_0(re)(R_+)$, and define its associated relation $\widehat{\Omega} = \widehat{\Omega}[\widehat{\chi}]$, as

$$(t, s \in R_+)\colon (t, s) \in \widehat{\Omega} \text{ iff } t \leq \widehat{\chi}(s).$$

The basic properties of this relation are concentrated in

Theorem 1.38. *Under the precise conventions, we have*

(73-a) $\widehat{\Omega}$ *is (zero first and zero second)*
(73-b) $\widehat{\Omega}$ *is upper diagonal*
(73-c) $\widehat{\Omega}$ *is first variable decreasing*
(73-d) $\widehat{\Omega}$ *is Matkowski admissible whenever* $\widehat{\chi}$ *is Matkowski admissible*
(73-e) $\widehat{\Omega}$ *is strongly Matkowski admissible whenever* $\widehat{\chi}$ *is strongly Matkowski admissible*

(73-f) $\widehat{\Omega}$ *is geometric Meir-Keeler whenever* $\widehat{\chi}$ *is Meir-Keeler admissible.*

Proof. i), ii), iii): Evident.

iv), v), vi): As $\widehat{\Omega}$ is (zero first and zero second), the properties of $\widehat{\Omega}$ involved there are identical with the corresponding properties of its restriction Ω. On the other hand, as precise, the properties of $\widehat{\chi}$ involved there are identical with the corresponding properties of its restriction χ. This ends the argument. □

VII-2) Let us say that $(\widehat{\psi}, \widehat{\varphi})$ is a *normal couple* on $\mathcal{F}_0(R_+)$, if

(nc-1) $\widehat{\psi}$ is increasing and reflexive sufficient ($\widehat{\psi}(t) = 0$ iff $t = 0$)
(nc-2) $\widehat{\varphi}$ is reflexive sufficient (whence: $\widehat{\varphi}(0) = 0$, $\widehat{\varphi}(R_+^0) \subseteq R_+^0$).

For any such couple, define the additional properties

(ad-pr-1) $\widehat{\varphi}$ is *asymptotic positive*: for each sequence $(t_n; n \geq 0)$ in R_+^0 and each $\varepsilon > 0$ with $t_n \to \varepsilon+$, we have $\limsup_n(\widehat{\varphi}(t_n)) > 0$
(ad-pr-2) $\widehat{\varphi}$ is *asymptotic expansive*:
for each strictly descending sequence $(t_n; n \geq 0)$ in R_+^0 with $\lim_n t_n = 0$ and $\sum_n \widehat{\varphi}(t_n) < \infty$, we have $\sum_n t_n < \infty$
(ad-pr-3) $\widehat{\varphi}$ is *asymptotic μ-dominating* (for some $\mu > 1$):
$\forall$ strictly descending sequence $(t_n; n \geq 0)$ in R_+^0 with $\lim_n t_n = 0$ and $\sum_n(\widehat{\varphi}(t_n))^\mu < \infty$, we have $\sum_n t_n < \infty$.

Given the normal couple $(\widehat{\psi}, \widehat{\varphi})$ in $\mathcal{F}_0(R_+)$, define the (complete) relation $\widehat{\Omega} = \widehat{\Omega}[\widehat{\psi}, \widehat{\varphi}]$ over R_+, as

$(t, s \in R_+)$: $(t, s) \in \widehat{\Omega}$ iff $\widehat{\psi}(t) \leq \widehat{\psi}(s) - \widehat{\varphi}(s)$.

The basic properties of this relation are described in

Theorem 1.39. *Under the precise conventions, we have*

(74-a) $\widehat{\Omega}$ *is zero second*
(74-b) $\widehat{\Omega}$ *is upper diagonal*
(74-c) $\widehat{\Omega}$ *is first variable decreasing*
(74-d) $\widehat{\Omega}$ *is second variable increasing if* $\widehat{\psi} - \widehat{\varphi}$ *is increasing*
(74-e) $\widehat{\Omega}$ *is geometric/asymptotic Meir-Keeler (hence, Matkowski admissible) whenever* $\widehat{\varphi}$ *is asymptotic positive*
(74-f) $\widehat{\Omega}$ *is strongly Matkowski admissible if* $\widehat{\varphi}$ *is asymptotic positive and asymptotic expansive*

(74-g) $\widehat{\Omega}$ *is strongly Matkowski admissible if*
$\widehat{\varphi}$ *is increasing and asymptotic* μ*-dominating, for some* $\mu > 1$.

Proof. i): Suppose that $(t,s) \in \widehat{\Omega}$ is taken with $s = 0$. Then,

$$0 \leq \widehat{\psi}(t) \leq \widehat{\psi}(0) - \widehat{\varphi}(0) = 0;\ \text{whence},\ \widehat{\psi}(t) = 0.$$

This, along with [$\widehat{\psi}$ = reflexive sufficient], gives $t = 0$.

ii): Suppose that $(t,s) \in \widehat{\Omega}$ is taken with $s > 0$. Then, $\widehat{\varphi}(s) > 0$; so that, by the admitted hypothesis, $\widehat{\psi}(t) < \widehat{\psi}(s)$. Combining with ($\widehat{\psi}$ = increasing), yields $t < s$.

iii): Evident, in view of ($\widehat{\psi}$ = increasing).

iv): Evident, by definition.

v): Suppose, by absurd, that $\widehat{\Omega}$ is not geometric/asymptotic Meir-Keeler: there are strictly descending sequences (t_n) and (s_n) in R^0_+ and elements ε in R^0_+, with $((t_n, s_n) \in \widehat{\Omega}, \forall n)$ and $(t_n \to \varepsilon+,\ s_n \to \varepsilon+)$. The former of these relations means

$$(\forall n): \widehat{\psi}(t_n) \leq \widehat{\psi}(s_n) - \widehat{\varphi}(s_n);\ \text{that is: } \widehat{\varphi}(s_n) \leq \widehat{\psi}(s_n) - \widehat{\psi}(t_n).$$

Passing to lim sup as $n \to \infty$, and noting that

$$\lim_n \widehat{\psi}(s_n) = \lim_n \widehat{\psi}(t_n) = \widehat{\psi}(\varepsilon + 0),$$

we necessarily derive

$$(0 <) \limsup_n \widehat{\varphi}(s_n) \leq 0:\ \text{contradiction}.$$

Hence, our working assumption cannot be accepted; and we are done.

vi): Let $(t_n; n \geq 0)$ be a sequence in R^0_+, with

$$(t_{n+1}, t_n) \in \widehat{\Omega}\ (\text{that is: } \widehat{\psi}(t_{n+1}) \leq \widehat{\psi}(t_n) - \widehat{\varphi}(t_n)),\ \text{for all } n.$$

By a re-arrangement in this relation, we get

$$\widehat{\varphi}(t_n) \leq \widehat{\psi}(t_n) - \widehat{\psi}(t_{n+1}),\ \text{for all } n.$$

From the upper diagonal property of $\widehat{\Omega}$, $(t_n; n \geq 0)$ is strictly descending; hence, $\tau := \lim_n t_n$ exists. We claim that

$$\tau = 0;\ \text{that is: } t_n \to 0 \text{ as } n \to \infty.$$

Assume by contradiction that $\tau > 0$. Taking the obtained relation into account, gives (via $\widehat{\psi}$ = increasing)

$$(0 <) \limsup_n \widehat{\varphi}(t_n) \leq \widehat{\psi}(\tau + 0) - \widehat{\psi}(\tau + 0) = 0;$$

in contradiction with $\widehat{\varphi}$ = asymptotic positive. On the other hand,

$$\sum_n [\widehat{\psi}(t_n) - \widehat{\psi}(t_{n+1})] < \infty; \text{ because } (\infty >)\widehat{\psi}(0+0) \geq \widehat{\psi}(0) = 0.$$

This, combined with the re-arranged inequality once again, gives

$$(t_n \to 0 \text{ and}) \sum_n \widehat{\varphi}(t_n) < \infty; \text{ whence, } \sum_n t_n < \infty,$$

if we remember that $\widehat{\varphi}$ is asymptotic expansive.

vii): Let $(t_n; n \geq 0)$ be a sequence in R^0_+, with

$$t_{n+1}\widehat{\Omega}t_n \text{ (that is: } \widehat{\psi}(t_{n+1}) \leq \widehat{\psi}(t_n) - \widehat{\varphi}(t_n)), \text{ for all } n.$$

By the upper diagonal property of $\widehat{\Omega}$, $(t_n; n \geq 0)$ is strictly descending; hence, $\tau := \lim_n t_n$ exists. We claim that

$$\tau = 0; \text{ and then: } t_n \to 0 \text{ as } n \to \infty.$$

In fact, by a simple re-arrangement in the preceding relation,

$$\widehat{\varphi}(t_n) \leq \widehat{\psi}(t_n) - \widehat{\psi}(t_{n+1}), \text{ for all } n.$$

Adding the first $(n+1)$ relations we derive (via $\widehat{\varphi}$ = increasing)

$$(n+1)\widehat{\varphi}(t_{n+1}) \leq (n+1)\widehat{\varphi}(t_n) \leq \widehat{\psi}(t_0) - \widehat{\psi}(t_{n+1}) \leq A, \text{ for all } n;$$
$$\text{where } A := 1 + \widehat{\psi}(t_0) - \widehat{\psi}(0+) \text{ (hence, } 1 \leq A < \infty).$$

Assume by contradiction that $\tau > 0$. From the above inequality,

$$(n+1)\varphi(\tau) \leq A, \text{ for all } n.$$

Passing to limsup as $n \to \infty$, we derive (via $\varphi(\tau) > 0$)

$$\infty \leq A < \infty; \text{ contradiction; hence the claim.}$$

Now, by an appropriate multiplication of the (additively generated) inequality above, one gets

$$(\widehat{\varphi}(t_{n+1}))^\mu \leq A^\mu/(n+1)^\mu, \text{ for all } n \geq 0.$$

As $\sum_n 1/(n+1)^\mu < \infty$, it follows that

$$\sum_n (\widehat{\varphi}(t_{n+1}))^\mu < \infty; \text{ whence, } \sum_n t_n < \infty,$$

as $\widehat{\varphi}$ is asymptotic μ-dominating. This proves our assertion. □

Note, finally, that some plane versions of such relations are available, by a simple inspection of our previous developments involving the corresponding relations over R^0_+. Further aspects will be delineated elsewhere.

1.6 Topological Structures

Abstract. Some preliminary facts involving the topological structures are discussed, with a special emphasis on gauge spaces.

Subject-Class 2020. 47H10 (Primary), 54H25 (Secondary).

Keywords. Topology and cotopology, Kuratowski closure, neighborhoods system and subsystem, basis and subbasis, continuous function, interior, adherence, compactness, gauge space.

1.6.1 *Topology and cotopology*

Let $N = \{0, 1, ...\}$ denote the set of natural numbers. Given $p \in N(1, \leq)$, each set M equivalent with $N(p, >) := \{0, ..., p-1\}$ (in the sense: there exists a bijection between M and $N(p, >)$) is called *p-finite*. For completeness, we accept that the empty set $\emptyset$ is 0-finite. Finally, let us say that the set M is *finite*, provided it is q-finite, for some $q \in N$.

Let E be a nonempty set. Remember that $\exp[E]$ denotes the class of all subsets in E. Given the subset $\mathcal{D} \subseteq \exp[E]$ (that is: a family of subsets in E), remember that its union $\cup\mathcal{D}$ is introduced as

$$z \in \cup\mathcal{D} \text{ iff } [(\exists H)\text{: } H \in \mathcal{D} \text{ and } z \in H]\text{; hence, in particular, } \cup\emptyset = \emptyset.$$

On the other hand, whenever $\mathcal{D}$ is nonempty, then $\cap\mathcal{D}$ is introduced as

$$z \in \cap\mathcal{D} \text{ iff } [(\forall H)\text{: } H \in \mathcal{D} \text{ implies } z \in H].$$

Note that, in this way,

$$\cap\emptyset = \text{the class of all sets in our ambient system;}$$

so, it cannot be viewed as a set. To avoid this drawback, one *formally* takes $\cap\emptyset = E$. We stress that this definition is entirely *locally*; precisely, it is restricted (at the level of our ambient set E) to the class $\exp[E]$.

Let X be a nonempty set. We say that the family $\mathcal{G} \subseteq \exp[X]$ is

semi-normal, if $\emptyset \in \mathcal{G}$; *normal*, if $\{\emptyset, X\} \subseteq \mathcal{G}$.

By a *topology* on X, we mean any normal family $\mathcal{T} \subseteq \exp[X]$, with the properties (cf. Bourbaki [33, Ch. I, Sec. 1])

(top-1) the union of any subset in $\mathcal{T}$ belongs to $\mathcal{T}$

(top-2) the intersection of any finite subset of $\mathcal{T}$ is in $\mathcal{T}$;

then, $(X, \mathcal{T})$ is referred to as a *topological space*. Any $D \in \mathcal{T}$ will be called *open*; and its complement $X \setminus D$ is termed *closed* (with respect to $\mathcal{T}$). An equivalent way of introducing such a concept is the *complementary* one. Let us say that the normal family $\mathcal{S} \subseteq \exp[X]$ is a *cotopology* on X, when

(cotop-1) the union of any finite subset of $\mathcal{S}$ is in $\mathcal{S}$
(cotop-2) the intersection of any subset in $\mathcal{S}$ belongs to $\mathcal{S}$.

In this case, the couple $(X, \mathcal{S})$ will be called a *cotopological space*.

Given the topology $\mathcal{T}$ on X, the (normal) family $\mathcal{S} \subseteq \exp[X]$ introduced as $[E \in \mathcal{S}$ iff $X \setminus E \in \mathcal{T}]$ is a cotopology on X. This will be referred to as the *cotopology induced* by the topology $\mathcal{T}$; denoted as: $\mathcal{S} = X \setminus \mathcal{T}$. Conversely, given the cotopology $\mathcal{S}$ on X, the (normal) family $\mathcal{T} \subseteq \exp[X]$ introduced as $[D \in \mathcal{T}$ iff $X \setminus D \in \mathcal{S}]$ is a topology on X. This will be referred to as the *topology induced* by the cotopology $\mathcal{S}$; denoted as: $\mathcal{T} = X \setminus \mathcal{S}$.

1.6.2 *Topological constructions*

Let X be a nonempty set. There are several ways of introducing a topology – or cotopology – on X; the basic ones are described below.

I) We say that the selfmap $A \mapsto \text{Klo}(A)$ of $\exp[X]$ is a *(Kuratowski) closure* over X, provided (cf. Kuratowski [155, Ch. I, Sec. 4])

(Klo-1) $\emptyset = \text{Klo}(\emptyset)$, $X = \text{Klo}(X)$
(Klo-2) $A \subseteq \text{Klo}(A)$, for each $A \in \exp[X]$
(Klo-3) $\text{Klo}(A \cup B) = \text{Klo}(A) \cup \text{Klo}(B)$, $\forall A, B \in \exp[X]$
(Klo-4) $\text{Klo}(\text{Klo}(A)) = \text{Klo}(A)$, for each $A \in \exp[X]$.

Note that, as a direct consequence of (Klo-3),

(Klo-incr) $A \subseteq B$ implies $\text{Klo}(A) \subseteq \text{Klo}(B)$.

In fact, let $A, B \in \exp[X]$ be such that $A \subseteq B$. As $B = A \cup (B \setminus A)$,

$\text{Klo}(B) = \text{Klo}(A) \cup \text{Klo}(B \setminus A) \supseteq \text{Klo}(A)$; and the assertion follows.

Let $\mathcal{T}$ be a topology over X. Define a selfmap $A \mapsto \text{cl}(A)$ of $\exp[X]$, as: for each $A \in \exp[X]$,

$\text{cl}(A) =$ the intersection of all closed subsets $G \in \exp[X]$ with $A \subseteq G$.

It is not hard to see that all properties (Klo-1)-(Klo-4) hold; whence, $A \mapsto \text{cl}(A)$ is a Kuratowski closure over X; referred to as the *closure (adherence) operator* induced by $\mathcal{T}$. Conversely, suppose that $A \mapsto \text{Klo}(A)$

is a Kuratowski closure over X. Then, the (normal) class $\mathcal{S} = \{E \in \exp[X]; E = \text{Klo}(E)\}$ is a cotopology over X (see above); so that, $\mathcal{T} = X \setminus \mathcal{S}$ is a topology over X. Moreover, with respect to this topology, we have

$$\text{cl}(A) = \text{Klo}(A), \text{ for each } A \in \exp[X].$$

II) Suppose that, to each $x \in X$ we attached a class $\mathcal{W}(x) \subseteq \exp[X]$ such that the family $(\mathcal{W}(x); x \in X)$ fulfills

(ns-1) $(\forall x \in X)$: for each $W \in \mathcal{W}(x)$, we have $x \in W$
(ns-2) $(\forall x \in X)$: $A \in \mathcal{W}(x)$ and $A \subseteq B$ imply $B \in \mathcal{W}(x)$;
hence, in particular, $X \in \mathcal{W}(x)$
(ns-3) $(\forall x \in X)$: $W_1, W_2 \in \mathcal{W}(x)$ imply $W_1 \cap W_2 \in \mathcal{W}(x)$
(ns-4) $(\forall x \in X)$: for each $A \in \mathcal{W}(x)$ there exists $D \in \mathcal{W}(x)$
such that $A \in \mathcal{W}(y)$, for each $y \in D$ (hence, $A \supseteq D$).

We then say that $(\mathcal{W}(x); x \in X)$ is a *neighborhoods system* over X; and, for $x \in X$, $\mathcal{W}(x)$ is a *neighborhoods system* of x. A concrete example is to be given as follows. Let $\mathcal{T}$ be a topology over X. Given $x \in X$, let us say that $Y \in \exp[X]$ is a $\mathcal{T}$-*neighborhood* of x, provided

$$x \in D \subseteq Y, \text{ for some } D \in \mathcal{T};$$

let $\mathcal{V}(x)$ be the class of all these. It is directly seen that $(\mathcal{V}(x); x \in X)$ fulfills (ns-1)–(ns-4); so, it is a neighborhoods system over X [referred to as the *neighborhoods system over X generated by $\mathcal{T}$*]; and, for $x \in X$, $\mathcal{V}(x)$ is a neighborhoods system of x [referred to as the *neighborhoods system of x generated by $\mathcal{T}$*]. Concerning the reciprocal question, the following statement is available (see, for instance, Costinescu [66, Ch. II, Sec. 1]).

Proposition 1.52. *Suppose that $(\mathcal{W}(x); x \in X)$ is a neighborhoods system over X; and put*

$$\mathcal{T} = \textit{the set of all } D \in \exp[X] \textit{ with } [(\forall x): x \in D \Longrightarrow D \in \mathcal{W}(x)].$$

Then,

(21-1) $\mathcal{T}$ is (a normal family and) a topology over X
(21-2) with respect to this topology, $(\mathcal{V}(x) = \mathcal{W}(x)$, for all $x \in X)$.

A *cofinal* variant of this construction is as follows. Suppose that, to each $x \in X$ we attached a class $\mathcal{W}^*(x) \subseteq \exp[X]$ such that the family $(\mathcal{W}^*(x); x \in X)$ fulfills

(nss-1) $(\forall x \in X)$: for each $W^* \in \mathcal{W}^*(x)$, we have $x \in W^*$
(nss-2) $(\forall x \in X)$: for each $W_1^*, W_2^* \in \mathcal{W}^*(x)$ there exists $W_3^* \in \mathcal{W}^*(x)$ such that $W_1^* \cap W_2^* \supseteq W_3^*$
(nss-3) $(\forall x \in X)$: for each $A^* \in \mathcal{W}^*(x)$ there exists $D^* \in \mathcal{W}^*(x)$ such that for each $y \in D^*$, there exists $B_y^* \in \mathcal{W}^*(y)$ with $A^* \supseteq B_y^*$.

We then say that $(\mathcal{W}^*(x); x \in X)$ is a *neighborhoods subsystem over* X; and, for $x \in X$, $\mathcal{W}^*(x)$ will be referred to as a *neighborhoods subsystem of* x. A concrete example is to be given as follows. Let $\mathcal{T}$ be a topology over X; and $(\mathcal{V}(x); x \in X)$ stand for the neighborhoods system over X generated by $\mathcal{T}$. Then, for each $x \in X$, let the subset $\mathcal{V}^*(x)$ of $\mathcal{V}(x)$ be taken as

for each $V \in \mathcal{V}(x)$ where exists $V^* \in \mathcal{V}^*(x)$ with $V \supseteq V^*$.

The family $(\mathcal{V}^*(x); x \in X)$ is a neighborhoods subsystem over X; referred to as: *the neighborhoods subsystem over* X *generated by* $\mathcal{T}$. For example, $(\mathcal{V}^*(x); x \in X)$ may be taken as the *open neighborhoods subsystem over* X *generated by* $\mathcal{T}$:

$$\mathcal{V}^*(x) = \{D \in \mathcal{T}; x \in D\},\ x \in X.$$

Conversely, let $(\mathcal{W}^*(x); x \in X)$ be a neighborhoods subsystem over X. Then, the class $(\mathcal{W}(x); x \in X)$ introduced as

$$(\forall x \in X)\text{: } W \in \mathcal{W}(x) \text{ iff } W \supseteq W^*, \text{ for some } W^* \in \mathcal{W}^*(x)$$

is a neighborhoods system over X. By a preceding statement, there exists a (uniquely determined) topology $\mathcal{T}$ over X such that $[\mathcal{W}(x) = \mathcal{V}(x),\ x \in X]$; so, $(\mathcal{W}(x); x \in X)$ is the neighborhoods system on X generated by $\mathcal{T}$. But then, $(\mathcal{W}^*(x); x \in X)$ is a neighborhoods subsystem on X generated by $\mathcal{T}$.

III) Let us say that the open sets family $\mathcal{B} \subseteq \mathcal{T}$ is a *basis* for $\mathcal{T}$, provided

$$(\forall D \in \mathcal{T})\text{: } D\text{=union of a subset in } \mathcal{B}.$$

Note that (as $X \in \mathcal{T}$), $\mathcal{B}$ is *total* $[X = \cup\mathcal{B}]$; whence, $\mathcal{B} \neq \emptyset$. A characterization of this concept is to be obtained by means of neighborhoods subsystem on X generated by $\mathcal{T}$. Let $\mathcal{B} \subseteq \mathcal{T}$ be a family of open sets. Denote

$$\mathcal{B}(x) = \{B \in \mathcal{B}; x \in B\},\ x \in X.$$

We have (see Costinescu [66, Ch. II, Sec. 2] for details)

Proposition 1.53. *Given* $\mathcal{B} \subseteq \mathcal{T}$, *the following are equivalent:*

(22-1) $\mathcal{B}$ *is a basis for the topology* $\mathcal{T}$

(22-2) $(\mathcal{B}(x); x \in X)$ *is a neighborhoods subsystem on* X *generated by* $\mathcal{T}$*: for each* $(x \in X,\ V \in \mathcal{V}(x))$ *there exists* $B \in \mathcal{B}(x)$ *with* $V \supseteq B$.

Let $\mathcal{B} \subseteq \exp[X]$ be a class of sets. We may ask whether there exists a topology $\mathcal{T}$ on X such that $\mathcal{B}$ is a basis for $\mathcal{T}$. Note that whenever this happens, then (by the properties of $(\mathcal{B}(x); x \in X)$)

(B-total) $\mathcal{B}$ is *total*: $X = \cup\mathcal{B}$ ($\forall x \in X$, $\exists B \in \mathcal{B}$ such that $x \in B$)
(B-s-dir) $\mathcal{B}$ is *strongly directed*: for each $B_1, B_2 \in \mathcal{B}$ and each $x \in B_1 \cap B_2$ there exists $B_3 \in \mathcal{B}$ such that $x \in B_3 \subseteq B_1 \cap B_2$.

Theorem 1.40. *Let* $\mathcal{B} \subseteq \exp[X]$ *be total, strongly directed; and put*

$$\mathcal{T} = \textit{the class of all } D \in \exp[X] \textit{ with } D = \textit{union of a subset in } \mathcal{B}.$$

Then,

(21-a) $\mathcal{T}$ *is (a normal family and) a topology on* X
(21-b) $\mathcal{B}$ *is a basis of the topology* $\mathcal{T}$.

Proof. (cf. Engelking [87, Ch. 1, Sec. 1.2]). There are two steps to pass.

Step 1. Clearly, $\mathcal{T}$ is normal, via $X = \cup\mathcal{B}$, $\emptyset = \cup\emptyset$, and $(\mathcal{B}, \emptyset \in \exp[\mathcal{B}])$. In addition, each union of elements in $\mathcal{T}$ belongs to $\mathcal{T}$. It remains to establish that the intersection of any finite subset of $\mathcal{T}$ belongs to $\mathcal{T}$; which amounts to: $D_1, D_2 \in \mathcal{T}$ imply $D_1 \cap D_2 \in \mathcal{T}$. Given $x \in D_1 \cap D_2$, there exist $B_1, B_2 \in \mathcal{B}$ such that

$$x \in B_1 \subseteq D_1,\ x \in B_2 \subseteq D_2;\ \text{hence, } x \in B_1 \cap B_2 \subseteq D_1 \cap D_2.$$

From the strongly directed property, there exists $B_3 \in \mathcal{B}$ such that

$$x \in B_3 \subseteq B_1 \cap B_2 \subseteq D_1 \cap D_2.$$

This, along with the arbitrariness of x in $D_1 \cap D_2$, tells us that

$$D_1 \cap D_2 = \text{union of a subset in } \mathcal{B};\ \text{whence, } D_1 \cap D_2 \in \mathcal{T};$$

so that, $\mathcal{T}$ is a topology on X.

Step 2. For each $B \in \mathcal{B}$, we have $B = \cup\{B\} \in \mathcal{T}$; whence, $\mathcal{B} \subseteq \mathcal{T}$; and this, along with definition of $\mathcal{T}$, gives the desired conclusion. □

IV) Given the topology $\mathcal{T}$ over X, let us say that the family $\mathcal{A} \subseteq \exp[X]$ is a *subbasis* of it, when

the class $\mathcal{B}$ of intersections of finite subsets in $\mathcal{A}$ is a basis for $\mathcal{T}$.

Clearly, $A = \cap\{A\} \in \mathcal{B}$, $\forall A \in \mathcal{A}$; so, $\mathcal{A} \subseteq \mathcal{B}$; wherefrom $\cup\mathcal{A} \subseteq \cup\mathcal{B}$. Moreover, ($\forall B \in \mathcal{B}$, $\exists A \in \mathcal{A}$: $B \subseteq A$); whence $\cup\mathcal{B} \subseteq \cup\mathcal{A}$. Combining these, yields $\cup\mathcal{B} = \cup\mathcal{A}$; so (via $\cup\mathcal{B} = X$), we have that $\mathcal{A}$ is *total* $[\cup\mathcal{A} = X]$.

Conversely, let $\mathcal{A} \subseteq \exp[X]$ be a class of sets. As before, we may ask whether there exists a topology $\mathcal{T}$ on X such that $\mathcal{A}$ is a subbasis for $\mathcal{T}$. By the above developments, this happens when

$\mathcal{B}$ = the class of intersections of finite subsets in $\mathcal{A}$

fulfills the couple of conditions

(B-total) $\mathcal{B}$ is *total*: $X = \cup\mathcal{B}$ ($\forall x \in X$, $\exists B \in \mathcal{B}$ such that $x \in B$)
(B-s-dir) $\mathcal{B}$ is *strongly directed*: for each $B_1, B_2 \in \mathcal{B}$ and each $x \in B_1 \cap B_2$ there exists $B_3 \in \mathcal{B}$ such that $x \in B_3 \subseteq B_1 \cap B_2$.

Concerning the former one, note that (by the same argument as before) $\cup\mathcal{B} = \cup\mathcal{A}$; so that, whenever $\mathcal{A}$ is total ($\cup\mathcal{A} = X$) we are done. The latter of these is evident (in our case), via: $B_1, B_2 \in \mathcal{B}$ implies $B_1 \cap B_2 \in \mathcal{B}$. Summing up, the following answer to the posed question is available.

Theorem 1.41. *Let the family $\mathcal{A} \subseteq \exp[X]$ be total ($\cup\mathcal{A} = X$); and put*

$$\mathcal{T} = \{D \in \exp[X]; D = \textit{union of intersections of finite subsets in } \mathcal{A}\}.$$

Then,

(22-a) $\mathcal{T}$ is (a normal family and) a topology on X
(22-b) $\mathcal{A}$ is a subbasis of the topology $\mathcal{T}$.

1.6.3 *Comparison criteria*

In the following, the comparison of topologies will be discussed.

Given the couple of topological spaces $(X, \mathcal{T})$ and $(Y, \mathcal{S})$, let us say that $f : X \to Y$ is *$(\mathcal{T}, \mathcal{S})$-continuous*, when

for each $G \in \mathcal{S}$ we have $f^{-1}(G) \in \mathcal{T}$.

Note that, an equivalent characterization of this property is by means of cotopologies; precisely, $f : X \to Y$ is $(\mathcal{T}, \mathcal{S})$-continuous, iff

for each $H \in Y \setminus \mathcal{S}$ we have $f^{-1}(H) \in X \setminus \mathcal{T}$.

This concept may be used towards the comparison of topologies. Some preliminaries are needed. Let X be a nonempty set. Given the topologies $\mathcal{T}$, $\mathcal{S}$ over X, let us introduce the relation

$\mathcal{T} \subseteq \mathcal{S}$: each $\mathcal{T}$-open set is $\mathcal{S}$-open.

This will be referred to as: $\mathcal{T}$ is *coarser* than $\mathcal{S}$; or: $\mathcal{S}$ is *finer* than $\mathcal{T}$.

Let $i_X : X \to X$ stand for the *identity selfmap* ($i_X(x) = x; x \in X$). The following result is now immediate; so, we do not give details.

Proposition 1.54. *For the topologies $\mathcal{T}$, $\mathcal{S}$ over X, the following assertions are equivalent:*

(31-1) $\mathcal{T}$ is coarser than $\mathcal{S}$ (or, equivalently: $\mathcal{S}$ is finer than $\mathcal{T}$)
(31-2) the identical application $i_X : X \to X$ is $(\mathcal{S}, \mathcal{T})$-continuous.

This characterization is a starting point for constructing topologies on a set. Given a nonempty index set Λ, let $((Y_\lambda, \mathcal{S}_\lambda); \lambda \in \Lambda)$ be a family of topological spaces, and $(f_\lambda : X \to Y_\lambda; \lambda \in \Lambda)$ be a family of maps. We are interested to determine the (minimal) topology $\mathcal{T}$ over X with

f_λ is $(\mathcal{T}, \mathcal{S}_\lambda)$-continuous, for each $\lambda \in \Lambda$.

As we will see, this topology is to be introduced by the subbase

$\mathcal{A} = \cup\{f_\lambda^{-1}(\mathcal{S}_\lambda); \lambda \in \Lambda\}$; where, by definition,
$f_\lambda^{-1}(\mathcal{S}_\lambda) = \{f_\lambda^{-1}(E_\lambda); E_\lambda \in \mathcal{S}_\lambda\}$, $\lambda \in \Lambda$.

In fact, $\mathcal{A}$ is total; because ($\forall \lambda \in \Lambda$: $X = f_\lambda^{-1}(Y_\lambda)$, $Y_\lambda \in \mathcal{S}_\lambda$). So, by a preceding statement, $\mathcal{A}$ is a subbase of a topology $\mathcal{T}$ over X, given as

$\mathcal{T} = \{D \in \exp[X]; D =$ union of intersections of finite subsets in $\mathcal{A}\}$.

The obtained topology has all needed properties. Precisely, the following result holds. (Since the verification is immediate, we do not give details).

Theorem 1.42. *Under the above conventions,*
(31-a) *f_λ is $(\mathcal{T}, \mathcal{S}_\lambda)$-continuous, for each $\lambda \in \Lambda$*
(31-b) *If the topology $\mathcal{Z}$ over X satisfies [f_λ is $(\mathcal{Z}, \mathcal{S}_\lambda)$-continuous, $\forall \lambda \in \Lambda$] then, necessarily, $\mathcal{T} \subseteq \mathcal{Z}$ ($\mathcal{T}$ is coarser than $\mathcal{Z}$).*

Remark 1.19.

The following alternate representation of this topology is useful in many concrete cases. For each $\lambda \in \Lambda$, let $\mathcal{S}_\lambda^*$ be a subbasis of $\mathcal{S}_\lambda$; note that, by definition, $\mathcal{S}_\lambda^*$ is total in Y_λ. Then, let us denote

$\mathcal{A}^* = \cup\{f_\lambda^{-1}(\mathcal{S}_\lambda^*); \lambda \in \Lambda\}$; where, by definition,
$f_\lambda^{-1}(\mathcal{S}_\lambda^*) = \{f_\lambda^{-1}(E_\lambda^*); E_\lambda^* \in \mathcal{S}_\lambda^*\}$, $\lambda \in \Lambda$.

Clearly, $\mathcal{A}^*$ is total in X; because $(\forall \lambda \in \Lambda$: $X = f_\lambda^{-1}(Y_\lambda) = f_\lambda^{-1}(\cup \mathcal{S}_\lambda^*))$. From the above observation, the formula

$$\mathcal{T}^* = \{D \in \exp[X]; D = \text{union of intersections of finite subsets in } \mathcal{A}^*\}$$

defines a topology over X. We claim that, necessarily,

$$\mathcal{T} = \mathcal{T}^*; \text{ meaning: } \mathcal{T} \subseteq \mathcal{T}^* \text{ and } \mathcal{T}^* \subseteq \mathcal{T}.$$

The latter inclusion is clear, in view of $\mathcal{A}^* \subseteq \mathcal{A}$; so, it remains to establish the former inclusion. This, in turn, reduces to

(f-int) each intersection of a finite subset in $\mathcal{A}$ belongs to $\mathcal{T}^*$;

for, in such a case (by the very definition of $\mathcal{T}$)

$$D \in \mathcal{T} \text{ implies } D = \text{union of a subset in } \mathcal{T}^*; \text{ whence } D \in \mathcal{T}^*.$$

Now, (f-int) means, ultimately: $A, B \in \mathcal{A}$ implies $A \cap B \in \mathcal{T}^*$. To verify this, let $A, B \in \mathcal{A}$; hence,

$$A = f_\lambda^{-1}(G_\lambda),\ B = f_\mu^{-1}(H_\mu), \text{ where } \lambda, \mu \in \Lambda,\ G_\lambda \in \mathcal{S}_\lambda,\ H_\mu \in \mathcal{S}_\mu.$$

Fix some $x \in A \cap B$. As $f_\lambda(x) \in G_\lambda$, there exists ($G_\lambda^*$ = intersection of a finite subset in $\mathcal{S}_\lambda^*$), such that $f_\lambda(x) \in G_\lambda^* \subseteq G_\lambda$. Likewise, as $f_\mu(x) \in H_\mu$, there exists (H_μ^* = intersection of a finite subset in $\mathcal{S}_\mu^*$), such that $f_\mu(x) \in H_\mu^* \subseteq H_\mu$. Combining these, yields

$$x \in f_\lambda^{-1}(G_\lambda^*) \cap f_\mu^{-1}(H_\mu^*) \subseteq A \cap B; \text{ with}$$
$$f_\lambda^{-1}(G_\lambda^*) \cap f_\mu^{-1}(H_\mu^*) = \text{intersection of a finite subset in } \mathcal{A}^*;$$

and this, by the arbitrariness of $x \in A \cap B$, gives

$$A \cap B = \text{union of intersections of finite subsets in } \mathcal{A}^*; \text{ so, } A \cap B \in \mathcal{T}^*.$$

A basic particular case of these developments corresponds to the choice

$$Y_\lambda = X,\ f_\lambda = i_X, \text{ for all } \lambda \in \Lambda.$$

Precisely, let Λ be a nonempty index set; and $(\mathcal{S}_\lambda; \lambda \in \Lambda)$ be a family of topologies over X. We have

Theorem 1.43. *Under the described setting, there exists a unique topology $\mathcal{T}$ over X having as subbase the class of sets $\mathcal{A} = \cup\{\mathcal{S}_\lambda; \lambda \in \Lambda\}$, and representable as follows*

$$\mathcal{T} = \{D \in \exp[X]; D = \textit{union of intersections of finite subsets in } \mathcal{A}\}.$$

Moreover, the following properties hold:

(32-a) $\mathcal{S}_\lambda \subseteq \mathcal{T}$, *for each* $\lambda \in \Lambda$
(32-b) *If the topology* $\mathcal{Z}$ *over* X *fulfills* ($\mathcal{S}_\lambda \subseteq \mathcal{Z}$, $\forall \lambda \in \Lambda$) *then,* $\mathcal{T} \subseteq \mathcal{Z}$.

In other words:

(32-c) $\mathcal{T} = \sup\{\mathcal{S}_\lambda; \lambda \in \Lambda\}$ *(in the inclusion sense).*

Remark 1.20.

The following alternate representation of this topology is useful in many concrete cases. For each $\lambda \in \Lambda$, let $\mathcal{S}_\lambda^*$ be a subbasis of $\mathcal{S}_\lambda$; note that, by definition, $\mathcal{S}_\lambda^*$ is total. Then, denote $\mathcal{A}^* = \cup\{\mathcal{S}_\lambda^*; \lambda \in \Lambda\}$. From the above observation, $\mathcal{A}^*$ is also total; so, the formula

$$\mathcal{T}^* = \{D \in \exp[X]; D = \text{union of intersections of finite subsets in } \mathcal{A}^*\}$$

defines a topology over X. We claim that, necessarily,

$$\mathcal{T} = \mathcal{T}^*; \text{ meaning: } \mathcal{T} \subseteq \mathcal{T}^* \text{ and } \mathcal{T}^* \subseteq \mathcal{T}.$$

The latter inclusion is clear, in view of $\mathcal{A}^* \subseteq \mathcal{A}$; so, it remains to establish the former inclusion. This, in turn, reduces to

(f-int-id) each intersection of a finite subset in $\mathcal{A}$ belongs to $\mathcal{T}^*$;

for, in such a case (by the very definition of $\mathcal{T}$)

$D \in \mathcal{T}$ implies D = union of a subset in $\mathcal{T}^*$; whence $D \in \mathcal{T}^*$.

Now, (f-int-id) means, ultimately: $A, B \in \mathcal{A}$ implies $A \cap B \in \mathcal{T}^*$. To verify this, let $A, B \in \mathcal{A}$; hence,

$$A = G_\lambda,\ B = H_\mu, \text{ where } \lambda, \mu \in \Lambda,\ G_\lambda \in \mathcal{S}_\lambda,\ H_\mu \in \mathcal{S}_\mu.$$

Let $x \in A \cap B$ be arbitrary fixed. As $x \in G_\lambda$, there exists (G_λ^* = intersection of a finite subset in $\mathcal{S}_\lambda^*$), such that $x \in G_\lambda^* \subseteq G_\lambda$. Likewise, as $x \in H_\mu$, there exists (H_μ^* = intersection of a finite subset in $\mathcal{S}_\mu^*$), such that $x \in H_\mu^* \subseteq H_\mu$. Combining these, yields the assertion

$x \in G_\lambda^* \cap H_\mu^* \subseteq A \cap B$; with
$G_\lambda^* \cap H_\mu^*$ = intersection of a finite subset in $\mathcal{A}^*$;

and this, by the arbitrariness of $x \in A \cap B$, gives

$A \cap B$ = union of intersections of finite subsets in $\mathcal{A}^*$; so, $A \cap B \in \mathcal{T}^*$.

1.6.4 *Basic concepts*

In the following, some completions of the above facts will be provided. Let $(X, \mathcal{T})$ be a topological space.

Given $A \in \exp[X]$, let us say that $x \in X$ is *interior* to A, when $A \in \mathcal{V}(x)$; clearly, $x \in A$. The class of all these points is denoted as $\text{int}(A)$ (the *interior* of A). From a global perspective, this set may be represented as

$$\text{int}(A) = \text{the union of all } D \in \mathcal{T} \text{ with } D \subseteq A.$$

Given $A \in \exp[X]$, let us say that $x \in X$ is *adherent* to A, when $V \cap A \neq \emptyset$, for all $V \in \mathcal{V}(x)$; clearly, $x \in A$ in not in general true. The class of all these points is just $\text{cl}(A)$ [the *closure* (or, *adherence*) of A]. From a global perspective, this set may be represented as

$$\text{cl}(A) = \text{the intersection of all } G \in X \setminus \mathcal{T} \text{ with } A \subseteq G.$$

It is worth noting that the operators $A \mapsto \text{int}(A)$ and $A \mapsto \text{cl}(A)$ (from $\exp[X]$ to itself) are *duals*, in the sense

$$X \setminus \text{int}(A) = \text{cl}(X \setminus A),\ X \setminus \text{cl}(A) = \text{int}(X \setminus A),\ A \in \exp[X].$$

This means that each property involving int(.) has a dual property involving cl(.), and vice versa; we do not give details.

Given $A \in \exp[X]$, let us say that the point $x \in X$ is *cluster* to A, when $V \cap (A \setminus \{x\}) \neq \emptyset$, for all $V \in \mathcal{V}(x)$. The class of all these writes $\text{clu}(A)$ (the *cluster* of A). Note that, by definition,

$$A \cup \text{clu}(A) = \text{cl}(A); \text{ hence } \text{clu}(A) = \emptyset \text{ implies } A = \text{closed}.$$

Given $A \in \exp[X]$, let us say that $x \in X$ is *boundary* to A, when $x \in \text{cl}(A) \cap \text{cl}(X \setminus A)$; as before, such a point need not be in A. The class of all these writes

$$\text{bd}(A) = \text{cl}(A) \cap \text{cl}(X \setminus A) = \text{cl}(A) \setminus \text{int}(A) \text{ (the } \textit{boundary} \text{ of } A).$$

As a consequence of this,

$$\text{cl}(A) = (\text{cl}(A) \setminus \text{int}(A)) \cup \text{int}(A) = \text{bd}(A) \cup \text{int}(A),\ A \in \exp[X].$$

Let again $(X, \mathcal{T})$ be a topological space. We say that $(X, \mathcal{T})$ is *(Hausdorff) separated*, when

$$\forall x, y \in X \text{ with } x \neq y, \text{ there exist } V \in \mathcal{V}(x),\ W \in \mathcal{V}(y) \text{ with } V \cap W = \emptyset.$$

Given the (Hausdorff) separated space $(X, \mathcal{T})$, call it *compact* when

each open cover of X admits a finite open sub-cover of X; that is:
$X = \cup\mathcal{G}$ for $\mathcal{G} \subseteq \mathcal{T}$ implies $X = \cup\mathcal{H}$, where $\mathcal{H}$ = finite part of $\mathcal{G}$.

Generally, for the (Hausdorff) separated space $(X, \mathcal{T})$, we say that the (nonempty) subset $P \subseteq X$ is *compact*, when

each open cover of P yields a finite open sub-cover of P; that is:
$P \subseteq \cup\mathcal{G}$ for $\mathcal{G} \subseteq \mathcal{T}$ implies $P \subseteq \cup\mathcal{H}$, where $\mathcal{H}$ = finite part of $\mathcal{G}$.

The following properties of these concepts will be useful.

Proposition 1.55. *Let $(X, \mathcal{T})$ be a Hausdorff space, and P be a nonempty subset of X. Then,*

(41-1) P is compact implies P is closed
(41-2) $(X, \mathcal{T})$ is compact and P = closed imply P = compact.

For a discussion of all these aspects, we refer to Costinescu [66, Ch. V].

1.6.5 *Metrical structures*

Some basic applications of these facts are given in a metrical context.

I) Let M be a nonempty set. Remember that, by a *pseudometric* over M, we mean any map $d : M \times M \to R_+$. Given such an object, let us consider the properties

(m-1) d is *reflexive*: $d(x, x) = 0$, for all $x \in X$
(m-2) d is *triangular*: $d(x, y) \le d(x, z) + d(z, y)$, $\forall x, y, z \in X$
(m-3) d is *symmetric*: $d(x, y) = d(y, x)$, for all $x, y \in X$;
(m-4) d is *sufficient*: $d(x, y) = 0$ implies $x = y$.

When (m-1)–(m-3) hold, we say that $d(.,.)$ is a *semimetric* on M; and (M, d) is called a *semimetric space*. On the other hand, when (m-1)–(m-4) hold, we say that $d(.,.)$ is a *metric* on M; and (M, d) is called a *metric space*.

Let (M, d) be a semimetric space. The topology to be considered here is generated by the family of open spheres in M. Precisely, given $a \in M$, $\rho > 0$, denote

$$M(a, \rho)(d) = \{x \in M; d(a, x) < \rho\},\ M[a, \rho](d) = \{x \in M; d(a, x) \le \rho\};$$

referred to as the *open* (respectively, *closed*) sphere with center $a \in M$ and radius $\rho > 0$. (Clearly, both these are nonempty; because $a \in M$ is an element of them).

Define the family of (nonempty) sets

$$\mathcal{W}^*(a)(d) = \{M(a,\rho)(d); \rho > 0\},\ a \in M.$$

It is not hard to see that $(\mathcal{W}^*(a)(d); a \in M)$ is a neighborhoods subsystem over M. Its attached family of sets $(\mathcal{W}(a)(d); a \in M)$ defined as

$$(\forall a \in M)\text{: } W \in \mathcal{W}(a)(d) \text{ iff } W \supseteq W^*, \text{ for some } W^* \in \mathcal{W}^*(a)(d)$$

is a neighborhoods system over M. By a preceding statement, there exists a unique topology $\mathcal{T}(d)$ over M such that, for each $a \in M$,

$$\mathcal{W}(a)(d) \text{ is just the } \mathcal{T}(d)\text{-neighborhoods system of } a, \text{ for each } a \in M.$$

According to its definition, we have, for each $D \in \mathcal{T}(d)$ and each $a \in D$,

$$D \in \mathcal{W}(a)(d); \text{ so, there exists } \rho > 0 \text{ such that } M(a,\rho)(d) \subseteq D.$$

Moreover, $\mathcal{A}^*(d) = \{\mathcal{W}^*(a)(d); a \in M\}$ is a subbasis of $\mathcal{T}(d)$; so that

$$(\forall D \in \mathcal{T}(d))\text{: } D = \text{union of intersections of finite subsets in } \mathcal{A}^*(d).$$

In particular, $\mathcal{A}^*(d) \subseteq \mathcal{T}(d)$; whence,

$$M(a,\rho)(d) \text{ is open, for each } a \in M,\ \rho > 0.$$

On the other hand, by the semimetrical properties of d,

$$M^e(a,\rho)(d) := \{x \in M; d(a,x) > \rho\} \text{ is open, for each } a \in M,\ \rho > 0;$$

and this yields

$$M[a,\rho](d) = M \setminus M^e(a,\rho)(d) \text{ is closed, for each } a \in M,\ \rho > 0.$$

Finally, when no confusion can arise, it will be convenient to drop the index (d) from all notations above.

II) Let (M,d) be a metric space. As precise, we denoted by $\mathcal{T}(d)$ the associated to d topology; for simplicity, any concept (modulo $\mathcal{T}(d)$) will be referred to as a concept (modulo d). Note that, by our preceding discussion,

> $(M,\mathcal{T}(d))$ is Hausdorff separated: for each $x,y \in M$ with $x \neq y$, there exist $\rho = \rho(x) > 0$, $\sigma = \sigma(y) > 0$, with $M(x,\rho) \cap M(y,\sigma) = \emptyset$.

Then, fix some $P \in \exp(M)$. Remember that

> P is compact (modulo d) when each open covering $P \subseteq \cup\{D_i; i \in I\}$ yields a finite open sub-covering: $P \subseteq \cup\{D_j; j \in J\}$, where $J =$ finite subset of I.

A metrical characterizations of this (topological) concept is contained in

Theorem 1.44. *For each $P \in \exp(M)$, the following are equivalent in (the complete system) (ZF)*

(51-a) P is compact (modulo d)
(51-b) P is Bolzano-Weierstrass (modulo d):
each infinite subset of P has cluster points in P (modulo d)
(51-c) P is sequentially compact (modulo d):
each sequence in P has a d-convergent in P subsequence.

A verification of this may be found in Kumaresan [154, Ch. 4, Sec. 4.3]; see also Costinescu [66, Ch. VII, Sec. 3]. However, for completeness reasons, we will provide its basic lines, with some modifications.

Proof. i): Suppose that P is compact (modulo d); hence (by the Hausdorff separation) P is closed in M. We have to establish that P is Bolzano-Weierstrass (modulo d): each infinite part A of P admits cluster points. Suppose not: $\text{clu}(A) = \emptyset$. Then (according to a preceding fact)

A is closed in M; hence, $B := M \setminus A$ is open in M.

Moreover, by the same working hypothesis,

for each $x \in M$ there exists $\rho(x) > 0$ with
$M(x, \rho(x)) \cap (A \setminus \{x\}) = \emptyset$; so that, $M(x, \rho(x)) \subseteq B \cup \{x\}$.

The family of sets $\{M(x, \rho(x)); x \in A\} \cup \{B\}$ is an open covering of P; hence, by compactness, there exists a finite open sub-covering for the same:

$P \subseteq \cup\{M(x_i, \rho(x_i)); i \in I\} \cup B$, where $I =$ finite set.

But then (according to the relation above)

$A \subset \cup\{B \cup \{x_i\}; i \in I\} = B \cup \{x_i; i \in I\}$; hence, $A \subseteq \{x_i; i \in I\}$;

in contradiction with $A =$ infinite. Hence, A must have cluster points; which, as $P =$ closed, must belong to P.

ii): Suppose that P is Bolzano-Weierstrass (modulo d); we claim that P is sequentially compact. Let (x_n) be a sequence in P. If $\{x_n; n \geq 0\}$ is finite, then it has a constant (hence, convergent in P) subsequence; and we are done. Suppose now that $\{x_n; n \geq 0\}$ is infinite. By the construction

$i(0) = 0$, $i(n) = \min\{h > i(n-1); x_h \notin \{x_{i(0)}, ..., x_{i(n-1)}\}\}$, $n \geq 1$, we get a subsequence $(y_n = x_{i(n)}; n \geq 0)$ of $(x_n; n \geq 0)$, with

$(y_n; n \geq 0)$ is *full*: $i \neq j$ implies $y_i \neq y_j$.

The subset $A = \{y_n; n \geq 0\}$ of P is infinite; hence $\text{clu}(A) \cap P$ has at least one point, z say. Denote $k = \min\{s \in N; n \geq s \Longrightarrow y_n \neq z\}$. From the iterative construction

$j(0)$ = the minimal rank $i \geq k$ with $d(y_i, z) < 2^{-0}$;
and, for each $n \geq 1$: $j(n)$ = the minimal rank $i > j(n-1)$ with
$d(y_i, z) < \min\{2^{-n}, d(y_{j(0)}, z), ..., d(y_{j(n-1)}, z)\}$,

it follows that $y_{j(n)} \xrightarrow{d} z$ as $n \to \infty$; and conclusion follows.

iii): Suppose that P is sequentially compact (modulo d); we have to establish that P is compact (modulo d). Let $\{D_i; i \in I\}$ be an open covering of P. Denote

$r(x) = \sup\{\sigma > 0; M(x, \sigma) \subseteq D_i, \text{ for some } i \in I\}, x \in P$;
clearly, $0 < r(x) \leq \infty$, for all $x \in P$.

Then, put $\varepsilon := \inf\{r(x); x \in P\}$; we claim that $\varepsilon > 0$. For, if $\varepsilon = 0$, there exists a sequence (x_n) in P such that $r(x_n) \to 0$ as $n \to \infty$. By hypothesis, (x_n) has a subsequence $(y_n := x_{i(n)}; n \geq 0)$ with $y_n \xrightarrow{d} z$ as $n \to \infty$, for some $z \in P$. From the covering hypothesis, $z \in D_j$ for some $j \in I$; so, there exists $\sigma > 0$ such that $M(z, \sigma) \subseteq D_j$. But then, there must be some index $n(\sigma)$, such that

$(\forall n \geq n(\sigma))$: $d(z, y_n) < \sigma/2$; hence, $M(y_n, \sigma/2) \subseteq M(z, \sigma) \subseteq D_j$;
so (by definition) $r(y_n) \geq \sigma/2$.

This, by a limit process, gives $0 \geq \sigma/2$; a contradiction; hence the claim.

Finally, take some $x_0 \in P$ and consider the iterative process

$$x_n \in P \setminus \cup\{M(x_i, \varepsilon/2); 0 \leq i \leq n-1\},\ n \geq 1.$$

This cannot be constructed indefinitely; that is: it ends at a certain rank. For, otherwise, in view of

$$d(x_i, x_j) \geq \varepsilon/2, \text{ whenever } i < j,$$

the sequence (x_n) is not d-Cauchy; so, it cannot admit convergent subsequences; in contradiction with the working hypothesis. Consequently,

$$P \subseteq \cup\{M(x_i, \varepsilon/2); 0 \leq i \leq k-1\}, \text{ for some } k \geq 1.$$

But then, according to the above definition,

$$\forall i \in \{0, ..., k-1\}, \ \exists p(i) \in I, \text{ such that } M(x_i, \varepsilon/2) \subseteq D_{p(i)}.$$

This tells us that $\{D_{p(i)}; 0 \leq i \leq k-1\}$ is a finite open cover of P. □

Finally, call the (nonempty) subset Q of M, *relatively compact* when

$P := \text{cl}(Q)$ is compact; or, equivalently (see above):
each sequence in P has a d-convergent (in P) subsequence.

The following characterization of this concept will be in effect for us.

Proposition 1.56. *The subset $Q \in \exp(M)$ is relatively compact iff*

Q is narrow relatively compact:
each sequence in Q has a convergent (in M) subsequence.

Proof. Clearly, (Q = relatively compact) implies (Q = narrow relatively compact). Conversely, suppose that Q is narrow relatively compact; and let (x_n) be a sequence in $P := \text{cl}(Q)$. By definition, there exists a sequence (y_n) in Q with

(eva) $d(x_n, y_n) \leq 2^{-n}$, for all n.

Combined with the admitted hypothesis, there exists a subsequence $(y_{i(n)}; n \geq 0)$ of $(y_n; n \geq 0)$, with

$$y_{i(n)} \xrightarrow{d} z \text{ as } n \to \infty, \text{ for some } z \in M \text{ (hence, } z \in P).$$

This, by the relation (eva), proves $x_{i(n)} \xrightarrow{d} z$; and we are done. □

III) Let (P, d) and (Q, e) be a couple of metric spaces; and $f : P \to Q$ be a mapping. Let us say that f is *sequentially continuous*, when

for each sequence (x_n) in P and each $x \in P$ with $x_n \xrightarrow{d} x$
we have $f(x_n) \xrightarrow{e} f(x)$.

Proposition 1.57. *We have, in (ZF-AC+DC),*

$f(.)$ is (topological) continuous iff $f(.)$ is sequentially continuous.

Proof. i): Suppose that f is (topological) continuous; and let the sequence (x_n) in P and the point $x \in P$ be such that $x_n \xrightarrow{d} x$. Let $\varepsilon > 0$ be arbitrary fixed. By the topological definition of continuity, there exists some $\delta > 0$ such that $f(P(x,\delta)(d)) \subseteq Q(f(x),\varepsilon)(e)$. Further, given this $\delta > 0$, there exists (by the imposed hypothesis), some index $n(\delta) \in N$ such that $n \geq n(\delta)$ implies $x_n \in P(x,\delta)(d)$. Combining these gives

$$n \geq n(\delta) \text{ implies } f(x_n) \in Q(f(x),\varepsilon)(e);$$

and this, by the arbitrariness of $\varepsilon > 0$, tells us that $f(x_n) \xrightarrow{e} f(x)$.

ii): Suppose that f is sequentially continuous; but [contrary to the desired conclusion] f is not (topological) continuous: there exists $\varepsilon > 0$ with

$$(\forall \delta > 0): H(\delta) := \{u \in P(x,\delta)(d); f(u) \notin Q(f(x),\varepsilon)(e)\} \text{ is nonempty.}$$

Let $(\delta_n; n \geq 0)$ be a strictly descending sequence in R^0_+ with $\delta_n \to 0$. (For, example, one may take $(\delta_n = 2^{-n}; n \geq 0)$; hence, no choice procedure is needed). By the Denumerable Axiom of Choice (AC(N)) (deductible in (ZF-AC+DC)), there may be determined a sequence (x_n) in P with

$(\forall n)$: $x_n \in H(\delta_n)$
(that is: $x_n \in P(x,\delta_n)(d)$ and $f(x_n) \notin Q(f(x),\varepsilon)(e)$).

By the first half of this, $x_n \xrightarrow{d} x$; so that, by the sequential continuity,

$f(x_n) \xrightarrow{e} f(x)$; wherefrom: there exists an index $n(\varepsilon) \geq 0$, with
$f(x_n) \in Q(f(x),\varepsilon)(e)$, $\forall n \geq n(\varepsilon)$.

This, however, is in contradiction with the second half of underlying relation; and then, the conclusion follows. □

IV) Let again (P,d) and (Q,e) be a couple of metric spaces; and $f : P \to Q$ be a mapping. We say that f is *uniformly continuous*, when

$$\forall \varepsilon > 0, \exists \delta > 0, \text{ such that } d(x,y) \leq \delta \text{ implies } e(f(x),f(y)) \leq \varepsilon.$$

In this case, we may define a mapping $\psi : R_+ \to R_+ \cup \{\infty\}$, according to

$$\psi(t) = \sup\{e(f(x),f(y)); d(x,y) \leq t\},\ t \in R_+.$$

This will be referred to as the *uniform continuity modulus* of f. Clearly,

(uc-1) ψ is *zero-continuous* [$\psi(0) = 0$ and ($\psi(t) \to 0$ as $t \to 0$)],
(uc-2) $e(f(x),f(y)) \leq \psi(d(x,y))$, $\forall x,y \in P$.

A basic example of this type is obtainable in the compactness context.

Theorem 1.45. *Suppose that (P, d) is compact. Then, the generic inclusion is valid, in (ZF-AC+DC):*

$(\forall f \in \mathcal{F}(P, Q))$: *$f$ is continuous implies f is uniformly continuous.*

Proof. Suppose that f is continuous, but not uniformly continuous. There exists then $\varepsilon > 0$, such that

$$C(\delta) := \{(x, y) \in P \times P; d(x, y) \leq \delta, e(f(x), f(y)) > \varepsilon\} \neq \emptyset, \forall \delta > 0.$$

Let $(\delta_n; n \geq 0)$ be a sequence in R^0_+ with $\delta_n \to 0$. (For example, we may take $(\delta_n = 2^{-n}; n \geq 0)$; hence, this is not depending on (AC).) From the Denumerable Axiom of Choice (AC(N)) (deductible, as precise, in (ZF-AC+DC)), we get a sequence $((x_n, y_n); n \geq 0)$ in $P \times P$, such that

$((x_n, y_n) \in C(\delta_n), \forall n)$; that is:
$d(x_n, y_n) \leq \delta_n$, $e(f(x_n), f(y_n)) > \varepsilon$, $\forall n$.

Then, by the sequential compactness of (P, d), there exist subsequences

$$(x^*_n := x_{i(n)}; n \geq 0), (y^*_n := y_{i(n)}; n \geq 0), (\delta^*_n := \delta_{i(n)}; n \geq 0)$$

of (x_n), (y_n) and (δ_n) respectively, as well as elements $x^*, y^* \in P$, such that

(p-1) $(\forall n)$: $d(x^*_n, y^*_n) \leq \delta^*_n$, $e(f(x^*_n), f(y^*_n)) > \varepsilon$
(p-2) $x^*_n \xrightarrow{d} x^*$, $y^*_n \xrightarrow{d} y^*$ as $n \to \infty$; hence, by continuity, $(f(x^*_n) \xrightarrow{e} f(x^*)$, $f(y^*_n) \xrightarrow{e} f(y^*))$, as $n \to \infty$.

By (p-1) (the first half), $x^* = y^*$. Combining with (p-1) (the second half) and (p-2), we get (by a limit process)

$0 = e(f(x^*), f(y^*)) \geq \varepsilon$; contradiction.

Hence, our working assumption is not acceptable; and we are done. □

1.6.6 *Gauge spaces*

Let X be a nonempty set; and Λ be some nonempty (index) set. For each $\lambda \in \Lambda$, let $d_\lambda : X \times X \to R_+$ be a semimetric over X_λ; and $\mathcal{T}(d_\lambda)$ stand for the associated to d_λ topology (see above).

Technically speaking, these data generate two basic structures on X.

I) The former of these is represented by the supremum topology of the family $\{\mathcal{T}(d_\lambda); \lambda \in \Lambda\}$. Precisely, note that for each $\lambda \in \Lambda$,

$\mathcal{A}_\lambda = \{X(x,\rho)(d_\lambda); x \in X, \rho > 0\}$ is a subbase of $\mathcal{T}(d_\lambda)$;

hence, in particular, $\mathcal{A}_\lambda$ is total. Denote further

$\mathcal{A} = \cup\{\mathcal{A}_\lambda; \lambda \in \Lambda\}$; clearly, $\mathcal{A}$ is total too.

By a preceding result, the formula

$\mathcal{T} = \{D \in \exp[X]; D =$ union of intersections of finite subsets in $\mathcal{A}\}$

defines a topology over X; which is just the supremum topology of the family $\{\mathcal{T}(d_\lambda); \lambda \in \Lambda\}$. Let int(.) and cl(.) stand for the associated interior and closure operator. By definition, we have for each subset Y of X

(g-int) $x \in \text{int}(Y)$ iff there exists an intersection W
of a finite subset in $\mathcal{A}$ with $x \in W \subseteq Y$
(g-clo) $x \in \text{cl}(Y)$ iff each intersection W
of a finite subset in $\mathcal{A}$ with $x \in W$ fulfills $W \cap Y \neq \emptyset$.

II) The latter of these is represented by the (sequential) conv-Cauchy structure induced by the family $D = (d_\lambda; \lambda \in \Lambda)$ of these semimetrics. Take a sequence (x_n) in X. Given $\lambda \in \Lambda$, the d_λ-*convergence* property of this sequence towards an $x \in X$ [written: $x_n \xrightarrow{d_\lambda} x$], means:

$d_\lambda(x_p, x) \to 0$ as $p \to \infty$; that is:
$\forall \varepsilon > 0$, $\exists i = i(\lambda, \varepsilon)$, such that $i \leq p \Longrightarrow d_\lambda(x_p, x) \leq \varepsilon$.

If this holds $\forall \lambda \in \Lambda$, then (x_n) is said to *D-converge* towards x [written as: $x_n \xrightarrow{D} x$]. The set of all such points x will be denoted as $D - \lim_n(x_n)$; when it is nonempty, then $(x_n; n \geq 0)$ is called *D-convergent*. On the other hand, given $\lambda \in \Lambda$, the d_λ-Cauchy property of $(x_n; n \geq 0)$ means:

$d_\lambda(x_p, x_q) \to 0$ as $p, q \to \infty$, $p \leq q$; that is:
$\forall \varepsilon > 0$, $\exists j := j(\lambda, \varepsilon)$, such that $j \leq p \leq q \Longrightarrow d_\lambda(x_p, x_q) \leq \varepsilon$.

If this holds $\forall \lambda \in \Lambda$, we say that (x_n) is *D-Cauchy*; the class of all these will be denoted as $Cauchy(D)$. By definition, the triple $(X, \Lambda; D)$ endowed with the conv-Cauchy structure $((\xrightarrow{D}), Cauchy(D))$ and the regularity condition

D is *sufficient*: $x, y \in X$ and $(d_\lambda(x,y) = 0, \forall \lambda \in \Lambda)$ imply $x = y$

is called a *gauge space*. Note that, in this setting,

any D-convergent sequence is D-Cauchy too;

the reciprocal is not in general valid.

Using the previous conventions, we may introduce a *D-closure* operator $A \mapsto \mathrm{Dcl}(A)$ from $\exp[X]$ to itself as: for each $A \in \exp[X]$,

$$y \in \mathrm{Dcl}(A) \text{ iff } y = D - \lim_n(x_n), \text{ for some sequence } (x_n) \text{ in } A.$$

It is not hard to see that this operator is a *semi-closure* over X, in the sense

(Dcl-1) (identity) $\emptyset = \mathrm{Dcl}(\emptyset)$, $X = \mathrm{Dcl}(X)$,
(Dcl-2) (progressiveness) $Y \subseteq \mathrm{Dcl}(Y)$, $\forall Y \in \exp[X]$,
(Dcl-3) (additivity) $\mathrm{Dcl}(U \cup V) = \mathrm{Dcl}(U) \cup \mathrm{Dcl}(V)$, $\forall U, V \in \exp[X]$.

Note that, as a direct consequence of these, we also have

(Dcl-4) (monotonicity) $Y \subseteq Z$ implies $\mathrm{Dcl}(Y) \subseteq \mathrm{Dcl}(Z)$.

When $Y = \mathrm{Dcl}(Y)$, we say that Y is *D-closed*; this, according to the above, may be characterized as

$$\text{if } (x_n; n \geq 0) \text{ is a sequence in } Y \text{ and } x_n \xrightarrow{D} x, \text{ then } x \in Y.$$

Denote, for simplicity

$$\mathcal{K}[D] = \text{the class of all } D\text{-closed subsets in } X.$$

Some basic properties of this class are listed below:

(Dclo-1) $\emptyset, X \in \mathcal{K}[D]$
(Dclo-2) $U, V \in \mathcal{K}[D] \Longrightarrow U \cup V \in \mathcal{K}[D]$
(Dclo-3) $\mathcal{A} \subseteq \mathcal{K}[D] \Longrightarrow \cap\mathcal{A} \in \mathcal{K}[D]$.

Having these precise, define

$$\mathcal{T}[D] = \{Z \in \exp[X]; X \setminus Z \in \mathcal{K}[D]\}.$$

By the result above, the following properties of this class are available:

(Dopen-1) $\emptyset, X \in \mathcal{T}[D]$
(Dopen-2) $U, V \in \mathcal{T}[D] \Longrightarrow U \cap V \in \mathcal{T}[D]$
(Dopen-3) $\mathcal{B} \subseteq \mathcal{T}[D] \Longrightarrow \cup\mathcal{B} \in \mathcal{T}[D]$.

This class is therefore a *topology* on X, according to Bourbaki [33, Ch. I, Sec. 1]. So, it has an associated closure operator $\mathrm{cl}_D(.)$, introduced as

$$(Y \in \exp[X]): x \in \mathrm{cl}_D(Y) \text{ iff } U \cap Y \neq \emptyset, \text{ for each } U \in \mathcal{T}[D] \text{ with } x \in U.$$

However, this closure operator $\text{cl}_D(.)$ does not coincide with the initial closure operator Dcl(.). An explanation of this bad property is due to the fact that – by the very choice of D-convergence concept – the closure operator Dcl(.) is not *involutive*; i.e., a property like

(Dcl-5) (*involution*) $\text{Dcl}(\text{Dcl}(Y)) = \text{Dcl}(Y)$, $\forall Y \in \exp[X]$

is not available here; so that, it cannot be viewed as a (genuine) Kuratowski closure operator [155, Ch. I, Sec. 4]. Further aspects may be found in Engelking [87, Ch. 1, Sec. 1.2].

Now, let us compare the topological closure cl(.) with the gauge closure Dcl(.). In this direction, we have

(p-1) $(\forall Y \in \exp[X])$: $\text{Dcl}(Y) \subseteq \text{cl}(Y)$
(p-2) $\mathcal{T} \subseteq \mathcal{T}[D]$ (Y = closed (open) implies Y = D-closed (D-open)).

In fact, letting $x \in \text{Dcl}(Y)$, there exists a sequence (x_n) in Y with

$$x_n \xrightarrow{D} x;\ \text{hence, } x_n \xrightarrow{d_\lambda} x, \text{ for each } \lambda \in \Lambda.$$

Let $W = \cap\{X(x, \rho_i)(d_{\lambda(i)}); i \in I\}$ be an intersection of open spheres centered at x; where I is a (nonempty) finite index set. By the above convergence property, there must be some index $n(x)$, such that

$$\{x_n; n \geq n(x)\} \subseteq W;\ \text{whence, } Y \cap W \neq \emptyset.$$

This yields $x \in \text{cl}(Y)$; and proves the first half of our assertion. The second half of the same is immediate, via

Y=closed $\Longrightarrow \text{cl}(Y) = Y \subseteq \text{Dcl}(Y) \subseteq \text{cl}(Y)$;
hence, $Y = \text{Dcl}(Y)$; proving that Y is D-closed.

Finally, we have the possibility of introducing another topology $\mathcal{S} = \mathcal{S}[D]$ associated to this structure, as follows. Define

$\mathcal{S}$ = the class of all $E \in \exp[X]$ with:
$x_n \xrightarrow{D} x \in E$ implies $\{x_n; n \geq n(x)\} \subseteq E$, for some $n(x) \in N$.

Then, $\mathcal{S}$ is a topology on X, as it can be directly seen. The properties of this topology as well as its relationships with $\mathcal{T}$ and $\mathcal{T}[D]$ seem to raise interesting questions; but, we do not go any further with this subject.

1.7 Banach Contractions and Relatives

Abstract. Some direct extensions of Banach fixed point theorem (B-fpt) including the related fixed point statements due to Meir and Keeler, Boyd and Wong, or Matkowski are being formulated. Then, (B-fpt) itself is considered, as well as its Bessaga type converse.

Subject-Class 2020. 47H10 (Primary), 54H25 (Secondary).

Keywords. Metric space, fixed point, Banach contraction principle, converse, (geometric) relation, Meir-Keeler contraction, Boyd-Wong and Matkowski theorems.

1.7.1 *Meir-Keeler fixed point principle*

Let X be a nonempty set. Call the subset Y of X, *almost singleton* (in short: *asingleton*) provided $[y_1, y_2 \in Y$ implies $y_1 = y_2]$; and *singleton* if, in addition, Y is nonempty; note that in this case $Y = \{y\}$, for some $y \in X$.

Further, let $d : X \times X \to R_+ := [0, \infty[$ be a *metric* over X; the couple (X, d) will be termed a *metric space*. Finally, let $T \in \mathcal{F}(X)$ be a selfmap of X. [Here, for each pair A, B of nonempty sets, $\mathcal{F}(A, B)$ stands for the class of all functions from A to B; when $A = B$, we write $\mathcal{F}(A)$ in place of $\mathcal{F}(A, A)$.] Denote $\text{Fix}(T) = \{x \in X; x = Tx\}$; each point of this set is referred to as *fixed* under T. In the metrical fixed point theory, such points are to be determined according to the context below, comparable with the one described in Rus [203, Ch. 2, Sec. 2.2]:

pic-0) We say that T is *fix-asingleton* if $\text{Fix}(T)$ is an asingleton; and *fix-singleton* if $\text{Fix}(T)$ is a singleton

pic-1) We say that T is a *Picard operator* (modulo d) if, for each $x \in X$, the iterative sequence $(T^n x; n \geq 0)$ is d-Cauchy

pic-2) We say that T is a *strong Picard operator* (modulo d) if, for each $x \in X$, $(T^n x; n \geq 0)$ is d-convergent with $\lim_n(T^n x) \in \text{Fix}(T)$.

The regularity conditions for such properties are being founded on *orbital full* (in short: (o-f)) concepts. Given the sequence $(z_n; n \geq 0)$ in X, define the properties

(z_n) is T-*orbital*: $(z_n = T^n x; n \geq 0)$, for some $x \in X$
(z_n) is *full*: $n \mapsto z_n$ is injective ($i \neq j$ implies $z_i \neq z_j$).

In this case, the underlying conditions write

reg-1) X is *(o-f,d)-complete*:
(for each (o-f)-sequence) d-Cauchy $\Longrightarrow$ d-convergent

reg-2) T is *(o-f,d)-continuous*:
((z_n) is (o-f)-sequence and $z_n \xrightarrow{d} z$) imply $Tz_n \xrightarrow{d} Tz$.

As an essential completion of these developments, we must now discuss the contractive type conditions to be used. Denote, for $x, y \in X$,

$$P_0(x,y) = d(Tx,Ty),\ P_1(x,y) = d(x,y);$$

and let $P : X \times X \to R_+$ be a map. According to Ćirić [59] and Matkowski [166], we say that $T \in \mathcal{F}(X)$ is *Meir-Keeler* $(d;P)$*-contractive*, if

(mk-1) $P(x,y) > 0$ implies $P_0(x,y) < P(x,y)$;
expressed as: T is *strictly nonexpansive* (modulo $(d;P)$)
(mk-2) $\forall \varepsilon > 0, \exists \delta > 0$: $[\varepsilon < P(x,y) < \varepsilon + \delta] \Longrightarrow P_0(x,y) \leq \varepsilon$;
expressed as: T has the *Meir-Keeler property* (modulo $(d;P)$).

Note that, by the former of these, the Meir-Keeler property writes

(mk-3) $\forall \varepsilon > 0, \exists \delta > 0$: $[0 < P(x,y) < \varepsilon + \delta] \Longrightarrow P_0(x,y) \leq \varepsilon$.

The following technical aspect of the introduced convention must be noted. Call $T \in \mathcal{F}(X)$, *original Meir-Keeler* $(d;P)$*-contractive* provided

(o-mk) $\forall \varepsilon > 0, \exists \delta > 0$: $[\varepsilon \leq P(x,y) < \varepsilon + \delta] \Longrightarrow P_0(x,y) < \varepsilon$.

The motivation of this comes from (o-mk) being the contractive condition in the 1969 paper by Meir and Keller [168] (under $P = P_1$). Its relationships with Meir-Keller contractive condition are described as follows.

Proposition 1.58. *The generic inclusion is valid (for $T \in \mathcal{F}(X)$):*

T is original Meir-Keeler $(d;P)$-contractive implies
T is Meir-Keeler $(d;P)$-contractive.

Proof. i): Let $x, y \in X$ be such that $P(x,y) > 0$. With $\varepsilon = P(x,y)$ (and $\delta > 0$ taken by the underlying property), $\varepsilon \leq P(x,y) < \varepsilon + \delta$; whence, $P_0(x,y) < \varepsilon = P(x,y)$; so that, T is strictly nonexpansive (modulo $(d;P)$).

ii): Let $\varepsilon > 0$ be given, and $\delta > 0$ be assured by the original Meir-Keeler $(d;P)$-contractive property. If $x, y \in X$ fulfill $\varepsilon < P(x,y) < \varepsilon + \delta$, then (as $\varepsilon \leq P(x,y) < \varepsilon + \delta$) we have (via underlying condition) $P_0(x,y) < \varepsilon$; hence $P_0(x,y) \leq \varepsilon$; so that, T has the Meir-Keeler property (modulo $(d;P)$). □

Remark 1.21.

The question of reciprocal inclusion being valid is open for the moment; some related aspects may be found in Lim [162].

A geometric version of these developments may be given along the lines below. Remember that udiag(R^0_+) stands for the class of all $\Omega \subseteq R^0_+ \times R^0_+$ with the *upper diagonal* property [$(t,s) \in \Omega$ implies $t < s$]. Let us introduce the geometric and asymptotic condition (over the class udiag(R^0_+))

(g-mk) Ω has the *geometric Meir-Keeler property*:
$\forall \varepsilon > 0, \exists \delta > 0$: $(t\Omega s,\ \varepsilon < s < \varepsilon + \delta) \Longrightarrow t \leq \varepsilon$
(a-mk) Ω is *asymptotic Meir-Keeler*: there are no strictly descending sequences (t_n) and (s_n) in R^0_+ and no elements ε in R^0_+, with $((t_n, s_n) \in \Omega, \forall n)$ and $(t_n \to \varepsilon+,\ s_n \to \varepsilon+)$.

As precise, these properties are equivalent to each other.

Given $\Omega \subseteq R^0_+ \times R^0_+$, let us say that T is $(d;P;\Omega)$*-contractive*, provided

(Om-contr) $(P_0(x,y), P(x,y)) \in \Omega$,
for all $x, y \in X$ with $P_0(x,y), P(x,y) > 0$.

Proposition 1.59. *Suppose that the selfmap T is $(d;P;\Omega)$-contractive, where $\Omega \subseteq R^0_+ \times R^0_+$ is upper diagonal and geometric Meir-Keeler. Then, T is Meir-Keeler $(d;P)$-contractive.*

Proof. i): Let $x, y \in X$ be such that $P(x,y) > 0$. If $P_0(x,y) = 0$, all is clear. Suppose now that $P_0(x,y) > 0$. As a consequence of this,

$$(t,s) \in \Omega;\ \text{where } t := P_0(x,y),\ s := P(x,y).$$

Combining with the upper diagonal property of Ω, one gets $t < s$; i.e.: $P_0(x,y) < P(x,y)$. Summing up, T is strictly contractive (modulo $(d;P)$).

ii): Given $\varepsilon > 0$, let $\delta > 0$ be assured by the geometric Meir-Keeler property of Ω. Further, let $x, y \in X$ be such that $\varepsilon < P(x,y) < \varepsilon + \delta$. If $P_0(x,y) = 0$, all is clear. Suppose now that $P_0(x,y) > 0$. By definition,

$$(t,s) \in \Omega;\ \text{where } t := P_0(x,y),\ s := P(x,y);$$

and this, along with $\varepsilon < s < \varepsilon + \delta$, gives (by the geometric Meir-Keeler property of Ω), $t \leq \varepsilon$; that is: $P_0(x,y) \leq \varepsilon$. Putting these together, it follows that T has the Meir-Keeler property (modulo $(d;P)$). □

In the following, a converse result is formulated. Let $\Omega := \Omega[P;T]$ stand for the associated relation over R^0_+:

$$\Omega = \{(P_0(x,y), P(x,y)); x, y \in X, P_0(x,y), P(x,y) > 0\}.$$

This, by definition, means that

$(t, s) \in R^0_+ \times R^0_+$ belongs to Ω iff
$(t = P_0(x, y),\ s = P(x, y))$, for some $(x, y) \in X \times X$.

Roughly speaking, the upper diagonal and geometric Meir-Keeler properties of the associated to T relation $\Omega[P;T]$ give, ultimately, a characterization of the Meir-Keeler contractive properties upon T. The following result will certify this.

Proposition 1.60. *Under these conventions, we have*

(13-1) If T is Meir-Keeler $(d;P)$-contractive, then the attached relation $\Omega := \Omega[P;T]$ is upper diagonal and geometric Meir-Keeler
(13-2) T is Meir-Keeler $(d;P)$-contractive if and only if the attached relation $\Omega := \Omega[P;T]$ is upper diagonal and geometric Meir-Keeler.

Proof. i): Suppose that T is Meir-Keeler $(d;P)$-contractive.

i-1): Let $(t, s) \in R^0_+ \times R^0_+$ be such that $(t, s) \in \Omega$; hence (by definition)

$$t = P_0(x, y),\ s = P(x, y), \text{ where } x, y \in X \text{ and } [P_0(x, y), P(x, y) > 0].$$

From the strict contractive property of T, we must have $P_0(x, y) < P(x, y)$; or, equivalently, $t < s$; which shows that Ω is upper diagonal.

i-2): Let $\varepsilon > 0$ be arbitrary fixed; and $\delta > 0$ be the number associated by the Meir-Keeler property of T. Further, let $(t, s) \in R^0_+ \times R^0_+$ be such that $(t, s) \in \Omega$ and $\varepsilon < s < \varepsilon + \delta$; hence (see above)

$$t = P_0(x, y),\ s = P(x, y), \text{ where } x, y \in X \text{ and}$$
$$(P_0(x, y), P(x, y) > 0),\ \varepsilon < P(x, y) < \varepsilon + \delta.$$

By the underlying Meir-Keeler-property for T, we get

$$P_0(x, y) \le \varepsilon; \text{ that is (under our notation): } t \le \varepsilon;$$

so that, Ω has the geometric Meir-Keeler property.

ii): Suppose that the associated relation $\Omega := \Omega[P;T]$ over R^0_+ is upper diagonal and has the geometric Meir-Keeler property. By the very definition of this object, T is $(d;P;\Omega)$-contractive. Combining with the preceding result, one derives that T appears as Meir-Keeler $(d;P)$-contractive. □

1.7.2 *Main result*

Let (X, d) be a metric space. Further, take a selfmap $T \in \mathcal{F}(X)$. As precise, we have to determine sufficient conditions for $\mathrm{Fix}(T) =$ nonempty. The specific directions under which this problem is to be solved were already listed; and the contractive framework has been settled. It remains now to list the extra hypotheses to be used. Denote, for $x, y \in X$,

$P_0(x,y) = d(Tx,Ty)$, $P_1(x,y) = d(x,y)$,
$M(x,y) = \text{diam}\{x,Tx,y,Ty\}$, $L_0(x,y) = \min\{P_0(x,y), P_1(x,y)\}$,
$L_1(x,y) = \max\{P_0(x,y), P_1(x,y)\}$.

Further, let $P : X \times X \to R_+$ be a mapping with

(pos) P is *positive*: $L_0(x,y) > 0$ implies $P(x,y) > 0$
(tb) P is *telescopic bounded*: $P(x,Tx) \leq L_1(x,Tx)$, $\forall x \in X$
(diam) P is *diametral*: $P(x,y) \leq M(x,y)$, $\forall x,y \in X$;

referred to as: P is (p-tb-d). In addition, we require that

(sg-asy) P is *orbitally singular asymptotic*: for each d-Cauchy (o-f)-sequence $(x_n = T^n x_0)$ in X, and $\forall z \in X$ with $x_n \xrightarrow{d} z$, ($L_0(x_n, z) > 0$, $\forall n$), and $d(z,Tz) > 0$, one has $\liminf_n P(x_n, z) < d(z,Tz)$.

We are now in position to formulate our main statement in the area (referred to as *Meir-Keeler fixed point theorem*; in short: (MK-fpt)).

Theorem 1.46. *Suppose that T is Meir-Keeler $(d;P)$-contractive, where $P : X \times X \to R_+$ is (p-tb-d). In addition, let X be (o-f,d)-complete. Then,*

(21-a) T is a strong Picard operator (modulo d) when, in addition, T is (o-f,d)-continuous
(21-b) T is a strong Picard operator (modulo d) when, in addition, P is orbitally singular asymptotic
(21-c) T is fix-asingleton;
hence, fix-singleton, under either of the posed conditions.

Proof. There are several parts to be passed.

Part 1. Let us firstly check the fix-asingleton property. Take a couple of points $z_1, z_2 \in \text{Fix}(T)$; and assume by contradiction that $z_1 \neq z_2$; whence $L_0(z_1, z_2) > 0$. As P is positive, we have $P(z_1, z_2) > 0$; so, from the contractive hypothesis, $d(z_1, z_2) = P_0(z_1, z_2) < P(z_1, z_2)$. On the other hand, as P is diametral, $P(z_1, z_2) \leq M(z_1, z_2) = d(z_1, z_2)$. Since the obtained facts are contradictory, the working assumption cannot be accepted; and our assertion follows.

Part 2. We claim that, under the conditions in the statement

(1-2-iter) $d(Tx, T^2x) < P(x,Tx) \leq d(x,Tx)$, when $L_0(x,Tx) > 0$.

In fact, let $x \in X$ be as in the premise above. For the moment, $P(x, Tx) > 0$ (as P is positive). Combining with the contractive condition, implies $d(Tx, T^2x) < P(x, Tx)$. On the other hand, as P is telescopic bounded,

$$(0 <)P(x, Tx) \leq L_1(x, Tx) = \max\{d(x, Tx), d(Tx, T^2x)\}.$$

This, along with the preceding relation, gives

$$d(Tx, T^2x) < \max\{d(x, Tx), d(Tx, T^2x)\}; \text{ so, necessarily,}$$
$$d(Tx, T^2x) < d(x, Tx),\ L_1(x, Tx) = d(x, Tx);$$

wherefrom the claim follows.

Having these precise, take some $x_0 \in X$; and put $(x_n = T^n x_0; n \geq 0)$; this is an orbital sequence. If $x_n = x_{n+1}$ for some $n \geq 0$, we are done; so, without loss, one may assume that the non-telescopic property holds

(n-tele) $(\forall n)$: $x_n \neq x_{n+1}$; hence, $\rho_n := d(x_n, x_{n+1}) > 0$.

Part 3. By the preceding relation (1-2-iter), we have

(iter) $\rho_{n+1} < P(x_n, x_{n+1}) \leq \rho_n$, $\forall n$;
so, the sequence $(\rho_n; n \geq 0)$ is strictly descending in R_+^0.

Two basic consequences of this fact are to be noted.

Conseq 1. Firstly, we claim that (x_n) is full ($n \mapsto x_n$ is injective). In fact, suppose by contradiction that

there exist $i, j \in N$ such that $i < j$, $x_i = x_j$.

This, by definition, yields $x_{i+1} = x_{j+1}$; whence $\rho_i = \rho_j$; in contradiction with the strict descending property of $(\rho_n; n \geq 0)$.

Conseq 2. Secondly, $\rho := \lim_n \rho_n$ exists as an element of R_+; we claim that $\rho = 0$. Assume by contradiction that $\rho > 0$; and let $\sigma > 0$ be the number given by the Meir-Keeler contractive property of T. By definition, there exists $n(\sigma) \geq 0$, such that $[n \geq n(\sigma)$ implies $\rho < \rho_n < \rho + \sigma]$. Taking (iter) into account, yields

$$(\forall n \geq n(\sigma)): \rho < \rho_{n+1} < P(x_n, x_{n+1}) \leq \rho_n < \rho + \sigma;$$

and this, combined with the underlying Meir-Keeler property, gives

$$(\forall n \geq n(\sigma)): \rho < \rho_{n+1} = P_0(x_n, x_{n+1}) \leq \rho;$$

a contradiction. Hence, $\rho = 0$; so that,

$\rho_n := d(x_n, x_{n+1}) \to 0$, as $n \to \infty$; meaning: (x_n) is d-asymptotic.

Part 4. Summing up, (x_n) is orbital, full, d-asymptotic. We claim that (x_n) is d-Cauchy. Let $\varepsilon > 0$ be arbitrary fixed; and $\delta > 0$ be associated by the Meir-Keeler property; without loss, one may assume that $\delta < \varepsilon$. From the d-asymptotic property, there exists a rank $n(\delta) \geq 0$ such that

(d-asy) $(\forall n \geq n(\delta))$: $d(x_n, x_{n+1}) < \delta/4$;
so (by the triangular inequality): $d(x_n, x_{n+i}) < \delta/2$, $\forall i \in \{1, 2\}$.

We claim, via ordinary induction, that the evaluation

(d-C;i) $d(x_n, x_{n+i}) < \varepsilon + \delta/2, \ \forall n \geq n(\delta)$

holds for all $i \geq 1$; wherefrom, all is clear. The case of $i \in \{1, 2\}$ is evident, by (d-asy). Assume that (d-C;i) holds for all $i \in \{1, ..., q\}$, where $q \geq 2$; we must establish that (d-C;q+1) holds as well. So, let $n \geq n(\delta)$ be arbitrary fixed. From the full property of (x_n),

$L_0(x_n, x_{n+q}) > 0$; hence, $P(x_n, x_{n+q}) > 0$

(as P is positive). On the other hand, from the d-asymptotic property, inductive hypothesis, and triangular inequality,

$d(x_n, x_{n+1}), d(x_{n+q}, x_{n+q+1}) < \delta/4 < \delta$,
$d(x_n, x_{n+q}), d(x_{n+1}, x_{n+q}), d(x_{n+1}, x_{n+q+1}) < \varepsilon + \delta/2$,
$d(x_n, x_{n+q+1}) \leq d(x_n, x_{n+q}) + d(x_{n+q}, x_{n+q+1}) <$
$\varepsilon + \delta/2 + \delta/4 = \varepsilon + 3\delta/4 < \varepsilon + \delta$; so that $M(x_n, x_{n+q}) < \varepsilon + \delta$;
whence, $P(x_n, x_{n+q}) < \varepsilon + \delta$ (as P = diametral).

Combining with the Meir-Keeler property of T, yields

$d(x_{n+1}, x_{n+q+1}) = P_0(x_n, x_{n+q}) \leq \varepsilon$; wherefrom,
$d(x_n, x_{n+q+1}) \leq d(x_n, x_{n+1}) + d(x_{n+1}, x_{n+q+1}) < \varepsilon + \delta/4 < \varepsilon + \delta/2$;

and the assertion follows.

Part 5. As X is (o-f,d)-complete, we have that $x_n \xrightarrow{d} z$, for some $z \in X$. There are two cases to discuss.

Case A. Suppose that T is (o-f,d)-continuous. Then $y_n := Tx_n \xrightarrow{d} Tz$, as $n \to \infty$. On the other hand, $(y_n = x_{n+1}; n \geq 0)$ is a subsequence of $(x_n; n \geq 0)$; whence $y_n \xrightarrow{d} z$; and this yields (as d is separated), $z = Tz$.

Case B. Suppose that P is orbitally singular asymptotic. We intend to show that the alternative $[d(z, Tz) > 0]$ yields a contradiction.

As $(x_n; n \geq 0)$ is full, the natural subset $E := \{n \geq 0; Tx_n = Tz\}$ is an asingleton; so, necessarily, there exists an index $h \geq 0$, with

$(\forall n \geq h)$: $x_n \neq z$, $Tx_n \neq Tz$; hence, $L_0(x_n, z) > 0$.

Without loss of generality, one may assume that $h = 0$; that is

(z-dist) $(\forall n)$: $x_n \neq z$, $Tx_n \neq Tz$; that is: $L_0(x_n, z) > 0$;

for, otherwise, passing to the subsequence

$(u_n = x_{h+n}; n \geq 0)$ (that is: $(u_n = T^n u_0; n \geq 0)$), where $u_0 = x_h$

this property holds, as well as the (o-f) one. As a consequence of (z-dist),

$(\forall n)$: $P(x_n, z) > 0$ (as P is positive).

The contractive condition is thus applicable to (x_n, z), for each n; and yields (by the strict nonexpansive property)

(z-nexp) $P_0(x_n, z) < P(x_n, z)$, for all n.

From these relations (and a metric property of d), one gets (as $n \to \infty$)

$$d(x_n, Tx_n), d(x_n, z), d(Tx_n, z) \to 0,\ d(x_n, Tz), d(Tx_n, Tz) \to d(z, Tz).$$

Passing to lim inf as $n \to \infty$ in (z-nexp), gives

$$d(z, Tz) = \lim_n P_0(x_n, z) = \liminf_n P_0(x_n, z) \leq \liminf_n P(x_n, z).$$

This, however, contradicts the orbital singular asymptotic property of P. Hence, $d(z, Tz) = 0$ (that is: $z = Tz$); and conclusion follows. □

As a consequence of this, the following statement (referred to as *Meir-Keeler geometric fixed point theorem*; in short: (MK-g-fpt)) is available.

Theorem 1.47. *Suppose that the selfmap $T : X \to X$ is $(d; P; \Omega)$-contractive, where $P : X \times X \to R_+$ is (p-tb-d) and $\Omega \subseteq R^0_+ \times R^0_+$ is upper diagonal and geometric Meir-Keeler. In addition, let X be (o-f,d)-complete. Then,*

(22-a) T is a strong Picard operator (modulo d) when, in addition, T is (o-f,d)-continuous
(22-b) T is a strong Picard operator (modulo d) when, in addition, P is orbitally singular asymptotic
(22-c) T is fix-asingleton;
hence, fix-singleton, under either of the posed conditions.

Note that, further enlargements of these facts are possible, over dislocated metric spaces taken as in Hitzler [113, Ch. 1, Sec. 1.2]. On the other hand, this result admits multivalued type versions, under Nadler's model [175]; see also Reich [194] and Turinici [232]. Further aspects may be found in Khojasteh et al. [146].

1.7.3 *Functional counterparts*

In the following, a lot of functional versions are provided for the geometric contraction principle above. Some preliminaries are needed.

Let $\varphi \in \mathcal{F}(R^0_+, R)$ and $s \in R^0_+$ be given. Denote, for each $\varepsilon > 0$,

$$\Phi_*(s+)(\varepsilon) = \inf \varphi(]s, s+\varepsilon[), \ \ \Phi^*(s+)(\varepsilon) = \sup \varphi(]s, s+\varepsilon[);$$

clearly, the case of these quantities having infinite values cannot be avoided. From this convention, $\varepsilon \mapsto \Phi_*(s+)(\varepsilon)$ is decreasing and $\varepsilon \mapsto \Phi^*(s+)(\varepsilon)$ is increasing. Further, by the same definition,

$$\Phi_*(s+)(\alpha) \leq \Phi^*(s+)(\beta), \ \forall \alpha, \beta \in R^0_+.$$

Having these precise, define (for φ and s as before)

$$\Lambda_+\varphi(s) = \sup_{\varepsilon>0} \Phi_*(s+)(\varepsilon), \ \ \Lambda^+\varphi(s) = \inf_{\varepsilon>0} \Phi^*(s+)(\varepsilon);$$

these will be referred to as: *right inferior/superior limit* of φ at s. As before, the alternative of $s \mapsto \Lambda_+\varphi(s)$ and $s \mapsto \Lambda^+\varphi(s)$ having infinite values cannot be avoided. By the preceding relation, we have

$$\Lambda_+\varphi(s) \leq \Lambda^+\varphi(s), \ \forall s \in R^0_+.$$

When equality holds, the common value of these quantities is denoted as

$$\varphi(s+0) = \lim_{t\to s+} \varphi(t) \text{ (the } \textit{right limit} \text{ of } \varphi \text{ at } s).$$

Under these notations, we may now return to our initial objectives. Let $P : X \times X \to R_+$ be a map. There are two cases to discuss.

Case 1. Given $\chi \in \mathcal{F}(R^0_+, R)$, let us say that the selfmap T is *Boyd-Wong $(d; P; \chi)$-contractive,* if

$$P_0(x, y) \leq \chi(P(x, y)), \ \forall x, y \in X, \ P_0(x, y), P(x, y) > 0.$$

To discuss it, we need some conditions upon this function. Let $\mathcal{F}(re)(R^0_+, R)$ stand for the subclass of all *regressive* $\chi \in \mathcal{F}(R^0_+, R)$ $[\chi(t) < t, \ \forall t \in R^0_+]$. Then, call $\chi \in \mathcal{F}(re)(R^0_+, R)$, *Meir-Keeler admissible* if

(mk-adm) $\forall \gamma > 0, \ \exists \beta > 0, \ \forall t$: $(\gamma < t < \gamma + \beta) \Longrightarrow \chi(t) \leq \gamma$.

Proposition 1.61. *Suppose that T is Boyd-Wong $(d;P;\chi)$-contractive, where $\chi \in \mathcal{F}(re)(R_+^0, R)$ is given. Then, the following assertions involving the associated relation $\Omega[P;T]$ are true in (ZF-AC+DC):*

(31-1) $\Omega[P;T]$ is upper diagonal
(31-2) $\Omega[P;T]$ is, in addition, geometric Meir-Keeler (so that: T is Meir-Keeler $(d;P)$-contractive) whenever χ is Meir-Keeler admissible.

Proof. Let $\Omega := \Omega[\chi]$ be the relation over R_+^0 introduced as

$(t, s \in R_+^0)$: $t\Omega s$ iff $t \leq \chi(s)$.

By definition,

T is Boyd-Wong $(d;P;\chi)$-contractive implies T is $(d;P;\Omega)$-contractive.

Moreover, by a previous fact,

(p1) Ω is upper diagonal (as χ is regressive)
(p2) Ω is geometric Meir-Keeler whenever χ is Meir-Keeler admissible.

This, along with $\Omega[P;T] \subseteq \Omega$ yields (see above) the written conclusions. □

Now, by simply combining these with Meir-Keeler geometric fixed point theorem (MK-g-fpt), one gets the following statement (referred to as: *Boyd-Wong-Matkowski fixed point theorem*; in short: (BWM-fpt)).

Theorem 1.48. *Suppose that T is Boyd-Wong $(d;P;\chi)$-contractive, where $P : X \times X \to R_+$ is (p-tb-d) and $\chi \in \mathcal{F}(re)(R_+^0, R)$ is Meir-Keeler admissible. In addition, let X be (o-f,d)-complete. Then,*

(31-a) T is a strong Picard operator (modulo d) when, in addition, T is (o-f,d)-continuous
(31-b) T is a strong Picard operator (modulo d) when, in addition, P is orbitally singular asymptotic
(31-c) T is fix-asingleton;
hence, fix-singleton, under either of the posed conditions.

Some basic examples of such functions are given below.
I) Call $\varphi \in \mathcal{F}(re)(R_+^0, R)$, *Boyd-Wong admissible* [35] if

(BW-adm) $\Lambda^+\varphi(s) < s$, for all $s > 0$.

In particular, $\varphi \in \mathcal{F}(re)(R_+^0, R)$ is Boyd-Wong admissible provided it is *upper semicontinuous at the right* on R_+^0: $\Lambda^+\varphi(s) \leq \varphi(s)$, for each $s \in R_+^0$.

This, e.g., is fulfilled when φ is *continuous at the right* on R^0_+; for, in such a case, $\Lambda^+\varphi(s) = \varphi(s)$, for each $s \in R^0_+$.

II) Let $\mathcal{F}(re, in)(R^0_+, R)$ stand for the class of all increasing $\varphi \in \mathcal{F}(re)(R^0_+, R)$; note that, for any such object,

$$\Lambda^+\varphi(s) = \varphi(s+0) := \lim_{t\to s+}\varphi(t) \text{ [the } \textit{right limit} \text{ of } \varphi \text{ at } s], \ s \in R^0_+.$$

Then, call $\varphi \in \mathcal{F}(re, in)(R^0_+, R)$, *Matkowski admissible* [164] provided

(M-adm) $(\forall t > 0)$: $\lim_n \varphi^n(t) = 0$, as long as $(\varphi^n(t); n \geq 0)$ exists.

Here, as usual, we denoted for each $t > 0$

$$\varphi^0(t) = t; \text{ and, for each } n \geq 0\text{: } \varphi^{n+1}(t) = \varphi(\varphi^n(t)).$$

Note that such a construction may be non-effective; because, for example, $\varphi^2(t) = \varphi(\varphi(t))$ is undefined whenever $\varphi(t) \leq 0$.

The relationships with the previous class of functions are described in

Proposition 1.62. *The following inclusions are valid:*

(32-1) each Boyd-Wong admissible function in $\mathcal{F}(re)(R^0_+, R)$ is Meir-Keeler admissible

(32-2) each Matkowski admissible function in $\mathcal{F}(re, in)(R^0_+, R)$ is Meir-Keeler admissible, and vice versa

(32-3) each Boyd-Wong admissible function in $\mathcal{F}(re, in)(R^0_+, R)$ is Matkowski admissible

(32-4) there exist Matkowski admissible functions in $\mathcal{F}(re, in)(R^0_+, R)$, that are not Boyd-Wong admissible.

Proof. i) (cf. Boyd and Wong [35]) Let $\varphi \in \mathcal{F}(re)(R^0_+, R)$ be Boyd-Wong admissible, and fix $\gamma > 0$; hence $\Lambda^+\varphi(\gamma) < \gamma$. By definition, $[(\exists\beta > 0)$: $\gamma < t < \gamma + \beta \Longrightarrow \varphi(t) < \gamma]$; proving that φ is Meir-Keeler admissible.

ii-1) (cf. Jachymski [120]) Assume that $\varphi \in \mathcal{F}(re, in)(R^0_+, R)$ is Matkowski admissible; we have to establish that it is Meir-Keeler admissible. If the underlying property fails, then (for some $\gamma > 0$):

$$\forall\beta > 0, \exists t \in]\gamma, \gamma + \beta[, \text{ such that } \varphi(t) \geq \gamma.$$

Combining with the increasing property of φ, one gets

$$(\forall t > \gamma)\text{: } \varphi(t) \geq \gamma \text{ [whence (by induction): } \varphi^n(t) \geq \gamma, \text{ for each } n].$$

Fixing some $t > \gamma$ and passing to limit, gives $0 \geq \gamma$; contradiction.

ii-2) Let $\varphi \in \mathcal{F}(re, in)(R^0_+, R)$ be Meir-Keeler admissible; we have to establish that it is Matkowski admissible. Given $s_0 > 0$, suppose that

the iterative sequence $(s_n = \varphi^n(s_0); n \geq 0)$ exists (in R^0_+).

As φ = regressive, (s_n) is strictly descending; hence, $s := \lim_n s_n$ exists, with $(s_n > s, \forall n)$. Suppose by contradiction that $s > 0$; and let $r > 0$ be the number assured by the Meir-Keeler admissible property of φ. By definition, there exists a rank $n(r) \geq 0$, such that $[n \geq n(r)$ implies $s < s_n < s + r]$. This, by the Meir-Keeler property, gives (for the same ranks) $[s < s_{n+1} = \varphi(s_n) \leq s]$; a contradiction. Hence, $s = 0$; and we are done.

iii) Let $\varphi \in \mathcal{F}(re, in)(R^0_+, R)$ be Boyd-Wong admissible

$\Lambda^+\varphi(s) < s$, $\forall s \in R^0_+$; that is: $\varphi(s+0) < s$, $\forall s \in R^0_+$.

Given $s_0 > 0$, assume that

the iterative sequence $(s_n = \varphi^n(s_0); n \geq 0)$ exists (in R^0_+).

From φ=regressive, (s_n) is strictly descending; so, $s := \lim_n s_n$ exists, with $(s_n > s, \forall n)$. Suppose by contradiction that $s > 0$. Combining with

$\varphi(s+0) = \lim_n \varphi(s_n) = \lim_n s_{n+1}$,

yields $\varphi(s+0) = s$; contradiction. Hence, $s = 0$; and we are done.

iv) (cf. Turinici [228]) Let the function $\varphi \in \mathcal{F}(re, in)(R^0_+, R)$, be taken as (for some $r > 0$):

$(\varphi(t) = t/2, \text{ if } t \leq r), (\varphi(t) = r, \text{ if } t > r)$.

Clearly, φ is Matkowski admissible. On the other hand,

$(\Lambda^+\varphi(r) =)\varphi(r+0) = r$; whence, φ is not Boyd-Wong admissible;

and this proves our claim. □

Finally, the Boyd-Wong-Matkowski fixed point theorem (BWM-fpt) includes the 1969 fixed point result in Boyd and Wong [35], when χ is Boyd-Wong admissible; as well as the 1975 fixed point result in Matkowski [164], when χ is Matkowski admissible. It also includes the 1977 fixed point result in Turinici [228]; we do not give details.

Case II. Given the functional couple (ψ, φ) over $\mathcal{F}(R^0_+, R)$, let us say that the mapping T is *Rhoades* $(d; P; \psi, \varphi)$*-contractive*, provided

$\psi(P_0(x,y)) \leq \psi(P(x,y)) - \varphi(P(x,y))$,
for all $x, y \in X$ with $P_0(x,y), P(x,y) > 0$.

To discuss it, remember that some compatible properties of the couple (ψ, φ) were introduced. First, let us assume that

(norm) (ψ, φ) is *normal*:
ψ is increasing, and φ is *strictly positive* ($\varphi(t) > 0$, $\forall t > 0$).

Further, let us introduce the conditions [upon the normal couple (ψ, φ)]:

(as-pos) (φ is *asymptotic positive*):
for each strictly descending sequence (t_n) in R_+^0 and each $\varepsilon > 0$ with $t_n \to \varepsilon+$, we must have $\limsup_n(\varphi(t_n)) > 0$
(r-loc-sp) (φ is *right locally strict positive*): $\Lambda_+\varphi(\varepsilon) > 0$, $\forall \varepsilon > 0$.

Remark 1.22.
The latter condition above is a particular case of the former one. In fact, assume that the premise of (r-loc-sp) is holding; and let the strictly descending sequence (t_n) in R_+^0 and the number $\varepsilon > 0$ be such that $t_n \to \varepsilon+$. From the accepted premise, there exists $\lambda > 0$ and $\delta > 0$, such that

$\varphi(t) \geq \lambda(> 0)$, for all $t \in]\varepsilon, \varepsilon + \delta[$.

But then, there must be some index $n(\delta)$, with

$\varphi(t_n) \geq \lambda(> 0)$, $\forall n \geq n(\delta)$; wherefrom, $\limsup_n \varphi(t_n) \geq \lambda(> 0)$;

and our claim follows.

Under these preliminaries, we may now formulate

Proposition 1.63. *Suppose that T is Rhoades $(d; P; \psi, \varphi)$-contractive, where the couple (ψ, φ) is normal. Then, the following assertions involving the associated relation $\Omega[P;T]$ are true in (ZF-AC+DC):*

(33-1) $\Omega[P;T]$ is upper diagonal
(33-2) $\Omega[P;T]$ is, in addition, geometric Meir-Keeler (whence: T is Meir-Keeler $(d;P)$-contractive) whenever φ is asymptotic positive or φ is right locally strict positive.

Proof. Let $\Omega := \Omega[\psi, \varphi]$ be the associated relation over R_+^0

$(t, s \in R_+^0)$: $t\Omega s$ iff $\psi(t) \leq \psi(s) - \varphi(s)$.

By definition,

T is Rhoades $(d;P;\psi,\varphi)$-contractive implies T is $(d;P;\Omega)$-contractive.

Moreover, by a previous fact,

(q1) Ω is upper diagonal (as φ is strictly positive)
(q2) Ω is geometric Meir-Keeler if φ is asymptotic positive or φ is right locally strict positive.

This, along with $\Omega[P;T] \subseteq \Omega$ yields (see above) the written conclusion. □

Now, by simply combining these with Meir-Keeler geometric fixed point theorem (MK-g-fpt), one gets the following statement (referred to as: *Rhoades-Dutta-Choudhury fixed point theorem*; in short: (RDC-fpt)).

Theorem 1.49. *Suppose that the selfmap T is Rhoades $(d;P;\psi,\varphi)$-contractive, where $P : X \times X \to R_+$ is (p-tb-d), and (ψ,φ) is a normal couple over $\mathcal{F}(R^0_+,R)$, with (φ = asymptotic positive) or (φ = right locally strict positive). In addition, let X be (o-f,d)-complete. Then,*

(32-a) T is a strong Picard operator (modulo d) when, in addition, T is (o-f,d)-continuous
(32-b) T is a strong Picard operator (modulo d) when, in addition, P is orbitally singular asymptotic
(32-c) T is fix-asingleton;
hence, fix-singleton, under either of the posed conditions.

In particular, when

$P = P_1$ and both ψ and φ are increasing, continuous,

the obtained statement is just the 2008 one in Dutta and Choudhury [84]; which, in turn extends the 2001 contribution due to Rhoades [197]. Further aspects may be found in Jleli and Samet [128]; see also Alsulami et al. [13].

In the following, two basic examples of such contractions are discussed.

Example 1. Let $F : R^0_+ \to R$ be a function; we say that it is *semi-Wardowski*, provided the following conditions hold

(sw-1) F is *strictly increasing*: $t < s$ implies $F(t) < F(s)$
(sw-2) F is *zero-vertical*: $t_n \to 0$ iff $F(t_n) \to -\infty$;

and a *Wardowski function* if, in addition,

(sw-3) $\lim_{t\to 0+} t^k F(t) = 0$, for some $k \in]0,1[$.

Given the function $F : R^0_+ \to R$ and the number $C > 0$, let us say that T is $(d; P; F, C)$-*contractive*, provided

(FC-con) $F(P_0(x,y)) \leq F(P(x,y)) - C$,
$\forall x, y \in X$, $P_0(x,y), P(x,y) > 0$.

Theorem 1.50. *Suppose that T is $(d; P; F, C)$-contractive, where $P : X \times X \to R_+$ is (p-tb-d), $F(.)$ is a Wardowski function and $C > 0$ is a constant. In addition, let X be (o-f,d)-complete. Then,*

(33-a) T is a strong Picard operator (modulo d) when, in addition, T is (o-f,d)-continuous
(33-b) T is a strong Picard operator (modulo d) when, in addition, P is orbitally singular asymptotic
(33-c) T is fix-asingleton;
hence, fix-singleton, under either of the posed conditions.

Proof. Take the couple (ψ, φ) over $\mathcal{F}(R^0_+, R)$ according to

$\psi(t) = F(t)$, $\varphi(t) = C$, $t \in R^0_+$.

Clearly, (ψ, φ) is normal; in addition, φ is right locally strict positive. Hence, Theorem 1.49 applies to these data; wherefrom, all is clear. □

Remark 1.23.

Technically speaking, the second and third property of a Wardowski function were ignored in this argument; while the first one may be replaced by its non-strict counterpart:

(sw-1a) F is increasing: $t \leq s$ implies $F(t) \leq F(s)$.

In other words: even in this larger setting, the initial Wardowski fixed point result [264] is retainable; see also Turinici [248].

Example 2. Let (F, θ) be a couple over $\mathcal{F}(R^0_+, R)$; we call it *Amini-Harandi*, provided the following conditions hold

(ah-1) F is semi-Wardowski and $[\theta(r) > 0, \ \forall r > 0]$
(ah-2) for each strictly descending (hence, convergent) sequence (r_n) in R^0_+, the relation $\theta(r_n) \to 0$ is impossible
(ah-3) for each strictly descending sequence (r_n) in R^0_+ with $r_n \to 0$ and $[\theta(r_n) \leq F(r_n) - F(r_{n+1}), \ \forall n]$, we have $\sum_n r_n < \infty$.

Given the couple (F, θ), let us say that T is $(d; P; F, \theta)$-*contractive*, provided

(F-th-con) $F(P_0(x,y)) \leq F(P(x,y)) - \theta(P(x,y))$,
for all $x, y \in X$, $P_0(x,y), P(x,y) > 0$.

Theorem 1.51. *Suppose that T is $(d; P; F, \theta)$-contractive, where $P : X \times X \to R_+$ is (p-tb-d), and (F, θ) is an Amini-Harandi couple. In addition, let X be (o-f,d)-complete. Then,*

(34-a) T is a strong Picard operator (modulo d) when, in addition, T is (o-f,d)-continuous
(34-b) T is a strong Picard operator (modulo d) when, in addition, P is orbitally singular asymptotic
(34-c) T is fix-asingleton;
hence, fix-singleton, under either of the posed conditions.

Proof. Take the couple (ψ, φ) over $\mathcal{F}(R^0_+, R)$, as

$$\psi(t) = F(t),\ \varphi(t) = \theta(t),\ t \in R^0_+.$$

Clearly, (ψ, φ) is normal, by the choice of (F, θ). Moreover, we claim that φ is right locally strict positive. Suppose not: there exists $\varepsilon > 0$, with

$$\inf \varphi(]\varepsilon, \varepsilon + \gamma[) = 0, \text{ for all } \gamma > 0.$$

Let $(\delta_n; n \geq 0)$ be a strictly descending sequence in R^0_+ with $\delta_n \to 0$. By the very definition of infimum, it follows that

$$\text{for each } s \in R^0_+, \text{ there exists } t \in R^0_+, \text{ such that } t < s/2,\ \varphi(\varepsilon + t) < \delta_n.$$

Consequently, for each $n \geq 0$, the relation $\mathcal{R}_n$ over R^0_+, introduced as

$$(s, t \in R^0_+)\text{: } s\mathcal{R}_n t \text{ iff } t < s/2 \text{ and } \varphi(\varepsilon + t) < \delta_n$$

is proper. But then, from the Diagonal Dependent Choice principle (DDC) [deductible in the reduced system (ZF-AC+DC)], it results that there must be a sequence $(t_n; n \geq 0)$ in R^0_+, with $t_0 = 1$ and

$$t_n \mathcal{R}_n t_{n+1} \text{ [that is: } t_{n+1} < t_n/2 \text{ and } (0 <)\ \varphi(\varepsilon + t_{n+1}) < \delta_n], \forall n.$$

The sequence $(r_n := \varepsilon + t_n; n \geq 0)$ in R^0_+ is strictly descending, with

$$r_n \to \varepsilon+, \text{ and } (\theta(r_n) =)\ \varphi(r_n) \to 0.$$

But then, the second Amini-Harandi condition (ah-2) is not fulfilled by the couple (F, θ); contradiction. Hence, φ is indeed right locally strict positive, as claimed; and Theorem 1.49 is applicable here. The proof is complete. □

Note that, from the properties of a semi-Wardowski function appearing in our first Amini-Harandi condition (ah-1), the second one was ignored; while, the first one was used in the non-strict sense we just precise. On the other hand, the third Amini-Harandi condition (ah-3) was ignored along this reasoning. In other words: with all these enlargements, the univalued version of the fixed point result in Amini-Harandi [16] (corresponding to $P = P_1$) is still remaining valid. However, it is questionable whether the multivalued version of the quoted result is available too with the same techniques; we conjecture that the answer is negative. Further aspects may be found in Abbas et al. [1], Batra and Vashistha [23], Sgroi and Vetro [213], Imdad et al. [118], or Wardowski and Dung [266]; see also Gupta [105].

1.7.4 *Banach contraction principle*

Let (X, d) be a metric space; and $T \in \mathcal{F}(X)$ be a selfmap of X. As precise, we look for existence and uniqueness conditions involving points of $\text{Fix}(T)$. The basic directions to be considered are described by the concepts below:

(conc-1) T is *strong Picard* (modulo d): for each $x \in X$,
$(T^n x; n \geq 0)$ is d-convergent with $\lim_n (T^n x) \in \text{Fix}(T)$
(conc-2) T is *fix-asingleton* (respectively, *fix-singleton*):
$\text{Fix}(T)$ is asingleton (respectively, singleton).

A natural extension of these is introduced by the iterative type concepts:

(i-conc-1) T is *iterative strong Picard* (modulo d):
for each $k \geq 1$, T^k is strong Picard (modulo d)
(i-conc-2) T is *iterative fix-asingleton* (*iterative fix-singleton*):
for each $k \geq 1$, T^k is fix-asingleton (fix-singleton).

Remark 1.24.
The strong Picard property is invariant to iterations; i.e.,

T is strong Picard (modulo d) implies
T is iterative strong Picard (modulo d).

This, however, is no longer valid for the fix-asingleton (respectively, fix-singleton) property; precisely, a relation like

T is fix-asingleton implies T is iterative fix-asingleton

is not valid, in general.

Now, to solve the posed problem in this iterative sense, a lot of special contractive conditions upon T are needed. The simplest one, referred to as *Banach fixed point theorem* (in short: (B-fpt)) may be stated as below. Given $\alpha \geq 0$, let us say that T is (d, α)-*contractive*, provided

(con) $d(Tx, Ty) \leq \alpha d(x, y)$, $\forall x, y \in X$.

Theorem 1.52. *Suppose that T is (d, α)-contractive, for some $\alpha \in [0, 1[$. Further, let X be d-complete. Then,*

(41-a) T is iterative strong Picard (modulo d)
(41-b) T is iterative fix-singleton:
$\mathrm{Fix}(T^k) = \mathrm{Fix}(T)$ = *singleton, for each $k \geq 1$.*

Proof. i): By the imposed contractive condition,

T is strict nonexpansive (modulo d):
$d(Tx, Ty) < d(x, y)$, $\forall x, y \in X$, $x \neq y$;

hence, in particular, T is fix-asingleton. As a consequence, we have

T is nonexpansive (modulo d): $d(Tx, Ty) \leq d(x, y)$, $\forall x, y \in X$, $x \neq y$;

which tells us that T is d-continuous.

ii): Fix some $x_0 \in X$; and put $(x_n = T^n x_0; n \geq 0)$. From the contractive condition once again,

$d(x_n, x_{n+1}) \leq \alpha^n d(x_0, x_1)$, $\forall n$; whence, (x_n) is a d-Cauchy sequence.

By completeness, $x_n \xrightarrow{d} z$ as $n \to \infty$, for some (uniquely determined) $z \in X$. This, by the continuity of T, yields

$y_n := x_{n+1} = Tx_n \xrightarrow{d} Tz$ as $n \to \infty$.

On the other hand, (y_n) is a subsequence of (x_n); whence $y_n \xrightarrow{d} z$; so that (as d is separated) $z = Tz$. As x_0 was arbitrarily chosen, T is strong Picard (modulo d).

iii): For each $k \geq 1$, the selfmap T^k is (d, α^k)-contractive, with $\alpha^k \in [0, 1[$. This, along with the preceding part, yields the desired conclusion. □

This result established in 1922 by Banach [22], found some basic applications to the operator equations theory; so, it was the subject of many extensions. According to the 1985 Deimling's monograph [71, Ch. 5], there were more that 5,000 extensions of (B-fpt) at that time; so – taking into account the tremendous development of this area – it is pretty realistic to presume that, at present, there must be more than 15,000 extensions of (B-fpt). A complete classification – hence, all the more, a complete unification – of all these is impossible. Perhaps, the most impressive example in this direction is Browder's attempt [41], [42] of including under a general scheme the family of all contractions known – at that moment – by the author. Unfortunately, as proved in Walter [263], his machinery needs very strong conditions to start; so that, Browder's initial objective cannot be attained in this way. Nevertheless, certain unifying efforts have been already done; see, for instance, Rhoades [195], Park [184], Piticari [186, Ch. II], Collaco and E Silva [64], or Kincses and Totik [148], to quote only a few survey type contributions in the area.

(B) Concerning all these results, the basic question is that of to what extent is an arbitrary member of it – Extended Banach theorem say – an effective generalization of the Banach theorem. This raises the question of determining the most general conditions under which the setting of Banach theorem is to be attained. An appropriate answer to it – in fact, a converse of the Banach theorem – may be stated along the lines below.

Theorem 1.53. *Let X be a nonempty set, and T be a selfmap of it, with*

$$T \text{ is iterative fix-singleton: } \mathrm{Fix}(T^k) = \mathrm{Fix}(T) = \text{singleton, for all } k \geq 1.$$

Further, take some $\lambda \in]0,1[$. There exists then a metric $d(.,.)$ on X, with

$$X \text{ is } d\text{-complete and } T \text{ is } (d,\lambda)\text{-contractive.}$$

Proof. (Sketch) By the imposed condition,

$$\mathrm{Fix}(T^k) = \{z\}, \text{ for all } k \geq 1 \text{ and some } z \in X.$$

Let $\mathcal{M}$ stand for the class of all pairs (A, α), where $[A \in \exp(X),\ \alpha \in \mathcal{F}(A, R_+)]$, with the compatible properties

$$\text{(comp)}\quad z \in A,\ T(A) \subseteq A,\ \alpha^{-1}(0) = \{z\},\ [\alpha(Tx) \leq \lambda\alpha(x),\ \forall x \in A].$$

Clearly, $\mathcal{M}$ is nonempty; because (A_0, α_0), where $(A_0 = \{z\},\ \alpha_0(z) = 0)$ is an element of it. Define a partial order $(\preceq)$ on $\mathcal{M}$ as

$$(A, \alpha) \preceq (B, \beta) \text{ iff } A \subseteq B \text{ and } \beta|_A = \alpha.$$

The ordered structure $(\mathcal{M}, \preceq)$ is inductive: any chain of $\mathcal{M}$ has an upper bound. So, by the Zorn Maximal Principle, there exists a $(\preceq)$-maximal element $(C, \gamma) \in \mathcal{M}$, in the sense

$$(C, \gamma) \preceq (D, \delta) \in \mathcal{M} \Longrightarrow (C = D, \gamma = \delta).$$

By these properties, we must have $C = X$. Moreover, the mapping

$$d(x, y) = \gamma(x) + \gamma(y), \text{ if } x \neq y;\ d(x, y) = 0, \text{ otherwise,}$$

is a metric on X such that X is d-complete and T is (d, λ)-contractive; wherefrom, all is clear. □

This line of proof, due to Jachymski [123], is a refinement of the one proposed by Deimling [71, Ch. 5, Sec. 17]; which, in turn, follows the developments in Wong [268]. An early "abstract" version of it was given in the 1959 paper due to Bessaga [30]; so, it is natural to call the obtained result as: Bessaga theorem.

Roughly speaking, the reduction of Extended Banach theorem (over the metric space (X, d)) to the Banach theorem (over X) runs as follows

Step 1. From the Extended Banach theorem, the selfmap T is iterative fix-singleton.

Step 2. By the Bessaga theorem, it results that, given the number $\lambda \in]0, 1[$, there exists a metric $e(.,.)$ on X, such that X is e-complete and T is (e, λ)-contractive.

Step 3. From the Banach theorem, it follows that T is iterative strong Picard (modulo e), and iterative fix-singleton.

Note that, establishing **Step 1** means that we practically arrived at the conclusion of Extended Banach theorem. However, suppose that this is a secondary question; i.e., we eventually reached the precise conclusion. By its very proof, **Step 2** needs the Zorn Maximal Principle, that holds in the (complete) system (ZF); hence, the reduction procedure above needs (ZF). On the other hand, the proof of most Extended Banach theorems needs the strongly reduced system (ZF-AC). This tells us that, from the "local" axiomatic perspective of (ZF-AC), the Bessaga theorem cannot help us to solve the reduction problem above; but, from the "global" axiomatic perspective of (ZF), it works in a remarkable way. Moreover, by these developments, one derives an important methodological fact: the question of establishing, in (ZF-AC), that Extended Banach theorem is reducible to (hence, equivalent with) the Banach theorem is not at all trivial. Further aspects of the discussed question may be found in the 2001 monograph by Rus [203, Ch. 2].

Chapter 2

CLASSES OF CONTRACTIVE MAPS

2.1 Anticipative Rhoades Contractions in bi-Relational Metric Spaces

Abstract. A lot of implicit methods is proposed for a class of anticipative Rhoades contractions on bi-relational metric spaces. The connections with some related statements obtained by Dung and Hang [Vietnam J. Math., 43 (2015), 743–755], Chandok et al. [J. Egyptian Math. Soc., 23 (2015), 95–101], and Wardowski [Proc. Amer. Math. Soc., 146 (2018), 1585–1598] are also being established.

Subject Class 2020. 47H10 (Primary), 54H25 (Secondary).

Keywords. Bi-relational metric space, fixed point, Picard operator, Cauchy and convergent sequence, Rhoades contraction.

2.1.1 *Dependent Choice principles*

See Ch. 1.

2.1.2 *Conv-Cauchy structures*

See Ch. 1.

2.1.3 *Admissible functions*

See Ch. 1.

2.1.4 *Statement of the problem*

Let X be a nonempty set. Take a *metric* $d : X \times X \to R_+$ on X; and let $\mathcal{G} \subseteq X \times X$ be a relation over X; the triple $(X, d, \mathcal{G})$ will be said to be a *relational metric space.* Given the relations $\mathcal{A}, \mathcal{B}$ on X, denote

$$(\mathcal{A}; \mathcal{B}; Z) = \mathcal{A} \cap \mathcal{B} \cap (Z \times Z),\ (\mathcal{A}; Z) = (\mathcal{A}; \mathcal{A}; Z),\ Z \in \exp(X);$$

in this perspective, $(\mathcal{A}; \mathcal{B}; X) := \mathcal{A} \cap \mathcal{B}$ is denoted as $(\mathcal{A}; \mathcal{B})$, and $(\mathcal{A}; X)$ is just $\mathcal{A}$. Then, given the sequence (z_n) in X and the point $z \in X$, put

$(z_n; n \geq 0)\mathcal{G}z$ iff $z_n\mathcal{G}z$, for all $n \geq 0$
$(z_n; n \geq 0)\mathcal{G}\mathcal{G}z$ iff there exists a subsequence $(w_n = z_{i(n)}; n \geq 0)$ of $(z_n; n \geq 0)$ such that $(w_n; n \geq 0)\mathcal{G}z$.

Let $T \in \mathcal{F}(X)$ be a selfmap of X. In the following, sufficient conditions are given for the existence and/or uniqueness of elements in $\mathrm{Fix}(T)$.

4-I) Concerning the uniqueness question, call the subset Y of X, *$\mathcal{G}$-almost-singleton* (in short: *$\mathcal{G}$-asingleton*) provided ($y_1, y_2 \in Y$, $y_1\mathcal{G}y_2 \Longrightarrow y_1 = y_2$); and *$\mathcal{G}$-singleton* when, in addition, Y is nonempty. Then, let us introduce the fix-uniqueness concepts

(fix-a-sg) T is *fix-$\mathcal{G}$-asingleton* (respectively, *fix-$\mathcal{G}$-singleton*) when $\mathrm{Fix}(T)$ is $\mathcal{G}$-asingleton (respectively, $\mathcal{G}$-singleton).

Some concrete examples of this type will be provided a bit further.

Passing to the existence question, the *sequential* way of solving it consists in the points of $\mathrm{Fix}(T)$ being chosen among the limit points (if any)

$T^\omega x_0 := \lim_n(T^n x_0)$, where $x_0 \in X$ is appropriately given.

(Here, ω is the first transfinite ordinal.) To do this, a lot of technical facts about *iterative processes* must be introduced and discussed.

4-I-1) Let x_0 be some point in X. By an *orbital* (or: *iterative*) sequence attached to x_0 (and T), we mean any sequence $X_0 := (x_n; n \geq 0)$ [or, simply, $X_0 = (x_n)$] defined as $(x_n = T^n x_0; n \geq 0)$. When x_0 is generic here, the resulting object will be referred to as an *orbital sequence* (in short *o-sequence*) on X.

Fix in the following such an object $X_0 = (x_n)$. Then, denote

$[X_0] = \{x_n; n \geq 0\}$ (the *trajectory* attached to $X_0 = (x_n)$)
$[[X_0]] := \mathrm{cl}([X_0])$ (the *complete trajectory* attached to $X_0 = (x_n)$).

Proposition 2.1. *Under these conventions,*

$[[X_0]] = [X_0] \cup \{z\}$, *whenever* $z := \lim_n(x_n)$ *exists.*

Proof. Denote $U_0 = [X_0]$, $V_0 = [[X_0]]$; hence, $V_0 = \mathrm{cl}(U_0)$. Clearly, $V_0 \supseteq U_0 \cup \{z\}$. Suppose that there exists $v \in V_0 \setminus (U_0 \cup \{z\})$. As $v \neq z$, there exists $\sigma > 0$ such that $X(v, \sigma) \cap X(z, \sigma) = \emptyset$; and this (by the choice of z) yields $X(v, \rho) \cap (U_0 \cup \{z\}) = \emptyset$, for some $\rho \in]0, \sigma[$. In particular, we derive that v cannot belong to $\mathrm{cl}(U_0) = V_0$; contradiction. Consequently, $V_0 \subseteq U_0 \cup \{z\}$; and the conclusion follows. □

4-I-2) Given the o-sequence $X_0 = (x_n)$, define the concept

X_0 is $\mathcal{G}$-*ascending* (in short: *ascending*): $x_n \mathcal{G} x_{n+1}$, $\forall n$.

When X_0 is generic, we will talk about *ascending orbital sequences* (in short: *(a-o)-sequences*) on X; the class of all these will be denoted as $\mathrm{ao}(T)$.

Further, let $X_0 = (x_n)$ be an (a-o) sequence in X. Given the natural number $h \geq 1$, let us say that $X_0 = (x_n)$ is $(\mathcal{G}, h)$-*uadmissible*, provided

$$\forall (m, n) \in (<; N),\ \exists p \in N, \text{ such that } n < p \leq n + h \text{ and } x_m \mathcal{G} x_p.$$

The set of all such elements h will be denoted as $\mathrm{uad}(X_0)$; when it is nonempty, we say that $X_0 = (x_n)$ is $\mathcal{G}$-*uadmissible* (in short: *uadmissible*). Finally, let $\mathrm{uadao}(T)$ stand for the class of all uadmissible, ascending orbital sequences $X_0 = (x_n)$ in X.

Concerning these concepts, let us consider the conditions

(a-reg) T is *ascending regular*: $\mathrm{ao}(T)$ is nonempty
(uad-a-reg) T is *uadmissible ascending regular*: $\mathrm{uadao}(T)$ is nonempty.

Under the precise framework, two basic problems occur.

Pro-1) The first one is to derive $\mathcal{G}$-ascending orbital sequences by starting from (global) conditions upon T. For example, this is available under

(s-pro) T is $\mathcal{G}$-*semi-progressive* $[X(T, \mathcal{G}) := \{x \in X; x\mathcal{G}Tx\} \neq \emptyset]$,
(incr) T is $\mathcal{G}$-*increasing* $[x\mathcal{G}y \Longrightarrow Tx\mathcal{G}Ty]$.

However, we do not effectively use these requirements in the sequel.

Pro-2) The second one is to get ($\mathcal{G}$-uadmissible, $\mathcal{G}$-ascending) o-sequences by starting from $\mathcal{G}$-ascending o-sequences. So, assume that

T is ascending regular ($\mathrm{ao}(T)$ is nonempty);
hence, T is $\mathcal{G}$-semi-progressive ($X(T, \mathcal{G})$ is nonempty).

Denote, for each $x \in X(T, \mathcal{G})$,

$$\mathrm{spec}(x) = \{i \in N(1, \leq); x\mathcal{G}T^i x\} \text{ [the } \textit{spectrum} \text{ of } x \text{ (modulo } (\mathcal{G}, T))].$$

Clearly, $1 \in \mathrm{spec}(x)$; but, the alternative $\mathrm{spec}(x) = \{1\}$ cannot be avoided.

The following particular answer to the underlying question is available. Letting the $\mathcal{G}$-ascending orbital sequence $X_0 = (x_n)$, define the property

(t-asc) $X_0 = (x_n)$ is translation ascending:
$x_i \mathcal{G} x_j$ implies $x_{i+s} \mathcal{G} x_{j+s}$, for all $s \in N$.

Clearly, this holds when T is $\mathcal{G}$-increasing. Given the natural number $h \geq 1$, call the subset M of $N(1, \leq)$, *h-admissible* provided

(M-adm) for each $n \in N$ there exists $r \in M$ with $n < r \leq n + h$.

When such numbers $h \geq 1$ exist, we say that M is *admissible*.

Proposition 2.2. *Let the $\mathcal{G}$-ascending orbital sequence $X_0 = (x_n)$ fulfill*

X_0 is translation ascending and $\text{spec}(x_0)$ *is admissible.*

Then, necessarily, $X_0 = (x_n)$ is $\mathcal{G}$-uadmissible.

Proof. Fix $(m, n) \in (<; N)$. By hypothesis, there exists $h \in N(1, \leq)$, with

$\text{spec}(x_0) = \{i \in N(1, \leq); x_0 \mathcal{G} x_i\}$ is h-admissible; wherefrom
there exists $r \in \text{spec}(x_0)$ (hence, $x_0 \mathcal{G} x_r$), with $n - m < r \leq n - m + h$.

In this case, under the notation $p = m + r$, we have $n < p \leq n + h$ and $x_m \mathcal{G} x_{m+r}$ (that is: $x_m \mathcal{G} x_p$), if we remember that $X_0 = (x_n)$ is translation ascending. Hence, $X_0 = (x_n)$ is $\mathcal{G}$-uadmissible; as claimed. □

In the following, some other example of such objects is given. The following auxiliary fact is our starting point.

Proposition 2.3. *Let the $\mathcal{G}$-ascending sequence $(z_n; n \geq 0)$ in X and the number $h \in N(2, \leq)$ be such that (under the notation $Z := \{z_n; n \geq 0\}$)*

$(\mathcal{G}; Z)$ is h-transitive: $(\mathcal{G}; Z)^h \subseteq (\mathcal{G}; Z)$.

Then, the regular property holds, for each $r \geq 0$

(Reg;r) $(z_i, z_{i+1+r(h-1)}) \in \mathcal{G}$, for all $i \geq 0$.

Proof. We will use the induction with respect to r. First, by the $\mathcal{G}$-ascending property of our sequence,

$(z_i, z_{i+1}) \in (\mathcal{G}; Z) \subseteq \mathcal{G}$, $\forall i \geq 0$; whence, the relation $(Reg; 0)$ holds.

This, by definition (and the h-transitive hypothesis), yields

$(z_i, z_{i+h}) \in (\mathcal{G}; Z)^h \subseteq (\mathcal{G}; Z) \subseteq \mathcal{G}$, $\forall i \geq 0$; hence, $(Reg; 1)$ holds too.

Suppose that the underlying relation (Reg;r) holds for $r \in \{0, ..., s\}$, where $s \geq 1$; we claim that it (Reg;s+1) holds as well. In fact, let $i \geq 0$ be arbitrary fixed. Again by the $\mathcal{G}$-ascending property of our sequence,

$$(z_{i+1+s(h-1)}, z_{i+1+(s+1)(h-1)}) \in (\mathcal{G}; Z)^{h-1};$$

so that, by the inductive hypothesis (and properties of relational product)

$$(z_i, z_{i+1+(s+1)(h-1)}) \in (\mathcal{G}; Z) \circ (\mathcal{G}; Z)^{h-1} = (\mathcal{G}; Z)^h \subseteq (\mathcal{G}; Z) \subseteq \mathcal{G};$$

hence the claim. The proof is thereby complete. □

Given the number $h \in N(2, \leq)$, call the $\mathcal{G}$-ascending orbital sequence $X_0 = (x_n)$ in X, $(\mathcal{G}, h)$-*transitive* when

the restriction $(\mathcal{G}; [X_0])$, where $[X_0] := \{x_n; n \geq 0\}$, is h-transitive.

When $h \in N(2, \leq)$ is generic in this convention, we then say that the $\mathcal{G}$-ascending orbital sequence $X_0 = (x_n)$ is $\mathcal{G}$-*transitive*.

Proposition 2.4. *Let the $\mathcal{G}$-ascending orbital sequence $X_0 = (x_n)$ be $\mathcal{G}$-transitive. Then, necessarily, $X_0 = (x_n)$ is $\mathcal{G}$-uadmissible.*

Proof. By definition, there exists $h \in N(2, \leq)$ such that $X_0 = (x_n)$ is $(\mathcal{G}, h)$-transitive. Let $(m, n) \in (<; N)$ be arbitrary fixed. By the preceding result, $[(x_m, x_{m+1+r(h-1)}) \in \mathcal{G}, \forall r \geq 0]$. As $m+1 \leq n$, there exists at least one $r \in N(1, \leq)$ with $n < m + 1 + r(h-1) \leq n + h$. It will suffice putting $p = m + 1 + r(h-1)$ to end the reasoning. □

4-III) We may now pass to the essential part of our developments. For each $x, y \in X$ define the non-anticipative and anticipative functions

(non-anti) $Q_1(x, y) = d(x, Tx)$, $Q_2(x, y) = d(x, y)$,
$Q_3(x, y) = d(x, Ty)$, $Q_4(x, y) = d(Tx, y)$,
$Q_5(x, y) = d(Tx, Ty)$, $Q_6(x, y) - d(y, Ty)$,
$\mathcal{Q}(x, y) = (Q_1(x, y), Q_2(x, y), Q_3(x, y), Q_4(x, y), Q_5(x, y), Q_6(x, y))$;
(anti) $K_1(x, y) = d(x, T^2x)$, $K_2(x, y) = d(Tx, T^2x)$,
$K_3(x, y) = d(T^2x, y)$, $K_4(x, y) = d(T^2x, Ty)$,
$\mathcal{K}(x, y) = (K_1(x, y), K_2(x, y), K_3(x, y), K_4(x, y))$.

Further, let us construct the family of functions [for $x, y \in X$]

$B_0(x, y) = \min\{Q_2, Q_5\}(x, y)$,
$B_1(x, y) = \min\{Q_1, Q_2, Q_5, Q_6; K_2, K_3\}(x, y)$,

$$B_2(x,y) = \min\{Q_2, Q_3, Q_4, Q_5, K_3, K_4\}(x,y),$$
$$B_3(x,y) = \min\{Q_1, Q_2, Q_3, Q_4, Q_5, Q_6, K_1, K_2, K_3, K_4\}(x,y);$$

as well as the associated relations (over X)

$$(B_i > 0) = \{(x,y) \in X \times X; B_i(x,y) > 0\},\ i \in \{0,1,2,3\}.$$

Take some $\Upsilon \in \exp(R_+^6 \times R_+^4)$; and let $\mathcal{S} = \mathcal{S}(\Upsilon)$ stand for the relation

$$\mathcal{S} = (\mathcal{Q}, \mathcal{K})^{-1}(\Upsilon);\ \text{hence, } (\mathcal{Q}, \mathcal{K})(\mathcal{S}) \subseteq \Upsilon.$$

Given the relation $\mathcal{R}$ over X, let us introduce the contractive property

(GR-contr) T is $(\mathcal{G};\mathcal{R};\Upsilon)$-*contractive*, provided $(\mathcal{G};\mathcal{R}) \subseteq \mathcal{S}$;
that is: $(\mathcal{Q},\mathcal{K})(x,y) \in \Upsilon$, $\forall (x,y) \in (\mathcal{G};\mathcal{R})$.

The starting point of our metrical fixed point theory for these objects is the construction of a $\mathcal{G}$-ascending orbital sequence $X_0 = (x_n)$, with

(R-asc) $X_0 = (x_n)$ is $\mathcal{R}$-ascending: $x_n \mathcal{R} x_{n+1}$, for all n.

A general answer to this question is rather difficult. Here, we will concentrate on a particular case of it – that, in fact, covers a large number of such contractions – based on the restrictive condition

(B1-adm) $\mathcal{R}$ is B_1-*admissible*: $(B_1 > 0) \subseteq \mathcal{R}$.

This tells us that our general contraction is reducible to

(B1-contr) T is $(\mathcal{G};(B_1 > 0);\Upsilon)$-*contractive*, provided
$(\mathcal{G}; B_1 > 0) \subseteq \mathcal{S}$; that is: $(\mathcal{Q},\mathcal{K})(x,y) \in \Upsilon$, $\forall (x,y) \in (\mathcal{G}; B_1 > 0)$.

Another condition to be used, for the uniqueness question, is

(B2-adm) $\mathcal{R}$ is B_2-admissible: $(B_2 > 0) \subseteq \mathcal{R}$;

with the associated (stronger) contractive condition

(B2-contr) T is $(\mathcal{G};(B_2 > 0);\Upsilon)$-*contractive*, provided
$(\mathcal{G}; B_2 > 0) \subseteq \mathcal{S}$; that is: $\mathcal{Q}(x,y) \in \Upsilon$, $\forall (x,y) \in (\mathcal{G}; B_2 > 0)$.

Note that both admissible conditions above hold under

(B0-adm) $\mathcal{R}$ is B_0-admissible: $(B_0 > 0) \subseteq \mathcal{R}$;

when both associated contractive conditions are fulfilled under

(B0-contr) T is $(\mathcal{G};(B_0>0);\Upsilon)$-*contractive*, provided $(\mathcal{G};B_0>0)\subseteq\mathcal{S}$; that is: $\mathcal{Q}(x,y)\in\Upsilon$, $\forall(x,y)\in(\mathcal{G};B_0>0)$.

4-III-1) Having these precise, let us return to the basic condition

(a-reg) T is *ascending regular*: ao(T) is nonempty.

By definition, there exist (a-o) sequences in X; let $X_0=(x_n)$ be one of these. Concerning its $\mathcal{R}$-ascending property, we stress that in the B_1-admissible setting, this is attainable in terms of the relation $(B_1>0)$. Precisely, call the (a-o) sequence $X_0=(x_n)$, $(B_1>0)$-*ascending* provided

$(B_1(x_n,x_{n+1})>0,\ \forall n)$; referred to as: $X_0=(x_n)$ is B-*ascending*.

We then say that the (a-o) sequence $X_0=(x_n)$ has the *B-property*; so that (putting these together) $X_0=(x_n)$ is (Ba-o). The possibility o reaching such a property is a consequence of the reasoning below. Given the ascending orbital sequence $X_0=(x_n)$, we have two alternatives.

Alt-1) The (a-o)-sequence X_0 is *telescopic*: there exists $h\geq 0$, such that $d(x_h,x_{h+1})=0$ [that is: $x_h=x_{h+1}$]. By the very choice of X_0,

$x_h=x_n$, for all $n\geq h$; whence, $z:=x_h$ is an element of Fix(T);

so that, this case is completely clarified from the fixed point perspective.

Alt-2) The ascending orbital sequence $X_0=(x_n)$ is *non-telescopic*: $[d(x_n,x_{n+1})>0$, for all $n]$. Note that, in this case,

$(B_1(x_n,x_{n+1})>0,\ \forall n)$; hence, $X_0=(x_n)$ is (Ba-o).

4-III-2) Having these precise, let us now pass to the effective condition to be used in the sequel

(uad-a-reg) T is *ascending uadmissible regular*: uadao(T) is nonempty.

According to it, there exist ($\mathcal{G}$-uadmissible, $\mathcal{G}$-ascending) orbital sequences (in short: (uad-a-o) sequences) in X; let $X_0=(x_n)$ be one of these. By the developments above, we may arrange for $X_0=(x_n)$ being, in addition, B-ascending; hence, a (uad-Ba-o) sequence. In this case, the basic directions under which the investigations be conducted are described in the list below.

(rpic-1) We say that the (uad-Ba-o) sequence $X_0=(x_n)$ is *semi Picard* (modulo $(\mathcal{G},\mathcal{R};T)$) when (x_n) is d-asymptotic ($d(x_n,x_{n+1})\to 0$ as $n\to\infty$)

(rpic-2) We say that the (uad-Ba-o) sequence $X_0 = (x_n)$ is *Picard* (modulo $(\mathcal{G}, \mathcal{R}; T)$) when (x_n) is d-Cauchy ($d(x_n, x_m) \to 0$ as $n, m \to \infty$)

(rpic-3) We say that the (uad-Ba-o) sequence $X_0 = (x_n)$ is *strongly Picard* (modulo $(\mathcal{G}, \mathcal{R}; T)$) when $x_\omega := \lim_n(x_n)$ exists, with $x_\omega \in \text{Fix}(T)$

(rpic-4) We say that the (uad-Ba-o) sequence $X_0 = (x_n)$ is *semi Bellman Picard* (modulo $(\mathcal{G}, \mathcal{R}; T)$) when $x_\omega := \lim_n(x_n)$ exists, with $x_\omega \in \text{Fix}(T)$, and $(x_n; n \geq 0)\mathcal{G}\mathcal{G}x_\omega$.

In particular, when T is $\mathcal{G}$-semi-progressive and $\mathcal{G}$-increasing, these conventions are comparable with the ones in Turinici [246]; which, in case of $\mathcal{G} = X \times X$ (the trivial relation over X) reduce to the ones in Rus [203, Ch. 2, Sec. 2.2]; because, in this setting, $X(T, \mathcal{G}) = X$.

Sufficient conditions for such properties are obtainable within the class of all uadmissible (Ba-o) sequences $X_0 = (x_n)$, with the additional property

$X_0 = (x_n)$ is *full*: $n \mapsto x_n$ is injective ($i \neq j$ implies $x_i \neq x_j$);

referred to as *uadmissible (Ba-o) full* [in short: (uad-Ba-o-f)] *sequences.*

reg-1) Call X, *(uad-Ba-o-f,d)-complete* provided (for each (uad-Ba-o-f) sequence) d-Cauchy $\Longrightarrow$ d-convergent

reg-2) Let us say that T is *(uad-Ba-o-f,d)-continuous*, if [(z_n) = (uad-Ba-o-f) sequence and $z_n \xrightarrow{d} z$] imply $Tz_n \xrightarrow{d} Tz$

reg-3) Call $\mathcal{G}$, *(uad-Ba-o-f,d)-almost-selfclosed* when [(z_n) = (uad-Ba-o-f) sequence and $z_n \xrightarrow{d} z$] imply $(z_n; n \geq 0)\mathcal{G}\mathcal{G}z$.

4-IV) As a basic completion of the developments above, we list the specific metrical conditions upon our data. So, let $X_0 = (x_n)$ be a (uad-Ba-o) sequence in X. Denote $[X_0] = \{x_n; n \geq 0\}$, $[[X_0]] = \text{cl}([X_0])$; and define the couple of sequences $(r_n := r_n(X_0))$ and $(p_n := p_n(X_0))$, as

$$(r_n = d(x_n, x_{n+1}); n \geq 0),\ (p_n = d(x_n, x_{n+2}); n \geq 0);$$

clearly, (r_n) and (p_n) and sequences in R_+^0 and R_+ respectively, with

$(\forall n)$: $(|p_n - r_n| \leq r_{n+1})$, and the couple $(E_n = E_n(X_0), F_n = F_n(X_0))$, where $(E_n := (r_n, r_n, p_n, 0, r_{n+1}, r_{n+1})$; $F_n := (p_n, r_{n+1}, r_{n+1}, 0))$ satisfies $(E_n, F_n) = (\mathcal{Q}, \mathcal{K})(x_n, x_{n+1}) \in (\mathcal{Q}, \mathcal{K})(\mathcal{G}; (B_1 > 0); [X_0])$;

these will be referred to as the *telescopic sequences* attached to X_0.

(I) The first condition is of *asymptotic* type. Two variants of it appear.

I-a) Call Υ, *asymptotic* on $[X_0]$ when

(asy) whenever the telescopic sequences $(r_n := r_n(X_0))$ and $(p_n := p_n(X_0))$ satisfy $((E_n, F_n) \in \Upsilon, \forall n)$, then $r_n \to 0$; hence, $p_n \to 0$.

I-b) Call Υ, *descending asymptotic* on $[X_0]$ when

(desc-asy) whenever the telescopic sequences ($r_n := r_n(X_0)$) and ($p_n := p_n(X_0)$) attached to X_0 satisfy ($(E_n, F_n) \in \Upsilon$, $\forall n$), then (r_n) = strictly descending and $r_n \to 0$; hence (via $0 < r_n - r_{n+1} \le p_n \le r_n + r_{n+1}$, $\forall n$), (p_n) is a sequence in R^0_+ with (in addition) $p_n \to 0$.

Clearly, the inclusion below holds

Υ is descending asymptotic on $[X_0]$ implies Υ is asymptotic on $[X_0]$.

However, as we will see, the immense majority of contractive conditions is based on the (stronger) descending asymptotic hypothesis.

(II) The second condition to be considered is related to *right* properties. Take some point $(e, f) \in R^6_+ \times R^4_+$, where $e = (e_1, ..., e_6)$, $f = (f_1, ..., f_4)$. We say that the sequence $((t^n, s^n); n \ge 0)$ in $R^6_+ \times R^4_+$, where $(t^n := (t^n_1, ..., t^n_6); n \ge 0)$, $(s^n := (s^n_1, ..., s^n_4); n \ge 0)$, is *right* at (e, f), if

(right-seq) $(t^n_i \to e_i+,\ s^n_j \to f_j+)$ as $n \to \infty$, $\forall i$, $\forall j$.

Given $b > 0$, let us say that Υ is *nright* at b on $[X_0]$, if

(nright-loc) $\forall$ sequence $((t^n, s^n); n \ge 0)$ in $(\mathcal{Q}, \mathcal{K})(\mathcal{G}; B_3 > 0; [X_0]) \cap \Upsilon$, the right property at $((0, b, b, b, b, 0); (0, 0, b, b))$ is not true;

the class of all these is denoted as $\text{nright}(\Upsilon; [X_0])$. In this case, we say that

(a-nright) Υ is *almost nright* on $[X_0]$, if $\Theta := \text{nright}(\Upsilon; [X_0])$ is $(>)$-cofinal in R^0_+ (for each $\varepsilon \in R^0_+$ there exists $\theta \in \Theta$ with $\varepsilon > \theta$)
(nright) Υ is *nright* on $[X_0]$, if $\text{nright}(\Upsilon; [X_0])$ is identical with R^0_+.

(III) The third condition involves *point* properties. Take some point $(e, f) \in R^6_+ \times R^4_+$, where $e = (e_1, ..., e_6)$, $f = (f_1, ..., f_4)$. We say that the sequence $((t^n, s^n); n \ge 0)$ in $R^6_+ \times R^4_+$, where $(t^n := (t^n_1, ..., t^n_6); n \ge 0)$, $(s^n := (s^n_1, ..., s^n_4); n \ge 0)$, is *point* at (e, f), if

(point-seq) $(t^n_i \to e_i,\ s^n_j \to f_j,\ \forall i,\ \forall j)$ and $[t^n_6 = e_6,\ \forall n]$.

Given $b > 0$, let us say that Υ is *npoint* at b on $[[X_0]]$, if

(npoint-loc) $\forall$ sequence $((t^n, s^n); n \ge 0)$ in $(\mathcal{Q}, \mathcal{K})(\mathcal{G}; B_3 > 0; [[X_0]]) \cap \Upsilon$, the point property at $((0, 0, b, 0, b, b); (0, 0, 0, b))$ is not true;

the class of all these will be denoted as $\text{npoint}(\Upsilon; [[X_0]])$. Then, define

(npoint) Υ is *npoint* on $[[X_0]]$, if $\text{npoint}(\Upsilon; [[X_0]])$ is identical with R^0_+.

(IV) The fourth condition is *n-normality*. Given $a > 0$, we say that

(nnorm-loc) Υ is *nnormal* at a, if $((0, a, a, a, a, 0); (0, 0, a, a)) \notin \Upsilon$.

The class of all nnormal $a > 0$ will be denoted as nnorm(Υ). Then, define

(nnormal) Υ is *nnormal*, if nnorm(Υ) is identical with R^0_+.

Some concrete examples of such objects will be given a bit further.

2.1.5 *Main result*

Let X be a nonempty set, $d : X \times X \to R_+$ be a metric on X, and $\mathcal{G}$ be a relation over X; the system $(X, d, \mathcal{G})$ will be referred to as a *relational metric space*. Further, take some selfmap $T \in \mathcal{F}(X)$, subject to the condition

(uad-a-reg) T is *uadmissible ascending regular*: uadao(T) is nonempty.

As a consequence, there exist ($\mathcal{G}$-uadmissible, $\mathcal{G}$-ascending) orbital sequences (in short: (uad-a-o) sequences) in X. Let $X_0 = (x_n)$ be one of these; remember that, without loss, one may assume that

X_0 is $(B_1 > 0)$-ascending; hence, a (uad-Ba-o) sequence.

Finally, denote $[X_0] = \{x_n; n \geq 0\}$, $[[X_0]] = \text{cl}([X_0])$.

The basic directions and regularity conditions relative to X_0 under which the problem of determining fixed points of T is to be solved were already listed; and the contractive type framework (involving X_0, $[X_0]$ and $[[X_0]]$) was settled. We may now state our main result in this exposition.

Theorem 2.1. *Assume that the selfmap T is $(\mathcal{G}; \mathcal{R}; \Upsilon)$-contractive, where the subset $\Upsilon \in \exp(R^6_+ \times R^4_+)$ and the relation $\mathcal{R}$ over X are introduced under the precise conventions, in conjunction with*

(51-i) $\mathcal{R}$ is B_1-admissible: $(B_1 > 0) \subseteq \mathcal{R}$
(51-ii) X is (uad-Ba-o-f,d)-complete.

Then, the following are valid

(51-a) *$X_0 = (x_n)$ is a (uad-Ba-o-f) sequence with the semi Picard property (modulo $(\mathcal{G}; \mathcal{R}; T)$) whenever Υ is descending asymptotic on $[X_0]$*

(51-b) *$X_0 = (x_n)$ is Picard (modulo $(\mathcal{G}; \mathcal{R}; T)$), if (in addition to the above) Υ is almost nright on $[X_0]$*

(51-c) *$X_0 = (x_n)$ is strongly Picard (modulo $(\mathcal{G}; \mathcal{R}; T)$) whenever (in addition to (51-a)+(51-b)) T is (uad-Ba-o-f,d)-continuous*

(51-d) $X_0 = (x_n)$ *is semi Bellman Picard (modulo* $(\mathcal{G};\mathcal{R};T)$*), whenever (in addition to the setting of (51-a)+(51-b))* $\mathcal{G}$ *is (uad-Ba-o-f,d)-almost-selfclosed and* Υ *is npoint on* $[[X_0]]$

(51-e) T *is fix-*$\mathcal{G}$*-asingleton (hence, fix-*$\mathcal{G}$*-singleton, under these extra requirements) when, in addition* $\mathcal{R}$ *is* B_2*-admissible and* Υ *is nnormal.*

Proof. There are several steps to be passed.

Step 1. Let $(r_n := r_n(X_0))$ and $(p_n := p_n(X_0))$ be the telescopic sequences attached to the (uad-Ba-o) sequence $X_0 = (x_n)$:

$$(r_n = d(x_n, x_{n+1}); n \geq 0),\ (p_n = d(x_n, x_{n+2}); n \geq 0);$$

as precise, these enjoy the property $(|p_n - r_n| \leq r_{n+1},\ \forall n)$ and

$(\forall n)$: $E_n := (r_n, r_n, p_n, 0, r_{n+1}, r_{n+1})$ and $F_n := (p_n, r_{n+1}, r_{n+1}, 0)$ satisfy $(E_n, F_n) = (\mathcal{Q}, \mathcal{K})(x_n, x_{n+1}) \in (\mathcal{Q}, \mathcal{K})(\mathcal{G}; (B_1 > 0); [X_0])$.

On the other hand, by the contractive property,

$$(\mathcal{Q}, \mathcal{K})(\mathcal{G}; (B_1 > 0); [X_0]) \subseteq \Upsilon; \text{ hence, } ((E_n, F_n) \in \Upsilon,\ \forall n).$$

By the descending asymptotic property of Υ at $[X_0]$, one derives

(sd-asy-1) (r_n) is strictly descending in R^0_+ with $r_n \to 0$ as $n \to \infty$
(sd-asy-2) (p_n) is a sequence in R^0_+ with $p_n \to 0$ as $n \to \infty$.

By the convergence part of the first relation, $X_0 = (x_n)$ is semi Picard (modulo $(\mathcal{G};\mathcal{R};T)$). On the other hand, by the strict descending part of the same, $X_0 = (x_n)$ is a full sequence. In fact, suppose that $x_i = x_j$, for some $i, j \in N$ with $i < j$. This, by the iterative construction of our sequence, gives $x_{i+1} = x_{j+1}$; whence, $r_i = r_j$; a contradiction.

Step 2. Summing up, $X_0 = (x_n)$ is (uad-Ba-o-f) and d-asymptotic. Moreover, as Υ is almost nright, $\Theta := \text{nright}(\Upsilon; [X_0])$ is $(>)$-cofinal in R^0_+. Finally, by the very choice of our sequence, there exists an index $h \geq 1$, such that X_0 is $(\mathcal{G}, h)$-uadmissible. We show that, under these conditions,

$$X_0 = (x_n) \text{ is } d\text{-Cauchy; hence, Picard (modulo } (\mathcal{G}, \mathcal{R}; T)).$$

Suppose that this is not true. As a consequence,

$$X_0 = (x_n) \text{ is } d\text{-asymptotic, } (\mathcal{G}, h)\text{-uadmissible, and not } d\text{-Cauchy.}$$

By a preliminary fact, there exist a number $\gamma \in \Theta$, a rank $J := J(\gamma, h) \geq 1$, and a couple of rank-sequences $(m(k); k \geq 0)$, $(n(k); k \geq 0)$, with

(p-1) $(\forall k)$: $J + k \leq m(k) < m(k) + 3h < n(k)$, and $x_{m(k)} \mathcal{G} x_{n(k)}$

(p-2) $\forall s,t \in N[0,3h]$: $V_k(s,t) := d(x_{m(k)+s}, x_{n(k)+t}) \to \gamma+$ as $k \to \infty$.

By the d-asymptotic and full properties of $X_0 = (x_n)$,

$(t_1^k := r_{m(k)}; k \geq 0)$, $(t_6^k := r_{n(k)}; k \geq 0)$,
$(s_1^k := p_{m(k)}; k \geq 0)$, $(s_2^k := r_{m(k)+1}; k \geq 0)$
are sequences in R_+^0 with $t_1^k, t_6^k, s_1^k, s_2^k \to 0$ as $k \to \infty$.

Moreover, taking (p-2) into account, yields

$(t_2^k := V_k(0,0); k \geq 0)$, $(t_3^k := V_k(0,1); k \geq 0)$, $(t_4^k := V_k(1,0); k \geq 0)$,
$(t_5^k := V_k(1,1); k \geq 0)$ $(s_3^k := V_k(2,0); k \geq 0)$, $(s_4^k := V_k(2,1); k \geq 0)$,
are sequences in R_+^0 with $t_2^k, t_3^k, t_4^k, t_5^k, s_3^k, s_4^k \to \gamma+$ as $k \to \infty$;

hence, putting these together,

(ts-right) the vectorial sequence $((t^k, s^k); k \geq 0)$, where
$(t^k = (t_1^k, ..., t_6^k); k \geq 0)$, $(s^k = (s_1^k, ..., s_4^k); k \geq 0)$,
is right at $((0,\gamma,\gamma,\gamma,\gamma,0); (0,0,\gamma,\gamma))$.

Concerning the localization of the object $((t^k, s^k); k \geq 0)$, note that by (p-1) and the full property of $X_0 = (x_n)$,

$(\forall k)$: $x_{m(k)} \mathcal{G} x_{n(k)}$, and
$B_3(x_{m(k)}, x_{n(k)}) = \min\{t_1^k, ..., t_6^k; s_1^k, ..., s_4^k\} > 0$; so that,
$(t^k, s^k) = (\mathcal{Q},\mathcal{K})(x_{m(k)}, x_{n(k)}) \in (\mathcal{Q},\mathcal{K})(\mathcal{G}; (B_3 > 0); [X_0]) = (\mathcal{Q},\mathcal{K})(\mathcal{G}; (B_3 > 0); [X_0]) \cap \Upsilon$;

where the last equality is a consequence of contractive property. This, via (ts-right), contradicts the choice of γ as element of $\Theta := \text{nright}(\Upsilon; [X_0])$. Hence, our working assumption is not acceptable; and the assertion follows.

Step 3. From these developments, $X_0 = (x_n)$ is (uad-Ba-o-f) and d-Cauchy. As X is (uad-Ba-o-f,d)-complete, we must have

X_0 is d-convergent: $x_n \xrightarrow{d} z_0$ as $n \to \infty$, for some $z_0 \in X$.

Two possible cases – treated in the steps below – are to be discussed.

Step 4. Suppose that T is (uad-Ba-o-f,d)-continuous at X_0. Then, $y_n := Tx_n \xrightarrow{d} Tz_0$ as $n \to \infty$. On the other hand, $(y_n = x_{n+1}; n \geq 0)$ is a subsequence of $(x_n; n \geq 0)$; whence $y_n \xrightarrow{d} z_0$ as $n \to \infty$. Combining with d = separated, yields $z_0 = Tz_0$; that is, $z_0 \in \text{Fix}(T)$.

Step 5. Suppose that $\mathcal{G}$ is (uad-Ba-o-f,d)-almost-selfclosed and Υ is npoint on $[[X_0]]$. By the full property of $X_0 = (x_n)$,

$H_1 := \{n \in N; Tx_n = Tz_0\}$, $H_2 := \{n \in N; x_n = Tz_0\}$, are asingletons;

and this tells us that there exists $i = i(z_0) \in N$, such that:

$m \geq i$ implies $Tx_m \neq Tz_0$ (hence, $x_m \neq z_0$) and $x_m \neq Tz_0$.

Denote for simplicity $(u_n = x_{n+i}; n \geq 0)$; clearly, by the preceding relation,

$(\forall n)$: $Tu_n \neq Tz_0$ (hence, $u_n \neq z_0$) and $u_n \neq Tz_0$.

Suppose by contradiction that

(posi-1) $b := d(z_0, Tz_0) > 0$ [whence, $Q_6(u_n, z_0) = b > 0$, $\forall n$].

From the preceding relations and the full property of $X_0 = (x_n)$,

(posi-2) $(Q_1^n := Q_1(u_n, z_0); n \geq 0)$, $(Q_2^n := Q_2(u_n, z_0); n \geq 0)$,
$(Q_3^n := Q_3(u_n, z_0); n \geq 0)$, $(Q_4^n := Q_4(u_n, z_0); n \geq 0)$,
$(Q_5^n := Q_5(u_n, z_0); n \geq 0)$, $(Q_6^n := Q_6(u_n, z_0); n \geq 0)$,
$(K_1^n := K_1(u_n, z_0); n \geq 0)$, $(K_2^n := K_2(u_n, z_0); n \geq 0)$,
$(K_3^n := K_3(u_n, z_0); n \geq 0)$, $(K_4^n := K_4(u_n, z_0); n \geq 0)$
are sequences in R_+^0 with $(Q_1^n, Q_2^n, Q_4^n \to 0)$, $(Q_3^n, Q_5^n \to b)$,
$(Q_6^n = b, \forall n)$, and $(K_1^n, K_2^n, K_3^n \to 0)$, $(K_4^n \to b)$.

As $\mathcal{G}$ is (uad-Ba-o-f,d)-almost-selfclosed, we have $(u_n; n \geq 0)\mathcal{G}\mathcal{G}z_0$. This, by definition (and a small arrangement in the rank sequence) tells us that

there exists a strictly ascending sequence of ranks $(j(n); n \geq 0)$, such that the subsequence $(v_n = u_{j(n)}; n \geq 0)$ of (u_n) fulfills $(v_n; n \geq 0)\mathcal{G}z_0$;

in addition since $(j(n))$ is strictly ascending, we must have $(j(n) \geq n, \forall n)$. Combining with the preceding relations, the full property of $X_0 = (x_n)$, and the metrical properties of $d(.,.)$, gives

(posi-3) $(q_1^n := Q_1(v_n, z); n \geq 0)$, $(q_2^n := Q_2(v_n, z); n \geq 0)$,
$(q_3^n := Q_3(v_n, z); n \geq 0)$, $(q_4^n := Q_4(v_n, z); n \geq 0)$,
$(q_5^n := Q_5(v_n, z); n \geq 0)$, $(q_6^n := Q_6(v_n, z) = b; n \geq 0)$,
$(k_1^n := K_1(v_n, z_0); n \geq 0)$, $(k_2^n := K_2(v_n, z_0); n \geq 0)$,
$(k_3^n := K_3(v_n, z_0); n \geq 0)$, $(k_4^n := K_4(v_n, z_0); n \geq 0)$
are sequences in R_+^0 with $(q_1^n, q_2^n, q_4^n \to 0)$, $(q_3^n, q_5^n \to b)$,
$(q_6^n = b, \forall n)$, and $(k_1^n, k_2^n, k_3^n \to 0)$, $(k_4^n \to b)$;

hence, putting these together,

(ts-point) the vectorial sequence $((q^n, k^n); n \geq 0)$, where $(q^n = (q_1^n, ..., q_6^n); n \geq 0)$, $(k^n = (k_1^n, ..., k_4^n); n \geq 0)$, is point at $((0, 0, b, 0, b, b); (0, 0, 0, b))$.

Concerning the localization of the same object $((q^n, k^n); n \geq 0)$, note that by the very choice of the subsequence $(v_n; n \geq 0)$ and the preceding relations

$(\forall n)$: $v_n \mathcal{G} z_0$, and $B_3(v_n, z_0) = \min\{q_1^n, ..., q_6^n; k_1^n, ..., k_4^n\} > 0$; whence,
$(q^n, k^n) = (\mathcal{Q}, \mathcal{K})(v_n, z_0) \in (\mathcal{Q}, \mathcal{K})(\mathcal{G}; (B_3 > 0); [[X_0]]) =$
$(\mathcal{Q}, \mathcal{K})(\mathcal{G}; (B_3 > 0); [[X_0]]) \cap \Upsilon$;

where the last inclusion is a consequence of contractive property. This, via (ts-point), contradicts the choice of b as element of $\Theta :=$ npoint$(\Upsilon; [[X_0]])$. Hence, our working assumption $b > 0$ is not acceptable; and then, $b = 0$; that is: $z_0 \in \text{Fix}(T)$.

Step 6. Letting $z_1, z_2 \in \text{Fix}(T)$, with $z_1 \mathcal{G} z_2$, assume that $z_1 \neq z_2$; hence, $a := d(z_1, z_2) > 0$. Then,

$B_2(z_1, z_2) = d(z_1, z_2) = a > 0$; hence, $(z_1, z_2) \in \mathcal{R}$
(as $\mathcal{R}$ is B_2-admissible).

The contractive condition applies to (z_1, z_2); and gives $(\mathcal{Q}, \mathcal{K})(z_1, z_2) \in \Upsilon$; wherefrom $((0, a, a, a, a, 0); (0, 0, a, a)) \in \Upsilon$; a contradiction with the nnormal property of Υ. Hence, $z_1 = z_2$; as desired. □

The obtained result extends some rational type fixed point statements in the area, including the ones in Cabrera et al. [44], Chouhan et al. [57], Dass and Gupta [70], Kumam et al. [153], or Saluja et al. [205], obtained via direct methods. On the other hand, further extensions of our main result are possible under the lines in Abtahi [4]. Finally, multivalued versions of Theorem 2.1 are possible, under the lines in Nadler [175]. Further aspects will be discussed elsewhere.

2.1.6 *Analytic methods in KSRDC-theory*

In the following, as a by-product of our main result, some analytic methods in the Rhoades contractions theory are being developed.

Let $(X, d, \mathcal{G})$ be a relational metric space; and $T \in \mathcal{F}(X)$ be such that

(uad-a-reg) T is *uadmissible ascending regular*: uadao(T) is nonempty.

As a consequence, there exist uadmissible (a-o)-sequences in X. Let $X_0 = (x_n)$ be such an object; remember that, without loss, one may assume that

$B_1(x_n, x_{n+1}) > 0$, $\forall n$; whence: $X_0 = (x_n)$ is a (uad-Ba-o) sequence.

As precise, we are interested to determine sufficient conditions for existence and uniqueness of elements in Fix(T), in a contractive type setting. The basic directions and regularity conditions relative to X_0 under which the problem of determining fixed points of T is to be solved were already listed; and the contractive type framework (involving $X_0 = (x_n; n \geq 0)$, $[X_0] = \{x_n; n \geq 0\}$, and $[[X_0]] = \mathrm{cl}([X_0])$ was settled. As a by-product of these, we established our main result, Theorem 2.1. It is our aim in the following to show that, starting from this principle, it is possible to sketch the basic lines for a kind of *analytical approach* in treating fixed points of certain contractions on relational metric spaces, extending the ones introduced by Krasnoselskii and Stetsenko [152], re-discovered by Rhoades [197], and refined by Dutta and Choudhury [84].

(A) Roughly speaking, the approach to be considered requires an appropriate description of contractive conditions. For each $x, y \in X$ define the non-anticipative and anticipative functions

(n-a-f) $Q_1(x,y) = d(x,Tx)$, $Q_2(x,y) = d(x,y)$, $Q_3(x,y) = d(x,Ty)$,
$Q_4(x,y) = d(Tx,y)$, $Q_5(x,y) = d(Tx,Ty)$, $Q_6(x,y) = d(y,Ty)$,
$\mathcal{Q}(x,y) = (Q_1(x,y), Q_2(x,y), Q_3(x,y), Q_4(x,y), Q_5(x,y), Q_6(x,y))$;
(a-f) $K_1(x,y) = d(x,T^2x)$, $K_2(x,y) = d(Tx,T^2x)$,
$K_3(x,y) = d(T^2x,y)$, $K_4(x,y) = d(T^2x,Ty)$,
$\mathcal{K}(x,y) = (K_1(x,y), K_2(x,y), K_3(x,y), K_4(x,y))$.

Then, let us construct the family of functions [for $x, y \in X$]

$$B_0(x,y) = \min\{Q_2, Q_5\}(x,y),$$
$$B_1(x,y) = \min\{Q_1, Q_2, Q_5, Q_6; K_2, K_3\}(x,y),$$
$$B_2(x,y) = \min\{Q_2, Q_3, Q_4, Q_5, K_3, K_4\}(x,y),$$
$$B_3(x,y) = \min\{Q_1, Q_2, Q_3, Q_4, Q_5, Q_6; K_1, K_2, K_3, K_4\}(x,y);$$

and put, for simplicity,

$$(B_i > 0) = \{(x,y) \in X \times X; B_i(x,y) > 0\},\ i \in \{0,1,2,3\}.$$

Let (ψ, φ) be a couple of functions over $\mathcal{F}(R)$. Further, letting (u, v, w) be a triple of functions over $\mathcal{F}(R^6_+ \times R^4_+, R_+)$, define the subsets

$$\Upsilon_0 = \{(t,s) \in R^6_+ \times R^4_+; u(t,s), v(t,s), w(t,s) > 0\},$$
$$\Upsilon = \{(t,s) \in \Upsilon_0; \psi(u(t,s)) \leq \psi(v(t,s)) - \varphi(w(t,s))\}.$$

Finally, define the couple of relations over X

$\mathcal{R} = (\mathcal{Q}, \mathcal{K})^{-1}(\Upsilon_0)$, $\mathcal{S} = (\mathcal{Q}, \mathcal{K})^{-1}(\Upsilon)$.

Under this setting, we may introduce two contractive conditions upon our data. The former of these is a functional contractive condition, written as:

(fct-contr) T is $(\mathcal{G}, \mathcal{R}; \psi, \varphi; u, v, w)$-*contractive*, if
$\psi(u \circ (\mathcal{Q}, \mathcal{K})(x, y)) \leq \psi(v \circ (\mathcal{Q}, \mathcal{K})(x, y)) - \varphi(w \circ (\mathcal{Q}, \mathcal{K})(x, y))$,
when $(x, y) \in (\mathcal{G}; \mathcal{R})$.

The latter of these appears as a set contractive condition and writes

(set-contr) T is $(\mathcal{G}, \mathcal{R}; \Upsilon)$-contractive:
$(\mathcal{Q}, \mathcal{K})(x, y) \in \Upsilon$, if $(x, y) \in (\mathcal{G}; \mathcal{R})$.

The connection between these is described in

Proposition 2.5. *Under the above developments, one has*

T *is* $(\mathcal{G}, \mathcal{R}; \psi, \varphi; u, v, w)$-*contractive iff* T *is* $(\mathcal{G}, \mathcal{R}; \Upsilon)$-*contractive.*

Proof. i): Suppose that (fct-contr) is holding; and let the couple $(x, y) \in (\mathcal{G}; \mathcal{R})$ be arbitrary fixed. For the moment,

$(x, y) \in (\mathcal{G}; \mathcal{R})$ implies $(t, s) := (\mathcal{Q}, \mathcal{K})(x, y) \in \Upsilon_0$
(that is: $u(t, s) > 0$, $v(t, s) > 0$, $w(t, s) > 0$).

Moreover, by the very concept of function contractive property,

$$\psi(u(t, s)) \leq \psi(v(t, s)) - \varphi(w(t, s)); \text{ that is: } (t, s) := (\mathcal{Q}, \mathcal{K})(x, y) \in \Upsilon.$$

This, by the arbitrariness of $(x, y) \in (\mathcal{G}; \mathcal{R})$, shows that the set contractive property (set-contr) holds; and the left to right inclusion follows.

ii): Suppose that the set contractive property (set-contr) is holding; and let the couple $(x, y) \in (\mathcal{G}; \mathcal{R})$ be arbitrary fixed. For the moment,

$$(t, s) := (\mathcal{Q}, \mathcal{K})(x, y) \in \Upsilon_0; \text{ that is: } u(t, s) > 0,\ v(t, s) > 0,\ w(t, s) > 0.$$

Moreover, note that we actually have

$$(t, s) := (\mathcal{Q}, \mathcal{K})(x, y) \in \Upsilon; \text{ whence, } \psi(u(t, s)) \leq \psi(v(t, s)) - \psi(w(t, s)).$$

This tells us that the couple (x, y) fulfills the functional contractive property (fct-contr); and proves the right to left inclusion. □

In other words, the functional contractive condition (fct-contr) is equivalent with the set-contractive condition (set-contr); to which the methods of the main result are applicable. Technically speaking, an effective application of the main result to the set contractive problem (set-contr) amounts to verifying the asymptotic, right, point, and normality properties upon Υ, in terms of the precise data: the pair of relations $(\mathcal{G}, \mathcal{R})$, the couple (ψ, φ) and the triple (u, v, w). Some preliminaries are needed.

(B) Let $\psi \in \mathcal{F}(R)$ and $s \in R$ be given. Define the *right superior/inferior limit* of ψ at s, as

$$\Lambda^+\psi(s) = \inf_{\varepsilon>0} \Psi^+(s+)(\varepsilon); \text{ where } \Psi^+(s+)(\varepsilon) = \sup \psi(]s, s+\varepsilon[)$$
$$\Lambda_+\psi(s) = \sup_{\varepsilon>0} \Psi_+(s+)(\varepsilon); \text{ where } \Psi_+(s+)(\varepsilon) = \inf \psi(]s, s+\varepsilon[);$$

and the *bilateral superior/inferior limit* of ψ at s, as

$$\Lambda^{\pm}\psi(s) = \inf_{\varepsilon>0} \Psi^{\pm}(s\pm)(\varepsilon); \text{ where } \Psi^{\pm}(s\pm)(\varepsilon) = \sup \psi(]s-\varepsilon, s+\varepsilon[)$$
$$\Lambda_{\pm}\psi(s) = \sup_{\varepsilon>0} \Psi_{\pm}(s\pm)(\varepsilon); \text{ where } \Psi_{\pm}(s\pm)(\varepsilon) = \inf \psi(]s-\varepsilon, s+\varepsilon[).$$

The following auxiliary fact will be useful.

Proposition 2.6. *Let $\psi \in \mathcal{F}(R)$ be given. Then, for each $s \in R$,*

(62-1) $\limsup_n(\psi(t_n)) \leq \Lambda^+\psi(s)$,
for each sequence (t_n) in R with $t_n \to s+$
(62-2) there exists a sequence (r_n) in R with
$r_n \to s+$ and $\psi(r_n) \to \Lambda^+\psi(s)$
(62-3) $\limsup_n(\psi(t_n)) \leq \Lambda^{\pm}\psi(s)$,
for each sequence (t_n) in R with $t_n \to s$
(62-4) there exists a sequence (r_n) in R with
$r_n \to s$ and $\psi(r_n) \to \Lambda^{\pm}\psi(s)$.

The verification is very similar with the one described in our preliminary section (involving admissible functions); so, we do not give further details.

(C) Having these precise, we introduce two basic concepts. Let $\psi \in \mathcal{F}(R)$ be a function; and $b, c \in R_+$, be a couple of points. Define the *right and point oscillation* of ψ at (b, c), as

(osc-r) $\text{osc}(+)(\psi; b, c) =$ the supremum of all $\limsup_n(\psi(t_n) - \psi(s_n))$ where (t_n), (s_n), are sequences in R_+^0 with $t_n \to c+$, $s_n \to b+$
(osc-p) $\text{osc}(\psi; b, c) =$ the supremum of all $\limsup_n(\psi(t_n) - \psi(s_n))$ where (t_n), (s_n), are sequences in R_+^0 with $t_n \to c$, $s_n \to b$.

A characterization of these oscillations is provided in the statement below.

Proposition 2.7. *For each $\psi \in \mathcal{F}(R)$ and each $b, c \in R_+$, we have (when the right member exists)*

(63-1) $\text{osc}(+)(\psi; b, c) = \Lambda^+\psi(c) - \Lambda_+\psi(b)$,
(63-2) $\text{osc}(\psi; b, c) \leq \Lambda^\pm\psi(c) - \Lambda_\pm\psi(b)$.

Proof. There are two parts to be verified.

(I) Let (t_n), (s_n) be sequences in R_+^0 with $t_n \to c+$, $s_n \to b+$. By a previous auxiliary fact involving extremal limits of real sequences,

$$\limsup_n[\psi(t_n) - \psi(s_n)] \leq$$
$$\limsup_n \psi(t_n) - \liminf_n \psi(s_n) \leq \Lambda^+\psi(c) - \Lambda_+\psi(b);$$

and this gives the left to right inequality. On the other hand, by the same fact, there are two sequences (u_n), (v_n) in R_+^0, with

$$v_n \to c+,\ \Lambda^+\psi(c) = \lim_n \psi(v_n);\ \text{as well as}$$
$$u_n \to b+,\ \Lambda^+(-\psi)(b) = \lim_n(-\psi(u_n));\ \text{so, } \Lambda_+\psi(b) = \lim_n \psi(u_n).$$

Adding these relations yields

$$\Lambda^+\psi(c) - \Lambda_+\psi(b) = \lim_n[\psi(v_n) - \psi(u_n)] \leq \text{osc}(+)(\psi; b, c);$$

and this gives the right to left inequality.

(II) Let (t_n), (s_n) be sequences in R_+^0 with $t_n \to c$, $s_n \to b$. By a previous auxiliary fact involving extremal limits of real sequences,

$$\limsup_n[\psi(t_n) - \psi(s_n)] \leq$$
$$\limsup_n \psi(t_n) - \liminf_n \psi(s_n) \leq \Lambda^\pm\psi(c) - \Lambda_\pm\psi(b);$$

and, from this, all is clear. □

Remark 2.1.

The inequality of the second part may be strict when $0 \in \{b, c\}$. To verify this, let $\psi : R \to R$ be defined as

$$\psi(t) = -3,\ t \leq 0;\ \psi(t) = 1,\ t > 0;$$

and let the couple (b, c) be identical with $(0, 2)$. Then, by definition,

$$\text{osc}(\psi; 0, 2) = 0,\ \Lambda^\pm\psi(2) = 1,\ \Lambda_\pm\psi(0) = -3;$$

and this ends the argument.

A simplified form of these representations is available under special regularity assumptions about ψ. Some preliminaries are needed.

For each function $\psi \in \mathcal{F}(R)$ and each $a \in R$, define

(rr-a) ψ is *right regulated* at a: $\psi(a+0) = \lim_{t\to a+} \psi(t)$
(the *right limit* of ψ at a) exists (in $R \cup \{-\infty, \infty\}$)
(lr-a) ψ is *left regulated* at a: $\psi(a-0) = \lim_{t\to a-} \psi(t)$
(the *left* limit of ψ at a) exists (in $R \cup \{-\infty, \infty\}$)
(r-a) ψ is *regulated* at a:
$\psi(a+0)$ and $\psi(a-0)$ exist (in $R \cup \{-\infty, \infty\}$)
(cont-a) ψ is *continuous* at a:
ψ is regulated at a and $\psi(a+0) = \psi(a-0) = \psi(a)$.

The class of all such points will be indicated, respectively, as

rreg(ψ) = the right regulated domain of ψ,
lreg(ψ) = the left regulated domain of ψ,
reg(ψ) = the regulated domain of ψ,
cont(ψ) = the continuous domain of ψ.

Proposition 2.8. *Given $\psi \in \mathcal{F}(R)$ and $b \in R$, we have*

(64-1) $\Lambda^+\psi(b) = \psi(b+0)$, *whenever* $b \in \text{rreg}(\psi)$
(64-2) $\Lambda^\pm\psi(b) = \max\{\psi(b+0), \psi(b), \psi(b-0)\}$, *if* $b \in \text{reg}(\psi)$.

Proof. i): By a previous observation, there exists a sequence (s_n) in R with $s_n \to b+$ and $\psi(s_n) \to \Lambda^+\psi(b)$. On the other hand, by the right regulated property, $\psi(s_n) \to \psi(b+0)$. Combining these, we are done.

ii): Let (t_n) be a sequence in R_+^0 with $t_n \to b$. By a previous observation,

$$\limsup_n \psi(t_n) \le \Lambda^\pm\psi(b); \text{ so, } \max\{\psi(b+0), \psi(b), \psi(b-0)\} \le \Lambda^\pm\psi(b);$$

by passing with (t_n) through the (sub-sequential type) alternatives

$$(t_n > b,\ \forall n),\ (t_n = b,\ \forall n),\ (t_n < b,\ \forall n).$$

In addition, there exists a sequence (r_n) in R with $r_n \to b$ and $\psi(r_n) \to \Lambda^\pm\psi(b)$. This, via $\{n \in N; r_n \nabla b\}$ = infinite for some $\nabla \in \{>, =, <\}$, gives $\Lambda^\pm\psi(b) \in \{\psi(b+0), \psi(b), \psi(b-0)\}$; wherefrom, all is clear. □

(D) In the following, some technical fact are developed for a class of binary operations to be used further.

Let $(a, b) \mapsto a * b$ be some operation over R_+; we call it *semi regular*, provided the following properties are valid

(sreg-1) $(a \le a * b)$ and $(c * c = c)$, $\forall a, b, c \in R_+$
(sreg-2) $(a, b) \mapsto a * b$ is continuous in its arguments.

Note that, by the former condition,

(sreg-3) $a, b > 0$ implies $a * b > 0$; so, $(*)$ is a binary operation on R_+^0.

Given such an object, call it

(reg) *regular*, when ($b < a * b$ implies $b < a$);
or, equivalently: ($a \leq b$ implies $a * b \leq b$)
(s-reg) *strongly regular*, when ($b \neq a * b$ implies $b < a$);
or, equivalently: ($a \leq b$ implies $a * b = b$).

A basic example of (strongly regular) operation is

$a \top b = \max\{a, b\} = a + \max\{0, b - a\}$, $a, b \in R_+$.

Some other examples are provided by the statement below. Let $\mathcal{F}(ci)(R_+)$ stand for the class of all continuous increasing $f \in \mathcal{F}(R_+)$. Given $f \in \mathcal{F}(ci)(R_+)$, define the binary operations

(op-1) $a(f)b = a + \max\{0, f(b) - f(a)\}$, $a, b \in R_+$
(op-2) $a[f]b = a + |f(b) - f(a)|$, $a, b \in R_+$.

Clearly, f and $f - \alpha$ (where $0 \leq \alpha \leq f(0)$) generate the same binary operations; so, without loss, one may assume that $f(0) = 0$. Let $\mathcal{F}_0(ci)(R_+)$ stand for the subclass of all $f \in \mathcal{F}(ci)(R_+)$ with this last property. For example, $I(.)$ (the unitary function of $\mathcal{F}(R_+)$) is an element of it.

Proposition 2.9. *Let $f \in \mathcal{F}_0(ci)(R_+)$ be given. Then,*

(65-1) *the binary operation (f) is regular when $I - f$ is increasing*

(65-2) *the binary operation $[f]$ is regular whenever $I - f$ is increasing*

(65-3) *the binary operation (f) is strongly regular iff $f = I$; moreover, in this case, $a \top b = a(f)b$, $\forall a, b \in R_+$.*

(65-4) *the binary operation $[f]$ is strongly regular iff $f = I$; moreover, in this case, $a \top b \leq a[f]b$, $\forall a, b \in R_+$.*

Proof. Clearly, both (f) and $[f]$ are semi regular.

i): Let $a, b \in R_+$ be such that $a \leq b$; hence, $f(b) \geq f(a)$; wherefrom $a(f)b = a + f(b) - f(a)$. The desired relation $a(f)b \leq b$ becomes $a - f(a) \leq b - f(b)$; evident, via $I - f$ = increasing.

ii): Let $a, b \in R_+$ be such that $a \leq b$; hence, $f(b) \geq f(a)$; wherefrom $a[f]b = a + f(b) - f(a)$. The desired relation $a[f]b \leq b$ becomes $a - f(a) \leq b - f(b)$; evident, via $I - f$ = increasing.

iii): Let $a, b \in R_+$ be such that $a \leq b$; hence, $f(b) \geq f(a)$; wherefrom $a(f)b = a + f(b) - f(a)$. The desired relation $a(f)b = b$ becomes: $a - f(a) =$

$b - f(b)$ (for all such (a, b)). But then (combining with $f(0) = 0$), $f = I$; and the assertion follows.

iv): Let $a, b \in R_+$ be such that $a \le b$; hence, $f(b) \ge f(a)$; wherefrom $a[f]b = a + f(b) - f(a)$. The desired relation $a[f]b = b$ becomes: $a - f(a) = b - f(b)$ (for all such (a, b)). But then (combining with $f(0) = 0$), $f = I$; and the assertion follows. □

Finally, one may ask if these operations are the only ones endowed with (strongly) regular properties. We conjecture that the answer is negative.

(E) The following completion of these will be useful. Let $L : R_+ \times R_+ \to R_+$ and $\lambda : R_+ \to R_+$ be a couple of maps, with

(posi-cp) (L, λ) is *positive*: $L(R_+^0 \times R_+^0) \subseteq R_+^0$, $\lambda(R_+^0) \subseteq R_+^0$.

Define (under this setting) the concept

(r-comp) (L, λ) is *right compatible*: for each sequence (r_n) in R_+^0 and each $r \in R_+^0$ with $r_n \to r+$ we have $L(r_n, r_{n+1}) \to \lambda(r)+$.

The following compatibility criterion is available. Denote, for $\alpha, \beta \in R$,

co(α, β)=the convex cover of $\{\alpha, \beta\}$;
that is: the interval $[\min\{\alpha, \beta\}, \max\{\alpha, \beta\}]$.

Proposition 2.10. *Let the positive couple (L, λ) be such that*

(66-I) L is λ-subordinated: $L(t, s) \in \text{co}(\lambda(t), \lambda(s))$, for each $t, s \in R_+^0$
(66-II) $\lambda(.)$ is continuous strictly increasing on R_+^0.

Then, necessarily, (L, λ) is right compatible.

Proof. Let the sequence (r_n) in R_+^0 and the point $r \in R_+^0$ be such that $r_n \to r+$. By the subordination type condition,

$(\forall n)$: $L(r_n, r_{n+1}) \in \text{co}(\lambda(r_n), \lambda(r_{n+1}))$; that is:
$\min\{\lambda(r_n), \lambda(r_{n+1})\} \le L(r_n, r_{n+1}) \le \max\{\lambda(r_n), \lambda(r_{n+1})\}$.

By the regularity conditions imposed upon λ, one has $\lambda(r_n) \to \lambda(r)+$ as $n \to \infty$; whence, $\min\{\lambda(r_n), \lambda(r_{n+1})\} \to r+$, $\max\{\lambda(r_n), \lambda(r_{n+1})\} \to r+$, as $n \to \infty$. Putting these together yields the desired fact. □

In particular, with λ taken as before, the positive mappings

$L_1(t, s) = a\lambda(t) + b\lambda(s)$, $t, s \in R_+$, where $a, b \in R_+$, $a + b = 1$,
$L_2(t, s) = \max\{\lambda(t), \lambda(s)\}$, $L_3(t, s) = \min\{\lambda(t), \lambda(s)\}$, $t, s \in R_+$

are λ-subordinated; hence, (L_i, λ) = right compatible, $i \in \{1, 2, 3\}$. For a distinct example, define $(L(t, s) = \mu(t) + |\mu(s) - \mu(t)|;\ t, s \in R_+)$, where $\mu \in \mathcal{F}(R_+)$ is positive, continuous, strictly increasing on R_+^0. Clearly, (L, μ) is right compatible; but this is not due to the preceding criterion, because

> L is not λ-subordinated, for each positive $\lambda \in \mathcal{F}(R_+)$ with
> λ = continuous, strictly increasing on R_+^0.

In fact, suppose that this would be true: there exists a positive $\lambda \in \mathcal{F}(R_+)$ with λ = continuous, strictly increasing on R_+^0, such that

> (sub) L is λ-*subordinated*: $L(t, s) \in \text{co}(\lambda(t), \lambda(s))$, for each $t, s \in R_+^0$;
> hence, $\lambda(s) \le 2\mu(t) - \mu(s) \le \lambda(t)$, for all $t, s \in R_+^0$ with $t > s$.

Passing to limit as $t \to s+$, yields

$$(\forall s > 0)\colon\ \lambda(s) \le \mu(s) \le \lambda(s);\ \text{hence, } \lambda(s) = \mu(s).$$

This, combined with (sub), gives

$$\mu(s) \le 2\mu(t) - \mu(s) \le \mu(t)\ (\text{hence, } \mu(t) = \mu(s)),\ \forall t, s \in R_+^0,\ t > s;$$

in contradiction with the strict increasing choice of μ; hence the claim.

(F) Let (ψ, φ) be a couple of functions over $\mathcal{F}(R)$. For each triple $(a \in R_+;\ b, c \in R_+)$, define the properties

> (n-r-osc) (ψ, φ) is *non right oscillating at* (b, c, a): for each sequence (τ_n) in R_+^0 with $\tau_n \to a+$, we have $\limsup_n \varphi(\tau_n) > \text{osc}(+)(\psi; b, c)$
> (n-p-osc) (ψ, φ) is *non point oscillating at* (b, c, a): for each sequence (τ_n) in R_+^0 with $\tau_n \to a$, we have $\limsup_n \varphi(\tau_n) > \text{osc}(\psi; b, c)$.

In particular, let (ξ, η, ζ) be a triple of functions in $\mathcal{F}(R_+)$; for example, one may take $\xi = I$ [the identity function $(I(t) = t; t \in R_+)$]; as well as $\eta = 0$ and/or $\zeta = 0$ [the identically zero function $(0(t) = 0; t \in R_+)$].

Call the system $(\psi, \varphi; \xi, \eta, \zeta)$, *non right oscillating* at $b > 0$ provided (ψ, φ) is *non right oscillating* at $(\xi(b), \eta(b), \zeta(b))$; the set of all such $b > 0$ will be denoted as $\text{nrosc}(\psi, \varphi; \xi, \eta, \zeta)$. Then, we say that

> (a-nrosc) $(\psi, \varphi; \xi, \eta, \zeta)$ is *almost non right oscillating*, if
> $\Theta := \text{nrosc}(\psi, \varphi; \xi, \eta, \zeta)$ is $(>)$-cofinal in R_+^0
> (nrosc) $(\psi, \varphi; \xi, \eta, \zeta)$ is *non right oscillating*, if
> $\Theta := \text{nrosc}(\psi, \varphi; \xi, \eta, \zeta)$ is identical with R_+^0.

Further, call the system $(\psi, \varphi; \xi, \eta, \zeta)$, *non point oscillating* at $b > 0$ provided (ψ, φ) is *non point oscillating* at $(\xi(b), \eta(b), \zeta(b))$. If this holds for all $b > 0$, then $(\psi, \varphi; \xi, \eta, \zeta)$ is called *non point oscillating.*

Finally, let us say that (u, v, w) is

(r-asy) *right asymptotic of type* (ξ, η, ζ) at $[X_0]$, when:
$\forall$ sequence $((t^n, s^n); n \geq 0)$ in $(\mathcal{Q}, \mathcal{K})(\mathcal{G}; (B_3 > 0); [X_0])$, and $\forall b > 0$:
$((t^n, s^n); n \geq 0)$ has the right property at $((0, b, b, b, b, 0); (0, 0, b, b))$
implies $u(t^n, s^n) \to \xi(b)+$, $v(t^n, s^n) \to \eta(b)+$, $w(t^n, s^n) \to \zeta(b)+$,
(pt-asy) *point asymptotic of type* (ξ, η, ζ) at $[[X_0]]$, when:
$\forall$ sequence $((t^n, s^n); n \geq 0)$ in $(\mathcal{Q}, \mathcal{K})(\mathcal{G}; (B_3 > 0); [[X_0]])$, and $\forall b > 0$:
$((t^n, s^n); n \geq 0)$ has the point property at $((0, 0, b, 0, b, b); (0, 0, 0, b))$
implies $u(t^n, s^n) \to \xi(b)$, $v(t^n, s^n) \to \eta(b)$, $w(t^n, s^n) \to \zeta(b)$.

(G) The following statement is an essential step towards our objective. Let (ψ, φ), (u, v, w), and $(\Upsilon_0, \Upsilon; \mathcal{R}, \mathcal{S})$ be taken as above.

Theorem 2.2. *Let the selfmap T be taken so as*

(61-i) T is $(\mathcal{G}, \mathcal{R}; \psi, \varphi; u, v, w)$-contractive
(or, equivalently: $(\mathcal{G}; \mathcal{R}; \Upsilon)$-contractive)
(61-ii) $\mathcal{R}$ is B_1-admissible, and φ is positive ($\varphi(R^0_+) \subseteq R^0_+$).

Further, let $X_0 = (x_n)$ be a (uad-Ba-o) sequence. Then,

61-a) *Υ is descending asymptotic on $[X_0]$ when, for the telescopic sequences $(r_n := d(x_n, x_{n+1}))$, $(p_n := d(x_n, x_{n+2}))$ attached to X_0, there exists a strongly regular binary operation $(*)$ and a positive right compatible pair (L, λ) (where $L \in \mathcal{F}(R_+ \times R_+, R_+)$, $\lambda \in \mathcal{F}(R_+)$), with*

(61-a-i) (u, v, w) is strongly $(*, L, \lambda)$-iterative, in the sense:
$(u^n := u(E_n, F_n) = r_{n+1}$, $v^n := v(E_n, F_n) = r_n * r_{n+1}$,
$w^n := w(E_n, F_n) = L(r_n, r_{n+1})$, for all n), where
$(E_n := (r_n, r_n, p_n, 0, r_{n+1}, r_{n+1}))$, $(F_n = (p_n, r_{n+1}, r_{n+1}, 0))$
(61-a-ii) $(\psi, \varphi; I, I, \lambda)$ is non right oscillating

61-b) *Υ is descending asymptotic on $[X_0]$ when ψ is increasing and, for the telescopic sequences $(r_n := d(x_n, x_{n+1}))$, $(p_n := d(x_n, x_{n+2}))$ attached to X_0, there exists a regular binary operation $(*)$ and a positive right compatible pair (L, λ) (where $L \in \mathcal{F}(R_+ \times R_+, R_+)$, $\lambda \in \mathcal{F}(R_+)$), with*

(61-b-i) (u, v, w) is weakly $(*, L, \lambda)$-iterative, in the sense:
$(u^n := u(E_n, F_n) \geq r_{n+1}$, $v^n := v(E_n, F_n) \leq r_n * r_{n+1}$,

$w^n := w(E_n, F_n) = L(r_n, r_{n+1})$, *for all* n), *where*
$(E_n := (r_n, r_n, p_n, 0, r_{n+1}, r_{n+1}))$, $(F_n = (p_n, r_{n+1}, r_{n+1}, 0))$
(61-b-ii) $(\psi, \varphi; I, I, \lambda)$ *is non right oscillating*

61-c) Υ *is almost nright on* $[X_0]$, *provided there exists* $\mu \in \mathcal{F}(R_+)$ *with*

(61-c-i) (u, v, w) *is right asymptotic of type* (I, I, μ) *at* $[X_0]$
(61-c-ii) $(\psi, \varphi; I, I, \mu)$ *is almost non right oscillating*

61-d) Υ *is npoint on* $[[X_0]]$, *if there is a triple* (ξ, η, ζ) *in* $\mathcal{F}(R_+)$ *with*

(61-d-i) (u, v, w) *is point asymptotic of type* (ξ, η, ζ) *at* $[[X_0]]$
(61-d-ii) $(\psi, \varphi; \xi, \eta, \zeta)$ *is non point oscillating*

61-e) Υ *is nnormal, provided*

(u, v, w) *is invariant, in the sense:* $\forall a > 0$, *one has* $u(E, F) = a$, $v(E, F) = a$, $w(E, F) = a$, *where* $E = (0, a, a, a, a, 0)$, $F = (0, 0, a, a)$.

Proof. The argument consists in a number of parts.

Part 1. [Υ is descending asymptotic on $[X_0]$ under the first lot of conditions.] Let the telescopic sequences $(r_n := d(x_n, x_{n+1}))$ and $(p_n := d(x_n, x_{n+2}))$ attached to $X_0 = (x_n)$ be given. As (u, v, w) is strongly $(*, L, \lambda)$-iterative,

$(\forall n)$: $u^n := u(E_n, F_n) = r_{n+1}$, $v^n := v(E_n, F_n) = r_n * r_{n+1}$,
$w^n := w(E_n, F_n) = L(r_n, r_{n+1})$.

By the very representation of Υ, we must have

(iter-1) $(\forall n)$: $\psi(r_{n+1}) \leq \psi(r_n * r_{n+1}) - \varphi(w^n)$.

From the positive condition upon φ (and $w^n = L(r_n . r_{n+1}) > 0$)

$(\forall n)$: $\psi(r_{n+1}) < \psi(r_n * r_{n+1})$; whence $r_{n+1} \neq r_n * r_{n+1}$;
which yields $r_{n+1} < r_n$ (as $(*)$ is strongly regular).

Consequently, (r_n) is strictly descending; so, $r := \lim_n(r_n)$ exists in R_+. Suppose by contradiction that $r > 0$. As (r_n) is strictly descending,

$r_n \to r+$; hence $r_n * r_{n+1} \to r+$, $w^n = L(r_n, r_{n+1}) \to \lambda(r)+$,

(cf. the choice of $(*)$ and (L, λ)). By (iter-1), we get

(iter-2) $\varphi(w^n) \leq \psi(r_n * r_{n+1}) - \psi(r_{n+1})$, $\forall n$.

Passing to lim sup as $n \to \infty$, one derives

$$\limsup_n \varphi(w^n) \leq \limsup_n [\psi(r_n * r_{n+1}) - \psi(r_{n+1})] \leq \text{osc}(+)(\psi; r, r);$$

in contradiction with $(\psi, \varphi; I, I, \lambda)$ being non right oscillating; hence, $r = 0$.

Part 2. [Υ is descending asymptotic on $[X_0]$ under the second lot of conditions.] Let the telescopic sequences $(r_n := d(x_n, x_{n+1}))$ and $(p_n := d(x_n, x_{n+2}))$ attached to $X_0 = (x_n)$ be given. As (u, v, w) is weakly $(*, L, \lambda)$-iterative,

$(\forall n)$: $u^n := u(E_n, F_n) \geq r_{n+1}$, $v^n := v(E_n, F_n) \leq r_n * r_{n+1}$,
$w^n := w(E_n, F_n) = L(r_n, r_{n+1})$.

From the very representation of Υ (and ψ = increasing)

(iter-3) $(\forall n)$: $\psi(r_{n+1}) \leq \psi(u^n) \leq \psi(v^n) - \varphi(w^n) \leq$
$\psi(r_n * r_{n+1}) - \varphi(w^n)$.

By the positive condition upon φ and ψ = increasing,

$(\forall n)$: $\psi(r_{n+1}) < \psi(r_n * r_{n+1})$; whence $r_{n+1} < r_n * r_{n+1}$;
proving that $r_{n+1} < r_n$ (as $(*)$ is regular);

so, (r_n) is strictly descending; wherefrom, $r := \lim_n(r_n)$ exists in R_+. Suppose by contradiction that $r > 0$. By the strict descending property,

$r_n \to r+$; hence $r_n * r_{n+1} \to r+$, $w^n = L(r_n, r_{n+1}) \to \lambda(r)+$

(cf. the choice of $(*)$ and (L, λ)). Further, by (iter-3), we get

(iter-4) $\varphi(w^n) \leq \psi(r_n * r_{n+1}) - \psi(r_{n+1})$, $\forall n$.

Passing to lim sup as $n \to \infty$, one derives

$$\limsup_n \varphi(w^n) \leq \limsup_n [\psi(r_n * r_{n+1}) - \psi(r_{n+1})] \leq \text{osc}(+)(\psi; r, r);$$

in contradiction with $(\psi, \varphi; I, I, \lambda)$ being non right oscillating; hence, $r = 0$.

Part 3. [Υ is almost nright on $[X_0]$, under the precise conditions.] Take some point $(c, f) \in R_+^6 \times R_+^4$, where $e = (e_1, ..., e_6)$, $f = (f_1, ..., f_4)$. We say that the sequence $((t^n, s^n); n \geq 0)$ in $R_+^6 \times R_+^4$, where $(t^n := (t_1^n, ..., t_6^n); n \geq 0)$, $(s^n := (s_1^n, ..., s_4^n); n \geq 0)$, is *right* at (e, f), if

(r-ef) $(t_i^n \to e_i+, s_j^n \to f_j+)$ as $n \to \infty$, $\forall i$, $\forall j$.

Let us say that Υ is *nright* at $b > 0$ on $[X_0]$, if

for each sequence $((t^n, s^n); n \geq 0)$ in $(\mathcal{Q}, \mathcal{K})(\mathcal{G}; (B_3 > 0); [X_0]) \cap \Upsilon$,
the right property at $((0, b, b, b, b, 0); (0, 0, b, b))$ is not true;

the class of all these is denoted as $\text{nright}(\Upsilon; [X_0])$. In this case, we say that

> Υ is *almost nright* (respectively: *nright*) on $[X_0]$, if $\text{nright}(\Upsilon; [X_0])$ is $(>)$-cofinal in R^0_+ (respectively: identical with R^0_+).

We have to establish that the former property is retainable for our data. To do this, we start by noting that

> $(\psi, \varphi; I, I, \mu)$ = almost non right oscillating implies
> $\Theta := \text{nrosc}(\psi, \varphi; I, I, \mu)$ is $(>)$-cofinal in R^0_+.

We now claim that

> $\Theta \subseteq \text{nright}(\Upsilon; [X_0])$; and then, $\text{nright}(\Upsilon; [X_0])$ is $(>)$-cofinal in R^0_+.

This amounts to establish that

> $\forall$ sequence $((t^n, s^n); n \geq 0)$ in $(\mathcal{Q}, \mathcal{K})(\mathcal{G}; (B_3 > 0); [X_0]) \cap \Upsilon$, and
> $\forall b \in \Theta$, the right property at $((0, b, b, b, b, 0); (0, 0, b, b))$ is not true.

Suppose by absurd that there exists a sequence $((t^n, s^n))$ in $R^6_+ \times R^4_+$ where $(t^n := (t^n_1, ..., t^n_6))$, $(s^n := (s^n_1, ..., s^n_4))$, and some $b \in \Theta$, with

> (right-b-1) $(\forall n)$: $(t^n, s^n) \in (\mathcal{Q}, \mathcal{K})(\mathcal{G}; (B_3 > 0); [X_0]) \cap \Upsilon$;
> whence, $\psi(u(t^n, s^n)) \leq \psi(v(t^n, s^n)) - \varphi(w(t^n, s^n))$
> (right-b-2) $((t^n, s^n); n \geq 0)$ has the right property at
> $((0, b, b, b, b, 0); (0, 0, b, b))$; that is: $(t^n_i \to b+,\ s^n_j \to b+,\ \forall i \in \{2, 3, 4, 5\}$, $\forall j \in \{3, 4\})$ and $(t^n_i \to 0,\ s^n_j \to 0,\ \forall i \in \{1, 6\},\ \forall j \in \{1, 2\})$.

(Note that, by the contractive condition, the intersection in the first half is just $(\mathcal{Q}, \mathcal{K})(\mathcal{G}; (B_3 > 0); [X_0])$.) By the former of these properties

$$(\forall n):\ \varphi(w(t^n, s^n)) \leq \psi(v(t^n, s^n)) - \psi(u(t^n, s^n)).$$

As (u, v, w) is right asymptotic of type (I, I, μ) at $[X_0]$,

$$u(t^n, s^n) \to b+,\ v(t^n, s^n) \to b+,\ w(t^n, s^n) \to \mu(b)+.$$

Passing to lim sup as $n \to \infty$ in this relation gives

$$\limsup_n \varphi(w(t^n, s^n)) \leq$$
$$\limsup_n [\psi(v(t^n, s^n)) - \psi(u(t^n, s^n))] \leq \text{osc}(+)(\psi, b, b);$$

in contradiction with $(\psi, \varphi; I, I, \mu)$ being non right oscillating at b.

Part 4. [Υ is npoint on $[[X_0]]$ under the described assumptions.] Take some point $(e, f) \in R^6_+ \times R^4_+$, where $e = (e_1, ..., e_6)$, $f = (f_1, ..., f_4)$. We say that the sequence $((t^n, s^n); n \geq 0)$ in $R_+^6 \times R_+^4$, where $(t^n := (t_1^n, ..., t_6^n); n \geq 0)$, $(s^n := (s_1^n, ..., s_4^n); n \geq 0)$, is *point* at (e, f), if

(p-ef) $(t_i^n \to e_i,\ s_j^n \to f_j,\ \forall i,\ \forall j)$ and $[t_6^n = e_6,\ \forall n]$.

Let us say that Υ is *npoint* at $b > 0$ on $[[X_0]]$, if

(npoint) $\forall$ sequence $((t^n, s^n); n \geq 0)$ in $(\mathcal{Q}, \mathcal{K})(\mathcal{G}; B_3 > 0; [[X_0]]) \cap \Upsilon$, the point property at $((0,0,b,0,b,b); (0,0,0,b))$ is not true;

the class of all these will be denoted as $\text{npoint}(\Upsilon; [[X_0]])$. Then, define

Υ is *npoint* on $[[X_0]]$, if $\text{npoint}(\Upsilon; [[X_0]])$ is identical with R_+^0.

We have to establish that this property holds; that is,

$\forall$ sequence $((t^n, s^n); n \geq 0)$ in $(\mathcal{Q}, \mathcal{K})(\mathcal{G}; (B_3 > 0); [[X_0]]) \cap \Upsilon$, and $\forall b > 0$, the point property at $((0,0,b,0,b,b); (0,0,0,b))$ is not true.

Assume by contradiction that this assertion is false: there exists a sequence $((t^n, s^n); n \geq 0)$ in $R_+^6 \times R_+^4$ where $(t^n := (t_1^n, ..., t_6^n); n \geq 0)$, $(s^n := (s_1^n, ..., s_4^n); n \geq 0)$, and some element $b > 0$, with

(point-b-1) $(\forall n)$: $(t^n, s^n) \in (\mathcal{Q}, \mathcal{K})(\mathcal{G}; (B_3 > 0); [[X_0]]) \cap \Upsilon$;
whence, $\psi(u(t^n, s^n)) \leq \psi(v(t^n, s^n)) - \varphi(w(t^n, s^n))$
(point-b-2) $((t^n, s^n))$ is point at $((0,0,b,0,b,b), (0,0,0,b))$; that is:
$(t_i^n \to 0,\ \forall i \in \{1,2,4\})$, $(t_i^n \to b,\ \forall i \in \{3,5,6\})$, $(t_6^n = b,\ \forall n)$, and
$(s_j^n \to 0,\ \forall j \in \{1,2,3\})$, $(s_4^n \to b)$.

(Note that, by the contractive condition, the intersection in the first half is just $(\mathcal{Q}, \mathcal{K})(\mathcal{G}; (B_3 > 0); [[X_0]])$.) By the former of these properties

$(\forall n)$: $\varphi(w(t^n, s^n)) \leq \psi(v(t^n, s^n)) - \psi(u(t^n, s^n))$.

As (u, v, w) is point asymptotic of type (ξ, η, ζ) at $[[X_0]]$,

$u(t^n, s^n) \to \xi(b)$, $v(t^n, s^n) \to \eta(b)$, $w(t^n, s^n) \to \zeta(b)$.

Passing to lim sup as $n \to \infty$ in this relation gives

$\limsup_n \varphi(w(t^n, s^n)) \leq$
$\limsup_n [\psi(v(t^n, s^n)) - \psi(u(t^n, s^n))] \leq \text{osc}(\psi, \xi(b), \eta(b))$;

in contradiction with $(\psi, \varphi; \xi, \eta, \zeta)$ being non point oscillating at b.

Part 5. [Υ is nnormal under the posed condition.]

Remember that, this property means:

$\forall a > 0$, one has that $(E, F) \in \Upsilon$ is impossible,
where $E = (0, a, a, a, a, 0)$, $F = (0, 0, a, a)$.

Suppose by contradiction that there exists $a > 0$ with $(E, F) \in \Upsilon$, where (E, F) is as before. In view of $(u, v, w) =$ invariant, we must have $u(E, F) = a$, $v(E, F) = a$, $w(E, F) = a$. This firstly means $(E, F) \in \Upsilon_0$ (because $a > 0$). Secondly, by the complete definition of Υ, $\psi(u(E, F)) \leq \psi(v(E, F)) - \varphi(w(E, F))$. So, again combining with $(u, v, w, z) =$ invariant, we must have $\psi(a) \leq \psi(a) - \varphi(a)$; that is: $\varphi(a) \leq 0$; absurd, by $\varphi =$ positive. □

Now, combining the obtained facts with Theorem 2.1, one gets a couple of fixed point statements with a practical value. The former of these statements (referred to as: Rhoades-Dutta-Choudhury principle for general functions; in short: (RDC-gen)) is based on a general framework upon ψ.

Theorem 2.3. *Let the selfmap T be taken so as*

(62-i) T is $(\mathcal{G}, \mathcal{R}; \psi, \varphi; u, v, w)$-contractive;
or, equivalently: $(\mathcal{G}; \mathcal{R}; \Upsilon)$-contractive,

where the relation $\mathcal{R}$ over X, the couple (ψ, φ) over $\mathcal{F}(R)$, and the triple (u, v, w) over $\mathcal{F}(R_+^6 \times R_+^4, R_+)$ fulfill

(62-ii) $\mathcal{R}$ is B_1-admissible, and φ is positive ($\varphi(R_+^0) \subseteq R_+^0$).

Further, let X be (uad-Ba-o-f,d)-complete; and $X_0 = (x_n)$ be a (uad-Ba-o) sequence. Then,

(62-a) *$X_0 = (x_n)$ is full (hence, a (uad-Ba-o-f) sequence) and semi Picard (modulo $(\mathcal{G}; \mathcal{R}; T)$), provided, for the telescopic sequences $(r_n := d(x_n, x_{n+1}); n \geq 0)$ and $(p_n := d(x_n, x_{n+2}; n \geq 0))$ attached to $X_0 = (x_n)$, there exists a strongly regular binary operation $(*)$ and a positive compatible pair (L, λ) (where $L \in \mathcal{F}(R_+ \times R_+, R_+)$, $\lambda \in \mathcal{F}(R_+)$), with*

(62-a-1) (u, v, w) is strongly $(*, L, \lambda)$-iterative, in the sense:
$(u^n := u(E_n, F_n) = r_{n+1}$, $v^n := v(E_n, F_n) = r_n * r_{n+1}$,
$w^n := w(E_n, F_n) = L(r_n, r_{n+1})$, for all n), where
$(E_n := (r_n, r_n, p_n, 0, r_{n+1}, r_{n+1}))$, $(F_n = (p_n, r_{n+1}, r_{n+1}, 0))$
(62-a-2) $(\psi, \varphi; I, I, \lambda)$ is non right oscillating

(62-b) *$X_0 = (x_n)$ is Picard (modulo $(\mathcal{G}; \mathcal{R}; T)$) when, in addition to the above, there exists $\mu \in \mathcal{F}(R_+)$ such that:*

(62-b-1) (u, v, w) is right asymptotic of type (I, I, μ) at $[X_0]$

(62-b-2) $(\psi,\varphi;I,I,\mu)$ *is almost non right oscillating*

(62-c) $X_0=(x_n)$ *is strongly Picard (modulo* $(\mathcal{G};\mathcal{R};T)$*), if (in addition to (62-a)+(62-b))* T *is (uad-Ba-o-f,d)-continuous*

(62-d) $X_0=(x_n)$ *is semi Bellman Picard (modulo* $(\mathcal{G};\mathcal{R};T)$*) if (in addition to (62-a)+(62-b))* $\mathcal{G}$ *is (uad-Ba-o-f,d)-almost-selfclosed and there exists a triple* (ξ,η,ζ) *in* $\mathcal{F}(R_+)$ *with*

(62-d-1) (u,v,w) *is point asymptotic of type* (ξ,η,ζ) *at* $[[X_0]]$
(62-d-2) $(\psi,\varphi;\xi,\eta,\zeta)$ *is non point oscillating*

(62-e) T *is fix-*$\mathcal{G}$*-asingleton (hence, fix-*$\mathcal{G}$*-singleton, under any of these extra requirements) when, in addition to the conditions above,* $\mathcal{R}$ *is* B_2*-admissible and* (u,v,w) *is invariant:* $\forall a>0$*, the couple* $(E=(0,a,a,a,a,0)$*,* $F=(0,0,a,a))$ *fulfills* $u(E,F)=v(E,F)=w(E,F)=a$*.*

The latter of these statements (referred to as: Rhoades-Dutta-Choudhury principle for increasing functions; in short: (RDC-incr)) is based on $\psi=$ increasing on R^0_+.

Theorem 2.4. *Let the selfmap* T *be taken so as*

(63-i) T *is* $(\mathcal{G},\mathcal{R};\psi,\varphi;u,v,w)$*-contractive;*
or, equivalently: $(\mathcal{G};\mathcal{R};\Upsilon)$*-contractive,*

where the relation $\mathcal{R}$ *over* X*, the couple* (ψ,φ) *over* $\mathcal{F}(R)$*, and the triple* (u,v,w) *over* $\mathcal{F}(R^6_+\times R^4_+,R_+)$ *fulfill (for the moment)*

(63-ii) $\mathcal{R}$ *is* B_1*-admissible,* ψ *is increasing on* R^0_+ *and* φ *is positive.*

Further, let X *be (uad-Ba-o-f,d)-complete; and* $X_0=(x_n)$ *be a (uad-Ba-o) sequence. Then,*

(63-a) $X_0=(x_n)$ *is full (hence, a (uad-Ba-o-f) sequence) and semi Picard (modulo* $(\mathcal{G};\mathcal{R};T)$*), provided, for the telescopic sequences* $(r_n:=d(x_n,x_{n+1});n\geq 0)$ *and* $(p_n:=d(x_n,x_{n+2});n\geq 0)$ *attached to* $X_0=(x_n)$ *there exists a regular binary operation* $(*)$ *and a positive compatible pair* (L,λ) *(where* $L\in\mathcal{F}(R_+\times R_+,R_+)$*,* $\lambda\in\mathcal{F}(R_+)$*), with*

(63-a-1) (u,v,w) *is weakly* $(*,L,\lambda)$*-iterative, in the sense:*
$(u^n:=u(E_n,F_n)\geq r_{n+1},\ v^n:=v(E_n,F_n)\leq r_n*r_{n+1},$
$w^n:=w(E_n,F_n)=L(r_n,r_{n+1})$*, for all* n*), where*
$(E_n:=(r_n,r_n,p_n,0,r_{n+1},r_{n+1}))$, $(F_n=(p_n,r_{n+1},r_{n+1},0))$
(63-a-2) $(\psi,\varphi;I,I,\lambda)$ *is non right oscillating*

(63-b) *$X_0 = (x_n)$ is Picard (modulo $(\mathcal{G}; \mathcal{R}; T)$) when, in addition to the above, there exists $\mu \in \mathcal{F}(R_+)$ such that:*

(63-b-1) (u, v, w) *is right asymptotic of type* (I, I, μ) *at* $[X_0]$
(63-b-2) $(\psi, \varphi; I, I, \mu)$ *is almost non right oscillating*

(63-c) *$X_0 = (x_n)$ is strongly Picard (modulo $(\mathcal{G}; \mathcal{R}; T)$), if (in addition to (63-a)+(63-b)) T is (uad-Ba-o-f,d)-continuous*

(63-d) *$X_0 = (x_n)$ is semi Bellman Picard (modulo $(\mathcal{G}; \mathcal{R}; T)$), if (in addition to (63-a)+(63-b)) $\mathcal{G}$ is (uad-Ba-o-f,d)-almost-selfclosed and there exists a triple (ξ, η, ζ) in $\mathcal{F}(R_+)$ with*

(63-d-1) (u, v, w) *is point asymptotic of type* (ξ, η, ζ) *at* $[[X_0]]$
(63-d-2) $(\psi, \varphi; \xi, \eta, \zeta)$ *is non point oscillating*

(63-e) *T is fix-$\mathcal{G}$-asingleton (hence, fix-$\mathcal{G}$-singleton, under any of these extra requirements) when, in addition to the conditions above, $\mathcal{R}$ is B_2-admissible and (u, v, w) is invariant: $\forall a > 0$, the couple ($E = (0, a, a, a, a, 0)$, $F = (0, 0, a, a)$) fulfills $u(E, F) = v(E, F) = w(E, F) = a$.*

It is worth noting that an extended setting of these developments is possible, under the lines described in the 2001 PhD Thesis by Hitzler [113, Ch. 1, Sec. 1.2]; see also Turinici [255].

2.1.7 *Particular aspects*

Let $(X, d, \mathcal{G})$ be a *relational metric space*. Further, take some selfmap $T \in \mathcal{F}(X)$, subject to the condition

(uad-a-reg) T is *uadmissible ascending regular*: uadao(T) is nonempty.

As a consequence, there exist uadmissible, (a-o)-sequences in X. Let $X_0 = (x_n)$ be such an object; remember that, without loss, one may assume that

$(B_1(x_n, x_{n+1}) > 0, \forall n)$; meaning: $X_0 = (x_n)$ is a (uad-Ba-o) sequence.

As precise, we are interested to determine conditions for existence and uniqueness of elements in Fix(T), via contractive type requirements involving the iterative process $X_0 = (x_n)$, and the associated sets $[X_0] = \{x_n; n \geq 0\}$, $[[X_0]] = \text{cl}([X_0])$. The basic directions and regularity conditions relative to X_0 under which the problem of determining fixed points of T is to be solved were already listed; and the contractive type framework (involving X_0, $[X_0]$, and $[[X_0]]$) was settled. As a by-product of these, we established our main result in this exposition, Theorem 2.1. Then, as an illustration, we gave a couple of practical variants of it – Theorem 2.3 and Theorem 2.4 – that appear as analytic approaches for solving the fixed point question involving a class of contractions introduced by Krasnoselskii and Stetsenko [152], re-discovered by Rhoades [197], and refined by Dutta and Choudhury [84]. It is our aim to show that, as particular cases of this approach, one gets a lot of fixed point statements extending certain well known contributions in the area.

(A) For the moment, some direct constructions will be proposed.

Constr-0) Define the auxiliary system of functions (e_0, e_1, e_2) and $(f_1, ..., f_6)$ over $\mathcal{F}(R^6_+ \times R^4_+, R_+)$, as: for each $t = (t_1, ..., t_6) \in R^6_+$, $s = (s_1, ..., s_4) \in R^4_+$,

$$
\begin{aligned}
&e_0(t,s) = t_2 + |t_1 - s_2|,\ e_1(t,s) = \max\{t_1, t_6\},\ e_2(t,s) = \max\{s_2, s_3, s_4\},\\
&f_1(t,s) = t_6(1 + t_1 + \theta s_2)/(1 + t_2 + \theta t_5) \text{ (where } \theta \geq 0 \text{ is a constant)},\\
&f_2(t,s) = t_4(1 + t_3)/(1 + t_2),\ f_3(t,s) = (1/2)[t_3 + t_4],\\
&f_4(t,s) = (1/2)[t_5 + t_6],\ f_5(t,s) = (1/2)[s_1 + s_4],\\
&f_6(t,s) = t_1 t_6/t_2, \text{ if } t_2 \neq 0, \text{ and } f_6(t,s) = 0, \text{ if } t_2 = 0.
\end{aligned}
$$

Constr-1) Further, let the system $(u_0, ..., u_\varepsilon)$ over $\mathcal{F}(R^6_+ \times R^4_+, R_+)$ be introduced as: for each $t = (t_1, ..., t_6) \in R^6_+$, $s = (s_1, ..., s_4) \in R^4_+$,

$$
\begin{aligned}
&\text{(u-def)}\ \ u_0(t,s) = \max\{t_5, f_1(t,s)\},\ u_1(t,s) = \max\{t_5, f_2(t,s)\},\\
&u_2(t,s) = \max\{t_5, f_4(t,s)\},\ u_3(t,s) = \max\{t_5, f_1(t,s), f_2(t,s)\},\\
&u_4(t,s) = \max\{t_5, f_1(t,s), f_4(t,s)\},\ u_5(t,s) = \max\{t_5, f_2(t,s), f_4(t,s)\},\\
&u_6(t,s) = \max\{t_5, f_1(t,s), f_2(t,s), f_4(t,s)\},\\
&u_7(t,s) = \max\{t_5, e_2(t,s), f_1(t,s)\},\ u_8(t,s) = \max\{t_5, e_2(t,s), f_2(t,s)\},\\
&u_9(t,s) = \max\{t_5, e_2(t,s), f_4(t,s)\},\\
&u_\alpha(t,s) = \max\{t_5, e_2(t,s), f_1(t,s), f_2(t,s)\},\\
&u_\beta(t,s) = \max\{t_5, e_2(t,s), f_2(t,s), f_4(t,s)\},\\
&u_\gamma(t,s) = \max\{t_5, f_6(t,s)\},\ u_\delta(t,s) = \max\{t_5, t_6\},\ u_\varepsilon(t,s) = t_5.
\end{aligned}
$$

Constr-2) Then, let us construct the system $(v_0, ..., v_\varepsilon)$ over $\mathcal{F}(R^6_+ \times R^4_+, R_+)$, as: for each $t = (t_1, ..., t_6) \in R^6_+$, $s = (s_1, ..., s_4) \in R^4_+$,

$$
\begin{aligned}
&\text{(v-def)}\ \ v_0(t,s) = e_0(t,s),\ v_1(t,s) = e_1(t,s),\ v_2(t,s) = \max\{t_2, e_1(t,s)\},\\
&v_3(t,s) = \max\{t_2, e_1(t,s), e_2(t,s)\},\ v_4(t,s) = \max\{t_2, e_1(t,s), f_3(t,s)\},\\
&v_5(t,s) = \max\{t_2, e_1(t,s), f_6(t,s)\},\ v_6(t,s) = \max\{t_2, f_1(t,s), f_2(t,s)\},\\
&v_7(t,s) = \max\{t_2, f_1(t,s)\},\ v_8(t,s) = \max\{t_2, f_2(t,s)\},\\
&v_9(t,s) = \max\{t_2, e_1(t,s), e_2(t,s), f_3(t,s)\},\\
&v_\alpha(t,s) = \max\{t_2, e_1(t,s), e_2(t,s), f_5(t,s)\},
\end{aligned}
$$

$v_\beta(t,s) = \max\{t_2, e_1(t,s), e_2(t,s), f_3(t,s), f_5(t,s)\}$,
$v_\gamma(t,s) = \max\{t_2, f_6(t,s)\}$, $v_\delta(t,s) = \max\{t_2, t_5\}$, $v_\varepsilon(t,s) = t_2$.

Constr-3) Further, define the system $(w_0, ..., w_\varepsilon)$ of functions over $\mathcal{F}(R_+^6 \times R_+^4, R_+)$, as: for each $t = (t_1, ..., t_6) \in R_+^6$, $s = (s_1, ..., s_4) \in R_+^4$,

(w-def) $w_k(t,s) = v_k(t,s)$, $k \in \{0, ..., \varepsilon\}$.

Constr-4) Finally, define the couple $(\top, *)$ of relations on R_+ as

$(a \top b = \max\{a, b\}, a * b = a + |b - a|; a, b \in R_+)$; as precise,
$(\top)$ and $(*)$ are strongly regular and $a \top b \le a * b$, $\forall a, b \in R_+$.

It is our aim to discuss of to what extent are the triples (u, v, w), where

$u \in \{u_0, ..., u_\varepsilon\}$, $v \in \{v_0, ..., v_\varepsilon\}$, $w \in \{w_0, ..., w_\varepsilon\}$,

compatible with the regularity conditions encountered in Theorem 2.3 and Theorem 2.4. The response to this question is to be precise as below.

R-1) Let $X_0 = (x_n)$ be a (uad-Ba-o) sequence, and $(r_n := d(x_n, x_{n+1}))$, $(p_n := d(x_n, x_{n+2}))$, stand for the attached telescopic sequences. Then, let $(E_n := (r_n, r_n, p_n, 0, r_{n+1}, r_{n+1}); n \ge 0)$, $(F_n := (p_n, r_{n+1}, r_{n+1}, 0); n \ge 0)$ denote the vectorial sequences generated by these. We have

$(\forall n)$: $e_0(E_n, F_n) = r_n * r_{n+1}$, $e_1(E_n, F_n) = r_n \top r_{n+1}$, $e_2(E_n, F_n) = r_{n+1}$, $f_1(E_n, F_n) = r_{n+1}$, $f_2(E_n, F_n) = 0$, $f_3(E_n, F_n) = (1/2)p_n \le r_n \top r_{n+1}$, $f_4(E_n, F_n) = r_{n+1}$, $f_5(E_n, F_n) = (1/2)p_n \le r_n \top r_{n+1}$, $f_6(E_n, F_n) = r_{n+1}$;

and this, by definition, gives

$(\forall n)$: $u_0(E_n, F_n) = r_{n+1}$, $u_1(E_n, F_n) = r_{n+1}$, $u_2(E_n, F_n) = r_{n+1}$,
$u_3(E_n, F_n) = r_{n+1}$, $u_4(E_n, F_n) = r_{n+1}$, $u_5(E_n, F_n) = r_{n+1}$,
$u_6(E_n, F_n) = r_{n+1}$, $u_7(E_n, F_n) = r_{n+1}$, $u_8(E_n, F_n) = r_{n+1}$,
$u_9(E_n, F_n) = r_{n+1}$, $u_\alpha(E_n, F_n) = r_{n+1}$, $u_\beta(E_n, F_n) = r_{n+1}$,
$u_\gamma(E_n, F_n) = r_{n+1}$, $u_\delta(E_n, F_n) = r_{n+1}$, $u_\varepsilon(E_n, F_n) = r_{n+1}$,
$v_0(E_n, F_n) = r_n * r_{n+1}$, $v_1(E_n, F_n) = r_n \top r_{n+1}$, $v_2(E_n, F_n) = r_n \top r_{n+1}$,
$v_3(E_n, F_n) = r_n \top r_{n+1}$, $v_4(E_n, F_n) = r_n \top r_{n+1}$, $v_5(E_n, F_n) = r_n \top r_{n+1}$,
$v_6(E_n, F_n) = r_n \top r_{n+1}$, $v_7(E_n, F_n) = r_n \top r_{n+1}$, $v_8(E_n, F_n) = r_n$,
$v_9(E_n, F_n) = r_n \top r_{n+1}$, $v_\alpha(E_n, F_n) = r_n \top r_{n+1}$, $v_\beta(E_n, F_n) = r_n \top r_{n+1}$, $v_\gamma(E_n, F_n) = r_n \top r_{n+1}$, $v_\delta(E_n, F_n) = r_n \top r_{n+1}$, $v_\varepsilon(E_n, F_n) = r_n$, $(w_k(E_n, F_n) = v_k(E_n, F_n)$, $k \in \{0, ..., \varepsilon\})$;

proving that, under the notations

$\nabla_j = (*)$, $j \in \{0\}$, $\nabla_j = (\top)$, for $j \in \{0, ..., \varepsilon\} \setminus \{0\}$,
$(L_k(t, s) = t * s,\ k \in \{0\})$, $(L_k(t, s) = t,\ k \in \{8, \varepsilon\})$,
$(L_k(t, s) = t \top s,\ k \in \{0, ..., \varepsilon\} \setminus \{0, 8, \varepsilon\})$, $t, s \in R_+^0$,

the following conclusions hold

for each $i \in \{0, ..., \varepsilon\}$, $j \in \{0, ..., \varepsilon\}$, $k \in \{0, ..., \varepsilon\}$:
$\{u_i, v_j, w_k\}$ is (strongly (∇_j, L_k, I)-iterative, if $j \in \{0, ..., \varepsilon\} \setminus \{8, \varepsilon\}$),
and (weakly (∇_j, L_k, I)-iterative, if $j \in \{8, \varepsilon\}$).

R-2) Let $X_0 = (x_n)$ be the given (uad-Ba-o) sequence, and (ξ, η, ζ) be a triple of functions in $\mathcal{F}(R_+)$. Remember that (u, v, w) is called *right asymptotic of type* (ξ, η, ζ) at $[X_0]$, when

for each sequence $((t^n, s^n); n \geq 0)$, where $(t^n := (t_1^n, ..., t_6^n); n \geq 0)$, $(s^n := (s_1^n, ..., s_4^n); n \geq 0)$, in $(\mathcal{Q}, \mathcal{K})(\mathcal{G}; (B_3 > 0); [X_0])$, and each $b > 0$: $((t^n, s^n); n \geq 0)$ has the right property at $((0, b, b, b, b, 0); (0, 0, b, b))$ (that is: $(t_i^n \to 0, \forall i \in \{1, 6\})$, $(t_i^n \to b+, \forall i \in \{2, 3, 4, 5\})$, and $(s_j^n \to 0, \forall j \in \{1, 2\})$, $(s_j^n \to b+, \forall j \in \{3, 4\})$), implies $u(t^n, s^n) \to \xi(b)+$, $v(t^n, s^n) \to \eta(b)+$, $w(t^n, s^n) \to \zeta(b)+$.

To verify this assertion upon our constructions, let the sequence $((t^n, s^n); n \geq 0)$ in $(\mathcal{Q}, \mathcal{K})(\mathcal{G}; (B_3 > 0); [X_0])$ and the point $b > 0$ be as before. We have (as $n \to \infty$)

$e_0(t^n, s^n) \to b$, $e_1(t^n, s^n) \to 0$, $e_2(t^n, s^n) \to b$,
$f_1(t^n, s^n) \to 0$, $f_2(t^n, s^n) \to b$, $f_3(t^n, s^n) \to b$,
$f_4(t^n, s^n) \to b/2$, $f_5(t^n, s^n) \to b/2$, $f_6(t^n, s^n) \to 0$;

and this, by definition, gives (again under $n \to \infty$)

$u_0(t^n, s^n) \to b+$, $u_1(t^n, s^n) \to b+$, $u_2(t^n, s^n) \to b+$,
$u_3(t^n, s^n) \to b+$, $u_4(t^n, s^n) \to b+$, $u_5(t^n, s^n) \to b+$,
$u_6(t^n, s^n) \to b+$, $u_7(t^n, s^n) \to b+$, $u_8(t^n, s^n) \to b+$,
$u_9(t^n, s^n) \to b+$, $u_\alpha(t^n, s^n) \to b+$, $u_\beta(t^n, s^n) \to b+$,
$u_\gamma(t^n, s^n) \to b+$, $u_\delta(t^n, s^n) \to b+$, $u_\varepsilon(t^n, s^n) \to b+$,
$v_0(t^n, s^n) \to b+$, $v_1(t^n, s^n) \to 0+$, $v_2(t^n, s^n) \to b+$,
$v_3(t^n, s^n) \to b+$, $v_4(t^n, s^n) \to b+$, $v_5(t^n, s^n) \to b+$,
$v_6(t^n, s^n) \to b+$, $v_7(t^n, s^n) \to b+$, $v_8(t^n, s^n) \to b+$,
$v_9(t^n, s^n) \to b+$, $v_\alpha(t^n, s^n) \to b+$, $v_\beta(t^n, s^n) \to b+$,
$v_\gamma(t^n, s^n) \to b+$, $v_\delta(t^n, s^n) \to b+$, $v_\varepsilon(t^n, s^n) \to b+$,
and $(\lim_n w_k(t^n, s^n) = \lim_n v_k(t^n, s_n),\ k \in \{0, ..., \varepsilon\})$.

Here, we used the following evident fact

$$(c_n \geq d_n, \forall n) \text{ and } (c_n \to b,\ d_n \to b+) \text{ imply } c_n \to b+.$$

As a consequence of these relations, we have that, under the conventions

$$(\mu_j = 0, \text{ for } j \in \{1\}),\ (\mu_j = I, \text{ for } j \in \{0, ..., \varepsilon\} \setminus \{1\}),$$
$$(\nu_k = 0, \text{ for } k \in \{1\}),\ (\nu_k = I, \text{ for } k \in \{0, ..., \varepsilon\} \setminus \{1\}),$$

one derives the conclusions

(for each $i \in \{1, ..., \varepsilon\}$, $j \in \{0, ..., \varepsilon\}$, $k \in \{0, ..., \varepsilon\}$):
(u_i, v_j, w_k) is right asymptotic of type (I, μ_j, ν_k) at $[X_0]$.

R-3) Let $X_0 = (x_n)$ be the given (uad-Ba-o) sequence, and (ξ, η, ζ) be a triple of functions in $\mathcal{F}(R_+)$. Remember that (u, v, w) is called *point asymptotic of type* (ξ, η, ζ) *at* $[[X_0]]$, when

for each sequence $((t^n, s^n))$, where $(t^n := (t_1^n, ..., t_6^n); n \geq 0)$, $(s^n := (s_1^n, ..., s_4^n); n \geq 0)$, in $(\mathcal{Q}, \mathcal{K})(\mathcal{G}; (B_3 > 0); [[X_0]])$, and each $b > 0$: $((t^n, s^n); n \geq 0)$ has the point property at $((0, 0, b, 0, b, b); (0, 0, 0, b))$ (that is: $(t_i^n \to 0, \forall i \in \{1, 2, 4\})$, $(t_i^n \to b, \forall i \in \{3, 5, 6\})$, $(t_6^n = b, \forall n)$, and $(s_j^n \to 0, \forall j \in \{1, 2, 3\})$, $(s_4^n \to b)$) implies $(u(t^n, s^n) \to \xi(b)$, $v(t^n, s^n) \to \eta(b)$, $w(t^n, s^n) \to \zeta(b))$.

To verify this upon our constructions, let the sequence $((t^n, s^n))$ (where $(t^n := (t_1^n, ..., t_6^n); n \geq 0)$, $(s^n := (s_1^n, ..., s_4^n); n \geq 0)$) in $(\mathcal{Q}, \mathcal{K})(\mathcal{G}; (B_3 > 0); [[X_0]])$ and the point $b > 0$ be as before. We have (as $n \to \infty$)

$e_0(t^n, s^n) \to 0$, $e_1(t^n, s^n) \to b$, $e_2(t^n, s^n) \to b$, $f_1(t^n, s^n) \to b/(1 + \theta b)$, $f_2(t^n, s^n) \to 0$, $f_3(t^n, s^n) \to b/2$, $f_4(t^n, s^n) \to b$, $f_5(t^n, s^n) \to b/2$, and $(f_6(t^n, s^n); n \geq 0)$ is not convergent;

and this, by definition, gives (again under $n \to \infty$)

$u_0(t^n, s^n) \to b$, $u_1(t^n, s^n) \to b$, $u_2(t^n, s^n) \to b$, $u_3(t^n, s^n) \to b$,
$u_4(t^n, s^n) \to b$, $u_5(t^n, s^n) \to b$, $u_6(t^n, s^n) \to b$, $u_7(t^n, s^n) \to b$,
$u_8(t^n, s^n) \to b$, $u_9(t^n, s^n) \to b$, $u_\alpha(t^n, s^n) \to b$, $u_\beta(t^n, s^n) \to b$,
$(u_\gamma(t^n, s^n); n \geq 0)$ is not convergent, $u_\delta(t^n, s^n) \to b$, $u_\varepsilon(t^n, s^n) \to b$,
$v_0(t^n, s^n) \to 0$, $v_1(t^n, s^n) \to b$, $v_2(t^n, s^n) \to b$, $v_3(t^n, s^n) \to b$,
$v_4(t^n, s^n) \to b$, $(v_5(t^n, s^n); n \geq 0)$ is not convergent,
$v_6(t^n, s^n) \to b/(1 + \theta b)$, $v_7(t^n, s^n) \to b/(1 + \theta b)$, $v_8(t^n, s^n) \to 0$,
$v_9(t^n, s^n) \to b$, $v_\alpha(t^n, s^n) \to b$, $v_\beta(t^n, s^n) \to b$,

$(v_\gamma(t^n, s^n); n \geq 0)$ is not convergent, $v_\delta(t^n, s^n) \to b$, $v_\varepsilon(t^n, s^n) \to 0$,
$\lim_n w_k(t^n, s^n) = \lim_n v_k(t^n, s^n)$, $k \in \{0, ..., \varepsilon\}$.

As a consequence of these relations, we have that under the conventions

$\eta_j = 0$, $j \in \{0, 8\}$, $\eta_j = \sigma$, $j \in \{6, 7\}$, where $(\sigma(t) = t/(1 + \theta t); t \in R_+)$,
$\eta_j = I$, $j \in \{0, ..., \varepsilon\} \setminus \{0, 6, 7, 8\}$; $\zeta_k = 0$, $k \in \{0, 8\}$, $\zeta_k = \sigma$, $k \in \{6, 7\}$,
where $(\sigma(t) = t/(1 + \theta t); t \in R_+)$, $\zeta_k = I$, $j \in \{0, ..., \varepsilon\} \setminus \{0, 6, 7, 8\}$,

one derives the conclusions

(for each $i \in \{0, ..., \varepsilon\} \setminus \{\gamma\}$, $j \in \{0, ..., \varepsilon\} \setminus \{5, \gamma\}$, $k \in \{0, ..., \varepsilon\} \setminus \{5, \gamma\}$):
(u_i, v_j, w_k) is point asymptotic of type (I, η_j, ζ_k) at $[[X_0]]$.

R-4) Finally, remember that the triple (u, v, w) is invariant, provided

$(\forall a > 0)$: $u(E, F) = v(E, F) = w(E, F) = a$,
where $E = (0, a, a, a, a, 0)$, $F = (0, 0, a, a)$.

To verify this upon our constructions, we have

$e_0(E, F) = a$, $e_1(E, F) = 0$, $e_2(E, F) = a$, $f_1(E, F) = 0$, $f_2(E, F) = a$,
$f_3(E, F) = a$, $f_4(E, F) = a/2$, $f_5(E, F) = a/2$, $f_6(E, F) = 0$;

and this, by definition, gives

$u_0(E, F) = a$, $u_1(E, F) = a$, $u_2(E, F) = a$,
$u_3(E, F) = a$, $u_4(E, F) = a$, $u_5(E, F) = a$,
$u_6(E, F) = a$, $u_7(E, F) = a$, $u_8(E, F) = a$,
$u_9(E, F) = a$, $u_\alpha(E, F) = a$, $u_\beta(E, F) = a$,
$u_\gamma(E, F) = a$, $u_\delta(E, F) = a$, $u_\varepsilon(E, F) = a$,
$v_0(E, F) = a$, $v_1(E, F) = 0$, $v_2(E, F) = a$,
$v_3(E, F) = a$, $v_4(E, F) = a$, $v_5(E, F) = a$,
$v_6(E, F) = a$, $v_7(E, F) = a$, $v_8(E, F) = a$,
$v_9(E, F) = a$, $v_\alpha(E, F) = a$, $v_\beta(E, F) = a$,
$v_\gamma(E, F) = a$, $v_\delta(E, F) = a$, $v_\varepsilon(E, F) = a$,
$w_k(E, F) = v_k(E, F)$, $k \in \{0, ..., \varepsilon\}$.

As a consequence, we have that

for each $i \in \{0, ..., \varepsilon\}$, $j \in \{0, ..., \varepsilon\} \setminus \{1\}$, $k \in \{0, ..., \varepsilon\} \setminus \{1\}$:
$\{u_i, v_j, w_k\}$ is invariant.

As a direct consequence of these facts and Rhoades-Dutta-Choudhury principles we just exposed, one may now derive a couple of fixed point statements with methodological value.

The former of these, starting from Rhoades-Dutta-Choudhury principle for general functions (RDC-gen) is based on a general framework upon ψ.

Theorem 2.5. *Let the selfmap T be taken so as*

> *(71-i) T is $(\mathcal{G}, \mathcal{R}; \psi, \varphi; u_i, v_j, w_k)$-contractive;*
> *or, equivalently: $(\mathcal{G}; \mathcal{R}; \Upsilon)$-contractive,*

where the relation $\mathcal{R}$ over X and the couple (ψ, φ) over $\mathcal{F}(R)$, fulfill

> *(71-ii) $\mathcal{R}$ is B_1-admissible and φ is positive ($\varphi(R^0_+) \subseteq R^0_+$),*

and (i, j, k) is a triple over $\{0, ..., \varepsilon\}$. Further, let X be (uad-Ba-o-f,d)-complete; and $X_0 = (x_n)$ be a (uad-Ba-o) sequence. Then,

(71-a) *$X_0 = (x_n)$ is full (hence, a (uad-Ba-o-f) sequence) and semi Picard (modulo $(\mathcal{G}; \mathcal{R}; T)$) if, in addition,*

> *(71-a-1) $i \in \{0, ..., \varepsilon\}$, $j \in \{0, ..., \varepsilon\} \setminus \{8, \varepsilon\}$, $k \in \{0, ..., \varepsilon\}$ and*
> *$(\psi, \varphi; I, I, I)$ is non right oscillating*

(71-b) *$X_0 = (x_n)$ is Picard (modulo $(\mathcal{G}; \mathcal{R}; T)$) if, in addition to (71-a),*

> *(71-b-1) $i \in \{0, ..., \varepsilon\}$, $j \in \{0, ..., \varepsilon\} \setminus \{1\}$, $k \in \{0, ..., \varepsilon\} \setminus \{1\}$*
> *(and (as accepted), $(\psi, \varphi; I, I, I)$ is non right oscillating)*

(71-c) *$X_0 = (x_n)$ is strongly Picard (modulo $(\mathcal{G}; \mathcal{R}; T)$), whenever (in addition to (71-a)+(71-b)) T is (uad-Ba-o-f,d)-continuous*

(71-d) *$X_0 = (x_n)$ is semi Bellman Picard (modulo $(\mathcal{G}; \mathcal{R}; T)$), provided (in addition to (71-a)+(71-b))*

> *(71-d-1) $\mathcal{G}$ is (uad-Ba-o-f,d)-almost-selfclosed*
> *(71-d-2) $i \in \{0, ..., \varepsilon\} \setminus \{\gamma\}$, $j \in \{0, ..., \varepsilon\} \setminus \{5, \gamma\}$, $k \in \{0, ..., \varepsilon\} \setminus \{5, \gamma\}$, and $(\psi, \varphi; I, \eta_j, \zeta_k)$ is non point oscillating*

(71-e) *T is fix-$\mathcal{G}$-asingleton (hence, fix-$\mathcal{G}$-singleton, under any of these extra requirements) when, in addition, $i \in \{0, ..., \varepsilon\}$, $j \in \{0, ..., \varepsilon\} \setminus \{1\}$, $k \in \{0, ..., \varepsilon\} \setminus \{1\}$, and $\mathcal{R}$ is B_2-admissible.*

The latter of these, starting from Rhoades-Dutta-Choudhury principle for increasing functions (RDC-incr) is based on ψ = increasing on R^0_+.

Theorem 2.6. *Let the selfmap T be taken so as*

(72-i) T *is* $(\mathcal{G}, \mathcal{R}; \psi, \varphi; u_i, v_j, w_k)$*-contractive;*
or, equivalently: $(\mathcal{G}; \mathcal{R}; \Upsilon)$*-contractive,*

where the relation $\mathcal{R}$ *over* X *and the couple* (ψ, φ) *over* $\mathcal{F}(R_+, R)$*, fulfill*

(72-ii) $\mathcal{R}$ *is* B_1*-admissible,* ψ *is increasing on* R^0_+ *and* φ *is positive*

and (i, j, k) *is a triple over* $\{0, ..., \varepsilon\}$*. Further, let* X *be (uad-Ba-o-f,d)-complete; and* $X_0 = (x_n)$ *be a (uad-Ba-o) sequence. Then,*

(72-a) $X_0 = (x_n)$ *is full (hence, a (uad-Ba-o-f) sequence) and semi Picard (modulo* $(\mathcal{G}; \mathcal{R}; T)$*) if, in addition,*

(72-a-1) (i, j, k) *is as before and* $(\psi, \varphi; I, I, I)$ *is non right oscillating*

(72-b) $X_0 = (x_n)$ *is Picard (modulo* $(\mathcal{G}; \mathcal{R}; T)$*), if (under (72-a))*

(72-b-1) $i \in \{0, ..., \varepsilon\}$, $j \in \{0, ..., \varepsilon\} \setminus \{1\}$, $k \in \{0, ..., \varepsilon\} \setminus \{1\}$
(and (as accepted), $(\psi, \varphi; I, I, I)$ *is non right oscillating)*

(72-c) $X_0 = (x_n)$ *is strongly Picard (modulo* $(\mathcal{G}; \mathcal{R}; T)$*), whenever (in addition to (71-a)+(71-b))* T *is (uad-Ba-o-f,d)-continuous*

(72-d) $X_0 = (x_n)$ *is semi Bellman Picard (modulo* $(\mathcal{G}; \mathcal{R}; T)$*), provided (in addition to (71-a)+(71-b))*

(72-d-1) $\mathcal{G}$ *is (uad-Ba-o-f,d)-almost-selfclosed*
(71-d-2) $i \in \{0, ..., \varepsilon\} \setminus \{\gamma\}$, $j \in \{0, ..., \varepsilon\} \setminus \{5, \gamma\}$, $k \in \{0, ..., \varepsilon\} \setminus \{5, \gamma\}$,
and $(\psi, \varphi; I, \eta_j, \zeta_k)$ *is non point oscillating*

(72-e) T *is fix-*$\mathcal{G}$*-asingleton (hence, fix-*$\mathcal{G}$*-singleton, under any of these extra requirements) when, in addition,* $i \in \{0, ..., \varepsilon\}$, $j \in \{0, ..., \varepsilon\} \setminus \{1\}$, $k \in \{0, ..., \varepsilon\} \setminus \{1\}$, *and* $\mathcal{R}$ *is* B_2*-admissible.*

Some remarks are in order.

Rem-1) A variant of these statements under the ambient relation $\mathcal{G}$ being a quasi-order on X is immediate; and yields the related developments in Turinici [254], [258, Paper-3-1].

Rem-2) Theorem 2.5 is a collection of (at most) $15 \times 12 \times 14 = 2520$ (anticipative or not) fixed point statements. Among these,

(I) the $(u_\varepsilon, v_4, w_4)$ variant of Theorem 2.5 extends to relational metric spaces the fixed point statements over (amorphous) metric spaces due to Xue [269], subsequently refined by Proinov [189]

(II) the $(u_\varepsilon, v_5, w_5)$ variant of Theorem 2.5 extends to relational metric spaces the fixed point statements over (amorphous) metric spaces obtained by Fulga [92].

The remaining variants appearing in this synthetic result seem to be new from the discussed perspective.

Rem-3) Theorem 2.6 is a collection of (at most) $15 \times 14 \times 14 = 2940$ (anticipative or not) fixed point statements. Among these

(I) the $(u_\varepsilon, v_6, w_7)$ variant of Theorem 2.6 includes from a technical perspective the fixed point statements obtained in Turinici [256, Paper-1-1]; which extend to the anticipative setting the non-anticipative fixed point statements over partially ordered metric spaces due to Chandok et al. [47]
(II) The $(u_\varepsilon, v_4, w_1)$ variant of Theorem 2.6 extends to relational metric spaces the non-anticipative fixed point statements over amorphous metric spaces due to Choudhury et al. [56]
(III) The $(u_\varepsilon, v_4, w_\varepsilon)$ variant of Theorem 2.6 extends to relational metric spaces the non-anticipative fixed point statements over amorphous metric spaces due to Vujaković et al. [261].

Rem-4) In particular, when $\varphi =$ constant

(I) The (u_ε, v_β) variant of Theorem 2.6 extends to relational metric spaces the fixed point statement over amorphous metric spaces in Dung and Hang [83]
(II) The $(u_\varepsilon, v_\varepsilon)$ version of Theorem 2.6 extends to relational metric spaces the fixed point statement in Wardowski [265]
(III) The (u_ε, v_0) variant of Theorem 2.6 extends to relational metric spaces the fixed point statement in Fulga and Proca [93].

Finally, note that Theorem 2.6 extends some other fixed point statements in the area, exposed in the survey paper by Karapinar et al. [140]. Further technical aspects of these extensions will be delineated elsewhere.

2.2 Functional Contractions in Local Branciari Ordered Metric Spaces

Abstract. The fixed point result in Fora et al. [Mat. Vesnik, 61 (2009), 203–208] is extended to a class of functional contractions over local Branciari ordered metric spaces.

Subject-Class 2020. 47H10 (Primary), 54H25 (Secondary).

Keywords. Local Branciari metric, Picard mapping, fixed point, quasi-order, Meir-Keeler contraction, admissible function, normal couple.

2.2.1 *Dependent Choice principles*

See Ch. 1.

2.2.2 *Conv-Cauchy structures*

See Ch. 1.

2.2.3 *Banach contractions and relatives*

See Ch. 1.

2.2.4 *Statement of the problem*

Let X be a nonempty set. Call the subset Y of X, *almost singleton* (in short: *asingleton*), provided $[y_1, y_2 \in Y$ implies $y_1 = y_2]$; and *singleton* if, in addition, Y is nonempty; note that in this case $Y = \{y\}$, for some $y \in X$.

Let the mapping $d : X \times X \to R_+$ be such that

(re-suf) d is *reflexive sufficient* $[x = y \iff d(x, y) = 0]$;
(symm) d is *symmetric* $[d(x, y) = d(y, x),\ \forall x, y \in X]$;

we then say that d is a *reflexive sufficient symmetric* (in short: *(r-s)-symmetric*); and (X, d) is a *(r-s)-symmetric space*. Suppose that we fixed such an object, $d(.,.)$; with, in addition,

(C-sep) d is Cauchy-separated:
$\lim_n(x_n)$ is an asingleton, for each d-Cauchy sequence in X;

we call it, a *Cauchy-separated (r-s)-symmetric* (in short: *(C-s;r-s)-symmetric*) over X; and (X, d), as a *Cauchy-separated (r-s)-symmetric space* (in short: *(C-s;r-s)-symmetric space*). Further, let $(\leq)$ be a *quasi-order* on X; the structure $(X, d, \leq)$ will be referred to as a *quasi-ordered*

(C-s;r-s)-symmetric space. We say that $Y \in \exp[X]$ is $(\leq)$-*almost-singleton* (in short: $(\leq)$-*asingleton*), if $[y_1, y_2 \in Y,\ y_1 \leq y_2$ imply $y_1 = y_2]$; and $(\leq)$-*singleton* if, in addition, $Y \in \exp(X)$.

Having these precise, take some $T \in \mathcal{F}(X)$; supposed to satisfy

(s-pro) T is semi-progressive $[X(T, \leq) := \{x \in X; x \leq Tx\} \neq \emptyset]$,
(incr) T is increasing $[x \leq y$ implies $Tx \leq Ty]$.

We have to determine circumstances under which $\mathrm{Fix}(T) := \{x \in X; x = Tx\}$ be nonempty. The following concepts establish the directions under which the investigations be conducted (cf. Turinici [255]):

opic-0) We say that T is *fix-$(\leq)$-asingleton*, when $\mathrm{Fix}(T)$ is $(\leq)$-asingleton; and *fix-$(\leq)$-singleton*, when $\mathrm{Fix}(T)$ is $(\leq)$-singleton

opic-1) We say that T is a *Picard operator* (modulo $(d, \leq)$) if, for each point $x \in X(T, \leq)$, $(T^n x; n \geq 0)$ is d-Cauchy, d-convergent [hence, $\lim_n(T^n x)$ is a singleton]

opic-2) We say that T is a *strong Picard operator* (modulo $d, \leq)$) when, for each $x \in X(T, \leq)$, $(T^n x; n \geq 0)$ is d-Cauchy, d-convergent, with $\lim_n(T^n x) \in \mathrm{Fix}(T)$

opic-3) We say that T is a *Bellman Picard operator* (modulo $(d, \leq)$) if, for each $x \in X(T, \leq)$, $(T^n x; n \geq 0)$ is d-Cauchy, d-convergent, with $\lim_n(T^n x) \in \mathrm{Fix}(T)$, and $T^n x \leq \lim_n(T^n x)$ for all n.

In particular, when $(\leq) = X \times X$, these notions are comparable with the ones in Rus [203, Ch. 2, Sec. 2.2]; because, in this case, $X(T, \leq) = X$.

The sufficient (regularity) conditions for such properties are being founded on *ascending orbital full* concepts (in short: (a-o-f)-concepts). Namely, call the sequence $(z_n; n \geq 0)$ in X,

(p-1) *ascending*, if $z_i \leq z_j$ for $i \leq j$;
(p-2) *orbital*, when $(z_n = T^n x; n \geq 0)$, for some $x \in X$;
(p-3) *full*, if the map $n \mapsto z_n$ is injective ($i \neq j$ implies $z_i \neq z_j$);

the intersection of these notions is just the precise one.

oreg-1) Call X, *(a-o-f,d)-complete*, provided (for each (a-o-f)-sequence in X) d-Cauchy $\Longrightarrow$ d-convergent

oreg-2) We say that the selfmap T is *(a-o-f,d)-Cauchy-continuous*, if $[(z_n)$ is d-Cauchy, (a-o-f)-sequence and $z_n \xrightarrow{d} z]$ imply $Tz_n \xrightarrow{d} Tz$

oreg-3) Call $(\leq)$, *(a-o-f,d)-Cauchy-selfclosed*, when $[(z_n)$ is d-Cauchy, (a-o-f)-sequence and $z_n \xrightarrow{d} z]$ imply $[z_n \leq z,\ \forall n]$.

Finally, when the orbital properties are ignored, these conventions may be written in the usual way; we do not give details.

As a completion of introduced concepts, we formulate the specific metrical contractive conditions upon our data. These, essentially, consist in a quasi-order variant of the Meir-Keeler condition [168]. Let $(<)$ stand for the relation $[x < y$ iff $x \leq y$ and $x \neq y]$; clearly, $(<)$ is irreflexive; but not transitive, in general. Denote, for $x, y \in X$:

$$A_1(x,y) = d(x,y),\ A_2(x,y) = (1/2)[d(x,Tx) + d(y,Ty)],$$
$$A_3(x,y) = \max\{d(x,Tx), d(y,Ty)\},\ A_4(x,y) = d(Tx,y);$$

and then, define the families of maps

$$\mathcal{A} = \{A_1, A_2, A_3, A_4\},\ \exp(A_1; \mathcal{A}) = \{\mathcal{K} \in \exp(\mathcal{A}); A_1 \in \mathcal{K}\},$$
$$\mathcal{G} = \{\max(\mathcal{K}); \mathcal{K} \in \exp(A_1; \mathcal{A})\},$$
$$\mathcal{G}_1 = \{\max(\mathcal{K}); \mathcal{K} \in \exp(A_1; \mathcal{A}), A_3 \notin \mathcal{K}\},$$
$$\mathcal{G}_2 = \{\max(\mathcal{K}); \mathcal{K} \in \exp(A_1; \mathcal{A}), A_3 \in \mathcal{K}\}.$$

Clearly, for each $G \in \mathcal{G}$ we have

$A_1(x,y) \leq G(x,y)$, $\forall x, y \in X$; so, in particular:
G is *separated* ($x, y \in X$, $x \neq y$ imply $G(x,y) > 0$).

Given $G \in \mathcal{G}$, we say that T is *Meir-Keeler* $(d, \leq; G)$*-contractive*, if

(mk-1) $x < y$ implies $d(Tx, Ty) < G(x,y)$;
expressed as: T is *strictly nonexpansive* (modulo $(d, \leq; G)$)
(mk-2) $\forall \varepsilon > 0$, $\exists \delta > 0$: $[x < y,\ \varepsilon < G(x,y) < \varepsilon + \delta] \Longrightarrow d(Tx, Ty) \leq \varepsilon$;
expressed as: T has the *Meir-Keeler property* (modulo $(d, \leq; G)$).

Note that, by the former of these, the Meir-Keeler property writes

(mk-3) $\forall \varepsilon > 0$, $\exists \delta > 0$: $[x < y,\ G(x,y) < \varepsilon + \delta] \Longrightarrow d(Tx, Ty) \leq \varepsilon$.

In the following, some basic examples of such contractions will be given.

(I) Let $\mathcal{F}(re)(R^0_+, R)$ be the class of all *regressive* $\varphi \in \mathcal{F}(R^0_+, R)$ $[\varphi(t) < t,\ \forall t \in R^0_+]$. We say that $\varphi \in \mathcal{F}(re)(R^0_+, R)$ is *Meir-Keeler admissible*, if

(MK-adm) $\forall \gamma > 0$, $\exists \beta > 0$, $(\forall t)$: $\gamma < t < \gamma + \beta \Longrightarrow \varphi(t) \leq \gamma$.

Given $(G \in \mathcal{G},\ \varphi \in \mathcal{F}(R^0_+, R))$, call T, $(d, \leq; G, \varphi)$*-contractive* if

(G-phi) $d(Tx, Ty) \leq \varphi(G(x,y))$, $\forall x, y \in X$, $x < y$.

Proposition 2.11. *Let T be $(d, \leq; G, \varphi)$-contractive, where $\varphi \in \mathcal{F}(re)(R^0_+, R)$ is Meir-Keeler admissible. Then, T is Meir-Keeler $(d, \leq; G)$-contractive, in the strongly reduced system (ZF-AC) (hence, all the more, in the reduced system (ZF-AC+DC)).*

Proof. i) Let $x, y \in X$ be such that $x < y$. The contractive condition and G = separated yield (by the regressiveness of φ), $d(Tx, Ty) < G(x, y)$; so that, T is strictly nonexpansive (modulo $(d, \leq; G)$).

ii) Let $\varepsilon > 0$ be arbitrary fixed; and $\delta > 0$ be the number assured by the Meir-Keeler admissible property of φ. Further, let $x, y \in X$ be such that $x < y$ and $\varepsilon < G(x, y) < \varepsilon + \delta$. By the imposed conditions, $[d(Tx, Ty) \leq \varphi(G(x, y)) \leq \varepsilon]$; so that, T has the Meir-Keeler property (modulo $(d, \leq; G)$). □

Some important classes of such functions are given below.

For any $\varphi \in \mathcal{F}(re)(R^0_+, R)$ and any $s \in R^0_+$, put

$\Lambda^+\varphi(s) = \inf_{\varepsilon>0} \Phi^*(s+)(\varepsilon)$; where $\Phi^*(s+)(\varepsilon) = \sup \varphi(]s, s + \varepsilon[)$;
clearly, $-\infty \leq \Lambda^+\varphi(s) \leq s$, for all $s \in R^0_+$.

Call $\varphi \in \mathcal{F}(re)(R^0_+, R)$, *Boyd-Wong admissible* [35] provided

(BW-adm) $\Lambda^+\varphi(s) < s$, for all $s \in R^0_+$.

In particular, $\varphi \in \mathcal{F}(re)(R^0_+, R)$ is Boyd-Wong admissible provided

φ is right-usc on R^0_+ $(\Lambda^+\varphi(s) \leq \varphi(s), \forall s \in R^0_+)$.

This, in particular, holds when φ is continuous at the right on R^0_+; for, in such a case, $[\Lambda^+\varphi(s) = \varphi(s), \forall s \in R^0_+]$.

Further, let $\mathcal{F}(re, in)(R^0_+, R)$ stand for the class of all regressive increasing $\varphi \in \mathcal{F}(R^0_+, R)$; note that, for each $s > 0$,

$\Lambda^+\varphi(s) = \varphi(s + 0) := \lim_{t \to s+} \varphi(t), \forall \varphi \in \mathcal{F}(re, in)(R_+)$.

Call $\varphi \in \mathcal{F}(re, in)(R^0_+, R)$, *Matkowski admissible* [164] provided

(M-adm) $(\forall t \in R^0_+)$: $\lim_n \varphi^n(t) = 0$, whenever $(\varphi^n(t); n \geq 0)$ exists.

The relationships between these classes of functions and the Meir-Keeler admissible one are now made precise in

Proposition 2.12. *Under the described conventions, we have*

(42-1) $(\forall \varphi \in \mathcal{F}(re)(R^0_+, R))$: *(BW-adm)* $\Longrightarrow$ *(MK-adm)*

(42-2) $(\forall\varphi \in \mathcal{F}(re,in)(R_+^0, R))$:
(BW-adm) $\Longrightarrow$ *(M-adm)* $\Longrightarrow$ *(MK-adm)*.

Proof. (Sketch) The former of these is an immediate consequence of definition. And the second one is to be found in Jachymski [120]. □

(II) Let (ψ, φ) be a couple of functions over $\mathcal{F}(R_+^0, R)$, with

(norm-cou) (ψ, φ) is *normal*:
ψ is increasing and φ is strictly positive ($\varphi(R_+^0) \subseteq R_+^0$).

The following extra conditions will be considered here:

(rs-ad) (φ is *right sequentially positive*)
$(\forall\varepsilon > 0)$: $\limsup_n \varphi(t_n) > 0$, whenever $t_n \to \varepsilon+$
(left-ad) ((ψ, φ) is *left admissible*)
$(\forall\varepsilon > 0)$: $\varphi(\varepsilon) > \psi(\varepsilon) - \psi(\varepsilon - 0)$.

Here, given the sequence $(r_n; n \geq 0)$ in R and the point $r \in R$, we denoted

$r_n \to r+$, if $r_n \to r$ and ($r_n > r$, for all $n \geq 0$).

Given $G \in \mathcal{G}$ and the couple (ψ, φ) of functions over $\mathcal{F}(R_+^0, R)$, let us say that T is *$(d, \leq; G, (\psi, \varphi))$-contractive*, provided

(G-psi-phi) $\psi(d(Tx, Ty)) \leq \psi(G(x, y)) - \varphi(G(x, y))$,
for all $x, y \in X$ with $x < y$, $Tx < Ty$.

Proposition 2.13. *Suppose that T is $(d, \leq; G, (\psi, \varphi))$-contractive, for a normal couple (ψ, φ) of functions over $\mathcal{F}(R_+^0, R)$, with φ = right sequentially positive. Then, T is Meir-Keeler $(d, \leq; G)$-contractive, in the reduced system (ZF-AC+DC).*

Proof. i) Let $x, y \in X$ be such that $x < y$. Then (via G = separated), $G(x, y) > 0$; moreover, as T is increasing, we have $Tx \leq Ty$. If $Tx = Ty$, then $d(Tx, Ty) = 0 < G(x, y)$. Assume now that $Tx \neq Ty$; whence, $Tx < Ty$, $d(Tx, Ty) > 0$. As φ = strictly positive, we have $\varphi(G(x, y)) > 0$; wherefrom (by the contractive condition) $\psi(d(Tx, Ty)) < \psi(G(x, y))$. This, via [$\psi$ = increasing], yields $d(Tx, Ty) < G(x, y)$; so that, T is strictly nonexpansive (modulo $(d, \leq; G)$).

ii) Assume by contradiction that T does not have the Meir-Keeler property (modulo $(d, \leq; G)$); that is, for some $\varepsilon > 0$,

$H(\delta) := \{(u,v) \in X \times X; u < v, \varepsilon < G(u,v) < \varepsilon + \delta,\ d(Tu,Tv) > \varepsilon\}$
is nonempty, for each $\delta > 0$.

Taking a zero converging sequence (δ_n) in R^0_+, we get by the Denumerable Axiom of Choice (AC(N)) [deductible, as precise, in (ZF-AC+DC)], a couple of sequences $(x_n; n \geq 0)$ and $(y_n; n \geq 0)$ in X, so as

$(\forall n)$: $(x_n, y_n) \in H(\delta_n)$;
or, equivalently (cf. the preceding step)
$x_n < y_n,\ \varepsilon < d(Tx_n, Ty_n) < G(x_n, y_n) < \varepsilon + \delta_n$.

As a first consequence of this relation,

$$t_n := G(x_n, y_n) \to \varepsilon+ \text{ and } s_n := d(Tx_n, Ty_n) \to \varepsilon+ \text{ as } n \to \infty$$

Then, as a second consequence, one gets (via contractive condition)

$$(0 <)\ \varphi(G(x_n, y_n)) \leq \psi(G(x_n, y_n)) - \psi(d(Tx_n, Ty_n)),\ \ \forall n;$$

so that, passing to lim sup as $n \to \infty$,

$$(0 \leq) \limsup_n \varphi(G(x_n, y_n)) \leq \psi(\varepsilon + 0) - \psi(\varepsilon + 0) = 0;$$

in contradiction with the right sequential positivity of φ. Hence, the Meir-Keeler property follows; and this ends the argument. □

A basic particular case of the obtained fact may be given along the following lines. For any $\varphi \in \mathcal{F}(R^0_+, R)$ and any $s \in R^0_+$, put

$$\Lambda_+\varphi(s) = \sup_{\varepsilon>0} \Phi_*(s+)(\varepsilon);\ \text{where } \Phi_*(s+)(\varepsilon) = \inf \varphi(]s, s+\varepsilon[).$$

By this definition (and some auxiliary facts)

$\limsup_n \varphi(t_n) \geq \liminf_n \varphi(t_n) \geq \Lambda_+\varphi(s)$,
for each sequence (t_n) in R^0_+ with $t_n \to s+$.

Further, define the property (over $\mathcal{F}(R^0_+, R)$)

(r-s-pos) φ is *right locally strict positive*: $\Lambda_+\varphi(s) > 0$, $\forall s \in R^0_+$.

Proposition 2.14. *Suppose that T is $(d, \leq; G, (\psi, \varphi))$-contractive, for a normal couple (ψ, φ) of functions over $\mathcal{F}(R^0_+, R)$, with φ = right locally strict positive. Then, T is Meir-Keeler $(d, \leq; G)$-contractive.*

Note that, by the previous relation involving superior/inferior limits, (for each $\varphi \in \mathcal{F}(R_+^0, R)$): φ is right locally strict positive implies φ is right sequentially positive;

so, this result is reducible to the preceding one. However, for completeness reasons, we provide an argument, patterned after Suzuki and Vetro [220].

Proof. i) See our preceding statement for the deduction of

T = strictly nonexpansive (modulo $(d, \leq; G)$).

ii) Let $\varepsilon > 0$ be arbitrary fixed. As φ is both strictly positive and right locally strict positive, there exists $\theta = \theta(\varepsilon) > 0$ such that $\eta := \inf \varphi[\varepsilon, \varepsilon + \theta] > 0$. On the other hand, by the characterization of right limit, we have

$\psi(\varepsilon + 0) = \inf\{\psi(\varepsilon + t); t > 0\}$; so that:
$\psi(\varepsilon + \delta) < \psi(\varepsilon + 0) + \eta$, for some $\delta \in]0, \theta[$.

Having these precise, let $x, y \in X$ be such that

$x < y$ and $\varepsilon < G(x, y) < \varepsilon + \delta$; hence, $G(x, y) \in]\varepsilon, \varepsilon + \theta[$.

Clearly, $Tx \leq Ty$. If $Tx = Ty$, then $d(Tx, Ty) = 0 < \varepsilon$. Assume now that $Tx \neq Ty$; whence, $Tx < Ty$, $d(Tx, Ty) > 0$. From the definition above, $\varphi(G(x, y)) \geq \eta > 0$; so, combining with the contractive condition, we get

$\psi(d(Tx, Ty)) \leq \psi(G(x, y)) - \varphi(G(x, y)) \leq \psi(\varepsilon + \delta) - \eta <$
$\psi(\varepsilon + 0) \leq \psi(\tau)$, for each $\tau > \varepsilon$.

This, via [ψ = increasing], yields

$d(Tx, Ty) < \tau$, $\forall \tau > \varepsilon$; whence, $d(Tx, Ty) \leq \varepsilon$;

which tells us that T has the Meir-Keeler property (modulo $(d, \leq; G)$). $\square$

The obtained facts, comparable with the ones in Zhu et al. [273], may be used for the construction of a fixed point theory over the described structures. Further aspects will be delineated elsewhere.

2.2.5 *Local Branciari metric spaces*

In the following, some technical facts about local Branciari metric spaces are being discussed.

(A) Let X be a nonempty set; and $d : X \times X \to R_+$ be a (r-s)-symmetric over it; the couple (X, d) will be referred to as a (r-s)-symmetric space. Given $k \geq 1$, any ordered system $C = (x_1, ..., x_k)$ in X^k will be called a *k-chain* of X; the class of all these is denoted as chain$(X;k)$. Given such an object, put $[C] = \{x_1, ..., x_k\}$ (the set of all points belonging to this k-chain); clearly, the alternative card$[C] < k$ (when $k > 1$) cannot be avoided. If card$[C] = k$, then C is called a *regular k-chain* (in X); denote the class of all these as rchain$(X;k)$. In particular, any point $a \in X$ may be identified with a regular 1-chain of X. For any $C \in$ chain$(X;k)$ where $k \geq 2$, denote

$$\Lambda(C) = d(x_1, x_2) + ... + d(x_{k-1}, x_k), \text{ if } C = (x_1, ..., x_k)$$

(the "length" of C). Given $h \geq 1$ and the h-chain $D = (y_1, ..., y_h)$ in X, let $(C; D)$ stand for the $(k+h)$-chain $E = (z_1, ..., z_{k+h})$ of X introduced as

$$z_i = x_i,\ 1 \leq i \leq k;\ \ z_{k+j} = y_j,\ 1 \leq j \leq h;$$

it will be referred to as the "product" between C and D. This operation may be extended to a finite family of such objects.

Having these precise, let us say that the (r-s)-symmetric d is a *local Branciari metric* when it has the "local" triangular property: for each effectively denumerable part $M \subseteq X$, there exists $k = k(M) \geq 1$ such that

(local-tri) $d(x, y) \leq \Lambda(x; C; y)$,
for all $x, y \in M$ with $x \neq y$,
and all $C \in$ rchain$(M;k)$, with $\{x, y\} \cap [C] = \emptyset$

(referred to as: the $(k+2)$-*triangular inequality*); in this case, (X, d) will be referred to as a *local Branciari metric space.* Clearly, any *standard* metric space is a global example of such a structure, corresponding to the 3-triangular inequality ($k = 1$). On the other hand, any Branciari metric space [36] is another global example of such a structure, corresponding to the 4-triangular inequality ($k = 2$). Finally, this local structure is not reducible to the above ones; because, in the described $(k+2)$-triangular inequality (involving the couple (k, M)) the index $k \geq 1$ depends essentially on the (effectively denumerable) subset M of X.

Suppose that we introduced such an object. Define a d-convergence $(\stackrel{d}{\longrightarrow})$ and d-Cauchy structure *Cauchy*(d) over X in the above discussed way. Namely, given the sequence (x_n) in X and the point $x \in X$, we say that (x_n), d-converges to x (written as: $x_n \stackrel{d}{\longrightarrow} x$) provided

$$d(x_n, x) \to 0 \text{ as } n \to \infty\ (\forall \varepsilon > 0,\ \exists i = i(\varepsilon)\text{: } i \leq n \Longrightarrow d(x_n, x) < \varepsilon).$$

The set of all such points x will be denoted $\lim_n(x_n)$; when it is nonempty, (x_n) is called *d-convergent*. Note that $(\stackrel{d}{\longrightarrow})$ is not (in general) separated;

precisely – cf. Samet [207] – there exist sequences (x_n) in X with $\lim_n(x_n)$ having at least two (distinct) points. Further, call the sequence (x_n), *d-Cauchy* when $d(x_m, x_n) \to 0$ as $m, n \to \infty$, $m < n$; that is,

$$\forall \varepsilon > 0,\ \exists j = j(\varepsilon):\ j \leq m < n \Longrightarrow d(x_m, x_n) < \varepsilon.$$

Then, let us say that the couple $((\stackrel{d}{\longrightarrow}), Cauchy(d))$ is a *conv-Cauchy structure* on X (attached to d). We stress that, by the adopted setting, this structure is, generally, not regular: there exist d-convergent sequences that are not d-Cauchy. Finally, call the sequence $(x_n; n \geq 0)$ in X

(d-asy) *d-asymptotic*, if $\lim_n d(x_n, x_{n+1}) = 0$,
(d-to-asy) *d-total-asymptotic*, if $\lim_n d(x_n, x_{n+i}) = 0$, $\forall i \in N(1, \leq)$.

Clearly, the following inclusions hold

(∀ sequence): d-Cauchy $\Longrightarrow$ d-total-asymptotic $\Longrightarrow$ d-asymptotic;

but the converse relations are not in general true.

(B) As already precise, the (nonempty) set of limit points for a convergent sequence is not in general a singleton. However, according to the described (metrical) fixed point methodology, the convergence property of this sequence comes from the d-Cauchy property of the same. So, we may ask whether this supplementary condition will suffice for such a property. The following positive answer to this is available. (See also Jleli and Samet [127], or Kirk and Shahzad [150], [151], for a related particular case.) Remember that the (r-s)-symmetric $d(.,.)$ is *Cauchy-separated*, if

(C-sep) for each (d-convergent, d-Cauchy) sequence (x_n) in X, we have that $\lim_n(x_n) = \{x \in X; x_n \stackrel{d}{\longrightarrow} x\}$ is a singleton.

Proposition 2.15. *Assume that $d(.,.)$ is a local Branciari metric (see above). Then, d is Cauchy-separated.*

Proof. Let (x_n) be a (d-convergent, d-Cauchy) sequence in X. Assume by contradiction that $\lim_n(x_n)$ has at least two distinct points:

$$\exists u, v \in X \text{ with } u \neq v, \text{ such that } x_n \stackrel{d}{\longrightarrow} u,\ x_n \stackrel{d}{\longrightarrow} v.$$

Step 1. Denote $A = \{n \in N; x_n = u\}$, $B = \{n \in N; x_n = v\}$. We claim that both A and B are finite. In fact, if A is effectively denumerable, then $A = \{i(n); n \geq 0\}$, where $(i(n); n \geq 0)$ is strictly ascending and $(x_{i(n)} = u$, $\forall n \geq 0)$. Since, on the other hand, $x_{i(n)} \stackrel{d}{\longrightarrow} v$ as $n \to \infty$, we must have $d(u, v) = 0$; so that, $u = v$, contradiction. An identical reasoning is applicable upon B; and the claim follows. As a consequence, there exists some rank $p \in N$, such that:

$$\{x_n; n \geq p\} \cap \{u, v\} = \emptyset \ \ (x_n \neq u \text{ and } x_n \neq v, \text{ for all } n \geq p).$$

Step 2. Put $h(0) = p$, and

$$S[0] = \{n \geq h(0); x_n = x_{h(0)}\}; \text{ hence, } h(0) \in S[0].$$

We claim that the set $S[0]$ is finite. For, otherwise, it has the representation $S[0] = \{j(n); n \geq 0\}$, where $(j(n); n \geq 0)$ is strictly ascending and $[x_{j(n)} = x_{h(0)}, \forall n]$. Combining with the working convergence hypothesis, one derives

$$x_{h(0)} = u, \ x_{h(0)} = v; \text{ wherefrom, } u = v, \text{ contradiction.}$$

Hence, $S[0]$ is indeed finite; wherefrom, $H(0) = \max S[0]$ exists; moreover, the rank $h(1) := H(0) + 1$ fulfills

$$h(0) < h(1), \ \{x_n; n \geq h(1)\} \cap \{x_{h(0)}\} = \emptyset.$$

Further, denote

$$S[1] = \{n \geq h(1); x_n = x_{h(1)}\}; \text{ hence, } h(1) \in S[1].$$

By a very similar argument, $S[1]$ is finite too; wherefrom, $H(1) = \max S[1]$ exists; moreover, the rank $h(2) := H(1) + 1$ fulfills

$$h(0) < h(1) < h(2), \ \{x_n; n \geq h(2)\} \cap \{x_{h(0)}, x_{h(1)}\} = \emptyset.$$

The procedure may continue indefinitely; it yields – inductively – a strictly ascending rank sequence $(h(n); n \geq 0)$ such that, the subsequence $(y_n := x_{h(n)}; n \geq 0)$ of $(x_n; n \geq 0)$ fulfills

$$(y_n) \text{ is full } (y_i \neq y_j, \text{ for } i \neq j) \text{ and } (y_n \xrightarrow{d} u, \ y_n \xrightarrow{d} v \text{ as } n \to \infty).$$

The subset $M = \{y_n; n \geq 0\} \cup \{u, v\}$ is effectively denumerable; let $k = k(M) \geq 1$ stand for the index assured by the local Branciari metric property of d. By the $(k+2)$-triangular inequality applied to the regular $(k+2)$-chain $C = (u; y_{n+1}, ..., y_{n+k}; v) \in \text{rchain}(M; k+2)$, we have

$$d(u, v) \leq d(u, y_{n+1}) + ... + d(y_{n+k}, v), \text{ for each } n.$$

Moreover, (y_n) is a d-Cauchy sequence; because, so is (x_n); hence $d(y_m, y_{m+1}) \to 0$ as $m \to \infty$. Passing to limit in the above relation gives

$d(u, v) = 0$; whence, $u = v$; contradiction. So, the working convergence condition above is not acceptable; and this concludes the argument. □

(C) Remember that, if $e : X \times X \to R_+$ is a metric on X, the mapping $(x, y) \mapsto e(x, y)$ is continuous, in the sense:

$$x_n \xrightarrow{e} x,\ y_n \xrightarrow{e} y \text{ imply } e(x_n, y_n) \to e(x, y).$$

Unfortunately, for the ambient local Branciari metric $d(.,.)$, this property is no longer valid. However, the following partial version of it is available.

Proposition 2.16. *Let $d(.,.)$ be a local Branciari metric on X. If the full d-asymptotic sequence (x_n) in X and the points $u, v \in X$ are such that*

$$\{x_n; n \geq 0\} \cap \{u, v\} = \emptyset \text{ (that is: } x_n \neq u \text{ and } x_n \neq v,\ \forall n),$$

then, necessarily,

$$x_n \xrightarrow{d} u \text{ implies } d(x_n, v) \to d(u, v).$$

Proof. The case $u = v$ is trivial; so, without loss, one may assume that $u \neq v$. Clearly, the subset $Y := \{x_n; n \geq 0\} \cup \{u, v\}$ is effectively denumerable; let $k = k(Y) \geq 1$ be the index attached to it, by the local Branciari property of d. For each $n \geq 0$, $A_n := (x_{n+1}, ..., x_{n+k})$ and $A_n^* := (x_{n+k}, ..., x_{n+1})$ are regular k-chains in X. By the $(k+2)$-triangular inequality applied to the regular $(k+2)$-chains $(u; A_n; v)$ and $(A_n^*; u; v)$, respectively, we have

$$d(u, v) \leq \Lambda(u; A_n) + d(x_{n+k}, v),\ \ d(x_{n+k}, v) \leq \Lambda(A_n^*; u) + d(u, v);$$
or, equivalently:
$$d(u, v) - \Lambda(u; A_n) \leq d(x_{n+k}, v) \leq d(u, v) + \Lambda(A_n^*; u).$$

On the other hand, from the d-asymptotic and d convergence hypotheses,

$$\Lambda(u; A_n) \to 0 \text{ and } \Lambda(A_n^*; u) \to 0 \text{ as } n \to \infty.$$

Combining these relations, we get $\lim_n d(x_{n+k}, v) = d(u, v)$; and this yields the desired conclusion. □

Note that these facts do not exhaust the subjects to be discussed. Further aspects may be found in Alamri et al. [11] and the references therein.

2.2.6 *Main result*

Let (X, d) be a local Branciari metric space; and $(\leq)$ be a quasi-order on X; the triplet $(X, d, \leq)$ will be called a *quasi-ordered local Branciari metric space*. Further, let T be a selfmap of X; supposed to be semi-progressive and increasing. The specific directions under which the problem of getting fixed points of T is to be solved, were already listed. Sufficient conditions for assuring such properties are being founded on the ascending orbital full concepts as well as specific Meir-Keeler contractive properties.

Our main result in this exposition is

Theorem 2.7. *Suppose that T is Meir-Keeler $(d, \leq; G)$-contractive, for some $G \in \mathcal{G}$. Further, let X be (a-o-f,d)-complete (see above). Then,*

(61-a) *T is fix-$(\leq)$-asingleton (Fix(T) is $(\leq)$-asingleton)*

(61-b) *T is strong Picard (modulo $(d, \leq)$), when (in addition) T is (a-o-f,d)-Cauchy-continuous*

(61-c) *T is Bellman Picard (modulo $(d, \leq)$), provided $(\leq)$ is (a-o-f,d)-Cauchy-selfclosed, and one of the extra conditions below holds:*

> *(61-c-1) $G(.,.)$ belongs to the subclass $\mathcal{G}_1$ (see above)*
> *(61-c-2) T is $(d, \leq; G, \varphi)$-contractive, for a certain Meir-Keeler admissible function $\varphi \in \mathcal{F}(re)(R^0_+, R)$*
> *(61-c-3) T is $(d, \leq; G, (\psi, \varphi))$-contractive, for a left admissible normal couple (ψ, φ) over $\mathcal{F}(R^0_+, R)$, with φ = right sequentially positive*
> *(61-c-4) T is $(d, \leq; G, (\psi, \varphi))$-contractive, for a left admissible normal couple (ψ, φ) over $\mathcal{F}(R^0_+, R)$, with φ = right locally strict positive.*

Proof. There are some steps to be passed.

Part 0. Let $z_1, z_2 \in$ Fix(T) be such that $z_1 \leq z_2$. By definition,

$A_1(z_1, z_2) = A_4(z_1, z_2) = d(z_1, z_2)$, $A_2(z_1, z_2) = A_3(z_1, z_2) = 0$;
wherefrom, $G(z_1, z_2) = d(z_1, z_2)$, for each $G \in \mathcal{G}$.

Assume by contradiction that $z_1 \neq z_2$; hence, $z_1 < z_2$. By the Meir-Keeler strict nonexpansive condition, we then get

$$d(z_1, z_2) = d(Tz_1, Tz_2) < G(z_1, z_2) = d(z_1, z_2);$$

contradiction; hence the claim.

It remains to prove that T has the strong/Bellman Picard property (modulo $(d, \leq)$). Fix some $x_0 \in X(T, \leq)$; and put $(x_n = T^n x_0; n \geq 0)$; clearly, (x_n) is ascending orbital. If $(x_n = x_{n+1}$ for some $n)$, we are done; so, without loss, one may assume that the *non-telescopic* property holds (n-tele) $x_n \neq x_{n+1}$, (hence, $x_n < x_{n+1}$), for all n.

Denote, for each $i \geq 0$,

$$D_i(n) = d(x_n, x_{n+i}),\ E_i(n) = \sup\{D_i(n), D_i(n+1), ...\},\ n \geq 0.$$

Note that $(E_i(n); n \geq 0)$ is a decreasing sequence; but, in general, the alternative of its terms being in $R_+ \cup \{\infty\}$ cannot be excluded.

Part 1. By the admitted hypothesis (and d being reflexive sufficient)

$$\rho_n := D_1(n)\ (= d(x_n, x_{n+1})) > 0, \text{ for all } n.$$

We now establish that

$$\lim_n(\rho_n) = 0; \text{ or, in other words: } (x_n; n \geq 0) \text{ is } d\text{-asymptotic}.$$

In fact, from the very definition of our functions one has, for each n,

$$A_1(x_n, x_{n+1}) = \rho_n \leq \max\{\rho_n, \rho_{n+1}\},$$
$$A_2(x_n, x_{n+1}) = (1/2)[\rho_n + \rho_{n+1}] \leq \max\{\rho_n, \rho_{n+1}\},$$
$$A_3(x_n, x_{n+1}) = \max\{\rho_n, \rho_{n+1}\},\ A_4(x_n, x_{n+1}) = 0;$$
whence (by the choice of G): $\rho_n \leq G(x_n, x_{n+1}) \leq \max\{\rho_n, \rho_{n+1}\}$.

On the other hand, by the imposed alternative upon (x_n), the Meir-Keeler contractive condition applies to (x_n, x_{n+1}), for each $n \geq 0$; and this, by the strict nonexpansive portion of it, gives

$$(\forall n)\text{: } \rho_{n+1} = d(Tx_n, Tx_{n+1}) < G(x_n, x_{n+1}) \leq \max\{\rho_n, \rho_{n+1}\};$$
so that: $\rho_{n+1} < \rho_n$, and (see above) $G(x_n, x_{n+1}) = \rho_n$.

From this strict descending property of $(\rho_n; n \geq 0)$, $\rho := \lim_n \rho_n$ exists in R_+; with, in addition, $(\rho_n > \rho, \forall n)$. Assume that $\rho > 0$; and let $\sigma > 0$ be given by the Meir-Keeler property of T. By definition, there exists some rank $n(\sigma)$ such that

$$n \geq n(\sigma) \Longrightarrow \rho < \rho_n < \rho + \sigma.$$

Combining with a previous relation gives

$$\rho < G(x_n, x_{n+1}) = \rho_n < \rho + \sigma, \text{ for all } n \geq n(\sigma).$$

This, along with the Meir-Keeler contractive condition, yields

$$(\rho <)\ \rho_{n+1} = d(Tx_n, x_{n+1}) \leq \rho,\ \forall n \geq n(\sigma);$$

contradiction. Hence, $\rho = 0$; and our claim follows.

Part 2. Suppose that

(non-full) there exists $i, j \in N$ such that $i < j$, $x_i = x_j$.

By the iterative construction of our sequence, we have $x_{i+1} = x_{j+1}$; which (by definition) yields $\rho_i = \rho_j$; in contradiction with (ρ_n) = strictly descending. Hence, our working hypothesis cannot hold; wherefrom,

$n \mapsto x_n$ is injective; so that: $(x_n; n \geq 0)$ is a full sequence.

Part 3. We now establish, inductively, that for each $i \geq 1$,

(t-asym;i) (x_n) is d-total-asymptotic: $D_i(n) \to 0$ as $n \to \infty$;

The case of $i = 1$ was already established. Assume that, for some $j \geq 1$,

(ind) (t-asym;i) holds, for all $i \in \{1, ..., j\}$;

we claim that, necessarily,

(t-asym;j+1) holds too, in the sense: $D_{j+1}(n) \to 0$, as $n \to \infty$.

In fact, by the inductive hypothesis, it follows that

$(\forall i \in \{1, ..., j\})$: $(E_i(n); n \geq 0)$ is well-defined and descending in R_+;
with in addition: $E_i(n) \to 0$ as $n \to \infty$.

Denote for simplicity $(\sigma_n = E_j(n); n \geq 0)$; hence (see above)

$(\sigma_n; n \geq 0)$ is descending and $\sigma_n \to 0$ as $n \to \infty$.

By the strict descending property of (ρ_n) one has, for each $n \geq 0$,

$$
\begin{aligned}
&A_1(x_n, x_{n+j+1}) = D_{j+1}(n),\\
&A_2(x_n, x_{n+j+1}) = (1/2)[\rho_n + \rho_{n+j+1}] \leq \max\{\rho_n, \rho_{n+j+1}\} = \rho_n,\\
&A_3(x_n, x_{n+j+1}) = \max\{\rho_n, \rho_{n+j+1}\} = \rho_n,\\
&A_4(x_n, x_{n+j+1}) = d(x_{n+1}, x_{n+j+1}) = D_j(n+1) \leq E_j(n) = \sigma_n.
\end{aligned}
$$

Combining with the Meir-Keeler strict nonexpansive property at (x_n, x_{n+j+1}) (possible, via $x_n < x_{n+j+1}$) gives

(for all n): $D_{j+1}(n+1) = d(Tx_n, Tx_{n+j+1}) <$
$G(x_n, x_{n+j+1}) \leq \max\{D_{j+1}(n), \rho_n, \sigma_n\}$.

Denote $(t_n = \max\{D_{j+1}(n), \rho_n, \sigma_n\}; n \geq 0)$; hence

$D_{j+1}(n+1) < t_n$, $\forall n$ [cf. the above relation].

On the other hand, the decreasing properties of our sequences give

$\rho_{n+1} < \rho_n \leq t_n$, $\sigma_{n+1} \leq \sigma_n \leq t_n$, $\forall n$;

so, by simply combining these,

$$t_{n+1} = \max\{D_{j+1}(n+1), \rho_{n+1}, \sigma_{n+1}\} \le t_n,\ \forall n.$$

The sequence $(t_n; n \ge 0)$ is therefore descending; so that

$$t := \lim_n t_n \text{ exists in } R_+; \text{ with, in addition, } (t_n \ge t, \text{ for all } n).$$

Assume that $t > 0$. As $(\rho_n \to 0, \sigma_n \to 0)$, there exists some rank $n(t)$, with

$$(\forall n \ge n(t)):\ \rho_n, \sigma_n < t \le t_n; \text{ whence, } t_n = D_{j+1}(n).$$

Combining with a previous relation, we get

$$(t_{n+1} =) D_{j+1}(n+1) < t_n = D_{j+1}(n),\ \forall n \ge n(t);$$

meaning: the sequence $(D_{j+1}(n); n \ge 0)$ is strictly descending, beyond the rank $n(t)$; with, in addition (see above)

$$\lim_n D_{j+1}(n) = \lim_n t_n = t\ (> 0), \text{ and } D_{j+1}(n) > t,\ \forall n \ge n(t).$$

Let $s > 0$ be the number associated to t, by the Meir-Keeler contractive condition. There exists some rank $n(t,s) > n(t)$, so that

$$\rho_n, \sigma_n < t < D_{j+1}(n) < t + s,\ \forall n \ge n(t,s).$$

On the other hand, for the same ranks,

$$\begin{aligned}
&A_1(x_n, x_{n+j+1}) = D_{j+1}(n) \in]t, t+s[,\\
&A_2(x_n, x_{n+j+1}) = (1/2)[\rho_n + \rho_{n+j+1}] \le \max\{\rho_n, \rho_{n+j+1}\} = \rho_n < t,\\
&A_3(x_n, x_{n+j+1}) = \max\{\rho_n, \rho_{n+j+1}\} = \rho_n < t,\\
&A_4(x_n, x_{n+j+1}) = D_j(n+1) \le E_j(n) = \sigma_n < t;
\end{aligned}$$

wherefrom (by definition)

$$t < A_1(x_n, x_{n+j+1}) \le G(x_n, x_{n+j+1}) < t + s,\ \forall n \ge n(t,s).$$

This, along with the Meir-Keeler property at (x_n, x_{n+j+1}) (possible, in view of $x_n < x_{n+j+1}$) yields, again for all $n \ge n(t,s)$,

$$t < t_{n+1} = D_{j+1}(n+1) = d(Tx_n, Tx_{n+j+1}) \le t;$$

contradiction. Hence, necessarily, $t = 0$; and our assertion follows.

Part 4. By a previous part, (x_n) is full; so that, $Y := \{x_n; n \ge 0\}$ is effectively denumerable; let $k = k(Y) \ge 1$ be the number attached to it, by the local Branciari property of d. Fix $\gamma > 0$; and let $\beta > 0$ be given by the Meir-Keeler property of T; without loss, one may assume $\beta < \gamma$. From the d-total-asymptotic property of (x_n), there exists $j(\beta) \ge 0$, such that (t-asym;i) $D_i(n) = d(x_n, x_{n+i}) < \beta/4k, \ \forall n \geq j(\beta)$

is retainable, for all $i \in \{1, ..., k+2\}$. We now claim, by induction, that

(d-C;p) $D_p(n) = d(x_n, x_{n+p}) < \gamma + \beta/2$, whenever $n \geq j(\beta)$

holds, for all $p \geq 1$; wherefrom, the d-Cauchy property for $(x_n; n \geq 0)$ follows. The case of $p \in \{1, ..., k+2\}$ is clear, by [(t-asym;i); $1 \leq i \leq k+2$]. Assume that, given $q > k+2$ (hence, $q \geq k+3$), (d-C;p) holds, for all $p < q$; we show that (d-C;q) holds as well. Let $n \geq j(\beta)$ be arbitrary fixed. By the inductive hypothesis and a preceding relation, we have

$A_1(x_{n+k}, x_{n+q-1}) = D_{q-k-1}(n+k) < \gamma + \beta/2,$
$A_2(x_{n+k}, x_{n+q-1}) \leq \rho_{n+k} = D_1(n+k) < \beta/4k < \beta,$
$A_3(x_{n+k}, x_{n+q-1}) = \rho_{n+k} = D_1(n+k) < \beta/4k < \beta,$
$A_4(x_{n+k}, x_{n+q-1}) = D_{q-k-2}(n+k+1) < \gamma + \beta/2;$
whence (as (x_n) = full) $0 < G(x_{n+k}, x_{n+q-1}) < \gamma + \beta$.

From the (variant of) Meir-Keeler property attached to (x_{n+k}, x_{n+q-1}) (possible, via $x_{n+k} < x_{n+q-1}$), one derives

$$d(x_{n+k+1}, x_{n+q}) = d(Tx_{n+k}, Tx_{n+q-1}) \leq \gamma.$$

Combining with the $(2+k)$-triangular inequality (applied to the regular k-chain $C = (x_{n+2}, ..., x_{n+k+1})$ in Y), we finally get

$$d(x_n, x_{n+q}) \leq \Lambda(x_n; C) + d(x_{n+k+1}, x_{n+q}) <$$
$$k\beta/4k + \gamma < \beta/2 + \gamma;$$

and the assertion follows.

Part 5. As X is (a-o-f,d)-complete, $x_n \xrightarrow{d} z$ as $n \to \infty$, for some $z \in X$; moreover, as $d(.,.)$ is Cauchy-separated, z is uniquely determined in this way. We claim that, necessarily, $z = Tz$; that is: $z \in \text{Fix}(T)$. Two situations occur.

Case 5-1. Assume that T is (a-o-f,d)-Cauchy-continuous. Then, the sequence $(y_n := Tx_n; n \geq 0)$, d-converges towards Tz. As $(y_n = x_{n+1}; n \geq 0)$ is a subsequence of (x_n), we must also have $y_n \xrightarrow{d} z$. Combining with (y_n) being d-Cauchy, gives (as $d(.,.)$ is Cauchy-separated), $z = Tz$.

Case 5-2. Assume that $(\leq)$ is (a-o-f,d)-Cauchy-selfclosed; note that, by the d-Cauchy and d-convergence properties above,

$x_n \leq z$, for all n (because (x_n) is a d-Cauchy, (a-o-f)-sequence).

We now show that the alternative

(non-fp) $z \neq Tz$; or, equivalently, $\delta := d(z, Tz) > 0$

leads to a contradiction; and this ends the reasoning. Denote $A = \{n \in N; x_n = z\}$, $B = \{n \in N; x_n = Tz\}$. As $(x_n; n \geq 0)$ is full, both A and B are asingletons; so that, the property below must be true

there exists $h \geq 0$, such that: $\{x_n; n \geq h\} \cap \{z, Tz\} = \emptyset$;
hence (by the above): $x_n < z$, for all $n \geq h$.

The subset $Y := \{x_n; n \geq h\} \cup \{z, Tz\}$ is therefore effectively denumerable; let $k = k(Y) \geq 1$ be the natural number attached to it, by the local Branciari property of d. From the very definition of involved functions,

$A_1(x_n, z) = d(x_n, z) \to 0$, $A_2(x_n, z) = (1/2)[\rho_n + \delta] \to \delta/2$,
$A_4(x_n, z) = d(x_{n+1}, z) \to 0$; wherefrom
($\exists$) $\lim_n G(x_n, z) \in \{0, \delta/2\}$, for each $G \in \mathcal{G}_1$.

On the other hand, by a previous auxiliary fact about full sequences,

$$d(x_{n+1}, Tz) \to d(z, Tz) = \delta, \text{ as } n \to \infty.$$

Finally, by the d-total-asymptotic and convergence properties of our ascending orbital full sequence (x_n), there exists some rank $j(\delta) \geq h$, with

$$n \geq j(\delta) \Longrightarrow (\rho_n =) d(x_n, x_{n+1}), d(x_n, z) < \delta/2.$$

As a consequence, we must have for each $n \geq j(\delta)$,

$A_1(x_n, z) = d(x_n, z) < \delta/2$, $A_2(x_n, z) = (1/2)[\rho_n + \delta] < 3\delta/4$,
$A_3(x_n, z) = \max\{\rho_n, \delta\} = \delta$, $A_4(x_n, z) = d(x_{n+1}, z) < \delta/2$;
so that: $G(x_n, z) = \delta$, whenever $G \in \mathcal{G}_2$.

Now, the following alternatives appear.

Alter 5-2-I. Suppose that $G \in \mathcal{G}_1$. For each $n \geq h$, the Meir-Keeler strict nonexpansive property applies to (x_n, z) (via $x_n < z$); and gives

$$d(x_{n+1}, Tz) = d(Tx_n, Tz) < G(x_n, z), \ \forall n \geq h.$$

Passing to limit as $n \to \infty$ in this relation, we get (by the above properties) $\delta \leq \delta/2 < \delta$; a contradiction. Hence, $\delta = 0$; that is, $z \in \text{Fix}(T)$.

Alter 5-2-II. Suppose that T is $(d, \leq; G, \varphi)$-contractive, for a certain Meir-Keeler admissible function $\varphi \in \mathcal{F}(re)(R_+)$. The case of $G \in \mathcal{G}_1$ was already clarified (via T = Meir-Keeler $(d, \leq; G)$-contractive); so, it remains to discuss the case of $G \in \mathcal{G}_2$. By the contractive property,

$$d(Tx_n, Tz) \leq \varphi(G(x_n, z)), \forall n \geq j(\delta);$$

or, equivalently (combining with our previous evaluation of G)

$$d(x_{n+1}, Tz) \leq \varphi(\delta), \ \forall n \geq j(\delta).$$

Passing to limit gives (see above): $\delta \leq \varphi(\delta) < \delta$; contradiction. Hence, necessarily, $\delta = 0$; or, equivalently: $z = Tz$.

Alter 5-2-III. Suppose that T is $(d, \leq; G, (\psi, \varphi))$-contractive, for a left admissible normal couple (ψ, φ) over $\mathcal{F}(R_+^0, R)$, with φ = right sequentially admissible. The case $G \in \mathcal{G}_1$ is clear (via T = Meir-Keeler $(d, \leq; G)$-contractive); so, assume that $G \in \mathcal{G}_2$. From this contractive condition,

$$\psi(d(T_n, Tz)) \leq \psi(G(x_n, z)) - \varphi(G(x_n, z)), \forall n \geq j(\delta);$$

or, equivalently (combining with our previous evaluation of $G \in \mathcal{G}_2$)

$$0 < \varphi(\delta) \leq \psi(\delta) - \psi(d(x_{n+1}, Tz)), \ \forall n \geq j(\delta).$$

Note that, as a consequence,

$$d(x_{n+1}, Tz) < \delta, \text{ for all } n \geq j(\delta).$$

Passing to limit as $n \to \infty$ yields (see above) $\varphi(\delta) \leq \psi(\delta) - \psi(\delta - 0)$. This, however, contradicts the left admissible property of the couple (ψ, φ); so, $\delta = 0$; that is: $z = Tz$.

Alter 5-2-IV. Suppose that T is $(d, \leq; G, (\psi, \varphi))$-contractive, for a left admissible normal couple (ψ, φ) over $\mathcal{F}(R_+^0, R)$ with φ = right locally strict positive (hence, φ = right sequentially admissible). As before, the case $G \in \mathcal{G}_1$ is clear (via T = Meir-Keeler $(d, \leq; G)$-contractive); so, it remains to discuss the case of $G \in \mathcal{G}_2$. This (by the remark above) is the case we just exposed in our preceding stage; and yields $\delta = 0$; that is: $z \in \text{Fix}(T)$. The proof is complete. □

Note that, further extensions of these facts are possible, in the realm of (non-reflexive) sufficient symmetric spaces, taken as in Hicks and Rhoades [112]; or, in the setting of reflexive sufficient partial symmetric spaces, introduced under the lines in Matthews [167]; we shall discuss them elsewhere.

2.2.7 *Particular cases*

Let $(X, d, \leq)$ be a quasi-ordered local Branciari metric space. Further, let T be a selfmap of X; supposed to be semi-progressive and increasing.

Technically speaking, our main result consists of three sub-statements; according to the alternatives of Theorem 2.7 we already listed. For both practical and theoretical reasons, it would be useful to highlight all of these; further aspects involving the obtained facts are also discussed.

Case 1. As a direct consequence of Theorem 2.7, we get

Theorem 2.8. *Suppose that T is Meir-Keeler $(d, \leq; G)$-contractive, for some $G \in \mathcal{G}$. Further, let X be (a-o-f,d)-complete. Then,*

(71-a) *T is fix-$(\leq)$-asingleton (*$\text{Fix}(T)$ *is $(\leq)$-asingleton)*

(71-b) *T is a strong Picard operator (modulo $(d, \leq)$), when (in addition to this) T is (a-o-f,d)-Cauchy-continuous.*

In particular, assume that $(\leq) = X \times X$ (the *trivial* quasi-order on X). Then, under $G = A_1$, the result above reduces to the one in Suzuki [218], proved under pretty different methods. Further aspects may be found in Jleli and Samet [129].

Case 2. As another consequence of Theorem 2.7, we have the following statement (with practical value):

Theorem 2.9. *Suppose that T is Meir-Keeler $(d, \leq; G)$-contractive, for some $G \in \mathcal{G}$. Further, let X be (a-o-f,d)-complete. Then,*

(72-a) *T is fix-$(\leq)$-asingleton (*$\text{Fix}(T)$ *is $(\leq)$-asingleton)*

(72-b) *T is Bellman Picard (modulo $(d, \leq)$), when (in addition to this) $(\leq)$ is (a-o-f,d)-Cauchy-selfclosed, and one of the extra conditions holds:*

(72-b-1) G belongs to the subclass $\mathcal{G}_1$ (see above)
(72-b-2) T is $(d, \leq; G, \varphi)$-contractive, for a certain Meir-Keeler admissible function $\varphi \in \mathcal{F}(re)(R^0_+, R)$.

In particular, suppose that $(\leq)$ is the trivial quasi-order on X (hence, $(\leq)$ is (a-o-f,d)-Cauchy-selfclosed) and the function $\varphi \in \mathcal{F}(re, in)(R^0_+, R)$ is Matkowski admissible. Then, in case of $G = \max\{A_1, A_3, A_4\}$, we have that Theorem 2.9 covers the one obtained by Fora et al. [90]; and, under the choice $G = \max\{A_1, A_2, A_4\}$, the same result includes the one due to Dey and Saha [72]; see also Das and Dey [69]. Note that, by the developments in Jachymski [126], it includes as well the related statements in Di Bari and Vetro [76]. Finally, we must underline that the fixed point statements in Chen and Sun [50] or Kikina et al. [147] are also deductible via these techniques. Further aspects may be found in Abtahi et al. [5].

Case 3. As a final consequence of Theorem 2.7, we have

Theorem 2.10. *Suppose that T is Meir-Keeler $(d, \leq; G)$-contractive, for some $G \in \mathcal{G}$. Further, let X be (a-o-f,d)-complete. Then,*

(73-a) *T is fix-$(\leq)$-asingleton (*Fix(T) *is $(\leq)$-asingleton)*

(73-b) *T is Bellman Picard (modulo $(d, \leq)$), provided (in addition to this) $(\leq)$ is (a-o-f,d)-Cauchy-selfclosed, and one of the extra conditions below holds:*

(73-b-1) G belongs to the subclass $\mathcal{G}_1$ (see above)
(73-b-2) T is $(d, \leq; G, (\psi, \varphi))$-contractive, for a left admissible normal couple (ψ, φ) over $\mathcal{F}(R^0_+, R)$, with $\varphi =$ right sequentially positive
(73-b-3) T is $(d, \leq; G, (\psi, \varphi))$-contractive, for a left admissible normal couple (ψ, φ) over $\mathcal{F}(R^0_+, R)$, with $\varphi =$ right locally strict positive.

The obtained result seems to be new, even if $(\leq)$ is the trivial quasi-order on X and $G = A_1$. Technically speaking, this is comparable with some contributions in Azam and Arshad [21]. Some other aspects of historical nature may be found in the survey paper by Kadelburg and Radenović [135].

2.3 Matkowski Regular Maps in Ordered Metric Spaces

Abstract. Some fixed point results are given for a class of Matkowski regular operators acting on quasi-ordered metric spaces. Technical connections with the related statements due to Chandok et al. [J. Egyptian Math. Soc., 23 (2015), 95–101] are also being discussed.

Subject-Class 2020. 47H10 (Primary), 54H25 (Secondary).

Keywords. Metric space, fixed point, strong Picard operator, quasi-order, Matkowski regular map, rational type contraction.

2.3.1 *Dependent Choice principles*

See Ch. 1.

2.3.2 *Conv-Cauchy structures*

See Ch. 1.

2.3.3 *Banach contractions and relatives*

See Ch. 1.

2.3.4 *Statement of the problem*

Let X be a nonempty set. Take a metric $d(.,.)$ on it; as well as a *quasi-order* $(\leq)$ over X; the triple $(X, d, \leq)$ will be referred to as a *quasi-ordered metric space.* Call the subset Y of X, $(\leq)$-*asingleton* if $[y_1, y_2 \in Y,\ y_1 \leq y_2$ imply $y_1 = y_2]$; and $(\leq)$-*singleton* if, in addition, Y is nonempty.

(A) Fix a selfmap $T \in \mathcal{F}(X)$; and assume in the following that

(s-prog) T is *semi-progressive* $[X(T, \leq) := \{x \in X; x \leq Tx\} \neq \emptyset]$,
(incr) T is *increasing* $[x \leq y$ implies $Tx \leq Ty]$.

We have to determine circumstances under which $\text{Fix}(T) := \{z \in X; z = Tz\}$ is nonempty. The basic directions of our investigations are described by the list below, comparable with the one in Turinici [246]:

Pic-0) We say that T is *fix-*$(\leq)$*-asingleton* when $\text{Fix}(T)$ is $(\leq)$-asingleton; and *fix-*$(\leq)$*-singleton*, provided $\text{Fix}(T)$ is $(\leq)$-singleton

Pic-1) Call T, a *semi Picard operator* (modulo $(d, \leq)$) if, for each $x \in X(T, \leq)$, $(T^n x; n \geq 0)$ is d-asymptotic $(d(T^n x, T^{n+1} x) \to 0$ as $n \to \infty)$

Pic-2) We say that T is a *Picard operator* (modulo $(d, \leq)$) if, for each $x \in X(T, \leq)$, $(T^n x)$ is d-Cauchy $(d(T^n x, T^m x) \to 0$ as $n, m \to \infty$, $n \leq m)$

Pic-3) Call T, a *strong Picard operator* (modulo $(d, \leq)$) when, for each $x \in X(T, \leq)$, $(T^n x; n \geq 0)$ is d-convergent with $\lim_n(T^n x) \in \text{Fix}(T)$

Pic-4) We say that T is a *Bellman Picard operator* (modulo $(d, \leq)$) if, for each $x \in X(T, \leq)$, the iterative sequence $(T^n x)$ is d-convergent with $T^n x \leq \lim_n(T^n x) \in \text{Fix}(T)$, $\forall n$.

In particular, when $(\leq) = X \times X$, these directions are identical with the ones in Rus [203, Ch. 2, Sec. 2.2]; because, in this case, $X(T, \leq) = X$.

The regularity conditions to be used are being founded on *ascending orbital full* concepts (in short: (a-o-f)-concepts). Namely, call the sequence $(z_n; n \geq 0)$ in X,

(def-1) $(\leq)$-*ascending*, whenever $z_i \leq z_j$ for $i \leq j$;
(def-2) T-*orbital*, when $(z_n = T^n x; n \geq 0)$, for some $x \in X$;
(def-3) *full*, if the map $n \mapsto z_n$ is injective ($i \neq j$ implies $z_i \neq z_j$);

the intersection of these notions is just the precise one.

Reg-1) Call the space X, *(a-o-f,d)-complete* when (for each (a-o-f)-sequence) d-Cauchy $\Longrightarrow$ d-convergent

Reg-2) We say that T is *(a-o-f,d)-continuous*, if ((z_n) is (a-o-f)-sequence and $z_n \xrightarrow{d} z$) imply $Tz_n \xrightarrow{d} Tz$

Reg-3) Call $(\leq)$, *(a-o-f,d)-selfclosed* provided ((z_n) is (a-o-f)-sequence and $z_n \xrightarrow{d} z$) imply ($z_n \leq z$, for all n).

As a completion of the developments above, we list the specific metrical conditions upon our data. These, essentially, consist in a quasi-order variant of the 1980 Matkowski regularity conditions [166]; which, in turn, extend the 1969 ones in Meir and Keeler [168]. Let $(<)$ stand for the relation

$x < y$ iff $x \leq y$ and $x \neq y$;

referred to as: the *irreflexive* part of $(\leq)$. Define the family of sets

$X(T, <; p) = \{x \in X; T^i x < T^{i+1} x, \forall i \in N[0, p]\}$, $p \in N$;
as well as $X(T, <; \omega) = \{x \in X; T^i x < T^{i+1} x, \forall i \in N\}$;
note that (by definition) $X(T, <; \omega) = \cap\{X(T, <; p); p \in N\}$.

Sometimes, it will be convenient to use the notations

$$X(T, <) = X(T, <; 0),\ X(T, T, <) = X(T, <; 1).$$

(MR-1) Let us say that T is *orbital strict* $(d, \leq)$-*nonexpansive*, if

(os-nex) $(\forall x)$: $x \in X(T, <; \omega)$ implies $d(Tx, T^2 x) < d(x, Tx)$.

(MR-2) Let us say that T is *orbital* $(d, \leq)$*-admissible*, provided

(o-admi-1) T is orbital strict $(d, \leq)$-nonexpansive (see above)
(o-admi-2) $\forall \varepsilon > 0$, $\exists \delta > 0$, $\forall x \in X(T, <; \omega)$: $\Pi(\varepsilon, \delta; x)$ is false;
where $\Pi(\varepsilon, \delta; x)$ means: $(\varepsilon < d(Tx, T^2x) < d(x, Tx) < \varepsilon + \delta)$.

A sufficient circumstance when this second condition holds is

(o-admi-2a) $\forall \varepsilon > 0$, $\exists \delta > 0$, such that $\forall x \in X(T, <; \omega)$:
$(\varepsilon < d(x, Tx) < \varepsilon + \delta)$ implies $d(Tx, T^2x) \leq \varepsilon$.

In fact, suppose by contradiction that the written condition is not true:

(o-admi-2-non) $\exists \varepsilon > 0$, $\forall \delta > 0$, $\exists x \in X(T, <; \omega)$: $\Pi(\varepsilon, \delta; x)$ is true.

Letting $\delta > 0$ be as in (o-admi-2a) and noting that (for $x \in X(T, <; \omega)$)

$$\Pi(\varepsilon, \delta; x) \text{ is true implies } (\varepsilon < d(x, Tx) < \varepsilon + \delta) \text{ and } (\varepsilon < d(Tx, T^2x)),$$

we contradict the underlying condition; wherefrom, all is clear.

(MR-3) Let us say that T is *Picard* $(d, \leq)$*-contractive*, provided

(le-contr) $\forall \varepsilon > 0$, $\exists \delta > 0$, such that, $\forall (x, y)$:
$(x, y \in X(T, <; \omega)$, $x < y$, $\varepsilon < d(x, y) < \varepsilon + \delta$, and
$0 < d(y, Ty) < d(x, Tx) < \delta)$ imply $d(Tx, Ty) \leq \varepsilon$.

(MR-4) Let us say that T is *Picard* $(d, \leq)$*-admissible*, provided

there exist no (a-o-f)-sequences $(x_n; n \geq 0)$ in $X(T, <; \omega)$ and no elements $z \in X$, with: $x_n \xrightarrow{d} z$, $(x_n < z, \forall n)$, and $d(z, Tz) > 0$.

When all these conditions (MR-1)–(MR-4) hold, the selfmap T is referred to as *Matkowski* $(d, \leq)$*-regular.*

(B) A basic example of such maps is given by the Meir-Keeler ones. Given $T \in \mathcal{F}(X)$, let us call it *Meir-Keeler* $(d, \leq)$*-contractive*, if

(mk-1) $x < y$ implies $d(Tx, Ty) < d(x, y)$;
expressed as: T is *strictly nonexpansive* (modulo $(d, \leq)$)
(mk-2) $\forall \varepsilon > 0$, $\exists \delta > 0$: $[x < y,\ \varepsilon < d(x, y) < \varepsilon + \delta] \Longrightarrow d(Tx, Ty) \leq \varepsilon$;
expressed as: T has the *Meir-Keeler property* (modulo $(d, \leq)$).

Note that, by the former of these, the Meir-Keeler property writes

(mk-3) $\forall \varepsilon > 0$, $\exists \delta > 0$: $[x < y,\ d(x, y) < \varepsilon + \delta] \Longrightarrow d(Tx, Ty) \leq \varepsilon$;
expressed as: T has the *complete Meir-Keeler property* (modulo $(d, \leq)$).

Proposition 2.17. *Suppose that $T \in \mathcal{F}(X)$ is Meir-Keeler $(d, \leq)$-contractive. Then, necessarily, T is Matkowski $(d, \leq)$-regular.*

Proof. I) [T is orbital strict $(d, \leq)$-nonexpansive.] Let $x \in X(T, <; \omega)$ be given; hence, in particular, $x < Tx$. By the strict nonexpansive property (modulo $(d, \leq)$), $d(Tx, T^2x) < d(x, Tx)$; and the desired fact follows.

II) [T is orbital $(d, \leq)$-admissible.] For the moment (cf. the preceding stage), T is orbital strict $(d, \leq)$-nonexpansive; whence, the first half of this property holds. Further, take some $\varepsilon > 0$; and let $\delta > 0$ be given by the Meir-Keeler property (modulo $(d, \leq)$) of T. Then

$$x \in X(T, <) \text{ and } \varepsilon < d(x, Tx) < \varepsilon + \delta \text{ imply } d(Tx, T^2x) \leq \varepsilon; \text{ wherefrom}$$
$$x \in X(T, <; \omega) \text{ and } \varepsilon < d(x, Tx) < \varepsilon + \delta \text{ imply } d(Tx, T^2x) \leq \varepsilon;$$

This, along with a previous remark, yields the second half of our property.

III) [T is Picard $(d, \leq)$-contractive.] Take some $\varepsilon > 0$; and let $\delta > 0$ be given by the Meir-Keeler property (modulo $(d, \leq)$) of T. Then,

$$(\forall x, y \in X)\text{: } x < y \text{ and } \varepsilon < d(x, y) < \varepsilon + \delta \text{ imply } d(Tx, Ty) \leq \varepsilon;$$

and, from this, the written property follows.

IV) [T is Picard $(d, \leq)$-admissible.] Suppose by contradiction that there exist an (a-o-f)-sequence (x_n) in $X(T, <; \omega)$ and some element $z \in X$, with

$$x_n \xrightarrow{d} z, \ (x_n < z, \text{ for all } n), \text{ and } d(z, Tz) > 0.$$

By the strict nonexpansive property (modulo $(d, \leq)$) of T, we have $[d(Tx_n, Tz) < d(x_n, z), \forall n]$; and this, by the imposed properties, yields $Tx_n \xrightarrow{d} Tz$ as $n \to \infty$. Combining with

$$Tx_n = x_{n+1} \xrightarrow{d} z \text{ as } n \to \infty \text{ (obtainable via } (x_n) \text{ being } T\text{-orbital)}$$

one gets (as $d =$ separated) $z = Tz$ [that is: $d(z, Tz) = 0$]; in contradiction with the choice of our data. Consequently, the working hypothesis cannot be accepted; and then, our claim is proved. □

Note that, as a consequence of these facts, the Matkowski regular maps include a pretty large number of (metrical) contractive operators; we do not give details. Some other (concrete) examples of such objects will be given a bit further.

2.3.5 *Main result*

Let $(X, d, \leq)$ be a quasi-ordered metric space; and T be a selfmap of X; supposed to be semi-progressive and increasing. The basic directions and regularity conditions under which the problem of determining elements in Fix(T) is to be solved, were already listed; and the contractive type framework was settled. Our main result in this exposition is

Theorem 2.11. *Assume that T is orbital strict $(d, \leq)$-nonexpansive, orbital $(d, \leq)$-admissible, and Picard $(d, \leq)$-contractive. In addition, let X be (a-o-f,d)-complete. Then, the following conclusions hold*

(51-a) T is a strong Picard operator (modulo $(d, \leq)$), when (in addition) T is (a-o-f,d)-continuous
(51-b) T is a Bellman Picard operator (modulo $(d, \leq)$), when (in addition) $(\leq)$ is (a-o-f,d)-selfclosed and T is Picard $(d, \leq)$-admissible.

Proof. There are several steps to be passed.

Part 1. Take some $x_0 \in X(T, \leq)$; and put $(x_n = T^n x_0; n \geq 0)$; clearly, (x_n) is ascending orbital. If $x_n = x_{n+1}$ for some n, we are done; so, it may be assumed (via T = increasing) that the non-telescopic condition holds

(n-tele) $(x_n < x_{n+1}$ [so that, $\rho_n := d(x_n, x_{n+1}) > 0$], $\forall n)$;
whence, in particular: (x_n) is a sequence in $X(T, <; \omega)$.

As T is orbital strict $(d, \leq)$-nonexpansive,

$\rho_{n+1} = d(Tx_n, Tx_{n+1}) < d(x_n, x_{n+1}) = \rho_n$, $\forall n$;
so, in particular, (ρ_n) is strictly descending.

Two consequences of this must be noted.

Conseq-1. Firstly, we claim that (x_n) is full; hence, an (a-o-f) sequence. Suppose not: there exist $i, j \in N$ such that $i < j$, $x_i = x_j$. By the iterative construction of our sequence, $x_{i+1} = x_{j+1}$; so that, $\rho_i = \rho_j$; in contradiction with (ρ_n) = strictly descending. Hence, our working hypothesis cannot hold; wherefrom the assertion follows.

Conseq-2. Secondly, $\rho := \lim_n \rho_n$ exists in R_+; with, in addition, $(\rho_n > \rho, \forall n)$. Assume by contradiction that $\rho > 0$; and let $\sigma > 0$ be given by the orbital $(d, \leq)$-admissible property of T. By the limit definition, there must be some rank $n(\sigma)$ such that

$(\forall n \geq n(\sigma))$: $\rho < \rho_{n+1} < \rho_n < \rho + \sigma$; or, equivalently,
$(x_n \in X(T, <; \omega)$ and$)$ $\rho < d(Tx_n, T^2 x_n) < d(x_n, Tx_n) < \rho + \sigma$.

But, from the underlying property, existence of such elements x_n (for $n \geq n(\sigma)$) is denied. So, necessarily, $\rho = 0$; that is

$\rho_n := d(x_n, x_{n+1}) \to 0$, as $n \to \infty$; whence: (x_n) is d-asymptotic; proving that T is semi Picard (modulo $(d, \leq)$).

Part 2. Let $\varepsilon > 0$ be arbitrary fixed; and $\delta > 0$ be given by the Picard $(d, \leq)$-contractive property; without loss, one may assume that $\delta < \varepsilon$. By the d-asymptotic property, there exists some rank $n(\delta)$, with

(d-asy) $d(x_n, x_{n+1}) < \delta$ $[< \varepsilon + \delta]$, for all $n \geq n(\delta)$.

We prove, by induction, that for all $i \geq 1$,

(d-C;i) $d(x_n, x_{n+i}) < \varepsilon + \delta$, $\forall n \geq n(\delta)$; whence, (x_n) is d-Cauchy; proving that T is Picard (modulo $(d, \leq)$).

The case $i = 1$ is evident, by (d-asy). Assume that (d-C;i) holds for all $i \in \{1, ..., j\}$, where $j \geq 1$; we must establish that (d-C;j+1) holds too. So, let $n \geq n(\delta)$ be arbitrary fixed. Two alternatives occur.

Alt 1. Suppose that

$d(x_n, x_{n+j}) \leq \varepsilon$; hence, $0 < d(x_n, x_{n+j}) \leq \varepsilon$ (as (x_n) = full).

By the triangular inequality (and our choice of $n(\delta)$)

$$d(x_n, x_{n+j+1}) \leq d(x_n, x_{n+j}) + d(x_{n+j}, x_{n+j+1}) < \varepsilon + \delta;$$

wherefrom, the desired relation follows.

Alt 2. Suppose that

$\varepsilon < d(x_n, x_{n+j})$; hence, $\varepsilon < d(x_n, x_{n+j}) < \varepsilon + \delta$ (cf. the inductive part).

By the strict descending property of (ρ_n) and our very choice of the rank $n(\delta)$, we also have (by a previous remark, and (x_n) being (a-o-f))

$x_n, x_{n+j} \in X(T, <; \omega)$, $x_n < x_{n+j}$, as well as
$0 < d(x_{n+j}, x_{n+j+1}) < d(x_n, x_{n+1}) < \delta$.

This, via posed alternative and the Picard $(d, \leq)$-contractive property, gives

$$d(x_{n+1}, x_{n+j+1}) = d(Tx_n, Tx_{n+j}) \leq \varepsilon;$$

so that, by the triangular inequality (and our choice of $n(\delta)$ once again)

$$d(x_n, x_{n+j+1}) \leq d(x_n, x_{n+1}) + d(x_{n+1}, x_{n+j+1}) < \varepsilon + \delta;$$

whence, the inductive relation follows; and our claim is proved.

Part 3. As X is (a-o-f,d)-complete, $x_n \overset{d}{\longrightarrow} z$, for some (uniquely determined) $z \in X$. By the full property of $(x_n; n \geq 0)$, we have that $A = \{n \in N; x_n = z\}$ is a singleton; so that, the following is true:

(z-dist) $(\exists h = h(z) \geq 0)$: $n \geq h \Longrightarrow x_n \neq z$.

Without loss, one may assume that

(z-dist-0) $h = 0$; that is: $x_n \neq z$, for all n;

for, otherwise, the translated (a-o-f)-sequence $(u_n := x_{h+n}; n \geq 0)$, fulfills this property, as well as the (a-o-f) one. Two cases occur.

Case 1. Suppose that T is (a-o-f,d)-continuous. Then $y_n := Tx_n \overset{d}{\longrightarrow} Tz$ as $n \to \infty$. On the other hand, $(y_n = x_{n+1}; n \geq 0)$ is a subsequence of $(x_n; n \geq 0)$; whence $y_n \overset{d}{\longrightarrow} z$; and this yields (as d is separated), $z = Tz$.

Case 2. Suppose now that $(\leq)$ is (a-o-f,d)-selfclosed and T is Picard $(d, \leq)$-admissible. We show that the alternative

(non-fp) $z \neq Tz$; that is: $d(z, Tz) > 0$

yields a contradiction. From the imposed condition and (z-dist-0),

$(\forall n)$: $x_n \leq z$; hence, $x_n < z$.

This, via (x_n) being an (a-o-f) sequence in $X(T, <; \omega)$, yields [under the limit property of z and (non-fp)], a negation of the imposed Picard $(d, \leq)$-admissible property of T. Hence, $z = Tz$; and the proof is complete. □

Technically speaking, this result is an existence – but, not a uniqueness – one. In other words: Theorem 2.11 cannot give us fix-$(\leq)$-asingleton properties for the underlined map T. However – by a careful examination of our concrete cases to be discussed – the considered property is sometimes accessible (see below).

Finally, note that further extensions of these facts are possible, in the realm of triangular symmetric spaces, taken as in Hicks and Rhoades [112]; or, in the setting of partial metric spaces, introduced under the lines in Matthews [167]; we will discuss them elsewhere.

2.3.6 *Altering rational contractions*

Let $(X, d, \leq)$ be a quasi-ordered metric space; and $T \in \mathcal{F}(X)$ be a selfmap of X; supposed to be semi-progressive and increasing.

The main result above is formulated in terms of Matkowski $(d, \leq)$-regular maps; so, it is natural getting concrete circumstances under which the properties (MR-1)–(MR-4) listed there are attained. It is our aim in the following to show that – along the reduced system (ZF-AC+DC) – a certain version of the main result involving *altering rational contractions* is available; this, in particular, yields a related fixed point statement on (partially) ordered metric spaces due to Chandok et al. [47].

Given the couple of functions (ψ, φ) over $\mathcal{F}(R^0_+, R)$, the regularity conditions below are to be considered

(reg-1) (ψ, φ) is *normal*:
ψ is *increasing* and φ is *strictly positive* $(\psi(R^0_+) \subseteq R^0_+)$
(reg-2) (ψ, φ) is *limit-bounded oscillating*: $\forall$ sequence (t_n) in R^0_+ and $\forall b > 0$ with $t_n \to b$, we have $\limsup_n(\varphi(t_n)) > \psi(b+0) - \psi(b-0)$.

Note that, as a direct consequence of these, we must have

(reg-3) φ is *asymptotic positive*: for each sequence (t_n) in R^0_+ and each $b > 0$ with $t_n \to b$, we have $\limsup_n(\varphi(t_n)) > 0$.

Further, denote for each function $\gamma \in \mathcal{F}(R^0_+, R_+)$ and each point $s > 0$

$$\Lambda^{\pm}\gamma(s) = \inf_{0<\varepsilon<s} \Gamma(s\pm)(\varepsilon); \text{ where } \Gamma(s\pm)(\varepsilon) = \sup \gamma(]s-\varepsilon, s+\varepsilon[).$$

Then, call the function $\gamma \in \mathcal{F}(R^0_+, R_+)$, *locally bounded* provided

$$\Lambda^{\pm}\gamma(s) < \infty \text{ (hence, } 0 \leq \Lambda^{\pm}\gamma(s) < \infty\text{), for each } s > 0;$$

the class of all these functions will be denoted as $\mathcal{F}(loc-bd)(R^0_+, R_+)$. The following characterization of its objects is available

$\gamma \in \mathcal{F}(loc-bd)(R^0_+, R_+)$ iff for each $s > 0$, there exists $\varepsilon \in]0, s[$, such that $\gamma([s-\varepsilon, s+\varepsilon])$ is bounded in R_+;

since the verification is immediate, we do not give details. Finally, denote

$$\mathcal{F}(loc-bd-1)(R^0_+, R_+) = \{\gamma \in \mathcal{F}(loc-bd)(R^0_+, R_+);\ \lim_{t\to 1} \gamma(t) = \gamma(1) = 1\}.$$

Note that, for any such function, we have (by the limit definition)

(p-1) $a_n\gamma(b_n) \to 0$, whenever $a_n \to 0$, $b_n \to b > 0$
(p-2) $a_n\gamma(b_n) \to a$, whenever $a_n \to a > 0$, $b_n \to 1$.

Now, let the quasi-ordered metric space $(X, d, \leq)$ and the selfmap T of X be taken according to the precise general conditions. Further, take a couple $f_1, f_2 \in \mathcal{F}(loc-bd-1)(R^0_+, R_+)$; and let us introduce the functions: for each $x, y \in X$,

$$A_1(x,y) = (1 + d(x,Tx))/(1 + d(x,y)),$$
$$A_2(x,y) = (1 + d(x,Ty))/(1 + d(x,y)),$$
$$P_1(x,y) = d(y,Ty)f_1(A_1(x,y)),\ \ P_2(x,y) = d(Tx,y)f_2(A_2(x,y)),$$
$$P(x,y) = \max\{d(x,y), P_1(x,y), P_2(x,y)\};$$
$$Q_1(x,y) = P_1(x,y),\ \ Q(x,y) = \max\{d(x,y), Q_1(x,y)\}.$$

Note that, by these very definitions,

$$A_1(x,y), A_2(x,y) > 0,\ \forall x,y \in X;$$

hence, the introduced maps $[P_1,\ P_2,\ P,\ Q_1,\ Q]$ are well-defined, as elements in $\mathcal{F}(X \times X, R_+)$. Moreover, again by definition,

$P(x,y), Q(x,y) \geq d(x,y)$, $\forall x,y \in X$; whence:
$P(x,y), Q(x,y) > 0$, for all $x,y \in X$ with $x < y$.

Given the couple of functions (φ, ψ) over $\mathcal{F}(R^0_+, R)$, let us say that the selfmap T of X is $(d, \leq; \psi, P; \varphi, Q)$-*contractive*, if

(PQ-con) $\psi(d(Tx,Ty)) \leq \psi(P(x,y)) - \varphi(Q(x,y))$,
for all $x, y \in X$ with $x < y$, $Tx < Ty$.

By the observations above, this convention is meaningful. We want to determine of to what extent is Theorem 2.11 applicable to the system $[(X, d, \leq)$, T, $(\psi, P; \varphi, Q)]$ involved at the level of our contractive condition. The following auxiliary statement – that is holding over the reduced system (ZF-AC+DC) – is a basic step towards our desired answer.

Proposition 2.18. *Suppose that the selfmap T of X is $(d, \leq; \psi, P; \varphi, Q)$-contractive, where (P, Q) is introduced as before, and (ψ, φ) is a couple of functions over $\mathcal{F}(R^0_+, R)$, with*

(ψ, φ) is normal and limit-bounded oscillating.

Then, the following conclusions hold

(61-1) T is orbital strict $(d, \leq)$-nonexpansive
(61-2) T is orbital $(d, \leq)$-admissible
(61-3) T is Picard $(d, \leq)$-contractive
(61-4) T is Picard $(d, \leq)$-admissible.

Proof. I) [T is orbital strict $(d, \leq)$-nonexpansive.] Take some point $x \in X(T, <; \omega)$; hence, in particular, $x < Tx$, $Tx < T^2x$. By definition,

$$A_1(x,Tx) = 1,\ A_2(x,Tx) = (1 + d(x,T^2x))/(1 + d(x,Tx)),$$
$$P_1(x,Tx) = d(Tx,T^2x)f_1(1) = d(Tx,T^2x),\ P_2(x,Tx) = 0,$$
$$Q_1(x,Tx) = P_1(x,Tx) = d(Tx,T^2x)f_1(1) = d(Tx,T^2x);$$

and this yields

$$P(x,Tx) = Q(x,Tx) = \max\{d(x,Tx), d(Tx,T^2x)\} > 0.$$

The contractive condition applies to the couple (x, Tx), because $x < Tx$, $Tx < T^2x$; and gives an evaluation like

$$\psi(d(Tx,T^2x)) \leq \psi(P(x,Tx)) - \varphi(Q(x,Tx));$$
or, equivalently,
$$\varphi(Q(x,Tx)) \leq \psi(P(x,Tx)) - \psi(d(Tx,T^2x)).$$

As $P(x,Tx) = Q(x,Tx) > 0$ and φ is strictly positive, $\varphi(Q(x,Tx)) > 0$. Combining with the previous relation, we derive $\psi(d(Tx,T^2x)) < \psi(P(x,Tx))$; and this, along with ψ = increasing, tells us that

$$d(Tx,T^2x) < P(x,Tx) = \max\{d(x,Tx), d(Tx,T^2x)\}.$$

But then, necessarily,

$d(Tx,T^2x) < d(x,Tx)$; and conclusion follows.

In addition, as a by-product of this,

(PQ-eva) $P(x,Tx) = Q(x,Tx) = d(x,Tx)$, for all $x \in X(T, <; \omega)$.

II) [T is orbital $(d, \leq)$-admissible.] Suppose by contradiction that T is not orbital $(d, \leq)$-admissible. This, according to a previous stage, means: there exists $\varepsilon > 0$ such that

$$H(\delta) := \{x \in X(T, <; \omega); \varepsilon < d(Tx,T^2x) < d(x,Tx) < \varepsilon + \delta\}$$
is nonempty, for each $\delta > 0$.

Taking a zero converging sequence $(\delta_n; n \geq 0)$ in R^0_+, we get by the Denumerable Axiom of Choice (AC(N)) [deductible, as precise, in (ZF-AC+DC)], a sequence $(x_n; n \geq 0)$ in X so as

$(\forall n)$: $x_n \in H(\delta_n)$; that is (by definition):
$x_n \in X(T, <; \omega)$ and $\varepsilon < d(Tx_n, T^2x_n) < d(x_n, Tx_n) < \varepsilon + \delta_n$.

Note that, as a direct consequence,

$$d(x_n, Tx_n),\ d(Tx_n, T^2x_n) \to \varepsilon+ \text{ as } n \to \infty.$$

Moreover, by our previous evaluation (PQ-eva) over $X(T, <; \omega)$, we have

$$P(x_n, Tx_n) = Q(x_n, Tx_n) = d(x_n, Tx_n),\ \forall n \geq 0.$$

The contractive condition applies to (x_n, Tx_n) (as $x_n < Tx_n$, $Tx_n < T^2x_n$), for each $n \geq 0$; and yields

$(\forall n)$: $\psi(d(Tx_n, T^2x_n)) \leq \psi(d(x_n, Tx_n)) - \varphi(d(x_n, Tx_n))$;
or equivalently
$(0 <)\varphi(d(x_n, Tx_n)) \leq \psi(d(x_n, Tx_n)) - \psi(d(Tx_n, T^2x_n))$.

Passing to lim sup as $n \to \infty$, gives (by a previous convergence relation)

$$0 \leq \limsup_n \varphi(d(x_n, Tx_n)) \leq \psi(\varepsilon + 0) - \psi(\varepsilon + 0) = 0;$$

in contradiction with the limit-bounded oscillating property of (ψ, φ). Hence, our working assumption is not acceptable; and the claim follows.

III) [T is Picard $(d, \leq)$-contractive.] Suppose, by contradiction, that T is not Picard $(d, \leq)$-contractive; that is (for some $\varepsilon > 0$)

$K(\delta) := \{(x, y) \in X \times X;$
$x, y \in X(T, <; \omega),\ x < y,\ \varepsilon < d(x, y) < \varepsilon + \delta,$
$0 < d(y, Ty) < d(x, Tx) < \delta,\ d(Tx, Ty) > \varepsilon\} \neq \emptyset$, for each $\delta > 0$.

Taking a zero converging sequence $(\delta_n; n \geq 0)$ in R^0_+, we get by the Denumerable Axiom of Choice (AC(N)) [deductible, as precise, in (ZF-AC+DC)], a sequence $((x_n, y_n); n \geq 0)$ in $X \times X$, so as

$(\forall n)$: $(x_n, y_n) \in K(\delta_n)$; or, equivalently (by definition)
$x_n, y_n \in X(T, <; \omega)$, $x_n < y_n$, $\varepsilon < d(x_n, y_n) < \varepsilon + \delta_n$,
$0 < d(y_n, Ty_n) < d(x_n, Tx_n) < \delta_n$, $d(Tx_n, Ty_n) > \varepsilon$.

Note that, as a direct consequence,

(p-3) $d(x_n, y_n) \to \varepsilon+$, $d(x_n, Tx_n) \to 0+$, $d(y_n, Ty_n) \to 0+$, as $n \to \infty$.

Moreover, the Lipschitz property of the metric $d(.,.)$ yields, for all n,

$|d(x_n, Ty_n) - d(x_n, y_n)| \leq d(y_n, Ty_n)$,
$|d(Tx_n, y_n) - d(x_n, y_n)| \leq d(x_n, Tx_n)$,
$|d(Tx_n, Ty_n) - d(x_n, y_n)| \leq d(x_n, Tx_n) + d(y_n, Ty_n)$;

wherefrom (as a completion of the preceding relations)

(p-4) $(d(x_n, Ty_n), d(Tx_n, y_n) \to \varepsilon)$, $(d(Tx_n, Ty_n) \to \varepsilon+)$, as $n \to \infty$.

From the definition of our functions and (p-3)+(p-4), one has

$$\begin{aligned}
&\lim_n A_1(x_n, y_n) = 1/(1+\varepsilon),\ \lim_n A_2(x_n, y_n) = 1,\\
&\lim_n P_1(x_n, y_n) = \lim_n d(y_n, Ty_n) f_1(A_1(x_n, y_n)) = 0,\\
&\lim_n P_2(x_n, y_n) = \lim_n d(Tx_n, y_n) f_2(A_2(x_n, y_n)) = \varepsilon,\\
&\lim_n Q_1(x_n, y_n) = \lim_n P_1(x_n, y_n) = 0;
\end{aligned}$$

wherefrom

$$\lim_n P(x_n, y_n) = \max\{\varepsilon, 0, \varepsilon\} = \varepsilon,\ \lim_n Q(x_n, y_n) = \max\{\varepsilon, 0\} = \varepsilon.$$

Moreover, again by definition (and a previous relation)

$$\begin{aligned}
&(P(x_n, y_n), Q(x_n, y_n) \geq d(x_n, y_n) > \varepsilon,\ \forall n);\ \text{which gives}\\
&P(x_n, y_n) \to \varepsilon+ \text{ and } Q(x_n, y_n) \to \varepsilon+,\ \text{as } n \to \infty.
\end{aligned}$$

Finally, the contractive condition applies to (x_n, y_n) (because $x_n < y_n$, $Tx_n < Ty_n$), for each n; and yields

$$\begin{aligned}
&(\forall n){:}\ \psi(d(Tx_n, Ty_n)) \leq \psi(P(x_n, y_n)) - \varphi(Q(x_n, y_n));\\
&\text{or, equivalently}\\
&(0<)\ \varphi(Q(x_n, y_n)) \leq \psi(P(x_n, y_n)) - \psi(d(Tx_n, Ty_n)).
\end{aligned}$$

Passing to lim sup as $n \to \infty$, gives (by our previous convergence relations)

$$0 \leq \limsup_n \varphi(Q(x_n, y_n)) \leq \psi(\varepsilon + 0) - \psi(\varepsilon + 0) = 0;$$

in contradiction with (ψ, φ) = limit-bounded oscillating. Hence, our working assumption is not acceptable; and the assertion follows.

IV) [T is Picard $(d, \leq)$-admissible.] Suppose by contradiction that there exist an (a-o-f)-sequence (x_n) in $X(T, <; \omega)$, and an element $z \in X$, with:

$$x_n \xrightarrow{d} z,\ (x_n < z, \text{ for all } n),\ \text{and } b := d(z, Tz) > 0.$$

Note that, as a direct consequence of these,

$$\text{(p-5)}\quad d(x_n, z) \to 0+,\ d(Tx_n, z) \to 0+,\ d(x_n, Tx_n) \to 0+ \text{ as } n \to \infty.$$

In addition, from the Lipschitz property of $d(.,.)$,

$$\text{(p-6)}\quad d(x_n, Tz) \to b,\ d(Tx_n, Tz) \to b \text{ as } n \to \infty;$$

and this (by the limit definition) yields some index $k = k(b) \geq 0$, such that

$$n \geq k \Longrightarrow d(Tx_n, Tz) \geq b/2 > 0;\ \text{hence (as } T = \text{increasing) } Tx_n < Tz.$$

From the definition of our functions and (p-5)+(p-6), one has

$\lim_n A_1(x_n, z) = 1$, $\lim_n A_2(x_n, z) = 1 + b$,
$\lim_n P_1(x_n, z) = \lim_n d(z, Tz) f_1(A_1(x_n, z)) = b$,
$\lim_n P_2(x_n, z) = \lim_n d(Tx_n, z) f_2(A_2(x_n, z)) = 0$,
$\lim_n Q_1(x_n, z) = \lim_n P_1(x_n, z)) = b$;

and this gives, directly,

$$\lim_n P(x_n, z) = \max\{0, b, 0\} = b,\ \lim_n Q(x_n, z) = \max\{0, b\} = b.$$

Moreover, note that (again by definition)

$$P(x_n, z), Q(x_n, z) \geq d(x_n, z) > 0, \text{ for all } n.$$

Finally, the contractive condition applies to (x_n, z) (because $x_n < z$, $Tx_n < Tz$), for each $n \geq k$; and yields

$(\forall n \geq k)$: $\psi(d(Tx_n, Tz)) \leq \psi(P(x_n, z)) - \varphi(Q(x_n, z))$;
or, equivalently
$(0 <)\varphi(Q(x_n, z)) \leq \psi(P(x_n, z)) - \psi(d(Tx_n, Tz))$.

Passing to lim sup as $n \to \infty$, gives (by a previous convergence relation)

$$0 \leq \limsup_n \varphi(Q(x_n, z)) \leq \psi(b + 0) - \psi(b - 0);$$

in contradiction with the limit-bounded oscillating property of (ψ, φ). Hence, our working assumption is not acceptable; and the claim follows. □

Now, by simply combining this with Theorem 2.11, one gets the following altering type fixed point result (based on rational expressions):

Theorem 2.12. *Suppose that T is $(d, \leq; \psi, P; \varphi, Q)$-contractive, where (P, Q) is introduced as before, and (ψ, φ) is a couple of functions over $\mathcal{F}(R^0_+, R)$, with*

(ψ, φ) is normal and limit-bounded oscillating.

In addition, let X be (a-o-f,d)-complete. Then,

(61-a) T is a strong Picard operator (modulo $(d, \leq)$),
whenever (in addition), T is (a-o-f,d)-continuous
(61-b) T is a Bellman Picard operator (modulo $(d, \leq)$),
whenever (in addition), $(\leq)$ is (a-o-f,d)-selfclosed
(61-c) T is fix-$(\leq)$-asingleton; hence, fix-$(\leq)$-singleton.

Proof. The first and second conclusion are clear – by means of our main result – from the auxiliary statement above; so, the only point to be clarified is that concerning our third conclusion. Let the points $z_1, z_2 \in \mathrm{Fix}(T)$ be such that $z_1 \leq z_2$; whence, $Tz_1 \leq Tz_2$. Suppose by contradiction that

$$z_1 \neq z_2; \text{ hence, } Tz_1 = z_1 < z_2 = Tz_2, \ \rho := d(z_1, z_2) > 0.$$

As a direct consequence, the contractive condition is applicable to (z_1, z_2). But, in view of the very definitions above

$$A_1(z_1, z_2) = 1/(1+\rho), \ A_2(z_1, z_2) = 1,$$
$$P_1(z_1, z_2) = Q_1(z_1, z_2) = 0, \ P_2(z_1, z_2) = \rho f_2(1) = \rho,$$
$$P(z_1, z_2) = \max\{\rho, 0, \rho\} = \rho, \ Q(z_1, z_2) = \max\{\rho, 0\} = \rho.$$

In this case, an effective application of contractive condition gives

$$\psi(\rho) \leq \psi(\rho) - \varphi(\rho); \text{ that is: } \varphi(\rho) \leq 0;$$

in contradiction with the strict positivity of φ. Hence, necessarily, $\rho = 0$ (that is: $z_1 = z_2$); and the proof is complete. □

In particular, when $(\leq)$ is a (partial) order on X, φ is continuous, $(f_1(t) = f_2(t) = t; t > 0)$, and X is d-complete, this result is just the one in Chandok et al. [47], proved by a direct method. Moreover, it also includes – in a partial way – some recent contributions due to Cabrera et al. [44], Dass and Gupta [70], Harjani et al. [109], Kumam et al. [153], Saluja et al. [205], or Samet et al. [210]. Further aspects will be delineated elsewhere.

2.4 Fixed Points via Multistep Type Iterations

Abstract. A functional extension is given for the diagonal fixed point result in Cirić and Presić [Acta Math. Comenianae, 76 (2007), 143–147] involving multistep type iterations.

Subject-Class 2020. 47H17 (Primary), 54H25 (Secondary).

Keywords. Metric space, diagonal fixed point, strong Presić operator, Cirić-Presić and Tasković contraction, Matkowski admissible function.

2.4.1 *Banach contractions and relatives*

See Ch. 1.

2.4.2 *Cirić-Presić result*

Let X be a nonempty set; and $d : X \times X \to R_+$ be a *metric* over it; the couple (X, d) will be referred to as a *metric space.*

Given the couple $p, q \in N(1, \leq)$ and the vectors $a = (a_1, ..., a_p) \in X^p$, $b = (b_1, ..., b_q) \in X^q$, define the *concatenation vector* $(a, b) \in X^{p+q}$ [also written as $(a, b) = (a_1, ..., a_p, b)$ or $(a, b) = (a, b_1, ..., b_q)$], according to

$$(a, b) = (c_1, ..., c_{p+q}), \text{ where}$$
$$(c_i = a_i, 1 \leq i \leq p), (c_{p+j} = b_j, 1 \leq j \leq q);$$

This operation may be extended to a finite number of vectors. In particular, for each natural $h \geq 1$ and each $u \in X$, we introduce the convention

$$u^h\text{=the vector } (z_0, ..., z_{h-1}) \in X^h, \text{ where } (z_j = u, 0 \leq j \leq h - 1).$$

Further, let $k \geq 1$ be a natural number; and $T : X^k \to X$ be a mapping. Denote for simplicity

$$Sx = T(x^k), x \in X \text{ (the } \textit{diagonal} \text{ map attached to } T);$$

clearly, $S(.)$ is a selfmap of X. Then, let us put

$$\text{Fixd}(T) = \text{Fix}(S)(= \{z \in X; z = Sz\});$$

each element of this set will be called a *diagonal fixed point* of T.

By this very definition, it follows that there are two complementary ways in handling the existence and/or uniqueness of these objects.

Way 1. The former of these is *traditional*; and its context is, practically, the one in Rus [203, Ch. 2, Sec. 2.2]:

pic-0) We say that S is *fix-asingleton*, provided Fix(S) is an asingleton; and *fix-singleton*, provided Fix(S) is a singleton

pic-1) Call $x_0 \in X$, a *Picard point* (modulo (d, S)), provided the associated iterative sequence $(x_n := S^n x_0; n \geq 0)$ is d-Cauchy. If each $x_0 \in X$ is a Picard point (modulo (d, S)), then S will be referred to as a *Picard operator* (modulo d).

pic-2) Call $x_0 \in X$, a *strong Picard point* (modulo (d, S)), provided the iterative sequence $(x_n := T^n x_0; n \geq 0)$ is d-convergent and $z := \lim_n(x_n)$ is an element of Fix(S). If each $x_0 \in X$ is a strong Picard point (modulo (d, S)), then S will be referred to as a *strong Picard operator* (modulo d).

The following 1922 fixed point principle due to Banach [22] is in effect here. Call $S(.)$, *Banach* $(d; \alpha)$*-contractive* (where $\alpha \geq 0$), if

(B-con) $d(Sx, Sy) \leq \alpha d(x, y)$, for all $x, y \in X$.

Theorem 2.13. *Assume that S is Banach $(d; \alpha)$-contractive, for some $\alpha \in [0, 1[$. In addition, let X be d-complete. Then,*

(21-a) S is fix-singleton: Fix$(S) = \{z\}$, *for some $z \in X$;*
hence, T is fixd-singleton: Fixd$(T) = \{z\}$, *for some $z \in X$*
(21-b) S is a strong Picard operator (modulo d): for each $x_0 \in X$, the iterative sequence $(x_n = T^n x_0; n \geq 0)$ fulfills $x_n \xrightarrow{d} z$ as $n \to \infty$.

Way 2. The latter of these is a *diagonal* version of the previous one. Given $U_0 := (u_0, ..., u_{k-1}) \in X^k$, the sequence $(u_n; n \geq 0)$ in X given as

$$(u_n = \text{the above one}, n \in N[0, k-1]), \ (u_n = T(u_{n-k}, ..., u_{n-1}), n \geq k)$$

will be called the *iterative process* generated by U_0; and denoted – for simplicity – as $(u_n = T^n U_0; n \geq 0)$.

pp-0) Let us say that T is *fixd-asingleton*, if Fixd(T) = Fix(S) is an asingleton; and *fixd-singleton*, provided Fixd(T) = Fix(S) is a singleton.

pp-1) Call $U_0 = (u_0, ..., u_{k-1}) \in X^k$, a *Presić point* (modulo (d, T)), provided the associated iterative sequence $(u_n := T^n U_0; n \geq 0)$ is d-Cauchy. If each $U_0 = (u_0, ..., u_{k-1}) \in X^k$ is a Presić point (modulo (d, T)), then T will be referred to as a *Presić operator* (modulo d).

pp-2) Call $U_0 = (u_0, ..., u_{k-1}) \in X^k$, a *strong Presić point* (modulo (d, T)), provided the associated iterative sequence $(u_n := T^n U_0; n \geq 0)$ is d-convergent and $z := \lim_n(u_n)$ is an element of Fixd(T) = Fix(S). If each $U_0 = (u_0, ..., u_{k-1}) \in X^k$ is a strong Presić point (modulo (d, T)), then T will be referred to as a *strong Presić operator* (modulo d).

Sufficient conditions for such properties involve metrical contractions. Given $\Gamma := (\gamma_0, ..., \gamma_{k-1}) \in R_+^k$, call T, *Presić $(d;\Gamma)$-contractive* if

(P-contr) $d(T(x_0, ...x_{k-1}), T(x_1, ..., x_k)) \leq$
$\gamma_0 d(x_0, x_1) + ... + \gamma_{k-1} d(x_{k-1}, x_k)$, for each $(x_0, ..., x_k) \in X^{k+1}$.

The regularity condition imposed upon the vector Γ writes

(norm-sub) Γ is *norm subunitary:* $\gamma := \gamma_0 + ... + \gamma_{k-1} < 1$.

The following 1965 result obtained by Presić [188] is our starting point:

Theorem 2.14. *Suppose that T is Presić $(d;\Gamma)$-contractive, where the vector $\Gamma = (\gamma_1, ..., \gamma_k) \in R_+^k$ is norm subunitary. In addition, let X be d-complete. Then,*

(22-a) *T is fixd-singleton:* $\text{Fixd}(T) = \text{Fix}(S) = \{z\}$, *for some $z \in X$*

(22-b) *T is strong Presić (modulo d): for each starting $U_0 := (u_0, ..., u_{k-1}) \in X^k$, the iterative sequence $(u_n = T^n U_0; n \geq 0)$ given as [u_n=the above ones, $0 \leq n \leq k-1$; $u_n = T(u_{n-k}, ..., u_{n-1})$, $n \geq k$] fulfills $u_n \xrightarrow{d} z$ as $n \to \infty$.*

Remark 2.2.

In particular, when $k = 1$, Theorem 2.14 is just the Banach contraction principle [22]. Actually, the existence and uniqueness portion of Theorem 2.14 is entirely reducible to the underlying principle. For, let $x, y \in X$ be arbitrary fixed. From the Presić contractive condition

$$\begin{aligned} &d(Sx, Sy) = d(T(x^k), T(y^k)) \leq \\ &d(T(x^k), T(x^{k-1}, y)) + \cdots + d(T(x, y^{k-1}), T(y^k)) \leq \\ &\gamma_0 d(x, y) + \cdots + \gamma_{k-1} d(x, y) = \gamma d(x, y); \end{aligned}$$

which tells us that S is Banach $(d;\gamma)$-contractive. But then, Theorem 2.13 is applicable here; and from this, we are done.

Summing up, the novelty of Theorem 2.14 consists in possibility of approximating the diagonal fixed point of T by means of diagonal iterative processes. From such a perspective, the natural question to be posed is that of extending the underlying result in such a way that its existence and/or uniqueness part be no longer depending on Theorem 2.13. A basic contribution in this area is the 2007 one due to Cirić and Presić [62]. Let us say that T is *Cirić-Presić $(d;\beta)$-contractive* (where $\beta \in R_+$), provided

(CP-contr) $d(T(x_0, ..., x_{k-1}), T(x_1, ..., x_k)) \leq$
$\beta \max\{d(x_0, x_1), ..., d(x_{k-1}, x_k)\}$, for each $(x_0, ..., x_k) \in X^{k+1}$.

Theorem 2.15. *Assume that T is Cirić-Presić $(d;\beta)$-contractive, for some $\beta \in [0,1[$. In addition, let X be d-complete. Then, T is strong Presić (modulo d): for each $U_0 := (u_0, ...u_{k-1}) \in X^k$, the iterative sequence $(u_n = T^n U_0; n \geq 0)$ given as [u_n=the above ones, $0 \leq n \leq k-1$; $u_n = T(u_{n-k}, ..., u_{n-1})$, $n \geq k$] fulfills $u_n \xrightarrow{d} z$ as $n \to \infty$, for some* $z \in \text{Fixd}(T) = \text{Fix}(S)$.

For a number of technical reasons, we provide a proof of this statement; which differs, in part, from the original one.

Proof. It will suffice considering the case of $\beta > 0$; hence, $\beta \in]0,1[$. There are several parts to be passed.

Part 1. Denote for simplicity

$$(\rho_n := d(u_n, u_{n+1}); n \geq 0),\ \theta = \beta^{1/k} \text{ (hence, } 0 < \theta < 1,\ \theta^k = \beta)$$
$$B = \max\{\rho_i/\theta^i; 0 \leq i \leq k-1\} \text{ (hence, } \rho_i \leq B\theta^i, \text{ for } 0 \leq i \leq k-1).$$

We prove, by induction, that the evaluation

$$\text{(eva;i)}\quad \rho_i \leq B\theta^i \text{ [where } \theta \in]0,1[\text{ is the above number]},$$

is available, for all $i \geq 0$. The case of $i \in \{0, ..., k-1\}$ is clear, by definition. Suppose that the assertion (eva;i) is true for all $i \in \{0, ..., j\}$, where $j \geq k-1$; we want to establish that (eva;j+1) holds as well. From the contractive property (and imposed convention), we have

$$\rho_{j+1} \leq \beta \max\{\rho_{j+1-k}, ..., \rho_j\} \leq$$
$$B\beta \max\{\theta^{j+1-k}, ..., \theta^j\} = B\beta\theta^{j+1-k} = B\theta^{j+1};$$

and the claim follows. As a consequence of this,

$$\textstyle\sum_n \rho_n \text{ converges; whence, } (u_n) \text{ is } d\text{-Cauchy.}$$

Combining with X being d-complete, one derives

$$x_n \xrightarrow{d} z \text{ as } n \to \infty, \text{ for some } z \in X.$$

Part 2. We now claim that the obtained limit z is a diagonal fixed point for T: $z = T(z^k)(= Sz)$. To do this, it will suffice proving that

$$u_n \xrightarrow{d} Sz(= T(z^k)), \text{ as } n \to \infty.$$

In fact, the triangle inequality and contractive condition yield

$(\forall n \geq k)$: $d(u_n, Sz) = d(T(u_{n-k}, ..., u_{n-1}), T(z^k)) \leq$
$d(T(u_{n-k}, u_{n-k+1}, ..., u_{n-1}), T(u_{n-k+1}, ..., u_{n-1}, z)) +$
$d(T(u_{n-k+1}, u_{n-k+2}, ..., u_{n-1}, z), T(u_{n-k+2}, ..., u_{n-1}, z^2))$
$+ \cdots + d(T(u_{n-1}, z^{k-1}), T(z^k)) \leq$
$\beta \max\{\rho_{n-k}, ..., \rho_{n-2}, d(u_{n-1}, z)\} +$
$\beta \max\{\rho_{n-k+1}, ..., \rho_{n-2}, d(u_{n-1}, z)\} + \cdots + \beta d(u_{n-1}, z)$.

Passing to limit as $n \to \infty$, gives

$\lim_n d(u_n, Sz) = 0$; whence (as d = separated), $z = Sz$;

and this yields the desired fact. The proof is thereby complete. □

Note, finally, that further extensions – to the coincidence point setting – of these facts are possible, under the lines in George and Khan [95], or Păcurar [181], [182]; these, however, will be not treated here. For direct applications of such facts to convergence questions involving real sequences, we refer to Chen [49] and the references therein.

2.4.3 *Previous Tasković approach*

In the following, we show that the 2007 Ćirić-Presić result [62] we just presented is *almost* equivalent with the 1976 statement in Tasković [223].

Given $k \geq 1$, let $f : R_+^k \to R_+$ be a function; and $\Gamma := (\gamma_0, ..., \gamma_{k-1})$ be an element in R_+^k. We say that (f, Γ) is *admissible*, provided

(adm-1) f is increasing (in its variables):
$(u_i \leq v_i,\ i \in \{0, ..., k-1\})$ imply $f(u_0, ..., u_{k-1}) \leq f(v_0, ..., v_{k-1})$
(adm-2) f is *semi-homogeneous*:
$f(\lambda x_0, ..., \lambda x_{k-1}) \leq \lambda f(x_0, ..., x_{k-1}),\ \forall (x_0, ..., x_{k-1}) \in R_+^k,\ \forall \lambda \geq 0$
(adm-3) the associated map $g : R_+ \to R_+$ introduced as
$(g(t) = f(\gamma_0 t, ..., \gamma_{k-1} t^k); t \in R_+)$ is continuous at $t = 1$.

[Note that, in the 2011 Tasković paper [224], the above imposed conditions (adm-1)–(adm-3) are being referred to as: f has the *M-property*.] Then, for an easy reference, let us introduce the couple of conditions

(sub) (f, Γ) is *subunitary*: $\beta := f(\gamma_0, ..., \gamma_{k-1}) < 1$
(tele-sub) (f, Γ) is *telescopic subunitary*:
$\alpha := f(\gamma_0, 0, ..., 0) + ... + f(0, ..., 0, \gamma_{k-1}) < 1$.

Having these precise, let us say that $T : X^k \to X$ is *Tasković $(d; f, \Gamma)$-contractive* (where $f \in \mathcal{F}(R_+^k, R_+)$, $\Gamma := (\gamma_0, ..., \gamma_{k-1}) \in R_+^k$), provided

(T-contr) $d(T(x_0,...,x_{k-1}),T(x_1,...,x_k)) \leq$
$f(\gamma_0 d(x_0,x_1),...,\gamma_{k-1}d(x_{k-1},x_k))$, for each $(x_0,...,x_k) \in X^{k+1}$.

The following 1976 statement in Tasković [223] is our starting point.

Theorem 2.16. *Assume that T is Tasković $(d; f, \Gamma)$-contractive, where $f \in \mathcal{F}(R_+^k, R_+)$ is a mapping and $\Gamma = (\gamma_0,...,\gamma_{k-1})$ is a vector in R_+^k. In addition, let X be d-complete. Then, the following conclusions hold:*

(31-a) *If the couple (f, Γ) is admissible and telescopic subunitary, then T is fixd-asingleton, in the sense:* $\text{Fixd}(T) = \text{Fix}(S)$ *is an asingleton*

(31-b) *If the couple (f, Γ) is admissible and subunitary, then T is strong Presić (modulo d): for each $U_0 := (u_0,...u_{k-1}) \in X^k$, the iterative sequence $(u_n = T^n U_0; n \geq 0)$ fulfills $u_n \xrightarrow{d} z$, for some $z \in$* $\text{Fixd}(T) = \text{Fix}(S)$.

Before passing to the discussion of this result, some technical remarks about the semi-homogeneous condition are in order. Given $f \in \mathcal{F}(R_+^k, R_+)$, call it *homogeneous*, provided

$$f(\lambda x_0,...,\lambda x_{k-1}) = \lambda f(x_0,...,x_{k-1}),\ \forall (x_0,...,x_{k-1}) \in R_+^k,\ \forall \lambda \geq 0.$$

Clearly, any such function is semi-homogeneous. But, the reciprocal inclusion is also true; as it results from

Proposition 2.19. *For each function $f \in \mathcal{F}(R_+^k, R_+)$, we have*

semi-homogeneous $\Longrightarrow$ homogeneous;
whence, semi-homogeneous $\Longleftrightarrow$ homogeneous.

Proof. Suppose that $f : R_+^k \to R_+$ is semi-homogeneous; that is:

$(0 \leq) f(\lambda x_0,...,\lambda x_{k-1}) \leq \lambda f(x_0,...,x_{k-1})$,
for all $(x_0,...,x_{k-1}) \in R_+^k$ and all $\lambda \geq 0$.

Putting $\lambda = 0$ in this relation gives

$$0 \leq f(0,...,0) \leq 0;\ \text{whence},\ f(0,...,0) = 0;$$

meaning: the homogeneous property is fulfilled in case of $\lambda = 0$. It remains then to verify the underlying property in case of $\lambda > 0$. Denote for simplicity $\mu := 1/\lambda$; and let $(x_0,...,x_{k-1}) \in R_+^k$ be arbitrary fixed. By the semi-homogeneous property, we have

$f(x_0,...,x_{k-1}) = f(\mu\lambda x_0,...,\mu\lambda x_{k-1}) \leq \mu f(\lambda x_0,...,\lambda x_{k-1})$;
or, equivalently:
$\lambda f(x_0,...,x_{k-1}) \leq f(\lambda x_0,...,\lambda x_{k-1})$.

This, along with the semi-homogeneous inequality, gives

$$f(\lambda x_0, ..., \lambda x_{k-1}) = \lambda f(x_0, ..., x_{k-1}), \ \forall (x_0, ..., x_{k-1}) \in R_+^k, \ \forall \lambda > 0;$$

and proves our assertion. □

In this perspective, the following couple of results is in effect here. Given $k \geq 1$, take some function $f : R_+^k \to R_+$ and some vector $\Gamma := (\gamma_0, ..., \gamma_{k-1}) \in R_+^k$. We say that f is *quasi admissible*, provided

(qadm-1) f is increasing (in its variables):
$(u_i \leq v_i, \ i \in \{0, ..., k-1\})$ imply $f(u_0, ..., u_{k-1}) \leq f(v_0, ..., v_{k-1})$
(qadm-2) f is homogeneous:
$f(\lambda x_0, ..., \lambda x_{k-1}) = \lambda f(x_0, ..., x_{k-1}), \ \forall (x_0, ..., x_{k-1}) \in R_+^k, \ \forall \lambda \geq 0.$

Theorem 2.17. *Assume that T is Tasković $(d; f, \Gamma)$-contractive, where $f \in \mathcal{F}(R_+^k, R_+)$ and $\Gamma = (\gamma_0, ..., \gamma_{k-1}) \in R_+^k$ are such that*

f is quasi admissible and (f, Γ) is telescopic subunitary.

In addition, let X be d-complete. Then, the following conclusions hold:

(32-a) S is $(d; \alpha)$-contractive [$d(Sx, Sy) \leq \alpha d(x, y), \ \forall x, y \in X$]; hence, S is fix-asingleton (or, equivalently: T is fixd-asingleton)
(32-b) S is strongly Picard (modulo d): for each $v_0 \in X$, the iterative sequence $(v_n = S^n v_0; n \geq 0)$ is d-convergent and $v := \lim_n(v_n)$ is a fixed point of S (or, equivalently: a diagonal fixed point of T).

Theorem 2.18. *Assume that T is Tasković $(d; f, \Gamma)$-contractive, where $f \in \mathcal{F}(R_+^k, R_+)$ and $\Gamma = (\gamma_0, ..., \gamma_{k-1}) \in R_+^k$ are such that*

f is quasi admissible and (f, Γ) is subunitary.

In addition, let X be d-complete. Then, T is strong Presić (modulo d): for each $U_0 := (u_0, ..., u_{k-1}) \in X^k$, the iterative sequence $(u_n = T^n U_0; n \geq 0)$ fulfills $u_n \xrightarrow{d} z$, for some diagonal fixed point $z \in \mathrm{Fixd}(T) = \mathrm{Fix}(S)$.

Concerning the relationships between these results and the previous ones the following synthetic answer is available.

Proposition 2.20. *Under the above conventions,*

(32-1) Theorem 2.13 $\Longrightarrow$ Theorem 2.17
$\Longrightarrow$ Theorem 2.16 (the first half)

(32-2) Theorem 2.15 $\Longrightarrow$ *Theorem 2.18*
$\Longrightarrow$ *Theorem 2.16 (the second half)* $\Longrightarrow$ *Theorem 2.15.*

Proof. i): Two inclusions must be verified.

i-1): Let the premises of Theorem 2.17 be fulfilled. In particular, as (f, Γ) is telescopic subunitary,

$$\alpha := f(\gamma_0, 0, ..., 0) + ... + f(0, ..., 0, \gamma_{k-1}) < 1.$$

We claim that (under the precise notations)

$$S \text{ is } (d; \alpha)\text{-contractive: } d(Sx, Sy) \leq \alpha d(x, y), \ \forall x, y \in X.$$

In fact, let $x, y \in X$ be arbitrary fixed. From the Tasković contractive condition (and f = homogeneous), one derives

$$\begin{aligned} &d(Sx, Sy) = d(T(x^k), T(y^k)) \leq \\ &d(T(x^k), T(x^{k-1}, y)) + \cdots + d(T(x, y^{k-1}), T(y^k)) \leq \\ &f(\gamma_0 d(x, y), 0, ..., 0) + \cdots + f(0, ..., 0, \gamma_{k-1} d(x, y)) = \alpha d(x, y); \end{aligned}$$

hence (32-a) follows. On the other hand, (32-b) is obtainable from this contractive condition, via Theorem 2.13; so that, the first inclusion follows.

i-2): Evident, via: admissible $\Longrightarrow$ quasi admissible.

ii): We have three inclusions to discuss.

ii-1): Let the premises of Theorem 2.18 be fulfilled. From the imposed properties, we have (under our notations)

$$f(\gamma_0 t_0, ..., \gamma_{k-1} t_{k-1}) \leq \beta \max\{t_0, ..., t_{k-1}\}, \text{ for each } (t_0, ..., t_{k-1}) \in R_+^k.$$

As a consequence, the mapping T is Cirić-Presić $(d; \beta)$-contractive; and this tells us that Theorem 2.15 implies Theorem 2.18.

ii-2): Clearly, Theorem 2.18 implies Theorem 2.16 (the second half).

ii-3): Under the premises of Theorem 2.15, take the couple (f, Γ) as

$$\begin{aligned} &f(t_0, ..., t_{k-1}) = \max\{t_0, ..., t_{k-1}\}, \ (t_0, ..., t_{k-1}) \in R_+^k; \\ &\Gamma = (\gamma_0, ..., \gamma_{k-1}) \in R_+^k, \text{ where } (\gamma_i = \beta; \ 0 \leq i \leq k-1). \end{aligned}$$

Clearly, T is Tasković $(d; f, \Gamma)$-contractive; with, in addition,

$$(f, \Gamma) \text{ is admissible and subunitary.}$$

Putting these together, it results that Theorem 2.16 (the second half) implies Theorem 2.15. The proof is thereby complete. □

Summing up these facts, we firstly derive

(eq-1) Theorem 2.16 (the second half)
is equivalent with Theorem 2.15;

which shows that the claim in Tasković [224] is retainable. On the other hand, by the same argument we do have

(eq-2) Theorem 2.16 (the second half)
is equivalent with Theorem 2.18.

Concerning this aspect, note that (by the involved concepts), the right to left inclusion is trivial. However, the left to right inclusion is not a trivial one; because, the third admissible condition upon (f, Γ)

the associated map $g : R_+ \to R_+$ introduced as
$(g(t) = f(\gamma_0 t, ..., \gamma_{k-1} t^k); t \in R_+)$ is continuous at $t = 1$

is not deductible from the quasi admissible condition upon f. Further aspects may be found in Tasković [224].

2.4.4 *Main result*

Let (X, d) be a metric space, and $k \in N(1, \leq)$ be a natural number. Further, let $T : X^k \to X$ be a mapping; and $(Sx = T(x^k); x \in X)$ be its associated diagonal map. Remember that we denoted $\text{Fixd}(T) = \text{Fix}(S)$; each element of it will be referred to as a *diagonal fixed point* of T. The basic setting of our problem was already sketched in a previous place. It is our aim in the following to give a functional version of the Ćirić-Presić fixed point result we already presented. Some conventions are in order. Given a function $\varphi \in \mathcal{F}(R_+)$, call it

(in) *increasing* provided $t_1 \leq t_2$ implies $\varphi(t_1) \leq \varphi(t_2)$;
(re) *regressive*, if $\varphi(0) = 0$ and $\varphi(t) < t$, for all $t \in R_+^0 :=]0, \infty[$;

the class of all these will be denoted as $\mathcal{F}(in)(R_+)$ and $\mathcal{F}(re)(R_+)$, respectively. Further, put

$$\mathcal{F}(in, re)(R_+) = \mathcal{F}(in)(R_+) \cap \mathcal{F}(re)(R_+);$$

each function in this class is both increasing and regressive. Finally, call the function $\varphi \in \mathcal{F}(in, re)(R_+)$,

(M-ad) *Matkowski admissible* if $\varphi^n(t) \to 0$ as $n \to \infty$, $\forall t \geq 0$;
(s-M-ad) *strongly Matkowski admissible*, if $\sum_n \varphi^n(t) < \infty$, $\forall t \geq 0$;

here, φ^n = the n-th iterate of φ, for each n. (These conventions are taken from Matkowski [164], [165] and Turinici [248], respectively.) Clearly,

(for each $\varphi \in \mathcal{F}(in, re)(R_+)$):
strongly Matkowski admissible implies Matkowski admissible;

but the reciprocal is not in general true.

Returning to our initial setting, denote for each $i, j \in N$ with $i < j$

$$A(z_i, ..., z_j) = \max\{d(z_i, z_{i+1}), ..., d(z_{j-1}, z_j)\},\ (z_i, ..., z_j) \in X^{j-i+1}.$$

Given $\varphi \in \mathcal{F}(R_+)$, let us say that T is (d, φ)-*contractive*, provided

(d-con) $d(T(y_0, ..., y_{k-1}), T(y_1, ..., y_k)) \le \varphi(A(y_0, ..., y_{k-1}, y_k))$,
for all $(y_0, ..., y_{k-1}, y_k) \in X^{k+1}$.

Our main result in this exposition is

Theorem 2.19. *Suppose that the selfmap T is (d, φ)-contractive, for some strongly Matkowski admissible $\varphi \in \mathcal{F}(in, re)(R_+)$. In addition, let X be d-complete. Then, T is strong Presić (modulo d): for each $U_0 := (u_0, ..., u_{k-1}) \in X^k$, the iterative sequence $(u_n := T^n U_0; n \ge 0)$ given as*

$$(u_n = \text{the above ones},\ n \in N[0, k-1]),\ (u_n = T(u_{n-k}, ..., u_{n-1}),\ n \ge k)$$

fulfills $u_n \xrightarrow{d} z$, for some diagonal fixed point $z \in \mathrm{Fixd}(T) = \mathrm{Fix}(S)$.

Proof. There are several stages to be passed.

Step 1. Put, for simplicity,

$$\rho_n = d(u_n, u_{n+1}),\ n \ge 0.$$

Note that, by a previous convention, we have for each i, j with $i < j$,

$$A(u_i, ..., u_j) = \max\{\rho_i, ..., \rho_{j-1}\}.$$

For technical reasons, it would be useful to denote, for each $n \ge k$,

$$B_n = A(u_{n-k}, ..., u_{n-1}, u_n);\ \text{that is: } B_n = \max\{\rho_{n-k}, ..., \rho_{n-1}\}.$$

Before passing to the effective part of our argument, two auxiliary facts are in order. The former of these may be viewed as a stationarity criterion involving the terms of our iterative process.

Proposition 2.21. *Suppose that*

(B-zero) $B_n = 0$, *for some* $n \ge k$.

Then

(41-1) *The subsequence* $(y_i := u_{n-k+i}; i \geq 0)$ *is constant:*

$$u_{n-k} = u_{n-k+1} = \cdots = a, \text{ for some } a \in X.$$

(41-2) *In addition, we have*

$$a = T(a^k) = Sa;$$

hence, $a \in X$ *is a diagonal fixed point of* T.

Proof. (Proposition 2.21) Let $n \geq k$ be such that

$$B_n = 0; \text{ that is: } \rho_{n-k} = \cdots = \rho_{n-1} = 0.$$

This, by definition, yields (for the moment)

$$u_{n-k} = \cdots = u_{n-1} = u_n = a, \text{ for some } a \in X.$$

Combining with the iterative procedure, we also get

$$a = u_n = T(u_{n-k}, ..., u_{n-1}) = T(a^k);$$

so that, $a \in X$ is a diagonal fixed point of T. Finally, by a (finite) induction argument, we prove that

(eq;i) $u_{n-k+i} = a$, for all $i \geq 0$;

and, from this, we are done. The case of $i \in \{0, ..., k\}$ is clear, by the above. Suppose that (eq;i) is true for all $i \in \{0, ..., j\}$, where $j \geq k$; we have to establish that (eq;j+1) holds too. But, from this inductive assumption,

$$u_{n-k+j+1} = T(u_{n-k+j-k+1}, ..., u_{n-k+j}) = T(a^k) = a;$$

and the proof is complete. □

As a consequence of these remarks, it follows that, whenever (B-zero) holds, we are done; so, without loss, one may assume that the positivity condition holds

(B-pos) $B_n > 0$, for all $n \geq k$.

The latter auxiliary fact concentrates the "deep" part of our argument.

Proposition 2.22. *Under these conditions, we have*

(42-1) $\rho_n \leq \varphi(B_n) < B_n$, *for each* $n \geq k$

(42-2) $B_{n+1} \leq B_n$, $\forall n \geq k$; *whence,* $(B_{k+i}; i \geq 0)$ *is descending*

(42-3) $\rho_{n+k} < B_{n+k} \leq \varphi(B_n) < B_n$, $\forall n \geq k$.

Proof. (Proposition 2.22)

i): Let $n \geq k$ be arbitrary fixed. By definition,

$$A(u_{n-k}, ..., u_{n-1}, u_n) = B_n = \max\{\rho_{n-k}, ..., \rho_{n-1}\}.$$

On the other hand, from our iterative construction,

$$\rho_n = d(u_n, u_{n+1}) = d(T(u_{n-k}, ..., u_{n-1}), T(u_{n-k+1}, ..., u_{n-1}, u_n));$$

and this, by the contractive property, gives (via (B-pos))

$$\rho_n \leq \varphi(A(u_{n-k}, ..., u_{n-1}, u_n)) = \varphi(B_n) < B_n;$$

proving the desired fact.

ii): From the representation above one has, for each $n \geq k$,

$$B_{n+1} = \max\{\rho_{n-k+1}, ..., \rho_{n-1}, \rho_n\} \leq$$
$$\max\{\rho_{n-k}, \rho_{n-k+1}, ..., \rho_{n-1}, \rho_n\} = \max\{B_n, \rho_n\};$$

and this, along with $\rho_n < B_n$, yields the written relation.

iii): By the obtained facts one gets (via φ = increasing), for each $n \geq k$

$$\rho_{n+j} \leq \varphi(B_{n+j}) \leq \varphi(B_n), \ \forall j \in \{0, ..., k-1\}.$$

This yields (by definition)

$$B_{n+k} = \max\{\rho_n, ..., \rho_{n+k-1}\} \leq \varphi(B_n), \text{ for all } n \geq k;$$

and proves our claim. □

Having these established, we may pass to the final part of our argument.

Step 1. From the evaluations above, one has

$$(B_{ik+k} =)\ B_{(i+1)k} \leq \varphi^i(B_k), \text{ for all } i \geq 0.$$

We now establish, inductively, that the inequality

$$\text{(ineq;i) } \rho_{ik} + \cdots + \rho_{ik+k-1} \leq kB_{ik+k} \leq k\varphi^i(B_k)$$

is holding for each $i \geq 0$. In fact, the case of $i = 0$ means

$$\rho_0 + \cdots + \rho_{k-1} \leq kB_k = k\varphi^0(B_k);\text{ evident, by the definition of } B_k.$$

Suppose that (ineq;i) holds, for all $i \in \{0, ..., j\}$, where $j \geq 0$; we claim that (ineq;j+1) holds too; that is

$$\rho_{(j+1)k} + \cdots + \rho_{(j+1)k+k-1} \leq kB_{(j+1)k+k} \leq k\varphi^{j+1}(B_k).$$

But, according to definition (and the preceding evaluation)

$\max\{\rho_{(j+1)k}, ..., \rho_{(j+1)k+k-1}\} = B_{(j+1)k+k} \leq \varphi^{j+1}(B_k)$;
hence,
$\rho_{(j+1)k} + \cdots + \rho_{(j+1)k+k-1} \leq kB_{(j+1)k+k} \leq k\varphi^{j+1}(B_k)$;

and, from this, the claim follows. Adding the inequalities in (ineq;i), gives (if we remember that φ is strongly Matkowski admissible)

$\sum_n \rho_n \leq k \sum_i B_{(i+1)k} \leq k \sum_i \varphi^i(B_k) < \infty$;

proving that the iterative sequence (u_n) is d-Cauchy. As X is d-complete,

$u_n \xrightarrow{d} z$ as $n \to \infty$, for some $z \in X$.

Step 2. We now claim that the obtained limit z is a diagonal fixed point for T, in the sense: $z = T(z^k)(= Sz)$. To verify this, it will suffice establishing that

$u_n \xrightarrow{d} Sz(= T(z^k))$, as $n \to \infty$.

But, from the triangle inequality, contractive condition, and regressiveness of φ, we have for each $n \geq k$

$d(u_n, Sz) = d(T(u_{n-k}, ..., u_{n-1}), T(z^k)) \leq$
$d(T(u_{n-k}, u_{n-k+1}, ..., u_{n-1}), T(u_{n-k+1}, ..., u_{n-1}, z)) +$
$d(T(u_{n-k+1}, u_{n-k+2}, ..., u_{n-1}, z), T(u_{n-k+2}, ..., u_{n-1}, z^2)) + \cdots +$
$d(T(u_{n-1}, z^{k-1}), T(z^k)) \leq \varphi(\max\{\rho_{n-k}, ..., \rho_{n-2}, d(u_{n-1}, z)\}) +$
$\varphi(\max\{\rho_{n-k+1}, ..., \rho_{n-2}, d(u_{n-1}, z)\}) + \cdots + \varphi(d(u_{n-1}, z)) \leq$
$\max\{\rho_{n-k}, ..., \rho_{n-2}, d(u_{n-1}, z)\} +$
$\max\{\rho_{n-k+1}, ..., \rho_{n-2}, d(u_{n-1}, z)\} + \cdots + d(u_{n-1}, z)$.

Passing to limit as $n \to \infty$, we get

$\lim_n d(u_n, Sz) = 0$; so that (by uniqueness), $z = Sz(= T(z^k))$;

and this gives us the desired fact. The proof is thereby complete. □

In particular, when φ is linear; that is,

$\varphi(t) = \beta t$, $t \in R_+$ (for some $\beta \in [0, 1[$),

the obtained result is just the one in Ćirić and Presić [62]. On the other hand, when φ is continuous, the corresponding version of our main result includes a related 1981 statement in Rus [202]; we will discuss this fact elsewhere. Further technical aspects may be found in Khan et al. [142]; see also Shahzad and Shukla [214].

2.4.5 Related Abbas-Ilić-Nazir statement

Let us now return to our initial setting. So, take a metric space (X, d); as well as a natural number $k \in N(1, \leq)$. Further, let $T : X^k \to X$ be a mapping; and $S : X \to X$ be its associated diagonal operator. Remember that, our main diagonal fixed point result is essentially based upon strongly Matkowski admissible functions. Precisely, denote for $i, j \in N$, $i < j$,

$$A(z_i, ..., z_j) = \max\{d(z_i, z_{i+1}), ..., d(z_{j-1}, z_j)\},\ (z_i, ..., z_j) \in X^{j-i+1}.$$

Given $\varphi \in \mathcal{F}(R_+)$, let us say that T is (d, φ)-*contractive*, provided

$$d(T(y_0, ..., y_{k-1}), T(y_1, ..., y_k)) \leq \varphi(A(y_0, ..., y_{k-1}, y_k)),$$
for all $(y_0, ..., y_{k-1}, y_k) \in X^{k+1}$.

The basic conditions imposed upon φ are

(in-re) $\varphi \in \mathcal{F}(in, re)(R_+)$ (φ is increasing and regressive)
(s-M-ad) φ is strongly Matkowski admissible:
$\sum_n \varphi^n(t) < \infty$, for all $t \geq 0$.

A natural question to be posed is that of conclusion in our main result remaining valid if the strong admissible property above is replaced by its weaker counterpart:

(M-ad) φ is Matkowski admissible:
$\varphi^n(t) \to 0$ as $n \to \infty$, for all $t \geq 0$.

Technically speaking, such extensions (to the Matkowski admissible case) are not in general possible; because, the contractive condition imposed upon T connects two *consecutive* members of our iterative process. However, in a recent paper, Abbas et al. [2] established a fixed point statement founded on Matkowski admissible property. To discuss it, a lot of preliminary facts involving such functions is needed. Given some regressive function $\psi \in \mathcal{F}(re)(R_+)$, denote

$$\chi(t) = \sup\{\psi(s); 0 \leq s \leq t\},\ \ t \in R_+.$$

As ψ = regressive, χ is well defined, as an element of $\mathcal{F}(R_+)$; it will be referred to as the *associated* to ψ function. Some other properties of it are collected in the statement below.

Proposition 2.23. *Let the regressive function $\psi \in \mathcal{F}(re)(R_+)$ be, in addition, continuous. Then, the associated function χ has the properties*
(51-1) *χ is increasing, $\chi(0) = 0$ and $\psi(t) \leq \chi(t)$, $\forall t \in R_+$*

(51-2) *$\chi(t) < t$, for all $t > 0$; hence χ is regressive*
(51-3) *$\chi(t+0) < t$, $\forall t > 0$; wherefrom, χ is Matkowski admissible.*

Proof. i): Evident, by definition (and the choice of ψ).

ii): Fix some $t > 0$. As ψ is continuous, we must have $\chi(t) = \psi(r)$, for some $r \in [0, t]$. If $r = 0$, then $\chi(t) = 0 < t$; and, if $r > 0$, then $\chi(t) = \psi(r) < r \leq t$.

iii): Let $s > 0$ be arbitrary fixed. Clearly, $\psi(t) \leq \chi(s) < s$, $\forall t \in [0, s]$. By the continuity (at s) of ψ, there exists $\delta > 0$ such that $\psi(r) < s$, $\forall r \in [s, s+\delta]$. Putting these together, yields

$$\psi(r) < s, \ \forall r \in [0, s+\delta];$$

whence, again via ψ = continuous, one gets (cf. the preceding step)

$$\chi(t) < s, \text{ for all } t \in [0, s+\delta].$$

This, by definition, gives

$$\chi(s+0) := \lim_{t \to s+} \chi(t) = \inf\{\chi(t); t > s\} < s, \ \forall s > 0;$$

wherefrom, all is clear. □

In particular, let Φ_1 be the family of all functions $\varphi \in \mathcal{F}(R_+)$ with

(phi-1-a) $\varphi(0) = 0 < \varphi(t) \leq t$, for all $t \in R_+^0$
(phi-1-b) φ is continuous on R_+ (in the usual sense).

For each $\varphi \in \Phi_1$, the complementary function $\psi \in \mathcal{F}(R_+)$ introduced as

$$\psi(t) = t - \varphi(t), \ t \in R_+$$

is continuous regressive; hence, its attached function χ (see above) fulfills

χ is increasing, regressive, and Matkowski admissible.

An extension of this functional class may be given along the lines below. Let Φ_2 be the class of all functions $\varphi \in \mathcal{F}(R_+)$ with

(phi-2-a) $\varphi(0) = 0$, $\varphi(t) > 0$, $\forall t \in R_+^0$
(phi-2-b) φ is lower semicontinuous on R_+.

Clearly, $\Phi_1 \subseteq \Phi_2$, as it can be directly seen. This functional class was used by Abbas et al. [2] to establish a related fixed point statement involving these data. Let the metric space (X, d), the number $k \in N(1, \leq)$ and the mapping $T : X^k \to X$ be given.

Theorem 2.20. *Suppose that there exists $\varphi \in \Phi_2$, such that*

$$d(T(x_1,...,x_k),T(x_2,...,x_{k+1})) \leq$$
$$\max\{d(x_i,x_{i+1});1 \leq i \leq k\} - \varphi(\max\{d(x_i,x_{i+1});1 \leq i \leq k\}),$$
for all $(x_1,...,x_{k+1}) \in X^{k+1}$.

In addition, let X be d-complete. Then, the following conclusions hold:

(51-a) *For each $(x_1,...,x_k) \in X^k$, the iterative sequence (x_n) introduced according to*

$$x_{n+k} = T(x_n,...,x_{n+k-1}),\ n \geq 1$$

converges to some $u \in X$ with $u = T(u,...,u)$

(51-b) *Moreover, if*

$$d(T(x,...,x),T(y,...,y)) \leq d(x,y) - \varphi(d(x,y)),$$
for each $x, y \in X$ with $x \neq y$,

then $u \in X$ is uniquely determined with such a property.

Proof. According to authors, the argument consists in the following.

Part 1. Under the imposed conditions in the statement, one has

$$\lim_n d(x_{n+k-1},x_{n+k}) = 0; \text{ or, equivalently: } \lim_n d(x_{n-1},x_n) = 0.$$

Part 2. Further, (x_n) is a Cauchy sequence. For, given any couple of ranks $n, m \in N$ with $n \leq m$, we have (using the contractive condition)

$$\begin{aligned}
&\text{(R1)}\ \ d(x_{k+n},x_{k+m}) = \\
&d(T(x_1,...,x_{k+n-1}),T(x_2,...,x_{k+m-1})) \leq \\
&d(T(x_1,...,x_{k+n-1}),T(x_2,...,x_{k+n})) + \\
&d(T(x_2,...,x_{k+n}),T(x_3,...,x_{k+n+1})) + \cdots + \\
&d(T(x_2,...,x_{k+n}),T(x_3,...,x_{k+m-1})) \leq \\
&\max\{d(x_{i+n},x_{i+n+1});1 \leq i \leq k-1\} - \\
&\varphi(\max\{d(x_{i+n},x_{i+n+1});1 \leq i \leq k-1\}) + \\
&\max\{d(x_{i+n},x_{i+n+1});1 \leq i \leq k\} - \\
&\varphi(\max\{d(x_{i+n},x_{i+n+1});1 \leq i \leq k\}) + \cdots + \\
&\max\{d(x_{i+n},x_{i+n+1});1 \leq i \leq k+m-1\} - \\
&\varphi(\max\{d(x_{i+n},x_{i+n+1});1 \leq i \leq k+m-1\}).
\end{aligned}$$

Taking the upper limit as $n, m \to \infty$, implies that

$$\text{(R2)}\ \ \lim_{n\to\infty} d(x_{k+n},x_{k+m}) = 0.$$

Hence, (x_n) is a Cauchy sequence.

Part 3. Since X is d-complete, there exists $u \in X$ such that

$$\lim_{n,m\to\infty} d(x_n, x_m) = \lim_{n\to\infty} d(x_n, u).$$

Using the contractive condition gives $u = T(u, ..., u)$; which means: $u \in X$ is a diagonal fixed point of T.

Part 4. Assume that there exists $v \neq u$ such that $v = T(v, ..., v)$. As $d(u, v) > 0$, we must have $\varphi(d(u, v)) > 0$. Combining with the imposed extra condition yields

$$d(u, v) = d(T(u, ..., u), T(v, ..., v)) \leq d(u, v) - \varphi(d(u, v)) < d(u, v);$$

a contradiction. Hence, $u = v$; and the last conclusion follows. □

At a first glance, all seems to be OK in the authors' reasoning. However, even a superficial analysis shows that the situation is exactly opposite. Two basic motivations will explain our assertion.

A) The first motivation concerns the evaluation of iterates x_{k+n} and x_{k+m} appearing in relation (R1)

The authors' evaluation of the former iterate

$$x_{k+n} = T(x_1, ..., x_{k+n-1})$$

is not correct; because it involves $k+n-1$ variables in the right-hand side; and this, along with

$$k + n - 1 > k, \text{ whenever } n \geq 2$$

is in contradiction with the (correct) definition of our iterative sequence

$$x_{k+n} = T(x_n, ..., x_{n+k-1}),\ n \geq 1;$$

which involves, at any stage, k variables in the right-hand side.

Likewise, the authors' evaluation of the latter iterate

$$x_{k+m} = T(x_2, ..., x_{k+m-1})$$

is again incorrect; because it involves $k+m-2$ variables in the right-hand side; and this, along with

$$k + m - 2 > k, \text{ whenever } m \geq 3$$

is in contradiction with the (correct) definition of our iterative sequence

$$x_{k+m} = T(x_m, ..., x_{m+k-1}),\ m \geq 1;$$

which involves, at any stage, k variables in the right-hand side.

B) The second motivation concerns the deduction of relation (R2). Precisely, it is not clear of to what extent is this relation obtainable, along the class Φ_2, by passing to the upper limit as $n, m \to \infty$ in relation (R1).

Summing up, the reasoning in Part 2 of authors' theorem is not acceptable; and this tells us that the authors' statement is – at this very moment – without proof. As a consequence, all corollaries based upon authors' result above are not acceptable in general. Further aspects may be found in Rao et al. [193]; see also Shukla et al. [215].

Chapter 3

MAXIMAL TECHNIQUES

3.1 Caristi-Kirk Results in Banach Lattices

Abstract. The fixed point result over Banach lattices due to Agarwal and Khamsi [Nonlin. Anal., 74 (2011), 141–145] is reducible to the (pseudometric) ordering principle in Turinici [Demonstr. Math., 22 (1989), 213–228].

Subject-Class 2020. 47H10 (Primary), 54F05 (Secondary)

Keywords. Metric space, lsc function, contraction, fixed point, Zermelo-Fraenkel system, Dependent Choice, quasi-order, pseudometric, maximal element, Cauchy and asymptotic sequence, metrizable topological vector space, regular and normal cone, Banach lattice.

3.1.1 *Dependent Choice principles*

See Ch. 1.

3.1.2 *Pseudometric maximal principles*

In the following, a lot of maximal principles over pseudometric structures is given. Their usefulness will become clear a bit further.

(A) Let X be some nonempty set. Take a *quasi-order* $(\leq)$ on X; as well as a function $\sigma : X \to R_+ \cup \{\infty\}$. The following *asymptotic* statement – deductible in the reduced system (ZF-AC+DC) – is our starting point. Given the sequence (x_n) in X, let us say that $y \in X$ is an *upper bound* of it, when $(x_n \leq y, \forall n)$; the class of all these will be denoted $\text{ubd}(x_n)$. Further, call (x_n), *σ-asymptotic* provided $\lim_n \sigma(x_n) = 0$.

Proposition 3.1. *Let the conventions above be accepted. Then*

(21-1) *Under the condition*

($(X, \leq)$ is almost regular (modulo σ))
$\forall \varepsilon > 0$, $\forall x \in X$, $\exists y = y(\varepsilon, x) \in X(x, \leq)$: $\sigma(y) \leq \varepsilon$

it follows that, for each $u \in X$ there exists at least one u-starting sequence (x_n) in X with the properties: ascending, σ-asymptotic

(21-2) *If, in addition,*

($(X, \leq)$ is sequentially inductive (modulo σ)): each ascending and σ-asymptotic sequence in X has an upper bound (modulo $(\leq)$))

the sequence (x_n) obtained as before fulfills also $\mathrm{ubd}(x_n) \neq \emptyset$.

Proof. I) For the arbitrary fixed $\varepsilon > 0$, let $\mathcal{R}(\varepsilon)$ be the relation (over X):

$x\mathcal{R}(\varepsilon)y$ iff $x \leq y$ and $\sigma(y) \leq \varepsilon$;

note that, from the almost regular condition, $\mathcal{R}(\varepsilon)$ is a proper relation. Now, let (ε_n) be a strictly descending sequence in $R^0_+ :=]0, \infty[$, with $\varepsilon_n \to 0$ as $n \to \infty$. (For instance, we may take $(\varepsilon_n = 2^{-n}; n \geq 0)$; so, no choice is needed.) From the remark above, $(\mathcal{R}_n := \mathcal{R}(\varepsilon_n); n \geq 0)$ is a sequence of proper relations. So, by the Diagonal Dependent Choice principle (DDC), it follows that, for the arbitrary fixed $u \in X$, there must be a sequence (x_n) in X with $x_0 = u$ and

$x_n \leq x_{n+1}$, $\sigma(x_{n+1}) \leq \varepsilon_n$, for all n;

so, (x_n) is ascending and σ-asymptotic.

II) Combining with the sequential inductivity (modulo σ) condition, the sequence (x_n) admits upper bounds ($\mathrm{ubd}(x_n) \neq \emptyset$); and we are done. □

Let $(X, \leq, \sigma)$ be as before. Call $z \in X$, $(\leq, \sigma)$-*maximal* provided

$(w \in X$, $z \leq w)$ imply $\sigma(z) = \sigma(w)$.

A characterization of this concept may be given along the lines below.

Proposition 3.2. *The following are valid (for a generic $z \in X$):*

(22-1) *If $(X, \leq)$ is almost regular (modulo σ), then*

z is $(\leq, \sigma)$-maximal implies $\sigma(z) = 0$

(22-2) *Under σ = decreasing, we necessarily have*

$\sigma(z) = 0$ implies z is $(\leq, \sigma)$-maximal

(22-3) *If $(X, \leq)$ is almost regular (modulo σ) and σ = decreasing, then*

z is $(\leq, \sigma)$-maximal iff $\sigma(z) = 0$.

Proof. I) Assume by contradiction that $\alpha := \sigma(z) > 0$; and let the number $\varepsilon > 0$ be taken according to $\varepsilon < \alpha$. By the imposed property, there must be some $w \in X(z, \leq)$ with $\sigma(w) \leq \varepsilon$; hence $\sigma(w) < \alpha = \sigma(z)$; in contradiction with the $(\leq, \sigma)$-maximal property of z; hence the claim.

II) Suppose that $\sigma(z) = 0$. By the decreasing property of σ, we have

$$z \leq w \text{ implies } 0 = \sigma(z) \geq \sigma(w) \geq 0; \text{ whence, } \sigma(z) = \sigma(w) = 0;$$

which tells us that z is $(\leq, \sigma)$-maximal.

III) Evident, by the preceding stages. □

Combining these, one gets an existence principle involving such elements (referred to as: Asymptotic Maximal Principle; in short: (MP-a)).

Theorem 3.1. *Let the conditions below hold*

(a-reg-fct) $(X, \leq)$ is almost regular (modulo σ)
(seq-ind-fct) $(X, \leq)$ is sequentially inductive (modulo σ)
(decr) σ is decreasing ($x \leq y$ implies $\sigma(x) \geq \sigma(y)$).

Then, for each $u \in X$, there exists some $v \in X$, with

$u \leq v$ and v is $(\leq, \sigma)$-maximal (or, equivalently: $\sigma(v) = 0$).

Proof. By the preceding statement, it follows that, for the given $u \in X$, there exists at least one sequence (x_n) in X with the properties:

$$x_0 = u, \ (x_n) \text{ is ascending, } \sigma\text{-asymptotic, and } \mathrm{ubd}(x_n) \neq \emptyset.$$

Let $v \in \mathrm{ubd}(x_n)$ be arbitrary fixed; we claim that it is our desired point. In fact, as $x_0 = u$, we must have $u \leq v$. Then, by $\sigma =$ decreasing,

$$\sigma(x_n) \geq \sigma(v) (\geq 0), \text{ for all } n;$$

wherefrom (passing to limit), one gets (by the asymptotic property)

$$\sigma(v) = 0; \text{ or, equivalently (see above): } v \text{ is } (\leq, \sigma)\text{-maximal.}$$

The proof is thereby complete. □

(B) A basic particular case of these facts is obtainable in the pseudometric setting. Let X be a nonempty set. By a *(generalized) pseudometric* over X we shall mean any map $d : X \times X \to R_+ \cup \{\infty\}$. Concerning such objects, the following conditions imposed upon them are of interest:

(ref) d is *reflexive*: $d(x,x) = 0$, for each $x \in X$
(suf) d is *sufficient*: $d(x,y) = 0$ implies $x = y$
(tri) d is *triangular*: $d(x,z) \leq d(x,y) + d(y,z)$, $\forall x,y,z \in X$
(sym) d is *symmetric*: $d(x,y) = d(y,x)$, $\forall x,y \in X$.

When the reflexive condition holds, we get a *reflexive pseudometric* (in short: *r-pseudometric*); and (X,d) is termed a *r-pseudometric space*. Likewise, if the reflexive and sufficient properties hold, then we talk about a *reflexive sufficient pseudometric* (in short: *rs-pseudometric*); and (X,d) is termed a *rs-pseudometric space*.

Let in the following (X,d) be a r-pseudometric space. Call the subset Y of X, *almost singleton* (in short: *asingleton*), provided $[y_1, y_2 \in Y$ implies $y_1 = y_2]$; and *singleton* if, in addition, Y is nonempty; note that in this case $Y = \{y\}$, for some $y \in X$. A natural way of introducing a conv-Cauchy structure over X is the one below. Given the sequence (x_n) in X and the point $x \in X$, we say that (x_n), *d-converges* to x (written as: $x_n \xrightarrow{d} x$) provided $d(x_n, x) \to 0$ as $n \to \infty$; that is,

$$\forall \varepsilon > 0, \exists i = i(\varepsilon): \ i \leq n \Longrightarrow d(x_n, x) < \varepsilon.$$

The set of all such points will be denoted $\lim_n(x_n)$; if it is nonempty, then (x_n) is called *d-convergent*. By this very definition (and reflexivity of d)

(d-conv-1) ($(\xrightarrow{d})$ is *hereditary*)
$x_n \xrightarrow{d} x$ implies $y_n \xrightarrow{d} x$, for each subsequence (y_n) of (x_n)
(d-conv-2) ($(\xrightarrow{d})$ is *reflexive*)
$\forall u \in X$, the constant sequence $(x_n = u; n \geq 0)$ fulfills $x_n \xrightarrow{d} u$;

so, $(\xrightarrow{d})$ is a *sequential convergence* on X, under the lines in Kasahara [141]. An important extra property to be discussed here writes

(d-conv-3) $(\xrightarrow{d})$ is *separated*
(referred to as: (X,d) is *separated*):
$\lim_n(x_n)$ is an asingleton, for each sequence (x_n) of X.

Concerning this aspect, note that (X,d) is not separated, in general. However, the underlying property holds, provided (in addition)

$d(.,.)$ is sufficient, triangular, and symmetric
(when $d(.,.)$ becomes a *(generalized) metric* on X).

Further, call (x_n), *d-Cauchy* when

$d(x_m, x_n) \to 0$ as $m, n \to \infty$, $m < n$; that is:
$\forall \varepsilon > 0$, $\exists j = j(\varepsilon)$: $j \leq m < n \Longrightarrow d(x_m, x_n) < \varepsilon$;

the class of all these is denoted as $Cauchy(d)$. By definition (and d = reflexive),

(d-Cauchy-1) ($Cauchy(d)$ is *hereditary*)
(x_n) is d-Cauchy implies (y_n) is d-Cauchy,
for each subsequence $(y_n; n \geq 0)$ of $(x_n; n \geq 0)$
(d-Cauchy-2) ($Cauchy(d)$ is *reflexive*)
$\forall u \in X$, the constant sequence $(x_n = u; n \geq 0)$ is d-Cauchy;

so that, $Cauchy(d)$ is a *Cauchy structure*, under the lines in Turinici [244]. Finally, call $((\xrightarrow{d}), Cauchy(d))$, the *conv-Cauchy structure* attached to d. The following extra properties are to be discussed here

(conv-Cauchy-1) (X, d) is *admissible*:
each d-convergent sequence is d-Cauchy
(conv-Cauchy-2) (X, d) is *complete*:
each d-Cauchy sequence is d-convergent.

Note that, by the imposed (upon d) conditions, (X, d) is not admissible, in general. But, when (in addition)

(tri-sym) d is triangular and symmetric (see above)

this happens. On the other hand, (X, d) is not complete in general. Sufficient conditions for this to be valid will be given a bit further.

Let in the following X be a nonempty set, $(\leq)$ be a quasi-order on X and $d : X \times X \to R_+$ be a r-pseudometric on X; the triple $(X, \leq, d)$ will be then called a *quasi-ordered r-pseudometric space*. Call $z \in X$, $(\leq, d)$-*maximal* if

$(u, v \in X$ and $z \leq u \leq v)$ imply $d(u, v) = 0$.

Note that, in terms of the associated function (from X to $R_+ \cup \{\infty\}$)

$\sigma_d(x) = \sup\{d(u, v); x \leq u \leq v\}, x \in X$,

this property may be characterized as: $\sigma_d(z) = 0$. So, a basic source for determining such elements is the Asymptotic Maximal Principle (MP-a), applied to $\sigma_d(.)$. To do this, we start from the obvious fact that σ_d is decreasing. Concerning the remaining properties, let us say that $(X, \leq)$ is *almost regular* (modulo d), when

$(X, \leq)$ is almost regular (modulo σ_d); that is,
$\forall \varepsilon > 0$, $\forall x \in X$, $\exists y = y(\varepsilon, x) \in X(x, \leq)$: $(y \leq u \leq v \Longrightarrow d(u,v) \leq \varepsilon)$.

Further, call the sequence (x_n), *strongly* $(\leq, d)$-*Cauchy* provided

(x_n) is σ_d-asymptotic: $\sigma_d(x_n) \to 0$ as $n \to \infty$.

Finally, call $(X, \leq)$, *weakly sequentially inductive* (modulo d) when

$\forall$ ascending strongly $(\leq, d)$-Cauchy sequence has an upper bound;
that is: $(X, \leq)$ is sequentially inductive (modulo σ_d).

Combining with (MP-a), one gets the following maximal result (referred to as: r-pseudometric Function Maximal Principle; in short: (MP-f-rp)).

Theorem 3.2. *Let the quasi-ordered r-pseudometric space $(X, \leq, d)$ fulfill*

(a-reg-ps) $(X, \leq)$ is almost regular (modulo d)
(w-s-ind-ps) $(X, \leq)$ is weakly sequentially inductive (modulo d).

Then, for each $u \in X$ there exists some $v \in X$, with

$u \leq v$ and v is $(\leq, d)$-maximal ($v \leq x \leq y$ implies $d(x, y) = 0$).

Now, evidently, the generic property holds

(for each (ascending) sequence (x_n) in X):
(x_n) is strongly $(\leq, d)$-Cauchy implies (x_n) is d-Cauchy.

For, given such a sequence (x_n), we have

$d(x_i, x_j) \leq \sigma_d(x_n)$, whenever $n \leq i \leq j$;

and this, along with $\lim_n \sigma_d(x_n) = 0$, proves our assertion. Hence,

$(X, \leq)$ is weakly sequentially inductive (modulo d) whenever
$(X, \leq)$ is *Cauchy sequentially inductive* (modulo d):
each ascending d-Cauchy sequence has an upper bound.

It remains to give conditions guaranteeing the almost regular property. Given the quasi-ordered r-pseudometric space $(X, \leq, d)$, define the concept

$(X, \leq)$ is *Cauchy regular (modulo d)*:
each ascending sequence in X is d-Cauchy.

The following auxiliary fact involving this notion will be useful for us.

Proposition 3.3. *For each quasi-ordered r-pseudometric space $(X, \leq, d)$,*

$(X, \leq)$ is Cauchy regular (modulo d) implies
$(X, \leq)$ is almost regular (modulo d).

Proof. Assume by contradiction that $(X, \leq)$ is not almost regular (modulo d); that is (for some $\varepsilon > 0$, $x \in X$):

$$\forall y \in X(x, \leq),\ \exists u, v \in X\colon [y \leq u \leq v,\ d(u, v) > \varepsilon].$$

This, by definition, yields (for the same (ε, x)):

$$\forall y \in X(x, \leq),\ \exists (u, v) \in (\leq)\colon\ y \leq u,\ d(u, v) > \varepsilon.$$

(Here, $(\leq) := \{(a, b) \in X \times X; a \leq b\}$ is just the original (quasi-order) relation over X.) Fix $y_0, y_1 \in X$ with $x \leq y_0 \leq y_1$; hence, $w_0 := (y_0, y_1) \in (\leq)$. Define a relation $\mathcal{R} = \mathcal{R}(\varepsilon)$ on $M := \{(a, b) \in (\leq); y_1 \leq a\}$ as:

$$(a_1, b_1)\mathcal{R}(a_2, b_2) \text{ if and only if } b_1 \leq a_2,\ d(a_2, b_2) > \varepsilon.$$

From the imposed condition, $M((a, b), \mathcal{R}) \neq \emptyset$, $\forall (a, b) \in M$. So, by the Dependent Choice principle (DC), there exists a sequence $(w_n := (y_{2n}, y_{2n+1}); n \geq 0)$ in M with

$$w_0 = (y_0, y_1) \text{ and } (w_n \mathcal{R} w_{n+1},\ \forall n);$$

hence, by definition,

$$(\forall n)\colon\ y_{2n} \leq y_{2n+1} \leq y_{2n+2} \leq y_{2n+3},\ d(y_{2n+2}, y_{2n+3}) > \varepsilon.$$

As a consequence, $(y_n; n \geq 0)$ is ascending and not d-Cauchy; in contradiction with the imposed hypothesis; hence the claim. □

Now, combining these facts, one gets the following maximal statement (referred to as: r-pseudometric maximal principle; in short: (MP-r-p)).

Theorem 3.3. *Let the quasi-ordered r-pseudometric space $(X, \leq, d)$ fulfill*

(C-reg-ps) $(X, \leq)$ is Cauchy regular (modulo d)
(C-seq-ind-ps) $(X, \leq)$ is Cauchy sequentially inductive (modulo d).

Then, for each $u \in X$ there exists some $v \in X$, with

$$u \leq v \text{ and } v \text{ is } (\leq, d)\text{-maximal } (v \leq x \leq y \text{ implies } d(x, y) = 0).$$

Proof. By the auxiliary fact we just exposed, our preceding maximal principle is applicable to these data; wherefrom, all is clear. □

Technically speaking, this result was established in Turinici [238], through a slightly different approach; see also Kang and Park [137]. In particular, the Cauchy sequential inductivity holds under

(d-com) $(X, \leq, d)$ is complete:
each ascending d-Cauchy sequence is d-convergent
(d-s-c) $(\leq)$ is d-selfclosed:
the d-limit of each ascending sequence is an upper bound of it.

The obtained variant of our maximal principle includes then the results in Conserva and Rizzo [65]; see also Granas and Horvath [104].

A basic particular case of these developments corresponds to the ambient r-pseudometric $d(.,.)$ being, in addition, sufficient. So, let $(X, \leq, d)$ be a quasi-ordered rs-pseudometric space. Given $z \in X$, the following equivalence property holds

z is $(\leq, d)$-maximal ($u, v \in X$, $z \leq u \leq v$ imply $d(u, v) = 0$)
if and only if
z is $(\leq)$-*maximal*: $w \in X$ and $z \leq w$ imply $z = w$.

In fact, the direct inclusion is clear, via ($z \leq w \Longrightarrow z \leq z \leq w$) and ($d$ = sufficient). For the converse one, let $u, v \in X$ be such that $z \leq u \leq v$. Then, (as $(\leq)$ is transitive), $z \leq u$, $z \leq v$; so that, $(z =)u = v$; hence, $d(u, v) = 0$.

As a completion of this, the following auxiliary fact involving our structure will be useful for us (cf. Hamel [107, Ch. 4, Sec. 4.1]):

Proposition 3.4. *Let the quasi-ordered rs-pseudometric space $(X, \leq, d)$ be such that $(X, \leq)$ is Cauchy regular (modulo d). Then, $(\leq)$ has the anti-symmetric property; hence, it is a (partial) order on X.*

Proof. Let $u, v \in X$ be such that $u \leq v$, $v \leq u$. The sequence (z_n) in X introduced as ($z_{2n} = u$, $z_{2n+1} = v; n \geq 0$) is ascending; so that, as $(X, \leq)$ is Cauchy regular (modulo d), we must have that (z_n) is d-Cauchy; wherefrom

$\lim_n d(z_{2n}, z_{2n+1}) = 0$; that is: $d(u, v) = 0$;

and this (by sufficiency) gives $u = v$. □

An application of the preceding maximal statement to these data yields the following Zorn type maximal result (referred to as: rs-pseudometric Maximal Principle; in short: (MP-rs-p)).

Theorem 3.4. *Let the quasi-ordered rs-pseudometric space* $(X, \leq, d)$ *fulfill*

> *(C-reg-ps)* $(X, \leq)$ *is Cauchy regular (modulo d);*
> *hence by the above,* $(\leq)$ *is an ordering*
> *(C-s-ind-ps)* $(X, \leq)$ *is Cauchy sequentially inductive (modulo d).*

Then, for each $u \in X$ *there exists some* $v \in X$, *with*

> $u \leq v$ *and* v *is* $(\leq)$*-maximal* ($v \leq x$ *implies* $v = x$).

(C) In the following, two applications of our maximal results are given.

Let X be a nonempty set. Take a quasi-order $(\leq)$ on X; and a function $\varphi : X \to R_+$. Call $z \in X$, $(\leq, \varphi)$-*maximal* when:

$$z \leq w \in X \text{ implies } \varphi(z) = \varphi(w).$$

A basic result about such points is the 1976 Brezis-Browder ordering principle [38] (in short: (BB)).

Theorem 3.5. *Suppose that*

> *(seq-ind)* $(X, \leq)$ *is sequentially inductive:*
> *each ascending sequence has an upper bound (modulo* $(\leq)$*)*
> *(decr)* φ *is* $(\leq)$*-decreasing* ($x \leq y \Longrightarrow \varphi(x) \geq \varphi(y)$).

Then, for each $u \in X$ *there exists some* $v \in X$, *with*

> $u \leq v$ *and* v *is* $(\leq, \varphi)$*-maximal* ($v \leq w$ *implies* $\varphi(v) = \varphi(w)$).

Proof. Define the pseudometric d over X according to

$$d(x, y) = |\varphi(x) - \varphi(y)|,\ x, y \in X.$$

Clearly, d is reflexive; in addition, we have

> d is *triangular*: $d(x, z) \leq d(x, y) + d(y, z),\ \forall x, y, z \in X$
> d is *symmetric*: $d(x, y) = d(y, x),\ \forall x, y \in X$.

In other words: d has all properties of a metric, except sufficiency; we then say that $d(.,.)$ is a *semimetric* on X; and (X, d) is a *semimetric space*. Having these precise, we show that the r-pseudometric Maximal Principle (MP-r-p) is applicable to the structure $(X, \leq, d)$.

I) Let (x_n) be an ascending sequence in X. As φ is decreasing, the sequence $(\varphi(x_n); n \geq 0)$ is decreasing (in R_+) – hence, Cauchy; and this tells us that (x_n) is d-Cauchy. As (x_n) was arbitrarily chosen with the described property, one derives that $(X, \leq)$ is Cauchy regular (modulo d).

II) As $(X, \leq)$ is sequentially inductive, it trivially follows that $(X, \leq)$ is Cauchy sequentially inductive (modulo d).

Summing up, the quoted maximal principle is indeed applicable here. From its conclusion, it then follows that, for each $u \in X$ there exists some other point $v \in X$, with

(MP-r-p-1) $u \leq v$ (that is: $v \in X(u, \leq)$)
(MP-r-p-2) v is $(\leq, d)$-maximal ($v \leq x \leq y$ implies $d(x, y) = 0$).

But, from the second conclusion above, it is clear that $v \in X$ appears as $(\leq, \varphi)$-maximal. The proof is thereby complete. □

This statement found some useful applications to convex as well as non-convex analysis. So, it was the subject of many extensions; see, for instance, Hyers et al. [117, Ch. 5]. These are interesting from a technical perspective; but, in all concrete situations when a (sequential) variational principle of this type ((VP), say) is to be applied, a substitution by the Brezis-Browder's is always possible. This raises the question of to what extent are the logical inclusions (BB) $\Longrightarrow$ (VP) effective. For an appropriate answer, we have to complete the list of maximal principles above with a Zorn-Bourbaki one, valid on standard metric spaces.

Let X be a nonempty set. We say that the mapping $d : X \times X \to R_+$ is a *(standard) metric* on X, when

(m-1) d is reflexive-sufficient: $x = y \Longleftrightarrow d(x, y) = 0$
(m-2) d is triangular: $d(x, z) \leq d(x, y) + d(y, z)$, $\forall x, y, z \in X$
(m-3) d is symmetric: $d(x, y) = d(y, x)$, $\forall x, y \in X$.

Fix in the following such an object, $d(., .)$; and let $\varphi : X \to R_+$ be some function. The following 1974 statement in Ekeland [85] (referred to as Ekeland variational principle; in short: (EVP)) is available.

Theorem 3.6. *Suppose that*

(lsc-std) φ *is d-lsc:* $\liminf_n \varphi(x_n) \geq \varphi(x)$, *whenever* $x_n \xrightarrow{d} x$.

In addition, let (X, d) *be complete. Then, for each* $u \in X$ *there exists some other point* $v \in X$, *with*

(26-a) $d(u, v) \leq \varphi(u) - \varphi(v)$ *(hence,* $\varphi(u) \geq \varphi(v)$*)*
(26-b) $d(v, x) > \varphi(v) - \varphi(x)$, *for all* $x \in X \setminus \{v\}$.

Proof. For technical reasons – to be discussed a bit further – it would be useful giving two proofs of this statement.

[First proof, via (MP-rs-p)] Let $(\leq)$ stand for the relation (over X)

$$x \leq y \text{ iff } d(x,y) \leq \varphi(x) - \varphi(y).$$

Clearly, $(\leq)$ is reflexive, transitive and antisymmetric; hence, a (partial) order on X. We now claim that conclusions of (EVP) are deductible over $(X, \leq, d)$ by means of the rs-pseudometric Maximal Principle (MP-rs-p).

MP-rs-p-1) Let (x_n) be an ascending sequence of X:

$$d(x_n, x_m) \leq \varphi(x_n) - \varphi(x_m), \text{ if } n \leq m.$$

The sequence $(\varphi(x_n); n \geq 0)$ is descending in R_+; hence, a Cauchy one. This, combined with our working hypothesis, tells us that (x_n) is a d-Cauchy sequence in X; whence, $(X, \leq)$ is Cauchy regular (modulo d).

MP-r-s-p-2) As (X, d) is complete, we have $x_n \xrightarrow{d} y$ as $n \to \infty$, for some $y \in X$. Moreover, by the d-lsc property of φ, one has (via $(\varphi(x_n))$ = descending) $\varphi(x_m) \geq \varphi(y)$, for all m. Combining with our working hypothesis gives

$$d(x_n, x_m) \leq \varphi(x_n) - \varphi(y), \text{ if } n \leq m;$$

and this finally yields (passing to limit as $m \to \infty$)

$d(x_n, y) \leq \varphi(x_n) - \varphi(y)$ (meaning: $x_n \leq y$), for all n; wherefrom:
$(X, \leq)$ is sequentially inductive; so, all the more,
$(X, \leq)$ is Cauchy sequentially inductive (modulo d).

Putting these together, it follows that the rs-pseudometric Maximal Principle (MP-rs-p) is indeed applicable to $(X, \leq, d)$; and, from its conclusions, all is clear.

[Second proof, via (BB)] Let $(\leq)$ stand for the relation we just introduced; that is,

$$x \leq y \text{ iff } d(x,y) \leq \varphi(x) - \varphi(y);$$

as precise, it is a (partial) order on X. We claim that conclusions of (EVP) are deductible over $(X, \leq, \varphi)$ by means of Brezis-Browder ordering principle (BB) we just formulated.

BB-1) Let (x_n) be an ascending sequence of X:

$$d(x_n, x_m) \leq \varphi(x_n) - \varphi(x_m), \text{ if } n \leq m.$$

The sequence $(\varphi(x_n); n \geq 0)$ is descending in R_+; hence, a Cauchy one. This, combined with our working hypothesis tells us that (x_n) is a d-Cauchy sequence in X. As (X, d) is complete, we have

$$x_n \xrightarrow{d} y \text{ as } n \to \infty, \text{ for some } y \in X.$$

Moreover, by the d-lsc property of φ, one has (via ($\varphi(x_n)$ = descending) $\varphi(x_m) \geq \varphi(y)$, for all m. Combining with our working hypothesis gives

$$d(x_n, x_m) \leq \varphi(x_n) - \varphi(y), \text{ if } n \leq m;$$

and this finally yields (passing to limit as $m \to \infty$)

$$d(x_n, y) \leq \varphi(x_n) - \varphi(y) \text{ (meaning: } x_n \leq y\text{), for all } n;$$

wherefrom: $(X, \leq)$ is sequentially inductive.

BB-2) By the very definition of our (partial) order, the function φ appears as $(\leq)$-decreasing.

Putting these together, it follows that the Brezis-Browder ordering principle (BB) is indeed applicable to $(X, \leq, \varphi)$. So, by its conclusion, we have that, for the starting $u \in X$ there exists some $v \in X$, with

$u \leq v$; that is: $d(u, v) \leq \varphi(u) - \varphi(v)$
$w \in X$, $v \leq w$ imply $\varphi(v) = \varphi(w)$; or, in other words:
$w \in X$, $d(v, w) \leq \varphi(v) - \varphi(w)$ imply $\varphi(v) = \varphi(w)$.

This, finally combined with

$$d(v, w) \leq \varphi(v) - \varphi(w) \text{ and } \varphi(v) = \varphi(w) \text{ imply } v = w$$

gives the desired conclusion. □

The obtained Ekeland variational principle (EVP) found some basic applications to control and optimization, generalized differential calculus, critical point theory and global analysis; we refer to the 1979 paper by Ekeland [86] for a survey of these. So, it cannot be surprising that, soon after its formulation, many extensions of (EVP) were proposed. For example, the dimensional way of extension refers to the ambient space (R) of $\varphi(X)$ being substituted by a (topological or not) vector space. An account of the results in this area is to be found in the 2003 monograph by Goepfert et al. [102, Ch. 3]; see also the contributions by Hamel [107, Ch. 4], Isac [119], Khanh [145], Nemeth [178], and Rozoveanu [200]. Further, the (pseudo) metrical one consists in the conditions imposed upon the ambient metric over X being relaxed. The basic result in this direction was obtained in 1992 by Tataru [225], via Ekeland type techniques; subsequent extensions of it may be found in the 1996 paper by Kada et al. [133]. Finally, we must add to this list the "functional" extension of (EVP) obtained in 1997 by Zhong [272]; see also Turinici [242].

(D) Summing up, all maximal/variational results we just listed may be viewed as extensions of (EVP). Precisely, note that, by the developments above, we have the chain of implications:

$$\text{(DC)} \Longrightarrow \text{(MP-a)} \Longrightarrow \text{(MP-f-rp)} \Longrightarrow \text{(MP-r-p)}$$
$$\text{(MP-f-rp)} \Longrightarrow \text{(MP-r-p)} \Longrightarrow \text{(MP-rs-p)} \Longrightarrow \text{(EVP)}$$
$$\text{(MP-f-rp)} \Longrightarrow \text{(MP-r-p)} \Longrightarrow \text{(BB)} \Longrightarrow \text{(EVP)}.$$

All these inclusions may be reversed as long as (EVP) $\Longrightarrow$ (DC) is retainable. An appropriate (positive) answer to this is available in the *strongly reduced* Zermelo-Fraenkel system (ZF-AC). Some preliminaries are needed.

Let X be a nonempty set; and $(\leq)$ be a (partial) order on it. We say that $(\leq)$ has the *inf-lattice* property, provided:

$$x \wedge y := \inf(x, y) \text{ exists, for all } x, y \in X.$$

Remember that $z \in X$ is a $(\leq)$-*maximal* element if $X(z, \leq) = \{z\}$; the class of all these points will be denoted as $\max(X, \leq)$. Then, call $(\leq)$ a *Zorn order*, when

$\max(X, \leq)$ is nonempty and *cofinal* in X
(for each $u \in X$ there exists a $(\leq)$-maximal $v \in X$ with $u \leq v$).

Further aspects are to be described in a metric setting. Let $d : X \times X \to R_+$ be a metric over X; and $\varphi : X \to R_+$ be some function. Then, the natural choice for $(\leq)$ above is

$$x \leq_{(d,\varphi)} y \text{ iff } d(x, y) \leq \varphi(x) - \varphi(y);$$

referred to as the *Brøndsted order* [39] attached to (d, φ). Denote

$$X(x, \rho) = \{u \in X; d(x, u) < \rho\}, \ x \in X, \ \rho > 0$$

[the *open sphere* with center x and radius ρ]. Call the ambient metric space (X, d), *discrete* when

for each $x \in X$ there exists $\rho = \rho(x) > 0$ such that $X(x, \rho) = \{x\}$.

Note that, under such an assumption, any function $\psi : X \to R$ is continuous over X. However, the *d-Lipschitz* property

$$|\psi(x) - \psi(y)| \leq Ld(x, y), \ x, y \in X, \text{ for some } L > 0$$

cannot be assured, in general; this is also true for the *d-nonexpansive* property of our function (characterized via $L = 1$).

Now, the statement below is a particular case of (EVP):

Theorem 3.7. *Let the metric space* (X, d) *and the function* $\varphi : X \to R_+$ *satisfy the conditions*

(27-i) (X, d) *is discrete bounded and complete*
(27-ii) $(\leq_{(d,\varphi)})$ *has the inf-lattice property*
(27-iii) φ *is d-nonexpansive (see above)*
(27-iv) the range $\varphi(X)$ *is (at most) countable.*

Then, the Brøndsted order $(\leq_{(d,\varphi)})$ *is a Zorn one.*

We shall refer to it as: the *discrete Lipschitz countable* version of (EVP) (in short: (EVP-dLc)). As precise, (EVP) $\Longrightarrow$ (EVP-dLc). The remarkable fact to be added is that this last (maximal) principle yields (DC); so, it completes the circle between all these.

Proposition 3.5. *We have (in the strongly reduced system (ZF-AC))*

$$((EVP) \Longrightarrow)\ (EVP\text{-}dLc) \Rightarrow (DC)\ (\Longrightarrow (EVP)).$$

So (by the above), the maximal/variational principles (MP-a), (MP-f-rp), (MP-r-p), (MP-rs-p), (BB), and (EVP) are all equivalent with (DC); hence, equivalent (to each other).

For a complete proof, see the 2014 paper by Turinici [252]. In particular, when the specific assumptions (27-ii) and (27-iii) are ignored in (EVP-dLc), Proposition 3.5 above reduces to the result in Brunner [43].

Summing up, all maximal/variational principles we just evoked are nothing but logical equivalents of (EVP). Finally, we may ask whether the abstract ordering principle in Qiu [190] is also comprised in this list. The answer to this is affirmative; further aspects will be delineated elsewhere. Some technical developments of the described facts may be found in Turinici [253].

3.1.3 *Vectorial Caristi-Kirk principles*

Let V be a (real) vector space. Given $Z \in \exp(V)$, denote

$$\mathrm{bal}(Z) = \{\lambda z; \lambda \in [-1, 1], z \in Z\} \text{ (the } \textit{balanced cover} \text{ of } Z).$$

If $Z = \mathrm{bal}(Z)$, then Z is called *balanced*; note that in such a case, $Z = -Z$ (hence, Z is *symmetric*) and $0 \in Z$.

(A) Let $\mathcal{T}$ be a linear topology over V; that is:

$\mathcal{T}$ is a topology on V and the vector space operations are $\mathcal{T}$-continuous;

in this case, $(V, \mathcal{T})$ will be referred to as a *(real) topological vector space*. Let $\mathcal{V}$ stand for the family of all zero $\mathcal{T}$-neighborhoods in V; note that

$\lambda \neq 0$, $V \in \mathcal{V}$ imply $\lambda V \in \mathcal{V}$ (as $x \mapsto \lambda x$ is a homeomorphism).

Let us show that $\mathcal{V}$ has a *balanced basis*, in the sense

$\mathcal{B} := \{V \in \mathcal{V}; V = \mathrm{bal}(V)\}$ is $(\supseteq)$-cofinal in $\mathcal{V}$
($\forall V \in \mathcal{V}$, there exists $B \in \mathcal{B}$ with $V \supseteq B$).

In fact, as $(\lambda, x) \mapsto \lambda x$ is continuous at $(0,0) \in R \times V$, it follows that

$\forall V \in \mathcal{V}$, $\exists \varepsilon > 0$, $\exists W \in \mathcal{V}$: $\alpha\varepsilon W \subseteq V$, if $|\alpha| \leq 1$;
hence, $B := \mathrm{bal}(\varepsilon W)$ is balanced and $B \subseteq V$;

and this, along with $B \in \mathcal{V}$ (because $\varepsilon W \subseteq B$) proves our claim. As a direct consequence, the basic properties hold

(bzn-1) $(\mathcal{B}, \supseteq)$ is *directed* ($\forall B_1, B_2 \in \mathcal{B}$, $\exists B_3 \in \mathcal{B}$: $B_1 \cap B_2 \supseteq B_3$)
(bzn-2) each $B \in \mathcal{B}$ is *absorbing* ($\forall v \in V$, $\exists \varepsilon > 0$: $[-\varepsilon, \varepsilon]v \subseteq B$)
(bzn-3) $\mathcal{B}$ is *quasi-additive* ($\forall B \in \mathcal{B}$, $\exists D \in \mathcal{B}$: $B \supseteq D + D$).

Conversely, let the family $\mathcal{B}$ of balanced sets in V fulfill (bzn-1)–(bzn-3). According to Cristescu [67, Ch. 1, Sec. 2], there exists a (uniquely determined) linear topology $\mathcal{T}$ over V, such that $\mathcal{B}$ is a balanced basis of the zero $\mathcal{T}$-neighborhoods family $\mathcal{V}$. This allows us expressing all (topological) $\mathcal{V}$-concepts in terms of $\mathcal{B}$, and vice versa.

(B) Let $(V, \mathcal{T})$ be a topological vector space; as precise, the family $\mathcal{V}$ of all zero $\mathcal{T}$-neighborhoods has a balanced basis, $\mathcal{B}$. Then, take the subset $V_+ \in \exp(V)$ as

(c-1) V_+ is a *(convex) cone* of V ($\alpha V_+ + \beta V_+ \subseteq V_+$, for all $\alpha, \beta \in R_+$)
(c-2) V_+ is *non-degenerated* ($V_+ \neq \{0\}$) and *proper* ($V_+ \neq V$).

In this case, the associated relation $(\preceq)$ (over V) introduced via

$x \preceq y$ iff $y - x \in V_+$

is reflexive and transitive; hence, a quasi-order on V; and $(V, \mathcal{T}, V_+)$ will be referred to as a *quasi-ordered (real) topological vector space*. Under this context, we are in position to introduce and discuss a lot of working conditions to be needed in the sequel.

I) The first condition refers to the structure $(V, \mathcal{T})$ only; and writes

(Cond-1) $(V, \mathcal{T})$ is *metrizable*;

which means the couple of (independent) conditions

(Cond-1-a) $\mathcal{T}$ is *separated*: $\cap \mathcal{B} = \{0\}$
(Cond-1-b) $\mathcal{T}$ is *semi-metrizable*: $\mathcal{B}$ has countable $(\supseteq)$-cofinal parts.

A basic consequence of this is the following (cf. Precupanu [187, Ch. 2, Sec. 2.1]). Call the mapping $x \mapsto |||x|||$ (from V to R_+) a *Fréchet norm* over V, provided the conditions below hold

(fn-1) $x = 0$ if and only if $|||x||| = 0$
(fn-2) $(\forall x, y \in V)$: $|||x + y||| \leq |||x||| + |||y|||$
(fn-3) $(\forall x \in V)$: $|||\lambda x||| \leq |||x|||$, whenever $|\lambda| \leq 1$
(hence, in particular: $|||x||| = |||-x|||$)
(fn-4) $(\forall x \in V)$: $|||2^{-n}x||| \to 0$ as $n \to \infty$.

Proposition 3.6. *Suppose that $(V, \mathcal{T})$ is metrizable (see above). Then, there exists a Fréchet norm $x \mapsto |||x|||$ over V such that, with the convention*

$$(G(\varepsilon) = \{v \in V; |||v||| < \varepsilon\},\ \varepsilon > 0),\ \mathcal{G} := \{G(\varepsilon); \varepsilon > 0\},$$

the families of (balanced) sets $\mathcal{B}$ and $\mathcal{G}$ are equivalent, in the sense:

(31-1) $\mathcal{B}$ is finer than $\mathcal{G}$ ($\forall G \in \mathcal{G}$, $\exists B \in \mathcal{B}$: $G \supseteq B$)
(31-2) $\mathcal{G}$ is finer than $\mathcal{B}$ ($\forall B \in \mathcal{B}$, $\exists G \in \mathcal{G}$: $B \supseteq G$).

As a consequence, all topological concepts [which are stated in terms of the system $\mathcal{B}$] may be also characterized in terms of the system $\mathcal{G}$; or, equivalently: in terms of the invariant to translations metric (over V)

$$e(x, y) = |||x - y|||,\ x, y \in V.$$

This is also valid for the convergence and Cauchy structures over V attached to the underlying topology. A few basic examples are listed below.

Ex-1) Given the sequence $(v_n; n \geq 0)$ in V and the point $v \in V$, we say that $v_n \xrightarrow{\mathcal{T}} v$ (and read: $(v_n; n \geq 0)$, *$\mathcal{T}$-converges* towards v), provided

$$\forall B \in \mathcal{B},\ \exists p = p(B)\text{: } p \leq n \Longrightarrow v_n - v \in B.$$

This, along with the stated equivalence result, writes

$\forall \varepsilon > 0$, $\exists p = p(\varepsilon)$: $p \leq n \Longrightarrow e(v_n, v) = |||v_n - v||| < \varepsilon$;

which is nothing else than the metrical convergence property $v_n \xrightarrow{e} v$. As $\mathcal{T}$ is separated, the class $\lim_n(v_n)$ of all limit points v is either empty or a singleton; in the latter case, we say that $(v_n; n \geq 0)$ is $\mathcal{T}$-convergent (or, equivalently: e-convergent). Further, given the (nonempty) subset Z of V, its $\mathcal{T}$-*adherence*, denoted as $\text{cl}(Z)$, is the class of all $\mathcal{T}$-*adherent* points $v \in V$, introduced as

$(v + B) \cap Z \neq \emptyset$ (meaning: $v \in Z + B$), for each $B \in \mathcal{B}$.

Note that, in particular, this yields (cf. Cristescu [67, Ch. 1, Sec. 2])

$$\text{cl}(Z) = \cap\{Z + B; B \in \mathcal{B}\}.$$

Moreover, by the denumerable and metrical character of our ambient linear topology, $\text{cl}(Z)$ is the class of all $v \in V$ fulfilling

$v_n \xrightarrow{\mathcal{T}} v$ (or, equivalently: $v_n \xrightarrow{e} v$), for some $(v_n; n \geq 0)$ in Z.

Ex-2) Given the sequence (x_n) in V, the $\mathcal{T}$-Cauchy property of it means

$\forall B \in \mathcal{B}$, $\exists q = q(B)$: $q \leq m \leq n \Longrightarrow v_m - v_n \in B$;
or, equivalently (by the quasi-additive property of $\mathcal{B}$)
$\forall B \in \mathcal{B}$, $\exists r = r(B)$: $r \leq n \Longrightarrow v_r - v_n \in B$.

This, along with the same equivalence result, may be written as

$\forall \varepsilon > 0$, $\exists q = q(\varepsilon)$: $q \leq m \leq n \Longrightarrow e(v_m, v_n) = |||v_m - v_n||| < \varepsilon$;
or, equivalently (by the properties of the metric $e(.,.)$)
$\forall \varepsilon > 0$, $\exists r = r(\varepsilon)$: $r \leq n \Longrightarrow e(v_r, v_n) = |||v_r - v_n||| < \varepsilon$;

which is nothing else than the e-Cauchy property of $(v_n; n \geq 0)$. Clearly, any $\mathcal{T}$-convergent (or, equivalently: e-convergent) sequence is $\mathcal{T}$-Cauchy (or, equivalently: e-Cauchy); the reciprocal is not in general true.

Ex-3) Call the sequence $(v_n; n \geq 0)$ in V, $\mathcal{T}$-*asymptotic* provided

$\forall B \in \mathcal{B}$, $\exists r = r(B)$: $r \leq n \Longrightarrow v_n - v_{n+1} \in B$.

By the underlying equivalence result, this may be also written as

$\forall \varepsilon > 0$, $\exists r = r(\varepsilon)$: $r \leq n \Longrightarrow e(v_n, v_{n+1}) = |||v_n - v_{n+1}||| < \varepsilon$;
or, equivalently: $\lim_n e(v_n, x_{n+1}) = 0$;

which, by definition, means: $(v_n; n \geq 0)$ is e-*asymptotic*.

Ex-4) Concerning the relationships with the previous concept, we note the obvious inclusion

(∀ descending sequence) $\mathcal{T}$-Cauchy $\Longrightarrow$ $\mathcal{T}$-asymptotic.

The reverse implication holds too, in a global sense. Precisely, the conditions below

(C-reg) V_+ is *Cauchy $\mathcal{T}$-regular*:
each descending sequence in V_+ is $\mathcal{T}$-Cauchy (that is: e-Cauchy)
(A-reg) V_+ is *asymptotic $\mathcal{T}$-regular*:
each descending sequence in V_+ is $\mathcal{T}$-asymptotic (that is: e-asymptotic)

are equivalent to each other, in (ZF-AC). In fact, (C-reg) $\Longrightarrow$ (A-reg) is clear, by the above. Conversely, suppose that (A-reg) holds. If, by absurd, (C-reg) is not true, there must be a descending sequence $(v_n; n \geq 0)$ in V_+, such that (according to a previous characterization of this concept)

(v_n) is not $\mathcal{T}$-Cauchy; that is (for some $B \in \mathcal{B}$):
for each r, there exists $s > r$, with $v_r - v_s \notin B$.

Define a relation F over N as

$F(r) = \{s \in N(r, <); v_r - v_s \notin B\}$, $r \in N$.

By the working condition, F is a proper relation; so that, its selection $(f(n) = \min F(n);\ n \in N)$ is well defined as a selfmap of N. [Clearly, no choice techniques are needed here; so that, the strongly reduced system (ZF-AC) is sufficient as a framework.] By the very definition above,

f is progressive ($n < f(n)$, $\forall n \in N$) and $x_n - x_{f(n)} \notin B$, $\forall n \in N$.

As a consequence of this, $(i(n) := f^n(0); n \geq 0)$ is strictly ascending, and the subsequence $(w_n := v_{i(n)}; n \geq 0)$ of $(v_n; n \geq 0)$ is descending (in V_+), with (in addition)

($w_n \in V_+$ and) $w_n - w_{n+1} \notin B$, for all n.

This contradicts the asymptotic regularity hypothesis; and proves the claim. By definition, either of these (equivalent) properties will be referred to as

(reg) V_+ is *$\mathcal{T}$-regular* (or, equivalently: *e-regular*).

The usefulness of this condition is underlined by the following vectorial form of an auxiliary fact we just encountered in a previous place:

Proposition 3.7. *Let the quasi-ordered topological vector space* $(V, \mathcal{T}, V_+)$ *be such that* $(V, \mathcal{T})$ *is metrizable and* V_+ *is* $\mathcal{T}$*-regular. Then,*

(32-1) V_+ *is pointed* $(V_+ \cap (-V_+) = \{0\})$

(32-2) *the associated quasi-order* $(\preceq)$ *is a (partial) order on* V.

Proof. Let $u, v \in X$ be such that $u \preceq v$, $v \preceq u$. The sequence (z_n) in V_+ $(z_{2n} = u - v,\ z_{2n+1} = v - u; n \geq 0)$ is descending; so that, as V_+ is $\mathcal{T}$-regular (or, equivalently: e-regular), we must have that (z_n) is $\mathcal{T}$-asymptotic (or, equivalently: e-asymptotic); wherefrom $2(u - v) \in \cap\mathcal{B}$. This, along the sufficiency of $\mathcal{B}$, gives $u = v$; as desired. □

II) The second condition about our structure $(V, \mathcal{T}, V_+)$ writes

(Cond-2) V_+ is $\mathcal{T}$-closed (or, equivalently: e-closed).

Note that, by a representation formula, this may be written as

$$V_+ = \mathrm{cl}(V_+) = \cap\{V_+ + B; B \in \mathcal{B}\}.$$

Moreover, under the linear character of $\mathcal{T}$, this condition may be also written in terms of $(-V_+)$; precisely,

$-V_+$ is $\mathcal{T}$-closed (or, equivalently: e-closed); meaning:
$-V_+ = \mathrm{cl}(-V_+) = \cap\{-V_+ + B; B \in \mathcal{B}\}$.

III) For the last condition about $(V, \mathcal{T}, V_+)$, we need some preliminaries. Given any pair $a, b \in V$ with $a \preceq b$, let us put

$[a, b; V, V_+] = \{x \in V; a \preceq x \preceq b\}$ (the *order segment* between a and b).

Further, take a nonempty part Z of V; we say that it is *full*, in case

$a, b \in Z$, $a \preceq b$, imply $[a, b; V, V_+] \subseteq Z$.

A unilateral version of this concept may be introduced as follows: given the balanced part Z of V (hence, $0 \in Z$), let us say that it is *0-full*, provided

$0 \preceq b \in Z$ implies $[0, b; V, V_+] \subseteq Z$; or, equivalently:
$b \in V_+ \cap Z$ and $0 \preceq x \preceq b$ imply $x \in Z$ (hence, $x \in V_+ \cap Z$).

Clearly, if the balanced part Z of V is full, then it is 0-full as well; but, the converse is not in general valid, as shown in

Example 3.1.

Let $V = R^2$, endowed with the linear topology $\mathcal{N}$ induced by Euclidean norm $||.||$ on V, and the (partial) ordering $(\preceq)$ attached to the standard cone

$V_+ = R^2_+$ in V. Further, take the open unitary sphere $Z = \{v \in V; ||v|| < 1\}$. Clearly, Z is balanced and 0-full, by the monotonicity of Euclidean norm over the cone V_+. On the other hand, given ε in $]0, 1 - 2^{-1/2}[$, the points $a = (-1+\varepsilon, 0)$, $b = (0, 1-\varepsilon)$ belong to Z, with $a \preceq b$. Moreover, the point $c = (-1+\varepsilon, 1-\varepsilon)$ belongs to $[a, b; V, V_+]$, as it can be directly seen. But, in view of $||c|| = (1-\varepsilon)2^{1/2} > 1$, this point does not belong to Z; hence Z is not full.

Under these preliminaries, we may write the announced third condition upon our structure $(V, \mathcal{T}, V_+)$:

(Cond-3) V_+ is $(\mathcal{T}, 0)$-normal: $\mathcal{B}_0 := \{C \in \mathcal{B}; C \text{ is 0-full}\}$ is $(\supseteq)$-cofinal in $\mathcal{B}$ ($\forall B \in \mathcal{B}$, $\exists C \in \mathcal{B}_0$: $B \supseteq C$).

[Note that, a corresponding version of this in terms of the associated metric $e(.,.)$ is not in general available; we do not give details.] By the above remark, it is clear that under this unilateral type assumption, we cannot get a global property like

V_+ is $\mathcal{T}$-normal: $\mathcal{B}$ admits $(\supseteq)$-cofinal parts consisting of full subsets.

However, a weak "transitive" form of it is ultimately accessible, as below.

Proposition 3.8. *Suppose that V_+ is $(\mathcal{T}, 0)$-normal. Then*

V_+ is transitively $\mathcal{T}$-normal:
for each $B \in \mathcal{B}$ there exists $D \in \mathcal{B}_0$ with $B \supseteq D + D \supseteq D$, such that ($a, b \in D$, $a \preceq b$, imply $[a, b; V, V_+] \subseteq B$).

Proof. Let $B \in \mathcal{B}$ be arbitrary fixed. By the quasi-additive property of $\mathcal{B}$ and the $(\mathcal{T}, 0)$-normality condition, there exists $C \in \mathcal{B}_0$ such that $B \supseteq C + C \supseteq C$. Further, by the same couple of properties, there exists $D \in \mathcal{B}_0$ with $C \supseteq D + D \supseteq D$; hence, combining these, $B \supseteq D + D \supseteq D$. We claim that D is our desired element so as to verify the precise assertion. In fact, let $a, b \in D$ be such that $a \preceq b$; and let $x \in V$ be taken according to $a \preceq x \preceq b$. By the relations above

$$0 \preceq x - a \preceq b - a \in D + D \subseteq C;$$

and since C is 0-full, we must have $x - a \in C$. This yields

$$x = a + x - a \in D + C \subseteq C + C \subseteq B;$$

and the proof is complete. □

(C) Having these precise, let $(V, \mathcal{T}, V_+)$ be a quasi-ordered metrizable topological vector space; and X be a nonempty set. By a V_+-*metric* on X, we shall mean any map $\Delta : X \times X \to V_+$ with

(vm-1) Δ is *reflexive sufficient* $[x = y \Longleftrightarrow \Delta(x, y) = 0]$,
(vm-2) Δ is *triangular* $[\Delta(x, z) \preceq \Delta(x, y) + \Delta(y, z), \forall x, y, z \in X]$,
(vm-3) Δ is *symmetric* $[\Delta(x, y) = \Delta(y, x), \forall x, y \in X]$;

then, (X, Δ) will be referred to as a V_+-*metric space.* Given this object, we may introduce a Δ-convergence and Δ-Cauchy structure on X as follows.

Given the sequence (x_n) in X and the point $x \in X$, we say that (x_n), Δ-*converges* to x (written as: $x_n \xrightarrow{\Delta} x$), provided

$\forall B \in \mathcal{B}$, $\exists p = p(B)$: $p \leq n \Longrightarrow \Delta(x_n, x) \in B$.

By this very definition (and reflexivity of Δ), we have the properties:

(De-conv-1) $((\xrightarrow{\Delta})$ is *hereditary*)
$x_n \xrightarrow{\Delta} x$ implies $y_n \xrightarrow{\Delta} x$, for each subsequence (y_n) of (x_n)
(De-conv-2) $((\xrightarrow{\Delta})$ is *reflexive*)
$\forall u \in X$, the constant sequence $(x_n = u; n \geq 0)$ fulfills $x_n \xrightarrow{\Delta} u$.

As a consequence, $(\xrightarrow{\Delta})$ is a sequential convergence on X, under the lines in Kasahara [141]. The set of all such limit points of (x_n) will be denoted $\Delta - \lim_n(x_n)$; if it is nonempty, then (x_n) is called Δ-*convergent.* As before, the following extra property is to be discussed here

(De-conv-3) $(\xrightarrow{\Delta})$ is *separated* (referred to as: (X, Δ) is *separated*):
$\Delta - \lim_n(x_n)$ is an asingleton, for each sequence (x_n) of X.

Concerning this aspect, note that (X, Δ) is not separated, in general. However, under some extra conditions upon our data, this happens. Precisely, the following result is available.

Proposition 3.9. *Suppose that $\mathcal{T}$ is separated and V_+ is $(\mathcal{T}, 0)$-normal. Then, the V_+-metric space (X, Δ) is separated.*

Proof. We start by noting that (as $(V, \mathcal{T})$ is metrizable)

$\mathcal{T}$ is separated $(\cap \mathcal{B} = \{0\})$.

Let (x_n) be a sequence in X with

$x_n \xrightarrow{\Delta} u$, $x_n \xrightarrow{\Delta} v$ (for some $u, v \in X$).

Given $B \in \mathcal{B}$, there exists $C \in \mathcal{B}_0$ with $B \supseteq C$; then, for this element, there must be some $D \in \mathcal{B}_0$ such that $C \supseteq D + D$. From the above convergence relation, there exists some index $r \geq 0$, such that

$$r \leq n \Longrightarrow \Delta(x_n, u), \Delta(x_n, v) \in D.$$

On the other hand, by the properties of Δ, one has for each $n \geq r$,

$$0 \preceq \Delta(u, v) \preceq \Delta(x_n, u) + \Delta(x_n, v) \in D + D \subseteq C;$$

and this, along with $C \in \mathcal{B}_0$, yields

$$\Delta(u, v) \in C \subseteq B; \text{ whence, } \Delta(u, v) \in B.$$

As $B \in \mathcal{B}$ was arbitrarily chosen, this yields (via $\mathcal{T}$ = separated)

$$\Delta(u, v) \in \cap\mathcal{B} = \{0\} \text{ (so that: } u = v);$$

and the conclusion follows. □

Further, call the sequence (x_n), Δ-*Cauchy* provided

$$\forall B \in \mathcal{B}, \exists p = p(B) \colon p \leq m \leq n \Longrightarrow \Delta(x_m, x_n) \in B;$$

the class of all these will be denoted as $Cauchy(\Delta)$. Clearly,

(De-Cauchy-1) ($Cauchy(\Delta)$ is *hereditary*)
(x_n) is Δ-Cauchy implies (y_n) is Δ-Cauchy,
for each subsequence (y_n) of (x_n)
(De-Cauchy-2) ($Cauchy(\Delta)$ is *reflexive*)
$\forall u \in X$, the constant sequence $(x_n = u; n \geq 0)$ is Δ-Cauchy;

so, the class $Cauchy(\Delta)$ is a Cauchy structure, under the lines in Turinici [244]. Finally, call $((\xrightarrow{\Delta}), Cauchy(\Delta))$, the *conv-Cauchy structure* attached to Δ. As before, the following extra properties are to be discussed here

(De-Cauchy-3) (X, Δ) is *admissible*:
each Δ-convergent sequence is Δ-Cauchy
(De-Cauchy-4) (X, Δ) is *complete*:
each Δ-Cauchy sequence is Δ-convergent.

Generally, (X, Δ) is not complete. Likewise, by the imposed conditions, (X, Δ) is not admissible, in general. But, under extra conditions upon our data, this happens. Precisely, we have

Proposition 3.10. *Let V_+ be $(\mathcal{T}, 0)$-normal. Then, (X, Δ) is admissible.*

Proof. Let (x_n) be a sequence in X with $x_n \xrightarrow{\Delta} u$, for some $u \in X$. Given $B \in \mathcal{B}$, there exists $C \in \mathcal{B}_0$ with $B \supseteq C$; then, for this element, there must be some $D \in \mathcal{B}_0$ such that $C \supseteq D + D$. Now, by the above convergence relation, there exists some index $p \geq 0$, such that $[p \leq n \Longrightarrow \Delta(x_n, u) \in D]$. Further, by the triangular property of Δ, one has for $p \leq n \leq m$,

$$0 \preceq \Delta(x_n, x_m) \preceq \Delta(x_n, u) + \Delta(x_m, u) \in D + D \subseteq C;$$

and this, along with $C \in \mathcal{B}_0$, yields

$$\Delta(x_n, x_m) \in C \subseteq B, \text{ when } p \leq n \leq m.$$

As $B \in \mathcal{B}$ was arbitrarily chosen, this gives the Δ-Cauchy property of (x_n); and conclusion follows. □

Finally, let $\varphi : X \to V_+$ be a function. The following condition is essential in what follows

(lsc-desc) φ is descending Δ-lsc:
$x_n \xrightarrow{\Delta} x$ and $(\varphi(x_n))$=descending imply $\varphi(x_n) \succeq \varphi(x)$, $\forall n$;

this may be viewed as a natural extension of (descending) lsc property from the standard case $[V = R,\ V_+ = R_+]$.

(D) Under the introduced concepts, we are now in position to state a variational principle involving these data [referred to as: *vectorial Ekeland Variational Principle*; in short: (EVP-v)].

Theorem 3.8. *Let the quasi-ordered metrizable topological vector space* $(V, \mathcal{T}, V_+)$ *be taken as*

(31-i) V_+ *is* $\mathcal{T}$*-regular (hence (see above),* V_+ *is pointed)*
(31-ii) V_+ *is both* $\mathcal{T}$*-closed and* $(\mathcal{T}, 0)$*-normal.*

Further, let (X, Δ) *be a complete* V_+*-metric space; and* $\varphi : X \to V_+$ *be descending* Δ*-lsc. Then, for each starting* $x \in X$*, there exists an element* $z \in X$*, with*
(31-a) $\Delta(x, z) \preceq \varphi(x) - \varphi(z)$ *(hence,* $\varphi(x) \succeq \varphi(z)$*)*
(31-b) $w \in X,\ \Delta(z, w) \preceq \varphi(z) - \varphi(w) \Longrightarrow z = w$.

Proof. Let $(\leq)$ stand for the relation (over X)

$$x \leq y \text{ iff } \Delta(x, y) \preceq \varphi(x) - \varphi(y).$$

For the moment, $(\leq)$ is a quasi-order on X. Moreover, as V_+ is pointed (see above), $(\leq)$ is antisymmetric as well; hence, it is a (partial) order on X. Further, letting $|||.|||$ be the Fréchet norm of V given by the metrizable character of $(V, \mathcal{T})$, define the pseudometric d over X as

$$d(x, y) = |||\Delta(x, y)|||,\ x, y \in X.$$

By the properties of our (Fréchet) norm (and the ones of Δ), it follows that

(ref) d is *reflexive*: $d(x, x) = 0$, for each $x \in X$
(suf) d is *sufficient*: $d(x, y) = 0$ implies $x = y$
(sym) d is *symmetric*: $d(x, y) = d(y, x)$, $\forall x, y \in X$.

So, according to our previous terminology, d is a symmetric rs-pseudometric on X; in which case, (X, d) is termed a symmetric rs-pseudometric space. Unfortunately, the triangular property

(tri) d is *triangular*: $d(x, z) \leq d(x, y) + d(y, z)$, $\forall x, y, z \in X$

is not holding here; hence, in general, $d(.,.)$ is not a metric on X.

Having these precise, we now claim that the rs-pseudometric Maximal Principle (MP-rs-p) is applicable to the triple $(X, \leq, d)$; wherefrom, all is clear. Let $(x_n; n \geq 0)$ be an ascending sequence (modulo $(\leq)$) in X:

$$(0 \preceq)\Delta(x_n, x_m) \preceq \varphi(x_n) - \varphi(x_m), \text{ if } n \leq m.$$

Part 1) The sequence $(\varphi(x_n); n \geq 0)$ is descending in V_+; hence, $\mathcal{T}$-Cauchy, as V_+ is $\mathcal{T}$-regular. This, along with V_+ being $(\mathcal{T}, 0)$-normal (and our previous convention), may be written as:

$$\forall B \in \mathcal{B}_0, \exists q(B)\colon\ q(B) \leq n \leq m \Longrightarrow \varphi(x_n) - \varphi(x_m) \in B.$$

Combining with all elements of $\mathcal{B}_0$ being 0-full, gives

$$\forall B \in \mathcal{B}_0, \exists q(B)\colon\ q(B) \leq n \leq m \Longrightarrow \Delta(x_n, x_m) \in B;$$

or, in other words (by the cofinal character of $\mathcal{B}_0$),

$$\forall B \in \mathcal{B},\ \exists q(B)\colon\ q(B) \leq n \leq m \Longrightarrow \Delta(x_n, x_m) \in B;$$
whence: $(x_n; n \geq 0)$ is Δ-Cauchy.

This in turn yields (from the equivalence between the systems $\mathcal{B}$ and $\mathcal{G}$)

$$\forall \varepsilon > 0,\ \exists q(\varepsilon)\colon\ q(\varepsilon) \leq n \leq m \Longrightarrow \Delta(x_n, x_m) \in G(\varepsilon);$$
that is: $d(x_n, x_m) = |||\Delta(x_n, x_m)||| < \varepsilon$;
whence: (x_n) is d-Cauchy;

which tells us that

$(X, \le)$ is Cauchy regular (modulo d):
each ascending sequence in X is d-Cauchy.

Part 2) As (X, Δ) is complete, $x_n \xrightarrow{\Delta} x$, for some $x \in X$; moreover, by the choice of φ, we have $[\varphi(x_n) \succeq \varphi(x)$, for all $n]$. Let $B \in \mathcal{B}$ be arbitrary fixed. By the Δ-convergence property, there exists $p(B) \ge 0$, such that

$$m \ge p(B) \Longrightarrow \Delta(x_m, x) \in B.$$

Let $k \ge 0$ be arbitrary fixed. By the obtained boundedness from below relation we have (along with triangular property)

$$\Delta(x_k, x) \preceq \Delta(x_k, x_m) + \Delta(x_m, x) \preceq \varphi(x_k) - \varphi(x_m) + \Delta(x_m, x)$$
$$\preceq \varphi(x_k) - \varphi(x) + \Delta(x_m, x), \text{ for all } m \ge \max\{k, p(B)\};$$

wherefrom (combining with the previous relation)

$$\Delta(x_k, x) - (\varphi(x_k) - \varphi(x)) \in (-V_+) + B.$$

This, by the arbitrariness of B, yields (by a representation formula)

$$\Delta(x_k, x) - (\varphi(x_k) - \varphi(x)) \in \text{cl}(-V_+) = -V_+, \text{ (meaning: } x_k \le x), \forall k.$$

Combined with the preceding step, we thus obtained that

each ascending sequence in X is bounded above (modulo $(\le)$);
or, equivalently: $(X, \le)$ is sequentially inductive.

Part 3) Summing up, the rs-pseudometric maximal principle (MP-rs-p) is applicable to this context. But then, for the starting $x \in X$, there exists some $z \in X$, such that

(p-1) $x \le z$; hence, $\Delta(x, z) \preceq \varphi(x) - \varphi(z)$
(p-2) $z \le w$ implies $z = w$; that is:
$\Delta(z, w) \preceq \varphi(z) - \varphi(w)$ implies $z = w$.

This gives us all desired facts; and concludes the argument. □

In particular, when $V = R$, $V_+ = R_+$, (EVP-v) is just Ekeland's variational principle [85], [86] (in short: (EVP)); this, among others, motivated our terminology.

As a direct application of these, one gets the following fixed point statement [referred to as: the *vectorial Caristi-Kirk theorem*; in short: (CK-v)].

Theorem 3.9. *Let the conditions of (EVP-v) hold; and the selfmap $T : X \to X$ be such that*

(contr) T is (Δ, φ)-contractive: $\Delta(x, Tx) \le \varphi(x) - \varphi(Tx)$, $\forall x \in X$.

Then, for each $x \in X$ there exists $z \in \mathrm{Fix}(T)$ with the properties
(32-a) $\Delta(x, z) \preceq \varphi(x) - \varphi(z)$ *(hence, $\varphi(x) \succeq \varphi(z)$)*
(32-2) $\Delta(z, w) \preceq \varphi(z) - \varphi(w) \Longrightarrow z = w$.

In particular, when $V = R$, $V_+ = R_+$, Theorem 3.9 is just the Caristi-Kirk result [45] (in short: (CK)); so, our terminology is motivated. Note that, in the quoted paper, the conclusions of original statement (CK) were obtained via transfinite induction; so, it is natural to ask whether this procedure is also working in our vectorial setting. The answer to this is positive; we do not give details. Further technical aspects aspects may be found in Jachymski [124], Turinici [245], or Kirk and Saliga [149]; see also Goepfert et al. [103].

3.1.4 *Application to normed lattices*

An interesting particular case of these developments refers to the lattice properties of our ambient (real) space V. Some preliminaries are needed.

Let $||.||$ be a norm on V. Clearly, the associated to norm metric over V

$$e(x, y) = ||x - y||,\ x, y \in V$$

is invariant to translations; moreover, as precise, the (normed) linear topology $\mathcal{N}$ over V is defined by means of the balanced convex sets family

$$\mathcal{G} = (G(\varepsilon); \varepsilon > 0);\ \text{where } (G(\varepsilon) = \{x \in V; ||x|| < \varepsilon\},\ \varepsilon > 0).$$

Finally, let V_+ be a (non-degenerated, proper) pointed convex cone of V; its associated (partial) order will be denoted as $(\preceq)$.

Clearly, an application of the variational and fixed point results above is possible when the convex cone V_+ fulfills (in addition)

(r-c-0-n) V_+ is $\mathcal{N}$-regular, $\mathcal{N}$-closed, $(\mathcal{N}, 0)$-normal.

A discussion of these will be carried out in a *normed lattice* setting.

Suppose in the following that

(NL-1) (V, V_+) is a lattice: $x \vee y$ (hence, $x \wedge y$) exists, for all $x, y \in V$;

here, for each (nonempty) part A of V, we put

$\vee A = \sup(A)$ (*supremum* of A), $\wedge A = \inf(A)$ (*infimum* of A).

The following property is almost immediate (so, we do not give details):

$x \vee y + x \wedge y = x + y, \ \forall x, y \in V.$

The lattice setting relative to (V, V_+) allows us introducing some basic maps in $\mathcal{F}(V, V_+)$. The former of these is

$|x| = x \vee (-x)$ (the *modulus* of x), $x \in V$;

note that, as a direct consequence of our conventions,

$|x \vee z - y \vee z| \preceq |x - y|, \ |x \wedge z - y \wedge z| \preceq |x - y|, \ \forall x, y, z \in V;$

see, for instance, Cristescu [67, Ch. 5, Sec. 2]. The next functions to be introduced are

$x_+ = x \vee 0, \ x_- = (-x) \vee 0$ (the *positive/negative part* of x), $x \in V$.

Clearly, the following relations are valid

(for each $x \in V$): $x = x_+ - x_-, \ |x| = x_+ + x_-$;
hence, $x_+ = (1/2)(|x| + x), \ x_- = (1/2)(|x| - x)$.

Having these precise, the second basic condition upon our data writes

(NL-2) V_+ is a $\mathcal{N}$-*solid* cone: $x, y \in V, \ |x| \preceq |y|$ imply $||x|| \le ||y||$.

When (NL-1)+(NL-2) holds, the triple $(V, \mathcal{N}, V_+)$ will be called a *normed lattice*. Note that, as a consequence of the last convention,

(p-1) $(\forall x \in V)$: $||x|| = ||y||$, where $y = |x|$
(p-2) $(\forall x, y \in V)$: $0 \preceq x \preceq y \Longrightarrow ||x|| \le ||y||$.

Proposition 3.11. *Assume that $(V, \mathcal{N}, V_+)$ is a normed lattice. Then,*
(41-1) *V_+ is a $\mathcal{N}$-closed cone*
(41-2) *V_+ is a $(\mathcal{N}, 0)$-normal cone (relative to $\mathcal{G}$)*
(41-3) *V_+ is not (in general) a $\mathcal{N}$-normal cone (relative to $\mathcal{G}$).*

Proof. I) Let us firstly prove the following property:

whenever (x_n) and (y_n) fulfill $x_n \xrightarrow{e} x, \ y_n \xrightarrow{e} y$,
we must have $x_n \vee y_n \xrightarrow{e} x \vee y, \ x_n \wedge y_n \xrightarrow{e} x \wedge y$.

In fact, by a previous formula, one gets

$$\text{(for each } n\text{): } |x_n \vee y_n - x \vee y| \preceq$$
$$|x_n \vee y_n - x \vee y_n| + |x \vee y_n - x \vee y| \preceq |x_n - x| + |y_n - y|.$$

Passing to norm, yields (from a consequence of normed lattice condition)

$$||x_n \vee y_n - x \vee y|| \leq ||x_n - x|| + ||y_n - y||, \ \forall n;$$

wherefrom, the first part follows. For the second part, it will suffice noting that the following reduction formula holds

$$x \wedge y = -((-x) \vee (-y)), \ \forall x, y \in V.$$

Now, let (x_n) be a sequence in V_+ with $[x_n \stackrel{e}{\longrightarrow} x$, for some $x \in V]$. Clearly,

$$(\forall n)\text{: } -x_n \preceq 0 \preceq x_n\text{; so that, } x_n = x_n \vee (-x_n) = |x_n|.$$

Passing to limit as $n \to \infty$ (and taking the preceding step into account) yields $x = |x|$; which tells us that $x \in V_+$.

II) By the very definition of $\mathcal{N}$-solid cone, we have for each $\varepsilon > 0$

$$0 \preceq a \preceq b \text{ and } b \in G(\varepsilon) \text{ imply } a \in G(\varepsilon);$$

hence the claim.

III) Let $\varepsilon > 0$ be arbitrary fixed. We have to establish of to what extent is the implication below available:

$$(a, b \in G(\varepsilon), a \preceq b) \Longrightarrow [a, b; V, V_+] \subseteq G(\varepsilon).$$

To discuss this, note that, given $g \in G(\varepsilon)$, one gets (see above)

$$||h|| = ||g|| < \varepsilon, \text{ when } h = |g|\text{; so that, } |g| \in G(\varepsilon);$$
$$||-g|| = ||g|| < \varepsilon\text{; wherefrom, } -g \in G(\varepsilon).$$

This, along with a couple of representations involving the positive/negative part, yields (in combination with $G(\varepsilon) =$ convex)

$$g_+ = (1/2)(|g| + g) \in G(\varepsilon), \ g_- = (1/2)(|g| - g) \in G(\varepsilon);$$

Now, take $a, b \in G(\varepsilon)$ with $a \preceq b$; and let $x \in V$ be such that

$$a \preceq x \preceq b \text{ (or, equivalently, } -b \preceq -x \preceq -a).$$

For the moment, we have by definition,

$$0 \preceq a_+ \preceq x_+ \preceq b_+, \ 0 \preceq b_- \preceq x_- \preceq a_-;$$

and this, combined with the preceding relations, yields (as $G(\varepsilon)$ is convex, balanced) $u := x_+ \in G(\varepsilon)$, $v := -x_- \in G(\varepsilon)$. But, from the obtained facts,

$(u + v =)x \in G(\varepsilon)$ is not in general true;

see a previous example concerning this question. □

We are now passing to the basic part of our developments. Let X be a nonempty set; and $\Delta : X \times X \to V_+$ be a V_+-metric over it. Further, let $\varphi : X \to V_+$ be a function. As an application of the vector Ekeland Variational Principle (EVP-v) and the vector Caristi-Kirk theorem (CK-v), one derives the following synthetic (variational and fixed point) result:

Theorem 3.10. *Assume that $(V, \mathcal{N}, V_+)$ is a normed lattice, V_+ is $\mathcal{N}$-regular, (X, Δ) is complete and*

(lsc-desc-De) φ is descending Δ-lsc:
$x_n \xrightarrow{\Delta} x$ *and* $(\varphi(x_n)) =$ *descending imply* $\varphi(x_n) \succeq \varphi(x)$, $\forall n$.

Then,

(41-a) *For each $x \in X$ there exists $z \in X$ with the properties*

(41-a-1) $\Delta(x, z) \le \varphi(x) - \varphi(z)$ *(hence* $\varphi(x) \succeq \varphi(z)$*)*
(41-a-2) $w \in X$, $\Delta(z, w) \le \varphi(z) - \varphi(w)$, *imply* $z = w$.

(41-b) *If, in addition, $T : X \to X$ fulfills*

$\Delta(x, Tx) \le \varphi(x) - \varphi(Tx)$, $\forall x \in X$,

we also have $z \in \mathrm{Fix}(T)$.

In particular, when $X = V$, a basic example of metric $\Delta(.,.)$ like before is provided by the modulus function

$\Delta(x, y) = |x - y|$ (hence, $||\Delta(x, y)|| = ||x - y||$), $x, y \in V$.

In this case, for each sequence $(x_n; n \ge 0)$ in V and each $x \in V$,

(conv-eq) $x_n \xrightarrow{\Delta} x$ iff $x_n \xrightarrow{e} x$
(Cauchy-eq) (x_n) is Δ-Cauchy iff (x_n) is e-Cauchy.

This tells us that (V, Δ) is complete, whenever so is (V, e); the triple $(V, \mathcal{N}, V_+)$ is then referred to as a *Banach lattice*. We therefore have

Theorem 3.11. *Assume that $(V, \mathcal{N}, V_+)$ is a Banach lattice, V_+ is $\mathcal{N}$-regular, and*

(lsc-desc-e) φ is descending e-lsc:
$x_n \xrightarrow{e} x$ *and* $(\varphi(x_n)) =$ *descending imply* $\varphi(x_n) \succeq \varphi(x)$, $\forall n$.

Then,

(42-a) *For each* $x \in V$, *there exists* $z \in V$, *such that*

(42-a-1) $|x - z| \preceq \varphi(x) - \varphi(z)$ *(hence,* $\varphi(x) \succeq \varphi(z)$*)*
(42-a-2) $w \in V,\ |z - w| \preceq \varphi(z) - \varphi(w) \Longrightarrow z = w$.

(42-b) *If, in addition,* $T : V \to V$ *fulfills*

$$|x - Tx| \preceq \varphi(x) - \varphi(Tx),\ \forall x \in V,$$

we also have $z = Tz$.

Finally, note that the ambient cone V_+ is $\mathcal{N}$-regular, whenever

(reg-seq-e) for each descending sequence $(y_n; n \geq 0)$ in V_+, there exists $y_\infty := \wedge\{y_n; n \geq 0\}$ and $y_n \xrightarrow{e} y_\infty$.

Note that, in such a case, the former of these results includes the related one due to Agarwal and Khamsi [7], proved via transfinite induction; and the latter one may be viewed as an extension of some contributions due to Zhao and Sun [271]. Further aspects may be found in Latif [160]; see also Zhu and Li [275].

3.2 Tarski Fixed Point Results over Separable Structures

Abstract. A simplified version is proposed – in the separable framework – for a multivalued Tarski fixed point result due to Zhu and Li [JOTA, 132 (2007), 493–507], over the class of Fang uniform spaces.

Subject-Class 2020. 54H25 (Primary), 06A06 (Secondary).

Keywords. Fang space, completeness, multivalued map, compactness, fixed point, chain, ordinal, cardinal, separable structure, quasi-uniformity, transfinite induction.

3.2.1 *Dependent Choice principles*

See Ch. 1.

3.2.2 *Statement of the problem*

Let X be a nonempty set. By a *pseudometric* over X we shall mean any map $d : X \times X \to R_+ := [0, \infty[$. If, in addition,

(ref) d is *reflexive* ($d(x, x) = 0, \forall x \in X$)
(sym) d is *symmetric* ($d(x, y) = d(y, x), \forall x, y \in X$)

we say that $d(.,.)$ is a *reflexive symmetric pseudometric* (in short: *rs-pseudometric*) over X. Put $J =]0, 1]$; note that $(J, \geq)$ has no maximal elements. Call the family $D = (d_\alpha; \alpha \in J)$ of rs-pseudometrics (over X), *J-normal* when

(fang-1) D is *monotone* ($d_\alpha(x, y) \leq d_\beta(x, y), \forall x, y \in X$, if $\alpha \geq \beta$)
(fang-2) D is *J-triangular* (for each $\lambda \in J$, there exists $\mu \in J(\lambda, \geq)$, with $d_\lambda(x, z) \leq d_\mu(x, y) + d_\mu(y, z), \forall x, y, z \in X$)
(fang-3) D is *sufficient* ($d_\alpha(x, y) = 0, \forall \alpha \in J$ imply $x = y$).

Equivalently, we say that D is a *Fang structure* on X; and the couple (X, D) is called a *Fang space*. The motivation of terminology comes from the fact that the first systematic study of such structures is due to Fang [88]. Note that (cf. the quoted paper) any *probabilistic metric space* may be represented in such a way; moreover, a similar conclusion is valid for the class of *fuzzy metric spaces*; see, for instance, Chang et al. [48]. Further, let $(\leq)$ be a *(partial) order* (reflexive, transitive, antisymmetric relation) over X; the triple $(X, D, \leq)$ will be then referred to as a *(partially) ordered Fang space*. Finally, let $T : X \to \exp(X)$ be a multivalued map, subjected to the (regularity) conditions

(s-pro) T is *semi-progressive* $[X(T;\leq) := \{u \in X; u \leq Tu\} \neq \emptyset]$,
(incr) T is *increasing* $[x \leq y \Longrightarrow Tx \leq Ty]$.

Here, given the couple of subsets $U, V \in \exp(X)$, we put

$U \leq V$ iff $[\forall u \in U, \exists v \in V\colon u \leq v]$.

Clearly, this relation is reflexive and transitive; but, not in general antisymmetric. Denote $\mathrm{Fix}(T) = \{z \in X; z \in Tz\}$; any point of it will be called *fixed* (under T). An outstanding way of getting such elements is the one involving image compactness and countable iterative compactness conditions upon T introduced as

(comp-val) T has *compact values* [Tx is compact, for each $x \in X$],
(comp-count-it) T has *compact countable iterations*:
each countable totally ordered part $C = \{x_i; i < \beta\}$ with
$C \subseteq \mathrm{cl}[\{x_0\} \cup T(C)]$, is relatively compact.

The following 2007 fixed point result in Zhu and Li [275, Theorem 4.1] (referred to as: Zhu-Li fixed point result; in short: (ZL-fpr)) is basic for our next developments.

Theorem 3.12. *Let the (partially) ordered Fang space $(X, D, \leq)$ and the multivalued map $T : X \to \exp(X)$ be such that (in addition)*

(con-1) T is semi-progressive, increasing
(con-2) T has compact values and compact countable iterations
(con-3) $(\leq)$ is D-right-closed:
$X(x, \leq) := \{y \in X; x \leq y\}$ is D-closed, for each $x \in X$.

Then, $\mathrm{Fix}(T)$ *is not empty [that is: T has at least one fixed point in X].*

(As a matter of fact, the underlying result involves multivalued maps $T : Y \to \exp(Y)$, where $Y \in \exp(X)$ is a D-closed subset; but this is not essential for us).

In particular, if T is univalued, the Zhu-Li fixed point result (ZL-fpr) is comparable with some related statements in Moench [169] and Turinici [229]; which, in turn, may be viewed as topological versions of the 1955 Tarski fixed point principle [222]; see also Abian and Brown [3]. Further aspects of the problem involving the weak topology setting were discussed in O'Regan [180] and Hong [114]. Note that all these require upon the (partial) order $(\leq)$ (over X) the stronger condition

(con-3a) $(\leq)$ is *D-bilateral-closed*:
both $X(x, \leq)$ and $X(x, \geq)$ are D-closed, for each $x \in X$.

This shows that – surprisingly – the framework of Theorem 3.12 is pretty large; so, a discussion of its basic lines is naturally coming here. Concerning this aspect, we stress that (according to authors) the proof of Zhu-Li fixed point result (ZL-fpr) consists in the stages below.

Proof. Denote $H = X(T; \leq)$ (hence $H \neq \emptyset$). Let ω and Ω stand for the first transfinite ordinal and, respectively, the first uncountable ordinal.

I) Define a relation $(\preceq)$ over H as:

$x \preceq y$ iff either $x = y$ or (in the opposite case) there exists a (countable) ordinal $\nu < \Omega$ and a transfinite sequence $(x_i; i < \nu + 1)$ in H with
(def-1) $x_0 = x$, $x_\nu = y$ and $x_i \leq x_{i+1} \in Tx_i$, $\forall i < \nu$
(def-2) if $\beta < \nu + 1$ is a (countable) limit ordinal then the transfinite sequence $(x_i; i < \beta)$ converges to x_β.

II) The relation $(\preceq)$ acts as a (partial) order on H; with, in addition,

(pro-0) $(\preceq)$ is *coarser* than $(\leq)$ on H
$(x, y \in H$, and $x \preceq y$ imply $x \leq y)$.

III) Further, the following property is available:

if the sequence $(x_n; n < \omega)$ in H is $(\preceq)$-ascending, then
(pro-1) $u := \lim_n(x_n)$ exists in H
(pro-2) $x_n \leq u$, $x_n \preceq u$, for all $n < \omega$.

To verify the first part, note that, by the very definition of (partial) order $(\preceq)$, there must be a sequence $(\Lambda_n; n < \omega)$ of (countable) ordinals, with

$$\text{(for each } n < \omega\text{)}: x_n = x_0^n,\ x_{n+1} = x_{\Lambda_n}^n = x_0^{n+1}.$$

Denote, for simplicity,

$$C = \cup\{C_n; n < \omega\}, \text{ where } C_n = \{x_i^n; i < \Lambda_n + 1\},\ n < \omega;$$

this is a countable set fulfilling (in addition) $C \subseteq \text{cl}[\{x_0\} \cup T(C)]$; so that [with (con-2)], C is relatively compact. As a consequence, $(x_n) \subseteq C$ must have at least one convergent subsequence $(x_{p(n)}; n < \omega)$; that is,

$$\text{(there exists } u \in X\text{)}: x_{p(n)} \xrightarrow{D} u, \text{ as } n \to \infty.$$

This, and the initial construction of (x_n), yields $x_n \xrightarrow{D} u$ as $n \to \infty$. In fact, if such a property fails, there must be a subsequence $(x_{q(n)}; n < \omega)$ of $(x_n; n < \omega)$ and some $v \in X$, with

$$x_{q(n)} \xrightarrow{D} v \text{ as } n \to \infty \text{ and } u \neq v.$$

Since (x_n) is $(\leq)$-ascending, we then have

for each n, there exists $a(n)$, with $x_{p(n)} \leq x_{q(a(n))}$;
for each n, there exists $b(n)$, with $x_{q(n)} \leq x_{p(b(n))}$.

Taking the limit in both sides, gives

(le-sym) $u \leq v$ and $v \leq u$; hence $u = v$ (as $(\leq)$ is an order);

a contradiction. Consequently, claim (pro-1) and the first half of claim (pro-2) (relative to $(\leq)$) hold; the second part of claim (pro-2) (relative to $(\preceq)$) holds too, by means of the compactness type condition (con-2).

IV) From the obtained facts and a maximal principle obtained by the quoted authors [275, Theorem 2.5], it follows that

$\forall u \in H$, $\exists v \in H$: v is $(\preceq)$-maximal; wherefrom, $v \in Tv$.

This ends the proof. □

Now, a simple inspection of the argument shows that (le-sym) cannot be obtained in the setting of right closeness requirement (con-3); but it is retainable when the bilateral closeness requirement (con-3a) is accepted. However, we must say that – even if the underlying bilateral condition corrects the precise drawback – the argument developed in the second part of this proof is rather complicated; so, it is natural to ask whether this may be removed. It is our aim in the following to show that all reasoning involved here may be simplified. The basic tool for our approach is the concept of *separable* structure; where a new separability criterion is proposed. And the specific tool of the same is essentially related to the convergence structure of Fang spaces. Finally, in the last part of this exposition, a purely topological version of the obtained result is considered.

3.2.3 *Separable structures*

In the following, the methodological basis of our exposition is the *naive (descriptive) Cantor set theory,* as developed by Natanson [176, Ch. 14].

To avoid contradictory sets in this theory, we restrict all constructions and arguments to a (non-contradictory) *Grothendieck universe* $\mathcal{G}$, taken as in Hasse and Michler [110, Ch. 1, Sec. 2]. For simplicity reasons, we drop any reference to this universe; so, by a *set*, one actually means a $\mathcal{G}$-admissible set with respect to a "sufficiently large" Grothendieck universe $\mathcal{G}$. In particular, this allows us treating in a rather unitary way some basic structural problems involving the concepts of *ordinal* and *cardinal*.

Concerning the proposed methodology, one might think that the objects we just precise may be very well handled in the *Zermelo-Fraenkel axiomatic system* (abbreviated: (ZF)), as described by Cohen [63, Ch. 2]; so, this recursion to the descriptive Cantor set theory is, ultimately, obsolete. However, the assertion is not true; because, in (ZF), the ordinals to be introduced cannot be organized a set; and this is valid as well for the (associated) cardinals. On the other hand, the Goedel-Bernays axiomatic system (in short: (GB)) – to be found, e.g., in Goedel [101] and Bernays [26, Part II] – is more appropriate for a complete structural description of such objects. However, from the perspective of our concrete facts to be developed, these axioms of (GB) are a little too technical. Hence, the only realistic alternative to follow is the one we just sketched.

(A) A basic construction of this theory is to be described as below. Let $N = \{0, 1, 2, ...\}$ denote the set of *natural* numbers, introduced as

$$0 := \emptyset,\ 1 := \{0\},\ 2 := \{0, 1\}, \text{ and so on.}$$

Technically speaking, the basic (algebraic and order) structures over N may be obtained via *Peano system techniques*, consisting in an *(immediate) successor* function $\text{suc} : N \to N$, and a set of *Peano axioms*, (as described in Halmos [106, Ch. 12]):

(pea-1) $0 \in N$ and $0 \notin \text{suc}(N)$
(pea-2) suc(.) is injective ($\text{suc}(n) = \text{suc}(m)$ implies $n = m$)
(pea-3) if $M \subseteq N$ fulfills ($0 \in M$ and $\text{suc}(M) \subseteq M$), then $M = N$.

In fact, starting from these properties, one may construct, in a recurrent way, an *addition* $(a, b) \mapsto a + b$ over N, according to

$$(\forall m \in N)\text{: } m + 0 = m;\ m + \text{suc}(n) = \text{suc}(m + n).$$

This makes possible the introduction of a (partial) order ($\leq$) over N, as

$$(m, n \in N)\text{: } m \leq n \text{ iff } m + p = n, \text{ for some } p \in N.$$

Concerning further properties of this structure, the basic one writes

$(N, \leq)$ is well-ordered:
any (nonempty) subset of N has a first element.

This tells us that *inductive* reasonings and constructions are allowed over this structure; we do not give details. Denote, for simplicity

$$N(r, \leq) = \{n \in N; r \leq n\} = \{r, r+1, ..., \},\ r \geq 0,$$
$$N(r, >) = \{n \in N; r > n\} = \{0, ..., r-1\},\ r \geq 1;$$

the latter one is referred to as the *initial interval* (in N) induced by r. Any set P with $P \sim N$ (in the sense: there exists a bijection from P to N) will be referred to as *effectively countable*. In addition, given some natural number $n \geq 1$, any (nonempty) set Q with $Q \sim N(n, >)$ will be said to be *n-finite*; when n is generic here, we say that Q is *finite*. Finally, the (nonempty) set Y is called (at most) *countable* iff it is either effectively countable or finite.

(B) Let W stand for the class of ordinal numbers, introduced in a "factorial" way; cf. Kuratowski and Mostowski [156, Ch. 7, Sec. 2]. Precisely, call the partially ordered structure $(P, \leq)$, *well ordered* if each (nonempty) part of P admits a first element. Given a couple $(P, \leq)$, $(Q, \preceq)$ of such objects, put

$$(P, \leq) \equiv (Q, \preceq) \text{ iff there exists a strictly increasing bijection in } \mathcal{F}(P, Q).$$

[Here, given the nonempty sets A and B, we let $\mathcal{F}(A, B)$ stand for the class of all functions $f : A \to B$; when $A = B$, one writes $\mathcal{F}(A, A)$ as $\mathcal{F}(A)$.] This is an equivalence relation; the order type of $(P, \leq)$ (denoted $\text{ord}(P, \leq)$) is just its equivalence class; also referred to as an *ordinal*.

Note that from the *Burali-Forti paradox*, W is not a set; cf. Sierpinski [216, Ch. 14, Sec. 2]. However, since we restricted all considerations to a *Grothendieck universe* $\mathcal{G}$ (see above), this contradictory character is removed for the class $W(\mathcal{G})$ of all *admissible* (modulo $\mathcal{G}$) ordinals (generated by (non-contradictory) well ordered parts of $\mathcal{G}$). As precise, any reference to this universe will be eliminated. So, by an *ordinal* in W, one actually means a $\mathcal{G}$-admissible ordinal with respect to a "sufficiently large" Grothendieck universe $\mathcal{G}$. Clearly,

$$\xi = \text{admissible ordinal and } \eta \leq \xi \text{ imply } \eta = \text{admissible ordinal}.$$

Hence, in the formulae (attached to some couple $\alpha, \beta \in W$)

$$W[\beta, \alpha[= \{\xi \in W; \beta \leq \xi < \alpha\},\ W(\alpha) = W[0, \alpha[,$$
$$W[\beta, \alpha] = \{\xi \in W; \beta \leq \xi \leq \alpha\},\ W[\alpha] = W[0, \alpha],$$

$$W]\beta,\alpha[= \{\xi \in W; \beta < \xi < \alpha\}, \ \ W]\beta,\alpha] = \{\xi \in W; \beta < \xi \le \alpha\},$$

the symbol W in the brackets is the "absolute" class of all ordinals.

Now, an enumeration of W is realized via the *immediate successor* map

$$\mathrm{suc}(M) = \min\{\xi \in W; M < \xi\}, \ M \in \exp(W);$$

where, by definition,

$$(\forall M \in \exp(W)): \ M < \xi \text{ iff } \lambda < \xi, \text{ for each } \lambda \in M.$$

It begins with the natural numbers $N = \{0, 1, ...\}$. Their immediate successor is $\omega = \mathrm{suc}(N)$ (the *first transfinite ordinal*); the next in this enumeration is $\omega+1$, and so on. Put $W_0 = W \setminus \{0\} (= \{\xi \in W; \xi > 0\})$. This set is composed of two disjoint classes of ordinals. The former of these, W_0^1, collects all *first kind* ordinals $\xi > 0$, in the sense:

$W(\xi)$ admits a last element $\max[W(\xi)] = \xi - 1$;
called the *predecessor* of ξ.

And the latter one, W_0^2, collects all *second kind* ordinals $\xi > 0$, in the sense:

$W(\xi)$ does not admit a last element;
or, equivalently: $\lambda < \xi \Longrightarrow \lambda + 1 < \xi$;

this is also referred to as $\xi > 0$ being a *limit* ordinal.

The basic operations with ordinals may be introduced in a synthetic way as follows. Let α, β be two ordinals; and $(A, \le)$, $(B, \le)$ be well ordered structures with $\mathrm{ord}(A, \le) = \alpha, \mathrm{ord}(B, \le) = \beta$, $A \cap B = \emptyset$. Then

Add) $\alpha + \beta = \mathrm{ord}(A \cup B, \le)$, where the associated strict order is given by the *concatenation* procedure:

$$x < y \text{ iff either } [x \in A, y \in B] \text{ or } [(x, y \in A, x < y), (x, y \in B, x < y)];$$

Multi) $\alpha \cdot \beta = \mathrm{ord}(A \times B, \le)$, where the associated strict order is defined by the *lexicographic* procedure:

$$(x, y) < (u, v) \text{ iff either } [y < v] \text{ or } [x < u, y = v].$$

Further details may be found in Kuratowski and Mostowski [156, Ch. 7, Sec. 5].

Let X be a nonempty set; and $\lambda > 0$ be an ordinal. Any mapping $a : W(\lambda) \to X$ will be referred to as a λ-*net*; denoted for simplicity as $(a(\xi); \xi < \lambda)$ or $(a_\xi; \xi < \lambda)$. If $\lambda > 0$ is generic here, this will be referred to as a *net*; and, if $\lambda = \omega$, the corresponding object is a *sequence*. Given the couple (λ, μ) of second kind ordinals with $\lambda \geq \mu$, the map $a : W(\lambda) \to X$ (meaning: the λ-net $(a_\xi; \xi < \lambda)$) and the mapping $i : W(\mu) \to W(\lambda)$ with

$$i(.) \text{ is strictly increasing and } \sup i(W(\mu)) = \lambda$$

the composed map in $\mathcal{F}(W(\mu), X)$ (that is: the μ-net)

$$b(\eta) = a(i(\eta)), \text{ (also written: } b_\eta = a_{i(\eta)}), \ \eta < \mu$$

will be referred to as a *μ-subnet of* $(a_\xi; \xi < \lambda)$. In particular (when $\lambda = \mu = \omega$), given the map $a : W(\omega) \to X$ (that is: the sequence $(a_n; n < \omega)$), and the mapping $i : W(\omega) \to W(\omega)$ with

$$i(.) \text{ is strictly increasing (hence, } \sup i(W(\omega)) = \omega),$$

the composed map in $\mathcal{F}(W(\omega), X)$ (that is: the sequence)

$$b(n) = a(i(n)), \text{ (also written: } b_n = a_{i(n)}), \ n < \omega$$

will be referred to as a *subsequence* of $(a_n; n < \omega)$.

Returning to the general setting above, it is to be stressed that – in parallel to this – we may (construct and) enumerate the class of all admissible cardinals. Letting P and Q be nonempty sets, put

$$P \preceq Q \ (P \sim Q) \text{ iff there exists an injection (bijection) in } \mathcal{F}(P, Q);$$

the former is a quasi-order, while the latter is an equivalence. Denote also

$$P \prec Q \text{ if and only if } P \preceq Q \text{ and } \neg(P \sim Q).$$

This relation is *irreflexive* and *transitive*; hence a *strict order*. Let $\alpha > 0$ be an (admissible) ordinal; we say that it is an (admissible) *cardinal* if

$$W(\xi) \prec W(\alpha), \text{ for each } \xi < \alpha;$$

the class of all these will be denoted by Z. Its enumeration is realized via the *immediate successor* (in Z) map

$$\text{SUC}(M) = \min\{\eta \in Z; M < \eta\}, \ \ M \in \exp(Z).$$

Precisely, this begins with the natural numbers $N = \{0, 1, ...\}$. The immediate successor (in Z) of all these is (again) $\omega = \text{SUC}(N)$ (the *first transfinite cardinal*). To describe the remaining ones, we may introduce via transfinite recursion the function $\lambda \mapsto \aleph_\lambda$ from W to Z as

$\aleph_0 = \omega$; and, for each $\lambda > 0$:
$\aleph_\lambda = \text{SUC}(\aleph_{\lambda-1})$, if $\lambda - 1$ exists
$\aleph_\lambda = \text{SUC}\{\aleph_\xi; \xi < \lambda\}$, if $\lambda - 1$ does not exist.

Note that, in such a case, the order structure of $Z(\omega, \leq) = \{\xi \in Z; \omega \leq \xi\}$ is reducible to the one of W; cf. Sierpinski [216, Ch. 15, Sec. 7].

Given the ordinal $\lambda > 0$, any nonempty part P with $P \prec W(\lambda)$ ($P \sim W(\lambda)$) is termed *λ-finite* (*λ-effectively countable*); the union of these ($P \preceq W(\omega)$) is referred to as P is *λ-countable.* In the case of $\lambda = \omega$, any ω-*concept* from the ones above will be referred to as a *concept.* Finally, when $P = W(\alpha)$ (where $\alpha \in W$), all these conventions will be transferred to α.

Now, the immediate successor in Z of $\omega = \aleph_0$ is $\Omega = \aleph_1$ (the first uncountable ordinal). The motivation of our convention comes from

ξ is countable, for each $\xi < \Omega$; but Ω is not countable.

A basic consequence of this is described by the statement in Alexandrov [12, Ch. 3, Sec. 4] (referred to as: *Alexandrov theorem*):

Proposition 3.12. *The following are valid:*

(31-1) *The ordinal Ω cannot be attained via limits of countable ordinals. That is: if $(\alpha_n; n < \omega)$ is an ascending sequence of countable ordinals, then*

$\alpha = \sup(\alpha_n)(= \lim_n(\alpha_n))$ *(the limit ordinal of (α_n))*

is countable too (hence, $\alpha < \Omega$)

(31-2) *Each second kind countable ordinal is attainable via such sequences. In other words: if $\alpha < \Omega$ is of second kind then, there exists a strictly ascending sequence $(\alpha_n; n < \omega)$ of countable ordinals with the limit property above.*

(C) Let M be a nonempty set; and $(\leq)$, some *(partial) order* (= antisymmetric quasi-order) over it. By a $(\leq)$-*chain* of M we shall mean any (nonempty) part A of M with $(A, \leq)$ being well ordered (see above). Note that any such object may be written as $A = \{a_\xi; \xi < \lambda\}$, where the net $\xi \mapsto a_\xi$ is strictly ascending ($\xi < \eta \Longrightarrow a_\xi < a_\eta$); the uniquely determined ordinal λ is just $\text{ord}(A, \leq)$. Now, by the remark above,

A is countable $\Longleftrightarrow \text{ord}(A, \leq) < \Omega$.

If, moreover, $\text{ord}(A, \leq) \leq \omega$, we say that A is *normally countable*; equivalently, this reads

$A = \{b_n; n < \omega\}$, where $n \mapsto b_n$ is ascending ($p < q \Longrightarrow b_p \leq b_q$).

Let P, Q be nonempty parts of M with $P \supseteq Q$. We say that P is *majorized* by Q (and write $P \propto Q$) provided

Q is *cofinal* in P ($\forall x \in P, \exists y \in Q$, with $x \leq y$).

Then, call the ($\leq$)-chain S of M, *upper countable* in case

(u-count) $S \propto T$, for some normally countable ($\leq$)-chain $T \subseteq S$.

Clearly, this happens if S is normally countable. The following essential completion of it is available.

Proposition 3.13. *Under these conventions,*

(32-1) *Each countable ($\leq$)-chain of M is necessarily upper countable*

(32-2) *The ($\leq$)-chain S of M is upper countable if and only if*

$S \propto T$, *for some countable ($\leq$)-chain $T \subseteq S$*

(32-3) *Any ($\leq$)-chain S of M with* $\mathrm{ord}(S, \leq) = \Omega$ *is not upper countable; hence, not countable as well*

(32-4) *Suppose that there exist ($\leq$)-chains S in M with* $\mathrm{ord}(S, \leq) > \Omega$. *Then, there exist upper countable ($\leq$)-chains U in M with* $\mathrm{ord}(U, \leq) > \Omega$ *(whence, U is not countable).*

Proof. I) Let $S = \{u(\xi); \xi < \lambda\}$ be the representation of this chain, where $\lambda := \mathrm{ord}(S, \leq) < \Omega$. If λ is a first kind ordinal, we are done; because $T = \{u(\lambda - 1)\}$ is normally countable and $S \propto T$. Assume now that λ is a second kind ordinal. By the Alexandrov theorem, there exists a strictly ascending sequence of countable ordinals (α_n) with $\lambda = \sup\{\alpha_n\}$. But then,

$T := \{u(\alpha_n); n < \omega\}$ is a normally countable ($\leq$)-chain of S
with $S \propto T$;

that is, we are again done.

II) The left to right inclusion is clear, via normally countable $\Longrightarrow$ countable. Conversely, suppose that $S \propto T$, for some countable ($\leq$)-chain $T \subseteq S$. By the previous part, $T \propto U$, for some normally countable ($\leq$)-chain $U \subseteq T$. Combining these, yields

$S \propto U$, for some normally countable ($\leq$)-chain $U \subseteq S$;

and, from this, all is clear.

III) Let $S = \{u(\xi); \xi < \Omega\}$ be a ($\leq$)-chain of M with $\mathrm{ord}(S, \leq) = \Omega$; hence, $\xi \mapsto u(\xi)$ is an order isomorphism between S and $W(\Omega)$. Suppose that S is upper countable. By definition, there exists then a strictly ascending sequence $(\beta_n; n < \omega)$ in $W(\Omega)$ such that

$S \propto T := \{u(\beta_n); n < \omega\}$ (= normally countable chain of S).

But then, necessarily, $\Omega = \sup\{\beta_n; n < \omega\}$; a contradiction. Hence, S is not upper countable; wherefrom (see above) S is not countable.

IV) Suppose that there exists a $(\leq)$-chain $S = \{u(\xi); \xi < \Theta\}$ in M, with $\Theta := \text{ord}(S, \leq) > \Omega$; whence, $\Theta \geq \Omega + 1$. As a consequence, $U = \{u(\xi); \xi < \Omega + 1\}$ is a $(\leq)$-chain of S, with $\text{ord}(U, \leq) = \Omega + 1 > \Omega$; whence, U is not countable. Moreover, U is upper countable; because $V = \{u(\Omega)\}$ is a normally countable $(\leq)$-chain in U with $U \propto V$. □

(D) Let us now return to our initial setting. Given the (partially) ordered structure $(M, \leq)$, let us introduce the concepts

(sep) $(M, \leq)$ is *separable* (cf. Zhu et al. [274]):
any $(\leq)$-chain of M is upper countable
(s-sep) $(M, \leq)$ is *strongly separable*: each $(\leq)$-chain of M is countable.

From the preceding facts, we have a relation like

strongly separable $\Longrightarrow$ separable.

But, the reciprocal holds too; so that, we may formulate

Proposition 3.14. *For each (partially) ordered structure $(M, \leq)$, we have*

separable $\Longrightarrow$ strongly separable; whence:
separable $\Longleftrightarrow$ strongly separable.

Proof. Assume that $(M, \leq)$ is separable; and let $S = \{s(\xi); \xi < \lambda\}$ be some upper countable $(\leq)$-chain of M, where $\lambda := \text{ord}(S, \leq)$. If, by absurd, S is not countable, we must have $\lambda \geq \Omega$. The initial segment $U = \{s(\xi); \xi < \Omega\}$ of S is a non-countable $(\leq)$-chain of M (see above). On the other hand, as S is upper countable, U is upper countable too; so, there exists a strictly ascending sequence $(\xi_n; n < \omega)$ of ranks in $W(\Omega)$ with

$U \propto \{s(\xi_n); n < \omega\}$; hence $\Omega = \sup_n(\xi_n)(= \lim_n(\xi_n))$.

But then, in view of Alexandrov theorem, Ω would be countable; contradiction. Hence, S is countable; and the proof is complete. □

A basic example of such structures may be given under the lines below. Let M be a nonempty set; and $\mathcal{I}(M) := \{(x, x); x \in M\}$ stand for the *identical relation* over M. By an *almost uniformity* (on M), we shall mean any (nonempty) family $\mathcal{U} \subseteq \exp[M \times M]$, with

$$\cap\mathcal{U} \supseteq \mathcal{I}(M) \text{ (hence, } \mathcal{U} \subseteq \exp(M \times M));$$

in this case, $(M,\mathcal{U})$ will be referred to as an *almost uniform space*. Let also $(\leq)$ be a (partial) order on M; then, $(M,\mathcal{U};\leq)$ will be referred to as a *(partially) ordered almost uniform space*. We say that the (ascending) net $(a_\xi; \xi < \lambda)$ in M, $\mathcal{U}$*-converges* towards the point $a \in M$ (and we write: $a_\xi \xrightarrow{\mathcal{U}} a$), provided

$$\forall U \in \mathcal{U}, \exists \mu = \mu(U) < \lambda, \text{ such that: } \mu \leq \xi < \lambda \Longrightarrow (a_\xi, a) \in U.$$

This will be also referred to as: a is the $\mathcal{U}$*-limit* of $(a_\xi; \xi < \lambda)$; the class of all such elements will be denoted as

$$\mathcal{U} - \lim_\xi(a_\xi; \xi < \lambda), \text{ or } \mathcal{U} - \lim_\xi(a_\xi) \text{ (if } \lambda \text{ is understood)};$$

when it is nonempty, we say that $(a_\xi; \xi < \lambda)$ is $\mathcal{U}$*-convergent*. Further, call the (ascending) net $(a_\xi; \xi < \lambda)$ in M, $\mathcal{U}$*-Cauchy* when:

$$\forall U \in \mathcal{U}, \exists \nu = \nu(U) < \lambda, \text{ such that: } \nu \leq \xi \leq \eta < \lambda \Longrightarrow (a_\xi, a_\eta) \in U.$$

Note that, by the imposed (upon $\mathcal{U}$) conditions, no subordination between these properties is possible. Finally, call the (ascending) sequence $(b_n; n < \omega)$ in M, $\mathcal{U}$*-asymptotic* provided:

$$\forall U \in \mathcal{U}, \exists k = k(U) < \omega, \text{ such that: } k \leq n < \omega \Longrightarrow (b_n, b_{n+1}) \in U.$$

Concerning the relationships between these concepts, we have

$$(\forall \text{ ascending sequence in } M)\text{: } \mathcal{U}\text{-Cauchy} \Longrightarrow \mathcal{U}\text{-asymptotic};$$

but the converse inclusion is not in general true.

Now, given the nonempty subset P of M without maximal elements (in the sense: $P(x,<) \neq \emptyset$, for each $x \in P$), let us consider the conditions

(a-reg) $(P,\leq)$ is *asymptotic regular* (modulo $\mathcal{U}$):
each ascending sequence (in P) is $\mathcal{U}$-asymptotic
(w-a-reg) $(P,\leq)$ is *weakly asymptotic regular* (modulo $\mathcal{U}$):
each strictly ascending sequence (in P) is $\mathcal{U}$-asymptotic.

Clearly, the former of these implies the latter; because, a strictly ascending sequence in P is ascending too. The converse implication is also true; as results from the statement below.

Proposition 3.15. *For each nonempty part P of M without maximal elements, the following properties hold*

(34-1) $(P, \leq)$ is weakly asymptotic regular (modulo $\mathcal{U}$) $\Longrightarrow$
$(P, \leq)$ is asymptotic regular (modulo $\mathcal{U}$); whence
(34-2) $(P, \leq)$ is weakly asymptotic regular (modulo $\mathcal{U}$) $\Longleftrightarrow$
$(P, \leq)$ is asymptotic regular (modulo $\mathcal{U}$).

Proof. Let $(P, \leq)$ be weakly asymptotic regular (modulo $\mathcal{U}$); and $(b_n; n < \omega)$ be an ascending sequence in P. If the alternative below holds

$(\exists k < \omega)$, (for all $h > k$): $b_k = b_h$,

we get (by the definition of $\mathcal{U}$)

$(b_j, b_{j+1}) = (b_k, b_k) \in \cap\mathcal{U}$, for each $j \geq k$;

whence, $(b_n; n < \omega)$ is $\mathcal{U}$-asymptotic. Suppose that the opposite alternative is valid

$(\forall k < \omega)$, (there exists $h > k$): $b_k \neq b_h$ (hence, $b_k < b_h$).

The mapping (from $W(\omega)$ to $\exp(W(\omega))$)

$\Psi(n) = \{h \in W]n, \omega[; b_n < b_h\}$, $n < \omega$

is therefore well defined; let $\psi : W(\omega) \to W(\omega)$ be its *minimal* selection

$\psi(n) = \min \Psi(n)$, $n < \omega$.

Note that, by the relations above, one necessarily has

$(\forall n < \omega)$: $n < \psi(n)$, $b_n < b_{\psi(n)}$, and ($b_n = b_k$, if $n \leq k < \psi(n)$).

The rank-sequence $(i(n) := \psi^n(0); n < \omega)$ has the property

$(i(n) < i(n+1), \forall n)$; hence, $(i(n))$ is strictly ascending;

in addition, the sequence $(c_n := b_{i(n)}; n < \omega)$ fulfills

(p-1) $(c_n < c_{n+1}, \forall n)$; hence, (c_n) is strictly ascending in P
(p-2) $(\forall n < \omega)$: $c_n = b_k$, whenever $i(n) \leq k < \psi(i(n))(= i(n+1))$.

From the imposed hypothesis, $(c_n; n < \omega)$ is $\mathcal{U}$-asymptotic; and this, via

$(b_k, b_{k+1}) = (c_m, c_m)$, when $i(m) \leq k < k+1 < i(m+1)$
$(b_k, b_{k+1}) = (c_m, c_{m+1})$, when $i(m) \leq k < k+1 = i(m+1)$

gives the asymptotic property of $(b_n; n < \omega)$. Having explored all possible alternatives, we are done. □

Returning to our initial setting, the following working conditions upon the almost uniformity $\mathcal{U}$ (over M) will be considered:

(suf) $\mathcal{U}$ is *sufficient*: $\cap\mathcal{U} = \mathcal{I}(M)$
(p-metr) $\mathcal{U}$ is *pseudo-metrizable*: there exists a countable subfamily $\mathcal{V} \subseteq \mathcal{U}$, cofinal in $(\mathcal{U}, \supseteq)$ $(\forall U \in \mathcal{U}, \exists V \in \mathcal{V}$: $U \supseteq V)$.

Proposition 3.16. *Suppose that the (partially) ordered almost uniform space $(M,\mathcal{U}; \leq)$ is such that $\mathcal{U}$ is sufficient, pseudo-metrizable. Further, let C be a (nonempty) $(\leq)$-chain of M without last ($=$ maximal) element. Then,*

(35-1) *If $(C, \leq)$ is weakly asymptotic regular (modulo $\mathcal{U}$), then C is upper countable (hence, countable)*

(35-2) *If C is not upper countable, then $(C, \leq)$ is not weakly asymptotic regular (modulo $\mathcal{U}$); so, at least one strictly ascending sequence in C is not $\mathcal{U}$-asymptotic.*

Proof. I) Let $\mathcal{V} = \{V_n; n < \omega\}$ be the representation of our countable family $\mathcal{V}$; note that, by the cofinal property,

$\mathcal{V}$ is sufficient: $\cap\mathcal{V} = \cap\{V_n; n < \omega\} = \mathcal{I}(M)$.

Let C be some $(\leq)$-chain in M, without last ($=$ maximal) element:

(non-last) for each $x \in C$ there exists $y \in C$ with $x < y$.

We claim that the following *diametral* property holds

$\forall x \in C, \forall n < \omega$, there exists $y = y(x, n) \in C(x, <)$
such that: $(p, q \in C,\ y < p < q) \Longrightarrow (p, q) \in V_n$.

In fact, suppose by absurd that this property fails; that is: for at least one couple $(x \in C, i < \omega)$,

for each $y \in C(x, <)$, there exists a couple of points (u, v) in $C \times C$, such that $y < u < v$, $(u, v) \notin V_i$.

Fix some point $y(0) \in C(x, <)$. By this assumption, there exists a couple of points $(y(1), y(2))$ in $C \times C$, with

$y(0) < y(1) < y(2)$, $(y(1), y(2)) \notin V_i$.

Further, given the point $y(2) \in C(x, <)$, there exists another couple of points $(y(3), y(4))$ in $C \times C$, with

$$y(2) < y(3) < y(4),\ (y(3), y(4)) \notin V_i;$$

and so on. By (ordinary) induction, we therefore get a strictly ascending sequence $(y(n); n < \omega)$ in $C(x, <)$, with

$$(y_{2n+1}, y_{2n+2}) \notin V_i, \text{ for all } n < \omega.$$

This, however, contradicts the weak asymptotic regularity of $(C, \leq)$ we just admitted; and the claim follows.

Fix some $x \in C$. By the diametral property above, it follows that, for the starting pair $(x, 0) \in C \times W(\omega)$, there must be some $x_0 \in C(x, <)$, with

$$p, q \in C, x_0 < p < q \Longrightarrow (p, q) \in V_0.$$

Further, by the same diametral property, it follows that, for the starting pair $(x_0, 1) \in C \times W(\omega)$, there must be some $x_1 \in C(x_0, <)$, with

$$p, q \in C, x_1 < p < q \Longrightarrow (p, q) \in V_1;$$
$$\text{whence } (p, q) \in V_0 \cap V_1;$$

and so on. The procedure may continue indefinitely; and yields a strictly ascending sequence $(x_n; n < \omega)$ in C with

$$\text{(for each } n < \omega\text{): } p, q \in C,\ x_n < p < q \Longrightarrow (p, q) \in \cap\{V_i; i \leq n\}.$$

We claim that C is majorized by $D := \{x_n; n < \omega\}$; wherefrom, the upper countable property of C follows. For, if this were not true, D admits an upper bound $u \in C$. Moreover, as C is without last element, there must be some $v \in C$ with $u < v$ (hence $u \neq v$). But then, the written property of (x_n) yields

$$(u, v) \in \cap\{V_i; i < \omega\} = \mathcal{I}(M) \text{ (hence } u = v\text{); contradiction.}$$

This proves our claim; and completes the first part.

II) The property in question is a simple converse of the previous one. □

Finally, note that further extensions of these facts may be obtained beyond the countable setting. Some technical aspects may be found in Turinici [243].

3.2.4 *Fang spaces*

Let X be some nonempty set. Remember that, by a *pseudometric* over X we mean any map $d : X \times X \to R_+$. Fix such an object, endowed with the additional properties below

(ref) d is *reflexive* ($d(x, x) = 0, \forall x \in X$)
(sym) d is *symmetric* ($d(x, y) = d(y, x), \forall x, y \in X$);

we then say that $d(.,.)$ is a *rs-pseudometric* over X. The convergence structure ($\stackrel{d}{\longrightarrow}$) attached to d is introduced as: given the net $(u_\xi; \xi < \mu)$ in X and the point $u \in X$, we write

$u_\xi \stackrel{d}{\longrightarrow} u$ when
$\forall \varepsilon > 0$, $\exists \lambda < \mu$: $\lambda \leq \xi < \mu \Longrightarrow d(u_\xi, u) < \varepsilon$.

This also reads: u is a d-limit of $(u_\xi; \xi < \mu)$; the set of all such elements will be denoted as

$$d - \lim_\xi(u_\xi; \xi < \mu), \text{ or } d - \lim_\xi(u_\xi) \text{ (if } \mu \text{ is understood);}$$

when it is nonempty, we say that $(u_\xi; \xi < \mu)$ is *d-convergent*. Further, the Cauchy structure $Cauchy(d)$ attached to d is introduced as:

the net $(u_\xi; \xi < \mu)$ is *d-Cauchy*, provided:
$\forall \varepsilon > 0$, $\exists \lambda < \mu$ such that $\lambda \leq \xi \leq \eta < \mu \Longrightarrow d(u_\xi, u_\eta) < \varepsilon$.

Clearly, by the imposed (upon d) conditions, no relationships between these concepts/properties are possible.

Having these precise, put $J =]0, 1]$; note that $(J, \geq)$ is without maximal elements. Further, take a family $D = (d_\alpha; \alpha \in J)$ of rs-pseudometrics over X with the J-normality properties:

(fang-1) D is *monotone* ($d_\alpha(x, y) \leq d_\beta(x, y), \forall x, y \in X$, if $\alpha \geq \beta$)
(fang-2) D is *J-triangular* (for each $\lambda \in J$, there exists $\mu \in J(\lambda, \geq)$, with $d_\lambda(x, z) \leq d_\mu(x, y) + d_\mu(y, z), \forall x, y, z \in X$)
(fang-3) D is *sufficient* ($d_\alpha(x, y) = 0, \forall \alpha \in J$ imply $x = y$).

We then say that D is a *Fang structure* on X; and (X, D) is referred to as a *Fang space*. Define the family of subsets in $X \times X$

$$\mathcal{V} = \{B(\alpha, \varepsilon) := X^2(d_\alpha; \varepsilon); \alpha \in J, \varepsilon > 0\}$$

where, by definition,

$$X^2(d_\alpha;\varepsilon) = \{(x,y) \in X^2; d_\alpha(x,y) < \varepsilon\},\ \alpha \in J, \varepsilon > 0.$$

This is an almost uniformity over X; which, in addition, fulfills

(au-1) $(\mathcal{V}, \supseteq)$ is directed $(\forall V_1, V_2 \in \mathcal{V},\ \exists V_3 \in \mathcal{V}\colon\ V_1 \cap V_2 \supseteq V_3)$
(au-2) each $V \in \mathcal{V}$ is symmetric $(V = V^{-1})$
(au-3) for each $V \in \mathcal{V}$ there exists $W \in \mathcal{V}$ with $V \supseteq W \circ W$.

In fact: the first property is a consequence of $D =$ monotone, the second one is deductible from ($d_\alpha =$ symmetric, $\forall \alpha \in J$), and the last one follows from the J-triangular condition. This shows that $\mathcal{V}$ is a fundamental system of entourages for a uniformity $\mathcal{U}$ over X constructed as (by taking super-sets)

$$\mathcal{U} = \{Z \in \exp[X \times X];\ Z \supseteq V, \text{ for some } V \in \mathcal{V}\};$$

see for instance Bourbaki [33, Ch. 2, Sec. 1.2]. As a rule, the "uniform" terminology refers to the family $\mathcal{U}$. However (cf. the definition), all $\mathcal{U}$-notions are in fact $\mathcal{V}$-notions. For example, the (attached to $\mathcal{U}$) convergence property is identical with its corresponding notion related to $\mathcal{V}$; a similar conclusion is valid for the Cauchy and asymptotic properties. On the other hand, by the very representation of this quasi-uniformity, the convergence/Cauchy notions attached to $\mathcal{V}$ may be attached to $D = (d_\alpha; \alpha \in J)$ as well. For example, given the net $(u_\xi; \xi < \mu)$ in X and the point $u \in X$,

(conv-net) $u_\xi \xrightarrow{\mathcal{V}} x$ means:
$u_\xi \xrightarrow{d_\alpha} u$, $\forall \alpha \in J$ (written as: $u_\xi \xrightarrow{D} u$);
and reads: u is a *D-limit* of $(u_\xi; \xi < \mu)$.

The set of all such elements writes

$$D - \lim_\xi(u_\xi; \xi < \mu), \text{ or } D - \lim_\xi(u_\xi) \text{ (if } \mu \text{ is understood)};$$

when it is nonempty, we say that $(u_\xi; \xi < \mu)$ is *D-convergent.* On the other hand, the $\mathcal{V}$-Cauchy property of $(u_\xi; \zeta < \mu)$ amounts to:

(Cauchy-net) $(u_\xi; \xi < \mu)$ is d_α-Cauchy, $\forall \alpha \in J$
(referred to as: $(u_\xi; \xi < \mu)$ is *D-Cauchy*);

the class of all these will be denoted as $Cauchy(D)$.

Concerning the topological properties of $\mathcal{V}$, we firstly note that, by the sufficiency of D,

(au-4) $\cap \mathcal{V} = \mathcal{I}(X)$; so, $\mathcal{V}$ is sufficient (as well as $\mathcal{U}$).

Secondly, we claim that $\mathcal{V}$ is pseudo-metrizable. In fact, let $(\alpha(n); n < \omega)$ be a strictly descending sequence of J with $\alpha(n) \to 0$; and $(\beta(n); n < \omega)$ be a strictly descending sequence in $R_+^0 :=]0, \infty[$ with $\beta(n) \to 0$. Denote

$$\mathcal{W} = \{B(\alpha(n), \beta(m)); n, m < \omega\};$$

this is a countable sub-family of $\mathcal{V}$. We then have

(au-5) $\mathcal{W}$ is cofinal in $\mathcal{V}$; hence $\mathcal{V}$ is pseudo-metrizable (as well as $\mathcal{U}$).

In fact, let $\alpha \in J$ and $\varepsilon > 0$ be arbitrary fixed. By the choice of $(\alpha(n); n < \omega)$ and $(\beta(n); n < \omega)$, there must be a couple of ranks (n, m) with

$$\alpha(n) < \alpha \text{ (hence, } d_{\alpha(n)}(.,.) \geq d_\alpha(.,.)\text{), and } \beta(m) < \varepsilon.$$

This gives $B(\alpha(n), \beta(m)) \subseteq B(\alpha, \varepsilon)$; wherefrom, all is clear.

As a consequence of these facts, it follows that the uniformity $\mathcal{U}$ is *metrizable*: there exists a metric $g : X \times X \to R_+$ for which the associated uniformity is just $\mathcal{U}$; see, for instance, Bourbaki [34, Ch. 9, Sec. 2.4]. This may be also formulated as:

$\mathcal{V}$ is equivalent with the quasi-uniformity
$\mathcal{Z} = \{A(\varepsilon) := X^2(g; \varepsilon); \varepsilon > 0\}$;

where the exact meaning of this the one below

(eq-1) $\mathcal{Z}$ is finer than $\mathcal{V}$ ($\forall V \in \mathcal{V}, \exists Z \in \mathcal{Z}$: $V \supseteq Z$)
(eq-2) $\mathcal{V}$ is finer than $\mathcal{Z}$ ($\forall Z \in \mathcal{Z}, \exists V \in \mathcal{V}$: $Z \supseteq V$).

Further aspects may be found in Turinici [236].

A natural problem to be solved here is that concerning the relationships between the convergence and Cauchy structure we just introduced. An appropriate answer is contained in

Proposition 3.17. *Under these conventions, we have*

(for each net): D-convergent $\Longrightarrow$ D-Cauchy.

Proof. Since – as precise – the convergence structure $(\xrightarrow{D})$ and the Cauchy structure $Cauchy(D)$ are equivalent with the ones induced by the associated metric $g(.,.)$, the assertion is clear. However, the construction of our metric is not in general accessible; so, a direct proof of this assertion – by means of the associated uniformity – would be welcomed. Suppose that the net $(u_\xi; \xi < \mu)$ is D-convergent; or, equivalently, $\mathcal{V}$-convergent (see above):

$\mathcal{V} - \lim_\xi(u_\xi; \xi < \mu) = u$, for some $u \in X$.

Let $V \in \mathcal{V}$ be arbitrary fixed. By the properties of this family, there may be found some $W \in \mathcal{V}$ such that $W \circ W \subseteq V$. From the convergence property, there exists some rank $\lambda = \lambda(W) < \mu$, such that

$(u_\xi, u) \in W$ (hence, $(u, u_\xi) \in W$), if $\lambda \leq \xi < \mu$.

But then, by definition,

$(u_\xi, u_\eta) \in W \circ W \subseteq V$, whenever $\lambda \leq \xi \leq \eta < \mu$;

which tells us that $(u_\xi; \xi < \mu)$ is $\mathcal{V}$-Cauchy; hence, D-Cauchy. □

Let again (X, D) be a Fang space. Take some (partial) order $(\leq)$ over X; then, $(X, D, \leq)$ is called a *(partially) ordered Fang space.* Each ascending D-convergent sequence is D-Cauchy; when the reciprocal holds too, we say that D is $(\leq)$-*complete*; or, equivalently: X is $(D, \leq)$-*complete.* For each part M of X, define

M is $(D, \leq)$-closed:
the D-limit of each ascending sequence in M belongs to M.

The following conditions involving this concept will be considered here:

(D-sclo) $(\leq)$ is D-*selfclosed:*
$X(u, \leq)$ is $(D, \leq)$-closed, for each $u \in X$
(D-a-sclo) $(\leq)$ is D-*anti-selfclosed:*
$X(u, \geq)$ is $(D, <)$-closed, for each $u \in X$
(D-bi-sclo) $(\leq)$ is D-*bilateral-selfclosed:*
$(\leq)$ is both D-selfclosed and D-anti-selfclosed.

The usefulness of these concepts is to be judged from the following simple property of convergent (ascending) sequences in our Fang space.

Proposition 3.18. *Assume that $(\leq)$ is D-bilateral-selfclosed. Then, for each ascending sequence $(a_n; n < \omega)$ in X and each $a \in X$, we have*

$$(\exists) D - \lim_n(a_n; n < \omega) = a \Longrightarrow (\exists) \sup\{a_n; n < \omega\} = a.$$

Proof. Let the premise of this implication holds: $D - \lim_n(a_n; n < \omega) = a$. As $(\leq)$ is D-selfclosed, $[a_n \leq a, \forall n < \omega]$. Moreover, let $b \in X$ be such that

$a_n \leq b, \forall n < \omega$; or, equivalently; $\{a_n; n < \omega\} \subseteq X(b, \geq)$.

Combining with the D-anti-selfclosed property of $(\leq)$, we derive $a \leq b$; wherefrom, all is clear. □

In the following, a few basic properties of (relatively) compact sets in our Fang structure are being discussed. Remember that a subset K of X is *compact* if, e.g., the condition below holds

(comp) every family of closed parts in K with the finite intersection property has a nonempty intersection;

see for instance Bourbaki [33, Ch. 1, Sec. 9.1]. Further, call the subset H of X *relatively compact* iff $\text{cl}(H)$ is compact. Note that, by the metrizable character of the associated to D uniformity (see above), this may be also written as:

(r-comp) each sequence in H admits a convergent (in X) subsequence;

cf. Bourbaki [34, Ch. 9, Sec. 2.9]. In particular, let us say that the net $(u_\xi; \xi < \mu)$ is relatively compact, when the attached subset $H = \{u_\xi; \xi < \mu\}$ is relatively compact.

In the following, some limit properties of such objects are given. Two alternatives are open before us.

I) As a direct consequence of our previous facts, the following property of relatively compact sequences is available.

Proposition 3.19. *Assume that $(\leq)$ is D-bilateral-selfclosed. Further, let the ascending sequence $(a_n; n < \omega)$ in X be relatively compact. Then,*

there exists one and only one point $a \in X$, with
$D - \lim_n(a_n) = a$, $\sup\{a_n; n < \omega\} = a$.

Proof. By the relative compactness hypothesis upon $A = \{a_n; n < \omega\}$, it follows that $(a_n; n < \omega)$ (= ascending sequence in A) admits (ascending) D-convergent subsequences. Let $(b_n := a_{i(n)}; n < \omega)$ be such an object; hence (by definition)

$$b_n \xrightarrow{D} a \text{ (as } n \to \infty\text{), for some } a \in X.$$

By the preceding statement (and cofinal property of $\{b_n; n < \omega\}$ in $\{a_n; n < \omega\}$), we must have

$$\sup\{b_n; n < \omega\} = a; \text{ whence, } \sup\{a_n; n < \omega\} = a;$$

and this, by the uniqueness of $\sup\{a_n; n < \omega\}$, tells us that each ascending convergent subsequence $(c_n; n < \omega)$ of $(a_n; n < \omega)$ fulfills $D - \lim_n(c_n) = a = \sup\{a_n; n < \omega\}$.

We now claim that

$a_n \xrightarrow{D} a$ (or, equivalently: $a_n \xrightarrow{\mathcal{V}} a$), as $n \to \infty$.

In fact, suppose that this is not true. There exists then $V \in \mathcal{V}$ and a subsequence $(u_n := a_{h(n)}; n < \omega)$ of $(a_n; n < \omega)$, such that

$(u_n, a) \notin V$, for all n.

On the other hand, by the relative compactness condition, there exists at least a convergent subsequence $(v_n := u_{k(n)}; n < \omega)$ of $(u_n; n < \omega)$ (hence, of $(a_n; n < \omega)$ as well); wherefrom (see above)

$v_n \xrightarrow{D} a$ (or, equivalently: $v_n \xrightarrow{\mathcal{V}} a$), as $n \to \infty$.

In particular, taking $V \in \mathcal{V}$ according to the previous requirements, there must be some rank $n(V)$, such that

$(v_n, a) \in V$ (that is: $(u_{k(n)}, a) \in V$), for all $n \geq n(V)$.

This, however, contradicts the preceding relation involving $[(u_n), a, V]$; so that, the convergence property follows. □

II) Denote, for each ordinal $\alpha > 0$,

$W_0^2(\alpha) = \{\xi \in W_0^2; \xi < \alpha\}$, $W_0^2[\alpha] = \{\xi \in W_0^2; \xi \leq \alpha\}$.

Let $\mu \in W_0^2(\Omega)$ be fixed in the sequel; hence, μ is a second kind countable ordinal. Given the ascending net $(u_\xi; \xi < \mu)$ in X, any (standard) sequence $(a_n := u_{\alpha(n)}; n < \omega)$, where $(\alpha(n); n < \omega)$ is a strictly ascending sequence of ranks in $W(\mu)$ with $\sup\{\alpha(n); n < \omega\} = \mu$, is referred to as an *ascending subsequence* of $(u_\xi; \xi < \mu)$. The following (countable) "transfinite" version of the obtained fact is available.

Proposition 3.20. *Assume that $(\leq)$ is D-bilateral-selfclosed. If the ascending net $(u_\xi; \xi < \mu)$ is relatively compact (in X), then*

there exists one and only one point $u \in X$, with
$D - \lim_\xi(u_\xi; \xi < \mu) = u$, $\sup\{u_\xi; \xi < \mu\} = u$.

Proof. Let $(\alpha(n); n < \omega)$ be a strictly ascending sequence of ranks in $W(\mu) \subseteq W(\Omega)$, with $\sup\{\alpha(n); n < \omega\} = \mu$ (possible, by the Alexandrov theorem). The ascending subsequence $(a_n := u_{\alpha(n)}; n < \omega)$ is relatively compact too; so, by the (countable type) statement above,

there exists one and only one point $u \in X$, with
$D - \lim_n(a_n) = u$, $\sup\{a_n; n < \omega\} = u$.

On the other hand, by the cofinality of $\{\alpha(n); n < \omega\}$ in $W(\mu)$,

$$(u =) \sup\{a_n; n < \omega\} = \sup\{u_\xi; \xi < \mu\}.$$

This tells us that all ascending subsequences of $(u_\xi; \xi < \mu)$ have the same limit, u; with, in addition, $\sup\{u_\xi; \xi < \mu\} = u$. It remains to establish that

$$D - \lim_\xi(u_\xi; \xi < \mu) = u \text{ (or, equivalently: } \mathcal{V} - \lim_\xi(u_\xi; \xi < \mu) = u).$$

Suppose that this is not true: hence, for some $V \in \mathcal{V}$

$$\forall \xi < \mu, \exists \eta \in W]\xi, \mu[: (u_\eta, u) \notin V.$$

Let $\Phi : W(\mu) \to \exp(W(\mu))$ be the associated multivalued map

$$\Phi(\xi) = \{\eta \in W]\xi, \mu[; (u_\eta, u) \notin V\}, \ \xi \in W(\mu);$$

and $\varphi : W(\mu) \to W(\mu)$ denote its (minimum) selection

$$\varphi(\xi) = \min \Phi(\xi), \ \xi \in W(\mu);$$

note that, by the relations above

$$(\forall \xi \in W(\mu)): \ \xi < \varphi(\xi) \text{ and } (u_{\varphi(\xi)}, u) \notin V.$$

Let $(A(n); n < \omega)$ be a (non-ascending, in general) sequence in $W(\mu)$ deductible from its cardinal representation $W(\omega) \sim W(\mu)$; hence,

$$\sup\{A(n); n < \omega\} = \sup W(\mu) = \mu \text{ (as } \mu \text{ is a limit ordinal).}$$

By the properties involving our selfmap $\varphi(.)$, the (non-ascending, in general) sequence $(B(n) = \varphi(A(n)); n < \omega)$ fulfills

(p-1) $A(n) < B(n)$, for each $n < \omega$; hence, $\sup\{B(n); n < \omega\} = \mu$
(p-2) $(u_{B(n)}, u) \notin V$, for each $n < \omega$.

From the cofinal property of $\{B(n); n < \omega\}$ in $W(\mu)$, there exists a strictly ascending sequence of ranks $(i(n); n < \omega)$ in $W(\omega)$, such that

$(C(n) = B(i(n)); n < \omega)$ is a strictly ascending sequence
of ordinals in $W(\mu)$, with $\sup\{C(n); n < \omega\} = \mu$.

This, according to a preceding convention, tells us that

(p-3) $(v_n := u_{C(n)}; n < \omega)$ is an ascending subsequence of $(u_\xi; \xi < \mu)$
(p-4) $(v_n, u) \notin V$, for all $n < \omega$.

On the other hand, by the previous discussion upon such objects

(p-5) $D - \lim_n(v_n) = u$; in contradiction with the above relation.

So, the working hypothesis is not acceptable; and the proof is complete. □

Finally, note that further extensions of these are possible, in terms of (sequential) convergence structures as in Kasahara [141]; we shall develop them elsewhere.

3.2.5 *Main result*

With this information at hand, we are now returning to the questions of our introductory section. Let (X, D) be a Fang space; here, $D = (d_\alpha; \alpha \in J)$ is a family of rs-pseudometrics over X fulfilling the J-normality conditions. Remember that the uniform structure of (X, D) is metrizable; so, it may be introduced by a metric over X. Further, let $(\leq)$ be a (partial) order on X; the triple $(X, D, \leq)$ will be then termed a *(partially) ordered Fang space*. Finally, let $T : X \to \exp(X)$ be a multivalued map.

The main result in this exposition is

Theorem 3.13. *Let the (partially) ordered Fang space $(X, D, \leq)$ and the multivalued map $T : X \to \exp(X)$ be such that*

(kon-1) T is semi progressive, increasing, compact valued
(kon-2) each strictly ascending net $(u_\xi; \xi < \beta)$ [where $\beta \in W_0^2(\Omega)$], fulfilling $\{u(\xi); 1 \leq \xi < \beta\} \subseteq \mathrm{cl}[T\{u_\xi; \xi < \beta\}]$, is relatively compact
(kon-3) $(\leq)$ is D-bilateral-closed:
both $X(x, \leq)$ and $X(x, \geq)$ are D-closed, for each $x \in X$.

Then, $\mathrm{Fix}(T)$ *is nonempty (that is: T has at least one fixed point in X).*

Proof. Suppose by contradiction that

(non-fp) $x \notin Tx$, for each $x \in X$.

As T is semi-progressive, there exists some $u(0) \in X$ with $u(0) \leq Tu(0)$; hence (by the working condition)

$u(0) < u(1)$, for some $u(1) \in Tu(0)$.

As T is increasing, we must have $Tu(0) \leq Tu(1)$; whence $u(1) \leq Tu(1)$; so that (again by the working condition) $u(1) < u(2)$, for some $u(2) \in Tu(1)$; and so on. Generally, assume that for the ordinal $\mu \in W[3, \Omega[$, we constructed a net $(u(\xi); \xi < \mu)$ in X so that: for each $\lambda < \mu$, the assertions $[(P_1(\lambda)), (P_2(\lambda)), (P_3(\lambda))]$ hold, where

$(P_1(\lambda))$ the net $(u(\xi); \xi \leq \lambda)$ is strictly ascending
$(P_2(\lambda))$ $u(\xi+1) \in Tu(\xi)$, if $(\xi <)\ \xi + 1 \leq \lambda$
$(P_3(\lambda))$ (the net $(u(\xi); \xi < \nu)$ is strictly ascending and) $D - \lim_\xi(u(\xi); \xi < \nu) = u(\nu)$, whenever $\nu \leq \lambda$ is of second kind.

We intend to establish that, by an appropriate choice of $u(\mu)$, the relations $(P_1(\mu))$–$(P_3(\mu))$ hold as well. Two alternatives are possible.

Case 1) Suppose that $\mu \in W[3, \Omega[$ is a first kind ordinal; that is, $\beta := \mu - 1$ exists (so, $\beta < \beta + 1 = \mu$). Two sub-cases must be considered.

S-case 1-a) Suppose that $\beta(< \mu)$ is a first kind ordinal; that is, $\alpha := \beta - 1$ exists (so, $\alpha < \alpha + 1 = \beta$). By $(P_1(\beta))$ and $(P_2(\beta))$, we have

$$u(\alpha) < u(\beta) \text{ and } u(\beta) = u(\alpha + 1) \in Tu(\alpha).$$

As T is increasing,

$$Tu(\alpha) \leq Tu(\beta); \text{ hence } u(\beta) \leq Tu(\beta);$$

so that (by the non fixed point hypothesis), there exists $u(\mu) \in Tu(\beta)$ with $u(\beta) < u(\mu)$. As a consequence, $(P_1(\mu))$ and $(P_2(\mu))$ are valid. On the other hand, $(P_3(\mu))$ is identical with $(P_3(\beta))$ – because μ is a first kind ordinal– so, it is valid as well.

S-case 1-b) Assume now that $\beta(< \mu)$ is a second kind (limit) ordinal; that is, $\beta - 1$ does not exist. We have, successively,

$u(\xi) < u(\beta)$ [via $(P_1(\beta))$], for all $\xi < \beta$
$u(\xi) < u(\xi + 1)$ [via $(P_1(\xi + 1))$], for all $\xi < \beta$ (as $\xi + 1 < \beta$)
$u(\xi + 1) \in Tu(\xi)$ [via $(P_2(\beta))$], for all $\xi < \beta$ (as $\xi + 1 < \beta$).

Combining with T = increasing, yields (for all $\xi < \beta$)

$$Tu(\xi) \leq Tu(\beta); \text{ wherefrom } u(\xi) < u(\xi + 1) \leq Tu(\beta).$$

The family of (nonempty) closed sets (in $Tu(\beta)$)

$$\{Tu(\beta)(u(\xi), \leq) = Tu(\beta) \cap X(u(\xi), \leq); \xi < \beta\}$$

has the finite intersection property; so, by the compactness of $Tu(\beta)$, we have (cf. a previous characterization) $\cap\{Tu(\beta)(u(\xi), \leq); \xi < \beta\} \neq \emptyset$. Let $u(\mu)$ be some point in this intersection. For the moment,

$$u(\mu) = u(\beta + 1) \in Tu(\beta); \text{ hence, } (P_2(\mu)) \text{ holds.}$$

In addition, we must have

$$u(\xi) \leq u(\mu), \forall \xi < \beta; \text{ whence: } \{u_\xi; \xi < \beta\} \subseteq X(u(\mu), \geq).$$

On the other hand, $(P_3(\beta))$ yields $D - \lim_\xi(u_\xi; \xi < \beta) = u(\beta)$. Combining these with the closeness of $X(u(\mu), \geq)$, gives $u(\beta) \leq u(\mu)$; hence

$$u(\beta) < u(\mu) \text{ [by the working hypothesis and } u(\mu) \in Tu(\beta)];$$

proving that $(P_1(\mu))$ holds too. Finally, $(P_3(\mu))$ is identical with $(P_3(\beta))$ – because μ is a first kind ordinal– so, it is valid as well.

Case 2) Suppose that $\mu \in W[3, \Omega[$ is a second kind (limit) ordinal; that is, $\mu - 1$ does not exist. We claim that

$$(\mathrm{Q}(\mu)) \quad u(\eta) \in \mathrm{cl}[T(C(\mu))], \forall \eta \in W[1, \mu[,$$
$$\text{where } C(\mu) := \{u(\xi); \xi < \mu\}.$$

In fact, take some $\eta \in W[1, \mu[$. If $\tau := \eta - 1$ exists, then, by $(P_2(\tau))$

$$u(\eta) = u(\tau + 1) \in Tu(\tau) \subseteq T(C(\mu)) \subseteq \mathrm{cl}[T(C(\mu))];$$

and the assertion follows. Suppose now that $\eta - 1$ does not exist. The (strictly ascending) net $(u(\xi); \xi < \eta)$, D-converges to $u(\eta)$, by $(P_3(\eta))$; this is also true for the net $(u(\xi + 1); \xi < \eta)$, because $\{\xi + 1; \xi < \eta\}$ is cofinal in $W(\eta)$. Combining with

$$u(\xi + 1) \in Tu(\xi) \subseteq T(C(\mu)) \subseteq \mathrm{cl}[T(C(\mu))], \text{ for all } \xi < \eta,$$

gives the relation $(\mathrm{Q}(\mu))$; hence the claim. Taking (kon-2) into account, one derives that $(u_\xi; \xi < \mu)$ is strictly ascending, relatively compact. By a preceding auxiliary statement and (kon-3), there exists $u(\mu) \in X$, with

$$\text{(q-1)} \quad D - \lim_\xi(u(\xi); \xi < \mu) = u(\mu)$$
$$\text{(q-2)} \quad \sup[C(\mu)] = u(\mu) \text{ [hence } u(\xi) < u(\mu), \text{ for all } \xi < \mu].$$

As a consequence, $(P_1(\mu))$–$(P_3(\mu))$ are holding in the described case.

By the transfinite induction principle, this constructive process is effective over all of $W(\Omega)$; and gives us a strictly ascending net $(u(\xi); \xi < \Omega)$ in X; hence, $C(\Omega) := \{u(\xi); \xi < \Omega\}$ is non-countable. From a previous result involving this concept, $C(\Omega)$ is not upper countable; so – by an asymptotic property of these objects – there must be a strictly ascending sequence of ranks $(\xi_n; n < \omega)$ in $W(\Omega)$ such that the sequence $(u_n := u(\xi_n); n < \omega)$ is not D-asymptotic. Denote $\lambda := \sup\{\xi_n; n < \omega\}$; it is a second kind ordinal in $W(\Omega)$. From $(P_3(\lambda))$ above,

$(u_\xi; \xi < \lambda)$ is D-convergent; hence, so is $(u_n; n < \omega)$.

Consequently, $(u_n; n < \omega)$ must be D-asymptotic; contrary to its choice. Hence, our initial working condition (non-fp) is not acceptable; and the conclusion follows. □

In particular, the relative compactness condition (kon-2) holds under

(kon-2a) each countable totally ordered part $C = \{x_i; i < \beta\}$ with $C \subseteq \text{cl}[\{x_0\} \cup T(C)]$, is relatively compact.

In this case, our main result includes the one in Zhu and Li [275]. Further technical aspects were discussed in O'Regan [180]; see also Hong [114].

3.2.6 *Cardinal extensions*

Concerning the obtained main result, a basic assumption imposed upon our ambient space X is the Fang character of its topology; so, it is natural to ask whether this can be removed in some way. For an appropriate answer to this, we note that – from a technical perspective – the basic motivation of relative compactness assumption appearing in our main result is to provide – via (ordinary) sequences techniques – the convergence of involved chains; so, if we impose *ab initio* such a property, the reasoning we just developed there is no longer depending on (ordinary) sequences – hence, ultimately, on the initial Fang space setting. In this case, the ending of the process is to be decided as follows. Denote

$\Gamma = \text{card}(X)$, $\Delta = \text{SUC}(\Gamma)$; that is: $\Delta = \aleph_{\gamma+1}$ if $\Gamma = \aleph_\gamma$.

Then, any $(\leq)$-chain C of $(X, \leq)$ fulfills

$\text{card}(C) \leq \text{card}(X)$; wherefrom $\text{ord}(C, \leq) < \Delta$;

which tells us that, by the iterative constructions like before, we cannot reach $(\leq)$-chains of order type Δ.

A precise formulation of these facts may be given under the lines below.

Let X be a nonempty set. By a *topology* over X we mean any family $\mathcal{T} \subseteq \exp[X]$, with the properties

(top-1) $\emptyset, X \in \mathcal{T}$ (meaning: $\{\emptyset, X\} \subseteq \mathcal{T}$)
(top-2) $\mathcal{T}$ is invariant to arbitrary unions and finite intersections;

in this case, $(X, \mathcal{T})$ is referred to as a *topological space*. Assume that we fixed such a structure. By definition, each subset $A \in \mathcal{T}$ is called $\mathcal{T}$*-open*; and each subset $B \in \exp[X]$ with $X \setminus B \in \mathcal{T}$ is referred to as $\mathcal{T}$*-closed.* Given $x \in X$, each subset V of X with

$$x \in A \subseteq V, \text{ for some } A \in \mathcal{T},$$

will be called a $\mathcal{T}$*-neighborhood* of x; the class of all these is the $\mathcal{T}$*-neighborhoods family* $\mathcal{V}(x)$. This allows us to introduce all topological concepts to be needed; the basic ones may be described as below.

I) Given the subset M of X, define its $\mathcal{T}$*-closure* as

$$\mathrm{cl}(M) = \{x \in X; V \cap M \neq \emptyset, \ \forall V \in \mathcal{V}(x)\}.$$

Note that a subset B of X is $\mathcal{T}$-closed if and only if $B = \mathrm{cl}(B)$.

II) Define the *Hausdorff separation property* of $(X, \mathcal{T})$, as

$$\forall x, y \in X, \ x \neq y, \ \exists P \in \mathcal{V}(x), \ \exists Q \in \mathcal{V}(y)\text{: } P \cap Q = \emptyset.$$

III) Given the ordinal $\lambda \geq 1$, let us say that the net $(a_\xi; \xi < \lambda)$ in X, $\mathcal{T}$*-converges* towards the point $a \in X$ (and we write: $a_\xi \xrightarrow{\mathcal{T}} a$), provided

$$\forall U \in \mathcal{V}(a), \exists \mu = \mu(U) < \lambda, \text{ such that: } \mu \leq \xi < \lambda \Longrightarrow a_\xi \in U.$$

This will be also referred to as: a is the $\mathcal{T}$-limit of $(a_\xi; \xi < \lambda)$; the class of all such elements will be denoted as

$$\mathcal{T} - \lim_\xi (a_\xi; \xi < \lambda) \ [\text{or } \mathcal{T} - \lim_\xi (a_\xi) \ (\text{if } \lambda \text{ is understood})];$$

when it is nonempty, we say that $(a_\xi; \xi < \lambda)$ is $\mathcal{T}$*-convergent.* Note that, if the Hausdorff property is admitted, then the set of all such $\mathcal{T}$-limits is a singleton, whenever it is nonempty.

Let $(X, \mathcal{T})$ be a topological space; and $(\leq)$ be a (partial) order on it; the triple $(X, \mathcal{T}, \leq)$ is then referred to as a *(partially) ordered topological space.* Further, let $T : X \to \exp(X)$ be some multivalued map. The following extension of our main result is then available.

Theorem 3.14. *Let the (partially) ordered topological space $(X, \mathcal{T}, \leq)$ and the multivalued map $T : X \to \exp(X)$ be such that*

(kon-t-1) $(X, \mathcal{T})$ is Hausdorff separated and
T is semi-progressive, increasing, compact valued

(kon-t-2) each strictly ascending net $(u_\xi; \xi < \beta)$ *[where* $\beta \in W_0^2(\Omega)$*], fulfilling* $\{u_\xi; 1 \le \xi < \beta\} \subseteq \mathrm{cl}[T\{u_\xi; \xi < \beta\}]$, *is* $\mathcal{T}$*-convergent*
(kon-t-3) $(\le)$ *is* $\mathcal{T}$*-bilateral-closed:*
both $X(x, \le)$ *and* $X(x, \ge)$ *are* $\mathcal{T}$*-closed, for each* $x \in X$.

Then, $\mathrm{Fix}(T)$ *is nonempty (that is:* T *has at least one fixed point in* X*).*

The argument is very similar to the one we just developed for the main result; however, for completeness reasons, we provide it.

Proof. Let the couple of cardinals (Γ, Δ) be introduced as before. Suppose by contradiction that

(non-fp) $x \notin Tx$, for each $x \in X$.

As T is semi-progressive, there exists some $u(0) \in X$ with $u(0) \le Tu(0)$; hence (by the working condition)

$u(0) < u(1)$, for some $u(1) \in Tu(0)$.

As T is increasing, we must have $Tu(0) \le Tu(1)$; whence $u(1) \le Tu(1)$; so that (again by the working condition) $u(1) < u(2)$, for some $u(2) \in Tu(1)$; and so on. Generally, assume that for the ordinal $\mu \in W[3, \Delta[$, we constructed a net $(u(\xi); \xi < \mu)$ in X so that: for each $\lambda < \mu$, the assertions $[(P_1(\lambda)), (P_2(\lambda)), (P_3(\lambda))]$ are holding, where

$(P_1(\lambda))$ the net $(u(\xi); \xi \le \lambda)$ is strictly ascending
$(P_2(\lambda))$ $u(\xi + 1) \in Tu(\xi)$, if $(\xi <)\ \xi + 1 \le \lambda$
$(P_3(\lambda))$ (the net $(u(\xi); \xi < \nu)$ is strictly ascending and)
$\mathcal{T} - \lim_\xi (u(\xi); \xi < \nu) = u(\nu)$, whenever $\nu \le \lambda$ is of second kind.

We intend to establish – via transfinite induction – that, by an appropriate choice of $u(\mu)$, the relations $(P_1(\mu))$–$(P_3(\mu))$ hold as well. Two alternatives are possible.

Case 1) Suppose that $\mu \in W[3, \Delta[$ is a first kind ordinal; that is, $\beta := \mu - 1$ exists (so, $\beta < \beta + 1 = \mu$). Two sub-cases must be considered.

S-case 1-A) Suppose that $\beta (< \mu)$ is a first kind ordinal; that is, $\alpha := \beta - 1$ exists (so, $\alpha < \alpha + 1 = \mu$). By $(P_1(\beta))$ and $(P_2(\beta))$, we have

$u(\alpha) < u(\beta)$ and $u(\beta) = u(\alpha + 1) \in Tu(\alpha)$.

As T is increasing,

$Tu(\alpha) \le Tu(\beta)$; hence $u(\beta) \le Tu(\beta)$;

so that (by the non fixed point hypothesis), there exists $u(\mu) \in Tu(\beta)$ with $u(\beta) < u(\mu)$. As a consequence, $(P_1(\mu))$ and $(P_2(\mu))$ are valid. On the other hand, $(P_3(\mu))$ is identical with $(P_3(\beta))$ – because μ is a first kind ordinal– so, it is valid as well.

S-case 1-B) Assume now that $\beta(<\mu)$ is a second kind (limit) ordinal; that is, $\beta - 1$ does not exist. We have, successively,

$u(\xi) < u(\beta)$ [via $(P_1(\beta))$], for all $\xi < \beta$
$u(\xi) < u(\xi+1)$ [via $(P_1(\xi+1))$], for all $\xi < \beta$ (as $\xi + 1 < \beta$)
$u(\xi+1) \in Tu(\xi)$ [via $(P_2(\beta))$], for all $\xi < \beta$ (as $\xi + 1 < \beta$).

Combining with $T =$ increasing, yields (for all $\xi < \beta$)

$Tu(\xi) \leq Tu(\beta)$; wherefrom $u(\xi) < u(\xi+1) \leq Tu(\beta)$.

The family of (nonempty) closed sets

$$\{Tu(\beta)(u(\xi), \leq) = Tu(\beta) \cap X(u(\xi), \leq); \xi < \beta\}$$

has the finite intersection property; so, by the compactness of $Tu(\beta)$, we have (cf. a previous characterization) $\cap\{Tu(\beta)(u(\xi), \leq); \xi < \beta\} \neq \emptyset$. Let $u(\mu)$ be some point in this intersection. For the moment,

$u(\mu) = u(\beta+1) \in Tu(\beta)$; hence, $(P_2(\mu))$ holds.

In addition, we must have

$u(\xi) \leq u(\mu)$, $\forall \xi < \beta$; whence: $\{u_\xi; \xi < \beta\} \subseteq X(u(\mu), \geq)$.

On the other hand, $(P_3(\beta))$ yields $\mathcal{T} - \lim_\xi (u_\xi; \xi < \beta) = u(\beta)$. Combining these with the closeness of $X(u(\mu), \geq)$, gives $u(\beta) \leq u(\mu)$; hence

$u(\beta) < u(\mu)$ [by the working hypothesis and $u(\mu) \in Tu(\beta)$];

proving that $(P_1(\mu))$ holds too. Finally, $(P_3(\mu))$ is identical with $(P_3(\beta))$ – because μ is a first kind ordinal– so, it is valid as well.

Case 2) Suppose that $\mu \in W[3, \Delta[$ is a second kind (limit) ordinal; that is, $\mu - 1$ does not exist. We claim that

$$(\mathrm{Q}(\mu)) \quad u(\eta) \in \mathrm{cl}[T(C(\mu))],\ \forall \eta \in W[1, \mu[, \text{ where } C(\mu) = \{u(\xi); \xi < \mu\}.$$

In fact, take some $\eta \in W[1, \mu[$. If $\tau := \eta - 1$ exists then, by $(P_2(\tau))$,

$$u(\eta) = u(\tau+1) \in Tu(\tau) \subseteq T(C(\mu)) \subseteq \mathrm{cl}[T(C(\mu))];$$

hence, $(Q(\mu))$ holds. Suppose now that $\eta - 1$ does not exist. The net $(u(\xi); \xi < \eta)$, $\mathcal{T}$-converges to $u(\eta)$, by $(P_3(\eta))$; this is also true for the net $(u(\xi+1); \xi < \eta)$, because $\{\xi+1; \xi < \eta\}$ is cofinal in $W(\eta)$. Combining with

$$u(\xi + 1) \in Tu(\xi) \subseteq T(C(\mu)) \subseteq \mathrm{cl}[T(C(\mu))], \text{ for all } \xi < \eta,$$

gives again $(Q(\mu))$; and the assertion follows. Taking (kon-t-2) into account, one derives that the net $(u_\xi; \xi < \mu)$, $\mathcal{T}$-converges to some $u(\mu) \in X$; with, in addition (by the property (kon-t-3))

$$\sup[C(\mu)] = u(\mu) \text{ [hence } u(\xi) < u(\mu), \text{ for all } \xi < \mu].$$

As a consequence, relations $(P_1(\mu))$-$(P_3(\mu))$ are holding in the above described case.

By the transfinite induction principle, this constructive process is effective over all of $W(\Delta)$; and gives us a strictly ascending net $(u(\xi); \xi < \Delta)$ in X. But then, from a previous cardinality remark involving such objects, this is not possible. Hence, the initial working condition is not acceptable; and the conclusion follows. □

In particular, when T is univalent, this result is comparable with the one in Turinici [240]. Further aspects and concrete applications to differential equations may be found in Heikkila [111] and Hong [115]; see also Agarwal et al. [9, Ch. 4].

3.3 Homotopic Metric Interval Maps in Gauge Spaces

Abstract. A Matkowski type version is given for the 1967 contraction mapping principle in Gheorghiu [Stud. Cerc. Mat., 19 (1967), 119–122]. As a by-product of this, an appropriate functional extension is proposed for the homotopic fixed point result in gauge spaces due to Frigon [L. Notes Nonlin. Anal., 16 (2017), 9–91].

Subject-Class 2020. 47H10 (Primary), 54H25 (Secondary).

Keywords. Gauge space, Matkowski type contraction, fixed point, homotopy, sequential completeness.

3.3.1 *Dependent Choice principles*

See Ch. 1.

3.3.2 *Conv-Cauchy structures*

See Ch. 1.

3.3.3 *Admissible functions*

See Ch. 1.

3.3.4 *Topological structures*

See Ch. 1.

3.3.5 *Banach contractions and relatives*

See Ch. 1.

3.3.6 *Gheorghiu functional contractions*

Let X be a nonempty set; and Λ be some nonempty (index) set. For each $\lambda \in \Lambda$, let $d_\lambda : X \times X \to R_+$ be a *semimetric* (that is: non-sufficient metric) over X; and $\mathcal{T}(d_\lambda)$ stand for the associated topology.

(A) Technically speaking, these data generate two structures on X.

I) The former of these is represented by the *supremum topology* of the family $\{\mathcal{T}(d_\lambda); \lambda \in \Lambda\}$. Precisely, note that for each $\lambda \in \Lambda$,

$$\mathcal{A}_\lambda = \{X(x, \rho)(d_\lambda); x \in X, \rho > 0\} \text{ is a subbase of } \mathcal{T}(d_\lambda);$$

hence, in particular, $\mathcal{A}_\lambda$ is total. Denote further

$\mathcal{A} = \cup\{\mathcal{A}_\lambda; \lambda \in \Lambda\}$; clearly, $\mathcal{A}$ is total too.

By a preceding result, the formula

$$\mathcal{T} = \{D \in \exp[X]; D = \text{union of finite intersections of subsets in } \mathcal{A}\}$$

defines a topology over X. This, as precise, is just the supremum (modulo $(\subseteq)$) topology of the family $\{\mathcal{T}(d_\lambda); \lambda \in \Lambda\}$. Let int(.) and cl(.) stand for the associated interior and closure operator. According to definition, we have for each subset Y of X

(g-int) $x \in \text{int}(Y)$ iff there exists an intersection W
of a finite subset in $\mathcal{A}$ with $x \in W \subseteq Y$
(g-clo) $x \in \text{cl}(Y)$ iff each intersection W
of a finite subset in $\mathcal{A}$ with $x \in W$ fulfills $W \cap Y \neq \emptyset$.

II) The latter of these is represented by the *(sequential) conv-Cauchy structure* induced by the family $D = (d_\lambda; \lambda \in \Lambda)$ of these semimetrics. This, essentially, may be described in the way below. Take an arbitrary sequence (x_n) in X. Given $\lambda \in \Lambda$, the d_λ*-convergence* of this sequence towards an $x \in X$ [depicted as: $x_n \xrightarrow{d_\lambda} x$], means:

$d_\lambda(x_n, x) \to 0$ as $n \to \infty$
(that is: $\forall \varepsilon > 0$, $\exists i := i(\lambda, \varepsilon)$, such that $i \leq p \Longrightarrow d_\lambda(x_p, x) \leq \varepsilon$).

If this holds for all $\lambda \in \Lambda$, then $(x_n; n \geq 0)$ is said to *D-converge* towards x [written as: $x_n \xrightarrow{D} x$]. The set of all such points x will be denoted as $D-\lim_n(x_n)$; when it is nonempty, then $(x_n; n \geq 0)$ is called *D-convergent.* Further, given $\lambda \in \Lambda$, the d_λ*-Cauchy* property of $(x_n; n \geq 0)$ means:

$d_\lambda(x_n, x_m) \to 0$ as $n, m \to \infty$, $n \leq m$
(that is: $\forall \varepsilon > 0$, $\exists j := j(\lambda, \varepsilon)$, such that $j \leq p \leq q \Longrightarrow d_\lambda(x_p, x_q) \leq \varepsilon$).

If this holds for each $\lambda \in \Lambda$, we say that $(x_n; n \geq 0)$ is *D-Cauchy.* Note that any D-convergent sequence is D-Cauchy too; the reciprocal is not in general valid. By definition, the triple $(X, \Lambda; D)$ endowed with the described conv-Cauchy structure and the regularity condition

D is *sufficient*: $x, y \in X$ and $(d_\lambda(x, y) = 0, \forall \lambda \in \Lambda)$ imply $x = y$

is called a *gauge space*.

(B) Let $(X, \Lambda; D)$ be a gauge space. Remember that the subset Y of X is called *asingleton*, if $[y_1, y_2 \in Y$ imply $y_1 = y_2]$; and *singleton* if, in addition, Y is nonempty; note that in this case $Y = \{y\}$, for some $y \in X$.

Further, let $T : X \to X$ be a selfmap of X. As usual, denote

$$\mathrm{Fix}(T) := \{z \in X; z = Tz\} \text{ (the set of all \textit{fixed points} of } T \text{ in } X).$$

These points are to be determined in the context below, comparable with the one in Rus [203, Ch. 2, Sec. 2.2]:

pic-1) We say that T is *fix-asingleton*, if $\mathrm{Fix}(T)$ is an asingleton; and *fix-singleton*, if $\mathrm{Fix}(T)$ is a singleton

pic-2) We say that $x \in X$ is a *Picard point* (modulo (D, T)) when $(T^n x; n \geq 0)$ is D-Cauchy. If this property holds for all $x \in X$, then T is called a *Picard operator* (modulo D)

pic-3) We say that $x \in X$ is a *strongly Picard point* (modulo (D, T)) when $(T^n x; n \geq 0)$ is D-convergent with $\lim_n (T^n x) \in \mathrm{Fix}(T)$. If this property holds for all $x \in X$, then T is called a *strongly Picard operator* (modulo D).

The sufficient (regularity) conditions for such properties are being founded on *orbital* concepts (in short: o-concepts). Namely, given the sequence $(z_n; n \geq 0)$ in X, define the property

$(z_n; n \geq 0)$ is T-*orbital*, when $(z_n = T^n x; n \geq 0)$, for some $x \in X$.

greg-1) Call X, *(o;D)-complete*, provided (for each o-sequence) D-Cauchy $\Longrightarrow$ D-convergent

greg-2) We say that T is *(o;D)-continuous*, if $[(z_n)$ is .o-sequence and $z_n \xrightarrow{D} z]$ imply $Tz_n \xrightarrow{D} Tz$.

As a completion of these, we must now formulate the metrical contractive conditions upon our data. Denote

(class-1) $\mathcal{F}_0(R_+)$ = the class of all $\omega \in \mathcal{F}(R_+)$ with $\omega(0) = 0$
(class-2) $\mathcal{F}_0(re)(R_+)$ = the class of all $\omega \in \mathcal{F}_0(R_+)$ with
the *regressive* property ($\omega(t) < t$, for all $t > 0$)
(class-3) $\mathcal{F}_0(re, in)(R_+)$ = the class of all increasing $\omega \in \mathcal{F}_0(re)(R_+)$.

Let $\Omega = (\omega_\lambda; \lambda \in \Lambda)$ be a family over $\mathcal{F}_0(R_+)$; and $\varphi : \Lambda \to \Lambda$ be a function. As usual, we denote the iterates of this function as

$$(\forall \lambda \in \Lambda): \ \varphi^0(\lambda) = \lambda, \ \varphi^n(\lambda) = \varphi \circ \varphi^{n-1}(\lambda), \ n \geq 1.$$

We say that T is (Ω, φ)-*contractive*, if

$$d_\lambda(Tx, Ty) \leq \omega_\lambda(d_{\varphi(\lambda)}(x, y)), \ \forall x, y \in X, \ \forall \lambda \in \Lambda.$$

Our first main result in this exposition is

Theorem 3.15. *Suppose that T is (Ω, φ)-contractive, where the family $\Omega = (\omega_\lambda; \lambda \in \Lambda)$ over $\mathcal{F}_0(re)(R_+)$ and the selfmap $\varphi : \Lambda \to \Lambda$ fulfill*

(61-i) $(\forall \lambda \in \Lambda)$: *$\omega_\lambda$ is increasing (hence, $\omega_\lambda \in \mathcal{F}_0(re, in)(R_+)$)*
(61-ii) $\sum_{n\geq 1} \omega_\lambda \circ ... \circ \omega_{\varphi^{n-1}(\lambda)}(d_{\varphi^n(\lambda)}(x,y)) < \infty$, $\forall x, y \in X$, $\forall \lambda \in \Lambda$.

In addition, let X be (o;D)-complete. Then,
(61-a) *T is fix-asingleton*
(61-b) *T is strongly Picard (modulo D).*

Proof. There are several steps to be passed.

Part 0) Let $x, y \in X$ be arbitrary fixed. By the contractive condition (and ω_λ = increasing, for each $\lambda \in \Lambda$),

$$d_\lambda(T^n x, T^n y) \leq \omega_\lambda \circ ... \circ \omega_{\varphi^{n-1}(\lambda)}(d_{\varphi^n(\lambda)}(x,y)),\ \forall n \geq 1,\ \forall \lambda \in \Lambda.$$

Denote for simplicity

$$B_\lambda(x,y) := \sum_{n\geq 0} d_\lambda(T^n x, T^n y),\ x, y \in X,\ \lambda \in \Lambda.$$

From the imposed convergence property, we have

(eva) $(\forall x, y \in X, \forall \lambda \in \Lambda)$: $B_\lambda(Tx, Ty) = \sum_{n\geq 1} d_\lambda(T^n x, T^n y) \leq$
$\sum_{n\geq 1} \omega_\lambda \circ ... \circ \omega_{\varphi^{n-1}(\lambda)}(d_{\varphi^n(\lambda)}(x,y)) < \infty$; so that,
$B_\lambda(x,y) = d_\lambda(x,y) + B_\lambda(Tx,Ty) < \infty$.

Part 1) We prove that T is fix-asingleton. Let $z_1, z_2 \in \text{Fix}(T)$ be arbitrary fixed. By the previous evaluation,

$$d_\lambda(z_1, z_2) = B_\lambda(z_1, z_2) - B_\lambda(z_1, z_2) = 0, \text{ for all } \lambda \in \Lambda.$$

This, along with D = sufficient, gives $z_1 = z_2$; whence the assertion.

Part 2) Take some $x_0 \in X$, and put $(x_n := T^n x_0; n \geq 0)$; clearly, (x_n) is orbital. By the preceding relation (eva),

$$d_\lambda(x_n, x_{n+1}) = B_\lambda(x_n, x_{n+1}) - B_\lambda(x_{n+1}, x_{n+2}),\ \forall n \geq 0,\ \forall \lambda \in \Lambda.$$

For each $\lambda \in \Lambda$, the (real) sequence $(B_\lambda(x_n, x_{n+1}); n \geq 0)$ is descending bounded from below in R_+; hence, a Cauchy sequence. Combining with the above, yields

$(x_n; n \geq 0)$ is d_λ-Cauchy, for each $\lambda \in \Lambda$; hence, D-Cauchy.

Summing up, $(x_n; n \geq 0)$ is an orbital D-Cauchy sequence in X. This, along with X being (o;D)-complete, gives

$$x_n \xrightarrow{D} z \text{ (meaning: } \lim_n d_\lambda(x_n, z) = 0, \forall \lambda \in \Lambda) \text{ for some } z \in X.$$

As a consequence of contractive condition (and regressive property)

$$d_\lambda(x_{n+1}, Tz) \le \omega_\lambda(d_{\varphi(\lambda)}(x_n, z)) \le d_{\varphi(\lambda)}(x_n, z), \forall n \ge 0, \forall \lambda \in \Lambda.$$

By the convergence relation above, $y_n := Tx_n \xrightarrow{D} Tz$. On the other hand, $(y_n = x_{n+1}; n \ge 0)$ is a subsequence of $(x_n; n \ge 0)$; whence $y_n \xrightarrow{D} z$. This, again combined with D=separated, yields $z = Tz$; meaning: $z \in \text{Fix}(T)$. The proof is complete. □

(C) A basic counterpart of Theorem 3.15 corresponds to the choice

$$\varphi = \text{identity map of } \mathcal{F}(\Lambda)\ (\varphi(\lambda) = \lambda,\ \lambda \in \Lambda).$$

Let $\Omega = (\omega_\lambda; \lambda \in \Lambda)$ be a family over $\mathcal{F}_0(R_+)$. We say that the selfmap T is *Ω-contractive*, if

$$d_\lambda(Tx, Ty) \le \omega_\lambda(d_\lambda(x, y)), \forall x, y \in X,\ \ \forall \lambda \in \Lambda.$$

The deep part of the next result is an auxiliary fact involving almost Picard properties over semimetric structures. Some preliminaries are needed.

Let X be a nonempty set; and $d : X \times X \to R_+$ be a semimetric over X; the couple (X, d) will be referred to as a *semimetric space*. Further, take some $T \in \mathcal{F}(X)$. In the following, sufficient conditions are given so that T be *almost Picard* (modulo d), in the sense:

(a-p-1) $(\forall x \in X)$: the iterative sequence $(T^n x; n \ge 0)$ is d-Cauchy
(a-p-2) $(\forall x \in X, \forall z \in X)$: $T^n x \xrightarrow{d} z$ implies $d(z, Tz) = 0$.

The conditions in question are to be introduced as follows. Given $\omega \in \mathcal{F}_0(R_+)$, let us say that T is *ω-contractive*, provided

$$d(Tx, Ty) \le \omega(d(x, y)), \forall x, y \in X.$$

Concerning the functional classes to be used, remember that the function $\omega \in \mathcal{F}_0(re)(R_+)$ is

(MK-adm) *Meir-Keeler admissible*, when $\forall \varepsilon > 0$, $\exists \delta > 0$, such that: $\varepsilon < t < \varepsilon + \delta \Longrightarrow \omega(t) \le \varepsilon$; (or, equivalently: $t < \varepsilon + \delta$ implies $\omega(t) \le \varepsilon$)
(M-adm) *Matkowski admissible*, when each sequence (t_n) in R_+^0 with $(t_{n+1} \le \omega(t_n), \forall n)$ fulfills $t_n \to 0$.

Note that, according to a previous auxiliary fact,

$(\forall \omega \in \mathcal{F}_0(re)(R_+))$: Meir-Keeler admissible $\Longleftrightarrow$ Matkowski admissible.

In particular, if ω is increasing, this last property writes

$\omega^n(t) \to 0$ as $n \to \infty$, for each $t \in R_+$.

Returning to the general setting, we already observed that

$(\forall \omega \in \mathcal{F}_0(re)(R_+))$: ω is Matkowski admissible whenever ω is *Boyd-Wong admissible*: $\Delta^+\omega(s) := \limsup_{t\to s+} \omega(t) < s$, $\forall s \in R^0_+$.

In particular, when $\omega \in \mathcal{F}_0(re)(R_+)$ is increasing,

$\Delta^+\omega(s) = \omega(s+0)(:= \lim_{t\to s+} \omega(t))$, for all $s > 0$

wherefrom, the assertion below holds over $\mathcal{F}_0(re, in)(R_+)$:

ω is Boyd-Wong admissible iff $\omega(s+0) < s$, $\forall s > 0$
(referred to as: ω is *strongly regressive*).

The following almost Picard type statement is available.

Theorem 3.16. *Suppose that T is ω-contractive, where $\omega \in \mathcal{F}_0(re)(R_+)$ is endowed with one of the properties*

(62-i) ω is Meir-Keeler admissible
(or, equivalently: Matkowski admissible)
(62-ii) ω is increasing Matkowski admissible
(62-iii) ω is Boyd-Wong admissible (hence, Matkowski admissible).

Then, necessarily, T is almost Picard (modulo d).

The case (62-i) was essentially proved in the 1969 paper by Meir and Keeler [168]. Further, the cases (62-ii) and (62-iii) are directly reducible to (62-i). However, for completeness reasons, we develop the argument for all these alternatives.

Proof. Fix some point $x_0 \in X$, and put $(x_n = T^n x_0; n \geq 0)$.

Step 1. [**Asymptotic property**] From the contraction hypothesis,

$d(x_{n+1}, x_{n+2}) \leq \omega(d(x_n, x_{n+1}))$, $\forall n$;

wherefrom (noting that, anyway, ω is Matkowski admissible)

(x_n) is d-asymptotic: $r_n := d(x_n, x_{n+1}) \to 0$ as $n \to \infty$.

Step 2. [**Cauchy property**] We have three alternatives to discuss.

Alter 1) Suppose that ω is Meir-Keeler admissible (or, equivalently: Matkowski admissible). Let $\varepsilon > 0$ be arbitrary fixed; and $\delta > 0$ be provided by the Meir-Keeler admissible property; without loss, one may assume that $\delta < \varepsilon$. By the d-asymptotic property, there exists a rank $n(\delta)$, such that

(d-asy) $d(x_n, x_{n+1}) < \delta/2, \forall n \geq n(\delta)$.

We claim, by induction, that the relation

(d-C;j) $d(x_n, x_{n+j}) < \varepsilon + \delta/2, \forall n \geq n(\delta)$

is holding for all $j \geq 1$; wherefrom, $(x_n; n \geq 0)$ is d-Cauchy. The case $j = 1$ is evident, by the evaluation (d-asy). Assume that (d-C;j) holds for all $j \in \{1, ..., p\}$, where $p \geq 1$; we must establish that (d-C;p+1) holds too. So, let $n \geq n(\delta)$ be arbitrary fixed. From the inductive hypothesis

$$d(x_n, x_{n+p}) < \varepsilon + \delta/2 < \varepsilon + \delta.$$

Combining with contractive hypothesis and Meir-Keeler property of ω, gives

$$d(x_{n+1}, x_{n+p+1}) = d(Tx_n, Tx_{n+p}) \leq \omega(d(x_n, x_{n+p})) \leq \varepsilon;$$

wherefrom (by the triangular inequality)

$$d(x_n, x_{n+p+1}) \leq d(x_n, x_{n+1}) + d(x_{n+1}, x_{n+p+1}) < \varepsilon + \delta/2;$$

and our claim follows.

Alter 2) Suppose that ω is increasing Matkowski admissible. Let $\varepsilon > 0$ be arbitrary fixed. By the d-asymptotic property, there exists some rank $m(\varepsilon) \geq 0$, such that (as ω is regressive)

(d-asym) $d(x_n, x_{n+1}) < \varepsilon - \omega(\varepsilon) \leq \varepsilon$, for all $n \geq m(\varepsilon)$.

We claim, via induction, that

(d-Ca;p) $d(x_n, x_{n+p}) < \varepsilon$, for all $n \geq m(\varepsilon)$

holds for all $p \geq 1$; wherefrom (by the arbitrariness of $\varepsilon > 0$), the d-Cauchy property of $(x_n; n \geq 0)$ follows. The case of $p = 1$ is clear, by the preceding relation (d-asym). Suppose that the evaluation (d-Ca;p) is true, for all $p \in \{1, ..., q\}$, where $q \geq 1$; we claim that (d-Ca;q+1) holds too. In fact, let $n \geq m(\varepsilon)$ be arbitrary fixed. By the inductive hypothesis and contractive condition, we have (as ω=increasing)

$$d(x_{n+1}, x_{n+q+1}) = d(Tx_n, Tx_{n+q}) \leq \omega(d(x_n, x_{n+q})) \leq \omega(\varepsilon);$$

so, by the triangle inequality (and our preceding relation (d-asym))

$$d(x_n, x_{n+q+1}) \leq d(x_n, x_{n+1}) + d(x_{n+1}, x_{n+q+1}) < \varepsilon - \omega(\varepsilon) + \omega(\varepsilon) = \varepsilon;$$

proving our claim. Hence, (x_n) is an orbital d-Cauchy sequence in X.

Alter 3) Suppose that ω is Boyd-Wong admissible; and let $(\pi(k))$ be a strictly ascending rank sequence in $N(1, \leq)$. If, by absurd, (x_n) is not d-Cauchy, then – by a previous auxiliary fact – there exist a number $\gamma > 0$ and a triple of rank sequences $[(J(k)), (m(k)), (n(k))]$ in $N(1, \leq)$, with

(q-1) $k + 1 \leq J(k) \leq m(k) < m(k) + 3\pi(k) < n(k), \forall k \geq 0$
(q-2) $V_k(s,t) := d(x_{m(k)+s}, x_{n(k)+t}) \to \gamma+$ as $k \to \infty$,
for all $s, t \in N[0, 3\pi(0)]$.

From the contractive condition,

$$V_k(1,1) \leq \omega(V_k(0,0)), \text{ for all } k.$$

Passing to lim sup as $k \to \infty$, one gets (by another previous fact)

$$\gamma \leq \Delta^+\omega(\gamma) < \gamma; \text{ contradiction.}$$

Hence, (x_n) is a d-Cauchy sequence in X, as claimed.

Step 3. [**Fixed point property**] Suppose that $z \in X$ is such that

$$x_n \xrightarrow{d} z \text{ (that is: } d(x_n, z) \to 0), \text{ as } n \to \infty.$$

From the contractive condition (and ω = regressive)

$$d(x_{n+1}, Tz) \leq \omega(d(x_n, z)) \leq d(x_n, z), \forall n.$$

This, by the convergence relation above, gives (passing to limit as $n \to \infty$)

$$\lim_n d(x_n, Tz) = 0, \text{ whence } x_n \xrightarrow{d} Tz.$$

Combining with d = semimetric, yields $d(z, Tz) = 0$; as desired. □

Under these preliminaries, we may now pass to our second main result.

Theorem 3.17. *Suppose that T is Ω-contractive, where $\Omega = (\omega_\lambda; \lambda \in \Lambda)$ is a family over $\mathcal{F}_0(re)(R_+)$, fulfilling*

($\forall\lambda \in \Lambda$): ω_λ is Meir-Keeler admissible
(or, equivalently: Matkowski admissible).

In addition, let X be (o;D)-complete. Then,

(63-a) *T is fix-asingleton*

(63-b) *T is strongly Picard (modulo D).*

Proof. There are several steps to be passed.

Part 1) We prove that T is fix-asingleton. Let $z_1, z_2 \in \mathrm{Fix}(T)$ be arbitrary fixed. By the contractive condition,

$$d_\lambda(z_1, z_2) \le \omega_\lambda(d_\lambda(z_1, z_2)),\ \lambda \in \Lambda.$$

This, by the regressive property of the functions in Ω, gives

$$d_\lambda(z_1, z_2) = 0,\ \forall \lambda \in \Lambda;\ \text{wherefrom (as } D \text{ is sufficient)},\ z_1 = z_2.$$

Part 2) Take some $x_0 \in X$, and put $(x_n := T^n x_0; n \ge 0)$; clearly, (x_n) is orbital. By the preceding semimetric statement (applied to $[(X, d_\lambda); \omega_\lambda; T]$ for each $\lambda \in \Lambda$), one derives that

(r-1) (x_n) is d_λ-Cauchy, $\forall \lambda \in \Lambda$; which means: (x_n) is D-Cauchy.

Part 3) Summing up, (x_n) is an orbital D-Cauchy sequence. As X is (o;D)-complete, there must be some (uniquely determined) $z \in X$, with

$$x_n \xrightarrow{D} z \text{ (hence, } x_n \xrightarrow{d_\lambda} z,\ \forall \lambda \in \Lambda).$$

Again by the preceding semimetric statement, we get

(r-2) $d_\lambda(z, Tz) = 0$, for each $\lambda \in \Lambda$.

This, via D = separated, gives $z = Tz$; and completes the argument. □

Some remarks concerning these results are in order.

Rem-1) In particular, when

$$(\forall \lambda \in \Lambda)\colon \omega_\lambda(t) = k_\lambda t,\ t \in R_+,\ \text{for some } k_\lambda \in [0, 1[,$$

our first main result extends directly the related statement in Gheorghiu [97]; see also Gheorghiu and Rotaru [99]. On the other hand, the result in question cannot extend the Maia type fixed point theorem in Gheorghiu [98]; so, we may ask whether this is possible in some way; further aspects will be discussed elsewhere.

Rem-2) When Λ is a singleton, our second main result covers the one in Matkowski [164]. Passing to the general case, note that the result in question includes in a partial way in the 2017 quasi-order fixed point theorem in Turinici [258, Paper 3-4]. In fact, it is possible to get a quasi-order version of our second main result so as to include the quoted statement; we do not give details.

Finally, some applications of these developments to integral equations may be found in Gheorghiu and Turinici [100] or Cherichi and Samet [51]; see also Angelov [17, Ch. II, Sec. 2.2].

3.3.7 *Homotopic fixed points*

Let X be a nonempty set; and Λ be some nonempty (index) set. For each $\lambda \in \Lambda$, let $d_\lambda : X \times X \to R_+$ be a semimetric over X; and $\mathcal{T}(d_\lambda)$ stand for the associated topology. There are two structures on X to be used further.

I) The former of these is represented by the *supremum topology* of the family $\{\mathcal{T}(d_\lambda); \lambda \in \Lambda\}$. Some facts about this object were already listed.

II) The latter one is represented by the *(sequential) conv-Cauchy structure* induced by the family $D = (d_\lambda; \lambda \in \Lambda)$ of these semimetrics. Precisely, let $((\xrightarrow{D}), Cauchy(D))$ be the conv-Cauchy structure attached to the underlying family. Remember that any D-convergent sequence is D-Cauchy too; the reciprocal is not in general valid. By definition, the triple $(X, \Lambda; D)$ endowed with this conv-Cauchy structure and the regularity condition

$$D \text{ is } \textit{separated}\text{: } (d_\lambda(x, y) = 0, \forall \lambda \in \Lambda) \Longleftrightarrow x = y$$

is called a *gauge space.*

Suppose that we fixed such a structure. Define a D-closure operator $A \mapsto \mathrm{Dcl}(A)$ from $\exp[X]$ to itself, as: for each $A \in \exp[X]$,

$$y \in \mathrm{Dcl}(A) \text{ iff } y = D - \lim_n(y_n), \text{ for some sequence } (y_n) \text{ in } A.$$

It is not hard to see that $A \mapsto \mathrm{Dcl}(A)$ is a *semi-closure* over X, in the sense

(Dcl-1) $\mathrm{Dcl}(\emptyset) = \emptyset$, $\mathrm{Dcl}(X) = X$,
(Dcl-2) $A \subseteq B$ implies $\mathrm{Dcl}(A) \subseteq \mathrm{Dcl}(B)$,
(Dcl-3) $\mathrm{Dcl}(A \cup B) = \mathrm{Dcl}(A) \cup \mathrm{Dcl}(B)$, $A, B \in \exp[X]$.

Unfortunately, $A \mapsto \mathrm{Dcl}(A)$ is not involutive; i.e., a property like

(Dcl-4) $\mathrm{Dcl}(\mathrm{Dcl}(A)) = \mathrm{Dcl}(A)$, for each $A \in \exp[X]$

is not in general true; so that, $A \mapsto \mathrm{Dcl}(A)$ is not a closure over X according to Kuratowski [155, Ch. I, Sec. 4]. Further, let us say that $A \in \exp[X]$ is *D-closed*, provided $A = \mathrm{Dcl}(A)$; note that, in this case,

the D-limit of each sequence in A belongs to A.

Having these precise, let (J, g) be a metric space; and $(\leq)$ be a (partial) order over J; the triple $(J, g, \leq)$ will be then referred to as an *ordered metric space*. Let $(<)$ stand for the *strict order* attached to $(\leq)$, in the sense

$(x, y \in X)$: $x < y$ iff $x \leq y$ and $x \neq y$.

Note that [by the very choice of $(\leq)$], $(<)$ is irreflexive and transitive; hence, the convention is meaningful. Call $r \in J$, $(\leq)$-*maximal* provided

$J(r, \leq) = \{r\}$; or, equivalently, $J(r, <) = \emptyset$;

the class of these will be denoted $\max(J, \leq)$. The opposite property (that is: $J(r, <) \neq \emptyset$) will be referred to as r is $(\leq)$-*nonmaximal*; let $\text{nmax}(J, \leq)$ be the class of all these. The following properties of $(J, g, \leq)$ are admitted:

(pro-1) $(J, g, \leq)$ is *conv-bd-regular*:
each ascending sequence (t_n) in J is g-convergent
(hence, g-Cauchy) in J, with $(t_n \leq \lim_n(t_n), \forall n)$
(pro-2) $(J, g, \leq)$ is *right accessible*: each $r \in \text{nmax}(J, \leq)$ is *cluster* to $J(r, <)$ (that is: $r = g - \lim_n(s_n)$ for some sequence (s_n) in $J(r, <)$).

Further, let the nonempty subset $U \in \exp(X)$ be fixed; and take a mapping $(t, x) \mapsto H(t, x) = H(t)x$ from $J \times U$ into X. Then, fix the extended functional family $\Psi = (\psi_\lambda; \lambda \in \Lambda)$ over $\mathcal{F}(R_+, R_+ \cup \{\infty\})$ and the functional family $\Omega = (\omega_\lambda; \lambda \in \Lambda)$ over $\mathcal{F}(R_+)$, according to

(hom-1) H is *first variable* Ψ-*contractive*:
$d_\lambda(H(t, x), H(s, x)) \leq \psi_\lambda(g(t, s))$, for all $t, s \in J$ and all $x \in U$
(hom-2) H is *second variable* Ω-*contractive*:
$d_\lambda(H(t, x), H(t, y)) \leq \omega_\lambda(d_\lambda(x, y))$, for all $t \in J$ and all $x, y \in U$.

We then say that H is a (Ψ, Ω)-*homotopic mapping*.

The extra regularity condition required for the elements in Ω may be expressed as follows. Let us say that $\omega \in \mathcal{F}_0(re)(R_+)$ is

(MK-adm) *Meir-Keeler admissible*, when $\forall \varepsilon > 0$, $\exists \delta > 0$,
such that: $\varepsilon < t < \varepsilon + \delta \Longrightarrow \omega(t) \leq \varepsilon$
(M-adm) *Matkowski admissible*, when each sequence (t_n) in R^0_+
with $(t_{n+1} \leq \omega(t_n), \forall n)$ fulfills $t_n \to 0$.

Note that, according to a previous auxiliary fact,

$(\forall \omega \in \mathcal{F}_0(re)(R_+))$: Meir-Keeler admissible $\Longleftrightarrow$ Matkowski admissible.

In particular, if ω is increasing, the last property writes

$\omega^n(t) \to 0$ as $n \to \infty$, for each $t \in R_+$.

Returning to the general setting, we already observed that

$(\forall \omega \in \mathcal{F}_0(re)(R_+))$: ω is Matkowski admissible whenever ω is *Boyd-Wong admissible*: $\Delta^+\omega(s) := \limsup_{t\to s+} \omega(t) < s$, $\forall s \in R_+^0$.

In particular, when $\omega \in \mathcal{F}_0(re)(R_+)$ is increasing,

$\Delta^+\omega(s) = \omega(s+0)(:= \lim_{t\to s+} \omega(t))$, for all $s > 0$

wherefrom, the assertion below holds over $\mathcal{F}_0(re, in)(R_+)$:

ω is Boyd-Wong admissible iff $\omega(s+0) < s$, $\forall s > 0$
(referred to as: ω is *strongly regressive*).

Finally, given $\omega \in \mathcal{F}_0(re, in)(R_+)$, let

$\omega^* = \mathrm{id} - \omega$ (that is: $(\omega^*(t) = t - \omega(t); t \in R_+)$)

denote its *complement*. We say that ω is *complementary coercive*, in case

ω^* is coercive: $\lim_{t\to\infty} \omega^*(t) = \infty$; or, equivalently:
$[\omega^* \le a] := \{t \in R_+; \omega^*(t) \le a\}$ is bounded, for each $a \ge 0$.

Denote for simplicity

$J(H) = \{t \in J; \mathrm{Fix}(H(t)) \neq \emptyset\}$.

We say that the homotopic map H is *fix-transversal*, provided

(fix-tr) $J(H) \cap \max(J, \le)$ is cofinal in $J(H)$:
for each $q \in J(H)$ there exists $q^* \in J(H) \cap \max(J, \le)$ with $q \le q^*$.

The following homotopic fixed point theorem is available, in the reduced system (ZF-AC+DC).

Theorem 3.18. *Let the mapping $(t, x) \mapsto H(t, x) = H(t)x$ in $\mathcal{F}(J \times U, X)$ be (Ψ, Ω)-homotopic, where the family $\Psi = (\psi_\lambda; \lambda \in \Lambda)$ over $\mathcal{F}(R_+, R_+ \cup \{\infty\})$ and the family $\Omega = (\omega_\lambda; \lambda \in \Lambda)$ over $\mathcal{F}_0(re, in)(R_+)$ fulfill*

(71-i) $(\forall \lambda \in \Lambda)$: ψ_λ is zero-continuous ($\psi_\lambda(t) \to 0 = \psi_\lambda(0)$ as $t \to 0$)
(71-ii) $(\forall \lambda \in \Lambda)$: ω_λ is strongly regressive ($\omega_\lambda(s+0) < s$, $\forall s \in R_+^0$) and complementary coercive ($\omega_\lambda^* := \mathrm{id} - \omega_\lambda$ is coercive).

In addition, suppose that

(71-iii) $(J, g, \le)$ is conv-bd-regular and right accessible

(71-iv) *U is D-closed, and X is D-complete*
(71-v) $\text{Fix}(H(t)) \cap (U \setminus \text{int}(U)) = \emptyset$, *for all* $t \in J$.

Then, the homotopic map H is fix-transversal.

Proof. The case of $J(H) = \emptyset$ is trivial; so, without loss, one may assume that $J(H) \neq \emptyset$. There are several steps to be passed.

Step 1. We claim that the following assertion holds

(c-bd-reg) the ordered metric subspace $(J(H), g, \leq)$ is conv-bd-regular.

To this end, let (r_n) be an ascending sequence in $J(H)$; hence

(r_n) is ascending in J and $(\text{Fix}(H(r_n))$ is nonempty, $\forall n)$.

By the Denumerable Axiom of Choice (deductible, as above said, in (ZF-AC+DC)), there exists a sequence (x_n) in U, with

$(\forall n)$: $x_n \in \text{Fix}(H(r_n))$; that is: $x_n = H(r_n, x_n)$.

As $(J, g, \leq)$ is conv-bd-regular, there exists a (unique) $r_\infty \in J$ with

$r_n \xrightarrow{g} r_\infty$ as $n \to \infty$, and $(r_n \leq r_\infty, \forall n)$.

We claim that $r_\infty \in J(H)$; and, from this, all is clear. For each $\lambda \in \Lambda$ and each couple of ranks (n, m) with $n \leq m$, we have (by the contractive requirements) the chain of relations

$$d_\lambda(x_n, x_m) = d_\lambda(H(r_n, x_n), H(r_m, x_m)) \leq$$
$$d_\lambda(H(r_n, x_n), H(r_m, x_n)) + d_\lambda(H(r_m, x_n), H(r_m, x_m)) \leq$$
$$\psi_\lambda(g(r_n, r_m)) + \omega_\lambda(d_\lambda(x_n, x_m)).$$

By the above convergence property, (r_n) is a g-Cauchy sequence. Combining with the functional couple $(\psi_\lambda, \omega_\lambda)$ having the precise properties for each $\lambda \in \Lambda$, one derives (by a previous result) that

(x_n) is D-Cauchy; wherefrom (by the hypotheses upon (X, U))
$x_n \xrightarrow{D} x_\infty$ as $n \to \infty$, for some $x_\infty \in U$.

For each $\lambda \in \Lambda$ and each n, we have (again by the contractive conditions)

$$d_\lambda(x_\infty, H(r_\infty, x_\infty)) - d_\lambda(x_\infty, x_n) \leq$$
$$d_\lambda(x_n, H(r_\infty, x_\infty)) = d_\lambda(H(r_n, x_n), H(r_\infty, x_\infty)) \leq$$
$$d_\lambda(H(r_n, x_n), H(r_\infty, x_n)) + d_\lambda(H(r_\infty, x_n), H(r_\infty, x_\infty)) \leq$$
$$\psi_\lambda(g(r_n, r_\infty)) + \omega_\lambda(d_\lambda(x_n, x_\infty)) \leq \psi_\lambda(g(r_n, r_\infty)) + d_\lambda(x_n, x_\infty).$$

Passing to limit as $n \to \infty$ in the extremal terms, yields

$d_\lambda(x_\infty, H(r_\infty, x_\infty)) = 0$, $\forall \lambda \in \Lambda$; whence (as $D =$ separated)
$x_\infty = H(r_\infty, x_\infty)$; which tells us that $r_\infty \in J(H)$.

Step 2. Summing up, the Zorn-Bourbaki metric maximality principle (ZB-m-mp) [to be discussed a bit further] applies to the ordered metric space $(J(H), g, \leq)$. Hence, for the fixed $q_0 \in J(H)$ (assured by hypothesis), there exists an ascending sequence (q_n) in $J(H)$, with

(ZB-1) $q^* := \lim_n(q_n)$ exists in $J(H)$ and $q_n \leq q^*$, for all n
(ZB-2) $q^* \in \max(J(H), \leq)$; that is: $q^* \leq r \in J(H)$ implies $q^* = r$.

Note that, as $q^* \in J(H)$, one has

$y^* = H(q^*, y^*)$, for some $y^* \in U$.

We show that, necessarily, $q^* \in \max(J, \leq)$; and this will complete the argument. Assume not: $J(q^*, <)$ is not empty. By the imposed hypotheses

there exists a sequence $(s_m; m \geq 0)$ in $J(q^*, <)$ with $\lim_m(s_m) = q^*$.

On the other hand (again by hypothesis)

$y^* \in U \cap \mathrm{Fix}(H(q^*))$ and $y^* \notin U \setminus \mathrm{int}(U)$; hence, $y^* \in \mathrm{int}(U)$.

From the definition of supremum topology, there exists a finite part Λ^* of Λ and a set of numbers $(\rho_\lambda; \lambda \in \Lambda^*)$ in R^0_+, such that

$V := \cap\{X[y^*, \rho_\lambda](d_\lambda); \lambda \in \Lambda^*\} \subseteq U$.

We now claim that there exists some index $i = i(V)$, such that

$H(s_m, y) \in V$, for all $m \geq i$, $y \in V$.

In fact, for each $\lambda \in \Lambda^*$, $m \in N$, $y \in V$ we have (in view of $\omega_\lambda =$ increasing)

$d_\lambda(y^*, H(s_m, y)) = d_\lambda(H(q^*, y^*), H(s_m, y)) \leq$
$d_\lambda(H(q^*, y^*), H(s_m, y^*)) + d_\lambda(H(s_m, y^*), H(s_m, y)) \leq$
$\psi_\lambda(g(q^*, s_m)) + \omega_\lambda(d_\lambda(y^*, y)) \leq \psi_\lambda(g(q^*, s_m)) + \omega_\lambda(\rho_\lambda)$.

On the other hand, from the very choice of (s_m),

$\lim_m g(q^*, s_m) = 0$; so that, $\lim_m \psi_\lambda(g(q^*, s_m)) = 0$, $\forall \lambda \in \Lambda$;

by the properties of our class $\Psi = (\psi_\lambda; \lambda \in \Lambda)$. Consequently, for each $\lambda \in \Lambda^*$ there must be an index $m(\lambda) \in N$, such that

$\psi_\lambda(g(q^*, s_m)) < \rho_\lambda - \omega_\lambda(\rho_\lambda)$, $\forall m \geq m(\lambda)$;
whence (combining with the above evaluation)
$d_\lambda(y^*, H(s_m, y)) < \rho_\lambda$, $\forall m \geq m(\lambda)$, $\forall y \in V$.

Putting $i := \max\{m(\lambda); \lambda \in \Lambda^*\}$, we then get

$(\forall \lambda \in \Lambda^*)$: $d_\lambda(y^*, H(s_m, y)) < \rho_\lambda$, for all $m \geq i$, $y \in V$;

whence the claim. As a consequence of this, $H(s_i)$ is a selfmap of the D-closed subset V of U. In addition, by the imposed conditions,

$d_\lambda(H(s_i)x, H(s_i)y) \leq \omega_\lambda(d_\lambda(x, y))$, $\forall \lambda \in \Lambda$, $\forall x, y \in V$;

proving that $H(s_i)$ is Ω-contractive. Finally, by a previous observation,

$(\forall \lambda \in \Lambda)$: ω_λ is Meir-Keeler admissible
(or, equivalently: Matkowski admissible).

Summing up, the second main result applies to (V, D) and $H(s_i) : V \to V$. So, for the starting point $y^* \in V$, there exists some point $z^* \in V \subseteq U$, with

$z^* = \lim_n H(s_i)^n y^*$, and $z^* = H(s_i)z^*$; whence, $s_i \in J(H)$.

This, in view of $q^* < s_i$, contradicts the maximality of q^* in $J(H)$. Hence, necessarily, $q^* \in \max(J, \leq)$; and the conclusion follows. □

A basic particular case of these developments corresponds to

$(\forall \lambda \in \Lambda)$: ω_λ is linear $[\omega_\lambda(t) = k_\lambda t,\ t \in R_+$, for some $k_\lambda \in R_+]$.

Precisely, let $U \in \exp(X)$ be fixed; and take a mapping $(t, x) \mapsto H(t, x) = H(t)x$ from $J \times U$ into X. Then, fix the family $\Psi = (\psi_\lambda; \lambda \in \Lambda)$ over $\mathcal{F}(R_+, R_+ \cup \{\infty\})$ and the family $k = (k_\lambda; \lambda \in \Lambda)$ over R_+, according to

(hom-1) H is first variable Ψ-contractive:
$d_\lambda(H(t, x), H(s, x)) \leq \psi_\lambda(g(t, s))$, for all $t, s \in J$ and all $x \in U$
(hom-2) H is second variable k-contractive:
$d_\lambda(H(t, x), H(t, y)) \leq k_\lambda d_\lambda(x, y)$, for all $t \in J$ and all $x, y \in U$.

We then say that H is a (Ψ, k)-homotopic mapping.

The following (linear) homotopic fixed point theorem is available, over the same reduced system (ZF-AC+DC).

Theorem 3.19. *Let the mapping $(t, x) \mapsto H(t, x) = H(t)x$ from $J \times U$ into X be (Ψ, k)-homotopic, where the family $\Psi = (\psi_\lambda; \lambda \in \Lambda)$ over $\mathcal{F}(R_+, R_+ \cup \{\infty\})$ and the family $k = (k_\lambda; \lambda \in \Lambda)$ over R_+ are such that*

(72-i) $(\forall \lambda \in \Lambda)$: *$\psi_\lambda$ is zero-continuous* $(\psi_\lambda(t) \to 0 = \psi(0)$ *as* $t \to 0)$
(72-ii) $(\forall \lambda \in \Lambda)$: *$k_\lambda$ is subunitary (in the sense:* $0 \leq k_\lambda < 1$*).*

In addition, suppose that

(72-iii) $(J, g, \leq)$ *is conv-bd-regular and right accessible*
(72-iv) *U is D-closed, and X is D-complete*
(72-v) $\mathrm{Fix}(H(t)) \cap (U \setminus \mathrm{int}(U)) = \emptyset$, *for all* $t \in J$.

Then, the homotopic map H is fix-transversal.

In particular, when the data (X, U) and $(J, g, \leq)$ are taken as

(q-1) X is D-complete and U is $\mathcal{T}$-closed
(q-2) $J = [0, 1]$, $(g, \leq) =$ the usual metric and ordering in R

our homotopic fixed point theorem is comparable with a similar one due to Ariza-Ruiz and Jimenez-Melado [19], [20]; and its linear version extends a related one in Frigon [91]. Concerning this last aspect, two remarks (of methodological nature) are in order.

I) The first variable contractive condition in the quoted paper is

$(\forall \lambda \in \Lambda)$: $d_\lambda(H(t, x), H(s, x)) \leq |\varphi_\lambda(t) - \varphi_\lambda(s)|$,
for all $t, s \in J$ and all $x \in U$, where $\varphi_\lambda : J \to R$ is continuous.

But, in view of any continuous function on the compact $J = [0, 1]$ being uniformly continuous, we must have $\forall \lambda \in \Lambda$

$|\varphi_\lambda(t) - \varphi_\lambda(s)| \leq \psi_\lambda(|t - s|)$, $t, s \in J$,
where $\psi_\lambda \in \mathcal{F}(R_+, R_+ \cup \{\infty\})$ is zero-continuous;

and, from this, we arrive at the posed first variable condition above.

II) On the other hand, the linear homotopic result above holds over the reduced system (ZF-AC+DC); while, the quoted result is holding over the complete system (ZF) (by means of the (general) Zorn-Bourbaki principle used by the author).

Finally, it is worth noting that the linear homotopy result we just formulated admits nonlinear counterparts. In fact, define the family of maps $\Omega = (\omega_\lambda; \lambda \in \Lambda)$ according to the requirements

$(\forall \lambda \in \Lambda)$: $\omega_\lambda \in \mathcal{F}_0(re, in)(R_+)$ is
right continuous and complementary coercive.

Clearly, all these functions are strongly regressive; hence, the general homotopy result is holding in such a context. Combining with the functions of this type being nonlinear in general, our assertion follows.

An interesting problem to be posed is that of these functional homotopy results covering the related statement in Chiş and Precup [52]. Further aspects will be discussed elsewhere.

3.3.8 *Kang-Park principles*

Let X be a nonempty set. By a *pseudometric* over X we mean any map $d : X \times X \to R_+$. Fix such a map, endowed with

(re) $d(.,.)$ is *reflexive*: $d(x, x) = 0, \forall x \in X$;

we then say that $d(.,.)$ is a *r-pseudometric* on X. Further, let $(\preceq)$ be a *quasi-order* (i.e.: reflexive and transitive relation) over X; the triple $(X, d, \preceq)$ will be referred to as a *quasi-ordered r-pseudometric space.*

Given the subset $M \in \exp(X)$, call $z \in X$,

(d-max) $(d, \preceq)$*-maximal* on M, if $(u, v \in M,\ z \preceq u \preceq v) \Longrightarrow d(u, v) = 0$
(max) $(\preceq)$*-maximal* on M, if $(w \in M,\ z \preceq w) \Longrightarrow z = w$.

It is our aim in the following to establish sufficient conditions for such properties. Then, an application is given to Ekeland variational principles.

(A) Call the sequence (x_n) in X, *d-Cauchy* when

$\forall \varepsilon > 0, \exists n(\varepsilon)$: $n(\varepsilon) \le p \le q \Longrightarrow d(x_p, x_q) \le \varepsilon$;
or, equivalently (as d=reflexive)
$(\forall \varepsilon > 0, \exists n(\varepsilon)$: $n(\varepsilon) < p < q \Longrightarrow d(x_p, x_q) \le \varepsilon$.

Then, call (x_n), *d-asymptotic* if $\lim_n d(x_n, x_{n+1}) = 0$; that is:

$\forall \varepsilon > 0, \exists m(\varepsilon)$: $m(\varepsilon) \le p \Longrightarrow d(x_p, x_{p+1}) \le \varepsilon$.

Given $M \in \exp(X)$, let us consider the global conditions

(C-reg) $(M, d, \preceq)$ is *Cauchy regular*:
each $(\preceq)$-ascending sequence in M is d-Cauchy
(A-reg) $(M, d, \preceq)$ is *asymptotic regular*:
each $(\preceq)$-ascending sequence in M is d-asymptotic.

Clearly, (C-reg) $\Longrightarrow$ (A-reg); because each [$(\preceq)$-ascending or not] d-Cauchy sequence is d-asymptotic. The reverse implication also holds, in the sense

Proposition 3.21. *We have, in (ZF-AC+DC),*

(A-reg) $\Longrightarrow$ *(C-reg); whence, (A-reg)* $\Longleftrightarrow$ *(C-reg).*

Proof. Suppose that (A-reg) holds; but (C-reg) does not hold: there exists a $(\preceq)$-ascending sequence (x_n) in M that is not d-Cauchy; this, by definition, means that (for some $\varepsilon > 0$)

$$C(n) = \{(p, q) \in N \times N; n < p < q, d(x_p, x_q) > \varepsilon\} \neq \emptyset, \forall n.$$

Denote, for simplicity

$$p(n) = \min \mathrm{Dom}[C(n)],\ q(n) = \min[C(n)(p(n))],\ n \in N.$$

Put $i(0) = 0$. By this assumption, there exist $[i(1) = p(i(0)),\ i(2) = q(i(0))]$ with $i(0) < i(1) < i(2)$, $d(x_{i(1)}, x_{i(2)}) > \varepsilon$. Further, given the rank $i(2)$, there exist $[(i(3) = p(i(2)),\ i(4) = q(i(2))]$ with $i(2) < i(3) < i(4)$, $d(x_{i(3)}, x_{i(4)}) > \varepsilon$; and so on. By induction, we get a subsequence $(y_n = x_{i(n)}; n \geq 0)$ of $(x_n; n \geq 0)$ with $d(y_{2n+1}, y_{2n+2}) > \varepsilon$, for all n. This contradicts (A-reg); hence the claim. □

By definition, either of these conditions will be referred to as: $(M, d, \preceq)$ is *regular*. Further consequences of this fact may be stated under the lines below. For each $\varepsilon > 0$, let $\mathcal{R}(\varepsilon)$ stand for the relation over M

$(x, y \in M)$: $x\mathcal{R}(\varepsilon)y$ iff $x \preceq y$ and
$((\forall u, v \in M)$: $y \preceq u \preceq v$ implies $d(u, v) \leq \varepsilon)$.

Proposition 3.22. *We have in (ZF-AC+DC), for each* $M \in \exp(X)$
(82-1) *If* $(M, d, \preceq)$ *is regular, then* $(M, d, \preceq)$ *is weakly regular:*

$(\forall \varepsilon > 0)$: $\mathcal{R}(\varepsilon)$ *is proper* $(\mathcal{R}(\varepsilon)(x) \neq \emptyset, \forall x \in M)$

(82-2) *If* $(M, d, \preceq)$ *is weakly regular, then,* $(M, d, \preceq)$ *is weakly sequentially regular: for each* $u \in M$*, and each strictly descending sequence* $(\varepsilon_n; n \geq 0)$ *in* R^0_+ *with* $\varepsilon_n \to 0$*, there exists a* $(\preceq)$*-ascending sequence* $(u_n; n \geq 0)$ *in* M*, with*

$u \preceq u_0$, *and* $[(\forall n \geq 1), (\forall y, z \in M)]$: $u_n \preceq y \preceq z \Longrightarrow d(y, z) \leq \varepsilon_n$.

Proof. There are two steps to be passed.
Step 1. Assume this would be false; i.e. (for some $\varepsilon > 0$ and $x_0 \in M$)

for each $y \in M(x_0, \preceq)$ there exist $u, v \in M$ with $y \preceq u \preceq v$, $d(u, v) > \varepsilon$.

This, by definition, yields (for the same (ε, x_0)):

$\forall y \in M(x_0, \preceq)$, $\exists (u,v) \in (\preceq)$: $y \preceq u$, $d(u,v) > \varepsilon$;

where $(\preceq) := \{(a,b) \in M \times M; a \preceq b\}$. Put $Q := \{(a,b) \in (\preceq); x_0 \preceq a\}$; and fix $(y_0, y_1) \in Q$ (for example, we may take $y_0 = y_1 = x_0$). Define a relation $\mathcal{S} = \mathcal{S}(\varepsilon)$ on Q as

$(a_1, b_1)\mathcal{S}(a_2, b_2)$ if and only if $b_1 \preceq a_2$, $d(a_2, b_2) > \varepsilon$.

From the imposed condition, $Q((a,b), \mathcal{S}) \neq \emptyset$, $\forall (a,b) \in Q$. So, by the Dependent Choice principle (DC), it follows that for the starting $w_0 = (y_0, y_1)$ in Q, there exists a sequence $(w_n := (y_{2n}, y_{2n+1}); n \geq 0)$ in Q, with

$w_n \mathcal{R} w_{n+1}$, for all n; hence, by definition,
$y_{2n+1} \preceq y_{2n+2}$, $d(y_{2n+2}, y_{2n+3}) > \varepsilon$, for all n.

As a consequence, $(y_n; n \geq 0)$ is $(\preceq)$-ascending and not d-asymptotic; in contradiction with the regularity of $(M, d, \preceq)$; hence the claim.

Step 2. Suppose that $(M, d, \preceq)$ is weakly regular; and let $(\varepsilon_n; n \geq 0)$ be a strictly descending sequence in R^0_+ with $\varepsilon_n \to 0$; clearly, the relational sequence $(\mathcal{R}(\varepsilon_n); n \geq 0)$ is proper. Fix some $u \in X$; as well as an element $u_0 \in X(u, \preceq)$. By the Diagonal Dependent Choice principle (DDC), we have that, there exists a $(\preceq)$-ascending sequence $(u_n; n \geq 0)$ in X with

$(u_n \mathcal{R}(\varepsilon_{n+1}) u_{n+1}$, for each $n)$; hence (in particular):
$((\forall n \geq 1), (\forall y, z \in X))$: $u_n \preceq y \preceq z \Longrightarrow d(y,z) \leq \varepsilon_n$;

and, from this, we are done. □

(B) Let again $(X, d, \preceq)$ be a quasi-ordered r-pseudometric space. Remember that we introduced a (sequential) convergence structure $(\xrightarrow{d})$ on X (under the lines in Kasahara [141]) as: given the sequence (x_n) in X and the point $x \in X$, let us say that (x_n), *d-converges* to x (written as: $x_n \xrightarrow{d} x$), provided

$\forall \varepsilon > 0$, $\exists i = i(\varepsilon)$: $i \leq n \Longrightarrow d(x_n, x) \leq \varepsilon$.

The string $x_n \xrightarrow{d} x$ is also referred to as: x is a *d-limit* of (x_n). The class of all these points will be denoted as $\lim_n(x_n)$; when it is nonempty, we say that (x_n) is *d-convergent*. Then, call $Y \in \exp[X]$, $(d, \preceq)$*-closed* provided

the d-limit of each ascending sequence in Y is included in Y.

The class of all these subsets will be denoted as $\mathcal{K}(d, \preceq)$; note that

$\mathcal{K}(d, \preceq)$ is invariant to arbitrary intersections and finite unions.

In this case, denote for each $M \in \exp[X]$

$$\text{ocl}(M) = \text{the intersection of all } Y \in \mathcal{K}(d, \preceq) \text{ with } M \subseteq Y.$$

The selfmap $Y \mapsto \text{ocl}(Y)$ of $\exp[X]$ is a *closure* over X, in the sense

(ocl-1) $\text{ocl}(\emptyset) = \emptyset$, $\text{ocl}(X) = X$,
(ocl-2) $U \subseteq V$ implies $\text{ocl}(U) \subseteq \text{ocl}(V)$,
(ocl-3) $\text{ocl}(U \cup V) = \text{ocl}(U) \cup \text{ocl}(V)$, $\forall\, U, V \in \exp[X]$,
(ocl-4) $Y \subseteq \text{ocl}(Y)$, $\forall Y \in \exp[X]$,
(ocl-5) $\text{ocl}(\text{ocl}(Y)) = \text{ocl}(Y)$, for each $Y \in \exp[X]$;

so, the selfmap $Y \mapsto \text{ocl}(Y)$ is complying with the general conventions in Kuratowski [155, Ch. I, Sec. 4].

As a completion of this, let us introduce a *quasi-order convergence* $\mathcal{B} = \mathcal{B}(\preceq)$ over X as follows: given the sequence (x_n) in X and the point $x \in X$,

$$x_n \xrightarrow{\mathcal{B}} x \text{ if } \exists m(x) \in N, \text{ such that } n \geq m(x) \text{ implies } x_n \preceq x.$$

Clearly, $(\xrightarrow{\mathcal{B}})$ is a (sequential) convergence on X, under the lines in Kasahara [141]; and $x_n \xrightarrow{\mathcal{B}} x$ is also referred to as: x is a $\mathcal{B}$-*limit* of (x_n). Denote the class of all these as $\mathcal{B} - \lim_n(x_n)$; when it is nonempty, we say that (x_n) is $\mathcal{B}$-*convergent.*

The following statement (referred to as: Kang-Park pseudometric maximal principle; in short: (KP-p-mp)), holding over the reduced system (ZF-AC+DC), will be our starting point in what follows.

Theorem 3.20. *Let the quasi-ordered r-pseudometric space* $(X, d, \preceq)$ *be taken according to the conditions*

(81-i) $(X, d, \preceq)$ *is regular:*
each $(\preceq)$*-ascending sequence in* X *is d-Cauchy*
(81-ii) $(X, d, \preceq)$ *is* $(d, \mathcal{B})$*-complete:*
each ascending d-Cauchy sequence (x_n) *in* X
is d-convergent (in X*) and* $\lim_n(x_n) \cap (\mathcal{B} - \lim_n(x_n)) \neq \emptyset$.

Then, the conclusions below are valid

(81-a) *for each* $u \in X$ *there exists* $v \in X$*, such that* $u \preceq v$ *and* v *is* $(d, \preceq)$*-maximal (over* X*)*

(81-b) *for each* $M, Y \in \exp(X)$ *with* $M \subseteq Y \in \mathcal{K}(d, \preceq)$ *and each* $u \in M$, *there exists* $v \in Y$, *with:* $u \preceq v$ *and* v *is* $(d, \preceq)$*-maximal over* Y *(hence,* a fortiori*:* $(d, \preceq)$*-maximal over* M*).*

Proof. i): From the auxiliary statement above, $(X, d, \preceq)$ is weakly regular; hence, weakly sequentially regular. So, given $u \in X$, one may construct a $(\preceq)$-ascending sequence (u_n) in X with

(p-1) $u \preceq u_0$, and $[(\forall n \geq 1), (\forall y, z \in X)]$:
$u_n \preceq y \preceq z \Longrightarrow d(y, z) \leq 2^{-n}$.

In particular, (u_n) is d-Cauchy (in X). Combining with the $(d, \mathcal{B})$-completeness hypothesis, there exists $v \in X$ such that $u_n \xrightarrow{d} v$ and $u_n \xrightarrow{\mathcal{B}} v$. From the last boundedness relation and (u_n) = ascending, we have $[u_n \preceq v, \forall n]$. This, along with the remaining properties of (u_n), tells us that v is $(d, \preceq)$-maximal over X.

ii): Let $M, Y \in \exp[X]$ be such that $M \subseteq Y \in \mathcal{K}(d, \preceq)$. Clearly, $(Y, d, \preceq)$ is regular; hence weakly regular; hence weakly sequentially regular. As a consequence of this, we have that, given $u \in M \subseteq Y$, one may construct a $(\preceq)$-ascending sequence (u_n) in Y with

(p-2) $u \preceq u_0$, and $[(\forall n \geq 1), (\forall y, z \in Y)]$:
$u_n \preceq y \preceq z \Longrightarrow d(y, z) \leq 2^{-n}$.

In particular, (u_n) is d-Cauchy (in Y); so, by completeness,

(u_n) is d-convergent in X and $\lim_n(u_n) \cap (\mathcal{B} - \lim_n(u_n)) \neq \emptyset$.

This, along with $\lim_n(u_n) \subseteq Y$ (see above), tells us that there exists $v \in Y$ with $u_n \xrightarrow{d} v$ and $u_n \xrightarrow{\mathcal{B}} v$. From the last boundedness relation and (u_n) = ascending, we have $[u_n \preceq v, \forall n]$. In addition, the remaining properties of (u_n) tell us that v is $(d, \preceq)$-maximal over Y, via

$$(x, y \in Y, v \preceq x \preceq y) \Longrightarrow d(x, y) = 0.$$

But then, v is $(d, \preceq)$-maximal over M as well; and the proof is complete. □

Formally, this result is comparable with the 1990 one in Kang and Park [137]. However, its basic lines were already set up in the papers by Turinici [233], [237]. Note that, both these results extend the 1976 Brezis-Browder ordering principle [38]. Further aspects may be found in Altman [15].

(C) An important particular case of these developments corresponds to $d(.,.)$ being a metric over X. The auxiliary fact below is to be needed.

Proposition 3.23. *Under the precise framework, we have*

$((X, d, \preceq) = \textit{regular})$ *implies* $(\preceq)$ *is antisymmetric; hence, an ordering.*

Proof. (see also Hamel [107, Ch. 4, Sec. 4.1]) Let $u, v \in X$ be such that $u \preceq v$ and $v \preceq u$. The sequence (y_n) given as $[y_{2n} = u,\ y_{2n+1} = v;\ n \geq 0]$ is ascending; hence, d-asymptotic by hypothesis. This yields $d(u, v) = 0$; whence (as d = metric), $u = v$. □

For each $M \in \exp(X)$, let $\mathrm{cl}(M)$ stand for its d-closure:

$z \in \mathrm{cl}(M)$ iff $z = \lim_n(x_n)$, for some (x_n) in M.

The following maximal statement (referred to as: Zorn-Bourbaki metric maximal principle; in short: (ZB-m-mp)) is available, over (ZF-AC+DC).

Theorem 3.21. *Let the ordered metric space* $(X, d, \preceq)$ *be such that*

$(X, d, \preceq)$ *is (conv,* $\mathcal{B}$*)-regular: each* $(\preceq)$*-ascending sequence* (x_n) *in* X *is* d*-convergent (in* X*) and* $(x_n \preceq \lim_n(x_n), \forall n)$*.*

Then, the conclusions below hold

(82-a) for each $u \in X$ *there exists* $v \in X$*, such that*
$u \preceq v$ *and* v *is* $(\preceq)$*-maximal (over* X*)*
(82-b) for each $M \in \exp(X)$ *and each* $u \in M$*, there exists* $v \in \mathrm{cl}(M)$*, with:* $u \preceq v$ *and* v *is* $(\preceq)$*-maximal over* $\mathrm{cl}(M)$ *(hence,* a fortiori*,* $(\preceq)$*-maximal over* M*).*

Proof. If $(X, d, \preceq)$ is (conv, $\mathcal{B}$)-regular, then $(X, d, \preceq)$ is regular, because any d-convergent sequence is d-Cauchy. Moreover, by the same premise, $(X, d, \preceq)$ is $(d, \mathcal{B})$-complete; hence, summing up, our preceding result is applicable here.

i): By the first conclusion of this result, it follows that, for the starting $u \in X$ there exists $v \in X$ such that $[u \preceq v$ and v is $(d, \preceq)$-maximal over $X]$. But then, v is $(\preceq)$-maximal in X, via

$$v \preceq x \in X \Longrightarrow (v, x \in X,\ v \preceq v \preceq x) \Longrightarrow d(v, x) = 0 \Longrightarrow v = x.$$

ii): Denote $Y = \mathrm{cl}(M)$; hence, $M \subseteq Y \in \mathcal{K}(d, \preceq)$. From the second conclusion of our preceding result we have that, for the starting $u \in M \subseteq Y$ there exists $v \in Y$ such that $[u \preceq v$ and v is $(d, \preceq)$-maximal over $Y]$. But then, v is necessarily $(\preceq)$-maximal in Y, via

$$v \preceq y \in Y \Longrightarrow (v, y \in Y,\ v \preceq v \preceq y) \Longrightarrow d(v, y) = 0 \Longrightarrow v = y;$$

and this in particular, tells us that v is $(\preceq)$-maximal in M. □

This result may be viewed as a metrical version of the Zorn-Bourbaki maximal principle [32]. Some early variants of it were formulated in Turinici [230], [231], [234]; see also Dancs et al. [68].

(D) A basic application of these facts is to (local) variational principles on metrical structures. Let (X, d) be a metric space. Then, let the function $\varphi : X \to R \cup \{\infty\}$ be such that the following admissible conditions hold:

(adm-1) φ is inf-proper ($\mathrm{Dom}(\varphi) \neq \emptyset$ and $\inf[\varphi(X)] > -\infty$)
(adm-2) φ is d-lsc on X: $\{x \in X; \varphi(x) \leq t\}$ is d-closed, $\forall t \in R$.

The following statement (referred to as: Ekeland variational principle on metric spaces; in short: (EVP-m)) is available over the reduced system (ZF-AC+DC).

Theorem 3.22. *Let the metric space (X, d) and the function $\varphi : X \to R \cup \{\infty\}$ be such that (X, d) is complete and φ is admissible. Further, take some $u \in \mathrm{Dom}(\varphi)$. Then, the following conclusions are valid*

(83-a) *There exists $v \in \mathrm{Dom}(\varphi)$, with*

(83-a1) $d(u, v) \leq \varphi(u) - \varphi(v)$ *(hence $\varphi(u) \geq \varphi(v)$)*
(83-a2) $d(v, x) > \varphi(v) - \varphi(x)$, *for each $x \in X \setminus \{v\}$*

(83-b) *Given $M \in \exp(X)$, $u \in M \cap \mathrm{Dom}(\varphi)$, there exists $v \in \mathrm{cl}(M) \cap \mathrm{Dom}(\varphi)$, with the variational properties*

(83-b1) $d(u, v) \leq \varphi(u) - \varphi(v)$ *(hence $\varphi(u) \geq \varphi(v)$)*
(83-b2) $d(v, x) > \varphi(v) - \varphi(x)$, *for each $x \in M \setminus \{v\}$.*

Proof. Denote $X[u] = \{x \in X; \varphi(u) \geq \varphi(x)\}$; note that, by the posed conditions, $X[u]$ is a nonempty d-closed subset of X, with $X[u] \subseteq \mathrm{Dom}(\varphi)$. Let $(\preceq)$ be the relation over X, defined as

(pre-def) $x \preceq y$ iff $d(x, y) + \varphi(y) \leq \varphi(x)$.

Clearly, $(\preceq)$ is a quasi-order on X; in addition, $(\preceq)$ is an order on $\mathrm{Dom}(\varphi)$; hence, all the more, on $X[u]$. We claim that the premises of Zorn-Bourbaki metric maximal principle (ZB-m-mp) are fulfilled over the ordered metric space $(X[u], d, \preceq)$. Let (x_n) be a $(\preceq)$-ascending sequence in $X[u]$:

(asc-seq) $d(x_n, x_m) \leq \varphi(x_n) - \varphi(x_m)$, if $n \leq m$.

The sequence $(\varphi(x_n))$ is descending bounded from below; hence a Cauchy one. This, along with the (asc-seq), shows that (x_n) is a $(\preceq)$-ascending d-Cauchy sequence in $X[u]$. By the completeness hypothesis (and closeness of $X[u]$), $x_\infty := \lim_n(x_n)$ exists in $X[u]$; moreover, from the d-lsc property, $[\varphi(x_n) \geq \varphi(x_\infty)$, for all $n]$. Finally, take some rank n. From (asc-seq) again (and the triangular property)

$$d(x_n, x_\infty) \leq d(x_n, x_m) + d(x_m, x_\infty) \leq \varphi(x_n) - \varphi(x_m) +$$
$$d(x_m, x_\infty) \leq \varphi(x_n) - \varphi(x_\infty) + d(x_m, x_\infty), \forall m \geq n;$$

wherefrom (passing to limit as $m \to \infty$)

$$d(x_n, x_\infty) \leq \varphi(x_n) - \varphi(x_\infty) \text{ (that is: } x_n \preceq x_\infty).$$

This, by the arbitrariness of n, gives $(x_n \preceq x_\infty, \forall n)$; and our claim follows.

i): By the first conclusion of Zorn-Bourbaki metric maximal principle (ZB-m-mp) relative to the triple $(X[u], d, \preceq)$, we have that, for the starting $u \in X[u]$, there exists some other point $v \in X[u]$, with

(maxi) $u \preceq v$ and v is $(\preceq)$-maximal over $X[u]$
(in the sense: $v \preceq w \in X[u]$ imply $v = w$).

Suppose, by absurd, that

v is not $(\preceq)$-maximal in X: $v \preceq y$, for some $y \in X \setminus \{v\}$.

As $u \preceq v$, one gets $u \preceq y$; hence, $y \in X[u]$. This yields $v \preceq y \in X[u]$; whence [by (maxi) above], $v = y$; contradiction; proving that v is $(\preceq)$-maximal in X.

ii): Let $M \in \exp(X)$ and $u \in M \cap \mathrm{Dom}(\varphi)$ be given; and put

(loc-u) $M_u := M \cap X[u]$; hence, $M_u \in \exp(X[u])$ (as $u \in M_u$).

By the second conclusion of Zorn-Bourbaki metric maximal principle (ZB-m-mp) relative to $(X[u], d, \preceq)$ and $(M_u \in \exp(X[u]), u \in M_u)$, it results that, for the starting $u \in M_u$, there exists some point $v \in \mathrm{cl}(M_u)$, with

(maxi-M-u) $u \preceq v$ and v is $(\preceq)$-maximal over $\mathrm{cl}(M_u)$;
hence, necessarily, $(\preceq)$-maximal over M_u.

Suppose, by absurd, that

v is not $(\preceq)$-maximal in M: $v \preceq y$, for some $y \in M \setminus \{v\}$.

As $u \preceq v$, one gets $u \preceq y$; hence, $y \in X[u]$; proving that $y \in M \cap X[u] = M_u$. This yields $v \preceq y \in M_u$; whence [by (maxi-M-u) above], $v = y$;

contradiction; so that v is $(\preceq)$-maximal in M_u. Finally (taking the relation (loc-u) into account)

$$\mathrm{cl}(M_u) \subseteq \mathrm{cl}(M) \cap \mathrm{cl}(X[u]) = \mathrm{cl}(M) \cap X[u] \subseteq \mathrm{cl}(M) \cap \mathrm{Dom}(\varphi);$$

which gives the desired localization of v; and the proof is complete. □

This principle, due to Ekeland [86], found some basic applications to control and optimization, generalized differential calculus, critical point theory and global analysis; we refer to the quoted paper for a survey of these. So, it cannot be surprising that, soon after its formulation, many extensions of (EVP-m) were proposed. For example, the *dimensional* way of extension refers to the ambient space (R) of $\varphi(X)$ being substituted by a (topological or not) vector space. An account of the results in this area is to be found in Goepfert et al. [102, Ch. 3]. The *metrical* extension of the same consists in conditions imposed upon our metric being relaxed. Some of these extensions were already stated; for the remaining ones, we refer to Hyers et al. [117, Ch. 5]; see also Turinici [239].

By the developments above, we therefore have the implications:

$$\text{(DC)} \Longrightarrow \text{(KP-p-mp)} \Longrightarrow \text{(ZB-m-mp)} \Longrightarrow \text{(EVP-m)}.$$

So, we may ask whether these may be reversed. Clearly, the natural setting for solving this problem is the *strongly reduced* Zermelo-Fraenkel system (ZF-AC). Some preliminaries are needed.

Let X be a nonempty set; and $(\leq)$ be a (partial) order on it. We say that $(\leq)$ has the *inf-lattice* property, provided

$$x \wedge y := \inf(x, y) \text{ exists, for all } x, y \in X.$$

Remember that $z \in X$ is a $(\leq)$-*maximal* element if $X(z, \leq) = \{z\}$; the class of all these will be denoted as $\max(X, \leq)$. Call $(\leq)$, a *Zorn order* when

$\max(X, \leq)$ is nonempty and *cofinal* in X
(for each $u \in X$ there exists a $(\leq)$-maximal $v \in X$ with $u \leq v$).

Further aspects are to be described in a metric setting. Let (X, d) be a metric space; and $\varphi : X \to R_+$ be some function. The natural choice for $(\leq)$ above is

$$x \leq_{(d,\varphi)} y \text{ iff } d(x, y) \leq \varphi(x) - \varphi(y);$$

referred to as the *Brøndsted order* [39] attached to (d, φ). Denote

$$X(x,\rho) = \{u \in X; d(x,u) < \rho\},\ x \in X,\ \rho > 0$$

[the *open* sphere with center x and radius ρ]. Call (X,d), *discrete* when

for each $x \in X$ there exists $\rho = \rho(x) > 0$ such that $X(x,\rho) = \{x\}$.

Note that, under such an assumption, any function $\psi : X \to R$ is continuous over X. However, the *d-Lipschitz property* of the same

$$|\psi(x) - \psi(y)| \leq Ld(x,y),\ x, y \in X, \text{ for some } L > 0$$

cannot be assured, in general; this is also true for the *d-nonexpansive property* (characterized by means of $L = 1$).

The statement below [referred to as: the discrete Lipschitz countable version of (EVP-m); (in short: (EVP-m-dLc)] enters into our discussion.

Theorem 3.23. *Let the metric space (X,d) and the function $\varphi : X \to R_+$ be taken according to*

(84-i) (X,d) is discrete bounded and complete
(84-ii) $(\leq_{(d,\varphi)})$ has the inf-lattice property
(84-iii) φ is d-nonexpansive and $\varphi(X)$ is countable.

Then, $(\leq_{(d,\varphi)})$ is a Zorn order.

Clearly, (EVP-m) $\Longrightarrow$ (EVP-m-dLc). The remarkable fact is that this last principle yields (DC); and completes the circle between all these.

Proposition 3.24. *The inclusion below holds*

(84-1) (EVP-m-dLc) $\Longrightarrow$ (DC)
(in the strongly reduced Zermelo-Fraenkel system (ZF-AC)).

So (by the above inclusions),

(84-2) the maximal/variational principles (KP-p-mp), (ZB-m-mp) and (EVP-m) are all equivalent with (DC); hence, mutually equivalent
(84-3) each maximal/variational statement (VP) with (DC) $\Longrightarrow$ (VP) $\Longrightarrow$ (EVP-m) is equivalent with both (DC) and (EVP-m).

For a complete proof, see Turinici [252]. In particular, when the inf-lattice and nonexpansive properties are ignored in (EVP-dLc), the last result above reduces to the one in Brunner [43]. Note that, under the same particular setting, a different proof of the underlying inclusion was provided in Dodu and Morillon [78]. Further aspects may be found in Schechter [211, Ch. 19, Sec. 19.51].

Chapter 4

VARIA

4.1 Geometric Functional Contractions in Ordered Quasi-Metric Spaces

Abstract. Some fixed point theorems are given for a class of functional contractions over quasi-ordered quasi-metric spaces. These may be viewed as (geometric type) *full* quasi-metric counterparts of some contributions in the area due to Jleli and Samet [Fixed Point Th. Appl., 2012, 2012:210].

Subject-Class 2020. 47H10 (Primary), 54H25 (Secondary).

Keywords. Quasi-ordered quasi-metric space, fixed point, strong/Bellman Picard operator, positive and almost semi-diametral property, functional/geometric Meir-Keeler contraction.

4.1.1 *Dependent Choice principles*

See Ch. 1.

4.1.2 *Conv-Cauchy structures*

See Ch. 1.

4.1.3 *Geometric-asymptotic relations*

See Ch. 1.

4.1.4 *Banach contractions and relatives*

See Ch. 1.

4.1.5 *Statement of the problem*

Let X be a nonempty set. Call the subset Y of X, *almost singleton* (in short: *asingleton*) when $[y_1, y_2 \in Y$ implies $y_1 = y_2]$; and *singleton* if, in addition, Y is nonempty; note that in this case $Y = \{y\}$, for some $y \in X$.

By a *quasi-metric* over X we shall mean any mapping $b : X \times X \to R_+$ with the properties

(qm-1) b is *reflexive*: $b(x, x) = 0$, for each $x \in X$
(qm-2) b is *triangular*: $b(x, z) \leq b(x, y) + b(y, z)$, $\forall x, y, z \in X$
(qm-3) b is *sufficient*: $b(x, y) = 0$ implies $x = y$;

the couple (X, b) will be referred to as a *quasi-metric space*. Note that, a quasi-metric has all properties of a metric, except symmetry; hence, any metric is a quasi-metric. The converse is not in general true; as shown by

Example 4.1.
Put $X = \{1, 2, 3\}$; and take the map $b : X \times X \to R_+$ as

$b(1, 1) = b(2, 2) = b(3, 3) = 0$;
$b(1, 2) = b(1, 3) = 1$, $b(2, 3) = 2$; $b(2, 1) = b(3, 1) = b(3, 2) = 1$.

Clearly, $b(.,.)$ is a quasi-metric on X. On the other hand, $b(.,.)$ is not a metric on X; because $b(2, 3) = 2$, $b(3, 2) = 1$.

Finally, note that, given a quasi-metric $b(.,.)$ on X, its *conjugate map*

$$(c : X \times X \to R_+): c(x, y) = b(y, x),\ x, y \in X$$

is a quasi-metric too. On the other hand, the associated (to b and c) mapping $d : X \times X \to R_+$, given as

$$d(x, y) = \max\{b(x, y), c(x, y)\},\ x, y \in X \text{ (in short: } d = \max(b, c))$$

is a symmetric quasi-metric; hence, a (standard) metric on X.

The concept of quasi-metric seems to have a pretty long tradition in metrical spaces theory. For example, in his 2001 PhD Thesis, Hitzler [113, Ch. 1, Sec. 1.2] introduced this notion as a tool for the topological study of logic semantics. Later, in a 2004 paper, Turinici [241] used the same concept [referred to as: *reflexive triangular sufficient pseudometric*] with the aim of establishing a Caristi-Kirk fixed point theorem over such structures. From a "purely" fixed point perspective, the quasi-metric techniques have been used in 2012 by Jleli and Samet [128]. Further aspects of this theory –

related to *simulation method* – were developed in a recent 2014 contribution due to Alsulami et al. [13].

Suppose that we introduced a quasi-metric $b(.,.)$ (on X). A natural convergence and Cauchy structure over X may be defined as below.

(A) We say that the sequence (x_n) in X, *b-converges* to $x \in X$ (and write: $x_n \xrightarrow{b} x$) iff $b(x_n, x) \to 0$ as $n \to \infty$; that is

$$\forall \varepsilon > 0, \exists p = p(\varepsilon), (\forall n): \ (p \le n \Longrightarrow b(x_n, x) < \varepsilon).$$

Then x is called a *b-limit* of (x_n); the set of all these will be denoted as $\lim_n(x_n)$; if such elements exist, we say that (x_n) is *b-convergent.*

Clearly, the introduced convergence $(\xrightarrow{b})$ is *hereditary* and *reflexive*; so, it fulfills the general requirements in Kasahara [141]. Concerning some other convergence properties to be used, the following ones are of interest.

I) Let us say that b is *separated*, when

(sepa) $(\xrightarrow{b})$ is *separated:*
$\lim_n(x_n)$ is an asingleton, for each sequence $(x_n; n \ge 0)$ in X.

In this case, given the b-convergent sequence (x_n), we must have

$$\lim_n(x_n) = \{z\} \text{ (written as: } \lim_n(x_n) = z\text{), for some } z \in X.$$

Moreover (under the framework of (qm-1)+(qm-2))

$$(b = \text{separated}) \text{ implies } (b = \text{sufficient}).$$

In fact, let $u, v \in X$ be such that $b(u, v) = 0$. The constant sequence $(x_n = u; n \ge 0)$ fulfills $x_n \xrightarrow{b} u$, $x_n \xrightarrow{b} v$; so that (by the separated property), $u = v$.

As a basic example of such objects, note that

$$b = \text{metric (hence, } b = d) \Longrightarrow b = \text{separated}.$$

For, let (x_n) be a sequence in X with $x_n \xrightarrow{d} u$, $x_n \xrightarrow{d} v$. Clearly

$$d(u, v) \le d(x_n, u) + d(x_n, v), \forall n \text{ (as } d \text{ is triangular and symmetric)}.$$

Passing to limit, gives $d(u, v) = 0$; hence, $u = v$ (as $d =$ sufficient).

II) Returning to the general case, define the concept

(1v-cont) b is *first variable continuous*: for each $y \in X$, the map $x \mapsto b(x, y)$ is continuous ($x_n \xrightarrow{b} x$ implies $b(x_n, y) \to b(x, y)$).

Note that if b = metric, this condition is fulfilled (cf. a previous result).

Remark 4.1.

Concerning the introduced convention, let us consider the following (apparently weaker) form of it:

(1v-lsc) b is *first variable lsc*: for each $y \in X$, the map $x \mapsto b(x, y)$ is b-lsc ($x_n \xrightarrow{b} x$ implies $\liminf_n b(x_n, y) \geq b(x, y)$).

Note that, in many concrete circumstances to be considered, this variant works; so, it is legitimate to ask of what is the reason of introducing its strong version. The answer to this is very simple: by the conditions imposed upon $b(.,.)$, the weaker form in question is equivalent with its initial (strong) counterpart. In fact, assume that b is first variable lsc; and fix some $y \in X$. Given the sequence (x_n) in X and the point $x \in X$ with $x_n \xrightarrow{b} x$, we have (by the triangular inequality)

$b(x_n, y) \leq b(x_n, x) + b(x, y)$, $\forall n$; so (as $\lim_n b(x_n, x) = 0$)
$\limsup_n b(x_n, y) \leq \limsup_n b(x_n, x) + b(x, y) = b(x, y)$.

Combining with the working assumption about b, gives

$b(x, y) \leq \liminf_n b(x_n, y) \leq \limsup_n b(x_n, y) \leq b(x, y)$;
or, equivalently: $\lim_n b(x_n, y) = b(x, y)$;

proving that $b(.,.)$ is first variable continuous. So, passing from (1v-cont) to (1v-lsc) does not bring any generality relative to the precise framework.

Finally, we show by an example – patterned after Alsulami et al. [13] – that such extra conditions imposed upon quasi-metrics are effective.

Example 4.2.

Let $X = R_+$; and $[g(x, y) = |x - y|;\ x, y \in X]$ stand for the usual metric over it. Then, fix $\beta \in]0, 1[$; and define a mapping $b : X \times X \to R_+$ as

$$(b(x, y) = \beta g(x, y), \text{ if } x < y), \quad (b(x, y) = g(x, y), \text{ if } x \geq y).$$

Clearly, $b(.,.)$ is a quasi-metric over X. However, it is not a metric; because, for example, $b(1, 2) = \beta < 1 = b(2, 1)$. Further, its convergence structure is equivalent with the one induced by the usual metric, in view of

$$\beta g(x, y) \leq b(x, y) \leq g(x, y), \ \forall x, y \in X.$$

In fact, as a direct consequence of this,

$(\forall(x_n), \forall x)$: $x_n \xrightarrow{b} x$ iff $x_n \xrightarrow{g} x$; wherefrom, b is separated.

Finally, for each $y \in X$, the partial map $\varphi_y := b(.,y)$ defined as

$$(\varphi_y(x) = \beta g(x,y), \text{ if } x < y), \quad (\varphi_y(x) = g(x,y), \text{ if } x \geq y)$$

is g-continuous over X; hence (by the above), b-continuous over the same; so that, $b(.,.)$ is first variable continuous.

Let us say that the sequence $(x_n; \geq 0)$ in X is *b-Cauchy*, when $b(x_m, x_n) \to 0$ as $m, n \to \infty$, $m < n$; that is

$$\forall \varepsilon > 0, \exists q = q(\varepsilon), \forall (m,n): \ (q \leq m < n \Longrightarrow b(x_m, x_n) < \varepsilon).$$

The class of all these, denoted as $Cauchy(b)$, is hereditary and reflexive; so, this concept fulfills the general requirements in Turinici [244].

Finally, call $((\xrightarrow{b}), Cauchy(b))$, the *conv-Cauchy structure* induced by b. The following regularity conditions about this structure are considered

(reg) b is *regular*: each b-convergent sequence in X is b-Cauchy
(comp) b is *complete*: each b-Cauchy sequence in X is b-convergent.

Note that, by the ambient quasi-metric setting, neither of these properties is attainable, in general.

(B) Let X be a nonempty set, $b : X \times X \to R_+$ be a quasi-metric over it, and $(\leq)$ be a *quasi-order* over X; then, $(X, b, \leq)$ will be called a *quasi-ordered quasi-metric space*. Let also $(<)$ be the *strict part* of $(\leq)$:

$x < y$ iff $x \leq y$ and $x \neq y$ (clearly, $(<)$ is *irreflexive*).

Call the subset Y of X, $(\leq)$-*asingleton* if $[y_1, y_2 \in Y, y_1 \leq y_2$ imply $y_1 = y_2]$; and $(\leq)$-*singleton* if, in addition, Y is nonempty.

Having these precise, assume that

(b-sepa) b is separated: $x_n \xrightarrow{b} u$ and $x_n \xrightarrow{b} v$ imply $u = v$
(b-1v-cont) b is first variable continuous:
for each $y \in X$, one has $(x_n \xrightarrow{b} x$ implies $b(x_n, y) \to b(x,y))$.

Further, take some selfmap $T \in \mathcal{F}(X)$; supposed to satisfy

(s-prog) T is *semi-progressive* $(X(T, \leq) := \{x \in X; x \leq Tx\} \neq \emptyset)$
(incr) T is *increasing* ($x \leq y$ implies $Tx \leq Ty$).

It is our objective to determine whether $\text{Fix}(T) \neq \emptyset$; and then, to establish whether $\text{Fix}(T)$ is $(\leq)$-singleton. The basic directions under which the investigations be conducted are shown in list below, comparable with the one in Turinici [246]:

opic-1) We say that T is *fix-(≤)-asingleton* (resp., *fix-(≤)-singleton*) if $\mathrm{Fix}(T)$ is (≤)-asingleton (resp., (≤)-singleton)

opic-2) We say that T is a *Picard operator* (modulo $(b, \le)$) when, for each $x \in X(T, \le)$, the iterative sequence $(T^n x; n \ge 0)$ is b-Cauchy

opic-3) We say that T is a *strong Picard operator* (modulo $(b, \le)$) when, for each $x \in X(T, \le)$, $(T^n x)$ is b-convergent with $\lim_n(T^n x) \in \mathrm{Fix}(T)$

opic-4) We say that T is a *Bellman Picard operator* (modulo $(b, \le)$) when, for each $x \in X(T, \le)$, $(T^n x)$ is b-convergent, $\lim_n(T^n x) \in \mathrm{Fix}(T)$, and (in addition) $T^n x \le \lim_n(T^n x)$, for all n.

In particular, when $(\le) = X \times X$ (the *trivial* quasi-order on X) the list of such notions is identical with the standard one in Rus [203, Ch. 2, Sec. 2.2]; because, in this case, $X(T, \le) = X$.

The sufficient (regularity) conditions to be used here are being founded on *ascending orbital full* concepts (in short: (a-o-f) concepts). Namely, let us say that the sequence $(z_n; n \ge 0)$ in X is

ascending, if $(\forall i, \forall j)$: $i \le j$ implies $z_i \le z_j$
orbital, when $(z_n = T^n x; n \ge 0)$, for some $x \in X$
full, provided $n \mapsto z_n$ is injective ($i \ne j$ implies $z_i \ne z_j$);

the intersection of these is just the precise concept.

oreg-1) Call X, *(a-o-f,b)-complete* when (for each (a-o-f)-sequence) b-Cauchy $\Longrightarrow$ b-convergent

oreg-2) We say that T is *(a-o-f,b)-continuous*, if [(z_n) is (a-o-f)-sequence and $z_n \xrightarrow{b} z$] imply $Tz_n \xrightarrow{b} Tz$

oreg-3) Call $(\le)$, *(a-o-f,b)-selfclosed* when [(z_n) is an (a-o-f)-sequence and $z_n \xrightarrow{b} z$] imply [$z_n \le z$, $\forall n$].

(B) As a completion of these, we must discuss the contractive type conditions to be used. Define the basic family of maps (for $x, y \in X$)

$$\begin{aligned}
&A_1(x,y) = b(x,y),\ A_2(x,y) = \max\{b(x,Tx), b(y,Ty)\},\\
&A_3(x,y) = (1/2)[b(x,Tx) + b(y,Ty)],\\
&A_4(x,y) = (1/2)[b(x,Ty) + b(Tx,y)],\\
&A_5(x,y) = \max\{(1/2)b(x,Ty), b(Tx,y)\},\\
&L(x,y) = \min\{b(x,Tx), b(y,Ty), b(x,y), b(Tx,Ty)\},\\
&M(x,y) := \mathrm{sdiam}(x,Tx,y,Ty).
\end{aligned}$$

Here, for each 4-tuple $(x_1, x_2, x_3, x_4) \in X^4$, we let

$$\text{sdiam}(x_1, x_2, x_3, x_4) = \max\{b(x_i, x_j); i, j \in \{1, 2, 3, 4\}, i < j\}$$

stand for its *semi-diameter.*

Having these precise, let $P \in \mathcal{F}(X \times X, R_+)$ be a map. In the following, we give the list of *orbital normality conditions* (to be optionally fulfilled by $P(.,.)$). There are three groups of such conditions.

I) The first one involves *positivity* and *telescopic boundedness* conditions:

(L-pos) P is *L-positive*: $x \le y$, $L(x, y) > 0$ imply $P(x, y) > 0$
(t-bd) P is *telescopic bounded*: $x \le Tx \implies P(x, Tx) \le A_2(x, Tx)$.

Some concrete examples will be given a bit further.

II) The second group concerns the *semi-diametral* concepts below

(a-s-diam) P is *almost semi-diametral*:
for each (ε, δ) with $\varepsilon > \delta > 0$, there exists $\gamma \in]0, \delta[$ such that
$(x \le y,\ b(x, y) \ge \varepsilon,\ A_2(x, y) < \gamma,\ M(x, y) < \varepsilon + \gamma) \implies P(x, y) < \varepsilon + \delta$
(s-diam) P is *semi-diametral*: $P(x, y) \le M(x, y)$, $\forall x, y \in X$, $x \le y$.

Concerning the introduced concepts, the following criterion is useful.

Proposition 4.1. *The following inclusions are valid*

(51-1) If the mapping $P : X \times X \to R_+$ *is semi-diametral, then it is almost semi-diametral*
(51-2) If the maps $P_1, P_2 : X \times X \to R_+$ *are almost semi-diametral, then* $P_3 := \max\{P_1, P_2\}$ *is almost semi-diametral.*

Proof. i): Let the couple (ε, δ) with $\varepsilon > \delta > 0$ be given; and fix some $\gamma \in]0, \delta[$. Given $x, y \in X$ with $x \le y$, $b(x, y) \ge \varepsilon$, $A_2(x, y) < \gamma$, $M(x, y) < \varepsilon + \gamma$, we have

$$P(x, y) \le M(x, y) < \varepsilon + \gamma < \varepsilon + \delta \text{ (as } P \text{ is semi-diametral);}$$

and the claim follows.

ii): Given (ε, δ) with $\varepsilon > \delta > 0$, let $\gamma_1 \in]0, \delta[$ and $\gamma_2 \in]0, \delta[$ be assured by the almost semi-diametral property of P_1 and P_2, respectively. Then, let us put $\gamma_3 = \min\{\gamma_1, \gamma_2\}$; we claim that the desired property of $P_3 = \max\{P_1, P_2\}$ is fulfilled with respect to the obtained number. In fact, given $x, y \in X$ with $x \le y$, $b(x, y) \ge \varepsilon$, $A_2(x, y) < \gamma_3$, $M(x, y) < \varepsilon + \gamma_3$, we have

$$x \le y,\ b(x, y) \ge \varepsilon,\ A_2(x, y) < \gamma_1,\ M(x, y) < \varepsilon + \gamma_1; \text{ so, } P_1(x, y) < \varepsilon + \delta;$$
$$x \le y,\ b(x, y) \ge \varepsilon,\ A_2(x, y) < \gamma_2,\ M(x, y) < \varepsilon + \gamma_2; \text{ so, } P_2(x, y) < \varepsilon + \delta.$$

Putting these together yields $P_3(x,y) < \varepsilon+\delta$; and the assertion follows. □

III) The third group involves certain *orbital* properties written as

(sg-asy) P is *orbitally singular asymptotic*: $\forall$ b-Cauchy (a-o-f)-sequence $(x_n = T^n x_0; n \geq 0)$ in $X(T,\leq)$, and $\forall z \in X$ with $x_n \xrightarrow{b} z$, $(x_n < z, \forall n)$, and $b(z,Tz) > 0$, we have $\liminf_n P(x_n,z) < b(z,Tz)$
(r-asy) P is *orbitally regular asymptotic*: $\forall$ b-Cauchy (a-o-f)-sequence $(x_n = T^n x_0; n \geq 0)$ in $X(T,\leq)$, and $\forall z \in X$ with $x_n \xrightarrow{b} z$, $(x_n < z, \forall n)$, and $b(z,Tz) > 0$, we have $P(x_n,z) \to b(z,Tz)$
(s-r-asy) P is *orbitally strongly regular asymptotic*: $\forall$ b-Cauchy (a-o-f)-sequence $(x_n = T^n x_0; n \geq 0)$ in $X(T,\leq)$, and $\forall z \in X$ with $x_n \xrightarrow{b} z$, $(x_n < z, \forall n)$, and $b(z,Tz) > 0$, we have $P(x_n,z) \to\to b(z,Tz)$.

Here, given the sequence $(r_n; n \geq 0)$ in R and the point $r \in R$, we denoted

$r_n \to\to r$, if $(r_n \to r)$ and there exists a subsequence
$(s_n = r_{i(n)}; n \geq 0)$ of $(r_n; n \geq 0)$ such that $[s_n = r, \forall n \geq 0]$.

(C) As a basic completion of these, we must introduce the contractive conditions to be used. Given the mapping $P : X \times X \to R_+$, let us say that T is *Meir-Keeler* $(b,\leq;P)$*-contractive*, if

(mk-1) $x \leq y$, $P(x,y) > 0$ imply $b(Tx,Ty) < P(x,y)$
(T is *strictly nonexpansive* (modulo $(b,\leq;P)$))
(mk-2) $\forall \varepsilon > 0, \exists \delta > 0$: $[x \leq y, \varepsilon < P(x,y) < \varepsilon+\delta] \Longrightarrow b(Tx,Ty) \leq \varepsilon$
(T has the *Meir-Keeler property* (modulo $(b,\leq;P)$)).

Note that, by the former of these, the Meir-Keeler property writes

(mk-2a) $\forall \varepsilon > 0, \exists \delta > 0$: $[x \leq y, 0 < P(x,y) < \varepsilon+\delta] \Longrightarrow b(Tx,Ty) \leq \varepsilon$.

In particular, if (b = metric, $P = A_1$), this convention is comparable with the one in Meir and Keeler [168]; cf. Matkowski [166] and Cirić [59].

A geometric version of the above concept may be given as below. Remember that the relation $\Omega \subseteq R_+^0 \times R_+^0$ is called *upper diagonal*, provided

(u-diag) $(t,s) \in \Omega$ implies $t < s$;

the class of all these will be denoted as $\mathrm{udiag}(R_+^0)$. Further, let us introduce the conditions (over the class $\mathrm{udiag}(R_+^0)$)

(g-mk) Ω has the *geometric Meir-Keeler property*:
$\forall \varepsilon > 0, \exists \delta > 0$: $t\Omega s$, $\varepsilon < s < \varepsilon+\delta \Longrightarrow t \leq \varepsilon$

(g-bila-s) Ω is *geometric bilateral separable*:
$\forall \beta > 0$, $\exists \gamma \in]0, \beta[$, $\forall (t, s)$: $t, s \in]\beta - \gamma, \beta + \gamma[\implies (t, s) \notin \Omega$
(g-left-s) Ω is *geometric left separable*:
$\forall \beta > 0$, $\exists \gamma \in]0, \beta[$, $(\forall t)$: $t \in]\beta - \gamma, \beta[\implies (t, \beta) \notin \Omega$.

Note that, any such concept is equivalent with its asymptotic counterpart.

Now, given the relation $\Omega \subseteq R^0_+ \times R^0_+$, let us say that the selfmap T of X is $(b, \leq; P; \Omega)$-*contractive*, provided

(Om-contr) $(b(Tx, Ty), P(x, y)) \in \Omega$,
$\forall x, y \in X$, $x \leq y$, $b(Tx, Ty) > 0$, $P(x, y) > 0$.

Proposition 4.2. *Suppose that the selfmap T is $(b, \leq; P; \Omega)$-contractive where $\Omega \in \exp(R^0_+ \times R^0_+)$ is upper diagonal and geometric Meir-Keeler. Then, T is Meir-Keeler $(b, \leq; P)$-contractive in the system (ZF-AC+DC).*

Proof. i): Let $x, y \in X$ be such that $x \leq y$ (hence, $Tx \leq Ty$) and $P(x, y) > 0$. If $Tx = Ty$, then $b(Tx, Ty) = 0 < P(x, y)$. Suppose now that $Tx < Ty$; hence, $b(Tx, Ty) > 0$. As a consequence of this,

$(t, s) \in \Omega$; where $t := b(Tx, Ty)$, $s := P(x, y)$.

This, by the upper diagonal property of Ω, gives $t < s$; that is: $b(Tx, Ty) < P(x, y)$. Summing up, T is strictly nonexpansive (modulo $(b, \leq; T)$).

ii): Let $\varepsilon > 0$ be arbitrary fixed; and $\delta > 0$ be the number assured by the geometric Meir-Keeler property for Ω. Further, let $x, y \in X$ be such that $x \leq y$ (hence, $Tx \leq Ty$), and $\varepsilon < s := P(x, y) < \varepsilon + \delta$. If $Tx = Ty$, then $b(Tx, Ty) = 0 < \varepsilon$. Suppose now that $Tx < Ty$; whence, $t := b(Tx, Ty) > 0$. By definition, we must have $(t, s) \in \Omega$; and this, along with $\varepsilon < s < \varepsilon + \delta$ (and the geometric Meir-Keeler property for Ω), gives $t \leq \varepsilon$; that is: $b(Tx, Ty) \leq \varepsilon$. Putting these alternatives together, it follows that T has the Meir-Keeler property (modulo $(b, \leq; T)$). □

In the following, a converse result is formulated. Given the selfmap T (endowed with the specified general properties) and the map $P : X \times X \to R_+$, let $\Omega = \Omega[b, \leq; P; T]$ stand for the associated relation over R^0_+:

$\Omega = \{(b(Tx, Ty), P(x, y)); x \leq y, b(Tx, Ty > 0, P(x, y) > 0\}$; so that
$(\forall (t, s) \in R^0_+ \times R^0_+)$: $(t, s) \in \Omega$ iff $t = b(Tx, Ty)$, $s = P(x, y)$,
for some $x, y \in X$ with $x \leq y$, $b(Tx, Ty) > 0$, $P(x, y) > 0$.

Proposition 4.3. *The following hold, over the system (ZF-AC+DC)*

(53-1) *If T is Meir-Keeler $(b, \leq; P)$-contractive, then the attached relation $\Omega = \Omega[b, \leq; P; T]$ over R^0_+ is upper diagonal and geometric Meir-Keeler*

(53-2) *T is Meir-Keeler $(b, \leq; P)$-contractive iff the attached relation $\Omega = \Omega[b, \leq; P; T]$ over R^0_+ is upper diagonal and geometric Meir-Keeler.*

Proof. i): Suppose that T is Meir-Keeler $(b, \leq; P)$-contractive. There are two steps to be passed.

i-1): Let $(t, s) \in R^0_+ \times R^0_+$ be such that $(t, s) \in \Omega$; hence (by definition)

$$t = b(Tx, Ty),\ s = P(x, y), \text{ where } x \leq y,\ b(Tx, Ty) > 0,\ P(x, y) > 0.$$

From the strict nonexpansive property of T, we must have $b(Tx, Ty) < P(x, y)$; or, equivalently, $t < s$; which shows that Ω is upper diagonal.

i-2): Let $\varepsilon > 0$ be arbitrary fixed; and $\delta > 0$ be given by the Meir-Keeler property for T. Further, let $(t, s) \in R^0_+ \times R^0_+$ be such that

$$(t, s) \in \Omega \text{ and } \varepsilon < s < \varepsilon + \delta.$$

From the former of these, we have

$$t = b(Tx, Ty),\ s = P(x, y), \text{ where } x \leq y,\ b(Tx, Ty) > 0,\ P(x, y) > 0;$$

so, combining with the latter

$$x \leq y,\ b(Tx, Ty) > 0, \text{ and } \varepsilon < P(x, y) < \varepsilon + \delta.$$

By the underlying Meir-Keeler-property for T, we get

$$b(Tx, Ty) \leq \varepsilon; \text{ that is (under our notation): } t \leq \varepsilon;$$

so that, Ω has the geometric Meir-Keeler property.

ii): Suppose that the associated relation $\Omega = \Omega[b, \leq; P; T]$ over R^0_+ is upper diagonal and has the geometric Meir-Keeler property. By the very definition of this object, T is $(b, \leq; P; \Omega)$-contractive. Combining with the preceding result, one derives that T is Meir-Keeler $(b, \leq; P)$-contractive; and the conclusion follows. □

As a consequence of this, it follows that the Meir-Keeler $(b, \leq; P)$-contractive property of T is finally reducible to the upper diagonal and geometric Meir-Keeler properties for the associated relation $\Omega[b, \leq; P; T]$. Concerning this aspect, remember that various examples of such objects were treated in a previous place. Precisely, two basic classes of such contractions are entering here.

(I) Given $\chi \in \mathcal{F}(R^0_+, R)$, call T, *Matkowski $(b, \leq; P; \chi)$-contractive* if

(M-contr) $b(Tx,Ty) \le \chi(P(x,y))$,
$\forall x,y \in X$, $x \le y$, $b(Tx,Ty) > 0$, $P(x,y) > 0$.

(This convention is taken from the developments in Matkowski [164].) The introduced functions may be taken as below. Let $\mathcal{F}(re)(R^0_+,R)$ stand for the class of all $\chi \in \mathcal{F}(R^0_+,R)$ with the *regressive* property: ($\chi(t) < t$, $\forall t \in R^0_+$). Given $\chi \in \mathcal{F}(re)(R^0_+,R)$, define the (geometric) concept

(MK-a) χ is *Meir-Keeler admissible*:
$\forall \varepsilon > 0$, $\exists \delta > 0$, such that ($\varepsilon < s < \varepsilon + \delta$) implies $\chi(s) \le \varepsilon$.

Proposition 4.4. *Assume that T is Matkowski $(b,\le;P;\chi)$-contractive, where $\chi \in \mathcal{F}(re)(R^0_+,R)$ is Meir-Keeler admissible. Then, necessarily, T is Meir-Keeler $(b,\le;P)$-contractive.*

Proof. By a previous result, the relation $\Omega := \Omega[\chi]$ introduced as [$t\Omega s$ iff $t \le \chi(s)$] is upper diagonal and geometric Meir-Keeler. This, along with

T is Matkowski $(b,\le;P;\chi)$-contractive iff T is $(b,\le;P;\Omega)$-contractive

gives the desired fact. □

(II) Given the couple (ψ,φ) of functions over $\mathcal{F}(R^0_+,R)$, let us say that T is *Rhoades $(b,\le;P;\psi,\varphi)$-contractive*, provided

(R-contr) $\psi(b(Tx,Ty)) \le \psi(P(x,y)) - \varphi(P(x,y))$,
$\forall x,y \in X$, $x \le y$, $b(Tx,Ty) > 0$, $P(x,y) > 0$.

(This convention is taken from the developments in Rhoades [197].) The first normality condition upon (ψ,φ) writes

(nc-1) (ψ,φ) has the *normal* property:
ψ is increasing and φ is *strictly positive* [$\varphi(t) > 0$, $\forall t > 0$].

Then, the second condition (involving our normal couple) writes

(nc-2) φ is asymptotic positive:
for each strictly descending sequence $(t_n; n \ge 0)$ in R^0_+
and each $\varepsilon > 0$ with $t_n \to \varepsilon+$, we have $\limsup_n(\varphi(t_n)) > 0$.

Proposition 4.5. *Suppose that T is Rhoades $(b,\le;P;\psi,\varphi)$-contractive, for a normal couple (ψ,φ) of functions in $\mathcal{F}(R^0_+,R)$, with φ=asymptotic positive. Then, T is Meir-Keeler $(b,\le;P)$-contractive.*

Proof. By a previous fact, the relation $\Omega := \Omega[\psi, \varphi]$ given as [$t\Omega s$ iff $\psi(t) \leq \psi(s) - \varphi(s)$] is upper diagonal and geometric Meir-Keeler. This, and

$$T \text{ is Rhoades } (b, \leq; P; \psi, \varphi)\text{-contractive iff } T \text{ is } (b, \leq; P; \Omega)\text{-contractive}$$

give the needed conclusion. □

Some concrete examples of such functions were given in a previous place; we do not give details. Further aspects may be found in Lim [162].

(D) The introduced concepts may be used to get some practical fix-($\leq$)-asingleton criteria concerning our selfmap T. Two situations occur.

Case 1. A first answer to the posed problem is contained in

Proposition 4.6. *Assume that the selfmap T is Meir-Keeler $(b, \leq; P)$-contractive, where the mapping $P : X \times X \to R_+$ fulfills*

(56-i) $P(.,.)$ is fix positive: $(x, y \in \mathrm{Fix}(T),\ x < y) \Longrightarrow P(x, y) > 0$
(56-ii) P is fix semi-diametral:
$(x, y \in \mathrm{Fix}(T),\ x < y) \Longrightarrow P(x, y) \leq M(x, y)(= b(x, y))$.

Then, T is fix-($\leq$)-asingleton.

Proof. Let $z_1, z_2 \in \mathrm{Fix}(T)$ be such that $z_1 \leq z_2$; and assume (by contradiction) that $z_1 \neq z_2$; hence $z_1 < z_2$. As P is fix positive, we must have $P(z_1, z_2) > 0$. This, by the Meir-Keeler contractive condition, yields $b(z_1, z_2) = b(Tz_1, Tz_2) < P(z_1, z_2)$. On the other hand, as P is fix semi-diametral, we must have $P(z_1, z_2) \leq M(z_1, z_2) = b(z_1, z_2)$. This, along with the preceding relation, yields a contradiction. Hence, $z_1 = z_2$; and the assertion is proved. □

In particular, $P = A_4$ is fix positive and semi-diametral, because

$$(x, y \in \mathrm{Fix}(T),\ x < y) \Longrightarrow A_4(x, y) = b(x, y) \in]0, b(x, y)];$$

so, the statement above is applicable to this map.

Case 2. It remains now to discuss the (opposite) alternative

(non-fix-pos) P is *non fix positive*:
there exist $(x, y \in \mathrm{Fix}(T),\ x < y)$, with $P(x, y) = 0$.

Two possible cases must be discussed.

Sub-case 2-a. Suppose that the global version of this condition holds

(fix-sing) P is *fix singular*: $(x, y \in \mathrm{Fix}(T),\ x < y) \Longrightarrow P(x, y) = 0$.

Note that – unlike the preceding case – the Meir-Keeler contractive condition cannot help us in deducing the fix-($\leq$)-asingleton property for T; because it is applicable only to points $x, y \in X$ with $(x \leq y,\ P(x, y) > 0)$. So, the only chance to get the fix-($\leq$)-asingleton property for T is that of *ab initio* imposing it – in a formal way – to the underlying selfmap.

Sub-case 2-b. Suppose that the strong global version of the preceding singularity condition holds

(s-fix-sing) P is *strongly fix singular*:
$(x, y \in \mathrm{Fix}(T),\ x \leq y) \Longrightarrow P(x, y) = 0.$

[For example, any $P \in \{A_2, A_3\}$ has such a property]. As before, the Meir-Keeler contractive condition cannot help us in deducing the fix-($\leq$)-asingleton property for T; because it is applicable only to points $x, y \in X$ with $(x \leq y,\ P(x, y) > 0)$. But – in compensation to this – certain extra conditions about our data may be used so as to solve our problem. An appropriate condition of this type writes

(s-comp) T is $(b, \leq; P)$-*strongly-compatible*:
$(x, y \in X,\ x \leq y,\ P(x, y) = 0) \Longrightarrow (Tx = Ty).$

We are now in position to formulate a second answer to the posed problem.

Proposition 4.7. *Suppose that P is strongly fix singular and T is $(b, \leq; P)$-strongly-compatible. Then, T is fix asingleton.*

Proof. Let $z_1, z_2 \in \mathrm{Fix}(T)$ be such that $z_1 \leq z_2$. As P is strongly fix singular, $P(z_1, z_2) = 0$; so, by the strong compatible condition, $Tz_1 = Tz_2$; or, equivalently, $z_1 = z_2$; This ends the argument. □

Remark 4.2.

Concerning these imposed facts, the following question is naturally arising here. Let us consider the condition

(comp) T is $(b, \leq; P)$-*compatible*:
$(x, y \in X,\ x < y,\ P(x, y) = 0) \Longrightarrow (Tx = Ty).$

We may ask whether this (weaker) compatible requirement upon T may produce (in conjunction with the strongly fix singular condition upon P) the conclusion in the preceding statement. Unfortunately, this is not possible. In fact, the $(b, \leq; P)$-compatible condition upon T is without object along the points $x, y \in \mathrm{Fix}(T)$ with $x < y$; for, in such a case, the strongly fix singular property of P yields $P(x, y) = 0$; and this, along with our compatible condition, yields $x = y$; impossible, via $x < y$. Hence, as already precise, the only chance to get the fix-($\leq$)-asingleton property for T is that of *ab initio* imposing it – in a formal way – to the selfmap T.

Finally, we show that the strong compatible condition (upon T) is sometimes obtainable from a strong contractive one. A concrete example in this direction may be constructed as follows. Denote

$\mathcal{F}_0(R_+)$ = the class of all $\chi \in \mathcal{F}(R_+)$ with $\chi(0) = 0$.

Given $\chi \in \mathcal{F}_0(R_+)$, call the selfmap T, $(b, \leq; P; \chi)$-*strongly-contractive* if

(s-con) $b(Tx, Ty) \leq \chi(P(x, y)), \forall x, y \in X, x \leq y$.

Proposition 4.8. *Assume that* T *is* $(b, \leq; P; \chi)$*-strongly-contractive, where* $\chi \in \mathcal{F}_0(R_+)$*. Then,* T *is* $(b, \leq; P)$*-strongly-compatible.*

Proof. Let $x, y \in X$ be such that $x \leq y$ and $P(x, y) = 0$. From the strong contractive condition, we have

$(0 \leq) b(Tx, Ty) \leq \chi(0) = 0$; whence, $Tx = Ty$;

and the conclusion follows. □

In particular, the linear case

$\chi(t) = \beta t$, $t \in R_+$, for some $\beta \in [0, 1[$

enters in this scheme; and this tells us that any selfmap T of X fulfilling a relational type Kannan contractive condition as (for some $P \in \{A_2, A_3\}$)

$b(Tx, Ty) \leq \beta P(x, y)$, $x, y \in X$, $x \leq y$,

must be fix-($\leq$)-asingleton. Further aspects will be delineated elsewhere.

4.1.6 *Main result*

Let $(X, b, \leq)$ be a quasi-ordered quasi-metric space; where

(s-1v-cont) b is separated and first variable continuous.

Further, take some selfmap $T \in \mathcal{F}(X)$; supposed to satisfy

(s-prog-incr) T is semi-progressive and increasing.

As precise, we have to determine sufficient conditions for $\text{Fix}(T) =$ nonempty. The directions under which this problem is to be solved were already listed. The ingredients for getting such properties involve *ascending orbital full concepts* (in short: *(a-o-f)-concepts*) and *contractive properties.*

Having these precise, we are now in position to state the main result in this exposition (referred to as: *Meir-Keeler geometric theorem on quasi-ordered quasi-metric spaces*; in short (MK-g-oqms)).

Theorem 4.1. *Let the selfmap T be (semi-progressive, increasing and) $(d, \leq; P; \Omega)$-contractive, where $P : X \times X \to R_+$ and $\Omega \subseteq R^0_+ \times R^0_+$ fulfill*

(61-i) P is L-positive, telescopic bounded, and almost semi-diametral
(61-ii) Ω is upper diagonal and geometric Meir-Keeler.

In addition, let X be (a-o-f,b)-complete. Then,

61-a) *T is a strong Picard operator (modulo $(b, \leq)$) when, in addition, T is (a-o-f,b)-continuous*

61-b) *T is a Bellman Picard operator (modulo $(b, \leq)$) when, in addition, $(\leq)$ is (a-o-f,b)-selfclosed and one of the extra conditions below holds:*

(61-b1) P is orbitally singular $(b, \leq; T)$-asymptotic
(61-b2) P is orbitally regular $(b, \leq; T)$-asymptotic, and Ω is (in addition) geometric/asymptotic bilateral separable
(61-b3) P is orbitally strongly regular $(b, \leq; T)$-asymptotic, and Ω is (in addition) geometric/asymptotic left separable

61-c) *T is fix-$(\leq)$-asingleton (hence, fix-$(\leq)$-singleton) whenever (in addition to thc above setting) P is fix positive and fix semi-diametral.*

Proof. There are several steps to be passed.

Part 1. Let us firstly check the fix-$(\leq)$-asingleton property. Take a couple of points $z_1, z_2 \in \text{Fix}(T)$ with $z_1 \leq z_2$; and assume by contradiction that $z_1 \neq z_2$; whence $z_1 < z_2$. As P is fix positive, we have $P(z_1, z_2) > 0$. In addition, by the fixed point property, $b(Tz_1, Tz_2) = b(z_1, z_2) > 0$. From the contractive hypothesis (and $\Omega =$ upper diagonal),

$$(b(Tz_1, Tz_2), P(z_1, z_2)) \in \Omega; \text{ so, } b(z_1, z_2) = b(Tz_1, Tz_2) < P(z_1, z_2).$$

On the other hand, as P is fix semi-diametral, $P(z_1, z_2) \leq M(z_1, z_2) = b(z_1, z_2)$. Since the obtained facts are contradictory, the working assumption cannot be accepted; and our assertion follows.

Part 2. We claim that, under the conditions in the statement

(1-2-iter) $(b(Tx, T^2x), P(x, Tx)) \in \Omega$, and
$b(Tx, T^2x) < P(x, Tx) \le b(x, Tx)$, whenever $x < Tx$, $Tx < T^2x$.

In fact, let $x \in X$ be as in the premise above; whence,

$$L(x, Tx) = \min\{b(x, Tx), b(Tx, T^2x)\} > 0; \text{ so that, } P(x, Tx) > 0$$

(as P is L-positive). Putting these together, gives $(b(Tx, T^2x), P(x, Tx)) \in \Omega$; and this, combined with the upper diagonal property of Ω, implies $b(Tx, T^2x) < P(x, Tx)$. On the other hand, as P is telescopic bounded,

$$(0 <)P(x, Tx) \le A_2(x, Tx) = \max\{b(x, Tx), b(Tx, T^2x)\}.$$

Combining with the preceding relation, gives

$b(Tx, T^2x) < \max\{b(x, Tx), b(Tx, T^2x)\}$;
so, necessarily, $b(Tx, T^2x) < b(x, Tx)$, $A_2(x, Tx) = b(x, Tx)$;

wherefrom the claim follows.

It remains to establish that T is strong/Bellman Picard (modulo $(b, \le)$). Take some $x_0 \in X(T, \le)$; and put $(x_n = T^n x_0; n \ge 0)$; this is an ascending orbital sequence. If $x_n = x_{n+1}$ for some $n \ge 0$, we are done; so, without loss, one may assume that the *non-telescopic* property holds

(n-tele) $(\forall n)$: $x_n \ne x_{n+1}$; hence, $x_n < x_{n+1}$, $\rho_n := b(x_n, x_{n+1}) > 0$.

Part 3. By the preceding relation (1-2-iter), we have

(iter) $(\rho_{n+1}, P(x_n, x_{n+1})) \in \Omega$ and $\rho_{n+1} < P(x_n, x_{n+1}) \le \rho_n$, $\forall n$;
so, the sequence $(\rho_n; n \ge 0)$ is strictly descending in R^0_+.

Two basic consequences of this fact are to be noted.

Conseq 1. Firstly, we claim that

(x_n) is strictly ascending: $i < j$ implies $x_i < x_j$;
hence, in particular, (x_n) is full ($n \mapsto x_n$ is injective).

In fact, suppose by contradiction that there exist $i, j \in N$ such that $i < j$, $x_i = x_j$. This, by definition, yields $x_{i+1} = x_{j+1}$; whence $\rho_i = \rho_j$; in contradiction with (ρ_n) = strictly descending; hence the assertion.

Conseq 2. Secondly, $\rho := \lim_n \rho_n$ exists in R_+; we claim that $\rho = 0$. Assume by contradiction that $\rho > 0$; and let $\sigma > 0$ be given by the Meir-Keeler property of Ω. By definition, there exists a rank $n(\sigma) \ge 0$, such that $[n \ge n(\sigma)$ implies $\rho < \rho_n < \rho + \sigma]$. Taking (iter) into account, yields

$$(\forall n \ge n(\sigma)): x_n < x_{n+1} \text{ and } \rho < \rho_{n+1} < P(x_n, x_{n+1}) \le \rho_n < \rho + \sigma;$$

and this, combined with the geometric Meir-Keeler property of Ω, gives

$$(\forall n \geq n(\sigma)) \colon \rho < \rho_{n+1} = b(Tx_n, Tx_{n+1}) \leq \rho;$$

a contradiction. Hence, $\rho = 0$; so that,

$$\rho_n := b(x_n, x_{n+1}) \to 0, \text{ as } n \to \infty; \text{ meaning: } (x_n) \text{ is } b\text{-asymptotic.}$$

Part 4. We now establish that $(x_n; n \geq 0)$ is a b-Cauchy sequence. Let $\varepsilon > 0$ be arbitrary fixed; and $\delta > 0$ be the number associated by the Meir-Keeler property of Ω; without loss, one may assume that $\delta < \varepsilon$. As P is almost semi-diametral, there exists $\gamma \in]0, \delta[$ with

$$(x \leq y,\ b(x,y) \geq \varepsilon,\ A_2(x,y) < \gamma,\ M(x,y) < \varepsilon + \gamma) \Longrightarrow P(x,y) < \varepsilon + \delta.$$

From the d-asymptotic property of (x_n), there exists $n(\gamma) \geq 0$ with

(b-asy) $(\forall n \geq n(\gamma))$: $b(x_n, x_{n+1}) < \gamma/4$;
so (by the triangular inequality): $b(x_n, x_{n+i}) < \gamma/2$, $\forall i \in \{1, 2\}$.

We now claim, by induction, that the evaluation

(b-C;p) $b(x_n, x_{n+p}) < \varepsilon + \gamma/2$, $\forall n \geq n(\gamma)$

holds for all $p \geq 1$; wherefrom, the desired property of (x_n) follows. The case $p \in \{1, 2\}$ is clear, via (b-asy). Assume that (b-C;p) holds, for $p \in \{1, ..., q\}$, where $q \geq 2$; we show that (b-C;q+1) holds too. So, let $n \geq n(\gamma)$ be arbitrary fixed. The expected relation is clear whenever $b(x_n, x_{n+q}) < \varepsilon$; because, in view of triangular inequality and (b-asy),

$$b(x_n, x_{n+q+1}) \leq b(x_n, x_{n+q}) + b(x_{n+q}, x_{n+q+1}) < \varepsilon + \gamma/4 < \varepsilon + \gamma/2;$$

whence, our assertion is true. So, in the following, there is no loss in generality by assuming that $b(x_n, x_{n+q}) \geq \varepsilon$. From the strict ascending property of (x_n),

(q-rela) $x_n < x_{n+1}$, $x_{n+q} < x_{n+q+1}$, $x_n < x_{n+q}$, $x_{n+1} < x_{n+q+1}$;
whence, $L(x_n, x_{n+q}) > 0$ and $P(x_n, x_{n+q}) > 0$ (as P is L-positive);
so, by definition, $(b(x_{n+1}, x_{n+q+1}), P(x_n, x_{n+q})) \in \Omega$.

On the other hand, from the b-asymptotic property, inductive hypothesis, and triangular property,

$$b(x_n, x_{n+1}), b(x_{n+q}, x_{n+q+1}) < \gamma/4 < \gamma,$$
$$b(x_n, x_{n+q}), b(x_{n+1}, x_{n+q}), b(x_{n+1}, x_{n+q+1}) < \varepsilon + \gamma/2,$$

$b(x_n, x_{n+q+1}) \le b(x_n, x_{n+q}) + b(x_{n+q}, x_{n+q+1}) <$
$\varepsilon + \gamma/2 + \gamma/4 = \varepsilon + 3\gamma/4 < \varepsilon + \gamma.$

Putting these together, gives

$x_n < x_{n+q}$, $b(x_n, x_{n+q}) \ge \varepsilon$, $A_2(x_n, x_{n+q}) < \gamma$, $M(x_n, x_{n+q}) < \varepsilon + \gamma$;
so that, necessarily, $(0 <)P(x_n, x_{n+q}) < \varepsilon + \delta$

by (q-rela) and the almost semi-diametral property of P. Combining with the Meir-Keeler property of Ω, yields

$b(x_{n+1}, x_{n+q+1}) \le \varepsilon$; so that, by the triangular inequality,
$b(x_n, x_{n+q+1}) \le b(x_n, x_{n+1}) + b(x_{n+1}, x_{n+q+1}) < \varepsilon + \gamma/4 < \varepsilon + \gamma/2$;

and the assertion follows.

Part 5. As X is (a-o-f,b)-complete, we have that $x_n \xrightarrow{b} z$, for some $z \in X$. There are two cases to discuss.

Case A. Suppose that T is (a-o-f,b)-continuous. Then $y_n := Tx_n \xrightarrow{b} Tz$, as $n \to \infty$. On the other hand, $(y_n = x_{n+1}; n \ge 0)$ is a subsequence of $(x_n; n \ge 0)$; whence $y_n \xrightarrow{b} z$; and this yields (as b is separated), $z = Tz$.

Case B. Suppose that $(\le)$ is (a-o-f,b)-selfclosed. For the moment,

$x_n \le z$ (hence, $Tx_n \le Tz$), for all n (cf. the choice of z).

As (x_n) is full, the subsets $E_1 := \{n \in N; x_n = z\}$ and $E_2 := \{n \in N; Tx_n = Tz\}$ are asingletons; so, there exists an index $h \ge 0$, with

$n \ge h \Longrightarrow x_n < z,\ Tx_n < Tz.$

Without loss of generality, one may assume that $h = 0$; that is

(z-dist) $(\forall n)$: $x_n < z$, $Tx_n < Tz$; hence, $b(x_n, z) > 0$, $b(Tx_n, Tz) > 0$;

for, otherwise, passing to the subsequence

$(u_n = x_{h+n}; n \ge 0)$; that is: $(u_n = T^n u_0; n \ge 0)$, where $u_0 = x_h$

this property holds, as well as the (a-o-f) one. We now intend to show that the alternative $z \ne Tz$ (hence $\beta := b(z, Tz) > 0$) yields a contradiction. The obtained relations give (in addition to the preceding ones)

$(\forall n)$: $x_n < Tx_n$, $b(z, Tz) > 0$, $x_n < z$, $Tx_n < Tz$; wherefrom
$L(x_n, z) > 0$; which, in turn, gives $P(u_n, z) > 0$ (as P is L-positive).

This tells us that for each n, our geometric contractive condition is applicable to (x_n, z); and yields

(contr) $(\forall n)$: $(b(Tx_n, Tz), P(x_n, z)) \in \Omega$; whence,
$b(Tx_n, Tz) < P(x_n, z)$ (as Ω is upper diagonal).

On the other hand, by the d-Cauchy, d-convergence, and the first variable continuous property of our quasi-metric b, one gets (as $n \to \infty$)

$$b(x_n, Tx_n), b(x_n, z), b(Tx_n, z) \to 0,\ b(x_n, Tz), b(Tx_n, Tz) \to \beta.$$

Now, three alternatives are to be discussed.

Alter 1. Suppose that P is orbitally singular asymptotic. Passing to $\liminf$ as $n \to \infty$ in the second half of contractive relation (contr) gives

$$\beta = \liminf_n b(x_{n+1}, Tz) \le \liminf P(x_n, z).$$

This, however, contradicts the orbital singular asymptotic property of P. Hence, necessarily, $\beta = 0$ (meaning: $z = Tz$); and conclusion follows.

Alter 2. Suppose that P is orbitally regular asymptotic, and Ω is geometric/asymptotic bilateral separable. By the previous convergence relations concerning (x_n) we have (via orbitally regular asymptotic property)

$$(b(x_{n+1}, Tz) \to \beta \text{ and) } P(x_n, z) \to \beta, \text{ as } n \to \infty.$$

This, by the first half of our contractive relation (contr) yields a contradiction with the choice of Ω. Hence, necessarily, $\beta = 0$; that is: $z = Tz$.

Alter 3. Suppose that P is orbitally strongly regular asymptotic, and Ω is (in addition) geometric/asymptotic left separable. From the convergence relations concerning (x_n) we have (by the imposed hypotheses) $P(x_n, z) \to\to \beta$, as $n \to \infty$. According to the definition of this relation, $(P(x_n, z) \to \beta$, as $n \to \infty)$ and there must be a subsequence $(v_n := x_{j(n)}; n \ge 0)$ of $(x_n; n \ge 0)$, such that

$$P(v_n, z) = \beta (> 0), \text{ for all } n \ge 0.$$

Denote for simplicity $(\xi_n = b(Tv_n, Tz); n \ge 0)$. From the contractive condition (contr), one gets

$$(\forall n)\text{: } (\xi_n, \beta) \in \Omega; \text{ hence, } \xi_n < \beta \text{ (as } \Omega = \text{upper diagonal).}$$

Moreover, combining with the previous convergence relations,

$$\xi_n \to \beta \text{ (hence, } \xi_n \to \beta-) \text{ as } n \to \infty.$$

By a previous auxiliary fact, there exists a subsequence $(\xi_n^* := \xi_{i(n)}; n \ge 0)$ of $(\xi_n; n \ge 0)$, with

$$(\xi_n^*; n \ge 0) \text{ is strictly ascending and } (\xi_n^* \to \beta-) \text{ as } n \to \infty.$$

This, via $[(\xi_n^*, \beta) \in \Omega$, for all $n]$, contradicts the asymptotic left separable property of Ω. Hence, $\beta = 0$; that is: $z = Tz$; and conclusion follows. □

A useful version of our main result corresponds to $(\leq) = X \times X$, $P = A_1$. Then, its conclusions are comparable with the related ones in Jleli and Samet [128]; see also Alsulami et al. [13]. Precisely, in the quoted results,

I) the (a-o-f,b)-completeness hypothesis upon X is substituted by

(d-com) X is d-complete (each d-Cauchy sequence is d-convergent)

where $d = \max(b, c)$ is the metric attached to the quasi-metric b and its conjugate $(c(x, y) = b(y, x); x, y \in X)$

II) the Picard type conclusions are also expressed in terms of d.

This forces us to ask whether a complete reduction of our main result to its (standard) metrical version is possible. A positive answer to this is available, under the lines in Turinici [257]; we do not give details.

Note that, further extensions of these facts are possible, to the realm of triangular symmetric spaces, taken as in Hicks and Rhoades [112], as well as the realm of partial metric spaces, introduced under the lines in Matthews [167]. On the other hand, multivalued extensions of this result are also possible, under Nadler's techniques [175]. Finally, some dislocated versions of our main result are available, by simply following the developments in Amini-Harandi [16]. We shall discuss all these facts elsewhere.

4.1.7 *Particular cases*

Let $(X, b, \leq)$ be a quasi-ordered quasi-metric space, where

(s-1v-cont) $b(.,.)$ is separated and first variable continuous.

Further, let T be a selfmap of X; supposed to be semi-progressive and increasing. We have to determine whether Fix(T) is nonempty; and, if this holds, to establish whether T is fix-$(\leq)$-singleton. The specific directions under which this problem is to be solved were already listed. Sufficient conditions for getting such properties are being founded on the ascending orbital full concepts (in short: (a-o-f) concepts) we just introduced. Finally, the specific contractive properties to be used have been described; and the main result incorporating all these is the already formulated one. It is our aim in the sequel to expose a certain particular case of it, with a technical relevance. Define a basic family of maps as (for $x, y \in X$)

$$M_1(x,y) = b(Tx,Ty),\ M_2(x,y) = b(x,y),\ M_3(x,y) = b(x,Tx),$$
$$M_4(x,y) = b(y,Ty),\ M_5(x,y) = b(x,Ty),\ M_6(x,y) = b(Tx,y).$$

By taking order-algebraic combinations between these, one gets a lot of functions to be used in our reasonings; the basic ones are (for $x, y \in X$)

$$A_1(x,y) = b(x,y),\ A_2(x,y) = \max\{b(x,Tx), b(y,Ty)\},$$
$$A_3(x,y) = (1/2)[b(x,Tx) + b(y,Ty)],$$
$$A_4(x,y) = (1/2)[b(x,Ty) + b(Tx,y)],$$
$$A_5(x,y) = \max\{(1/2)b(x,Ty), b(Tx,y)\},$$
$$L(x,y) = \min\{b(x,Tx), b(y,Ty), b(x,y), b(Tx,Ty)\},$$
$$M(x,y) := \text{sdiam}(x, Tx, y, Ty).$$

Here, for each 4-tuple $(x_1, x_2, x_3, x_4) \in X^4$, we let

$$\text{sdiam}(x_1,x_2,x_3,x_4) = \max\{b(x_i,x_j); i,j \in \{1,2,3,4\}, i < j\}$$

stand for its *semi-diameter*. Then, again via order-algebraic operations, we may define some other functions of this type; the following ones will be taken as concrete examples in our developments. Let us introduce the diagonal type subset of R_+^2

$$\Delta = \{(\lambda,\mu) \in R_+ \times R_+^0; \lambda \le \mu\}.$$

This set is decomposed in two (mutually disjoint) parts, expressed as

$\Delta_s = \{(\lambda,\mu) \in \Delta; \lambda < \mu\}$ (the *singular* part)
$\Delta_r = \{(\lambda,\mu) \in \Delta; \lambda = \mu\} = \{(\nu,\nu); \nu \in R_+^0\}$ (the *regular* part).

For each $(\lambda,\mu) \in \Delta$, define the map $B := B[\lambda,\mu] : X \times X \to R_+$, as

$$B(x,y) = b(y,Ty)[\lambda + b(x,Tx)]/[\mu + b(x,y)],\ x,y \in X.$$

Further, let us define

A_s = one of the maps $B[\lambda,\mu]$ with $(\lambda,\mu) \in \Delta_s$;
A_r = one of the maps $B[\lambda,\mu]$ with $(\lambda,\mu) \subset \Delta_r$
(or, equivalently: A_r = one of the maps $B[\nu,\nu]$ with $\nu \in R_+^0$);

the reason of such a splitting will become clear a bit further. Finally, given $(\alpha,\beta) \in \Delta$, let us introduce the map $C := C[\alpha,\beta] : X \times X \to R_+$, as

$$C(x,y) = b(Tx,y)[\alpha + b(x,Ty)]/[\beta + b(x,y)],\ x,y \in X;$$

and then, for technical reasons, define

A_u = one of the maps $C[\alpha,\beta]$ with $(\alpha,\beta) \in \Delta$.

Having these precise, fix in the following the couples $(\lambda, \mu) \in \Delta_s$, $(\nu, \nu) \in \Delta_r$, $(\alpha, \beta) \in \Delta$; and (according to the previous conventions), denote

$$Z = \{1, 2, 3, 4, 5, s, r, u\},\ Z_a = \{1, 2, 3, 4, 5\},\ Z_b = \{s, r, u\}.$$

For each subset $\Theta \in \exp(Z)$, let $E(\Theta) \in \mathcal{F}(X \times X, R_+)$ be the mapping

$$E(\Theta)(x, y) = \max\{A_i(x, y); i \in \Theta\},\ x, y \in X.$$

Now, the maps $P : X \times X \to R_+$ to be considered are of the form

$$P = E(\Theta);\ \text{where } \Theta \in \exp(Z).$$

So, we must establish of to what extent is this family compatible with the specific conditions required by our main result. To do this, denote

$$Z_1^* = \{\Theta \in \exp(Z); 1 \in \Theta\},\ Z_2^* = \{\Theta \in \exp(Z); 2 \in \Theta\},\ Z^* = Z_1^* \cup Z_2^*.$$

I) The first property to be checked is

(L-pos) P is *L-positive*: $x, y \in X$, $x \le y$, $L(x, y) > 0$ imply $P(x, y) > 0$.

Proposition 4.9. *All maps $P = E(\Theta)$, where $\Theta \in Z^*$ are L-positive.*

Proof. Let $x, y \in X$ be such that $x \le y$, $L(x, y) > 0$. Then,

$$P \in Z_1^* \text{ implies } P(x, y) \ge A_1(x, y) \ge L(x, y) > 0,$$
$$P \in Z_2^* \text{ implies } P(x, y) \ge A_2(x, y) \ge L(x, y) > 0;$$

and from this, all is clear, by the definition of our class. □

II) The second property to be discussed is

(t-bd) P is *telescopic bounded*: $x \le Tx \Longrightarrow P(x, Tx) \le A_2(x, Tx)$.

Proposition 4.10. *All maps $P = E(\Theta)$, where $\Theta \in Z^*$ are necessarily telescopic bounded.*

Proof. First, we prove that any function $Q \in \{A_i; i \in Z\}$ has such a property. Given the arbitrary point $x \in X$ with $x \le Tx$, we have

$$A_1(x, Tx) = b(x, Tx) \le \max\{b(x, Tx), b(Tx, T^2x)\} = A_2(x, Tx),$$
$$A_3(x, Tx) = (1/2)[b(x, Tx) + b(Tx, T^2x)] \le A_2(x, Tx),$$
$$A_4(x, Tx) = A_5(x, Tx) = (1/2)b(x, T^2x) \le A_3(x, Tx) \le A_2(x, Tx);$$
$$B(x, Tx) = b(Tx, T^2x)[\lambda + b(x, Tx)]/[\mu + b(x, Tx)] \le$$
$$b(Tx, T^2x) \le A_2(x, Tx),\ C(x, Tx) = 0 \le A_2(x, Tx);$$

where, $B = B[\lambda, \mu]$ with $(\lambda, \mu) \in \Delta$, and $C = C[\alpha, \beta]$, for $(\alpha, \beta) \in \Delta$; and this, along with any map A_s or A_r having the form $B = B[\lambda, \mu]$, and any map A_u having the form $C = C[\alpha, \beta]$, ends the argument. Now, the property in question is now transferable to all $P = E(\Theta)$, where $\Theta \in Z^*$; and, from this, all is clear. □

III) The third couple of properties to be checked is

(a-s-diam) P is *almost semi-diametral*:
for each (ε, δ) with $\varepsilon > \delta > 0$, there exists $\gamma \in]0, \delta[$ such that
$x \le y$, $b(x, y) \ge \varepsilon$, $A_2(x, y) < \gamma$, $M(x, y) < \varepsilon + \gamma$ imply $P(x, y) < \varepsilon + \delta$
(s-diam) P is *semi-diametral*:
for each $x, y \in X$ with $x \le y$, we have $P(x, y) \le M(x, y)$.

Proposition 4.11. *All maps $P = E(\Theta)$ where $\Theta \in Z^*$, are endowed with the almost semi-diametral property.*

Proof. The argument consists in two steps.

Step 1. Let us first establish that all maps $Q \in \{A_i; i \in Z\}$ have such a property. There are two cases to be discussed.

Case 1. $Q \in \{A_1, A_2, A_3, A_4, A_5\}$. By definition, we have for $1 \le i \le 5$,

$$A_i(x, y) \le M(x, y), \text{ for all } x, y \in X;$$

whence, all these maps are semi-diametral; and this, along with a previous fact, assures us that the underlying maps are almost semi-diametral.

Case 2. $Q \in \{A_s, A_r, A_u\}$; or, equivalently

$Q \in \{B, C\}$; where $B = B[\lambda, \mu]$, for some $(\lambda, \mu) \in \Delta$ and
$C = C[\alpha, \beta]$, for some $(\alpha, \beta) \in \Delta$.

Let (ε, δ) with $\varepsilon > \delta > 0$ be given; and $\gamma \in]0, \delta[$ be arbitrary for the moment. Let $x, y \in X$ be such that $x \le y$, $b(x, y) \ge \varepsilon$, $A_2(x, y) < \gamma$, $M(x, y) < \varepsilon + \gamma$; we intend to show that $P(x, y) < \varepsilon + \delta$. By definition,

$$\begin{aligned}
&B(x, y) = b(y, Ty)[\lambda + b(x, Tx)]/[\mu + b(x, y)] \le \gamma[\lambda + \gamma]/[\mu + \varepsilon], \\
&C(x, y) = b(Tx, y)[\alpha + b(x, Ty)]/[\beta + b(x, y)] \le \\
&M(x, y)[\alpha + M(x, y)]/[\beta + b(x, y)] \le (\varepsilon + \gamma)[\alpha + \varepsilon + \gamma]/[\beta + \varepsilon].
\end{aligned}$$

Denote, for simplicity,

$$\Phi(\gamma) = \gamma[\lambda + \gamma]/(\mu + \varepsilon), \ \Psi(\gamma) = (\varepsilon + \gamma)[\alpha + \varepsilon + \gamma]/[\beta + \varepsilon], \ 0 < \gamma < \delta.$$

By the above evaluations, we have

$B(x,y) \leq \Phi(\gamma)$, $C(x,y) \leq \Psi(\gamma)$, for all such γ.

On the other hand,

$$\lim_{\gamma\to 0+} \Phi(\gamma) = 0 < \varepsilon + \delta,$$
$$\lim_{\gamma\to 0+} \Psi(\gamma) = \varepsilon[\alpha+\varepsilon]/[\beta+\varepsilon] \leq \varepsilon < \varepsilon + \delta.$$

This tells us that, if $\gamma \in]0,\delta[$ is small enough, we have

$$\Phi(\gamma) < \varepsilon + \delta; \text{ hence, } B(x,y) < \varepsilon + \delta,$$
$$\Psi(\gamma) < \varepsilon + \delta; \text{ hence, } C(x,y) < \varepsilon + \delta.$$

Putting these together, one gets the desired assertion involving the underlying family of maps $Q \in \{A_i; i \in Z\}$.

Step 2. The final conclusion relative to the maps $P = E(\Theta)$, where $\Theta \in Z^*$ is now clear – by a previous auxiliary fact – via components of P being endowed with the almost semi-diametral property. □

IV) The fourth group of properties to be discussed is that concerning the orbital asymptotic ones. In this direction, the synthetic answer below is available.

Proposition 4.12. *Under the above conventions,*

(74-1) Each mapping $P = E(\Theta)$, where $\Theta \in Z^*$ fulfills $\{2,r\} \cap \Theta = \emptyset$, is orbitally singular asymptotic
(74-2) Each mapping $P = E(\Theta)$, where $\Theta \in Z^*$ fulfills $r \in \Theta$ is orbitally regular asymptotic
(74-3) Each mapping $P = E(\Theta)$, where $\Theta \in Z^*$ fulfills $2 \in \Theta$, $r \notin \Theta$ is orbitally strongly regular asymptotic.

Proof. First, we have to discuss the orbital asymptotic properties of the maps $Q \in \{A_1, A_2, A_3, A_4, A_5\}$; and then, the same properties of the maps $Q \in \{A_s, A_r, A_u\}$.

Step 1. $Q \in \{A_1, A_2, A_3, A_4, A_5\}$.

Let the b-Cauchy (a-o-f)-sequence $(x_n = T^n x_0; n \geq 0)$ in $X(T, \leq)$ and the point $z \in X$ be such that

$$x_n \xrightarrow{b} z, \ (x_n < z, \forall n), \text{ and } \beta := b(z, Tz) > 0.$$

From these, one gets (taking the d-Cauchy, d-convergence, and the metrical properties of $b(.,.)$ into account)

$$b(x_n, z),\ b(Tx_n, z) \to 0;\ b(x_n, Tx_n) \to 0;\ b(x_n, Tz), b(Tx_n, Tz) \to \beta.$$

This, by definition, gives (as $n \to \infty$)

$$A_1(x_n,z) \to 0,\ A_2(x_n,z) \to \beta,\ A_3(x_n,z) \to \beta/2,$$
$$A_4(x_n,z) \to \beta/2,\ A_5(x_n,z) \to \beta/2;$$

whence, any map $Q \in \{A_1, A_3, A_4, A_5\}$ is orbitally singular asymptotic. Moreover, the same convergence properties of $(x_n; n \geq 0)$ tell us that, for a certain rank $n(z) \geq 0$, we must have,

$$\text{(for all } n \geq n(z)\text{)}: b(x_n,z), b(Tx_n,z) < \beta/2,\ b(x_n,Tx_n) < \beta/2.$$

This, by the triangular inequality, gives for all $n \geq n(z)$,

$$b(x_n,Tz) \leq b(x_n,z) + \beta < 3\beta/2,\ b(Tx_n,Tz) \leq b(Tx_n,z) + \beta < 3\beta/2.$$

Combining these, yields, for all $n \geq n(z)$

$$A_1(x_n,z) < \beta/2,\ A_2(x_n,z) = \beta,\ A_3(x_n,z) < 3\beta/4,$$
$$A_4(x_n,z) < \beta,\ A_5(x_n,z) < 3\beta/4;$$

which, in particular, tells us that A_2 is orbitally strongly regular asymptotic.

Step 2. $Q \in \{A_s, A_r, A_u\}$; that is:

$$Q = B[\lambda,\mu], \text{ where } (\lambda,\mu) \in \Delta, \text{ or } Q = C[\alpha,\beta], \text{ where } (\alpha,\beta) \in \Delta.$$

Let the b-Cauchy (a-o-f)-sequence $(x_n = T^n x_0; n \geq 0)$ in $X(T, \leq)$, and the point $z \in X$ be such that

$$x_n \xrightarrow{b} z,\ (x_n < z,\ \forall n), \text{ and } \beta := b(z,Tz) > 0.$$

By definition, we have (under the notation $(\rho_n := b(x_n, x_{n+1}); n \geq 0)$)

$$B(x_n,z) = \beta[\lambda + \rho_n]/[\mu + b(x_n,z)],\ \forall n; \text{ so, } \lim_n B(x_n,z) = \beta\lambda/\mu.$$

This, along with the above conventions, means

$$\lim_n A_s(x_n,z) = \beta\lambda/\mu < \beta;\ \text{ so, } A_s \text{ is orbitally singular asymptotic;}$$
$$\lim_n A_r(x_n,z) = \beta\zeta/\zeta = \beta;\ \text{so, } A_r \text{ is orbitally regular asymptotic.}$$

Likewise, again by definition,

$$C(x_n,z) = b(x_{n+1},z)[\alpha + b(x_n,Tz)]/[\beta + b(x_n,z)],\ \forall n; \text{ so that,}$$
$$\lim_n C(x_n,z) = 0 < \beta; \text{ proving that } C \text{ is orbitally singular asymptotic.}$$

Step 3. By the above discussion and the very definition of the maps $P = E(\Theta)$, it is clear that our conclusion follows. □

V) Finally, the fifth group of properties to be considered is

(fix-pos) P is *fix-positive*: $x, y \in \text{Fix}(T)$ and $x < y$ imply $P(x, y) > 0$
(fix-s-diam) P is *fix semi-diametral*, provided:
$x, y \in \text{Fix}(T)$ and $x < y$ imply $P(x, y) \leq M(x, y)(= b(x, y))$.

Proposition 4.13. *The following assertions hold:*

(75-1) Each mapping $P = E(\Theta)$ where $\Theta \in Z^*$ fulfills $\{1, 4, 5, u\} \cap \Theta \neq \emptyset$ is fix positive
(75-2) Each mapping $P = E(\Theta)$ where $\Theta \in Z^*$, is fix semi-diametral.

Proof. Let $x, y \in \text{Fix}(T)$ be such that $x < y$; hence, $b(x, y) > 0$. Then (putting $B = B[\lambda, \mu]$, $C = C[\alpha, \beta]$, where $(\lambda, \mu) \in \Delta$, $(\alpha, \beta) \in \Delta$)

$$A_1(x, y) = A_4(x, y) = A_5(x, y) = b(x, y); \text{ hence,}$$
$$0 < A_1(x, y) = A_4(x, y) = A_5(x, y) = M(x, y);$$
$$A_2(x, y) = A_3(x, y) = B(x, y) = 0 \leq b(x, y);$$
$$0 < C(x, y) = b(x, y)[\alpha + b(x, y)]/[\beta + b(x, y)] \leq b(x, y) = M(x, y);$$

and, from this, we are done. □

Now, by simply combining this with Meir-Keeler geometric theorem on quasi-ordered quasi-metric spaces (MK-g-oqms) one gets the fixed point result below (referred to as: *Meir-Keeler geometric rational function theorem on quasi-ordered quasi-metric spaces*: in short: (MK-g-ra-oqms)).

Theorem 4.2. *Assume that the selfmap T is (semi-progressive, increasing and) $(b, \leq; E(\Theta); \Omega)$-contractive, where*

$$\Theta \in Z^* \text{ and } \Omega \subseteq R^0_+ \times R^0_+ \text{ is upper diagonal, geometric Meir-Keeler.}$$

In addition, let X be (a-o-f,b)-complete. Then,

71-a) *T is a strong Picard operator (modulo $(b, \leq)$) when, in addition, T is (a-o-f,b)-continuous*

71-b) *T is a Bellman Picard operator (modulo $(b, \leq)$) when, in addition, $(\leq)$ is (a-o-f,b)-selfclosed and one of the extra conditions below holds:*

(71-b1) $\{2, r\}$ is disjoint from Θ
(71-b2) $r \in \Theta$ and Ω is (in addition to the above) geometric/asymptotic bilateral separable
(71-b3) $2 \in \Theta$, $r \notin \Theta$, and Ω is (in addition) geometric/asymptotic left separable

71-c) *T is fix-(≤)-asingleton (hence, fix-(≤)-singleton) whenever (in addition to the above setting)* $\{1, 4, 5, u\} \cap \Theta \neq \emptyset$.

Technically speaking, (MK-g-ra-oqms) may be viewed as a *full* geometric type quasi-metric version of a related statement in Jleli and Samet [128]; see also Alsulami et al. [13]. In particular, when

$b = d$ (the associated metric)

the obtained result extends in a direct way some contributions belonging to Dutta and Choudhury [84]. In addition, the underlying variant may be used as well in deducing coupled fixed point theorems like in Samet [206]. Note that, further extensions of these facts are possible – in the context of convergence structures – under the lines in Yeh [270] and Turinici [251].

Now, concrete examples of relations appearing in these contractive properties were already given. This yields the following practical version of our fixed point result above (referred to as: *Meir-Keeler Rational Function theorem on quasi-ordered quasi-metric spaces*; in short: (MK-ra-oqms)).

Theorem 4.3. *Let the (semi-progressive, increasing) selfmap T and the subset* $\Theta \in Z^*$ *be such that one of the following conditions holds*

(chi) T *is* $(d, \leq; E(\Theta); \chi)$*-contractive,*
where $\chi \in \mathcal{F}(re)(R^0_+, R)$ *is Meir-Keeler admissible*
(psi-vphi) T *is* $(d, \leq; E(\Theta); \psi, \varphi)$*-contractive, for a normal couple* (ψ, φ) *over* $\mathcal{F}(R^0_+, R)$*, such that* φ *is asymptotic positive.*

In addition, let X be (a-o-f,b)-complete. Then

(72-a) *T is a strong Picard operator (modulo* $(b, \leq)$*), provided the selfmap T is (a-o-f,b)-continuous*

(72-b) *T is a Bellman Picard operator (modulo* $(b, \leq)$*), provided* $(\leq)$ *is (a-o-f,b)-selfclosed and one of the following extra conditions holds*

(72-b1) $\{2, r\}$ *is disjoint from* Θ
(72-b2) $r \in \Theta$*, and the (chi)-alternative holds, where* $\chi \in \mathcal{F}(re)(R^0_+, R)$ *is (in addition) bilateral Boyd-Wong admissible*
(72-b3) $2 \in \Theta$*,* $r \notin \Theta$*, and the (chi)-alternative holds*
(72-b4) $r \in \Theta$*, and the (psi-vphi)-alternative holds, where the couple* (ψ, φ) *over* $\mathcal{F}(R^0_+, R)$ *is (in addition) limit-bounded oscillating*
(72-b5) $2 \in \Theta$*,* $r \notin \Theta$*, and the (psi-vphi)-alternative holds, where the couple* (ψ, φ) *over* $\mathcal{F}(R^0_+, R)$ *is (in addition) bounded left oscillating*

(72-c) *T is fix-($\leq$)-asingleton (hence, fix-($\leq$)-singleton) whenever (in addition to the above setting)* $\{1, 4, 5, u\} \cap \Theta \neq \emptyset$.

Some particular cases of this result may be described as below.

Case 1. Suppose that ($\leq$) is the trivial quasi-order on X and $b = d$ (the associated metric). Then, our χ-version of (MK-ra-oqms) includes in a direct way the statements in Boyd and Wong [35], Matkowski [164], and Leader [161]. Likewise, our (ψ, φ)-version of (MK-ra-oqms) includes some related statements in Dutta and Choudhury [84], Wardowski [264], or Chuasuk and Kaewcharoen [58], to name only a few.

Case 2. Suppose that ($\leq$) is, in addition, antisymmetric (hence, a partial order) and $b = d$ (the associated metric). Then, the χ-version of (MK-ra-oqms) includes the related statements in Agarwal et al. [6] or Cabrera et al. [44]; but the (ψ, φ)-version of (MK-ra-oqms) seems to be new. Further aspects may be given in Saluja et al. [205], or Khojasteh et al. [146]; see also Turinici [249].

Finally, it is worth noting that, by the used techniques, we cannot get the statement in Chandok et al. [47]. However, if one starts from a certain bi-dimensional refinement of our developments, this inclusion holds. Moreover, the statements in Harjani et al. [109] may be also derived in this way; we will discuss these elsewhere.

4.2 Factorial Jungck Contractions in Ordered Metric Spaces

Abstract. The main coincidence point result (over ordered metric spaces) in Cirić et al. [Fixed Point Th. Appl., Volume 2008, Article ID 131294] is deductible – via factorial techniques – from (a variant of) the fixed point statement obtained by Agarwal et al. [Appl. Anal., 87 (2008), 109–116].

Subject-Class 2020. 47H17 (Primary), 54H25 (Secondary).

Keywords. Quasi-ordered metric space, Leader contraction, (common) fixed point, Matkowski admissible function, contractive pair, factor space.

4.2.1 *Dependent Choice principles*

See Ch. 1.

4.2.2 *Conv-Cauchy structures*

See Ch. 1.

4.2.3 *Admissible functions*

See Ch. 1.

4.2.4 *Banach contractions and relatives*

See Ch. 1.

4.2.5 *Leader contractions*

Let X be a nonempty set. Take a *metric* $d : X \times X \to R_+$ over it; and let $(\leq)$ be a *quasi-order* on X; then, $(X, d, \leq)$ will be referred to as a *quasi-ordered metric space.* Let us say that $Y \in \exp[X]$ is $(\leq)$-*almost-singleton* (in short: $(\leq)$-*asingleton*), provided $[y_1, y_2 \in Y,\ y_1 \leq y_2$ implies $y_1 = y_2]$; and $(\leq)$-*singleton* if, in addition, $Y \in \exp(X)$.

(A) Take some $T \in \mathcal{F}(X)$, supposed to satisfy

(s-prog) T is $(\leq)$-semi-progressive ($X(T, \leq) := \{x \in X; x \leq Tx\} \neq \emptyset$)
(incr) T is $(\leq)$-increasing ($x \leq y$ implies $Tx \leq Ty$).

We have to determine circumstances under which $\mathrm{Fix}(T) := \{z \in X; z = Tz\}$ is nonempty. The basic directions under which our investigations be conducted are described by the list below, comparable with the one in Turinici [246]:

opic-0) We say that T is *fix-*$(\leq)$*-asingleton*, provided $\text{Fix}(T)$ is $(\leq)$-asingleton; and *fix-*$(\leq)$*-singleton*, whenever $\text{Fix}(T)$ is $(\leq)$-singleton

opic-1) We say that T is a *Picard operator* (modulo $(d, \leq)$) if, for each $x \in X(T, \leq)$, $(T^n x; n \geq 0)$ is d-Cauchy

opic-2) We say that T is a *strong Picard operator* (modulo $(d, \leq)$) when, for each $x \in X(T, \leq)$, the iterative sequence $(T^n x; n \geq 0)$ is d-convergent with $\lim_n(T^n x) \in \text{Fix}(T)$

opic-3) We say that T is a *Bellman Picard operator* (modulo $(d, \leq)$) when, for each $x \in X(T, \leq)$, $(T^n x; n \geq 0)$ is d-convergent with $\lim_n(T^n x) \in \text{Fix}(T)$ and $[T^n x \leq \lim_n(T^n x), \forall n]$.

In particular, if $(\leq) := X \times X$ (the *trivial quasi-order* on X) the list of such notions is identical with the standard one in Rus [203, Ch. 2, Sec. 2.2]; because, in this case, $X(T, \leq) = X$.

The regularity conditions for such properties are being founded on *ascending* concepts. Given the sequence $(z_n; n \geq 0)$ in X, define the property

$$(z_n; n \geq 0) \text{ is } (\leq)\textit{-ascending: } i \leq j \Longrightarrow z_i \leq z_j.$$

reg-1) Let us say that X is $(d, \leq)$*-complete*, provided (for each $(\leq)$-ascending sequence) d-Cauchy $\Longrightarrow$ d-convergent

reg-2) Call T, $(d, \leq)$*-continuous* if $[(z_n)$ is $(\leq)$-ascending and $z_n \xrightarrow{d} z]$ imply $Tz_n \xrightarrow{d} Tz$

reg-3) We say that $(\leq)$ is $(d, \leq)$*-selfclosed*, when $[(z_n)$ is $(\leq)$-ascending and $z_n \xrightarrow{d} z]$ imply $[z_n \leq z, \forall n]$.

As a basic completion of these, we have to precise the metrical contractive setting imposed upon our data. Denote, for $x, y \in X$,

$$Q_0(x,y) = d(Tx,Ty),\ Q_1(x,y) = d(x,y),\ Q_2(x,y) = d(x,Tx),$$
$$Q_3(x,y) = d(y,Ty),\ \ Q_4(x,y) = d(x,Ty),\ Q_5(x,y) = d(Tx,y).$$

Further, let us construct the family of functions

$$A_1 = Q_1, A_2 = (1/2)[Q_2 + Q_3],$$
$$A_3 = \max\{Q_2, Q_3\}, A_4 = (1/2)[Q_4 + Q_5];$$

or, explicitly, for $x, y \in X$

$$A_1(x,y) = d(x,y),\ A_2(x,y) = (1/2)[d(x,Tx) + d(y,Ty)],$$
$$A_3(x,y) = \max\{d(x,Tx), d(y,Ty)\},$$
$$A_4(x,y) = (1/2)[d(x,Ty) + d(Tx,y)].$$

Under the notation

$$\exp(\{1;;2,3,4\}) = \{\Theta \in \exp(\{1,2,3,4\}); 1 \in \Theta\},$$

define the functions (from $X \times X$ to R_+)

$$B(\Theta) = \max\{A_i; i \in \Theta\},\ \Theta \in \exp(\{1;;2,3,4\});$$

and let $\mathcal{B} = \{B(\Theta); \Theta \in \exp(\{1;;2,3,4\})\}$ denote the family of all these. Clearly, $\mathcal{B}$ has $2^3 = 8$ elements; but, in view of $A_2 \leq A_3$, some of these are identical. Then, define the classes of functions

$\mathcal{F}_0(R_+)$ = the class of all $\varphi \in \mathcal{F}(R_+)$ with $\varphi(0) = 0$
$\mathcal{F}_0(re)(R_+)$ = the subclass of all *regressive* $\varphi \in \mathcal{F}_0(R_+)$
(in the sense: $\varphi(t) < t$, for each $t \in R_+^0$)
$\mathcal{F}_0(in)(R_+)$ = the class of all increasing $\varphi \in \mathcal{F}_0(R_+)$
$\mathcal{F}_0(re,in)(R_+) := \mathcal{F}_0(re)(R_+) \cap \mathcal{F}_0(in)(R_+)$.

For each function $\varphi \in \mathcal{F}_0(re)(R_+)$, define

(MK-a) φ is *Meir-Keeler admissible*:
$\forall \varepsilon > 0, \exists \delta > 0$, such that $(\varepsilon < s < \varepsilon + \delta)$ implies $\varphi(s) \leq \varepsilon$,
(M-a) φ is *Matkowski admissible*:
for each (t_n) in R_+^0 with $(t_{n+1} \leq \varphi(t_n), \forall n)$ we have $\lim_n t_n = 0$.

Remember that, by an auxiliary statement,

(MK-M) (for each function $\varphi \in \mathcal{F}_0(re)(R_+)$):
φ is Meir-Keeler admissible iff φ is Matkowski admissible.

In addition, note that (according to Matkowski [164]):

$\varphi \in \mathcal{F}_0(re,in)(R_+)$ is Matkowski admissible iff:
$\varphi^n(t) \to 0$, as $n \to \infty$, for all $t \geq 0$;

here, for each $n \geq 0$, φ^n stands for the n-th *iterate* of φ. In the same context, we stress that (cf. a previous auxiliary fact)

$\varphi \in \mathcal{F}_0(re,in)(R_+)$ is Matkowski admissible, whenever
it is *strongly regressive*: $\varphi(s+0) < s$, for each $s > 0$.

Finally, given $\Theta \in \exp(\{1;;2,3,4\})$, $\varphi \in \mathcal{F}(R_+)$, let us say that the selfmap T of X is $(d, \leq; B(\Theta); \varphi)$-*contractive*, provided

(con) $d(Tx,Ty) \leq \varphi(B(\Theta)(x,y))$, for all $x, y \in X$, $x \leq y$.

Our starting point is the following statement (referred to as: Leader fixed point theorem on ordered metric spaces; in short: (L-fp-oms)).

Theorem 4.4. *Suppose that $T \in \mathcal{F}(X)$ fulfills*

(51-i) T is $(\leq)$-semi-progressive, $(\leq)$-increasing
(51-ii) T is $(d, \leq; B(\Theta); \varphi)$-contractive, where $\Theta \in \exp(\{1;;2,3,4\})$ and $\varphi \in \mathcal{F}_0(re, in)(R_+)$ is Meir-Keeler admissible.

Further, let X be $(d, \leq)$-complete. Then
(51-a) *T is fix-$(\leq)$-asingleton*
(51-b) *If (in addition) T is $(d, \leq)$-continuous, then T is a strong Picard operator (modulo $(d, \leq)$)*
(51-c) *If (in addition) $(\leq)$ is $(d, \leq)$-selfclosed, then T is a Bellman Picard operator (modulo $(d, \leq)$).*

Proof. As $A_2 \leq A_3$ and φ=increasing, the existence part of this result follows from the particular case $\Theta = \{1, 3, 4\}$. However, for an appropriate analysis of the remaining aspects, it will be useful for us to consider the general case above. Denote for simplicity (with Θ as before)

$$K = B(\Theta); \text{ that is: } K(x, y) = \max\{A_i(x, y); i \in \Theta\},\ x, y \in X.$$

Part 0. Let us establish that T is fix-$(\leq)$-asingleton. Take a couple of points $z_1, z_2 \in \mathrm{Fix}(T)$, according to $z_1 \leq z_2$. By definition,

$$A_1(z_1, z_2) = A_4(z_1, z_2) = d(z_1, z_2),\ A_2(z_1, z_2) = A_3(z_1, z_2) = 0;$$

whence, $K(z_1, z_2) = d(z_1, z_2)$. By the contractive condition,

$$d(z_1, z_2) = d(Tz_1, Tz_2) \leq \varphi(d(z_1, z_2));$$

wherefrom, $d(z_1, z_2) = 0$; and this gives $z_1 = z_2$.

It remains to show that T is strong/Bellman Picard (modulo $(d, \leq)$). To do this, we start by noting that, in view of triangular inequality,

$$A_4(x, Tx) = (1/2)d(x, T^2x) \leq A_2(x, Tx) \leq A_3(x, Tx),\ x \in X;$$

so that, we necessarily have

$$\text{(K-A3)}\quad K(x, Tx) \leq A_3(x, Tx) = \max\{d(x, Tx), d(Tx, T^2x)\},\ x \in X.$$

Take some point $x_0 \in X(T, \leq)$; and put $(x_n = T^n x_0; n \geq 0)$; it is an ascending sequence in X. If $x_n = x_{n+1}$ for some $n \geq 0$, we are done; so, without loss, one may assume that the *non-telescopic* relation holds (non-tele) $x_n \neq x_{n+1}$ (hence, $\rho_n := d(x_n, x_{n+1}) > 0$), for all $n \geq 0$.

Part 1. By the contractive condition and (K-A3),

$d(x_{n+1}, x_{n+2}) \leq \varphi(K(x_n, x_{n+1})) \leq \varphi(A_3(x_n, x_{n+1}))$, $\forall n$;
that is (cf. our notations) $\rho_{n+1} \leq \varphi(\max\{\rho_n, \rho_{n+1}\})$, $\forall n$.

Combining with the strict positivity of $(\rho_n; n \geq 0)$ (see above), gives

$\rho_{n+1} \leq \varphi(\rho_n) < \rho_n$, for each n.

This tells us that $(\rho_n; n \geq 0)$ is a strictly descending sequence in R^0_+; so that $\rho := \lim_n \rho_n$ exists in R_+, with $[\rho_n > \rho, \forall n]$. Assume that $\rho > 0$; and let $\sigma > 0$ be the number given by the Meir-Keeler admissible property of φ. By definition, there exists a rank $n(\sigma)$ such that

$(\forall n \geq n(\sigma))$: $\rho < \rho_n = d(x_n, x_{n+1}) < \rho + \sigma$.

This, along with our Meir-Keeler admissible property, gives

$(\forall n \geq n(\sigma))$: $(\rho <)\ \rho_{n+1} \leq \varphi(\rho_n) \leq \rho$;

a contradiction. Hence, $\rho = 0$; so that, (x_n) is *d-asymptotic*, in the sense

$(\rho_n =) d(x_n, x_{n+1}) \to 0$, as $n \to \infty$.

Part 2. Suppose that there exist $i, j \in N$ such that $i < j$, $x_i = x_j$. This, by the very definition of our iterates, yields yields $x_{i+1} = x_{j+1}$; whence, $\rho_i = \rho_j$; in contradiction with the strict descending property of (ρ_n). Hence, our working hypothesis cannot hold; wherefrom

$i \neq j$ implies $x_i \neq x_j$; so that, $(x_n; n \geq 0)$ is a *full* sequence.

Part 3. We claim that (x_n) is d-Cauchy in X. Let $\gamma > 0$ be arbitrary fixed; and $\beta > 0$ be the number given by the Meir-Keeler admissible property of φ; without loss, one may assume that $\beta < \gamma$. By the preceding step, there exists an index $k(\beta)$ such that (by the triangular inequality)

(d-asy) $d(x_n, x_{n+1}) < \beta/4$ (hence, $d(x_n, x_{n+2}) < \beta/2$), $\forall n \geq k(\beta)$.

We show, by induction, that the evaluation

(d-C;i) $d(x_n, x_{n+i}) < \gamma + \beta/2$, for each $n \geq k(\beta)$

holds for all $i \geq 0$; wherefrom, the d-Cauchy property of (x_n) follows. The case of $i = 0$ is trivial; and the case of $i \in \{1, 2\}$ is clear, in view of (d-asy). Assume that (d-C; i) is true, for all $i \in \{0, ..., j\}$ (where $j \geq 2$); we want to establish that (d-C;j+1) holds too. So, let $n \geq k(\beta)$ be arbitrary fixed. By the inductive hypothesis and our choice of $k(\beta)$ [expressed via (d-asy)]

$$A_1(x_n, x_{n+j}) = d(x_n, x_{n+j}) < \gamma + \beta/2 < \gamma + \beta,$$
$$A_2(x_n, x_{n+j}) \leq A_3(x_n, x_{n+j}) =$$
$$\max\{d(x_n, x_{n+1}), d(x_{n+j}, x_{n+j+1})\} < \beta/4 < \gamma + \beta.$$

Moreover, the same premises (and triangular property) give

$$d(x_n, x_{n+j+1}) \leq d(x_n, x_{n+j}) + d(x_{n+j}, x_{n+j+1}) < \gamma + \beta,$$
$$d(x_{n+1}, x_{n+j}) = d(x_{n+1}, x_{n+1+j-1}) < \gamma + \beta/2 < \gamma + \beta;$$

so that, by definition

$$A_4(x_n, x_{n+j}) = (1/2)[d(x_n, x_{n+j+1}) + d(x_{n+1}, x_{n+j})] < \gamma + \beta.$$

In addition, by the full property of (x_n), one has $A_1(x_n, x_{n+j}) = d(x_n, x_{n+j}) > 0$. Putting these together, yields (by definition)

$$0 < A_1(x_n, x_{n+j}) \leq K(x_n, x_{n+j}) < \gamma + \beta.$$

Combining with our contractive condition and the choice of β, one derives

$$d(x_{n+1}, x_{n+j+1}) = d(Tx_n, Tx_{n+j}) \leq \varphi(K(x_n, x_{n+j})) \leq \gamma;$$

and this, along with triangular property, gives

$$d(x_n, x_{n+j+1}) \leq d(x_n, x_{n+1}) + d(x_{n+1}, x_{n+j+1}) < \gamma + \beta/2;$$

wherefrom, our claim follows.

Part 4. As X is $(d, \leq)$-complete, $x_n \xrightarrow{d} z$ for some $z \in X$. There are two cases to be discussed.

Case 1. Suppose that T is $(d, \leq)$-continuous. Then $y_n := Tx_n \xrightarrow{d} Tz$ as $n \to \infty$. On the other hand, $(y_n = x_{n+1}; n \geq 0)$ is a subsequence of (x_n); whence $y_n \xrightarrow{d} z$; and this yields (as d is separated), $z = Tz$.

Case 2. Suppose that $(\leq)$ is $(d, \leq)$-selfclosed. By the convergence property above, $(x_n \leq z, \forall n)$. We now show that the alternative

$$z \neq Tz \text{ (or, equivalently: } b := d(z, Tz) > 0)$$

cannot be accepted. In fact, by the convergence property of $(x_n; n \geq 0)$ and the continuity of $d(.,.)$, one gets

$$d(x_n, Tx_n), d(x_n, z), d(Tx_n, z) \to 0; \ d(x_n, Tz), d(Tx_n, Tz) \to b.$$

This yields some rank $n(b) \geq 0$, such that

$$d(x_n, Tx_n), d(x_n, z), d(Tx_n, z) < b/2, \text{ for all } n \geq n(b).$$

Moreover, by the triangular property, we have

$$d(x_n, Tz) \leq d(x_n, z) + b < 3b/2, \text{ for all } n \geq n(b).$$

Combining these yields (again for the same ranks n)

$A_1(x_n, z) = d(x_n, z) < b/2 < b$
$A_2(x_n, z) = (1/2)[d(x_n, Tx_n) + b] < 3b/4 < b$
$A_3(x_n, z) = \max\{d(x_n, Tx_n), b\} = b$
$A_4(x_n, z) = (1/2)[d(x_n, Tz) + d(Tx_n, z)] < b$;
whence (by definition), $K(x_n, z) \leq b$.

Taking the contractive property into account, we derive

$$d(x_{n+1}, Tz) \leq \varphi(K(x_n, z)) \leq \varphi(b), \ \forall n \geq n(b).$$

Passing to limit as $n \to \infty$ gives (by the relation above) $b \leq \varphi(b)$; contradiction. Hence, $z = Tz$; and the proof is complete. □

(B) Concerning the obtained fixed point result (L-fp-oms), two basic variants of it (concerning the ambient quasi-order $(\leq)$) are of interest.

(I) Suppose that $(\leq) = X \times X$ (the *trivial quasi-order* on X). Then,

(I-0) $Y \in \exp[X]$ is $(\leq)$-asingleton is expressed as Y is *asingleton* (see above); and T is fix-$(\leq)$-asingleton is referred to as T is *fix-asingleton*
(I-1) the $(\leq)$-semi-progressive and $(\leq)$-increasing conditions upon T are holding (in a formal way)
(I-2) the strong Picard property (modulo $(d, \leq)$) and Bellman Picard property (modulo $(d, \leq)$) relative to T are identical.
(I-3) the $(d, \leq)$-selfclosed property upon $(\leq)$ is fulfilled.

In addition, the remaining conventions (modulo $(d, \leq)$) will be referred to as *conventions* (modulo d). These, in particular, include the contraction property, that writes according to the lines below:

Given $\Theta \in \exp(\{1;;2,3,4\})$, $\varphi \in \mathcal{F}(R_+)$, let us say that T is $(d; B(\Theta); \varphi)$-*contractive*, provided

(con) $d(Tx, Ty) \leq \varphi(B(\Theta)(x, y))$, for all $x, y \in X$.

The following (amorphous type) fixed point statement (referred to as: Leader fixed point theorem on metric spaces; in short: (L-fp-ms)) is now available (for a later use).

Theorem 4.5. *Suppose that* $T \in \mathcal{F}(X)$ *fulfills*

> T *is* $(d; B(\Theta); \varphi)$*-contractive, where* $\Theta \in \exp(\{1;;2,3,4\})$ *and* $\varphi \in \mathcal{F}_0(re, in)(R_+)$ *is Meir-Keeler admissible.*

Further, let X *be d-complete. Then*

(52-a) T *is fix-asingleton*

(52-b) T *is a strong Picard operator (modulo d).*

Concerning the obtained facts, the following are valid

(I-1) the $B(\{1,3,4\})$-variant of (L-fp-ms) is just the 1979 statement in Leader [161], obtained via similar methods

(I-2) the $B(\{1\})$-variant of (L-fp-ms) is Matkowski's result [164].

(II) Suppose that $(\leq)$ is *antisymmetric*; hence, an ordering on X. Then, all concepts and properties (modulo $(\leq)$) may be written, formally, in the same way as the one described in the quasi-order case. This yields the following statement (referred to as: Leader fixed point theorem on partially ordered metric spaces; in short: (L-fp-poms)).

Theorem 4.6. *Suppose that* $T \in \mathcal{F}(X)$ *fulfills*

> *(53-i)* T *is* $(\leq)$*-semi-progressive,* $(\leq)$*-increasing*
>
> *(53-ii)* T *is* $(d, \leq; B(\Theta); \varphi)$*-contractive, where* $\Theta \in \exp(\{1;;2,3,4\})$ *and* $\varphi \in \mathcal{F}_0(re, in)(R_+)$ *is Meir-Keeler admissible.*

Further, let X *be* $(d, \leq)$*-complete. Then*

(53-a) T *is fix-*$(\leq)$*-asingleton*

(53-b) *If (in addition)* T *is* $(d, \leq)$*-continuous, then* T *is a strong Picard operator (modulo* $(d, \leq)$*)*

(53-c) *If (in addition)* $(\leq)$ *is* $(d, \leq)$*-selfclosed, then* T *is a Bellman Picard operator (modulo* $(d, \leq)$*).*

Concerning the obtained fact, it is worth noting that

(II-1) the $B(\{1,3,4\})$-variant of (L-fp-poms) includes the related statement in Agarwal et al. [6]

(II-2) the $B(\{1\})$-variant of (L-fp-poms) is comparable with the 1986 result in Turinici [236].

We have to stress that explicit type relationships between these two particular variants of (L-fp-oms) are not possible. That is:

(L-fp-ms) is not obtainable from (L-fp-poms) and vice versa.

This, among others, motivated us to state (and prove) the present quasi-order fixed point statement (L-fp-oms).

Finally, note that, further extensions of the obtained result are possible; to do this, it will suffice following the 1977 paper by Rhoades [195]; where – in the trivial quasi-order case – a large list of contractive conditions is proposed. For some concrete results of this type, see Turinici [246] and the references therein.

4.2.6 *Jungck-Matkowski maps*

Let $(X, d, \leq)$ be a quasi-ordered metric space.

(A) Given $T \in \mathcal{F}(X)$, any fixed point of T is a common fixed point of the couple (T, G), where $G \in \mathcal{F}(X)$ is taken as

$G = I$ (= the identical selfmap of X);

so, it is natural to see whether Leader fixed point theorem on ordered metric spaces (L-fp-oms) is extendable to a common fixed point result for couples (T, G), where (in general) $G \neq I$. Some early answers to this were proposed, in the trivial quasi-order framework ($(\leq) = X \times X$), by Jungck [131], [132]. Recently, an extension of these developments to the (partial) order setting was obtained by Cirić et al. [61]. It is our aim to show that, ultimately, the quoted statement is just a particular case of Leader fixed point theorem (L-fp-poms). In fact, this reduction scheme comprises the quasi-order setting as well. To expose it, some preliminaries are needed.

Let $(X, d, \leq)$ be a quasi-ordered metric space; and $T, G \in \mathcal{F}(X)$ be a couple of selfmaps. Call $Y \in \exp[X]$, $(\leq, G)$-*asingleton* if

$$y_1, y_2 \in Y,\ Gy_1 \leq Gy_2 \Longrightarrow Gy_1 = Gy_2;$$

and $(\leq, G)$-*singleton* if, in addition, $Y \in \exp(X)$. We say that $z \in X$ is a *coincidence point* of (T, G) provided $Tz = Gz$; the set of all these will be denoted as $\mathrm{cp}(T, G)$. Then, $z \in \mathrm{cp}(T, G)$ is called a *common fixed point* of (T, G), when $Tz = Gz = z$; the set of all these will be denoted as $\mathrm{cfp}(T, G)$.

To get existence and uniqueness results for such points, some general assumptions upon our data must be imposed; namely,

(n-as-1) (T, G) is *admissible*: $T(X) \subseteq G(X)$

(n-as-2) (T, G) is *compatible*: $TGz = GTz$, whenever $Tz = Gz$

(n-as-3) (T, G) is *similar*: $Gx = Gy$ implies $Tx = Ty$;

these will be referred to as: (T, G) is *normal*.

In addition to this, some specific assumptions are to be added, as

(s-as-1) (T, G) is $(\leq)$-*semi-progressive*
$(X(T; \leq; G) := \{x \in X; Gx \leq Tx\}$ is nonempty)
(s-as-2) (T, G) is $(\leq)$-*increasing* ($Gx \leq Gy$ implies $Tx \leq Ty$).

A direct consequence of these is to be described as follows. Let $\mathcal{R}$ be the relation (over X)

$(x, y \in X)$: $x\mathcal{R}y$ iff $Tx = Gy$.

Note that, by the admissible condition,

$\mathcal{R}$ is *proper* (that is: $\mathcal{R}(x) \neq \emptyset$, for each $x \in X$).

So, by the Dependent Choice principle (DC), we have that for the arbitrary fixed $x_0 \in X$, there may be constructed a sequence (x_n) in X, with

$x_n \mathcal{R} x_{n+1}$ (that is: $Tx_n = Gx_{n+1}$), for all n.

This will be referred to as a $(x_0; T, G)$-*iterative process* (in X); note that, if $x_0 \in X(T, \leq; G)$, both sequences $(Tx_n; n \geq 0)$ and $(Gx_n; n \geq 0)$ are ascending; because (cf. the increasing property)

$Gx_0 \leq Tx_0(= Gx_1) \leq Tx_1(= Gx_2) \leq Tx_2(= Gx_3) \leq ...;$

and the claim is proved.

(B) Having these precise, it is our aim in the following to determine circumstances (involving such iterative processes) under which $\mathrm{cp}(T, G)$ and/or $\mathrm{cfp}(T, G)$ is nonempty. The basic directions under which the investigations be conducted are described as below:

cpic-0) Call (T, G), *cp-$(\leq, G)$-asingleton* if $\mathrm{cp}(T, G)$ is $(\leq, G)$-asingleton; and *cp-$(\leq, G)$-singleton*, if $\mathrm{cp}(T, G)$ is $(\leq, G)$-singleton

cpic-1) We say that (T, G) is a *cp-Picard couple* (modulo $(d, \leq)$) if, for each starting point $x_0 \in X(T; \leq; G)$ and each $(x_0; T, G)$-iterative process $(x_n; n \geq 0)$, the sequences $(Gx_n; n \geq 0)$ and $(Tx_n; n \geq 0)$ are d-Cauchy

cpic-2) We say that (T, G) is a *strong cp-Picard couple* (modulo $(d, \leq)$) if, for each starting point $x_0 \in X(T; \leq; G)$ and each $(x_0; T, G)$-iterative process $(x_n; n \geq 0)$, there exists $z \in \mathrm{cp}(T, G)$ with $\lim_n Gx_n (= \lim_n Tx_n) = Gz(= Tz)$; and a *strong cfp-cp-Picard couple* (modulo $(d, \leq)$) if, in addition, $w := Gz \in \mathrm{cfp}(T, G)$

cpic-3) Call (T, G), a *Bellman cp-Picard couple* (modulo $(d, \leq)$) if, for each $x_0 \in X(T; \leq; G)$ and each $(x_0; T, G)$-iterative process $(x_n; n \geq 0)$, there exists $z \in \text{cp}(T, G)$ with $\lim_n Gx_n (= \lim_n Tx_n) = Gz (= Tz)$, $(Gx_n \leq Gz$, for all $n)$; and a *Bellman cfp-cp-Picard couple* (modulo $(d, \leq)$) if, in addition, $w := Gz \in \text{cfp}(T, G)$.

In addition to these, the basic regularity conditions to be used are the standard ones (modulo G), completed with some special requirements upon G; precisely

creg-1) Call X, $(d, \leq; G)$-*complete* if: for each sequence (x_n) in X with $[(Gx_n) = \text{ascending}, d\text{-Cauchy}]$, there exists $x \in X$ with $Gx_n \xrightarrow{d} Gx$

creg-2) We say that T is $(d, \leq; G)$-*continuous*, if for each sequence (x_n) in X and each $x \in X$: $[(Gx_n) = \text{ascending}, Gx_n \xrightarrow{d} Gx]$ imply $Tx_n \xrightarrow{d} Tx$

creg-3) Call $(\leq)$, $(d, \leq; G)$-*selfclosed* when, for each sequence (x_n) in X and each $x \in X$: $[(Gx_n) = \text{ascending}, Gx_n \xrightarrow{d} Gx]$ imply $[Gx_n \leq Gx$, for all $n]$

creg-4) Let us say that G is $\text{cp}(T, G)$-*involutive*, provided $Gz = GGz$, for each $z \in \text{cp}(T, G)$; and $\text{cp}(T, G)$-$(\leq)$-*self-progressive*, when $Gz \leq GGz$, for each $z \in \text{cp}(T, G)$.

As a completion of these, we must introduce the contractive properties to be used. Denote, for $x, y \in X$,

$$Q_0(G)(x, y) = d(Tx, Ty),\ Q_1(G)(x, y) = d(Gx, Gy),$$
$$Q_2(G)(x, y) = d(Gx, Tx),\ Q_3(G)(x, y) = d(Gy, Ty),$$
$$Q_4(G)(x, y) = d(Gx, Ty),\ Q_5(G)(x, y) = d(Tx, Gy).$$

Further, let us construct the family of functions

$$A_1(G) = Q_1(G),\ A_2(G) = (1/2)[Q_2(G) + Q_3(G)],$$
$$A_3(G) = \max\{Q_2(G), Q_3(G)\},\ A_4(G) = (1/2)[Q_4(G) + Q_5(G)];$$

or, explicitly, for $x, y \in X$

$$A_1(G)(x, y) = d(Gx, Gy),$$
$$A_2(G)(x, y) = (1/2)[d(Gx, Tx) + d(Gy, Ty)],$$
$$A_3(G)(x, y) = \max\{d(Gx, Tx), d(Gy, Ty)\},$$
$$A_4(G)(x, y) = (1/2)[d(Gx, Ty) + d(Tx, Gy)].$$

Finally, define the maximum functions

$$B(\Theta; G) = \max\{A_i(G); i \in \Theta\},\ \Theta \in \exp(\{1;; 2, 3, 4\});$$

and let $\mathcal{B}(G)$ stand for the class of such functions. As before, $\mathcal{B}(G)$ has $2^3 = 8$ elements; but, as $A_2(G) \leq A_3(G)$, some of these are identical.

Given $\Theta \in \exp(\{1;;2,3,4\})$, $\varphi \in \mathcal{F}(R_+)$, let us say that the couple (T, G) (taken as before) is $(d, \leq; B(\Theta; G); \varphi)$*-contractive*, provided

$$\text{(G-con)} \quad d(Tx, Ty) \leq \varphi(B(\Theta; G)(x, y)), \quad \text{for all } x, y \in X,\ Gx \leq Gy.$$

The properties of φ to be considered here are the already introduced ones.

We are now in position to formulate our second main result in this exposition (referred to as: Jungck-Matkowski common fixed point theorem on ordered metric spaces; in short: (JM-cp-oms)).

Theorem 4.7. *Suppose that the couple (T, G) over $\mathcal{F}(X)$ fulfills*

(61-i) (T, G) is normal, $(\leq)$-semi-progressive, $(\leq)$-increasing
(61-ii) the couple (T, G) is $(d, \leq; B(\Theta; G); \varphi)$-contractive, with $\Theta \in$ $\exp(\{1;;2,3,4\})$ *and $\varphi \in \mathcal{F}_0(re, in)(R_+)$ being Meir-Keeler admissible.*

In addition, let X be $(d, \leq; G)$-complete. Then,

(61-a) *(T, G) is cp-$(\leq, G)$-asingleton*

(61-b) *If in addition T is $(d, \leq; G)$-continuous, then (T, G) is a strong cp-Picard couple (modulo $(d, \leq)$); and if (in addition to this) G is* cp(T, G)-*involutive, then (T, G) is a strong cfp-cp-Picard couple (modulo $(d, \leq)$)*

(61-c) *If in addition $(\leq)$ is $(d, \leq; G)$-selfclosed, then (T, G) is a Bellman cp-Picard couple (modulo $(d, \leq)$); and if (in addition to this) G is* cp(T, G)-*$(\leq)$-self-progressive, then, necessarily, (T, G) is a Bellman cfp-cp-Picard couple (modulo $(d, \leq)$).*

(C) It is our aim in the following to prove that this coincidence point result (JM-cp-oms) is equivalent with the fixed point result (L-fp-oms). This will be done by a two steps argument, developed below. The first part of it is contained in

Proposition 4.14. *Under the precise conditions, one has in (ZF-AC+DC)*

$$\textit{(JM-cp-oms)} \Longrightarrow \textit{(L-fp-oms)}.$$

In other words: Leader fixed point result (L-fp-oms) is obtainable from the Jungck-Matkowski coincidence point result (JM-cp-oms).

Proof. Assume that (JM-cp-oms) holds, with respect to its data; and let the premises of (L-fp-oms) be accepted. Several steps must be passed.

Step 1. Clearly, the following absolute assertions hold

(I-a1) (T, I) is normal (meaning: admissible, compatible, similar)
(I-a2) I is cp(T, I)-involutive and cp(T, I)-$(\leq)$-self-progressive.

Step 2. On the other hand, the following relative assertions are valid

(I-b1) (T, I) is $(\leq)$-semi-progressive (as T is $(\leq)$-semi-progressive)
(I-b2) (T, I) is $(\leq)$-increasing (as T is $(\leq)$-increasing)
(I-b3) (with Θ and φ as before):
(T, I) is $(d, \leq; B(\Theta; I); \varphi)$-contractive
(as T is $(d, \leq; B(\Theta); \varphi)$-contractive)
(I-b4) X is $(d, \leq; I)$-complete if X is $(d, \leq)$-complete
(I-b5) T is $(d, \leq; I)$-continuous if T is $(d, \leq)$-continuous
(I-b6) $(\leq)$ is $(d, \leq; I)$-selfclosed if $(\leq)$ is $(d, \leq)$-selfclosed.

Summing up, the general conditions of (JM-cp-oms) are fulfilled by the quasi-ordered metric space $(X, d, \leq)$, the couple (T, I) over $\mathcal{F}(X)$, the subset Θ in $\exp(\{1; ; 2, 3, 4\})$, and the Meir-Keeler admissible function $\varphi \in \mathcal{F}_0(re, in)(R_+)$. Consequently, the conclusions of (JM-cp-oms) are retainable (with respect to the underlying data).

Step 3. By the first conclusion of (JM-cp-oms), one derives that

(I-c1) (T, I) is cp-$(\leq, I)$-asingleton;
which means: T is fix-$(\leq)$-asingleton.

Step 4. Further, let $x_0 \in X(T, \leq)$ be fixed. By definition, we have $x_0 \in X(T; \leq; I)$; in addition, the $(x_0; T, I)$-iterative process (x_n) is just the iterative process $(x_n = T^n x_0; n \geq 0)$. Two alternatives occur.

Alter 1. Suppose, in addition, that T is $(d; \leq)$-continuous. Then (see above), the selfmap T is $(d; \leq; I)$-continuous; so that, by the second conclusion of (JM-cp-oms),

(T, I) is strong cp-Picard (modulo $(d, \leq)$);
which means:
T is strong Picard (modulo $(d, \leq)$).

Alter 2. Suppose, in addition, that $(\leq)$ is $(d; \leq)$-selfclosed. Then (see above), the quasi-order $(\leq)$ is $(d; \leq; I)$-selfclosed; so that, by the third conclusion of (JM-cp-oms),

(T, I) is Bellman cp-Picard (modulo $(d, \leq)$);
which means:
T is Bellman Picard (modulo $(d, \leq)$).

Combining these, one gets that all conclusions in (J-fp-oms) follow. □

The second part of our argument is contained in

Proposition 4.15. *Under the precise conditions, one has in the reduced system (ZF-AC+DC)*

(L-fp-oms) $\Longrightarrow$ *(JM-cp-oms); hence, (L-fp-oms)* $\Longleftrightarrow$ *(JM-cp-oms).*

In other words: Leader fixed point result (L-fp-oms) is equivalent with Jungck-Matkowski coincidence point result (JM-cp-oms).

Proof. Assume that (L-fp-oms) holds; and let the premises of (JM-cp-oms) be accepted. Before passing to the essential part of our discussion, we need some factorial type constructions and auxiliary facts involving these data.

Let $(X, d, \leq)$ be a quasi-ordered metric space; and (T, G) be a couple of selfmaps like in (JM-cp-oms); hence, in particular

(hyp-1) (T, G) is normal (i.e.: admissible, compatible, similar)
(hyp-2) (T, G) is $(\leq)$-semi-progressive and $(\leq)$-increasing
(hyp-3) (T, G) is $(d, \leq; B(\Theta; G); \varphi)$-contractive (see above).

Part 1. Let us introduce an *equivalence relation* $(\sim)$ on X as:

(defi-1) $(x, y \in X)$: $x \sim y$ iff $Gx = Gy$.

Define the equivalence classes and the attached factor space as

(defi-2) for $x \in X$, let $x^* := G^{-1}(Gx)(= \{u \in X; x \sim u\})$
stand for its *equivalence class* (modulo $(\sim)$)
(defi-3) $X^* = X|_\sim$ = the collection of all these
(that is: the *factor space* attached to this relation).

Clearly, by this very definition we have

(rela-1) $(Gx^* = \{Gx\}, \forall x^* \in X^*)$;
hence, $G(.)$ may be viewed as an element of $\mathcal{F}(X^*, X)$
(rela-2) $G : X^* \to X$ is injective: $G(x^*) = G(y^*)$ implies $x^* = y^*$
(rela-3) $G : X^* \to G(X)$ is surjective (hence, bijective).

Note that, as a direct consequence of this,

$(x^* = G^{-1}(Gx^*), \forall x^* \in X^*)$, and
$(G(G^{-1}y) = Gx^*$, whenever $Gx = y)$.

The obtained facts tell us that a standard metrical and quasi-order structure over X^* may be derived as a copy of the corresponding ones in the quasi-ordered metric space $(G(X), d, \leq)$. Precisely, define a pseudometric Δ over X^* according to

$$\Delta(x^*, y^*) = d(Gx^*, Gy^*) \ [= d(Gx, Gy)],\ x^*, y^* \in X^*.$$

Note that, the singleton property (rela-1) assures the consistency of our definition; as well as the metrical character of Δ. In addition, let us introduce a relation $(\preceq)$ on X^*, as

$$x^* \preceq y^* \text{ iff } Gx^* \leq Gy^* \text{ [or, equivalently: } Gx \leq Gy].$$

As before, the singleton property (rela-1) tells us that this convention is meaningful; and yields the quasi-order character of $(\preceq)$.

Putting these together we finally constructed a quasi-ordered metric space $(X^*, \Delta, \preceq)$. Note that, as a consequence of the bijective property, the sequential concepts over this structure are direct translations of the ones attached to the quasi-ordered metric space $(G(X), d, \leq)$, as: for each sequence (x_n^*) in X^* and each $x^* \in X^*$,

(seq-1) $x_n^* \xrightarrow{\Delta} x^*$ iff $Gx_n^* \xrightarrow{d} Gx^*$ in $G(X)$
(seq-2) (x_n^*) is Δ-Cauchy iff (Gx_n^*) is d-Cauchy in $G(X)$
(seq-3) (x_n^*) is $(\preceq)$-ascending iff (Gx_n^*) is $(\leq)$-ascending in $G(X)$.

Finally, by the similarity condition, we have

(Rela-1) $(Tx^* = \{Tx\}, \forall x^* \in X^*)$;
hence, $T(.)$ may be viewed as an element of $\mathcal{F}(X^*, X)$.

This allows us to introduce a mapping $T^* : X^* \to X^*$ as: for each $x^* \in X^*$,

(Rela-2) $T^*x^* = G^{-1}(Tx^*)$ $[= G^{-1}(Tx)]$;
which means:
T^*x^* = the (unique) equivalence class $u^* \in X^*$ with $Gu^* = Tx^*$.

In fact, the second line of definition is consistent, via (T, G)=admissible; and the first line of the same is a consequence of

$(\forall x^* \in X^*)$ $u^* := G^{-1}(Tx^*)$
is an equivalence class with $Gu^* = Tx^*$.

Moreover, by these very facts, the following local relations are valid:

(Rela-3) $(\forall x^* \in X^*)$: $GT^*x^* = Tx^*$,

(Rela-4) $(\forall x, y \in X)$: $Tx = Gy \Longrightarrow Tx^* = Gy^*$ ($\Longleftrightarrow T^*x^* = y^*$).

Part 2. Passing to the regularity conditions fulfilled by the quasi-ordered metric space $(X^*, \Delta, \preceq)$ and the selfmap T^* of X^*, the following (groups of) properties are to be noted.

GProp 1. The first group of such properties is

(II-a1) T^* is $(\preceq)$-semi-progressive if (T, G) is $(\leq)$-semi-progressive
(II-a2) T^* is $(\preceq)$-increasing if (T, G) is $(\leq)$-increasing.

The verification is as follows:

a1): Clearly, the inclusion chain is holding

$$(Gx \leq Tx) \Longleftrightarrow (Gx^* \leq Tx^* = GT^*x^*) \Longleftrightarrow (x^* \preceq T^*x^*);$$

and the first assertion follows.

a2): On the other hand,

$$x^* \preceq y^* \Longrightarrow Gx^* \leq Gy^* \Longrightarrow Tx^* \leq Ty^* \Longrightarrow GT^*x^* \leq GT^*y^*;$$
which means: $T^*x^* \preceq T^*y^*$;

and the second assertion holds too.

GProp 2. The second group of such properties is

(II-b1) X^* is $(\Delta, \preceq)$-complete (as X is $(d, \leq; G)$-complete)
(II-b2) T^* is $(\Delta, \preceq)$-continuous (as T is $(d, \leq; G)$-continuous)
(II-b3) $(\preceq)$ is $(\Delta, \preceq)$-selfclosed (as $(\leq)$ is $(d, \leq; G)$-selfclosed).

The verification of this is done according to the lines below.

b1): If (x_n^*) is $(\preceq)$-ascending, Δ-Cauchy, we have that (Gx_n^*) is $(\leq)$-ascending, d-Cauchy; so, by the completeness assumption,

there exists some $x^* \in X^*$ with $Gx_n^* \xrightarrow{d} Gx^*$; that is, $x_n^* \xrightarrow{\Delta} x^*$.

b2): If (x_n^*) and x^* are such that (x_n^*) is $(\preceq)$-ascending and $x_n^* \xrightarrow{\Delta} x^*$ then (Gx_n^*) is $(\leq)$-ascending, and $Gx_n^* \xrightarrow{d} Gx^*$; so, by continuity,

$$Tx_n^* \xrightarrow{d} Tx^*; \text{ that is: } GT^*x_n^* \xrightarrow{d} GT^*x^*;$$
which means: $T^*x_n^* \xrightarrow{\Delta} T^*x^*$.

b3): If (x_n^*) and x^* are such that (x_n^*) is $(\preceq)$-ascending and $x_n \xrightarrow{\Delta} x^*$ then (Gx_n^*) is $(\leq)$-ascending, and $Gx_n^* \xrightarrow{d} Gx^*$; so, by self-closeness,

$$(Gx_n^* \leq Gx^*, \forall n); \text{ that is: } (x_n^* \preceq x^*, \forall n).$$

GProp 3. Passing to the contractive properties, denote for $x^*, y^* \in X^*$

$$Q_0^*(x^*, y^*) = \Delta(T^*x^*, T^*y^*),\ Q_1^*(x^*, y^*) = \Delta(x^*, y^*),$$
$$Q_2^*(x, y) = \Delta(x^*, T^*x^*),\ Q_3^*(x^*, y^*) = \Delta(y^*, T^*y^*),$$
$$Q_4^*(x^*, y^*) = \Delta(x^*, T^*y^*),\ Q_5^*(x^*, y^*) = \Delta(T^*x^*, y^*);$$

and let us construct the family of functions

$A_1^* = Q_1^*,\ A_2^* = (1/2)[Q_2^* + Q_3^*],$
$A_3^* = \max\{Q_2^*, Q_3^*\},\ A_4^* = (1/2)[Q_4^* + Q_5^*];$
that is (for $x^*, y^* \in X^*$):
$A_1^*(x^*, y^*) = \Delta(x^*, y^*),$
$A_2^*(x^*, y^*) = (1/2)[\Delta(x^*, T^*x^*) + \Delta(y^*, T^*y^*)],$
$A_3^*(x^*, y^*) = \max\{\Delta(x^*, T^*x^*), \Delta(y^*, T^*y^*)\},$
$A_4^*(x^*, y^*) = (1/2)[\Delta(x^*, T^*y^*) + \Delta(T^*x^*, y^*)].$

Finally, define

$$B^*(\Theta) = \max\{A_i^*; i \in \Theta\},\ \Theta \in \exp(\{1;;2,3,4\});$$

and let $\mathcal{B}^*$ stand for the class of such functions. As before, $\mathcal{B}^*$ has $2^3 = 8$ elements; but, as $A_2^* \le A_3^*$, some of these are identical. Given $\Theta \in \exp(\{1;;2,3,4\})$, $\varphi \in \mathcal{F}(R_+)$, let us say that the selfmap T^* is $(\Delta, \preceq; B^*(\Theta); \varphi)$-*contractive*, provided

(f-contr) $\Delta(T^*x^*, T^*y^*) \le \varphi(B^*(\Theta)(x^*, y^*)),$
for all $x^*, y^* \in X^*$, $x^* \preceq y^*$.

The functions φ to be considered here are the ones of introductory part.

Having these precise, note that for $x, y \in X$

$$Q_0(G)(x, y) = d(Tx^*, Ty^*) = \Delta(T^*x^*, T^*y^*) = Q_0^*(x^*, y^*),$$
$$Q_1(G)(x, y) = d(Gx^*, Gy^*) = \Delta(x^*, y^*) = Q_1^*(x^*, y^*),$$
$$Q_2(G)(x, y) = d(Gx^*, Tx^*) = \Delta(x^*, T^*x^*) = Q_2^*(x^*, y^*),$$
$$Q_3(G)(x, y) = d(Gy^*, Ty^*) = \Delta(y^*, T^*y^*) = Q_3^*(x^*, y^*),$$
$$Q_4(G)(x, y) = d(Gx^*, Ty^*) = \Delta(x^*, T^*y^*) = Q_4^*(x^*, y^*),$$
$$Q_5(G)(x, y) = d(Tx^*, Gy^*) = \Delta(T^*x^*, y^*) = Q_5^*(x^*, y^*).$$

This yields (for the same data)

$A_1(G)(x, y) = A_1^*(x^*, y^*),\ \ A_2(G)(x, y) = A_2^*(x^*, y^*),$
$A_3(G)(x, y) = A_3^*(x^*, y^*),\ \ A_4(G)(x, y) = A_4^*(x^*, y^*);$
whence,
$B(\Theta; G)(x, y) = B^*(\Theta)(x^*, y^*),\ \forall \Theta \in \exp(\{1;;2,3,4\}).$

As a direct consequence, it follows that

(II-b4) (for each $\Theta \in \exp(\{1;;2,3,4\})$ and $\varphi \in \mathcal{F}(R_+)$):
the selfmap T^* is $(\Delta, \preceq; B^*(\Theta); \varphi)$-contractive,
whenever the couple (T, G) is $(d, \leq; B(\Theta; G); \varphi)$-contractive.

Summing up, the general conditions of (L-fp-oms) are fulfilled by the quasi-ordered metric space $(X^*, \Delta, \preceq)$, the selfmap $T^* \in \mathcal{F}(X^*)$, the subset $\Theta \in \exp(\{1;;2,3,4\})$, and the Meir-Keeler admissible function $\varphi \in \mathcal{F}_0(re, in)(R_+)$. Consequently, the conclusions of (L-fp-oms) are retainable (with respect to the underlying data). It remains to determine that, under these, the conclusions of (JM-cp-oms) are holding.

Conclu 1. We start by noting that the following is true:

(II-c1) $\mathrm{cp}(T, G)$ is $(\leq; G)$-asingleton,
under the first closing assertion of (L-fp-oms);

and, from this, the first conclusion in (JM-cp-oms) is retainable. In fact, let $w_1, w_2 \in \mathrm{cp}(T, G)$ be such that

$$Gw_1 \leq Gw_2; \text{ that is: } Gw_1 = Tw_1,\ Gw_2 = Tw_2,\ Gw_1 \leq Gw_2.$$

This, by definition, yields

$$w_1^* = T^* w_1^*, w_2^* = T^* w_2^*,\ w_1^* \preceq w_2^*;$$

so that (as $\mathrm{Fix}(T^*)$ is $(\preceq)$-asingleton), $w_1^* = w_2^*$; whence $Gw_1 = Gw_2$.

Conclu 2. Further, we establish that

(II-c2) (T, G) is strong/Bellman cp-Picard (modulo $(d, \leq)$)
under the basic conditions in (JM-cp-oms)
(II-c3) (T, G) is strong/Bellman cfp-cp-Picard (modulo $(d, \leq)$),
under the supplementary conditions in (JM-cp-oms);

and, from this, the second and third conclusion of (JM-cp-oms) hold. Two possible alternatives occur.

Case 1) [T is $(d, \leq; G)$-continuous.]

Two sub-cases must be analyzed.

Sub-case 1-a) From the previous remarks, the associated selfmap T^* is $(\Delta, \preceq)$-continuous; whence, (L-fp-oms) (the second closing assertion) applies to this framework; so that,

T^* is a strong Picard operator (modulo $(\Delta, \preceq)$).

Given $x_0 \in X(T, \leq; G)$ (hence, $Gx_0 \leq Tx_0$), let $(x_n; n \geq 0)$ be a (T, G)-iterative sequence generated by this element; that is: $[Tx_n = Gx_{n+1}; n \geq 0]$. This tells us that $x_0^* \preceq T^* x_0^*$; and $(x_n^*; n \geq 0)$ is the T^*-iterative sequence generated by x_0^* [in the sense: $(x_{n+1}^* = T^* x_n^*; n \geq 0)$]. By the strong Picard property above, this iterative sequence is $(\preceq)$-ascending, Δ-convergent; and $z^* := \Delta - \lim_n x_n^*$ is an element of $\mathrm{Fix}(T^*)$; which means:

$$z^* = T^* z^*; \text{ hence } Gz^* = Tz^*.$$

Let $z \in z^*$ be fixed in the sequel. From the previous relation, $Gz = Tz$; whence $z \in \mathrm{cp}(T, G)$. Moreover, by the very definition of our (T, G)-iterative sequence (and the limit property above)

$$(Gx_n; n \geq 0) \text{ is ascending and } \lim_n Gx_n (= \lim_n Tx_n) = Gz (= Tz);$$

proving that (T, G) is a strong cp-Picard couple (modulo $(d, \leq)$).

Sub-case 1-b) Let in addition G be $\mathrm{cp}(T, G)$-involutive. Denote for simplicity $w := Gz (= Tz)$. As (T, G) is compatible, we have $Gw = GTz = TGz = Tw$. Combining with the involutive property, we must have

$$Gz = GGz (= GTz); \text{ whence } w = GTz = TGz = Tw (= Gw);$$

which tells us that $w \in \mathrm{cfp}(T, G)$.

Case 2) $[(\leq)$ is $(d, \leq; G)$-selfclosed.$]$

As before, two sub-cases must be analyzed.

Sub-case 2-a) By the preceding remark, the factor quasi-order $(\preceq)$ is $(\Delta, \preceq)$-selfclosed; whence, (L-fp-oms) (the third closing assertion) applies to this framework; so that

$$T^* \text{ is a Bellman Picard operator (modulo } (\Delta, \preceq)).$$

Given $x_0 \in X(T, \leq; G)$ (hence, $Gx_0 \leq Tx_0$), let $(x_n; n \geq 0)$ be a (T, G)-iterative sequence generated by this element; that is: $[Tx_n = Gx_{n+1}; n \geq 0]$. This tells us that $x_0^* \preceq T^* x_0^*$; and $(x_n^*; n \geq 0)$ is the T^*-iterative sequence generated by x_0^* (in the sense: $(x_{n+1}^* = T^* x_n^*; n \geq 0)$. By the Bellman Picard property above, this iterative sequence is $(\preceq)$-ascending, Δ-convergent; and $z^* := \Delta - \lim_n x_n^*$ fulfills

$$z^* = T^* z^* \text{ (hence, } Gz^* = Tz^*\text{), and } (x_n^* \preceq z^*, \forall n).$$

Let $z \in z^*$ be fixed in the sequel. From the previous relation, $Gz = Tz$; whence $z \in \mathrm{cp}(T, G)$. Moreover, by the very definition of our (T, G)-iterative sequence (and the limit property above)

$(Gx_n; n \geq 0)$ is ascending, $\lim_n Gx_n (= \lim_n Tx_n) = Gz (= Tz)$,
and $(Gx_n \leq Gz$, for all $n)$;

proving that (T, G) is a Bellman cp-Picard couple (modulo $(d, \leq)$).

Sub-case 2-b) Let in addition G be cp(T, G)-$(\leq)$-self-progressive. Denote for simplicity $w := Gz (= Tz)$. As (T, G) is compatible, we have $[Gw = GTz = TGz = Tw]$; moreover, by the self-progressive property, $[Gz \leq GGz (= Gw)]$. This tells us that the contractive condition upon (T, G) is applicable at (z, w); and yields (by the properties we just listed)

$$d(Tz, Tw) \leq \varphi(B(\Theta; G)(z, w)).$$

On the other hand, the relations above give

$$d(Tz, Tw) = d(w, Tw),\ d(Gz, Gw) = d(w, Gw) = d(w, Tw),$$
$$d(Gz, Tz) = 0,\ d(Gw, Tw) = 0,$$
$$d(Gz, Tw) = d(w, Tw),\ d(Tz, Gw) = d(w, Gw) = d(w, Tw).$$

Replacing in the contractive relation yields

$d(w, Tw) \leq \varphi(d(w, Tw))$; wherefrom $d(w, Tw) = 0$;
so that (as d=metric), $w = Tw (= Gw)$;

telling us that $w \in \text{cfp}(T, G)$. The proof is thereby complete. □

Summing up, the meta-equivalence deductible from our developed argument may be simply expressed as:

(meta-eq) the common fixed point result (JM-cp-oms)
is deductible from (hence, equivalent with)
the fixed point result (L-fp-oms).

This is also true for some other statements in this area; such as the ones in Chandok [46] and Radenović et al. [191]; we shall discuss these elsewhere.

4.2.7 *Particular cases*

Let $(X, d, \leq)$ be a quasi-ordered metric space; and $T, G : X \to X$ be self-maps of X. In the following, two basic particular cases of the result (JM-cp-oms) are considered.

(I) Suppose that $(\leq) = X \times X$ (the *trivial quasi-order* on X). Then

(I-0) the $(\leq, G)$-asingleton property of a subset Y in X means
$(y_1, y_2 \in Y \Longrightarrow Gy_1 = Gy_2)$; and is referred to as: Y is *G-asingleton*

(I-1) the ($\leq$)-semi-progressive and ($\leq$)-increasing conditions upon (T, G) are holding (in a formal way)
(I-2) the strong cp-Picard property (modulo $(d, \leq)$) and Bellman cp-Picard property (modulo $(d, \leq)$) relative to (T, G) are identical
(I-3) the $(d, \leq; G)$-selfclosed property upon ($\leq$) is fulfilled
(I-4) G is cp(T, G)-($\leq$)-self-progressive.

Finally, let the remaining conventions (modulo $(d, \leq)$) be referred to as conventions (modulo d). In particular, the contraction property writes:

Given $\Theta \in \exp(\{1;;2,3,4\})$, $\varphi \in \mathcal{F}(R_+)$, let us say that the couple (T, G) is $(d; B(\Theta; G); \varphi)$-*contractive*, provided

(G-con) $d(Tx, Ty) \leq \varphi(B(\Theta; G)(x, y))$, for all $x, y \in X$.

The following common fixed point statement (referred to as: Jungck-Matkowski common fixed point theorem on metric spaces; in short: (JM-cp-ms)) is holding.

Theorem 4.8. *Suppose that the couple (T, G) over $\mathcal{F}(X)$ fulfills*

(71-i) (T, G) is normal (i.e.: admissible, compatible, similar)
(71-ii) (T, G) is $(d; B(\Theta; G); \varphi)$-contractive, where $\Theta \in \exp(\{1;;2,3,4\})$ and $\varphi \in \mathcal{F}_0(re, in)(R_+)$ is endowed with the Meir-Keeler admissible property.

In addition, let X be $(d; G)$-complete. Then,

(71-a) (T, G) is cp-G-asingleton (see above)
(71-b) (T, G) is a strong cfp-cp-Picard couple (modulo d).

In particular, the $B(\{1\}; G)$-variant of this result based on contractive conditions for the couple (T, G) like

$$d(Tx, Ty) \leq \varphi(A_1(G)(x, y)) = \varphi(d(Gx, Gy)), \text{ for all } x, y \in X,$$

includes the one due to Jungck [132], proved by a direct method. Note that, in such a (particular) case, the similarity condition upon (T, G) is deductible from the underlying contractive property, in view of $\varphi(0) = 0$. Further aspects may be found in Rhoades [196], [197].

Concerning the deduction of this result, the following is valid

Proposition 4.16. *We have, in (ZF-AC+DC)*

(JM-cp-ms) $\Longrightarrow$ (L-fp-ms) $\Longrightarrow$ (JM-cp-ms).

In other words: Jungck-Matkowski coincidence point result (JM-cp-ms) is equivalent with Leader fixed point result (L-fp-ms).

To verify this, it will suffice taking as comparison term in our previous equivalence results, the Bellman part of (JM-cp-oms) corresponding to the trivial quasi-order; we do not give details.

(II) Suppose now that $(\leq)$ is antisymmetric; hence, a (partial) order on X. The basic facts to be stressed are

(II-1) the similarity condition upon (T, G) is a consequence of increasing property for the same. In fact, if $Gx = Gy$ then $Gx \leq Gy$, $Gy \leq Gx$; wherefrom (by the increasing condition) $Tx \leq Ty$, $Ty \leq Tx$; and this yields $Tx = Ty$. However, for technical reasons, it will be useful to consider it as a basic one.

(II-2) the associated quasi-order $(\preceq)$ over X^* is antisymmetric; hence, a (partial) order.

Putting these together, one gets the following common fixed point statement (referred to as: Jungck-Matkowski common fixed point theorem on partially ordered metric spaces; in short: (JM-cp-poms)).

Theorem 4.9. *Suppose that the couple (T, G) over $\mathcal{F}(X)$ fulfills*

(72-i) (T, G) is normal, $(\leq)$-semi-progressive, $(\leq)$-increasing

(72-ii) the couple (T, G) is $(d, \leq; B(\Theta; G); \varphi)$-contractive, where $\Theta \in \exp(\{1; ; 2, 3, 4\})$ and $\varphi \in \mathcal{F}_0(re, in)(R_+)$ is Meir-Keeler admissible.

In addition, let X be $(d, \leq; G)$-complete. Then,

(72-a) *(T, G) is cp-$(\leq, G)$-asingleton*

(72-b) *If in addition T is $(d, \leq; G)$-continuous, then (T, G) is a strong cp-Picard couple (modulo $(d, \leq)$); and if (in addition to this) G is* cp(T, G)-*involutive, then (T, G) is a strong cfp-cp-Picard couple (modulo $(d, \leq)$)*

(72-c) *If in addition $(\leq)$ is $(d, \leq; G)$-selfclosed then (T, G) is a Bellman cfp-cp-Picard couple (modulo $(d, \leq)$); and if (in addition) G is* cp(T, G)-*$(\leq)$-self-progressive, then, necessarily, (T, G) is a Bellman cfp-cp-Picard couple (modulo $(d, \leq)$).*

Concerning the deduction of this result, the following is valid

Proposition 4.17. *Under the precise conditions, one has in (ZF-AC+DC)*

(JM-cp-poms) $\Longrightarrow$ *(L-fp-poms)* $\Longrightarrow$ *(JM-cp-poms);*

In other words: Jungck-Matkowski coincidence point result (JM-cp-poms) is equivalent with Leader fixed point result (L-fp-poms).

The proof mimics the corresponding one proposed over the quasi-ordered case; so, further details are not necessary. Note that a direct method for getting these conclusions is valid, under the lines in Jachymski [120]. Further aspects may be found in Sessa et al. [212]; see also Di Bari and Vetro [77].

4.2.8 *Related approach*

Let $(X, d, \leq)$ be a (partially) ordered metric space; and $T, G : X \to X$, be a couple of selfmaps. Remember that

(cp-def) $z \in X$ is a *coincidence point* of (T, G), provided $Tz = Gz$; the set of all these will be denoted as $\text{cp}(T, G)$
(cfp-def) z is a *common fixed point* of (T, G), when $Tz = Gz = z$; the set of all these will be denoted $\text{cfp}(T, G)$.

Sufficient conditions for getting such points were described in Jungck-Matkowski common fixed point theorem on partially ordered metric spaces (JM-cp-poms). Recently, a different approach in the area was proposed by Ćirić et al. [61]. It is our aim in the following to study its relationships between our theorem and the statement we just evoked.

(A) Let the couple (T, G) be such that

(cond-1) (T, G) is normal (i.e.: admissible, compatible, similar)
(cond-2) (T, G) is $(\leq)$-semi-progressive
(cond-3) (T, G) is $(\leq)$-increasing.

[As already precise, the similarity condition upon (T, G) follows from the increasing property of the same (since $(\leq)$ is a (partial) order); but this is not essential for us.] The general setting under which the common fixed point problem is to be solved was described in a previous place. It remains now to discuss the contractive condition to be used.

Remember that, the following notations were introduced for $x, y \in X$

$$Q_0(G)(x, y) = d(Tx, Ty),\ Q_1(G)(x, y) = d(Gx, Gy),$$
$$Q_2(G)(x, y) = d(Gx, Tx),\ Q_3(G)(x, y) = d(Gy, Ty),$$
$$Q_4(G)(x, y) = d(Gx, Ty),\ Q_5(G)(x, y) = d(Tx, Gy).$$

Further, as a by-product of this, we constructed the family of functions

$$A_1(G) = Q_1(G),\ A_2(G) = (1/2)[Q_2(G) + Q_3(G)],$$
$$A_3(G) = \max\{Q_2(G), Q_3(G)\},\ A_4(G) = (1/2)[Q_4(G) + Q_5(G)];$$

or, explicitly, for $x, y \in X$

$$A_1(G)(x, y) = d(Gx, Gy),$$
$$A_2(G)(x, y) = (1/2)[d(Gx, Tx) + d(Gy, Ty)],$$
$$A_3(G)(x, y) = \max\{d(Gx, Tx), d(Gy, Ty)\},$$
$$A_4(G)(x, y) = (1/2)[d(Gx, Ty) + d(Tx, Gy)].$$

Finally, denote

$$\mathcal{M}(G) = \{A_1(G), Q_2(G), Q_3(G), A_4(G)\}.$$

Given $\psi \in \mathcal{F}(R_+)$, let us say that (T, G) is $(d, \leq; \mathcal{M}(G); \psi)$-*contractive*, if

(M-con) $d(Tx, Ty) \leq \max\{\psi(M(x, y)); M \in \mathcal{M}(G)\}$,
for all $x, y \in X$, with $Gx \leq Gy$.

The following common fixed point result in Cirić et al. [61] (referred to as: Cirić-Cakić-Rajović-Ume coincidence point result on partially ordered metric spaces; in short: (CCRU-cp-poms)) is our starting point.

Theorem 4.10. *Suppose that the couple (T, G) over $\mathcal{F}(X)$ fulfills*

(81-i) (T, G) is normal, $(\leq)$-semi-progressive, $(\leq)$-increasing
(81-ii) (T, G) is $(d, \leq; \mathcal{M}(G); \psi)$-contractive, where the function $\psi \in \mathcal{F}_0(re)(R_+)$ is continuous.

In addition, suppose that

(81-iii) X is d-complete and $G(X)$ is d-closed
(81-iv) $(\leq)$ is $(d, \leq; G)$-selfclosed and G is cp(T, G)*-self-progressive.*

Then, cfp(T, G) *is nonempty.*

B) For the moment, (CCRU-poms) is not comparable with Jungck-Matkowski common fixed point theorem on partially ordered metric spaces (JM-cp-poms). This forces us to get an appropriate form of it, so as to realize this comparison process. Some preliminaries are needed. For each $\psi \in \mathcal{F}_0(re)(R_+)$, denote

(phi-def) $\varphi(t) = \sup\{\psi(s); 0 \leq s \leq t\},\ t \in R_+$.

By the imposed regressive property, the resulting object is an element of $\mathcal{F}_0(R_+)$. The remaining properties of it are concentrated in

Proposition 4.18. *Assume (in addition) that $\psi \in \mathcal{F}_0(re)(R_+)$ is continuous. Then, its associated function φ has the properties*

(81-1) φ *is increasing,* $\varphi(0) = 0$ *and* $\psi(t) \leq \varphi(t)$, $\forall t \in R_+$
(81-2) $\varphi(t) < t$, $\forall t > 0$; *hence* $\varphi \in \mathcal{F}_0(re, in)(R_+)$
(81-3) φ *is strongly regressive* [$\varphi(t+0) < t$, $\forall t > 0$]; *hence,*
φ *is Matkowski admissible (or, equivalently: Meir-Keeler admissible).*

Proof. i) Evident, by definition (and $\psi(0) = 0$).

ii) Fix some $t > 0$. As ψ is continuous, we must have

$$\varphi(t) = \psi(r), \text{ for some } r \in [0, t].$$

If $r = 0$, then $\varphi(t) = \psi(0) = 0 < t$; and, if $r > 0$, then $\varphi(t) = \psi(r) < r \leq t$.

iii) Let $t > 0$ be arbitrary fixed. By the above,

$$\psi(s) \leq \varphi(t) < t, \ \forall s \in [0, t].$$

Combining with the continuity of ψ, there exists $\delta > 0$ such that $\psi(s) < t$, $\forall s \in [0, t+\delta]$. This (again via $\psi =$ continuous) gives (by means of a previous representation formula)

$$\varphi(t+\delta) < t; \text{ whence, } \varphi(t+0) = \inf\{\varphi(t+\eta); \eta > 0\} < t;$$

and the proof is complete. □

Having these precise, we are now in position to give the announced answer to our posed problem. Let the couple of selfmaps (T, G) be arbitrary for the moment.

Proposition 4.19. *Suppose that (T, G) is $(d, \leq; \mathcal{M}(G); \psi)$-contractive, where $\psi \in \mathcal{F}_0(re)(R_+)$ is continuous. Then,*

(T, G) is $(d, \leq; K(G); \varphi)$-contractive, where $K(G) := B(\{1, 3, 4\}; G)$ and $\varphi \in \mathcal{F}_0(re, in)(R_+)$ is given as before.

Proof. Let $x, y \in X$ be such that $Gx \leq Gy$. By the definitions above,

$$\psi(A_1(G)(x,y)) \leq \varphi(A_1(G)(x,y)) \leq \varphi(K(G)(x,y)),$$
$$\psi(Q_2(G)(x,y)) \leq \varphi(Q_2(G)(x,y)) \leq \varphi(K(G)(x,y)),$$
$$\psi(Q_3(G)(x,y)) \leq \varphi(Q_3(G)(x,y)) \leq \varphi(K(G)(x,y)),$$
$$\psi(A_4(G)(x,y)) \leq \varphi(A_4(G)(x,y)) \leq \varphi(K(G)(x,y));$$

and, from this, we are done. □

The following consequence of this is now holding.

Proposition 4.20. *We have, in the reduced system (ZF-AC+DC)*

$$\text{(L-fp-poms)} \Longrightarrow \text{(JM-cp-poms)} \Longrightarrow \text{(CCRU-cp-poms)}.$$

In other words: the Ćirić-Cakić-Rajović-Ume coincidence point result on partially ordered metric spaces (CCRU-cp-poms) is but a variant of Leader fixed point result on partially ordered metric spaces (L-fp-poms).

Proof. The first inclusion is clear, by a previous remark. For the second one, let the premises of (CCRU-cp-poms) be admitted. To apply (JM-cp-poms) upon this setting, the only point to be clarified is that of our ambient space X being $(d, \leq; G)$-complete. Let (x_n) be a sequence in X, with

$$(Gx_n) \text{ is ascending and } d\text{-Cauchy (in } G(X)).$$

As $G(X)$ is d-complete, there exists some $x \in X$, with $Gx_n \xrightarrow{d} Gx \in G(X)$; and, from this, all is clear. □

Summing up, the common fixed point result in Ćirić et al. [61] is reducible, via factorial techniques, to the fixed point statement in Leader [161]. Note that similar reductions to corresponding fixed point statements are to be derived for the coupled coincidence point results in Lakshmikantham and Ćirić [158] or Kadelburg et al. [134]. Finally, the conclusion is also retainable for multiple coupled points introduced under the lines in Mustafa et al. [174]. Further aspects may be found in Cho et al. [54].

4.3 Kannan Spectral Contractive Maps in Relational Metric Spaces

Abstract. Some fixed point results are given for a class of Meir-Keeler contractive maps with scattered spectrum acting on relational metric spaces. Technical connections with the corresponding statements in this area established by Choudhury and Kundu [Demonstr. Math., 46 (2013), 327–334] are also being discussed.

Subject-Class 2020. 47H10 (Primary), 54H25 (Secondary).

Keywords. Metric space, relation, fixed point, strong and almost Bellman Picard operator, scattered set, spectrum, Meir-Keeler contraction.

4.3.1 *Dependent Choice principles*

See Ch. 1.

4.3.2 *Conv-Cauchy structures*

See Ch. 1.

4.3.3 *Classes of natural progressions*

See Ch. 1.

4.3.4 *Geometric-asymptotic relations*

See Ch. 1.

4.3.5 *Banach contractions and relatives*

See Ch. 1.

4.3.6 *Statement of the problem*

Let X be a nonempty set. Then, let $d : X \times X \to R_+$ be a metric and $\mathcal{R}$ be a relation over X; the triple $(X, d, \mathcal{R})$ will be called a *relational metric space.* Call $Y \in \exp[X]$, *$\mathcal{R}$-almost singleton* (in short: *$\mathcal{R}$-asingleton*) provided $[y_1, y_2 \in Y,\ y_1 \mathcal{R} y_2 \Longrightarrow y_1 = y_2]$; and *$\mathcal{R}$-singleton* when, in addition, $Y \in \exp(X)$. Further, take some $T \in \mathcal{F}(X)$, and put $\text{Fix}(T) := \{z \in X; z = Tz\}$; each point of this set will be referred to as *fixed* under T. In the following, existence and uniqueness results involving such points will be established, over this framework, by starting from the basic hypotheses

(s-prog) T is *$\mathcal{R}$-semi-progressive* ($X(T,\mathcal{R}) := \{x \in X; x\mathcal{R}Tx\} \neq \emptyset$)
(incr) T is *$\mathcal{R}$-increasing* ($x\mathcal{R}y$ implies $Tx\mathcal{R}Ty$).

The directions under which the investigations be conducted are described by the list below (comparable with the one in Turinici [246]):

rpic-0) We say that T is *fix-$\mathcal{R}$-asingleton* if Fix(T) is $\mathcal{R}$-asingleton; and *fix-$\mathcal{R}$-singleton*, provided Fix(T) is $\mathcal{R}$-singleton

rpic-1) We say that T is a *Picard operator* (modulo $(d,\mathcal{R})$) if, for each $x \in X(T,\mathcal{R})$, $(T^n x; n \geq 0)$ is d-Cauchy

rpic-2) We say that T is a *strong Picard operator* (modulo $(d,\mathcal{R})$) if, for each $x \in X(T,\mathcal{R})$, $(T^n x)$ is d-convergent with $\lim_n(T^n x) \in$ Fix(T)

rpic-3) We say that T is an *almost Bellman Picard operator* (modulo $(d,\mathcal{R})$) if, for each $x \in X(T,\mathcal{R})$, $(T^n x; n \geq 0)$ is d-convergent with $\lim_n(T^n x)$ belonging to Fix(T) and $(T^n x; n \geq 0)\mathcal{R}\mathcal{R}\lim_n(T^n x)$.

Here, given the sequence (z_n) in X and the point $z \in X$, we put

(def-1) $(z_n; n \geq 0)\mathcal{R}z$, if $z_n\mathcal{R}z$, for all n
(def-2) $(z_n; n \geq 0)\mathcal{R}\mathcal{R}z$, if there exists a subsequence
$(w_n = z_{i(n)}; n \geq 0)$ of $(z_n; n \geq 0)$, with $(w_n; n \geq 0)\mathcal{R}z$.

The sufficient conditions for such properties involve *ascending orbital full* concepts (in short: (a-o-f)-concepts). Call the sequence (z_n) in X,

$\mathcal{R}$-ascending, if $z_i\mathcal{R}z_{i+1}$ for all $i \geq 0$;
T-orbital, when $(z_n = T^n x; n \geq 0)$, for some $x \in X$;
full, provided $n \mapsto z_n$ is *injective* ($i \neq j$ implies $z_i \neq z_j$);

the intersection of these notions is just the precise one.

rreg-1) Call the ambient space X, *(a-o-f,d)-complete* provided (for each (a-o-f)-sequence) d-Cauchy $\Longrightarrow$ d-convergent

rreg-2) We say that T is *(a-o-f,d)-continuous*, if ((z_n) is (a-o-f)-sequence and $z_n \xrightarrow{d} z$) imply $Tz_n \xrightarrow{d} Tz$

rreg-3) Call $\mathcal{R}$, *(a-o-f,d)-almost-selfclosed*, if: [$(z_n; n \geq 0)$ is (a-o-f)-sequence and $z_n \xrightarrow{d} z$] imply $(z_n; n \geq 0)\mathcal{R}\mathcal{R}z$.

As a completion of these facts, we must formulate the specific requirements and metrical contractive conditions upon our data. Denote

$$\widetilde{\mathcal{R}} := \mathcal{R} \setminus \mathcal{I} \text{ (the \textit{irreflexive} part of } \mathcal{R}).$$

Further, let us introduce the family of maps (for $x, y \in X$)

$$A_1(x,y) = d(x,y),\ A_2(x,y) = \max\{d(x,Tx), d(y,Ty)\},$$

$A_3(x,y) = (1/2)[d(x,Tx) + d(y,Ty)]$,
$A_4(x,y) = (1/2)[d(x,Ty) + d(Tx,y)]$,
$A_5(x,y) = \max\{(1/2)d(x,Ty), d(Tx,y)\}$,
$L(x,y) = \min\{d(x,Tx), d(y,Ty), d(x,y), d(Tx,Ty)\}$,
$M(x,y) := \text{sdiam}(x,Tx,y,Ty)$.

Here, for each 4-tuple $(x_1,x_2,x_3,x_4) \in X^4$, we let

$$\text{sdiam}(x_1,x_2,x_3,x_4) = \max\{d(x_i,x_j); i,j \in \{1,2,3,4\}, i < j\}$$

stand for its *semi-diameter*.

(I) Having these precise, let $P \in \mathcal{F}(X \times X, R_+)$ be a map. There are three groups of specific requirements (to be optionally fulfilled by $P(.,.)$).

I-1) The first group consists of *positivity and boundedness conditions*

(L-pos) P is *L-positive*: $x\mathcal{R}y$, $L(x,y) > 0$ imply $P(x,y) > 0$;
or, equivalently: $x\widetilde{\mathcal{R}}y$, $L(x,y) > 0$ imply $P(x,y) > 0$
(pos) P is *positive*: $x\mathcal{R}y$, $A_1(x,y) > 0$ imply $P(x,y) > 0$;
or, equivalently: $x\widetilde{\mathcal{R}}y$ implies $P(x,y) > 0$
(t-bd) P is *telescopic bounded*: $x\widetilde{\mathcal{R}}Tx \implies P(x,Tx) \le A_2(x,Tx)$.

It is evident, by definition, that

(pos-L-pos) (P is positive) implies (P is L-positive);

but, the reciprocal inclusion is not in general true.

I-2) The second group consists of *semi-diametral* concepts, as

(a-s-diam) P is *almost semi-diametral*:
for each (ε,δ) with $\varepsilon > \delta > 0$, there exists $\gamma \in]0,\delta[$ such that
($x\mathcal{R}y$, $d(x,y) \ge \varepsilon$, $A_2(x,y) < \gamma$, $M(x,y) < \varepsilon+\gamma$) imply $P(x,y) < \varepsilon+\delta$
(s-diam) P is *semi-diametral*:
for each $x,y \in X$ with $x\mathcal{R}y$, we have $P(x,y) \le M(x,y)$.

Concerning these, the following criterion is useful.

Proposition 4.21. *Under the above conventions,*

(61-1) If the mapping $P : X \times X \to R_+$ is semi-diametral, then it is almost semi-diametral
(61-2) If the maps $P_1, P_2 \in \mathcal{F}(X \times X, R_+)$ are almost semi-diametral, then $P_3 := \max\{P_1,P_2\}$ is almost semi-diametral too.

Proof. i): Let (ε,δ) with $\varepsilon > \delta > 0$ be given; and fix some $\gamma \in]0,\delta[$. Given $x,y \in X$ with $x\mathcal{R}y$, $d(x,y) \ge \varepsilon$, $A_2(x,y) < \gamma$, $M(x,y) < \varepsilon+\gamma$, we have

$$P(x,y) \le M(x,y) < \varepsilon + \gamma < \varepsilon + \delta \text{ (as } P \text{ is semi-diametral);}$$

and the claim follows.

ii): Given (ε, δ) with $\varepsilon > \delta > 0$, let $\gamma_1 \in]0, \delta[$ and $\gamma_2 \in]0, \delta[$ be assured by the almost semi-diametral property of P_1 and P_2, respectively. Then, put $\gamma_3 = \min\{\gamma_1, \gamma_2\}$; we claim that the desired property of $P_3 = \max\{P_1, P_2\}$ is fulfilled with respect to this number. In fact, given $x, y \in X$ with

$$x\mathcal{R}y,\ d(x,y) \ge \varepsilon,\ A_2(x,y) < \gamma_3,\ M(x,y) < \varepsilon + \gamma_3,$$

we have, by the very choice of P_1 and P_2,

$$x\mathcal{R}y,\ d(x,y) \ge \varepsilon,\ A_2(x,y) < \gamma_1,\ M(x,y) < \varepsilon + \gamma_1;\ \text{so, } P_1(x,y) < \varepsilon + \delta;$$
$$x\mathcal{R}y,\ d(x,y) \ge \varepsilon,\ A_2(x,y) < \gamma_2,\ M(x,y) < \varepsilon + \gamma_2;\ \text{so, } P_2(x,y) < \varepsilon + \delta.$$

Putting these together yields $P_3(x,y) < \varepsilon + \delta$; and the assertion follows. □

I-3) The last group consists of *orbital asymptotic properties*, as

(sg-asy) P is *orbitally singular asymptotic*: for each (a-o-f) sequence $(x_n; n \ge 0)$ in $X(T, \mathcal{R})$ with $[x_n \xrightarrow{d} z,\ d(z, Tz) > 0]$, and each subsequence $(u_n := x_{i(n)}; n \ge 0)$ of $(x_n; n \ge 0)$ with $[u_n \widetilde{\mathcal{R}} z,\ Tu_n \widetilde{\mathcal{R}} Tz,\ \forall n]$, we have $\liminf_n P(u_n, z) < d(z, Tz)$

(r-asy) P is *orbitally regular asymptotic*: for each (a-o-f) sequence $(x_n; n \ge 0)$ in $X(T, \mathcal{R})$ with $[x_n \xrightarrow{d} z,\ d(z, Tz) > 0]$, and each subsequence $(u_n := x_{i(n)}; n \ge 0)$ of $(x_n; n \ge 0)$ with $[u_n \widetilde{\mathcal{R}} z,\ Tu_n \widetilde{\mathcal{R}} Tz,\ \forall n]$, we have $\lim_n P(u_n, z) = d(z, Tz)$

(s-r-asy) P is *orbitally strongly regular asymptotic*: for each (a-o-f) sequence $(x_n; n \ge 0)$ in $X(T, \mathcal{R})$ with $[x_n \xrightarrow{d} z,\ d(z, Tz) > 0]$, and each subsequence $(u_n := x_{i(n)}; n \ge 0)$ of $(x_n; n \ge 0)$ with $[u_n \widetilde{\mathcal{R}} z,\ Tu_n \widetilde{\mathcal{R}} Tz,\ \forall n]$, we have $P(u_n, z) \to\to d(z, Tz)$.

Here, given the sequence $(r_n; n \ge 0)$ in R and the point $r \in R$, we denoted

$r_n \to\to r$, when $(r_n \to r)$ and there exists a subsequence
$(s_n = r_{i(n)}; n \ge 0)$ of $(r_n; n \ge 0)$ such that $(s_n = r$, for all $n)$.

(II) We are now passing to the contractive properties to be used. These consist in a "relational" variant of the Meir-Keeler condition [168].

Given the mapping $P : X \times X \to R_+$, let us say that T is *Meir-Keeler* $(d, \mathcal{R}; P)$*-contractive*, in case

(mk-1) $[x\mathcal{R}y,\ P(x,y)>0]$ imply $d(Tx,Ty)<P(x,y)$;
referred to as: T is *strictly nonexpansive* (modulo $(d,\mathcal{R};P)$)
(mk-2) $\forall \varepsilon>0,\ \exists\delta>0$: $[x\mathcal{R}y,\ \varepsilon<P(x,y)<\varepsilon+\delta] \Longrightarrow d(Tx,Ty)\le\varepsilon$;
expressed as: T has the *Meir-Keeler property* (modulo $(d,\mathcal{R};P)$).

Note that, by the former of these, the Meir-Keeler property writes

(mk-2a) $\forall \varepsilon>0,\ \exists\delta>0$: $[x\mathcal{R}y,\ 0<P(x,y)<\varepsilon+\delta] \Longrightarrow d(Tx,Ty)\le\varepsilon$.

In particular, when $\mathcal{R}=X\times X$ and $P=A_1$, this convention is comparable with the one due to Meir and Keeler [168].

A geometric version of the above concept may be given as below. Remember that the relation $\Omega\subseteq R^0_+\times R^0_+$ is called *upper diagonal*, provided

(u-diag) $(t,s)\in\Omega$ implies $t<s$;

the class of all these will be denoted as udiag(R^0_+). Further, let us introduce the conditions (over the class udiag(R^0_+))

(g-mk) Ω has the *geometric Meir-Keeler property*:
$\forall\varepsilon>0,\ \exists\delta>0$: $t\Omega s,\ \varepsilon<s<\varepsilon+\delta \Longrightarrow t\le\varepsilon$
(g-bila-s) Ω is *geometric bilateral separable*:
$\forall\beta>0,\ \exists\gamma\in]0,\beta[,\ \forall(t,s)$: $t,s\in]\beta-\gamma,\beta+\gamma[\Longrightarrow (t,s)\notin\Omega$
(g-left-s) Ω is *geometric left separable*:
$\forall\beta>0,\ \exists\gamma\in]0,\beta[,\ (\forall t)$: $t\in]\beta-\gamma,\beta[\Longrightarrow (t,\beta)\notin\Omega$.

The asymptotic (sequential) counterparts of these may be written as

(a-mk) Ω is *asymptotic Meir-Keeler*:
there are no strictly descending sequences (t_n) and (s_n) in R^0_+ and no elements ε in R^0_+, with $((t_n,s_n)\in\Omega,\ \forall n)$ and $(t_n\to\varepsilon+,\ s_n\to\varepsilon+)$
(a-bila-s) Ω is *asymptotic bilateral separable*:
there are no sequences $(t_n;n\ge0)$ and $(s_n;n\ge0)$ in R^0_+ and no elements β in R^0_+, with $((t_n,s_n)\in\Omega,\ \forall n)$ and $(t_n\to\beta,\ s_n\to\beta)$
(a-left-s) Ω is *asymptotic left separable*:
there are no strictly ascending sequences (t_n) in R^0_+ and no elements β in R^0_+, with $((t_n,\beta)\in\Omega,\ \forall n)$ and $(t_n\to\beta-)$.

Here, for each sequence (r_n) in R and each element $r\in R$, we denoted

$r_n\to r+$ (resp., $r_n\to r-$), if $r_n\to r$ and $[r_n>r$ (resp., $r_n<r$), $\forall n]$.

By a previous auxiliary fact, we have that any geometric concept is equivalent with its asymptotic counterpart.

Now, given the relation $\Omega \subseteq R_+^0 \times R_+^0$, let us say that the selfmap T of X is $(d, \mathcal{R}; P; \Omega)$-*contractive*, provided

$$(d(Tx, Ty), P(x, y)) \in \Omega,$$
$$\forall x, y \in X,\ x\mathcal{R}y,\ d(Tx, Ty) > 0,\ P(x, y) > 0.$$

Proposition 4.22. *Suppose that the selfmap T is $(d, \mathcal{R}; P; \Omega)$-contractive where $\Omega \in \exp(R_+^0 \times R_+^0)$ is upper diagonal and geometric Meir-Keeler. Then, T is Meir-Keeler $(d, \mathcal{R}; P)$-contractive.*

Proof. i): Let $x, y \in X$ be such that $x\mathcal{R}y$, $P(x, y) > 0$. If $Tx = Ty$, then $d(Tx, Ty) = 0 < P(x, y)$. Suppose that $Tx \neq Ty$; hence, $d(Tx, Ty) > 0$. Then (by definition)

$$(t, s) \in \Omega; \text{ where } t := d(Tx, Ty),\ s := P(x, y).$$

As $\Omega =$ upper diagonal, this yields $t < s$; that is: $d(Tx, Ty) < P(x, y)$.

ii): Let $\varepsilon > 0$ be arbitrary fixed; and $\delta > 0$ be the number assured by the geometric Meir-Keeler property for Ω. Further, let $x, y \in X$ be such that $x\mathcal{R}y$ and $\varepsilon < s := P(x, y) < \varepsilon + \delta$. If $Tx = Ty$, then $d(Tx, Ty) = 0 < \varepsilon$. Suppose now that $Tx \neq Ty$; whence, $t := d(Tx, Ty) > 0$. By definition, we must have $(t, s) \in \Omega$; and this, along with $\varepsilon < s < \varepsilon + \delta$ (and the geometric Meir-Keeler property for Ω), gives $t \leq \varepsilon$; that is: $d(Tx, Ty) \leq \varepsilon$. □

In the following, a kind of reciprocal for this result is formulated. Given the selfmap T (endowed with the general properties we already specified) and the mapping $P : X \times X \to R_+$, let $\Omega = \Omega[d, \mathcal{R}; P; T]$ be the associated relation over R_+^0:

$$\Omega = \{(d(Tx, Ty), P(x, y)); x\mathcal{R}y, d(Tx, Ty > 0, P(x, y) > 0\}; \text{ so that}$$
$$(\forall (t, s) \in R_+^0 \times R_+^0)\text{: } (t, s) \in \Omega \text{ iff } t = d(Tx, Ty),\ s = P(x, y),$$
$$\text{for some } x, y \in X \text{ with } x\mathcal{R}y,\ d(Tx, Ty) > 0,\ P(x, y) > 0.$$

Proposition 4.23. *Under these conventions,*

(63-1) *If T is Meir-Keeler $(d, \mathcal{R}; P)$-contractive, then the relation $\Omega = \Omega[d, \mathcal{R}; P; T]$ over R_+^0 is upper diagonal and geometric Meir-Keeler*

(63-2) *T is Meir-Keeler $(d, \mathcal{R}; P)$-contractive iff the attached relation $\Omega = \Omega[d, \mathcal{R}; P; T]$ over R_+^0 is upper diagonal and geometric Meir-Keeler.*

Proof. i): Suppose that T is Meir-Keeler $(d, \mathcal{R}; P)$-contractive. There are two steps to be passed.

i-1): Let $(t, s) \in R_+^0 \times R_+^0$ be such that $(t, s) \in \Omega$; hence (by definition)

$t = d(Tx, Ty),\ s = P(x, y)$, where $x\mathcal{R}y,\ d(Tx, Ty) > 0,\ P(x, y) > 0$.

From the strict nonexpansive property of T, we must have $d(Tx, Ty) < P(x, y)$; or, equivalently, $t < s$; which shows that Ω is upper diagonal.

i-2): Let $\varepsilon > 0$ be arbitrary fixed; and $\delta > 0$ be associated by the Meir-Keeler property for T. Further, let $(t, s) \in R^0_+ \times R^0_+$ be such that

$(t, s) \in \Omega$ and $\varepsilon < s < \varepsilon + \delta$.

From the former of these, we have

$t = d(Tx, Ty),\ s = P(x, y)$, where $x\mathcal{R}y,\ d(Tx, Ty) > 0,\ P(x, y) > 0$;

so, combining with the latter

$x\mathcal{R}y,\ d(Tx, Ty) > 0$, and $\varepsilon < P(x, y) < \varepsilon + \delta$.

By the underlying Meir-Keeler-property for T, we get

$d(Tx, Ty) \le \varepsilon$; that is (under our notation): $t \le \varepsilon$;

so that, Ω has the geometric Meir-Keeler property.

ii): Suppose that the associated relation $\Omega = \Omega[d, \mathcal{R}; P; T]$ over R^0_+ is upper diagonal and has the geometric Meir-Keeler property. By the definition of this object, T is $(d, \mathcal{R}; P; \Omega)$-contractive. Combining with the preceding result, one derives that, necessarily, the selfmap T is Meir-Keeler $(d, \mathcal{R}; P)$-contractive. □

As a consequence of this, it follows that the Meir-Keeler $(d, \mathcal{R}; P)$-contractive property of T is finally reducible to the upper diagonal and geometric Meir-Keeler properties for the associated relation $\Omega[d, \mathcal{R}; P; T]$. Concerning this aspect, remember that various examples of such objects were treated in a previous place. Precisely, two basic groups of such contractions may be constructed.

(A) Given $\chi \in \mathcal{F}(R^0_+, R)$, call the selfmap T, $(d, \mathcal{R}; P; \chi)$-*contractive* if

(M-con) $d(Tx, Ty) \le \chi(P(x, y))$,
$\forall x, y \in X,\ x\mathcal{R}y,\ d(Tx, Ty) > 0,\ P(x, y) > 0$.

(This convention is taken from the developments in Matkowski [164].) The basic condition to be imposed upon χ is depicted as

χ is *regressive*: $\chi(t) < t,\ \forall t > 0$;

let $\mathcal{F}(re)(R^0_+, R)$ be the class of all these functions. Then, the extra condition to be imposed upon the class $\mathcal{F}(re)(R^0_+, R)$ writes

χ is *Meir-Keeler admissible*:
$\forall \gamma > 0$, $\exists \beta > 0$: $\gamma < t < \gamma + \beta \Longrightarrow \chi(t) \leq \gamma$.

Proposition 4.24. *Assume that T is $(d, \mathcal{R}; P; \chi)$-contractive, where $\chi \in \mathcal{F}(re)(R^0_+, R)$ is Meir-Keeler admissible. Then, T is necessarily endowed with the Meir-Keeler $(d, \mathcal{R}; P)$-contractive property.*

Proof. We develop two arguments for proving this fact.

Geometric approach. By a previous result, the relation $\Omega := \Omega[\chi]$ introduced as $[t\Omega s$ iff $t \leq \chi(s)]$ is upper diagonal and geometric Meir-Keeler. This, along with

T is $(d, \mathcal{R}; P; \chi)$-contractive iff T is $(d, \mathcal{R}; P; \Omega)$-contractive

gives the desired fact.

Analytic approach. i) Let $x, y \in X$ be such that $x\mathcal{R}y$ and $P(x, y) > 0$. If $Tx = Ty$ then $d(Tx, Ty) = 0 < P(x, y)$. Suppose now that $Tx \neq Ty$; hence $d(Tx, Ty) > 0$. The contractive condition yields $d(Tx, Ty) < P(x, y)$. Summing up, T is strictly nonexpansive (modulo $(d, \mathcal{R}; P)$).

ii) Let $\varepsilon > 0$ be arbitrary fixed; and $\delta > 0$ be the number assured by the Meir-Keeler admissible property of χ. Further, let $x, y \in X$ be such that $x\mathcal{R}y$ and $\varepsilon < P(x, y) < \varepsilon + \delta$. If $Tx = Ty$ then $d(Tx, Ty) = 0 < \varepsilon$. Suppose now that $Tx \neq Ty$; hence $d(Tx, Ty) > 0$. By the contractive condition and admissible property, $d(Tx, Ty) \leq \varphi(P(x, y)) \leq \varepsilon$. Putting these together, one derives that T has the Meir-Keeler property (modulo $(d, \mathcal{R}; P)$). □

(B) Let (ψ, φ) be a couple of functions over $\mathcal{F}(R^0_+, R)$. We say that the selfmap T is $(d, \mathcal{R}; P; (\psi, \varphi))$-*contractive*, provided

$\psi(d(Tx, Ty)) \leq \psi(P(x, y)) - \varphi(P(x, y))$,
for all $x, y \in X$ with $x\mathcal{R}y$, $d(Tx, Ty) > 0$, $P(x, y) > 0$.

The following extra conditions involving this couple will be considered here

(ec-0) (ψ, φ) is a *normal couple*:
ψ is increasing and φ is *strictly positive* ($\varphi(t) > 0$, $\forall t > 0$)
(ec-1) φ is *asymptotic positive*:
for each strictly descending sequence $(t_n; n \geq 0)$ in R^0_+
and each $\varepsilon > 0$ with $t_n \to \varepsilon+$, we have $\limsup_n(\varphi(t_n)) > 0$.

Proposition 4.25. *Suppose that T is $(d, \mathcal{R}; P; (\psi, \varphi))$-contractive, for a normal couple (ψ, φ) of functions over $\mathcal{F}(R^0_+, R)$, with φ = asymptotic positive. Then, T is Meir-Keeler $(d, \mathcal{R}; P)$-contractive, in (ZF-AC+DC).*

Proof. As before, we develop two arguments for proving this fact.

Geometric approach. By a previous result, the associated relation $\Omega := \Omega[\psi, \varphi]$ over R^0_+ introduced as $[t\Omega s$ iff $\psi(t) \leq \psi(s) - \varphi(s)]$ is upper diagonal and geometric Meir-Keeler. This, and

T is $(d, \mathcal{R}; P; (\psi, \varphi))$-contractive iff T is $(d, \mathcal{R}; P; \Omega)$-contractive

give the desired fact.

Analytic approach. i) Let $x, y \in X$ be such that $x\mathcal{R}y$, $P(x, y) > 0$. If $Tx = Ty$, then $d(Tx, Ty) = 0 < P(x, y)$; and we are done. Suppose now that $Tx \neq Ty$; whence, $d(Tx, Ty) > 0$. By the normality of our couple, $\varphi(P(x, y)) > 0$. This, combined with the contractive condition, yields $\psi(d(Tx, Ty)) < \psi(P(x, y))$; wherefrom (via ψ = increasing), $d(Tx, Ty) < P(x, y)$; which tells us that T is strictly nonexpansive (modulo $(d, \mathcal{R}; P)$).

ii) Assume by contradiction that T does not have the Meir-Keeler property (modulo $(d, \mathcal{R}; P)$); that is (for some $\varepsilon > 0$)

$K(\delta) := \{(u, v) \in X \times X; u\mathcal{R}v, \varepsilon < P(u, v) < \varepsilon + \delta,\ d(Tu, Tv) > \varepsilon\}$
is nonempty, for each $\delta > 0$.

Taking a zero converging sequence (δ_n) in R^0_+, we get by the Denumerable Axiom of Choice (AC(N)) [deductible, as precise, in (ZF-AC+DC)], a couple of sequences $(x_n; n \geq 0)$ and $(y_n; n \geq 0)$ in X, so as

$(\forall n)$: $(x_n, y_n) \in K(\delta_n)$; or, equivalently (by the preceding step)
$x_n\mathcal{R}y_n$, $\varepsilon < d(Tx_n, Ty_n) < P(x_n, y_n) < \varepsilon + \delta_n$.

From the contractive condition, we get

$$(0 <)\ \varphi(P(x_n, y_n)) \leq \psi(P(x_n, y_n)) - \psi(d(Tx_n, Ty_n)),\quad \forall n.$$

On the other hand, by the previous relation,

$$t_n := d(Tx_n, Ty_n) \to \varepsilon+,\ s_n := P(x_n, y_n) \to \varepsilon+.$$

Combined with the preceding relation yields (passing to lim sup as $n \to \infty$),

$$(0 <) \limsup_n \varphi(s_n) \leq \psi(\varepsilon+0) - \psi(\varepsilon+0) = 0;\ \text{whence, } \lim_n \varphi(s_n) = 0.$$

On the other hand, by a previous auxiliary fact, there must be a subsequence $(s^*_n := s_{i(n)}; n \geq 0)$ of (s_n), with

$(s_n^*; n \geq 0)$ is strictly descending and $s_n^* \to \varepsilon+$;
with, in addition (by the above) $\lim_n \varphi(s_n^*) = 0$.

This cannot be true, in view of $\varphi =$ asymptotic positive. Hence, the working hypothesis is not acceptable; so that, the Meir-Keeler property follows. □

The introduced concepts may be used to get some practical fix-$\mathcal{R}$-asingleton criteria concerning our selfmap T. Two possible situations occur.

Case 1. A first answer to the posed problem is contained in

Proposition 4.26. *Let T be Meir-Keeler $(d, \mathcal{R}; P)$-contractive, where*

(66-I) $P(.,.)$ is fix positive: $(x, y \in \mathrm{Fix}(T),\ x\widetilde{\mathcal{R}}y) \Longrightarrow P(x, y) > 0$
(66-II) P is fix semi-diametral:
$(x, y \in \mathrm{Fix}(T),\ x\widetilde{\mathcal{R}}y) \Longrightarrow P(x, y) \leq M(x, y)(= d(x, y))$.

Then, T is fix-$\mathcal{R}$-asingleton.

Proof. Let $z_1, z_2 \in \mathrm{Fix}(T)$ be such that $z_1\mathcal{R}z_2$; and assume (by contradiction) that $z_1 \neq z_2$; hence $z_1\widetilde{\mathcal{R}}z_2$. As P is fix positive, we must have $P(z_1, z_2) > 0$. This, by the Meir-Keeler contractive condition, yields

$$d(z_1, z_2) = d(Tz_1, Tz_2) < P(z_1, z_2).$$

On the other hand, as P is fix semi-diametral, we must have

$$P(z_1, z_2) \leq M(z_1, z_2)(= d(z_1, z_2)).$$

This, along with the preceding relation, yields a contradiction. Hence, $z_1 = z_2$; and the assertion is proved. □

In particular, $P(.,.)$ is fix positive, whenever

(pos) P is *positive*: $x\widetilde{\mathcal{R}}y$ implies $P(x, y) > 0$.

For example, $P = A_4$ is fix positive; because

$$x, y \in \mathrm{Fix}(T),\ x\widetilde{\mathcal{R}}y \Longrightarrow A_4(x, y) = d(x, y) > 0.$$

Moreover, by the same relation, $P = A_4$ is semi-diametral; so, the statement above is applicable to this choice.

Case 2. It remains now to discuss the (opposite) alternative

(non-fix-sep) P is *non fix separated*:
there exist $(x, y \in \mathrm{Fix}(T),\ x\widetilde{\mathcal{R}}y)$, with $P(x, y) = 0$.

Two basic sub-cases must be discussed.

Sub-case A. Suppose that the global version of this condition holds

(fix-sing) P is *fix singular*: $(x, y \in \text{Fix}(T),\ x\widetilde{\mathcal{R}}y) \Longrightarrow P(x, y) = 0$.

Note that – unlike the preceding case – the Meir-Keeler contractive condition is of no use in deducing the fix-$\mathcal{R}$-asingleton property for T; because it is applicable only to points $x, y \in X$ with $(x\mathcal{R}y,\ P(x, y) > 0)$. So, the only chance to get the underlying property is that of imposing it – in a formal way – to the selfmap T.

Sub-case B. Suppose that the strong global version of the preceding singularity condition holds

(fix-s-sing) P is *strongly fix singular*:
$(x, y \in \text{Fix}(T),\ x\mathcal{R}y) \Longrightarrow P(x, y) = 0$.

[Clearly, any $P \in \{A_2, A_3\}$ has such a property.] Note that – exactly as before – the Meir-Keeler contractive condition is of no use in deducing the fix-$\mathcal{R}$-asingleton property for T; because it is applicable only to points $x, y \in X$ with $(x\mathcal{R}y,\ P(x, y) > 0)$. But – in compensation to this – certain extra conditions about our data may be used so as to solve our problem. An appropriate example of this type is to be obtained by the condition

(s-comp) T is $(d, \mathcal{R}; P)$*-strongly-compatible*:
$(x, y \in X,\ x\mathcal{R}y,\ P(x, y) = 0) \Longrightarrow (Tx = Ty)$.

We are now in position to formulate a second answer to the posed problem.

Proposition 4.27. *Suppose that P is strongly fix singular and T is $(d, \mathcal{R}; P)$-strongly-compatible. Then, T is fix-$\mathcal{R}$-asingleton.*

Proof. Let $z_1, z_2 \in \text{Fix}(T)$ be such that $z_1\mathcal{R}z_2$. As P is strongly fix singular, we have $P(z_1, z_2) = 0$; so, by the strong compatible condition

$Tz_1 = Tz_2$; or, equivalently, $z_1 = z_2$;

hence the claim. □

Remark 4.3.

Concerning these imposed facts, the following question is naturally arising here. Given $P : X \times X \to R_+$, let us consider the condition

(comp) T is $(d, \mathcal{R}; P)$*-compatible*:
$(x, y \in X,\ x\widetilde{\mathcal{R}}y,\ P(x, y) = 0) \Longrightarrow (Tx = Ty)$.

Formally, we may ask whether this (weaker) compatible requirement upon T may produce (in conjunction with the strongly fix singular condition upon P) the conclusion in the preceding statement. Unfortunately, this is not possible. In fact, the $(d, \mathcal{R}; P)$-compatible condition upon T is without object along the points $x, y \in \text{Fix}(T)$ with $x\widetilde{\mathcal{R}}y$; for, in such a case, the strongly fix singular property of P yields $P(x, y) = 0$; whence [by (comp) above] $Tx = Ty$ (or, equivalently, $x = y$); impossible, via $x\widetilde{\mathcal{R}}y$. Hence – as already precise – the only chance to get the underlying property is that of imposing it – in a formal way – to the selfmap T.

Finally, we show that the strong compatible condition (upon T) is sometimes obtainable from a strong contractive one. A concrete example in this direction may be constructed as follows. Denote

$\mathcal{F}_0(R_+) =$ the class of all $\chi \in \mathcal{F}(R_+)$ with $\chi(0) = 0$.

Given $\chi \in \mathcal{F}_0(R_+)$, call the selfmap T, $(d, \mathcal{R}; P; \chi)$-*strongly-contractive* if

$$d(Tx, Ty) \le \chi(P(x, y)),\ \forall x, y \in X,\ x\mathcal{R}y.$$

Proposition 4.28. *Assume that T is $(d, \mathcal{R}; P; \chi)$-strongly-contractive, where $\chi \in \mathcal{F}_0(R_+)$. Then, T is $(d, \mathcal{R}; P)$-strongly-compatible.*

Proof. Let $x, y \in X$ be such that $x\mathcal{R}y$ and $P(x, y) = 0$. From the strong contractive condition, we have

$$(0 \le) d(Tx, Ty) \le \chi(0) = 0;\ \text{whence, } Tx = Ty;$$

and the conclusion follows. □

In particular, the linear case

$$\chi(t) = \beta t,\ t \in R_+,\ \text{for some } \beta \in [0, 1[$$

enters in this scheme; and this tells us that any selfmap T of X fulfilling a relational type Kannan contractive condition as (for some $P \in \{A_2, A_3\}$)

$$d(Tx, Ty) \le \beta P(x, y),\ x, y \in X,\ x\mathcal{R}y,$$

must be fix-$\mathcal{R}$-asingleton. Further aspects will be delineated elsewhere.

4.3.7 *Main result*

Let $(X, d, \mathcal{R})$ be a *relational metric space.* Further, take a selfmap T of X, in accordance with the conditions

(s-prog) T is *$\mathcal{R}$-semi-progressive* $(X(T,\mathcal{R}) := \{x \in X; x\mathcal{R}Tx\} \neq \emptyset)$
(incr) T is *$\mathcal{R}$-increasing* ($x\mathcal{R}y$ implies $Tx\mathcal{R}Ty$).

Note that, by the imposed properties,

$T(X(T,\mathcal{R})) \subseteq X(T,\mathcal{R})$; meaning: $X(T,\mathcal{R})$ is T-invariant.

The basic directions and regularity conditions under which the problem of determining elements in $\text{Fix}(T)$ is to be solved were already listed; and the contractive type framework was settled. For the last specific condition to be added, we need some preliminaries. Let M be an *unbounded chain* of N, in the sense: $\sup(M) = \infty$. Given the number $k \in N(1, \leq)$, we say that M is *k-scattered*, provided

for each $n \in N$, there exists $m \in M$ with $|n - m| \leq k$;

the class of these writes $\text{sca}(M)$ (the *scattering domain* of M). Clearly,

$\text{sca}(M)$ is *hereditary*: $k \in \text{sca}(M)$, $q \geq k \Longrightarrow q \in \text{sca}(M)$;

if $\text{sca}(M)$ is nonempty, then M is called *scattered* (in N).

A related concept may be given along the lines below. Let M be an unbounded chain of M; clearly, it may be written as $M = \{a_n; n \geq 0\}$, where the sequence $(a_n; n \geq 0)$ in N is strictly ascending. Given $h \in N(1, \leq)$, let us say that M is *h-sub-aprogression* provided $(a_{n+1} - a_n \leq h$, $\forall n)$. The class $\text{sapr}(M)$ of all such $h \in N(1, \leq)$ will be referred to as the *sub-aprogression domain* of M. Clearly,

$\text{sapr}(M)$ is hereditary: $h \in \text{sapr}(M)$, $p \geq h \Longrightarrow p \in \text{sapr}(M)$;

if $\text{sapr}(M)$ is nonempty, we say that M is *sub-aprogression.* Concerning the relationships between these concepts, we have (cf. a previous result)

(sca-sapr) (for each unbounded chain M of N):
M is scattered iff M is sub-aprogression.

Under these preliminaries, we may now formulate the underlying condition. Given $x \in X(T,\mathcal{R})$, the iterative sequence $(T^n x; n \in N)$ is $\mathcal{R}$-ascending and T-orbital. Denote, for simplicity

$\text{spec}(x) = \{i \in N(1, \leq); x\mathcal{R}T^i x\}$ [the *spectrum* of x (modulo $(\mathcal{R}, T)$)].

Note that, by the choice of our element, we have $1 \in \text{spec}(x)$; whence, $\text{spec}(x)$ is not empty. Moreover, by the admitted hypotheses,

$$\text{spec}(x) \subseteq \text{spec}(Tx); \text{ hence, } \text{spec}(x) \subseteq \text{spec}(T^j x), \forall j \geq 0.$$

Call $x \in X(T, \mathcal{R})$, *fully orbital* provided

$$(T^n x; n \geq 0) \text{ is } \textit{full } (n \mapsto T^n x \text{ is an injective map from } N \text{ to } X).$$

The condition to be added writes:

(g-sc) T is *$\mathcal{R}$-globally scattered*: for each fully orbital $x \in X(T, \mathcal{R})$, $\text{spec}(x)$ is (unbounded chain and) scattered.

Note that, by the above equivalence relation (sc-sa), this also writes:

(g-sap) T is $\mathcal{R}$-globally sub-aprogression: for each fully orbital point $x \in X(T, \mathcal{R})$, $\text{spec}(x)$ is (unbounded chain and) sub-aprogression.

To get concrete examples of such couples $(\mathcal{R}, T)$, the following auxiliary fact is useful (cf. Turinici [255]):

Proposition 4.29. *Let the sequence (z_n) in X and the index $k \geq 2$ fulfill*

(z_n) is $\mathcal{R}$-ascending and $\mathcal{R}$ is k-transitive on $Z := \{z_n; n \geq 0\}$
(in the sense: $\mathcal{R}_Z^k \subseteq \mathcal{R}_Z$, where $\mathcal{R}_Z = \mathcal{R} \cap (Z \times Z)$).

Then, necessarily,

$$(z_i, z_{i+1+n(k-1)}) \in \mathcal{R}_Z \subseteq \mathcal{R}, \forall i \geq 0, \forall n \geq 0.$$

Having these precise, the following answer to the posed question is now available. Let us consider the condition

(g-o-tr) T is *$\mathcal{R}$-globally orbital transitive*: for each fully orbital $x \in X(T, \mathcal{R})$, there exists $k(x) \in N(2, \leq)$, such that the restriction of $\mathcal{R}$ to the associated T-orbit $\{T^n x; n \geq 0\}$ is $k(x)$-transitive.

By the above auxiliary fact, we therefore get

$\forall$ fully orbital $x \in X(T, \mathcal{R})$: $\{1 + n(k(x) - 1); n \geq 0\} \subseteq \text{spec}(x)$;
whence, $\text{spec}(x)$ is (unbounded chain and) sub-aprogression;

so that, the global sub-aprogression property is necessarily holding.

We are now in position to state the main result of this exposition (referred to as: *Meir-Keeler geometric theorem on relational metric spaces*; in short (MK-g-rms)).

Theorem 4.11. *Assume that T is ($\mathcal{R}$-semi-progressive, $\mathcal{R}$-increasing, as well as) $\mathcal{R}$-globally sub-aprogression and $(d,\mathcal{R};P;\Omega)$-contractive, where $P: X\times X\to R_+$ and $\Omega\subseteq R^0_+\times R^0_+$ fulfill*

(71-i) P is L-positive, telescopic bounded, and almost semi-diametral
(71-ii) Ω is upper diagonal and geometric Meir-Keeler.

In addition, let X be (a-o-f,d)-complete. Then

(71-a) *T is fix-$\mathcal{R}$-asingleton, provided*

P is fix-positive and fix semi-diametral

(71-b) *T is a strong Picard operator (modulo $(d,\mathcal{R})$), provided (in addition) T is (a-o-f,d)-continuous*

(71-c) *T is almost Bellman Picard (modulo $(d,\mathcal{R})$), if (in addition) $\mathcal{R}$ is (a-o-f,d)-almost-selfclosed and one of the extra conditions holds*

(71-c-i) P is orbitally singular asymptotic
(71-c-ii) P is orbitally regular asymptotic,
and Ω is (in addition) geometric/asymptotic bilateral separable
(71-c-iii) P is orbitally strongly regular asymptotic,
and Ω is (in addition) geometric/asymptotic left separable.

Proof. The first conclusion is clear, by a previous result; so, we have to establish the remaining ones. There are several steps to be passed.

Part 1. We claim that, under the posed conditions,

(2-iter-con) $(d(Tx,T^2x),P(x,Tx))\in\Omega$ and
$d(Tx,T^2x)<P(x,Tx)\le d(x,Tx)$, whenever $x\widetilde{\mathcal{R}}Tx, Tx\widetilde{\mathcal{R}}T^2x$.

In fact, let $x\in X(T,\mathcal{R})$ be as in the premise above. Then, by definition,

$L(x,Tx)=\min\{d(x,Tx),d(Tx,T^2x)\}>0$;
hence $P(x,Tx)>0$ (as P is L-positive);

which tells us that the first half of (2-iter-con) follows. By the upper diagonal property of Ω we must have $d(Tx,T^2x)<P(x,Tx)$. On the other hand, as P is telescopic bounded,

$P(x,Tx)\le A_2(x,Tx)=\max\{d(x,Tx),d(Tx,T^2x)\}$.

Combining with the preceding relation, gives

$d(Tx,T^2x)<\max\{d(x,Tx),d(Tx,T^2x)\}$;
wherefrom, $d(Tx,T^2x)<d(x,Tx)$, $A_2(x,Tx)=d(x,Tx)$;

and the second half of (2-iter-con) follows as well.

Part 2. Take some $x_0 \in X(T, \mathcal{R})$; and put $(x_n = T^n x_0; n \geq 0)$; clearly, this is ascending orbital. If $x_n = x_{n+1}$ for some $n \geq 0$, we are done; so, without loss, one may assume that the non-telescopic relation holds

(non-tele) $(\forall n)$: $x_n \neq x_{n+1}$; hence, $x_n \widetilde{\mathcal{R}} x_{n+1}$, $\rho_n := d(x_n, x_{n+1}) > 0$.

From the preceding part, we derive the iterative contractive relation

(iter-con) $(\forall n)$: $(\rho_{n+1}, P(x_n, x_{n+1})) \in \Omega$, $\rho_{n+1} < P(x_n, x_{n+1}) \leq \rho_n$.

Two basic consequences of this are holding.

Conseq 1. By this relation, the sequence $(\rho_n; n \geq 0)$ is strictly descending; hence, $\rho := \lim_n \rho_n$ exists as an element of R_+ and $(\rho_n > \rho, \forall n)$. Assume by contradiction that $\rho > 0$; and let $\delta > 0$ be the number given by the Meir-Keeler property of Ω. By definition, there exists a rank $n(\delta)$ such that $n \geq n(\delta)$ implies $\rho < \rho_n < \rho + \delta$; hence (by a previous relation)

$(\forall n \geq n(\delta))$: $(\rho_{n+1}, P(x_n, x_{n+1})) \in \Omega$ and
$\rho < \rho_{n+1} < P(x_n, x_{n+1}) \leq \rho_n < \rho + \delta$.

Combining with the Meir-Keeler property we just quoted, yields

$(\forall n \geq n(\delta))$: $\rho < \rho_{n+1} = d(Tx_n, Tx_{n+1}) \leq \rho$; contradiction.

Hence, $\rho = 0$; so that,

$\rho_n := d(x_n, x_{n+1}) \to 0$, as $n \to \infty$; meaning: (x_n) is d-*asymptotic.*

Conseq 2. From the same relation (iter-con), we have

$(x_n; n \geq 0)$ is an (ascending, orbital and) full sequence.

In fact, suppose that this is not true: there exist $i, j \in N$ such that $i < j$, $x_i = x_j$. As (x_n) is iterative, we get $x_{i+1} = x_{j+1}$; hence, $\rho_i = \rho_j$; in contradiction with the strict descending property of $(\rho_n; n \geq 0)$.

Part 3. In particular, $x_0 \in X(T, \mathcal{R})$ is fully orbital. As $(\mathcal{R}, T)$ is globally sub-aprogression, $\text{spec}(x_0) = \{m \in N(1, \leq); x_0 \mathcal{R} x_m\}$ is an unbounded chain and sub-aprogression (in N); so, $\text{spec}(x_0) = \{a(i); i \geq 0\}$, where

(p1) $a(0) = 1$ and $(a(i))$ is strictly ascending ($i < j$ implies $a(i) < a(j)$)
(p2) there exists $k = k(x) \in N(1, \leq)$, such that $(a(i))$ is k-sub-aprogression: $a(i+1) - a(i) \leq k$, for all $i \geq 0$
(p3) in addition (as T is $\mathcal{R}$-increasing) $x_n \mathcal{R} x_{n+a(i)}$, $\forall n \geq 0$, $\forall i \geq 0$.

Having these precise, let $\varepsilon > 0$ be arbitrary fixed; and $\delta > 0$ be assured by the Meir-Keeler property of Ω; without loss, one may assume that $\delta < \varepsilon$. As P is almost semi-diametral, there exists $\gamma \in]0, \delta[$ with

$$(x\mathcal{R}y,\ d(x,y) \geq \varepsilon,\ A_2(x,y) < \gamma,\ M(x,y) < \varepsilon+\gamma) \text{ imply } P(x,y) < \varepsilon+\delta.$$

By the d-asymptotic property, there exists a rank $n(\gamma) \geq 0$, such that

(d-asy) $(\forall n \geq n(\gamma))$: $d(x_n, x_{n+1}) < \gamma/4k$;
whence $d(x_n, x_{n+h}) < h\gamma/4k \leq \gamma/4$, $\forall h \in \{1, ..., k\}$.

We claim, via ordinary induction, that the evaluation

(d-C;j) $d(x_n, x_{n+j}) < \varepsilon + \gamma/2$, $\forall n \geq n(\gamma)$

holds for all $j \geq 1$; wherefrom, $(x_n; n \geq 0)$ is d-Cauchy. The case of $j \in \{1, ..., k\}$ is evident, by the preceding evaluation (d-asy). Assume that (d-C;j) is true for all $j \in \{1, ..., p\}$, where $p \geq k$; we must establish that (d-C;p+1) is also true. So, let $n \geq n(\gamma)$ be arbitrary fixed. As $\text{spec}(x_0) = \{a(i); i \geq 0\}$ and $a(0) = 1$, $p \geq k \geq 1$, there exists a uniquely determined rank $i \geq 0$, such that $a(i) \leq p < a(i+1)$. Denote for simplicity $q = a(i)$, $r = a(i+1)$; we then have

$$q, r \in \text{spec}(x_0),\ 1 \leq q \leq p < r,\ p - q < r - q \leq k.$$

The desired relation holds whenever $d(x_n, x_{n+q}) < \varepsilon$; because, in view of (d-asy) and the direct evaluation $p + 1 \leq q + k$,

$d(x_n, x_{n+p+1}) \leq d(x_n, x_{n+q}) + d(x_{n+q}, x_{n+p+1}) <$
$\varepsilon + k\gamma/4k \leq \varepsilon + \gamma/4 < \varepsilon + \gamma/2$; whence, our assertion is true.

So, in the following, there is no loss in generality by assuming that $d(x_n, x_{n+q}) \geq \varepsilon$. According to a preceding relation, $x_n \widetilde{\mathcal{R}} x_{n+q}$; in addition, from the inductive hypothesis and the asymptotic relation (d-asy),

$d(x_n, x_{n+q}), d(x_{n+1}, x_{n+q}), d(x_{n+1}, x_{n+q+1}) < \varepsilon + \gamma/2$,
$d(x_n, x_{n+1}), d(x_{n+q}, x_{n+q+1}) < \gamma/4k \leq \gamma/4$.

This, along with the triangular inequality, gives us

$d(x_n, x_{n+q+1}) \leq d(x_n, x_{n+q}) + d(x_{n+q}, x_{n+q+1}) <$
$\varepsilon + \gamma/2 + \gamma/4k \leq \varepsilon + \gamma/2 + \gamma/4 = \varepsilon + 3\gamma/4$.

The obtained relations yield, on the one hand

$x_n \widetilde{\mathcal{R}} x_{n+q}$, $d(x_n, x_{n+q}) \geq \varepsilon$, $A_2(x_n, x_{n+q}) < \gamma/4 < \gamma$,
$M(x_n, x_{n+q}) < \varepsilon + \gamma$; whence, $P(x_n, x_{n+q}) < \varepsilon + \delta$
(if we remember that P is almost semi-diametral).

On the other hand, as (x_n) is full,

$L(x_n, x_{n+q}) = \min\{\rho_n, \rho_{n+q}, d(x_n, x_{n+q}), d(x_{n+1}, x_{n+q+1})\} > 0$;
so that $P(x_n, x_{n+q}) > 0$ (in view of P being L-positive).

Putting these together, gives

$(d(x_{n+1}, x_{n+q+1}), P(x_n, x_{n+q})) \in \Omega$ (cf. the contractive condition);

and this in turn yields

$d(x_{n+1}, x_{n+q+1}) \leq \varepsilon$ (by the Meir-Keeler property for Ω).

Combining with triangular inequality and (d-asy), one derives

$$d(x_n, x_{n+p+1}) \leq d(x_n, x_{n+1}) + d(x_{n+1}, x_{n+q+1}) + d(x_{n+q+1}, x_{n+p+1}) \\ < \varepsilon + \gamma/4k + k\gamma/4k \leq \varepsilon + \gamma/4 + \gamma/4 = \varepsilon + \gamma/2;$$

whence, our claim follows.

Part 4. As X is (a-o-f,d)-complete, $x_n \xrightarrow{d} z$, for some (uniquely determined) $z \in X$. By the full property of $(x_n; n \geq 0)$,

$E := \{n \in N; (x_{n+1} =)Tx_n = Tz\}$ is asingleton;

so that, the following assertion is true:

(h-dist) $\exists h = h(z) \geq 0$: $n \geq h \Longrightarrow x_n \neq z$ and $Tx_n \neq Tz$.

There are several cases to discuss.

Case 1. Suppose that T is (a-o-f,d)-continuous. Then $y_n := Tx_n \xrightarrow{d} Tz$ as $n \to \infty$. On the other hand, $(y_n = x_{n+1}; n \geq 0)$ is a subsequence of $(x_n; n \geq 0)$; whence $y_n \xrightarrow{d} z$; and this yields (as d is separated), $z = Tz$.

Case 2. Suppose that $\mathcal{R}$ is (a-o-f,d)-almost-selfclosed. We show that $\beta := d(z, Tz) > 0$ yields a contradiction.

By definition, there exists a subsequence $(u_n := x_{j(n)}; n \geq 0)$ of $(x_n; n \geq 0)$, such that $(u_n \mathcal{R} z, \forall n)$. As $\lim_n j(n) = \infty$, one may arrange for $[j(n) \geq h, \forall n]$; so that, from the working condition above (and T being $\mathcal{R}$-increasing)

(u-z-R) $u_n \widetilde{\mathcal{R}} z$ and $Tu_n \widetilde{\mathcal{R}} Tz$, for all $n \geq 0$.

Further, as $(Tu_n = x_{j(n)+1}; n \geq 0)$ is also a subsequence of $(x_n; n \geq 0)$, we get (in conjunction with a metrical property of d) the limit relations

(lim-rela) $d(u_n, Tu_n) \to 0$, $d(u_n, z)$, $d(Tu_n, z) \to 0$,
$d(u_n, Tz), d(Tu_n, Tz) \to \beta$; whence, $M(u_n, z) \to \beta$.

Finally, by the non-telescopic condition involving $(x_n; n \geq 0)$,

$d(u_n, Tu_n) = d(x_{j(n)}, x_{j(n)+1}) = \rho_{j(n)} > 0$, for all n;

and this, combined with (u-z-R), yields (for the same ranks)

$L(u_n, z) = \min\{d(u_n, Tu_n), \beta, d(u_n, z), d(Tu_n, Tz)\} > 0$;
whence, $P(u_n, z) > 0$ (as P is L-positive).

The geometric contractive condition applies to (u_n, z), $\forall n$; and yields

(contr) $(\forall n)$: $(d(Tu_n, Tz), P(u_n, z)) \in \Omega$; whence,
$d(Tu_n, Tz) < P(u_n, z)$ [via Ω=upper diagonal].

There are several cases to discuss.

Alter 1. Assume that P is orbitally singular asymptotic. From the preceding limit relations, we get

$\beta = \lim_n d(Tu_n, Tz) = \liminf_n d(Tu_n, Tz) \leq \liminf_n P(u_n, z)$;

in contradiction with the imposed upon P assumption.

Alter 2. Suppose that P is orbitally regular asymptotic, and Ω is geometric/asymptotic bilateral separable. From (lim-rela), we have

$d(Tu_n, Tz) \to \beta$ and $P(u_n, z) \to \beta$, as $n \to \infty$.

This, along with the first half of (contr), is not in agreement with the asymptotic bilateral separable property of Ω.

Alter 3. Suppose that P is orbitally strongly regular asymptotic, and Ω is (in addition) geometric/asymptotic left separable. From the previous relations involving (u_n) and z, we have (by the admitted hypothesis)

(rela) $P(u_n, z) \to\to \beta$, as $n \to \infty$;

which, according to definition, means the couple of relations

(rela-1) $P(u_n, z) \to \beta$ as $n \to \infty$; as well as
(rela-2) there must be a subsequence $(v_n := u_{g(n)}; n \geq 0)$
of $(u_n; n \geq 0)$, such that $P(v_n, z) = \beta (> 0)$, for all $n \geq 0$.

Denote $(\alpha_n = d(Tv_n, Tz); n \geq 0)$. As a direct consequence of (contr),

$((\alpha_n, \beta) \in \Omega, \forall n)$; whence, in particular, $(\alpha_n < \beta, \forall n)$;
so that, $\alpha_n \to \beta-$ as $n \to \infty$ (cf. our limit relations).

By a previous auxiliary fact, there exists a subsequence $(\alpha_n^* := \alpha_{f(n)}; n \geq 0)$ of $(\alpha_n; n \geq 0)$, with

$$(\alpha_n^*; n \geq 0) \text{ is strictly ascending, and } \alpha_n^* \to \beta- \text{ as } n \to \infty.$$

On the other hand,

$$(\alpha_n^*, \beta) \in \Omega, \text{ for all } n \text{ (by the subsequential contractive relation).}$$

These, however, are not compatible with the imposed property of Ω.

Summing up, all these alternatives are impossible. This, via their starting point, tells us that the assumption $\beta := d(z, Tz) > 0$ cannot be accepted. Hence, $\beta = 0$; that is: $z = Tz$; and conclusion follows. □

Note that further extensions of these facts are possible, in the realm of triangular symmetric spaces, taken as in Hicks and Rhoades [112]; or, in the setting of partial metric spaces, introduced as in Matthews [167]. On the other hand, multivalued extensions of this result are also possible, under the directions in Nadler [175]; and some dislocated versions of our main statement are available by following the lines in Amini-Harandi [16]; see also Jleli and Samet [128]. Finally, we stress that one may extend all these facts to the setting of relational Branciari metric spaces taken as in Turinici [247]; further details will be given elsewhere.

4.3.8 *Particular cases*

Let $(X, d, \mathcal{R})$ be a relational metric space. Further, let T be a selfmap of X; supposed to be $\mathcal{R}$-semi-progressive and $\mathcal{R}$-increasing. As precise, we have to determine conditions under which Fix(T) is nonempty; and, if this holds, to establish whether Fix(T) is $\mathcal{R}$-singleton. The specific directions under which this problem is to be solved were already listed. Sufficient conditions for getting such properties are being founded on the ascending orbital full concepts we just introduced. Finally, the contractive properties to be used have been described; and the main result incorporating all these is Meir-Keeler geometric theorem on relational metric spaces; in short: (MK-g-rms). It is our aim in the sequel to expose a lot of particular cases of it, with some technical relevance. To do this, remember that we defined a basic family of maps, as: for each $x, y \in X$,

$$A_1(x, y) = d(x, y),\ A_2(x, y) = \max\{d(x, Tx), d(y, Ty)\},$$
$$A_3(x, y) = (1/2)[d(x, Tx) + d(y, Ty)],$$
$$A_4(x, y) = (1/2)[d(x, Ty) + d(Tx, y)],$$

$$A_5(x,y) = \max\{(1/2)d(x,Ty), d(Tx,y)\},$$
$$L(x,y) = \min\{d(x,Tx), d(y,Ty), d(x,y), d(Tx,Ty)\},$$
$$M(x,y) := \mathrm{sdiam}(x,Tx,y,Ty).$$

Here, for each 4-tuple $(x_1,x_2,x_3,x_4) \in X^4$, we let

$$\mathrm{sdiam}(x_1,x_2,x_3,x_4) = \max\{d(x_i,x_j); i,j \in \{1,2,3,4\}, i < j\}$$

stand for its *semi-diameter*. In addition to this, for an appropriate illustration of the alternatives encountered in (MK-g-rms), it would be useful for us to introduce some other parametric functions of this type. To do this, let us define the diagonal type subset of $R_+ \times R_+$

$$\Delta = \{(\lambda,\mu) \in R_+ \times R_+^0; \lambda \leq \mu\}.$$

This set will be written as a union of two disjoint parts, expressed as

$\Delta_s = \{(\lambda,\mu) \in \Delta; \lambda < \mu\}$ (the *singular* part)
$\Delta_r = \{(\lambda,\mu) \in \Delta; \lambda = \mu\} = \{(\nu,\nu); \nu \in R_+^0\}$ (the *regular* part).

For each $(\lambda,\mu) \in \Delta$, let us define the map $B := B[\lambda,\mu] : X \times X \to R_+$, as

$$B(x,y) = d(y,Ty)[\lambda + d(x,Tx)]/[\mu + d(x,y)],\ x,y \in X.$$

Further, let us define

A_s = one of the maps $B[\lambda,\mu]$ with $(\lambda,\mu) \in \Delta_s$;
A_r = one of the maps $B[\lambda,\mu]$ with $(\lambda,\mu) \in \Delta_r$
(or, equivalently: A_r = one of the maps $B[\nu,\nu]$ with $\nu \in R_+^0$);

the reason of such a splitting will become clear a bit further. Finally, given $(\alpha,\beta) \in \Delta$, let us introduce the map $C := C[\alpha,\beta] : X \times X \to R_+$, as

$$C(x,y) = d(Tx,y)[\alpha + d(x,Ty)]/[\beta + d(x,y)],\ x,y \in X.$$

Then, let us define

A_u = one of the maps $C[\alpha,\beta]$ with $(\alpha,\beta) \in \Delta$.

Having these precise, fix in the following the couples $(\lambda,\mu) \in \Delta_s$, $(\nu,\nu) \in \Delta_r$, $(\alpha,\beta) \in \Delta$; and (according to the previous conventions), denote

$$Z = \{1,2,3,4,5,s,r,u\},\ Z_0^* = \exp(Z).$$

For each subset $\Theta \in Z_0^*$, let $E(\Theta) \in \mathcal{F}(X \times X, R_+)$ be the mapping

$$E(\Theta)(x,y) = \max\{A_i(x,y); i \in \Theta\},\ x,y \in X.$$

Now, the maps $P : X \times X \to R_+$ to be considered are of the form

$$P = E(\Theta);\ \text{where } \Theta \in Z_0^*.$$

So, it remains to establish of to what extent is this functional family compatible with the positivity, boundedness, semi-diametral and orbital asymptotic conditions required by our main result. To do this, denote

$$Z_1^* = \{\Theta \in Z_0^*; 1 \in \Theta\},\ Z_2^* = \{\Theta \in Z_0^*; 2 \in \Theta\},\ Z^* = Z_1^* \cup Z_2^*.$$

I) According to a previous convention, P is called

(L-pos) *L-positive*, provided $x\widetilde{\mathcal{R}}y$, $L(x, y) > 0$ imply $P(x, y) > 0$
(pos) *positive*, provided $x\widetilde{\mathcal{R}}y$ implies $P(x, y) > 0$.

Proposition 4.30. *The following assertions hold*

(81-1) The maps $P = E(\Theta)$, where $\Theta \in Z^*$ are L-positive
(81-2) The maps $P = E(\Theta)$, where $\Theta \in Z_1^*$ are positive.

Proof. i): Let $x, y \in X$ be such that $x\widetilde{\mathcal{R}}y$, $L(x, y) > 0$. Then,

$$P \in Z_1^* \text{ implies } P(x, y) \geq A_1(x, y) \geq L(x, y) > 0,$$
$$P \in Z_2^* \text{ implies } P(x, y) \geq A_2(x, y) \geq L(x, y) > 0;$$

and from this, all is clear, by the definition of our class.

ii): Let $x, y \in X$ be such that $x\widetilde{\mathcal{R}}y$; hence, $d(x, y) > 0$. Then,

$$P \in Z_1^* \text{ implies } P(x, y) \geq A_1(x, y) = d(x, y) > 0;$$

and the assertion follows. □

II) Remember that $P : X \times X \to R_+$ is *telescopic bounded*, provided

(t-bd) $x\widetilde{\mathcal{R}}Tx$ implies $P(x, Tx) \leq A_2(x, Tx)$.

Proposition 4.31. *All $P = E(\Theta)$ where $\Theta \in Z^*$, are telescopic bounded.*

Proof. First, we prove that any map $Q \in \{A_i; i \in Z\}$ has such a property. Given the arbitrary point $x \in X$ with $x\widetilde{\mathcal{R}}Tx$, we have

$$A_1(x, Tx) = d(x, Tx) \leq \max\{d(x, Tx), d(Tx, T^2x)\} = A_2(x, Tx),$$
$$A_3(x, Tx) = (1/2)[d(x, Tx) + d(Tx, T^2x)] \leq A_2(x, Tx),$$
$$A_4(x, Tx) = A_5(x, Tx) = (1/2)d(x, T^2x) \leq A_3(x, Tx) \leq A_2(x, Tx),$$
$$B(x, Tx) = d(Tx, T^2x)[\lambda + d(x, Tx)]/[\mu + d(x, Tx)] \leq$$
$$d(Tx, T^2x) \leq \max\{d(x, Tx), d(Tx, T^2x)\} = A_2(x, Tx),$$
$$C(x, Tx) = 0 \leq A_2(x, Tx);$$

where, $B = B[\lambda, \mu]$ with $(\lambda, \mu) \in \Delta$, and $C = C[\alpha, \beta]$ for $(\alpha, \beta) \in \Delta$; wherefrom, combining with any map A_s or A_r having the form $B = B[\lambda, \mu]$, and any map A_u having the form $C = C[\alpha, \beta]$, we are done. Now, the property in question is transferable to all $P = E(\Theta)$, where $\Theta \in Z^*$; and, from this, all is clear. □

III) Given the mapping $P \in \mathcal{F}(X \times X, R_+)$, define the concepts

(a-s-diam) P is *almost semi-diametral*:
for each (ε, δ) with $\varepsilon > \delta > 0$, there exists $\gamma \in]0, \delta[$ such that
$x\mathcal{R}y$, $d(x, y) \geq \varepsilon$, $A_2(x, y) < \gamma$, $M(x, y) < \varepsilon + \gamma$ imply $P(x, y) < \varepsilon + \delta$
(s-diam) P is *semi-diametral*:
for each $x, y \in X$ with $x\mathcal{R}y$, we have $P(x, y) \leq M(x, y)$.

Proposition 4.32. *All maps $P = E(\Theta)$ where $\Theta \in Z^*$, are necessarily almost semi-diametral.*

Proof. The argument consists in two steps.

Step 1. Let us first establish that all maps $Q \in \{A_i; i \in Z\}$ have such a property. There are two cases to be discussed.

Case 1. $Q \in \{A_1, A_2, A_3, A_4, A_5\}$. By definition, we have for $1 \leq i \leq 5$,

$$A_i(x, y) \leq M(x, y), \text{ for all } x, y \in X;$$

whence, the maps in $\{A_1, A_2, A_3, A_4, A_5\}$ are semi-diametral; this, along with a previous auxiliary fact, assures us that the underlying maps are almost semi-diametral.

Case 2. $Q \in \{A_s, A_r, A_u\}$; or, equivalently

$Q \in \{B, C\}$; where $B = B[\lambda, \mu]$ for some $(\lambda, \mu) \in \Delta$,
and $C = C[\alpha, \beta]$ for some $(\alpha, \beta) \in \Delta$.

Let (ε, δ) with $\varepsilon > \delta > 0$ be given; and $\gamma \in]0, \delta[$ be arbitrary for the moment. Let $x, y \in X$ be such that

$$x\mathcal{R}y,\ d(x, y) \geq \varepsilon,\ A_2(x, y) < \gamma,\ M(x, y) < \varepsilon + \gamma;$$

we intend to show that $P(x, y) < \varepsilon + \delta$. By definition,

$$B(x, y) = d(y, Ty)[\lambda + d(x, Tx)]/[\mu + d(x, y)] < \gamma[\lambda + \gamma]/[\mu + \varepsilon],$$
$$C(x, y) = d(Tx, y)[\alpha + d(x, Ty)]/[\beta + d(x, y)] < (\varepsilon + \gamma)[\alpha + \varepsilon + \gamma]/[\beta + \varepsilon].$$

Denote, for simplicity,

$$\Phi(\gamma) = \gamma[\lambda + \gamma]/[\mu + \varepsilon],\ \Psi(\gamma) = (\varepsilon + \gamma)[\alpha + \varepsilon + \gamma]/[\beta + \varepsilon],\ 0 < \gamma < \delta.$$

By the above evaluations, we have

$$B(x, y) < \Phi(\gamma),\ C(x, y) < \Psi(\gamma), \text{ for all such } \gamma.$$

On the other hand,

$$\lim_{\gamma\to 0+} \Phi(\gamma) = 0 < \varepsilon + \gamma < \varepsilon + \delta,$$
$$\lim_{\gamma\to 0+} \Psi(\gamma) = \varepsilon[\alpha + \varepsilon]/[\beta + \varepsilon] \leq \varepsilon < \varepsilon + \gamma < \varepsilon + \delta.$$

This tells us that, if $\gamma \in]0, \delta[$ is small enough, we have

$$\Phi(\gamma) < \varepsilon + \delta; \text{ hence, } B(x, y) < \varepsilon + \delta,$$
$$\Psi(\gamma) < \varepsilon + \delta; \text{ hence, } C(x, y) < \varepsilon + \delta.$$

Putting these together, one gets the desired assertion involving the underlying family of maps $Q \in \{A_i; i \in Z\}$.

Step 2. The final conclusion relative to the maps $P = E(\Theta)$, where $\Theta \in Z^*$ is now clear – by a previous auxiliary fact – via components of P being endowed with the almost semi-diametral property. □

IV) Concerning the orbital asymptotic properties, the situation is a little bit complicated. Precisely, the following synthetic answer is available.

Proposition 4.33. *Under the above conventions,*

(84-1) *Each mapping $P = E(\Theta)$, where $\Theta \in Z^*$ fulfills $\{2, r\} \cap \Theta = \emptyset$ is orbitally singular asymptotic*

(84-2) *Each mapping $P = E(\Theta)$, where $\Theta \in Z^*$ fulfills $r \in \Theta$ is orbitally regular asymptotic*

(84-3) *Each mapping $P = E(\Theta)$, where $\Theta \in Z^*$ fulfills $2 \in \Theta$, $r \notin \Theta$ is orbitally strongly regular asymptotic.*

Proof. First, we have to discuss the orbital asymptotic properties of the maps $Q \in \{A_1, A_2, A_3, A_4, A_5\}$; and then, the properties of the maps $Q \in \{A_s, A_r, A_u\}$.

Step 1. $Q \in \{A_1, A_2, A_3, A_4, A_5\}$.

Let the (a-o-f) sequence (x_n) in $X(T, \mathcal{R})$ with $[x_n \stackrel{d}{\longrightarrow} z, \beta := d(z, Tz) > 0]$, and the subsequence $(u_n := x_{i(n)}; n \geq 0)$ of $(x_n; n \geq 0)$ with $[u_n \widetilde{\mathcal{R}} z, Tu_n \widetilde{\mathcal{R}} Tz, \forall n]$ be given. From these, one gets (taking the convergence relation and a metrical property of $d(.,.)$ into account)

$$d(u_n, z),\ d(Tu_n, z) \to 0,\ d(u_n, Tu_n) \to 0;\ d(u_n, Tz), d(Tu_n, Tz) \to \beta.$$

This, by definition, gives (as $n \to \infty$)

$A_1(u_n, z) \to 0$, $A_2(u_n, z) \to \beta$, $A_3(u_n, z) \to \beta/2$,
$A_4(u_n, z) \to \beta/2$, $A_5(u_n, z) \to \beta/2$;

whence, any map $Q \in \{A_1, A_3, A_4, A_5\}$ is orbitally singular asymptotic. Moreover, the same convergence properties of $(u_n; n \geq 0)$ and $(Tu_n; n \geq 0)$ tell us that, for a certain rank $n(z) \geq 0$, we must have

$$(\forall n \geq n(z)){:}\ d(u_n, z), d(Tu_n, z) < \beta/2,\ d(u_n, Tu_n) < \beta/2.$$

This, by the triangular inequality, gives for all $n \geq n(z)$,

$$d(u_n, Tz) \leq d(u_n, z) + \beta < 3\beta/2,\ d(Tu_n, Tz) \leq d(Tu_n, z) + \beta < 3\beta/2.$$

Combining these, yields, for all $n \geq n(z)$

$A_1(u_n, z) < \beta/2$, $A_2(u_n, z) = \beta$, $A_3(u_n, z) < 3\beta/4$,
$A_4(u_n, z) < \beta$, $A_5(u_n, z) < 3\beta/4$;

which, in particular, tells us that A_2 is orbitally strongly regular asymptotic.

Step 2. $Q \in \{A_s, A_r, A_u\}$; that is:

$$Q = B[\lambda, \mu],\ \text{where } (\lambda, \mu) \in \Delta,\ \text{or } Q = C[\alpha, \beta],\ \text{where } (\alpha, \beta) \in \Delta.$$

Let the (a-o-f) sequence (x_n) in $X(T, \mathcal{R})$ with $[x_n \xrightarrow{d} z, \beta := d(z, Tz) > 0]$, and the subsequence $(u_n := x_{i(n)}; n \geq 0)$ of (x_n) with $[u_n \widetilde{\mathcal{R}} z, Tu_n \widetilde{\mathcal{R}} Tz, \forall n]$ be given. By definition, we have

$$\lim_n B(u_n, z) = \lim_n \beta[\lambda + d(u_n, Tu_n)]/[\mu + d(u_n, z)] = \beta\lambda/\mu.$$

This, along with the above conventions, means

$\lim_n A_s(u_n, z) = \beta\lambda/\mu < \beta$; so, A_s is orbitally singular asymptotic;
$\lim_n A_r(u_n, z) = \beta\nu/\nu = \beta$; so, A_r is orbitally regular asymptotic.

Likewise, again by definition,

$\lim_n C(u_n, z) = \lim_n d(Tu_n, z)[\alpha + d(u_n, Tz)]/[\beta + d(u_n, z)] = 0 < \beta$;
so that: A_u is orbitally singular asymptotic.

Step 3. Putting these together, our conclusion follows. □

V) Finally, remember that the mapping $P(.,.)$ is

(fix-pos) *fix-positive*, if: $x, y \in \text{Fix}(T)$ and $x\widetilde{\mathcal{R}}y$ imply $P(x, y) > 0$
(fix-s-diam) *fix semi-diametral*, provided:
$x, y \in \text{Fix}(T)$ and $x\widetilde{\mathcal{R}}y$ imply $P(x, y) \leq M(x, y)(= d(x, y))$.

Proposition 4.34. *The following assertions hold:*

(85-1) Each mapping $P = E(\Theta)$ *where* $\Theta \in Z^*$ *fulfills* $\{1, 4, 5, u\} \cap \Theta \neq \emptyset$ *is fix positive*
(85-2) Each mapping $P = E(\Theta)$ *where* $\Theta \in Z^*$, *is fix semi-diametral.*

Proof. Let $x, y \in \text{Fix}(T)$ be such that $x\widetilde{\mathcal{R}}y$. Then (under the convention $B = B[\lambda, \mu]$, $C = C[\alpha, \beta]$, where $(\lambda, \mu) \in \Delta$, $(\alpha, \beta) \in \Delta$)

$A_1(x, y) = A_4(x, y) = A_5(x, y) = d(x, y)$;
hence, $0 < A_1(x, y) = A_4(x, y) = A_5(x, y) = d(x, y) = M(x, y)$;
$A_2(x, y) = A_3(x, y) = B(x, y) = 0 \leq d(x, y) = M(x, y)$;
$0 < C(x, y) = d(x, y)[\alpha + d(x, y)]/[\beta + d(x, y)] \leq d(x, y) = M(x, y)$;

and, from this, we are done. □

Now, by simply combining this with Meir-Keeler geometric theorem on relational metric spaces (MK-g-rms), one gets the fixed point result below (referred to as: *Meir-Keeler geometric rational function theorem on relational metric spaces*; in short: (MK-g-ra-rms)).

Theorem 4.12. *Assume that* T *is* $\mathcal{R}$*-globally sub-aprogression as well as* $(d, \mathcal{R}; E(\Theta); \Omega)$*-contractive, where* $\Theta \in Z_0^*$ *and* $\Omega \subseteq R_+^0 \times R_+^0$ *fulfill*

$\Theta \in Z^*$ *and* Ω *is upper diagonal, geometric Meir-Keeler.*

In addition, let X *be (a-o-f,d)-complete. Then,*
81-a) T *is fix-*$\mathcal{R}$*-asingleton if (in addition)* $\{1, 4, 5, u\} \cap \Theta \neq \emptyset$
81-b) T *is a strong Picard operator (modulo* $(d, \mathcal{R})$*) when, in addition,* T *is (a-o-f,d)-continuous*
81-c) T *is almost Bellman Picard (modulo* $(d, \mathcal{R})$*) when, in addition,* $\mathcal{R}$ *is (a-o-f,d)-almost-selfclosed and one of the extra conditions below holds:*

(81-c-1) $\{2, r\}$ *is disjoint from* Θ
(81-c-2) $r \in \Theta$ *and* Ω *is (in addition) geometric/asymptotic bilateral separable*
(81-c-3) $2 \in \Theta$, $r \notin \Theta$, *and* Ω *is (in addition) geometric/asymptotic left separable.*

Now, concrete examples of relations appearing in these contractive properties were given in a previous place. This yields the following practical version of our fixed point result above (referred to as: *Meir-Keeler rational function theorem on relational metric spaces*; in short: (MK-fct-ra-rms)).

Theorem 4.13. *Let the $\mathcal{R}$-globally sub-aprogression selfmap T of X and the subset $\Theta \in Z^*$ be such that and one of the following conditions holds*

(chi) T is $(d, \mathcal{R}; E(\Theta); \chi)$-contractive, where $\chi \in \mathcal{F}(re)(R^0_+, R)$ is Meir-Keeler admissible
(psi-vphi) T is $(d, \mathcal{R}; E(\Theta); (\psi, \varphi))$-contractive, where (ψ, φ) is a normal couple over $\mathcal{F}(R^0_+, R)$, such that φ is asymptotic positive.

In addition, let X be (a-o-f,d)-complete. Then

(82-a) *T is fix-$\mathcal{R}$-asingleton if (in addition) $\{1, 4, 5, u\} \cap \Theta \neq \emptyset$*

(82-b) *T is a strong Picard operator (modulo $(d, \mathcal{R})$) when, in addition, T is (a-o-f,d)-continuous*

(82-c) *T is almost Bellman Picard operator (modulo $(d, \mathcal{R})$) when, in addition, $\mathcal{R}$ is (a-o-f,d)-almost-selfclosed and one of the following extra conditions holds*

(82-c-1) $\{2, r\}$ is disjoint from Θ
(82-c-2) $r \in \Theta$, and the (chi)-alternative holds, where $\chi \in \mathcal{F}(re)(R^0_+, R)$ is bilateral Boyd-Wong admissible
(82-c-3) $2 \in \Theta$, $r \notin \Theta$, and the (chi)-alternative holds
(82-c-4) $r \in \Theta$, and the (psi-vphi)-alternative holds, where the couple (ψ, φ) over $\mathcal{F}(R^0_+, R)$ is (in addition) limit-bounded oscillating
(82-c-5) $2 \in \Theta$, $r \notin \Theta$, and the (psi-vphi)-alternative holds, where the couple (ψ, φ) over $\mathcal{F}(R^0_+, R)$ is (in addition) bounded left oscillating.

Concerning this last result, some technical remarks are in order.

I) According to our preceding developments, the Boyd-Wong admissible functions φ are allowed here. In particular, as established there, any Geraghty admissible function [96] is endowed with such a property. Note that, in this particular setting, our Meir-Keeler Rational Function theorem on relational metric spaces (MK-fct-ra-rms) includes a lot or related statements in Alam and Imdad [10]; see also Jachymski [125] and Bojor [31]. On the other hand, in the particular case of

$\mathcal{R}$ is reflexive transitive (hence, a quasi-order),

the globally sub-aprogression (or, equivalently: the globally scattered) condition upon $(\mathcal{R}, T)$ is trivially fulfilled; because, in this setting,

$\text{spec}(x) = N(1, \leq)$, for each $x \in X(T, \mathcal{R})$.

As a consequence, (MK-fct-ra-rms) includes the related statement in Choudhury and Kundu [55]; which, in turn, includes the pioneering work due to Kannan [138]. Concerning this aspect, we stress that the uniqueness property stated in Theorem 2.2 of the quoted paper is not true; because, given two comparable elements $z, y \in \mathrm{Fix}(T)$, the relation in Case 1 (of that result)

$$d(z,y) = d(T^n z, T^n y) \leq \beta(U_n(z,y))U_n(z,y) = 0,$$

where (by definition)

$$U_n(z,y) = A_2(T^{n-1}z, T^n z),\ n \geq 1,\ z, y \in X,$$

is not acceptable; because $\beta(0)$ is not defined. Finally, the 1986 statement in Turinici [235] cannot be obtained in this way; we do not give details.

II) Let (α, β) be a couple of functions in $\mathcal{F}(X \times X, R_+)$; and $[\alpha \leq 1]$, $[\beta \geq 1]$, stand for the associated relations

$$x[\alpha \leq 1]y \text{ iff } \alpha(x,y) \leq 1;\ x[\beta \geq 1]y \text{ iff } \beta(x,y) \geq 1.$$

Then, under the choice $\mathcal{R} := [\alpha \leq 1] \cap [\beta \geq 1]$, (MK-fct-ra-rms) includes a related statement in Berzig et al. [29]. But, as precise there, this last result includes the one due to Karapinar and Berzig [139], based on global contractive conditions like

$$\psi(d(Tx,Ty)) \leq \alpha(x,y)\psi(d(x,y)) - \beta(x,y)\varphi(d(x,y)),\ \forall x, y \in X;$$

hence, so does (MK-fct-ra-rms). On the other hand, the quoted statement extends as well some related contributions in Samet et al. [209], Mohammadi et al. [170], Moradi and Farajzadeh [172], or Salimi et al. [204]; so that, (MK-fct-ra-rms) may be also viewed as a common extension of all such statements. It is to be stressed that none of these results is to be deemed as a genuine extension for the fixed point statement in Samet and Turinici [208]; because, in the quoted result, the ambient relation $\mathcal{R}$ is not subjected to any kind of transitive type requirements. Further aspects (involving the same general setting) may be found in Berzig [28].

4.4 PPF Dependent Fixed Points and the Unexpected Way to Nirvana

Abstract. A metric version is given for the local and global PPF dependent fixed point result in Bernfeld et al. [Applicable Anal., 6 (1977), 271–280], over the class of continuous real variable functions. Strange enough, the methods to be used reduce all developments to the constant functions of this class; which, ultimately, means a reduction to Nirvana.

Subject-Class 2020. 47H10 (Primary), 54H25 (Secondary).

Keywords. Metric space, Picard operator, fixed and coincidence point, Meir-Keeler admissible and Matkowski admissible function, Razumikhin relation, chain, standard and isometric systems, constant functions class.

4.4.1 *Dependent Choice principles*

See Ch. 1.

4.4.2 *Banach contractions and relatives*

See Ch. 1.

4.4.3 *Meir-Keeler contractions*

Let (X, d) be a metric space. Call the subset Y of X, *almost singleton* (in short: *asingleton*) provided $[y_1, y_2 \in Y$ implies $y_1 = y_2]$; and *singleton* if, in addition, Y is nonempty; note that in this case $Y = \{y\}$, for some $y \in X$. Further, let $T \in \mathcal{F}(X)$ be a selfmap of X. We have to determine circumstances under which $\text{Fix}(T) \neq \emptyset$; and, if this holds, to establish whether T is fix-asingleton; or, equivalently: T is fix-singleton. The basic directions under which the investigations be conducted were already shown. The regularity conditions for such properties are being founded on *orbital-full* concepts (in short: (o-f)-concepts). Call the sequence (z_n) in X,

T-orbital, when $(z_n = T^n x; n \geq 0)$, for some $x \in X$;
full, when $n \mapsto z_n$ is *injective* ($i \neq j$ implies $z_i \neq z_j$);

the intersection of these notions is just the precise one.

reg-1) Call X, *(o-f,d)-complete* provided (for each (o-f)-sequence) d-Cauchy $\Longrightarrow$ d-convergent

reg-2) Let us say that T is *(o-f,d)-continuous*, if $[(z_n)$ is (o-f)-sequence and $z_n \xrightarrow{d} z]$ imply $Tz_n \xrightarrow{d} Tz$.

As a basic completion of these, we have to discuss the contractive type conditions to be used. Let $P : X \times X \to R_+$ be a map. We say that T is *Meir-Keeler (d, P)-contractive*, if

(mk-1) $P(x, y) > 0$ imply $d(Tx, Ty) < P(x, y)$;
referred to as: T is *strictly nonexpansive* (modulo (d, P))
(mk-2) $\forall \varepsilon > 0$, $\exists \delta > 0$: $[\varepsilon < P(x, y) < \varepsilon + \delta] \Longrightarrow d(Tx, Ty) \leq \varepsilon$;
referred to as: T has the *Meir-Keeler property* (modulo (d, P)).

Note that, by the former of these, the Meir-Keeler property writes

(mk-2a) $\forall \varepsilon > 0$, $\exists \delta > 0$: $[0 < P(x, y) < \varepsilon + \delta] \Longrightarrow d(Tx, Ty) \leq \varepsilon$.

Remark 4.4.

This convention – comparable with the ones in Cirić [59], Jachymski [121], and Matkowski [166] – is a little larger than the original one in Meir and Keeler [168]. Precisely, given the mapping $P : X \times X \to R_+$, let us say that T is *original Meir-Keeler (d, P)-contractive*, if

(mk-orig) $\forall \varepsilon > 0$, $\exists \delta > 0$: $[\varepsilon \leq P(x, y) < \varepsilon + \delta] \Longrightarrow d(Tx, Ty) < \varepsilon$.

The relationships between these are expressed as below [in (ZF-AC+DC)]:

T is original Meir-Keeler (d, P)-contractive implies
T is Meir-Keeler (d, P)-contractive.

For a direct proof, we refer to the survey paper by Turinici [257].

The following asymptotic version of Meir-Keeler contractive property is useful in applications. Call T, *asymptotic Meir-Keeler (d, P)-contractive* if

(a-mk-1) $P(x, y) > 0$ implies $d(Tx, Ty) < P(x, y)$;
referred to as: T is *strictly nonexpansive* (modulo (d, P))
(a-mk-2) there are no sequences (u_n), (v_n) in X and no elements $\varepsilon > 0$ with $d(Tu_n, Tv_n) \to \varepsilon+$, $P(u_n, v_n) \to \varepsilon+$; referred to as:
T has the *asymptotic Meir-Keeler property* (modulo (d, P)).

Theorem 4.14. *We have, in (ZF-AC+DC),*

T is Meir-Keeler (d, P)-contractive iff
T is asymptotic Meir-Keeler (d, P)-contractive.

Proof. i): Suppose that T is Meir-Keeler (d, P)-contractive; but [in contradiction with the conclusion] T is not endowed with the asymptotic Meir-Keeler property (modulo (d, P)); that is:

there are sequences (u_n), (v_n) in X and elements $\varepsilon > 0$ with
$d(Tu_n, Tv_n) \to \varepsilon+$, $P(u_n, v_n) \to \varepsilon+$.

Given this $\varepsilon > 0$, let $\delta > 0$ be the number associated to it, by the Meir-Keeler (d, P)-contractive property. By the convergence relations above, there exists some rank $n(\delta)$ with

(rela) $\varepsilon < d(Tu_n, Tv_n) < \varepsilon + \delta$, $\varepsilon < P(u_n, v_n) < \varepsilon + \delta$, $\forall n \geq n(\delta)$.

But, according to the second part of (rela), we must have

$d(Tu_n, Tv_n) \leq \varepsilon$, for all $n \geq n(\delta)$.

This, however, contradicts the first part of (rela). Hence, our working assumption cannot be accepted; and the assertion follows.

ii): Suppose that T is asymptotic Meir-Keeler (d, P)-contractive; but [in contradiction with the conclusion] T is not endowed with the Meir-Keeler property (modulo (d, P)); that is (for some $\varepsilon > 0$)

$A(\delta) := \{(u, v) \in X \times X; \varepsilon < P(u, v) < \varepsilon + \delta,\ d(Tu, Tv) > \varepsilon\}$
is nonempty, for each $\delta > 0$.

Taking a zero converging sequence (δ_n) in R^0_+, we get by the Denumerable Axiom of Choice (AC(N)) [deductible, as precise, in (ZF-AC+DC)], a couple of sequences $(x_n; n \geq 0)$ and $(y_n; n \geq 0)$ in X, so as

$(\forall n)$: $(x_n, y_n) \in A(\delta_n)$; that is (by the strict nonexpansive condition)
$\varepsilon < d(Tx_n, Ty_n) < P(x_n, y_n) < \varepsilon + \delta_n$.

As a direct consequence of this relation,

$d(Tx_n, Ty_n) \to \varepsilon+$ and $P(x_n, y_n) \to \varepsilon+$, as $n \to \infty$.

This contradicts the asymptotic Meir-Keeler property (modulo (d, P)) of T. Hence, the Meir-Keeler property (modulo (d, P)) of T follows. □

In the following, two examples of such contractions are constructed.

Ex-I) Denote, for simplicity,

$\mathcal{F}_0(R_+)$= the class of all $\chi \in \mathcal{F}(R_+)$ with $\chi(0) = 0$.

Given the mapping $P : X \times X \to R_+$ and the function $\chi \in \mathcal{F}_0(R_+)$, let us say that T is *$(d; P; \chi)$-contractive*, if

(d-P-chi) $d(Tx, Ty) \leq \chi(P(x, y))$, $\forall x, y \in X$.

The functions χ to be considered may described as below. Let $\mathcal{F}_0(re)(R_+)$ stand for the subclass of all $\chi \in \mathcal{F}_0(R_+)$, with

χ is *regressive*: $\chi(t) < t$, $\forall t > 0$.

Call $\chi \in \mathcal{F}_0(re)(R_+)$, *Meir-Keeler admissible* if

(mk-adm) $\forall \gamma > 0, \exists \beta > 0, (\forall t)$: $\gamma < t < \gamma + \beta \Longrightarrow \chi(t) \leq \gamma$;

and *Matkowski admissible*, provided

(M-adm) for each sequence $(t_n; n \geq 0)$ in R_+^0 with $(t_{n+1} \leq \chi(t_n), \forall n)$, we have $\lim_n t_n = 0$.

The relationships between these notions are made precise in

Theorem 4.15. *For each $\chi \in \mathcal{F}_0(re)(R_+)$, we have*

χ is Meir-Keeler admissible iff χ is Matkowski admissible.

For a complete proof, see the 2020 book by Turinici [259, Sec. 1.3]. Some partial aspects of the problem can be found in Jachymski [120].

In particular, letting $\mathcal{F}_0(re, in)(R_+)$ stand for the class of all increasing functions in $\mathcal{F}_0(re)(R_+)$, we have (cf. Matkowski [164])

$\chi \in \mathcal{F}_0(re, in)(R_+)$ is *Matkowski admissible* iff $\lim_n \chi^n(t) = 0$, for each $t \geq 0$.

Here, as usual, χ^n stands for the n-th iterate of φ, for each $n \in N$.

To get concrete examples of such functions, the constructions below are in effect. Let $\chi \in \mathcal{F}_0(re)(R_+)$ and $s \in R_+^0$ be given. Denote, for each $\varepsilon > 0$,

$\Lambda^+\chi(s) = \inf_{\varepsilon>0} \chi^*(s+)(\varepsilon)$; where $\chi^*(s+)(\varepsilon) = \sup \chi]s, s+\varepsilon[$;

this will be referred to as: *right superior limit* of χ at s. From the regressive property of φ, we have

$0 \leq \Lambda^+\chi(s) \leq s, \quad \forall s \in R_+^0$;

but the alternative of the extremal terms being attained cannot be avoided.

Call $\chi \in \mathcal{F}_0(re)(R_+)$, *Boyd-Wong admissible* if

(bw-adm) $\Lambda^+\chi(s) < s$, for all $s > 0$.

(This convention is related to the developments in Boyd and Wong [35].) In particular, $\varphi \in \mathcal{F}_0(re)(R_+, R)$ is Boyd-Wong admissible provided it is upper semicontinuous at the right on R_+^0:

$\Lambda^+\chi(s) \le \chi(s)$, $\forall s \in R_+^0$.

This, e.g., holds when χ is continuous at the right on R_+^0; because

$\Lambda^+\chi(s) = \chi(s)$, for each $s \in R_+^0$.

The following auxiliary fact will be useful.

Proposition 4.35. *Each Boyd-Wong admissible function in $\mathcal{F}_0(re)(R_+)$ is Meir-Keeler admissible (or, equivalently: Matkowski admissible).*

Proof. Suppose that $\chi \in \mathcal{F}_0(re)(R_+)$ is Boyd-Wong admissible, and fix $\gamma > 0$; hence $\Lambda^+\chi(\gamma) < \gamma$. By definition, there exists $\beta = \beta(\gamma) > 0$ such that $\gamma < t < \gamma + \beta$ implies $\chi(t) < \gamma$; and we are done. □

Having these precise, we may establish the connection between this contractive property and the Meir-Keeler one.

Proposition 4.36. *Let T be $(d; P; \chi)$-contractive, where $\chi \in \mathcal{F}_0(re)(R_+)$ is Meir-Keeler admissible. Then, T is Meir-Keeler (d, P)-contractive.*

Proof. i) Let $x, y \in X$ be such that $P(x, y) > 0$. By the contractive condition [and χ = regressive], one has $d(Tx, Ty) < P(x, y)$; so that, T is strictly nonexpansive (modulo (d, P)).

ii) Let $\varepsilon > 0$ be arbitrary fixed; and $\delta > 0$ be the number assured by the Meir-Keeler admissible property of φ. Further, let $x, y \in X$ be such that $\varepsilon < P(x, y) < \varepsilon + \delta$. By the contractive condition and admissible property,

$d(Tx, Ty) \le \chi(P(x, y)) \le \varepsilon$;

so that, T has the Meir-Keeler property (modulo (d, P)). □

Ex-II) Let (ψ, φ) be a couple of functions over $\mathcal{F}_0(R_+)$, endowed with

(norm) (ψ, φ) is *normal:*
ψ is increasing and φ is *strictly positive* [$\varphi(t) > 0$, $\forall t > 0$].

(This concept is related with the developments in Dutta and Choudhury [84].) The following extra condition will be considered here:

(ex-cond) (φ is *right sequentially positive*)
for each sequence (t_n) in R_+^0 and each $\varepsilon > 0$ with $t_n \to \varepsilon+$,
the relation $\lim_n \varphi(t_n) = 0$ is impossible.

Given the mapping $P : X \times X \to R_+$ and the couple (ψ, φ) of functions over $\mathcal{F}_0(R_+)$, let us say that T is $(d; P; (\psi, \varphi))$*-contractive*, provided

$$\psi(d(Tx,Ty)) \le \psi(P(x,y)) - \varphi(P(x,y)),\ \forall x,y \in X.$$

The following auxiliary fact establishes the necessary connection with the concept of Meir-Keeler contractions.

Proposition 4.37. *Suppose that T is $(d;P;(\psi,\varphi))$-contractive, for a normal couple (ψ,φ) over $\mathcal{F}_0(R_+)$ with $\varphi =$ right sequentially positive. Then, we necessarily have in (ZF-AC+DC),*

T is asymptotic Meir-Keeler (d,P)-contractive;
or, equivalently: Meir-Keeler (d,P)-contractive.

Proof. i) Let $x,y \in X$ be such that $P(x,y) > 0$. If $Tx = Ty$, then $d(Tx,Ty) = 0 < P(x,y)$. Suppose now that $Tx \neq Ty$; whence: $d(Tx,Ty) > 0$. As $P(x,y) > 0$, and $\varphi =$ strictly positive, one has $\varphi(P(x,y)) > 0$; wherefrom

$$\psi(d(Tx,Ty)) < \psi(P(x,y)) \text{ (by the contractive condition);}$$

and this, via [$\psi =$ increasing], yields $d(Tx,Ty) < P(x,y)$. Putting these together, one derives that T is strictly nonexpansive (modulo (d,P)).

ii) We have to establish that T is endowed with the asymptotic Meir-Keeler property (modulo (d,P)). Suppose not:

there are sequences (u_n), (v_n) in X and elements $\varepsilon > 0$
with $d(Tu_n,Tv_n) \to \varepsilon+$, $P(u_n,v_n) \to \varepsilon+$.

By the contractive property,

$(\forall n)$: $\psi(d(Tu_n,Tv_n)) \le \psi(P(u_n,v_n)) - \varphi(P(u_n,v_n))$;
or, equivalently (along with $\varphi =$ strictly positive):
$(0 <)\varphi(P(u_n,v_n)) \le \psi(P(u_n,v_n)) - \psi(d(Tu_n,Tv_n))$.

Passing to $\limsup$ as $n \to \infty$, gives

$0 \le \limsup \varphi(P(u_n,v_n)) \le \psi(\varepsilon+0) - \psi(\varepsilon+0) = 0$;
whence, $\lim_n \varphi(P(u_n,v_n)) = 0$;

in contradiction with the choice of φ. Hence, the underlying asymptotic property holds; and we are done. □

Note that, by the same techniques, one may recover the related statements in Suzuki and Vetro [220], and Wardowski [264]. Since, however, this is beyond the scope of our exposition, we do not give further details.

4.4.4 *Metrical fixed point principles*

Let (X, d) be a metric space; and $T \in \mathcal{F}(X)$ be a selfmap of X. We have to determine whether Fix(T) is nonempty; and, if this holds, to establish whether T is fix-asingleton; or, equivalently: T is fix-singleton. The basic directions under which this problem is to be solved were already listed; as well as the sufficient conditions for getting such properties. In addition, the contractive Meir-Keeler framework, expressed in terms of a certain mapping $P \in \mathcal{F}(X \times X, R_+)$, is being settled. It remains now to discuss the specific regularity conditions upon P to be used. Denote for each $x, y \in X$,

$$
\begin{aligned}
&A_1(x,y) = d(x,y),\ A_2(x,y) = \max\{d(x,Tx), d(y,Ty)\},\\
&L_0(x,y) = \min\{d(x,y), d(Tx,Ty)\},\\
&L_1(x,y) = \min\{d(x,y), d(Tx,Ty), d(x,Tx), d(y,Ty)\},\\
&M(x,y) = \text{diam}\{x, Tx, y, Ty\}.
\end{aligned}
$$

(Con-1) The first condition upon P writes

P is *L-positive* (where $L \in \{L_0, L_1\}$): $L(x,y) > 0$ implies $P(x,y) > 0$.

It has the role of working with our iterative sequences (under $L = L_1$) as well as to assure the fix-asingleton property (under $L = L_0$). Note that

P is L_0-positive implies P is L_1-positive;

but the reciprocal is not in general true.

(Con-2) The second condition upon P is expressed as

P is *orbitally bounded*: $P(x, Tx) \le A_2(x, Tx)$, for all $x \in X$;

it allows us deducing the d-asymptotic and full properties for the iterative sequences to be considered (see below).

(Con-3) The third condition upon P writes

P is *diametral*: $P(x,y) \le M(x,y)$, for all $x, y \in X$.

It has the role of obtaining (in conjunction with Meir-Keeler contractive assumption) the d-Cauchy property for the iterative sequence in question.

(Con-4) The fourth condition upon the mapping P is a couple of orbitally singular asymptotic requirements, formulated as

(o-sg-asy) P is *orbitally singular asymptotic*:
for each (o-f) sequence (x_n) in X and each $z \in X$ with $x_n \xrightarrow{d} z$ and $(L_1(x_n, z) > 0, \forall n)$, we have $\liminf_n P(x_n, z) < d(z, Tz)$

(s-o-sg-asy) P is *strongly orbitally singular asymptotic*:
for each (o-f) sequence (x_n) in X and each $z \in X$ with $x_n \xrightarrow{d} z$ and $(L_1(x_n, z) > 0, \forall n)$, we have $(\exists) \lim_n P(x_n, z) < d(z, Tz)$.

Clearly, the former of these is weaker than the latter one. Note that the imposed conditions allow us getting (in addition to the Meir-Keeler contractive assumption) the fixed point property for the limit of underlying iterative sequence.

We are in position to state a basic fixed point result for a latter use.

Theorem 4.16. *Suppose that T is Meir-Keeler (d, P)-contractive, (or, equivalently: asymptotic Meir-Keeler (d, P)-contractive), where the mapping $P : X \times X \to R_+$ is orbitally bounded and diametral. In addition, let X be (o-f,d)-complete. Then,*

(41-a) *T is a Picard operator (modulo d) when, in addition to this, P is L_1-positive*

(41-b) *T is a strong Picard operator (modulo d) when, in addition to the setting of (41-a), T is (o-f,d)-continuous*

(41-c) *T is a strong Picard operator (modulo d) when, in addition to the setting of (41-a), P is orbitally singular asymptotic*

(41-d) *T has the fix-asingleton property when (in addition to these requirements) P is L_0-positive.*

Proof. Let us firstly establish the fix-asingleton property. Take some couple $z_1, z_2 \in \text{Fix}(T)$; and suppose by contradiction that $z_1 \neq z_2$; hence, $L_0(z_1, z_2) = d(z_1, z_2) > 0$. As P is L_0-positive, we must have $P(z_1, z_2) > 0$; so, that $d(z_1, z_2) = d(Tz_1, Tz_2) < P(z_1, z_2)$ [because T is strictly nonexpansive (modulo (d, P))]. On the other hand, as P is diametral, $P(z_1, z_2) \leq M(z_1, z_2) = d(z_1, z_2)$. Since the obtained relations are contradictory, our working assumption cannot be accepted; wherefrom, the assertion follows.

It remains to establish that T is a strong Picard operator (modulo d). Fix some $x_0 \in X$; and put $(x_n = T^n x_0; n \geq 0)$; it is an orbital sequence. If $x_n = x_{n+1}$ for some $n \geq 0$, we are done; so, without loss, one may assume that the *non-telescopic* property holds

(n-tele) $x_n \neq x_{n+1}$ (that is: $\rho_n := d(x_n, x_{n+1}) > 0$), $\forall n$.

The argument will be divided into several parts.

Part 1. By the imposed condition,

$(\forall n)$: $L_1(x_n, x_{n+1}) = \min\{\rho_n, \rho_{n+1}\} > 0$; hence $P(x_n, x_{n+1}) > 0$

if we remember that P is L_1-positive. The Meir-Keeler (d, P)-contractive condition applies to (x_n, x_{n+1}), for each n; and yields (by the strict nonexpansive condition upon T and the orbital boundedness of P)

(iter-1) $(\forall n)$: $\rho_{n+1} = d(Tx_n, Tx_{n+1}) < P(x_n, x_{n+1}) \leq$
$A_2(x_n, x_{n+1}) = \max\{\rho_n, \rho_{n+1}\}$.

From the strict inequality between the extremal members of this relation,

(iter-2) $(\forall n)$: $\rho_{n+1} < \rho_n$; wherefrom, $\rho_{n+1} < P(x_n, x_{n+1}) \leq \rho_n$.

Two consequences of this fact are retainable.

Conseq 1. By the first half of (iter-2), (ρ_n) is strictly descending. We claim that, in this case,

(x_n) is *full*: $i \neq j$ implies $x_i \neq x_j$ (whence, $d(x_i, x_j) > 0$).

In fact, suppose by contradiction that there exist $i, j \in N$ with $i < j$, $x_i = x_j$. Then, by definition, $x_{i+1} = x_{j+1}$; so that $\rho_i = \rho_j$; in contradiction with $\rho_i > \rho_j$; and the assertion follows.

Conseq 2. By the same strict descending property of $(\rho_n; n \geq 0)$, we have that $\rho := \lim_n \rho_n$ exists in R_+; and $[\rho_n > \rho, \forall n]$. Assume that $\rho > 0$; and let $\sigma > 0$ be the number given by the Meir-Keeler property (modulo (d, P)) of T. From the convergence relation, there exists a rank $n(\sigma)$ with

$(\forall n \geq n(\sigma))$: $\rho < \rho_n < \rho + \sigma$; hence, by (iter-2),
$\rho < \rho_{n+1} < P(x_n, x_{n+1}) \leq \rho_n < \rho + \sigma$.

This, by the quoted condition, yields (for the same n),

$(\rho <)\ \rho_{n+1} = d(Tx_n, Tx_{n+1}) \leq \rho$;

a contradiction. Hence, $\rho = 0$; so that, (x_n) is a d-asymptotic sequence.

Part 2. Summing up, (x_n) is orbital, full, and d-asymptotic. We claim that, under the precise conditions, (x_n) is d-Cauchy. Let $\varepsilon > 0$ be arbitrary fixed; and $\delta > 0$ be the number associated by the Meir-Keeler property; without loss, one may assume that $\delta < \varepsilon$. From the obtained d-asymptotic property, there exists a rank $n(\delta) \geq 0$, such that

(d-asy) $(\forall n \geq n(\delta))$: $d(x_n, x_{n+1}) < \delta/4$; hence, $d(x_n, x_{n+2}) < \delta/2$.

We prove, inductively, that for each $i \geq 1$, the property below holds

(d-C;i) $d(x_n, x_{n+i}) < \varepsilon + \delta/2$, $\forall n \geq n(\delta)$;

wherefrom, the d-Cauchy property of $(x_n; n \geq 0)$ is clear. The case of $i \in \{1, 2\}$ is evident, by (d-asy). Assume that, for a certain $k \geq 2$, (d-C;i) holds for all $i \in \{1, ..., k\}$; we must establish that it holds for $i = k + 1$. So, let $n \geq n(\delta)$ be arbitrary fixed. From the inductive hypothesis,

$$d(x_n, x_{n+k}) < \varepsilon + \delta/2; \text{ as well as}$$
$$d(x_{n+1}, x_{n+k}) < \varepsilon + \delta/2,\ d(x_{n+1}, x_{n+k+1}) < \varepsilon + \delta/2.$$

On the other hand, by the asymptotic property,

$$d(x_n, x_{n+1}) < \delta/4,\ d(x_{n+k}, x_{n+k+1}) < \delta/4.$$

Finally, the triangular inequality gives (in a direct way)

$$d(x_n, x_{n+k+1}) \leq d(x_n, x_{n+k}) + d(x_{n+k}, x_{n+k+1}) < \varepsilon + \delta/2 + \delta/4 < \varepsilon + \delta.$$

Putting these together, gives

$$M(x_n, x_{n+k}) < \varepsilon + \delta; \text{ whence, } P(x_n, x_{n+k}) < \varepsilon + \delta;$$

if we remember that P is diametral. On the other hand, as (x_n) is full

$$L_1(x_n, x_{n+k}) = \min\{\rho_n, \rho_{n+k}, d(x_n, x_{n+k}), d(x_{n+1}, x_{n+k+1})\} > 0;$$
$$\text{whence } P(x_n, x_{n+k}) > 0;$$

if we take the L_1-positive property of P into account. Combining these with the complete Meir-Keeler property (modulo (d, P)), one derives

$$d(x_{n+1}, x_{n+k+1}) = d(Tx_n, Tx_{n+k}) \leq \varepsilon.$$

This, along with the triangular inequality, yields

$$d(x_n, x_{n+k+1}) \leq d(x_n, x_{n+1}) + d(x_{n+1}, x_{n+k+1}) < \varepsilon + \delta/2;$$

and the assertion is retainable.

Part 3. Since X is (a-o-f,d)-complete,

$$\text{there exists (a unique) } z \in X \text{ with } x_n \xrightarrow{d} z.$$

Two basic alternatives are to be discussed.

Alter 1. Suppose that T is (o-f,d)-continuous. Then, $u_n := Tx_n \xrightarrow{d} Tz$ as $n \to \infty$. On the other hand, $(u_n = x_{n+1}; n \geq 0)$ is a subsequence of $(x_n; n \geq 0)$; so, $u_n \xrightarrow{d} z$ as $n \to \infty$. Combining with d=separated, yields $z = Tz$; which tells us that T is strongly Picard (modulo d).

Alter 2. Suppose that P is orbitally singular asymptotic. As (x_n) is an (o-f)-sequence, we have that

$H_1 := \{n \in N; x_n = Tz\}$, $H_2 := \{n \in N; Tx_n = Tz\}$ are asingletons;

so that, the following separation property holds:

(sepa) $\exists k = k(z) \geq 0$, such that $n \geq k$ implies
$d(x_n.Tz) > 0$, and $d(Tx_n, Tz) > 0$ (hence, $d(x_n, z) > 0$).

Denote for simplicity $(u_n = x_{n+k}; n \geq 0)$. By the preceding relations,

$(\forall n)$: $d(u_n, z) > 0$, $d(u_n, Tz) > 0$, $d(Tu_n, Tz) > 0$; as well as
$d(u_n, Tu_n) > 0$ (via (u_n)=full), $d(Tu_n, z) = d(u_{n+1}, z) > 0$.

Suppose by contradiction that $d(z, Tz) > 0$. By the relations above,

$(\forall n)$: $L_1(u_n, z) > 0$; hence, $P(u_n, z) > 0$ (as P is L_1-positive).

The contractive condition is therefore applicable to (u_n, z), for each n; and yields (by the strict nonexpansive (modulo (d, P)) property of T)

(s-ineq) $d(Tu_n, Tz) < P(u_n, z)$, for all n.

Moreover, from the d-Cauchy and convergence relations, one gets (taking a metrical property of $d(.,.)$ into account)

(conv) $d(u_n, z)$, $d(Tu_n, z), d(u_n, Tu_n) \to 0$;
$d(u_n, Tz)$, $d(Tu_n, Tz) \to d(z, Tz)$.

By the strict inequality relation (s-ineq) we get, passing to (inferior) limit as $n \to \infty$ (and remembering that P is orbitally singular asymptotic)

$$d(z, Tz) = \liminf_n d(Tu_n, Tz) \leq \liminf_n P(u_n, z) < d(z, Tz);$$

a contradiction. Summing up, the working hypothesis $d(z, Tz) > 0$ cannot be true; so that, $z = Tz$. The proof is complete. □

The obtained statement includes a related contribution due to Găvruţa et al. [94], obtained via similar methods. Note that, coincidence type versions of these facts are available, by means of related techniques in Roldán et al. [198]. On the other hand, all these developments may be extended to quasi-metric structures, by following the methods in Turinici [255]. Finally, multivalued extensions of these facts are possible under the lines in Nadler [175]. We will discuss these elsewhere.

In the following two basic particular cases of this result are given.

(A) Remember that $\chi \in \mathcal{F}_0(re)(R_+)$ is *Meir-Keeler admissible*, if

(mk-adm) $\forall \varepsilon > 0$, $\exists \delta > 0$, $(\forall t)$: $\varepsilon < t < \varepsilon + \delta \Longrightarrow \chi(t) \leq \varepsilon$.

Given the mapping $P : X \times X \to R_+$ and the function $\chi \in \mathcal{F}_0(R_+)$, let us say that T is $(d; P; \chi)$*-contractive*, provided

$$d(Tx, Ty) \leq \chi(P(x, y)), \forall x, y \in X.$$

As precise, any such contraction is necessarily Meir-Keeler $(d; P)$-contractive, whenever $\chi \in \mathcal{F}_0(re)(R_+)$ is Meir-Keeler admissible. As a direct consequence of this, we have (by means of our main result)

Theorem 4.17. *Suppose that* T *is* $(d; P; \chi)$*-contractive, where the mapping* $P : X \times X \to R_+$ *and the function* $\chi \in \mathcal{F}_0(re)(R_+)$ *are such that*

(P-chi) P *is orbitally bounded, diametral, and*
χ *is Meir-Keeler admissible (or, equivalently: Matkowski admissible).*

In addition, let X *be (o-f,d)-complete. Then,*

(42-a) T *is a Picard operator (modulo d) when, in addition, the mapping* P *is* L_1*-positive*

(42-b) T *is a strong Picard operator (modulo d) when, in addition to the setting of (42-a),* T *is (o-f,d)-continuous*

(42-c) T *is a strong Picard operator (modulo d) when, in addition to the setting of (42-a),* P *is orbitally singular asymptotic*

(42-d) T *has the fix-asingleton property if (in addition to this) the mapping* P *is* L_0*-positive.*

In particular, the regularity condition (P-chi) holds under

(P-chi-1) $P = A_1$ and χ is Boyd-Wong admissible.

In this case, the corresponding version of Theorem 4.17 includes directly the related statements in Boyd and Wong [35] or Matkowski [164].

(B) Let (ψ, φ) be a pair of functions over $\mathcal{F}_0(R_+)$, with

(norm) (ψ, φ) is normal:
ψ is increasing and φ is strictly positive ($\varphi(t) > 0$, $\forall t \in R_+^0$).

The following extra condition will be considered here:

(r-s-pos) φ is right sequentially positive:
for each sequence $(t_n; n \geq 0)$ in R_+^0 and each element $\varepsilon > 0$
with $t_n \to \varepsilon+$, the relation $\lim_n \varphi(t_n) = 0$ is impossible.

Given the mapping $P : X \times X \to R_+$ and the couple (ψ, φ) of functions over $\mathcal{F}_0(R_+)$, let us say that T is $(d; P; (\psi, \varphi))$*-contractive*, provided

$$\psi(d(Tx,Ty)) \le \psi(P(x,y)) - \varphi(P(x,y)),\ \forall x,y \in X.$$

By a previous result, any such contraction is Meir-Keeler $(d;P)$-contractive, whenever (ψ,φ) is normal and φ is right sequentially positive. As a direct consequence of this, we have (by means of our main result)

Theorem 4.18. *Suppose that T is $(d;P;(\psi,\varphi))$-contractive, where the mapping $P: X\times X \to R_+$ and the couple (ψ,φ) of functions over $\mathcal{F}_0(R_+)$ are taken so as*

(P-psi-phi) P is orbitally bounded, diametral,
(ψ,φ) is normal, and φ is right sequentially positive.

In addition, let X be (o-f,d)-complete. Then,

(43-a) *T is a Picard operator (modulo d) when, in addition, the mapping P is L_1-positive*

(43-b) *T is a strong Picard operator (modulo d) when, in addition to the setting of (43-a), T is (o-f,d)-continuous*

(43-c) *T is a strong Picard operator (modulo d) when, in addition to the setting of (43-a), P is orbitally singular asymptotic*

(43-d) *T has the fix-asingleton property when (in addition to this) the mapping P is L_0-positive.*

The obtained result extends some fixed point principles started by Krasnoselskii and Stetsenko [152], re-discovered by Rhoades [197], and further developed by Dutta and Choudhury [84]; but, it cannot include the related statements in Wardowski [265] and Vujaković et al. [261]. However, by using the implicit techniques in Turinici [254], an effective inclusion is possible for these, as well as many other ones described in the survey paper by Karapinar et al. [140]. Further aspects will be delineated elsewhere.

4.4.5 *Razumikhin chains*

Let (E,d) be a complete metric space. Further, let $I=[a,b]$ be a bounded closed real interval; and $E_0 := C(I,E)$ stand for the class of all continuous functions in $\mathcal{F}(I,E)$, endowed with the supremum distance

(s-dist) $D(\varphi,\xi) = \sup\{d(\varphi(t),\xi(t)); t \in I\}$, $\varphi,\xi \in E_0$;

clearly, (E_0,D) is a complete metric space.

Given an operator $\mathcal{S}: E_0 \to E$, define the relations $(\perp)$ and $(\models)$ on E_0

(perp) $\varphi \perp \xi$ iff $D(\varphi,\xi) \le d(\mathcal{S}\varphi,\mathcal{S}\xi)$;

(models) $\varphi \models \xi$ iff $D(\varphi, \xi) = d(\mathcal{S}\varphi, \mathcal{S}\xi)$;

referred to as: the *Razumikhin standard relation* and *Razumikhin isometric relation* attached to $\mathcal{S}$.

Remark 4.5.

Clearly, the following relationship between these is valid

(mod-perp) $\varphi \models \xi$ implies $\varphi \perp \xi$.

Moreover, if $\mathcal{S}$ is taken as

(nonexp) $\mathcal{S}$ is (D, d)-nonexpansive: $d(\mathcal{S}\varphi, \mathcal{S}\xi) \leq D(\varphi, \xi)$, $\varphi, \xi \in E_0$

the reciprocal is also possible; i.e.,

(perp-mod-perp) $\varphi \perp \xi$ iff $\varphi \models \xi$.

In the following, some auxiliary facts about the Razumikhin standard relation $(\perp)$ will be given; but, these are valid as well for the Razumikhin isometric relation $(\models)$. For the moment, it is clear that

(rr-1) $(\perp)$ is *reflexive*: $\varphi \perp \varphi$, $\forall \varphi \in E_0$
(rr-2) $(\perp)$ is *symmetric*: $\varphi \perp \xi$ iff $\xi \perp \varphi$;

however, $(\perp)$ is not in general transitive, as simple examples show. Note that, by the very definition above, the injective property holds

(inj) $(\varphi \perp \xi, \mathcal{S}\varphi = \mathcal{S}\xi) \Longrightarrow \varphi = \xi$.

Having these precise, denote for each $\varphi \in E_0$

(perp-sect) $E_0(\varphi, \perp) = \{\xi \in E_0; \varphi \perp \xi\}$;

referred to as: the *section* of $(\perp)$ at φ. A natural problem is to determine those elements $\varphi \in E_0$ with "small" sections. In this direction, we have

Proposition 4.38. *Let $\varphi \in E_0$ and the subset M_0 of E_0 be such that*

$\mathcal{S}\varphi \in \mathcal{S}(M_0)$, $E_0(\varphi, \perp) \supseteq M_0$.

Then, necessarily, $\varphi \in M_0$.

Proof. As $\mathcal{S}\varphi \in \mathcal{S}(M_0)$, there exists $\xi \in M_0$, such that $\mathcal{S}\varphi = \mathcal{S}\xi$. On the other hand, by the imposed hypothesis, $\xi \in E(\varphi, \perp)$; whence, $\varphi \perp \xi$. Combining these, gives (by injective property) $\varphi = \xi$; so that, $\varphi \in M_0$. □

Define the concept (for the subset $M_0 \subseteq E_0$):

M_0 is *full*, provided $\mathcal{S}(M_0) = E$.

As a consequence of the above result, we have that

$$(M_0 = \text{full and } \varphi \in E_0 \setminus M_0) \text{ imply } [E_0(\varphi, \perp) \supseteq M_0 \text{ is false}];$$

hence, for any $\varphi \in E_0 \setminus M_0$, the section $E_0(\varphi, \perp)$ is not very large.

The following "global" consequence of this fact is to be noted. Call the subset L_0 of E_0, a *Razumikhin chain*, provided

$$\text{(raz-ch)} \quad \varphi, \xi \in L_0 \Longrightarrow \varphi \perp \xi \text{ (that is: } D(\varphi, \xi) \le d(\mathcal{S}\varphi, \mathcal{S}\xi));$$

and a *Razumikhin iso-chain*, provided

$$\text{(raz-i-ch)} \quad \varphi, \xi \in L_0 \Longrightarrow \varphi \models \xi \text{ (that is: } D(\varphi, \xi) = d(\mathcal{S}\varphi, \mathcal{S}\xi)).$$

This convention comes from $\mathcal{S}$ being an isometry between L_0 and $\mathcal{S}(L_0)$.

Remark 4.6.

Clearly, the following holds

$$L_0 = \text{Razumikhin iso-chain implies } L_0 = \text{Razumikhin chain}.$$

In particular, if $\mathcal{S}$ is (D, d)-nonexpansive, these concepts are identical:

$$L_0 = \text{Razumikhin iso-chain iff } L_0 = \text{Razumikhin chain}.$$

Returning to the general setting, the following fact about such objects – referred to as *Reduction statement* – is available.

Proposition 4.39. *Under the precise framework, we have*

(52-1) *If the Razumikhin chain L_0 of E_0 and the (nonempty) subset M_0 of E_0 are such that $[\mathcal{S}(L_0) = \mathcal{S}(M_0),\ L_0 \supseteq M_0]$, then, $L_0 = M_0$*

(52-2) *Let L_0 and M_0 be a couple of full Razumikhin chains of E_0. Then, either $[L_0 = M_0]$, or $[L_0$ and M_0 are not inclusion comparable$]$.*

Proof. i) Suppose that L_0 and M_0 are like in the premise above; and let $\varphi \in L_0$ be arbitrary fixed; hence,

$$\mathcal{S}\varphi \in \mathcal{S}(L_0) = \mathcal{S}(M_0).$$

As $L_0 =$ Razumikhin chain, we get (by the imposed hypothesis) $E_0(\varphi, \perp) \supseteq L_0 \supseteq M_0$; so that, from the previous result, $\varphi \in M_0$. As φ was arbitrarily chosen in L_0, we necessarily have $L_0 \subseteq M_0$; whence, $L_0 = M_0$.

ii) Evident, by $\mathcal{S}(L_0) = \mathcal{S}(M_0) = E$ and the preceding step. □

Further information about these objects is to be obtained under

(tran) $(\perp)$ is transitive: $\varphi \perp \xi$ and $\xi \perp \lambda$ implies $\varphi \perp \lambda$; referred to as: $\mathcal{S}$ is *transitive*;

clearly, $(\perp)$ becomes an *equivalence* over E_0 in this case. Denote, for each $\varphi \in E_0$,

(eq-cls) $\widetilde{\varphi} := \{\xi \in E_0; \varphi \perp \xi\}$ (the *equivalence class* of φ);

clearly, $\widetilde{\varphi}$ is a (nonempty) Razumikhin chain of E_0.

As a completion of the above statement, we then have

Proposition 4.40. *Suppose that $\mathcal{S}$ is transitive. Then,*

(53-1) *Each full Razumikhin chain M_0 of E_0 is an equivalence class: $M_0 = \widetilde{\varphi}$, where φ is some/any element of M_0*

(53-2) *For any couple M_0, L_0 of full Razumikhin chains, we have either $M_0 = L_0$ or $M_0 \cap L_0 = \emptyset$.*

Proof. i) Let M_0 be a full Razumikhin chain; and $\varphi \in M_0$ be arbitrary fixed. By the chain property (and definition of equivalence class), $M_0 \subseteq \widetilde{\varphi}$. On the other hand, the full property of M_0 yields

$$E = \mathcal{S}(M_0) \subseteq \mathcal{S}(\widetilde{\varphi}) \subseteq E; \text{ whence, } E = \mathcal{S}(M_0) = \mathcal{S}(\widetilde{\varphi});$$

which tells us that both M_0 and $\widetilde{\varphi}$ are full Razumikhin chains. Combining these, gives (by the previous result) $M_0 = \widetilde{\varphi}$; and the claim follows.

ii) Evident, by the properties of equivalence classes. □

As a consequence of this, the full Razumikhin chains must be sought among the equivalence classes of relation $(\perp)$. Some concrete examples of such objects will be given a bit further.

4.4.6 *Conditional BLR theorem*

Let again (E, d) be a complete metric space. Further, let $I = [a, b]$ be a bounded closed real interval; and $E_0 := C(I, E)$ stand for the class of all continuous functions in $\mathcal{F}(I, E)$, endowed with the supremum distance D; as precise, (E_0, D) is a complete metric space.

Let $\mathcal{S} : E_0 \to E$ be an operator with

(Dd-cont) $\mathcal{S}$ is (D, d)-continuous [$\varphi_n \xrightarrow{D} \varphi$ implies $\mathcal{S}\varphi_n \xrightarrow{d} \mathcal{S}\varphi$].

Remember that we introduced the *Razumikhin standard relation* $(\perp)$ and the *Razumikhin isometric relation* $(\models)$ attached to $\mathcal{S}$, as

$$\varphi \perp \xi \text{ iff } D(\varphi, \xi) \le d(\mathcal{S}\varphi, \mathcal{S}\xi); \ \varphi \models \xi \text{ iff } D(\varphi, \xi) = d(\mathcal{S}\varphi, \mathcal{S}\xi).$$

Remark 4.7.

In particular, S is (D, d)-continuous whenever

(Dd-nexp) $\mathcal{S}$ is (D, d)-nonexpansive: $d(\mathcal{S}\varphi, \mathcal{S}\xi) \le D(\varphi, \xi)$, $\forall \varphi, \xi \in E_0$.

Note that, under such a condition, the associated Razumikhin relations $(\perp)$ and $(\models)$ are identical (see above).

A basic example of such objects is given as below. Let $c \in I$ be arbitrary fixed. Then, the operator $\mathcal{S}_c : E_0 \to E$ introduced as

$$\mathcal{S}_c(\varphi) = \varphi(c), \ \varphi \in E_0,$$

is (D, d)-nonexpansive; hence, (D, d)-continuous. Consequently, the Razumikhin relations $(\perp_c)$ and $(\models_c)$ are identical and have the representation

$$\varphi \perp_c \xi \text{ iff } D(\varphi, \xi) = d(\mathcal{S}_c(\varphi), \mathcal{S}_c(\xi)).$$

Returning to the general case, let $\mathcal{T} : E_0 \to E$ be a mapping. We say that $\varphi \in E_0$ is a *coincidence point* of $\mathcal{S}$ and $\mathcal{T}$, when $\mathcal{S}\varphi = \mathcal{T}\varphi$; the class of all these will be denoted as $\text{Cop}(\mathcal{S}, \mathcal{T})$. In particular, when $\mathcal{S} = \mathcal{S}_c$, the coincidence point relation for the couple $(\mathcal{S}_c, \mathcal{T})$ becomes: $\varphi(c) = \mathcal{T}\varphi$; and φ is then called a *PPF dependent fixed point* of T.

To state our main result concerning such elements, the following concepts and constructions are needed.

CC-1) Given $k \ge 0$, we say that $\mathcal{T}$ is *k-contractive*, provided

(contr) $d(\mathcal{T}\varphi, \mathcal{T}\xi) \le kD(\varphi, \xi)$, for all $\varphi, \xi \in E_0$.

Note that, as a consequence of this,

$$\mathcal{T} \text{ is } (D, d)\text{-continuous: } \varphi_n \xrightarrow{D} \varphi \text{ implies } \mathcal{T}\varphi_n \xrightarrow{d} \mathcal{T}\varphi.$$

CC-2) Let us introduce a relation (∇) over E_0, according to:

(nab) $\varphi \nabla \xi$ iff $\mathcal{T}\varphi = \mathcal{S}\xi$ and $\varphi \perp \xi$.

CC-3) Finally, given the starting element $\varphi_0 \in E_0$, we say that the sequence $(\varphi_n; n \ge 0)$ in E_0 is (φ_0, ∇)-*iterative*, in case $(\varphi_n \nabla \varphi_{n+1}, \forall n)$.

The following coincidence point statement (referred to as: Conditional BLR theorem) is now available.

Theorem 4.19. *Let $\mathcal{S}$ be a (D, d)-continuous operator. Further, suppose that $\mathcal{T}$ is k-contractive, for some $k \in [0, 1[$. Then,*

(61-a) *Given the starting point $\varphi_0 \in E_0$, any (φ_0, ∇)-iterative sequence $(\varphi_n; n \geq 0)$ in E_0 is D-convergent, and $\varphi := \lim_n \varphi_n$ is an element of* $\text{Cop}(\mathcal{S}, \mathcal{T})$.

(61-b) *Let $\varphi_0, \xi_0 \in E_0$ be a couple of starting points, and $(\varphi_n; n \geq 0)$, $(\xi_n; n \geq 0)$ be a (φ_0, ∇)-iterative sequence and (ξ_0, ∇)-iterative sequence, respectively. Then,*

$$D(\varphi_n, \xi_n) \leq (1/(1-k))[D(\varphi_0, \varphi_1) + D(\xi_0, \xi_1)] + D(\varphi_0, \xi_0), \forall n.$$

Moreover, if $\varphi_n \perp \xi_n$ for all $n \geq 0$, then $\lim_n \varphi_n = \lim_n \xi_n$.

(61-c) *Let $\varphi_0 = \xi_0 \in E_0$ be a starting point, and $(\varphi_n; n \geq 0)$, $(\xi_n; n \geq 0)$ be a couple of (φ_0, ∇)-iterative sequence and $(\xi_0 = \varphi_0, \nabla)$-iterative sequence in E_0, respectively. Then,*

$$D(\varphi_n, \xi_n) \leq (2/(1-k))d(\mathcal{S}\varphi_0, \mathcal{T}\varphi_0), \text{ for all } n \geq 0.$$

In particular, if $\mathcal{S}$ is (D, d)-nonexpansive, the above relation writes

$$D(\varphi_n, \xi_n) \leq (2/(1-k))D(\varphi_0, \varphi_1) = (2/(1-k))D(\xi_0, \xi_1), \forall n.$$

(61-d) *If φ^*, ξ^* are two elements in* $\text{Cop}(\mathcal{S}, \mathcal{T})$ *with $\varphi^* \perp \xi^*$, then, necessarily, $\varphi^* = \xi^*$.*

Proof. There are several steps to be considered.

Part 1. Take a starting point $\varphi_0 \in E_0$, and let $(\varphi_n; n \geq 0)$ in E_0 be some (φ_0, ∇)-iterative sequence. By definition, $\varphi_0 \nabla \varphi_1$; which means

$$\mathcal{T}\varphi_0 = \mathcal{S}\varphi_1, \ D(\varphi_0, \varphi_1) \leq d(\mathcal{S}\varphi_0, \mathcal{S}\varphi_1).$$

Further, $\varphi_1 \nabla \varphi_2$; which means

$$\mathcal{T}\varphi_1 = \mathcal{S}\varphi_2, \ D(\varphi_1, \varphi_2) \leq d(\mathcal{S}\varphi_1, \mathcal{S}\varphi_2).$$

As a combination of these, we have (by the contractive property of T)

$$D(\varphi_1, \varphi_2) \leq d(\mathcal{S}\varphi_1, \mathcal{S}\varphi_2) = d(\mathcal{T}\varphi_0, \mathcal{T}\varphi_1) \leq kD(\varphi_0, \varphi_1).$$

This procedure may continue indefinitely; and gives the iterative relations

$$D(\varphi_{n+1}, \varphi_{n+2}) \leq kD(\varphi_n, \varphi_{n+1}), \forall n \geq 0.$$

By a finite induction argument, one gets the geometric evaluations

$$D(\varphi_n, \varphi_{n+1}) \leq k^n D(\varphi_0, \varphi_1), \forall n \geq 0.$$

Since $\sum_n k^n$ converges, $(\varphi_n; n \geq 0)$ is D-Cauchy. As (E_0, D) is complete,

$$\varphi_n \xrightarrow{D} \varphi^* \text{ as } n \to \infty, \text{ for some } \varphi^* \in E_0;$$

and the first part of conclusion **(61-a)** holds. Moreover, by the very definition of our (φ_0, ∇)-iterative sequence (φ_n), one has

$$(\forall n)\colon \varphi_n \nabla \varphi_{n+1} \text{ (whence, } T\varphi_n = S\varphi_{n+1}, \varphi_n \perp \varphi_{n+1}).$$

Passing to limit as $n \to \infty$ in the operator part of this relation, we get (as both $\mathcal{S}$ and $\mathcal{T}$ are (D, d)-continuous), $\mathcal{T}\varphi^* = \mathcal{S}\varphi^*$; or, in other words: $\varphi^* \in \mathrm{Cop}(\mathcal{S}, \mathcal{T})$; and the second part of conclusion **(61-a)** holds too.

Part 2. Take a couple of starting points $\varphi_0, \xi_0 \in E_0$, and let $(\varphi_n; n \geq 0)$, $(\xi_n; n \geq 0)$, be a (φ_0, ∇)-iterative sequence and (ξ_0, ∇)-iterative sequence, respectively. By the preceding part, $(\varphi_n; n \geq 0)$ fulfills the iterative and geometric relations above. Likewise, $(\xi_n; n \geq 0)$ fulfills the iterative relations

$$D(\xi_{n+1}, \xi_{n+2}) \leq kD(\xi_n, \xi_{n+1}), \forall n \geq 0;$$

wherefrom, the geometric evaluation holds

$$D(\xi_n, \xi_{n+1}) \leq k^n D(\xi_0, \xi_1), \forall n \geq 0.$$

Taking the triangle inequality into account one gets, for all $n \geq 1$,

$$D(\varphi_n, \xi_n) \leq D(\varphi_{n-1}, \varphi_n) + D(\xi_{n-1}, \xi_n) + D(\varphi_{n-1}, \xi_{n-1}) \leq$$
$$k^{n-1}[D(\varphi_0, \varphi_1) + D(\xi_0, \xi_1)] + D(\varphi_{n-1}, \xi_{n-1});$$

wherefrom – by repeating $(n-1)$ times this procedure – the first part of conclusion **(61-b)** follows. Further, suppose that (in addition) $\varphi_n \perp \xi_n$, $\forall n$. By the very definition of these points and the working condition,

$$D(\varphi_n, \xi_n) \leq d(\mathcal{S}\varphi_n, \mathcal{S}\xi_n) = d(\mathcal{T}\varphi_{n-1}, \mathcal{T}\xi_{n-1}) \leq kD(\varphi_{n-1}, \xi_{n-1}),$$
$$\text{for all } n \geq 1.$$

This, by a finite induction argument, gives

$$D(\varphi_n, \xi_n) \leq k^n D(\varphi_0, \xi_0), \forall n.$$

Passing to limit as $n \to \infty$, and noting that both (φ_n) and (ξ_n) are D-convergent (see above), the second part of conclusion **(61-b)** is established.

Part 3. Let $\varphi_0 = \xi_0 \in E_0$ be a starting point, and (φ_n), (ξ_n) be a (φ_0, ∇)-iterative sequence and $(\xi_0 = \varphi_0, \nabla)$-iterative sequence in E_0, respectively. Then,

$$D(\varphi_0, \varphi_1) \leq d(\mathcal{S}\varphi_0, \mathcal{S}\varphi_1) = d(\mathcal{S}\varphi_0, \mathcal{T}\varphi_0),$$
$$D(\xi_0, \xi_1) \leq d(\mathcal{S}\xi_0, \mathcal{S}\xi_1) = d(\mathcal{S}\xi_0, \mathcal{T}\xi_0) = d(\mathcal{S}\varphi_0, \mathcal{T}\varphi_0);$$

and so, via preceding part, the first half of conclusion **(61-c)** follows. Moreover, assume that $\mathcal{S}$ is (D, d)-nonexpansive. Then, from $\varphi_0 = \xi_0$, we get $\mathcal{T}\varphi_0 = \mathcal{T}\xi_0$; whence $\mathcal{S}\varphi_1 = \mathcal{S}\xi_1$. This (along with the representation of $(\perp)$), gives

$$D(\varphi_0, \varphi_1) = d(\mathcal{S}\varphi_0, \mathcal{S}\varphi_1) = d(\mathcal{S}\varphi_0, \mathcal{T}\varphi_0),$$
$$D(\xi_0, \xi_1) = d(\mathcal{S}\xi_0, \mathcal{S}\xi_1) = d(\mathcal{S}\varphi_0, \mathcal{S}\varphi_1) = D(\mathcal{S}\varphi_0, \mathcal{T}\varphi_0);$$

wherefrom, the second half of conclusion **(61-c)** follows too.

Part 4. Let φ^*, ξ^* be two elements in $\text{Cop}(\mathcal{S}, \mathcal{T})$ taken as precise. By the working condition and contractive property of T,

$$D(\varphi^*, \xi^*) \leq d(\mathcal{S}\varphi^*, \mathcal{S}\xi^*) = d(\mathcal{T}\varphi^*, \mathcal{T}\xi^*) \leq kD(\varphi^*, \xi^*);$$

and, this yields (as $D(.,.)$ is a metric), $\varphi^* = \xi^*$. The proof is complete. □

In particular, let $(E, ||.||)$ be a Banach space; and $(E_0, ||.||_0)$ stand for the associated Banach functional space. Further, letting $c \in I$, define

$$(\mathcal{S}_c : E_0 \to E): \ \mathcal{S}_c(\varphi) = \varphi(c), \ \varphi \in E_0,$$

It is immediately seen that

(p1) $\mathcal{S}_c$ is a linear operator (additive and homogeneous)
(p2) $\mathcal{S}_c$ is nonexpansive: $||\mathcal{S}_c(\varphi)|| \leq ||\varphi||_0, \ \varphi \in E_0$.

Further, define the *unilateral Razumikhin set* $\text{razu}(\mathcal{S}_c)$ and *unilateral Razumikhin isometric set* $\text{razu}(iso)(\mathcal{S}_c)$ attached to $\mathcal{S}_c$, as

$$\text{razu}(\mathcal{S}_c) = \{\chi \in E_0; ||\chi||_0 \leq ||\mathcal{S}_c(\chi)||\},$$
$$\text{razu}(iso)(\mathcal{S}_c) = \{\chi \in E_0; ||\chi||_0 = ||\mathcal{S}_c(\chi)||\};$$

clearly, these are identical here. The associated to $\mathcal{S}_c$ relation $(\perp_c)$ writes

$(\varphi, \xi \in E_0)$: $\varphi \perp_c \xi$ iff $\varphi - \xi \in \text{razu}(\mathcal{S}_c)$;
or, equivalently: $\varphi \perp_c \xi$ iff $\varphi - \xi \in \text{razu}(iso)(\mathcal{S}_c)$.

The corresponding version of Conditional BLR theorem is just the result in Bernfeld et al. [27] proved via "normed" type methods. Further extensions of this statement may be found in Pathak [185] and Som [217].

4.4.7 Standard systems

Technically speaking, the result we just exposed is, ultimately, a *conditional* one. Precisely, the existence of coincidence points appearing there is essentially related to the possibility of effectively constructing iterative sequences according to the described conventions. So, it is natural to ask whether such iterative sequences exist. It is our aim in the following to show that, if we restrict our developments to certain classes of operators and Razumikhin chains, this always happens.

Let (E, d) be a complete metric space. Further, let $I = [a, b]$ be a bounded closed real interval; and $E_0 := C(I, E)$ stand for the class of all continuous functions in $\mathcal{F}(I, E)$, endowed with the supremum distance D; as noticed, (E_0, D) is a complete metric space.

Let in the following $\mathcal{S} : E_0 \to E$ be an operator with

(Dd-cont) $\mathcal{S}$ is (D, d)-continuous [$\varphi_n \xrightarrow{D} \varphi$ implies $\mathcal{S}\varphi_n \xrightarrow{d} \mathcal{S}\varphi$].

(A) Remember that, the *standard* Razumikhin relation $(\perp)$ attached to $\mathcal{S}$ is introduced as

$$\varphi \perp \xi \text{ iff } D(\varphi, \xi) \leq d(\mathcal{S}\varphi, \mathcal{S}\xi).$$

Further, by a *standard Razumikhin chain* we mean any nonempty part L_0 of E_0, with respect to which $(\perp)$ is *total*; that is

$$\varphi, \xi \in L_0 \Longrightarrow \varphi \perp \xi \text{ (any two elements are comparable);}$$

the class of all these will be denoted as $\mathrm{Rch}(\mathcal{S})$.

The following auxiliary facts involving these concepts will be useful for us. Let cl and Cl stand for the closure operators in E and E_0 respectively.

Proposition 4.41. *The following are valid, for each* $L_0 \in \mathrm{Rch}(\mathcal{S})$

(71-1) $K_0 := \mathrm{Cl}(L_0)$ *is in* $\mathrm{Rch}(\mathcal{S})$

(71-2) the restriction $\mathcal{S} : L_0 \to E$ *is injective*

(71-3) L_0 *is D-closed iff* $G := \mathcal{S}(L_0)$ *is d-closed.*

Proof. There are several steps to be passed.

Step 1. Let φ, ξ be two elements in K_0; that is,

there exist sequences (φ_n) and (ξ_n) in L_0 with $\varphi_n \xrightarrow{D} \varphi$, $\xi_n \xrightarrow{D} \xi$.

According to the choice of L_0, we have

$(\forall n)$: $\varphi_n \perp \xi_n$; that is, $D(\varphi_n, \xi_n) \leq d(\mathcal{S}\varphi_n, \mathcal{S}\xi_n)$.

Passing to limit as $n \to \infty$ in this relation yields (via $\mathcal{S}$ = continuous) $\varphi \perp \xi$; and this, by the arbitrariness of these data, proves that $K_0 \in \mathrm{Rch}(\mathcal{S})$.

Step 2. Let $\varphi, \xi \in L_0$ be such that $\mathcal{S}\varphi = \mathcal{S}\xi$. By the (standard) Razumikhin chain property of L_0 we have $\varphi \perp \xi$. Combining these facts yields (see above) $\varphi = \xi$; proving the assertion.

Step 3. There are two parts to be clarified.

i): Suppose that $L_0 \in \mathrm{Rch}(\mathcal{S})$ is D-closed; we have to deduce that $G := \mathcal{S}(L_0)$ is d-closed. Let the sequence $(\varphi_n; n \geq 0)$ in L_0 and the point $z \in E$ be such that $\mathcal{S}\varphi_n \xrightarrow{d} z$. As a direct consequence of this, $(\mathcal{S}\varphi_n; n \geq 0)$ is a d-Cauchy sequence in E; wherefrom (as $\mathcal{S}$ is expansive over L_0), $(\varphi_n; n \geq 0)$ is D-Cauchy in L_0. By completeness, $\varphi_n \xrightarrow{D} \varphi$, for some $\varphi \in L_0$; and this gives (as $\mathcal{S}$ is continuous) $\mathcal{S}\varphi_n \xrightarrow{d} \mathcal{S}\varphi$. Combining with the above, gives $z = \mathcal{S}\varphi$; wherefrom, $z \in G$.

ii): Conversely, let $G = \mathcal{S}(L_0)$ be d-closed; we have to establish that

$L_0 \in \mathrm{Rch}(\mathcal{S})$ is D-closed; that is: $L_0 = K_0$, where $K_0 := \mathrm{Cl}(L_0)$.

Let (φ_n) be a sequence in L_0 and φ be an element in E_0 with $\varphi_n \xrightarrow{D} \varphi$; clearly, $\varphi \in K_0$. Since $\mathcal{S}$ continuous, we must have (in view of $G = \mathcal{S}(L_0)$)

$(g_n := \mathcal{S}\varphi_n; n \geq 0)$ is a sequence in G with $g_n \xrightarrow{d} \mathcal{S}\varphi$ as $n \to \infty$.

Combining with the d-closed property of G, we get

$g := \mathcal{S}\varphi$ is in $G = \mathcal{S}(L_0)$; whence, $g = \mathcal{S}\xi$, for some $\xi \in L_0$.

This yields $\mathcal{S}\varphi = \mathcal{S}\xi$; where $\varphi \in K_0$, $\xi \in L_0 \subseteq K_0$. This, via preceding stage (modulo K_0), yields $\varphi = \xi \in L_0$; and the conclusion follows. □

Given the continuous map $\mathcal{S} : E_0 \to E$ and the nonempty part L_0 of E_0, define

$(\mathcal{S}, L_0)$ is a *standard system*, if L_0 is a D-closed Razumikhin chain in E_0; (hence (see above) $G := \mathcal{S}(L_0)$ is a d-closed subset in E).

Suppose that we introduced such an object. By the previous auxiliary fact, $\mathcal{S}$ is a bijection between L_0 and $G := \mathcal{S}(L_0)$. As a consequence,

the inverse operator $U := \mathcal{S}^{-1}$ exists as a bijection in $\mathcal{F}(G, L_0)$, and $(\mathcal{S}Ug = g, \forall g \in G)$, $(U\mathcal{S}\varphi = \varphi, \forall \varphi \in L_0)$.

Moreover, from the definition of Razumikhin relation $(\perp)$,

(non-exp) U is (d, D)-nonexpansive: $D(Ux, Uy) \leq d(x, y)$, $\forall x, y \in G$.

In the following, a basic particular case when such systems exist will be described. Before doing this, we need an auxiliary fact.

Proposition 4.42. *Let $P : E \to E$ be a continuous operator, with*

P is expansive: $d(u,v) \le d(Pu,Pv)$, $\forall u,v \in E$.

Then, $P(H)$ is closed, for each d-closed part H of E.

Proof. Let the sequence $(x_n; n \ge 0)$ in H and the point $z \in E$ be such that $Px_n \xrightarrow{d} z$. As a direct consequence of this, $(Px_n; n \ge 0)$ is a d-Cauchy sequence; wherefrom (via P = expansive), $(x_n; n \ge 0)$ is d-Cauchy sequence in H. By H = complete (in view of H being d-closed), $x_n \xrightarrow{d} x$, for some $x \in H$; and this gives (via P = continuous), $Px_n \xrightarrow{d} Px$. Combining with the above, gives $z = Px$; wherefrom, $z \in P(H)$. □

Given the continuous $\mathcal{S} : E_0 \to E$, the (nonempty) d-closed part H of E, the map $W : H \to E_0$, and the continuous expansive $P : E \to E$, let us say that the quadruple $(\mathcal{S}, H, W; P)$ is *admissible*, provided

(adm-1) $(\mathcal{S}, W)$ is *P-invertible*: $\mathcal{S}Wh = Ph$, $\forall h \in H$
(adm-2) W is (d,D)-nonexpansive: $D(Wx,Wy) \le d(x,y)$, $\forall x,y \in H$.

Proposition 4.43. *Let the quadruple $(\mathcal{S}, H, W; P)$ be admissible. Then,*

(73-1) $L_0 := W(H)$ is a D-closed Razumikhin chain
(73-2) $(\mathcal{S}, L_0)$ is a standard system.

Proof. Let $\varphi = Wx$, $\xi = Wy$ (where $x,y \in H$) be a couple of elements in L_0. By the imposed conditions,

$$D(\varphi,\xi) = D(Wx,Wy) \le d(x,y) \le d(Px,Py) = \\ d(\mathcal{S}Wx, \mathcal{S}Wy) = d(\mathcal{S}\varphi, \mathcal{S}\xi);$$

wherefrom, by the arbitrariness of our starting elements, L_0 is a Razumikhin chain in E_0. On the other hand, by definition,

$$\mathcal{S}(L_0) = \{\mathcal{S}Wx; x \in H\} = \{Px; x \in H\} = P(H);$$

and this, along with $P(H)$ being d-closed tells us that so is $\mathcal{S}(L_0)$. Combining with a previous result, one derives that L_0 is D-closed; and concludes our argument. □

(B) Returning to the general case, let $(\mathcal{S}, L_0)$ be a standard system; so that (see above)

> $L_0 \in \mathrm{Rch}(\mathcal{S})$ is D-closed, $G := \mathcal{S}(L_0)$ is d-closed, and $\mathcal{S} : L_0 \to G$ is a bijection; hence, $U := \mathcal{S}^{-1}$ exists as a bijection over $\mathcal{F}(G, L_0)$ (in the sense: $(\mathcal{S}Ug = g, \forall g \in G)$, $(U\mathcal{S}\varphi = \varphi, \forall \varphi \in L_0)$), with the nonexpansive property $(D(Ux, Uy) \leq d(x, y),\ x, y \in G)$.

Then, let $\mathcal{T} : E_0 \to E$ be a mapping with

> (subord) T is subordinated to $\mathcal{S}$ on L_0: $\mathcal{T}(L_0) \subseteq \mathcal{S}(L_0)(= G)$.

Remember that $\varphi \in E_0$ is a *coincidence point* of $\mathcal{S}$ and $\mathcal{T}$, when $\mathcal{S}\varphi = \mathcal{T}\varphi$. To determine such points, the following construction will be useful. Let $T : G \to G$ be the selfmap of G introduced as

$$Tg = \mathcal{T}(Ug),\ g \in G \text{ [where, as precise, } U := \mathcal{S}^{-1} \in \mathcal{F}(G, L_0)].$$

Proposition 4.44. *The following assertions are valid:*

(74-1) *If $x \in G$ is a fixed point of T, then $\xi := Ux \in L_0$ is a coincidence point of $\mathcal{S}$ and $\mathcal{T}$*

(74-2) *Conversely, if $\zeta := Uz \in L_0$ is a coincidence point of $\mathcal{S}$ and $\mathcal{T}$, then $z \in G$ is a fixed point of T.*

Proof. i) If $x \in G$ is a fixed point of T, then $\mathcal{S}(Ux) = x = Tx = \mathcal{T}(Ux)$; or, equivalently, $\mathcal{S}\xi = \mathcal{T}\xi$, where $\xi := Ux \in L_0$; which tells us that $\xi \in L_0$ is a coincidence point of $\mathcal{S}$ and $\mathcal{T}$.

ii) Suppose that $\zeta := Uz \in L_0$ is a coincidence point of $\mathcal{S}$ and $\mathcal{T}$; that is: $\mathcal{S}\zeta = \mathcal{T}\zeta$. This yields (by the properties of $\mathcal{S}$) $z = \mathcal{S}(Uz) = \mathcal{T}(Uz) = Tz$; so that, $z \in G$ is a fixed point of T. □

Remark 4.8.

The following completions of these facts are to be noted.

Given $\varphi_0 = Ux_0 \in L_0$, we say that the sequence $(\varphi_n = Ux_n; n \geq 0)$ in L_0 is $(\varphi_0, \mathcal{S}, \mathcal{T})$*-iterative*, provided $(\mathcal{T}\varphi_n = \mathcal{S}\varphi_{n+1}$, for all $n)$. Note that, the family of such sequences is nonempty, by the subordination hypothesis.

com-1): Given the $(\varphi_0, \mathcal{S}, \mathcal{T})$-iterative process $(\varphi_n = Ux_n; n \geq 0)$ in L_0, we have (cf. our previous constructions)

$$Tx_n = \mathcal{T}(Ux_n) = \mathcal{T}\varphi_n = \mathcal{S}\varphi_{n+1} = \mathcal{S}(Ux_{n+1}) = x_{n+1},\ \forall n \geq 0;$$

so that the supporting sequence (x_n) in G is a (x_0, T)-iterative one, according to $(Tx_n = x_{n+1}; n \geq 0)$. Consequently, any $(\varphi_0 = Ux_0, \mathcal{S}, \mathcal{T})$- iterative sequence in L_0 has the unique representation $(\varphi_n = UT^n x_0; n \geq 0)$.

com-2): Conversely, given the (x_0, T)-iterative sequence $(x_n = T^n x_0; n \geq 0)$ in G, the sequence $(\varphi_n := Ux_n; n \geq 0)$ in L_0 is $(\varphi_0, \mathcal{S}, \mathcal{T})$-iterative, in view of

$$\mathcal{T}\varphi_n = \mathcal{T}(Ux_n) = Tx_n = x_{n+1} = \mathcal{S}(Ux_{n+1}) = \mathcal{S}\varphi_{n+1}, \forall n \geq 0.$$

(com-3) Summing up, each $(\varphi_0 = Ux_0, \mathcal{S}, \mathcal{T})$-iterative process in L_0 is the (unique) U-image of its associated (x_0, T)-iterative process in G.

Under these preliminary facts, we may now proceed to the formulation of our announced results. The following basic concepts appear.

I) Let $\Phi \in \mathcal{F}_0(R_+)$ be given.

I-1) We say that $(\mathcal{T}, \mathcal{S})$ is Φ-*contractive*, provided

(con-1) $d(\mathcal{T}\varphi, \mathcal{T}\xi) \leq \Phi(d(\mathcal{S}\varphi, \mathcal{S}\xi))$, for all $\varphi, \xi \in E_0$.

This concept is comparable with the one in Dhage [73], [74]. In particular, under $\varphi = Ux$, $\xi = Uy$ (where $x, y \in G$), the relation above becomes

(con-1a) $d(Tx, Ty) \leq \Phi(d(x, y))$, for all $x, y \in G$;

or, in other words: the selfmap $T : G \to G$ is $(d; A_1; \Phi)$-contractive.

I-2) We say that $\mathcal{T}$ is Φ-*contractive*, provided

(con-2) $d(\mathcal{T}\varphi, \mathcal{T}\xi) \leq \Phi(D(\varphi, \xi))$, for all $\varphi, \xi \in E_0$.

This class of contractions is a direct extension of the one appearing in Bernfeld et al. [27]. In particular, when

$$\Phi = \text{increasing (hence, } \Phi \in \mathcal{F}_0(in)(R_+))$$

then, letting $\varphi = Ux$, $\xi = Uy$ (where $x, y \in G$) in this relation, we have (taking into account that U is nonexpansive)

(con-2a) $d(Tx, Ty) \leq \Phi(D(Ux, Uy)) \leq \Phi(d(x, y))$, for all $x, y \in G$;

or, in other words: the selfmap $T : G \to G$ is $(d; A_1; \Phi)$-contractive.

II) Let $\Psi, \Phi \in \mathcal{F}_0(R_+)$ be given.

II-1) We say that $(\mathcal{T}, \mathcal{S})$ is (Ψ, Φ)-*contractive*, provided

(con-3) $\Psi(d(\mathcal{T}\varphi, \mathcal{T}\xi)) \leq \Psi(d(\mathcal{S}\varphi, \mathcal{S}\xi)) - \Phi(d(\mathcal{S}\varphi, \mathcal{S}\xi))$, for all $\varphi, \xi \in E_0$.

In particular, with $\varphi = Ux$, $\xi = Uy$ (where $x, y \in G$), this relation becomes

(con-3a) $\Psi(d(Tx, Ty)) \leq \Psi(d(x, y)) - \Phi(d(x, y))$, for all $x, y \in G$;

or, in other words: the selfmap $T : G \to G$ is $(d; A_1; (\Psi, \Phi))$-contractive.

II-2) We say that $\mathcal{T}$ is (Ψ, Φ)-*contractive*, provided

(con-4) $\Psi(d(\mathcal{T}\varphi, \mathcal{T}\xi)) \leq \Psi(D(\varphi, \xi)) - \Phi(D(\varphi, \xi))$, for all $\varphi, \xi \in E_0$.

In particular, when

$\Psi - \Phi$ is increasing (hence, $\Psi - \Phi \in \mathcal{F}_0(in)(R_+)$)

then, letting $\varphi = Ux$, $\xi = Uy$ (where $x, y \in G$) in this relation, we have (again taking into account that U is nonexpansive)

(con-4a) $\Psi(d(Tx, Ty)) \leq \Psi(D(Ux, Uy)) - \Phi(d(Ux, Uy)) \leq$
$\Psi(d(x, y)) - \Phi(d(x, y))$, for all $x, y \in G$;

or, in other words: the selfmap $T : G \to G$ is $(d; A_1; (\Psi, \Phi))$-contractive.

Now, as G is closed and $T : G \to G$ is either $(d; A_1; \Phi)$-contractive (in the cases (I-1) or (I-2)) or $(d; A_1; (\Psi, \Phi))$-contractive (in the cases (II-1) or (II-2)) our previous fixed point principles are applicable here; and yield the following synthetic coincidence point result involving our data (referred to as: Standard BLR theorem):

Theorem 4.20. *Let $(\mathcal{S}, L_0)$ be a standard system, and the mapping $\mathcal{T} : E_0 \to E$ be subordinated to $\mathcal{S}$ on L_0. Further, suppose that one of the conditions below is being fulfilled*

(71-i) $(\mathcal{T}, \mathcal{S})$ is Φ-contractive, where $\Phi \in \mathcal{F}_0(re)(R_+)$ is Meir-Keeler admissible (or, equivalently: Matkowski admissible)
(71-ii) $\mathcal{T}$ is Φ-contractive, where $\Phi \in \mathcal{F}_0(re, in)(R_+)$ is Meir-Keeler admissible (or, equivalently: Matkowski admissible)
(71-iii) $(\mathcal{T}, \mathcal{S})$ is (Ψ, Φ)-contractive, where $\Psi, \Phi \in \mathcal{F}_0(R_+)$ are such that (Ψ, Φ) is a normal couple with Φ = right sequentially positive
(71-iv) $\mathcal{T}$ is (Ψ, Φ)-contractive, where $\Psi, \Phi \in \mathcal{F}_0(R_+)$ are such that: (Ψ, Φ) is a normal couple with $\Psi - \Phi \in \mathcal{F}_0(in)(R_+)$, and Φ = right sequentially positive.

Then,

(71-a) *$\mathcal{S}$ and $\mathcal{T}$ have (in L_0) a unique coincidence point $\varphi^* \in L_0$*

(71-b) *for the arbitrary fixed $\varphi_0 \in L_0$, the unique $(\varphi_0, \mathcal{S}, \mathcal{T})$-iterative sequence $(\varphi_n; n \geq 0)$ in L_0 is D-convergent towards φ^*.*

Proof. Let $T \in \mathcal{F}(G)$ be defined as $(Tg = \mathcal{T}(Ug); g \in G)$. By the previous metric fixed point results we have the conclusions

(con-1) $\text{Fix}(T) = \{x \in G; x = Tx\}$ is a singleton, $\{x^*\}$
(con-2) for each $x_0 \in G$, the (x_0, T)-iterative process (x_n)
introduced as $(x_{n+1} = Tx_n; n \geq 0)$ fulfills $x_n \xrightarrow{d} x^*$ as $n \to \infty$.

The former of these tells us [by means of a preceding auxiliary fact] that $\varphi^* := Ux^*$ is the unique coincidence point of $\mathcal{S}$ and $\mathcal{T}$ in L_0. Further, note that $\varphi_0 = Ux_0$, where $x_0 = \mathcal{S}\varphi_0$. Let $(\varphi_n = Ux_n; n \geq 0)$ be the $(\varphi_0, \mathcal{S}, \mathcal{T})$-iterative sequence in L_0 (generated by φ_0). Then (see above) $(x_n; n \geq 0)$ is a (x_0, T)-iterative process (generated by x_0); so (by the latter conclusion above) $x_n \xrightarrow{d} x^*$ [that is: $d(x_n, x^*) \to 0$] as $n \to \infty$. On the other hand, as U is nonexpansive,

$$D(\varphi_n, \varphi^*) = D(Ux_n, Ux^*) \leq d(x_n, x^*), \text{ for all } n;$$

so, by combining these, $\varphi_n \xrightarrow{D} \varphi^*$ [via $D(\varphi_n, \varphi^*) \to 0$] as $n \to \infty$. □

Remark 4.9.
Note that, unlike the Conditional BLR theorem, the iterative sequences above are effectively constructible, by means of the selfmap $T : G \to G$.

(C) In the following, a basic example of standard system is given. For each $u \in E$, let $Ku \in E_0$ be defined as

$$Ku(t) = u,\ t \in I \text{ (the constant function of } E_0\text{, attached to } u).$$

Further, let us put, for simplicity

$$\mathcal{K} = \{Ku; u \in E\} \text{ (referred to as : the } \textit{constant class} \text{ of } E_0).$$

The following properties of $\mathcal{K}$ are immediate; so, we do not give details.

Proposition 4.45. *Under the above conventions,*
(75-1) $D(Ku, Kv) = d(u, v),\ \forall u, v \in E$
(75-2) *the mapping $u \mapsto Ku$ is an isometry (hence, necessarily, a topological isomorphism) between E and $\mathcal{K}$*
(75-3) *$\mathcal{K}$ is D-closed (hence, D-complete) in E_0.*

Having these precise, let $\mathcal{S} : E_0 \to E$ be a continuous operator, and $P : E \to E$ be a continuous expansive operator, taken as

(S-K-P) $(\mathcal{S}, K)$ is *P-invertible*: $\mathcal{S}Ku = Pu,\ \forall u \in E$.

Proposition 4.46. *Under these notations,*

(76-1) *the quadruple* $(\mathcal{S}, E, K; P)$ *is admissible*
(76-2) $\mathcal{K} = K(E)$ *is a D-closed Razumikhin chain*
(76-3) $(\mathcal{S}, \mathcal{K})$ *is a standard system.*

The verification is immediate, by the developments above.

Remark 4.10.
Let $P : E \to E$ be a continuous expansive map; and $c \in I$ be arbitrary fixed. Then, the operator $\mathcal{S}_c^P : E_0 \to E$ introduced as

$$\mathcal{S}_c^P(\varphi) = P(\varphi(c)),\ \varphi \in E_0,$$

is (D, d)-continuous and $(\mathcal{S}_c^P, K)$ is P-invertible ; hence, by the above, $(\mathcal{S}_c^P, \mathcal{K})$ is a standard system.

Finally, a corresponding version of Standard BLR theorem with respect to this setting is available. The obtained results seem to be new.

4.4.8 *Isometric systems*

The Standard BLR theorem we just exposed repairs the existential drawbacks of the Conditional BLR theorem. However, the classes of Φ-contractive maps (defined via (I-2)) and (Ψ, Φ)-contractive maps (defined via (II-2)) necessitate extra conditions upon Φ and (Ψ, Φ) respectively, to apply the related metrical fixed point principles above. Technically, this is due to the nonexpansiveness of inverse mapping $U = \mathcal{S}^{-1}$; which, in turn, follows from the properties of the standard system $(\mathcal{S}, L_0)$. It is natural then to ask of under which additional properties upon our data is this removable; a positive answer will be given along the lines below.

Let (E, d) be a complete metric space. Further, let $I = [a, b]$ be a bounded closed real interval; and $E_0 := C(I, E)$ stand for the class of all continuous functions in $\mathcal{F}(I, E)$, endowed with the supremum distance D; as precise, (E_0, D) is a complete metric space.

Let in the following $\mathcal{S} : E_0 \to E$ be an operator with

(Dd-cont) $\mathcal{S}$ is (D, d)-continuous $[\varphi_n \xrightarrow{D} \varphi$ implies $\mathcal{S}\varphi_n \xrightarrow{d} \mathcal{S}\varphi]$.

(A) Remember that, the standard and isometric Razumikhin relations $(\perp)$ and $(\models)$ attached to $\mathcal{S}$ write

$$(\varphi \perp \xi \text{ iff } D(\varphi, \xi) \le d(\mathcal{S}\varphi, \mathcal{S}\xi)),\ (\varphi \models \xi \text{ iff } D(\varphi, \xi) = d(\mathcal{S}\varphi, \mathcal{S}\xi)).$$

Further, by a standard (respectively: isometric) *Razumikhin chain* we mean any nonempty part L_0 of E_0, with the property

$$\varphi, \xi \in L_0 \Longrightarrow \varphi \perp \xi \quad \text{(respectively: } \varphi, \xi \in L_0 \Longrightarrow \varphi \models \xi\text{)};$$

the class of all these will be denoted as $\mathrm{Rch}(\mathcal{S})$ and $\mathrm{Rch}(iso)(\mathcal{S})$, respectively. Note that, under the regularity condition

(Dd-nexp) $\mathcal{S}$ is (D, d)-nonexpansive: $d(\mathcal{S}\varphi, \mathcal{S}\xi) \leq D(\varphi, \xi)$, $\forall \varphi, \xi \in E_0$,

we necessarily have

($\perp$) and ($\models$) are identical; hence, so are $\mathrm{Rch}(\mathcal{S})$ and $\mathrm{Rch}(iso)(\mathcal{S})$.

The following statement involving these data is our starting point. Let cl and Cl stand for the closure operators in E and E_0 respectively.

Proposition 4.47. *The following are valid, for each* $L_0 \in \mathrm{Rch}(iso)(\mathcal{S})$,

(81-1) $K_0 := \mathrm{Cl}(L_0)$ *is in* $\mathrm{Rch}(iso)(\mathcal{S})$
(81-2) the restriction $\mathcal{S} : L_0 \to E$ *is injective*
(81-3) L_0 *is D-closed iff* $G := \mathcal{S}(L_0)$ *is d-closed.*

Proof. There are three steps to be passed.

Step 1. Let φ, ξ be two elements in K_0; that is,

there exist sequences (φ_n) and (ξ_n) in L_0 with $\varphi_n \xrightarrow{D} \varphi$, $\xi_n \xrightarrow{D} \xi$.

According to the choice of L_0, we have

$(\forall n)$: $\varphi_n \models \xi_n$; that is, $D(\varphi_n, \xi_n) - d(\mathcal{S}\varphi_n, \mathcal{S}\xi_n)$.

Passing to limit as $n \to \infty$ in this relation yields $\varphi \models \xi$; and this, by the arbitrariness of these data, proves that K_0 is an element of $\mathrm{Rch}(iso)(\mathcal{S})$.

Steps 2-3. Evident, by the corresponding result concerning the class $\mathrm{Rch}(\mathcal{S})$, in view of $\mathrm{Rch}(iso)(\mathcal{S})$ being a subset of it. $\square$

Given the continuous map $\mathcal{S} : E_0 \to E$ and the nonempty part L_0 of E, let us introduce the concept

$(\mathcal{S}, L_0)$ is an *isometric system*, if L_0 is a D-closed isometric Razumikhin chain in E_0 (hence (see above) $G := \mathcal{S}(L_0)$ is a d-closed subset in E).

Remark 4.11.

Note that, under the same (D, d)-nonexpansive condition upon $\mathcal{S}$,

$(\mathcal{S}, L_0)$ is a standard system iff $(\mathcal{S}, L_0)$ is an isometric system.

Returning to the general setting, fix an isometric system $(\mathcal{S}, L_0)$. Clearly, $\mathcal{S}$ is a bijection between L_0 and $G := \mathcal{S}(L_0)$. As a consequence,

the inverse operator $U := \mathcal{S}^{-1}$ exists as a bijection in $\mathcal{F}(G, L_0)$; hence, $(\mathcal{S}Ug = g, \forall g \in G)$, $(U\mathcal{S}\varphi = \varphi, \forall \varphi \in L_0)$.

Moreover, by the definition of isometric Razumikhin relation ($\models$),

U is (D, d)-isometric: $D(Ux, Uy) = d(x, y)$, for each $x, y \in G$.

A basic particular case when such systems exist may be described as below. Given the continuous map $\mathcal{S} : E_0 \to E$, the (nonempty) d-closed part G of E and the map $W : G \to E_0$, let us say that $(\mathcal{S}, G, W)$ is *strongly admissible*, provided

(s-adm-1) $(\mathcal{S}, W)$ is *invertible*: $\mathcal{S}Wg = g$, $\forall g \in G$
(s-adm-2) W is (d, D)-isometric: $D(Wx, Wy) = d(x, y)$, $\forall x, y \in G$.

Proposition 4.48. *Let the triple $(\mathcal{S}, G, W)$ (taken under the general setting above) be strongly admissible; and put $L_0 := W(G)$. Then,*

(82-1) the restriction $\mathcal{S} : L_0 \to G$ is a bijection over $\mathcal{F}(L_0, G)$
(82-2) L_0 is a D-closed isometric Razumikhin chain
(82-3) $(\mathcal{S}, L_0)$ is an isometric system.

Proof. There are several steps to be passed.

Part 1. For the moment, $\mathcal{S} : L_0 \to G$ is surjective; because,

$$\mathcal{S}(L_0) = \{\mathcal{S}\varphi; \varphi \in L_0\} = \{\mathcal{S}Wg; g \in G\} = G.$$

On the other hand, $\mathcal{S} : L_0 \to G$ is injective, in view of

$\mathcal{S}\varphi = \mathcal{S}\xi$ and $(\varphi = Wa, \xi = Wb)$ imply $a = b$ (hence, $\varphi = \xi$).

Part 2. As a consequence, $U := \mathcal{S}^{-1}$ exists in $\mathcal{F}(G, L_0)$, and

$(\mathcal{S}Ug = g, \forall g \in G)$, $(U\mathcal{S}\varphi = \varphi, \forall \varphi \in L_0)$.

But then, from the invertible property, we must have (under $\mathcal{S}$ = injective)

$(Ug = Wg, \forall g \in G)$ (hence, $U = W$), and $(W\mathcal{S}\varphi = \varphi, \forall \varphi \in L_0)$.

Part 3. Further, we claim that L_0 is a D-closed isometric Razumikhin chain. In fact, by the isometric property and the preceding relation, one has (via $W^{-1} = \mathcal{S}$)

$$D(\varphi,\xi) = D(Wx,Wy) = d(x,y) = d(\mathcal{S}\varphi,\mathcal{S}\xi), \text{ if } \varphi = Wx,\ \xi = Wy;$$

so, L_0 is an isometric Razumikhin chain in E_0. This, via d-closed property of $G = \mathcal{S}(L_0)$, tells us (by a previous result) that L_0 is D-closed. □

(B) Returning to the general case, let $(\mathcal{S}, L_0)$ be an isometric system:

$L_0 \in \text{Rch}(iso)(\mathcal{S})$ is D-closed, $G := \mathcal{S}(L_0)$ is d-closed, and $\mathcal{S} : L_0 \to G$ is a bijection; hence, $U := \mathcal{S}^{-1}$ exists as a bijection over $\mathcal{F}(G, L_0)$ (in the sense: $(\mathcal{S}Ug = g, \forall g \in G)$, $(U\mathcal{S}\varphi = \varphi, \forall \varphi \in L_0)$), with the isometric property $(D(Ux, Uy) = d(x, y),\ x, y \in G)$.

Then, let $\mathcal{T} : E_0 \to E$ be a mapping with

(subord) T is subordinated to $\mathcal{S}$ on L_0: $\mathcal{T}(L_0) \subseteq \mathcal{S}(L_0)(= G)$.

Remember that $\varphi \in E_0$ is a *coincidence point* of $\mathcal{S}$ and $\mathcal{T}$, when $\mathcal{S}\varphi = \mathcal{T}\varphi$. To determine such points, a previous construction will be considered. Let $T : G \to G$ be the selfmap of G introduced as

(T-def) $Tx = \mathcal{T}(Ux),\ x \in G$.

By a preceding statement, the question of working with coincidence points relative to $(\mathcal{S}, \mathcal{T})$ is reducible to the question of working with fixed points generated by the selfmap T. Precisely, we have

Proposition 4.49. *The following assertions are valid:*
(83-1) *If $x \in G$ is a fixed point of T, then $\xi := Ux \in L_0$ is a coincidence point of $\mathcal{S}$ and $\mathcal{T}$*
(83-2) *Conversely, if $\zeta := Uz \in L_0$ is a coincidence point of $\mathcal{S}$ and $\mathcal{T}$, then $z \in G$ is a fixed point of T.*

The proof of this is very similar with the one of standard case; so, we do not give details.

Remark 4.12.
The following completions of these facts are to be noted.
Given $\varphi_0 = Ux_0 \in L_0$, we say that the sequence $(\varphi_n = Ux_n; n \geq 0)$ in L_0 is $(\varphi_0, \mathcal{S}, \mathcal{T})$-*iterative*, provided $(\mathcal{T}\varphi_n = \mathcal{S}\varphi_{n+1}$, for all $n)$. Note that, the family of such sequences is nonempty, by the subordination hypothesis.
com-1): Given the $(\varphi_0, \mathcal{S}, \mathcal{T})$-iterative process $(\varphi_n = Ux_n; n \geq 0)$ in L_0, we have (cf. our previous constructions)

$$Tx_n = \mathcal{T}(Ux_n) = \mathcal{T}\varphi_n = \mathcal{S}\varphi_{n+1} = \mathcal{S}(Ux_{n+1}) = x_{n+1},\ \forall n \geq 0;$$

so that the supporting sequence (x_n) in G is a (x_0, T)-iterative one, according to $(Tx_n = x_{n+1}; n \geq 0)$. Consequently, any $(\varphi_0 = Ux_0, \mathcal{S}, \mathcal{T})$- iterative sequence in L_0 has the unique representation $(\varphi_n = UT^n x_0; n \geq 0)$.

com-2): Conversely, given the (x_0, T)-iterative sequence $(x_n = T^n x_0; n \geq 0)$ in G, the sequence $(\varphi_n := Ux_n; n \geq 0)$ in L_0 is $(\varphi_0, \mathcal{S}, \mathcal{T})$-iterative, in view of

$$\mathcal{T}\varphi_n = \mathcal{T}(Ux_n) = Tx_n = x_{n+1} = \mathcal{S}(Ux_{n+1}) = \mathcal{S}\varphi_{n+1}, \forall n \geq 0.$$

(com-3) Summing up, each $(\varphi_0 = Ux_0, \mathcal{S}, \mathcal{T})$-iterative process in L_0 is the (unique) U-image of its associated (x_0, T)-iterative process in G.

Under these preliminary facts, we may now proceed to the formulation of our announced results. The following basic concepts appear.

I) Let $\Phi \in \mathcal{F}_0(R_+)$ be given.

I-1) We say that $(\mathcal{T}, \mathcal{S})$ is Φ-*contractive*, provided

(con-1) $d(\mathcal{T}\varphi, \mathcal{T}\xi) \leq \Phi(d(\mathcal{S}\varphi, \mathcal{S}\xi))$, for all $\varphi, \xi \in E_0$.

This concept is comparable with the one in Dhage [73], [74]. In particular, under $\varphi = Ux$, $\xi = Uy$ (where $x, y \in G$), the relation above becomes

(con-1a) $d(Tx, Ty) \leq \Phi(d(x, y))$, for all $x, y \in G$;

or, in other words: the selfmap $T : G \to G$ is $(d; A_1; \Phi)$-contractive.

I-2) We say that $\mathcal{T}$ is Φ-*contractive*, provided

(con-2) $d(\mathcal{T}\varphi, \mathcal{T}\xi) \leq \Phi(D(\varphi, \xi))$, for all $\varphi, \xi \in E_0$.

This class of contractions is a direct extension of the one appearing in Bernfeld et al. [27]. In particular, letting $\varphi = Ux$, $\xi = Uy$ (where $x, y \in G$) in this relation, we have (in view of U = isometric)

(con-2a) $d(Tx, Ty) \leq \Phi(D(Ux, Uy)) = \Phi(d(x, y))$, for all $x, y \in G$;

or, in other words: the selfmap $T : G \to G$ is $(d; A_1; \Phi)$-contractive.

II) Let $\Psi, \Phi \in \mathcal{F}_0(R_+)$ be given.

II-1) We say that $(\mathcal{T}, \mathcal{S})$ is (Ψ, Φ)-*contractive*, provided

(con-3) $\Psi(d(\mathcal{T}\varphi, \mathcal{T}\xi)) \leq \Psi(d(\mathcal{S}\varphi, \mathcal{S}\xi)) - \Phi(d(\mathcal{S}\varphi, \mathcal{S}\xi))$, for all $\varphi, \xi \in E_0$.

Letting $\varphi = Ux$, $\xi = Uy$ (where $x, y \in G$) in this relation, we have

(con-3a) $\Psi(d(Tx, Ty)) \leq \Psi(d(x, y)) - \Phi(d(x, y))$, for all $x, y \in G$;

or, in other words: the selfmap $T : G \to G$ is $(d; A_1; (\Psi, \Phi))$-contractive.

II-2) We say that $\mathcal{T}$ is (Ψ, Φ)-*contractive*, provided

(con-4) $\Psi(d(\mathcal{T}\varphi, \mathcal{T}\xi)) \leq \Psi(D(\varphi, \xi)) - \Phi(D(\varphi, \xi))$, for all $\varphi, \xi \in E_0$.

In particular, letting $\varphi = Ux$, $\xi = Uy$ (where $x, y \in G$) in this relation, we have (in view of U = isometric)

(con-4a) $\Psi(d(Tx, Ty)) \leq \Psi(D(Ux, Uy)) - \Phi(D(Ux, Uy)) =$
$\Psi(d(x, y)) - \Phi(d(x, y))$, for all $x, y \in G$;

or, in other words: the selfmap $T : G \to G$ is $(d; A_1; (\Psi, \Phi))$-contractive.

Now, as G is closed and $T : G \to G$ is either $(d; A_1; \Phi)$-contractive (in the cases (I-1) or (I-2)) or $(d; A_1; (\Psi, \Phi))$-contractive (in the cases (II-1) or (II-2)) our previous fixed point principles are applicable here; and yield the following synthetic coincidence point result involving our data (referred to as: Isometric BLR theorem):

Theorem 4.21. *Let $(\mathcal{S}, L_0)$ be an isometric system, and $\mathcal{T} : E_0 \to E$ be subordinated to $\mathcal{S}$ on L_0. Further, suppose that one of next conditions below is accepted*

(81-i) $(\mathcal{T}, \mathcal{S})$ is Φ-contractive, where $\Phi \in \mathcal{F}_0(re)(R_+)$ is Meir-Keeler admissible (or, equivalently: Matkowski admissible)
(81-ii) $\mathcal{T}$ is Φ-contractive, where $\Phi \in \mathcal{F}_0(re)(R_+)$ is Meir-Keeler admissible (or, equivalently: Matkowski admissible)
(81-iii) $(\mathcal{T}, \mathcal{S})$ is (Ψ, Φ)-contractive, where $\Psi, \Phi \in \mathcal{F}_0(R_+)$ are such that (Ψ, Φ) is a normal couple with Φ = right sequentially positive
(81-iv) $\mathcal{T}$ is (Ψ, Φ)-contractive, where $\Psi, \Phi \in \mathcal{F}_0(R_+)$ fulfill: (Ψ, Φ) is a normal couple with Φ = right sequentially positive.

Then,

(81-a) *$\mathcal{S}$ and $\mathcal{T}$ have (in L_0) a unique coincidence point $\varphi^* \in L_0$*

(81-b) *for the arbitrary fixed $\varphi_0 \in L_0$, the unique $(\varphi_0, \mathcal{S}, \mathcal{T})$-iterative sequence $(\varphi_n; n \geq 0)$ in L_0 is D-convergent towards φ^*.*

Proof. Let $T \in \mathcal{F}(G)$ be defined as $(Tg = \mathcal{T}(Ug); g \in G)$. By the previous fixed point results, we have the conclusions

(con-1) $\text{Fix}(T) = \{x \in G; x = Tx\}$ is a singleton, $\{x^*\}$
(con-2) for each $x_0 \in G$, the (x_0, T)-iterative process (x_n)
introduced as $(x_{n+1} = Tx_n; n \geq 0)$ fulfills $x_n \xrightarrow{d} x^*$ as $n \to \infty$.

The former of these tells us [by means of a preceding auxiliary fact] that $\varphi^* := Ux^*$ is the unique coincidence point of $\mathcal{S}$ and $\mathcal{T}$ in L_0. Further, note that $\varphi_0 = Ux_0$, where $x_0 = \mathcal{S}\varphi_0$. Let $(\varphi_n = Ux_n; n \geq 0)$ be the $(\varphi_0, \mathcal{S}, \mathcal{T})$-iterative sequence in L_0 (generated by φ_0). Then (see above) (x_n) is a (x_0, T)-iterative process (generated by x_0); so (by the latter conclusion above) $x_n \xrightarrow{d} x^*$ [that is: $d(x_n, x^*) \to 0$] as $n \to \infty$. On the other hand,

$$D(\varphi_n, \varphi^*) = d(x_n, x^*), \text{ for all } n \text{ (as } U \text{ is isometric);}$$

so, by combining these, $\varphi_n \xrightarrow{D} \varphi^*$ [via $D(\varphi_n, \varphi^*) \to 0$] as $n \to \infty$. □

The obtained coincidence point result may be given as a completion of the previous one, appearing in Standard BLR theorem.

(D) In the following, a basic example of isometric system is given. For each $u \in E$, let $Ku \in E_0$ be defined as

$$Ku(t) = u,\ t \in I \text{ (the constant function of } E_0 \text{, attached to } u).$$

Further, let us put, for simplicity

$$\mathcal{K} = \{Ku; u \in E\} \text{ (referred to as : the \textit{constant class} of } E_0).$$

The following properties of this constant class $\mathcal{K}$ are almost immediate; so, we do not give details.

Proposition 4.50. *Under the above conventions,*

(84-1) $D(Ku, Kv) = d(u, v),\ \forall u, v \in E$

(84-2) *the mapping $u \mapsto Ku$ is an isometry (hence, necessarily, a topological isomorphism) between E and $\mathcal{K}$*

(84-3) *$\mathcal{K}$ is D-closed (hence, D-complete) in E_0.*

Having these precise, let $\mathcal{S} : E_0 \to E$ be a continuous operator, with

(cfs) $(\mathcal{S}, K)$ is *invertible*: $\mathcal{S}Ku = u,\ \forall u \in E$.

Proposition 4.51. *Under these notations,*

(85-1) the triple $(\mathcal{S}, E, K)$ is strongly admissible

(85-2) $\mathcal{K} = K(E)$ is a D-closed isometric Razumikhin chain

(85-3) $(\mathcal{S}, \mathcal{K})$ is an isometric system.

The verification is immediate, by the developments above.

Remark 4.13.

Let $c \in I$ be fixed. Then, the operator $\mathcal{S}_c : E_0 \to E$ introduced as

$\mathcal{S}_c(\varphi) = \varphi(c)$, $\varphi \in E_0$,

is (D, d)-nonexpansive (hence, (D, d)-continuous), and $(\mathcal{S}_c, K)$ is invertible; so that, by the above, $(\mathcal{S}_c, \mathcal{K})$ is an isometric system.

We close these developments with a methodological question. Denote

Isometric Constant BLR theorem: =
Isometric BLR theorem generated by the isometric system $(\mathcal{S}, \mathcal{K})$.

Note that, by a previous fact and $\mathcal{S}(\mathcal{K}) = E$,

L_0 = full isometric Razumikhin chain ($\mathcal{S}(L_0) = E$) and
L_0 is inclusion comparable with $\mathcal{K}$ imply $L_0 = \mathcal{K}$;

so that the following assertion is holding:

all Isometric BLR theorems constructed under isometric systems $(\mathcal{S}, L_0)$ where the full isometric Razumikhin chain L_0 is inclusion comparable with $\mathcal{K}$, are identical with Isometric Constant BLR theorem.

Therefore, an Isometric BLR theorem is distinct from Isometric Constant BLR theorem when its attached isometric system $(\mathcal{S}, L_0)$ is taken as

L_0 = full isometric Razumikhin chain that is not inclusion comparable with $\mathcal{K}$ (both $L_0 \subseteq \mathcal{K}$ and $\mathcal{K} \subseteq L_0$ are false).

Concerning this aspect note that, under

(tr) $(\perp)$ is transitive (hence, an equivalence)

we have a conclusion like

L_0 = full isometric Razumikhin chain of E_0 implies
L_0 is an equivalence class: $L_0 = \widetilde{\varphi}$, for some/all $\varphi \in L_0$.

This, in particular, is valid for the constant class $\mathcal{K}$; and yields

for each full isometric Razumikhin chain L_0, we have:
$(L_0 \cap \mathcal{K} = \emptyset)$ iff (L_0 is not inclusion comparable with $\mathcal{K}$);

which, along the described alternative, reduces the determination of full isometric Razumikhin chains to a disjoint intersection property with the constant class $\mathcal{K}$. Some related aspects will be delineated elsewhere.

4.4.9 *Linear case*

In the following, the linear version of these developments will be discussed.

Let $(E, ||.||)$ be a Banach space; and $d(.,.)$ be the induced metric $[d(x,y) = ||x-y||,\ x,y \in E]$; hence, (E,d) is a complete metric space. Further, let $I = [a,b]$ be a real interval; and $E_0 := C(I,E)$ be the class of all continuous functions in $\mathcal{F}(I,E)$, endowed with the supremum norm

$$||\varphi||_0 = \sup\{||\varphi(t)||; t \in I\},\ \varphi \in E_0.$$

Clearly, $(E_0, ||.||_0)$ is a Banach space; and, if D is the induced metric $[D(\varphi,\xi) = ||\varphi - \xi||_0,\ \varphi,\xi \in E_0]$, then (E_0, D) is a complete metric space.

For each $u \in E$, let Ku denote the function of E_0, given as

(K-def) $Ku(t) = u,\ t \in I$ (the constant function generated by u);

and $\mathcal{K}$ stand for the collection of all these:

$\mathcal{K} := K(E) = \{Ku; u \in E\}$ (referred to as: the *constant class* of E_0).

The basic properties of our objects are described below. (Since the verification is immediate, we do not give details).

Proposition 4.52. *Under the precise conventions, we have*

(91-1) $K(u+v) = Ku + Kv,\ K(\lambda u) = \lambda Ku,\ \forall u,v \in E,\ \forall \lambda \in R$;
hence, K *is linear, and* $\mathcal{K} = K(E)$ *is a linear subspace of* E_0
(91-2) $||u|| = ||Ku||_0,\ \forall u \in E$; *hence (by the above)* $u \mapsto Ku$ *is an algebraic and topological isomorphism between* $(E, ||.||)$ *and* $(\mathcal{K}, ||.||_0)$.

(A) Let in the following $\mathcal{S} : E_0 \to E$ be a mapping with

(m-1) $\mathcal{S}$ is (D,d)-continuous ($\varphi_n \xrightarrow{D} \varphi$ implies $\mathcal{S}\varphi_b \xrightarrow{d} \mathcal{S}\varphi$)
(m-2) $\mathcal{S}$ is *linear*, in the sense:
$\mathcal{S}$ is additive $(\mathcal{S}(\varphi+\xi) = \mathcal{S}(\varphi) + \mathcal{S}(\xi),\ \forall \varphi,\xi \in E_0)$,
$\mathcal{S}$ is homogeneous $(\mathcal{S}(\lambda\varphi) = \lambda\mathcal{S}(\varphi),\ \forall \lambda \in R,\ \forall \varphi \in E_0)$.

The *unilateral Razumikhin set* and *unilateral Razumikhin isometric set* attached to this operator are defined according to

(r-cl) $\text{razu}(\mathcal{S}) = \{\varphi \in E_0; ||\varphi||_0 \le ||\mathcal{S}\varphi||\}$
(r-iso-cl) $\text{razu}(iso)(\mathcal{S}) = \{\varphi \in E_0; ||\varphi||_0 = ||\mathcal{S}\varphi||\}$.

Note that both razu($\mathcal{S}$) and razu(iso)($\mathcal{S}$) are nonempty; because they include $K(0)$ (the null element of E_0). The following properties of these classes are useful. (See also Turinici [250] for a number of related facts.)

Proposition 4.53. *Let M_0 stand for either of the unilateral Razumikhin sets* razu($\mathcal{S}$) *or* razu(iso)($\mathcal{S}$). *Then*

M_0 is homogeneous ($M_0 = \lambda M_0$, $\forall \lambda \in R \setminus \{0\}$);
hence, in particular, symmetric ($M_0 = -M_0$).

Proof. We discuss the case of $M_0 = \text{razu}(\mathcal{S})$; the remaining one is carried out in a similar way. Let $\lambda \in R \setminus \{0\}$ be arbitrary fixed. Given $\mu \in \{\lambda, 1/\lambda\}$ and $\varphi \in \text{razu}(\mathcal{S})$, denote $\xi = \mu\varphi$. Clearly, $||\xi||_0 = |\mu| \cdot ||\varphi||_0$, $||\mathcal{S}(\xi)|| = |\mu| \cdot ||\mathcal{S}(\varphi)||$. This, along with the choice of φ, gives $||\xi||_0 \leq ||\mathcal{S}(\xi)||$; that is: $\xi \in \text{razu}(\mathcal{S})$. As a consequence, the global inclusions hold: $\lambda\text{razu}(\mathcal{S}) \subseteq \text{razu}(\mathcal{S})$, $(1/\lambda)\text{razu}(\mathcal{S}) \subseteq \text{razu}(\mathcal{S})$; and, from this, we are done. □

The introduction of this class is motivated by the fact that, under our linear setting, the standard and isometric Razumikhin relations ($\perp$) and ($\models$) associated to $\mathcal{S}$ may be written as

$$(\varphi \perp \xi \text{ iff } \varphi - \xi \in \text{razu}(\mathcal{S})); \ (\varphi \models \xi \text{ iff } \varphi - \xi \in \text{razu}(iso)(\mathcal{S})).$$

This tells us that, by definition, the (nonempty) subset L_0 of E_0 is a standard (respectively, isometric) Razumikhin chain, whenever

$$\varphi, \xi \in L_0 \Longrightarrow \varphi - \xi \in \text{razu}(\mathcal{S})$$
$$\text{(respectively: } \varphi, \xi \in L_0 \Longrightarrow \varphi - \xi \in \text{razu}(iso)(\mathcal{S}));$$

the class of all these will be denoted as Rch($\mathcal{S}$) and Rch(iso)($\mathcal{S}$), respectively. Note that, under the regularity condition

(Dd-nexp) $\mathcal{S}$ is (D, d)-nonexpansive ($||\mathcal{S}\varphi|| \leq ||\varphi||_0$, $\forall \varphi \in E_0$)

we necessarily have

razu($\mathcal{S}$) = razu(iso)($\mathcal{S}$); wherefrom (see above)
($\varphi \perp \xi$ iff $\varphi \models \xi$) and Rch($\mathcal{S}$) = Rch(iso)($\mathcal{S}$).

Finally, remember that the following auxiliary fact is valid (under the precise framework involving the operator $\mathcal{S}$).

Proposition 4.54. *For each L_0 in* Rch($\mathcal{S}$) *(resp.,* Rch(iso)($\mathcal{S}$)*),*

(93-1) $K_0 = \text{Cl}(L_0)$ *is in* Rch($\mathcal{S}$) *(resp.,* Rch(iso)($\mathcal{S}$)*)*

(93-2) the restriction $\mathcal{S} : L_0 \to E$ is injective
(93-3) L_0 is D-closed iff $G := \mathcal{S}(L_0)$ is d-closed.

(B) Let again $\mathcal{S} : E_0 \to E$ be a continuous linear operator. In the following, two examples of standard and isometric systems are constructed by starting from the couple $(\mathcal{S}, K)$, where $K : E \to E_0$ is the precise one.

Case 1. Let us start from the basic condition

$(\mathcal{S}, \mathcal{K})$ is a *standard system*, in the sense: $\mathcal{K}$ is a D-closed element of $\mathrm{Rch}(\mathcal{S})$ (hence (see above): $G := \mathcal{S}(\mathcal{K})$ is d-closed in E).

The following particular case is useful in practice. Let $P : E \to E$ be a linear operator with

P is continuous and expansive ($||P(x)|| \geq ||x||$, $x \in E$);

and take the continuous linear operator $\mathcal{S} : E_0 \to E$ according to

$(\mathcal{S}, K)$ is P-invertible: $\mathcal{S}Ku = Pu$, $\forall u \in E$.

As K is (d, D)-isometric, we have that (cf. a previous construction)

$(\mathcal{S}, E, K; P)$ is admissible; wherefrom $(\mathcal{S}, \mathcal{K})$ is a standard system; hence, $G := \mathcal{S}(\mathcal{K}) = P(E)$ is a d-closed part of E.

In addition, as a direct consequence of these conclusions,

$\mathcal{S}$ is a bijection between $\mathcal{K}$ and $G = \mathcal{S}(\mathcal{K}) = P(E)$;
hence, $U := \mathcal{S}^{-1}$ exists as a bijection of $\mathcal{F}(G, \mathcal{K})$,
with the nonexpansive property ($D(Ux, Uy) \leq d(x, y)$, $x, y \in G$).

Having these precise, let $\mathcal{T} : E_0 \to E$ be a mapping with

(subord) $\mathcal{T}$ is subordinated to $\mathcal{S}$ on $\mathcal{K}$: $\mathcal{T}(\mathcal{K}) \subseteq \mathcal{S}(\mathcal{K}) = G(= P(E))$.

Then, let $T : G \to G$ be the selfmap of G introduced as

(T-def) $Tx = \mathcal{T}(Ux)$, $x \in G$.

Under these preliminary facts, we may now proceed to the formulation of our coincidence point results. The following basic concepts appear.

I) Let $\Phi \in \mathcal{F}_0(R_+)$ be given.

I-1) We say that $(\mathcal{T}, \mathcal{S})$ is Φ-*contractive*, provided

(con-1) $||\mathcal{T}\varphi - \mathcal{T}\xi|| \leq \Phi(||\mathcal{S}\varphi - \mathcal{S}\xi||)$, for all $\varphi, \xi \in E_0$.

These contractions are comparable with the ones in Dhage [73], [74]. In particular, with $\varphi = Ux$, $\xi = Uy$ (where $x, y \in G$), our definition becomes

(con-1a) $d(Tx, Ty) \leq \Phi(d(x, y))$, for all $x, y \in G$;

or, in other words: the selfmap $T : G \to G$ is $(d; A_1; \Phi)$-contractive.

I-2) We say that $\mathcal{T}$ is Φ-*contractive*, provided

(con-2) $||\mathcal{T}\varphi - \mathcal{T}\xi|| \leq \Phi(||\varphi - \xi)||_0)$, for all $\varphi, \xi \in E_0$.

These contractions extend the ones in Bernfeld et al. [27]. In particular, when Φ = increasing (hence, $\Phi \in \mathcal{F}_0(in)(R_+)$) then, letting $\varphi = Ux$, $\xi = Uy$ (where $x, y \in G$) in this definition, we have (combining with the nonexpansive property of U)

(con-2a) $d(Tx, Ty) \leq \Phi(D(Ux, Uy)) \leq \Phi(d(x, y))$, for all $x, y \in G$;

or, in other words: the selfmap $T : G \to G$ is $(d; A_1; \Phi)$-contractive.

II) Let $\Psi, \Phi \in \mathcal{F}_0(R_+)$ be given.

II-1) We say that $(\mathcal{T}, \mathcal{S})$ is (Ψ, Φ)-*contractive*, provided

(con-3) $\Psi(||\mathcal{T}\varphi - \mathcal{T}\xi||) \leq \Psi(||\mathcal{S}\varphi - \mathcal{S}\xi||) - \Phi(||\mathcal{S}\varphi - \mathcal{S}\xi||)$, $\forall \varphi, \xi \in E_0$.

Letting $\varphi = Ux$, $\xi = Uy$ (where $x, y \in G$) in this relation, gives

(con-3a) $\Psi(d(Tx, Ty)) \leq \Psi(d(x, y)) - \Phi(d(x, y))$, for all $x, y \in G$;

or, in other words: the selfmap $T : G \to G$ is $(d; A_1; (\Psi, \Phi))$-contractive.

II-2) We say that $\mathcal{T}$ is (Ψ, Φ)-*contractive*, provided

(con-4) $\Psi(||\mathcal{T}\varphi - \mathcal{T}\xi||) \leq \Psi(||\varphi - \xi||_0) - \Phi(||\varphi - \xi||_0)$, $\forall \varphi, \xi \in E_0$.

In particular, when $\Psi - \Phi$ is increasing (hence, $\Psi - \Phi \in \mathcal{F}_0(in)(R)$) then, letting $\varphi = Ux$, $\xi = Uy$ (where $x, y \in G$) in this relation, we have (combining with the nonexpansive property of U)

(con-4a) $\Psi(d(Tx, Ty)) \leq \Psi(D(Ux, Uy)) - \Phi(D(Ux, Uy)) \leq$
$\Psi(d(x, y)) - \Phi(d(x, y))$, for all $x, y \in G$;

or, in other words: the selfmap $T : G \to G$ is $(d; A_1; (\Psi, \Phi))$–contractive.

Combining with the Standard BLR theorem, one gets the following coincidence point result (referred to as Standard Constant BLR theorem).

Theorem 4.22. *Let $(\mathcal{S}, \mathcal{K})$ be the standard system generated by the admissible $(\mathcal{S}, E, K; P)$ (see above), and the mapping $\mathcal{T} : E_0 \to E$ be subordinated to $\mathcal{S}$ on $\mathcal{K}$. Further, suppose that one of the next conditions holds*

(91-i) $(\mathcal{T},\mathcal{S})$ *is* Φ*-contractive, where* $\Phi \in \mathcal{F}_0(re)(R_+)$ *is Meir-Keeler admissible (or, equivalently: Matkowski admissible)*
(91-ii) $\mathcal{T}$ *is* Φ*-contractive, where* $\Phi \in \mathcal{F}_0(re,in)(R_+)$ *is Meir-Keeler admissible (or, equivalently: Matkowski admissible)*
(91-iii) $(\mathcal{T},\mathcal{S})$ *is* (Ψ,Φ)*-contractive, where* (Ψ,Φ) *is a normal couple over* $\mathcal{F}_0(R_+)$ *with* Φ *= right sequentially positive*
(91-iv) $\mathcal{T}$ *is* (Ψ,Φ)*-contractive, where* (Ψ,Φ) *is a normal couple over* $\mathcal{F}_0(R_+)$ *with* $\Psi - \Phi$ *= increasing and* Φ *= right sequentially positive.*

Then,

(91-a) $\mathcal{S}$ *and* $\mathcal{T}$ *have (in* $\mathcal{K}$*) a unique coincidence point* $\varphi^* \in \mathcal{K}$
(91-b) *for the arbitrary fixed* $\varphi_0 \in \mathcal{K}$*, the unique* $(\varphi_0,\mathcal{S},\mathcal{T})$*-iterative sequence* $(\varphi_n; n \ge 0)$ *in* $\mathcal{K}$ *is D-convergent towards* φ^**.*

In the following, some basic examples of such standard systems are given. Let $P : E \to E$ be a continuous expansive linear operator.

Example 4.3.
Let $c \in I$ be some point. The linear operator

$$(\mathcal{S}_c^P : E_0 \to E)\text{: } \mathcal{S}_c^P(\varphi) = P(\varphi(c)),\ \varphi \in E_0$$

is continuous, and $(\mathcal{S}_c^P, K)$ is P-invertible; so, $(\mathcal{S}_c^P, \mathcal{K})$ is a standard system. The Standard Constant BLR theorem is thus applicable to these data.

Example 4.4.
Let $C = (c_n; n \ge 0)$ be a sequence in I, and $\Lambda = (\lambda_n; n \ge 0)$ be a sequence in R_+ with $\sum_n \lambda_n = 1$. The linear operator

$$(\mathcal{S}_{(C,\Lambda)}^P : E_0 \to E)\text{: } \mathcal{S}_{(C,\Lambda)}^P(\varphi) = P(\textstyle\sum_n \lambda_n \varphi(c_n)),\ \varphi \in E_0$$

is continuous, and $(\mathcal{S}_{(C,\Lambda)}^P, K)$ is P-invertible; hence, by the above, $(\mathcal{S}_{(C,\Lambda)}^P, \mathcal{K})$ is a standard system. As a consequence, the Standard Constant BLR theorem is applicable to these data. In particular, when the sequence (c_n) is constant and $(\lambda_0 = 1; \lambda_n = 0,\ \forall n \ge 1)$, this operator reduces to the preceding one.

Example 4.5.
The linear operator

$$(\mathcal{S}_\int^P : E_0 \to E)\text{: } \mathcal{S}_\int^P(\varphi) = P((1/(b-a)) \textstyle\int_a^b \varphi(t)dt),\ \varphi \in E_0$$

(where $\int$ stands for the Bochner integral) is continuous, and $(\mathcal{S}^P_\int, K)$ is P-invertible; hence, $(\mathcal{S}^P_\int, \mathcal{K})$ is a standard system. As a consequence, the Standard Constant BLR theorem is applicable to these data.

Case 2. Let us start from the basic condition

$(\mathcal{S}, \mathcal{K})$ is an isometric system, in the sense: $\mathcal{K}$ is a D-closed element of $\mathrm{Rch}(iso)(\mathcal{S})$ (hence (see above) $G := \mathcal{S}(\mathcal{K})$ is d-closed in E).

A sufficient condition for this may be formulated as below. Take the continuous linear operator $\mathcal{S} : E_0 \to E$ according to

$(\mathcal{S}, K)$ is invertible: $\mathcal{S}(Ku) = u$, $\forall u \in E$; hence, $\mathcal{S}(\mathcal{K}) = E$.

As K is (d, D)-isometric, we have that (cf. a previous construction)

$(\mathcal{S}, E, K)$ is strongly admissible; so, $(\mathcal{S}, \mathcal{K})$ is an isometric system.

In addition, as a direct consequence of these conclusions,

$\mathcal{S}$ is a bijection between $\mathcal{K}$ and $E = \mathcal{S}(\mathcal{K})$;
hence, $U := \mathcal{S}^{-1}$ exists as a bijection of $\mathcal{F}(E, \mathcal{K})$,
with the isometric property $(D(Ux, Uy) = d(x, y)$, $x, y \in E)$.

In fact, by the invertible property, we necessarily have $U = K$.

Having these precise, let $\mathcal{T} : E_0 \to E$ be a mapping; note that

(subord) $\mathcal{T}$ is subordinated to $\mathcal{S}$ on $\mathcal{K}$, in view of $\mathcal{S}(\mathcal{K}) = E$.

Then, let $T : E \to E$ be the selfmap of E introduced as

(T-defi) $Tx = \mathcal{T}(Kx)$, $x \in E$.

Under these preliminary facts, we may now proceed to the formulation of our coincidence point results. The following basic concepts appear.

I) Let $\Phi \in \mathcal{F}_0(R_+)$ be given.

I-1) We say that $(\mathcal{T}, \mathcal{S})$ is Φ-*contractive*, provided

(con-1) $||\mathcal{T}\varphi - \mathcal{T}\xi|| \leq \Phi(||\mathcal{S}\varphi - \mathcal{S}\xi||)$, for all $\varphi, \xi \in E_0$.

These contractions are comparable with the ones in Dhage [73], [74]. In particular, with $\varphi = Kx$, $\xi = Ky$ (where $x, y \in E$), our definition becomes

(con-1a) $d(Tx, Ty) \leq \Phi(d(x, y))$, for all $x, y \in E$;

or, in other words: the selfmap $T : E \to E$ is $(d; A_1; \Phi)$-contractive.

I-2) We say that $\mathcal{T}$ is Φ-*contractive*, provided (con-2) $||\mathcal{T}\varphi - \mathcal{T}\xi|| \leq \Phi(||\varphi - \xi)||_0)$, for all $\varphi, \xi \in E_0$.

These contractions extend the ones in Bernfeld et al. [27]. In particular, with $\varphi = Kx$, $\xi = Ky$ (where $x, y \in E$), this relation becomes (combining with the isometric property of K)

(con-2a) $d(Tx, Ty) \leq \Phi(D(Kx, Ky)) = \Phi(d(x, y))$, for all $x, y \in E$;

or, in other words: the selfmap $T : E \to E$ is $(d; A_1; \Phi)$-contractive.

II) Let $\Psi, \Phi \in \mathcal{F}_0(R_+)$ be given.

II-1) We say that $(\mathcal{T}, \mathcal{S})$ is (Ψ, Φ)-*contractive*, provided

(con-3) $\Psi(||\mathcal{T}\varphi - \mathcal{T}\xi||) \leq \Psi(||\mathcal{S}\varphi - \mathcal{S}\xi||) - \Phi(||\mathcal{S}\varphi - \mathcal{S}\xi||)$, $\forall \varphi, \xi \in E_0$.

In particular, with $\varphi = Kx$, $\xi = Ky$ (where $x, y \in E$) this relation becomes

(con-3a) $\Psi(d(Tx, Ty)) \leq \Psi(d(x, y)) - \Phi(d(x, y))$, for all $x, y \in E$;

or, in other words: the selfmap $T : E \to E$ is $(d; A_1; (\Psi, \Phi))$–contractive.

II-2) We say that $\mathcal{T}$ is (Ψ, Φ)-*contractive*, provided

(con-4) $\Psi(||\mathcal{T}\varphi - \mathcal{T}\xi||) \leq \Psi(||\varphi - \xi||_0) - \Phi(||\varphi - \xi||_0)$, $\forall \varphi, \xi \in E_0$.

In particular, with $\varphi = Kx$, $\xi = Ky$ (where $x, y \in E$), this relation becomes (combining with the isometric property of K)

(con-4a) $\Psi(d(Tx, Ty)) \leq \Psi(D(Kx, Ky)) - \Phi(D(Kx, Ky)) =$
$\Psi(d(x, y)) - \Phi(d(x, y))$, for all $x, y \in E$;

or, in other words: the selfmap $T : E \to E$ is $(d; A_1; (\Psi, \Phi))$–contractive.

Combining with the Isometric BLR theorem, one get the following coincidence point result (referred to as Isometric Constant BLR theorem).

Theorem 4.23. *Let $(\mathcal{S}, \mathcal{K})$ be the isometric system generated by the strongly admissible $(\mathcal{S}, E, K)$ (see above), and the mapping $\mathcal{T} : E_0 \to E$ be given. Further, suppose that one of the next conditions holds*

(92-i) $(\mathcal{T}, \mathcal{S})$ is Φ-contractive, where $\Phi \in \mathcal{F}_0(re)(R_+)$ is Meir-Keeler admissible (or, equivalently: Matkowski admissible)
(92-ii) $\mathcal{T}$ is Φ-contractive, where $\Phi \in \mathcal{F}_0(re)(R_+)$ is Meir-Keeler admissible (or, equivalently: Matkowski admissible)
(92-iii) $(\mathcal{T}, \mathcal{S})$ is (Ψ, Φ)-contractive, where (Ψ, Φ) is a normal couple over $\mathcal{F}_0(R_+)$ with Φ = right sequentially positive
(92-iv) $\mathcal{T}$ is (Ψ, Φ)-contractive, where (Ψ, Φ) is a normal couple over $\mathcal{F}_0(R_+)$ with Φ = right sequentially positive.

Then,

(92-a) *$\mathcal{S}$ and $\mathcal{T}$ have (in $\mathcal{K}$) a unique coincidence point $\varphi^* \in \mathcal{K}$*

(92-b) *for the arbitrary fixed $\varphi_0 \in \mathcal{K}$, the unique $(\varphi_0, \mathcal{S}, \mathcal{T})$-iterative sequence $(\varphi_n; n \geq 0)$ in $\mathcal{K}$ is D-convergent towards φ^*.*

Now, as before, some basic examples of such isometric systems are given.

Example 4.6.

Let $c \in I$ be some point. The linear operator

$$(\mathcal{S}_c : E_0 \to E)\colon \mathcal{S}_c(\varphi) = \varphi(c),\ \varphi \in E_0,$$

is nonexpansive (hence, continuous) and $(\mathcal{S}_c, K)$ is invertible; so, by the above, $(\mathcal{S}_c, \mathcal{K})$ is an isometric system. As a consequence, the Isometric Constant BLR theorem is applicable to these data.

Example 4.7.

Let $C = (c_n)$ be a sequence in I; and $\Lambda = (\lambda_n)$ be a sequence in R_+ with $\sum_n \lambda_n = 1$. The linear operator

$$(\mathcal{S}_{(C,\Lambda)} : E_0 \to E)\colon \mathcal{S}_{(C,\Lambda)}(\varphi) = \textstyle\sum_n \lambda_n \varphi(c_n),\ \varphi \in E_0$$

is well defined, because φ is bounded on I. In addition, $\mathcal{S}_{(C,\Lambda)}$ is nonexpansive (hence, continuous) and $(\mathcal{S}_{(C,\Lambda)}, K)$ is invertible; so, by the above, $(\mathcal{S}_{(C,\Lambda)}, \mathcal{K})$ is an isometric system. As a consequence, the Isometric Constant BLR theorem is applicable to these data. In particular, when the sequence (c_n) is constant and $(\lambda_0 = 1; \lambda_n = 0,\ \forall n \geq 1)$, this operator reduces to the preceding one.

Example 4.8.

Define the linear operator $\mathcal{S}_\int : E_0 \to E$ as

$$\mathcal{S}_\int(\varphi) = (1/(b-a)) \textstyle\int_a^b \varphi(t)dt,\ \varphi \in E_0;$$

(where $\int$ stands for the Bochner integral). By the elementary properties of this object, $\mathcal{S}_\int$ is nonexpansive (hence, continuous) and $(\mathcal{S}_\int, K)$ is invertible; so, by the above, $(\mathcal{S}_\int, \mathcal{K})$ is an isometric system. As a consequence, the Isometric Constant BLR theorem is applicable to these data.

Remark 4.14.

Concerning the relationships between these results, the following aspect is useful. Let the linear mapping $\mathcal{S} : E_0 \to E$ be such that (n-exp) $\mathcal{S}$ is *nonexpansive* ($||S\varphi|| \le ||\varphi||_0$, for each $\varphi \in E_0$); hence, in particular, $\mathcal{S}$ is continuous.

The general setting (modulo $\mathcal{S}$) of these developments is thus fulfilled. As direct consequence, one has

(q1) the unilateral Razumikhin sets $\text{razu}(\mathcal{S})$, $\text{razu}(iso)(\mathcal{S})$ are identical
(q2) the Razumikhin relations $(\perp)$ and $(\models)$ are identical
(q3) Razumikhin chain is identical with isometric Razumikhin chain
(q4) the Razumikhin classes $\text{Rch}(\mathcal{S})$, $\text{Rch}(iso)(\mathcal{S})$ are identical.

In particular, this tells us that Standard Constant BLR theorem is reducible to Isometric Constant BLR theorem.

(C) Returning to the general setting, suppose (in addition) that

(tr) $\mathcal{S}$ is *transitive*: the Razumikhin relation $(\perp)$ is transitive.

Proposition 4.55. *The following conditions are equivalent:*

(94-1) $\mathcal{S}$ *is transitive (in the sense:* $(\perp)$ *is transitive)*
(94-2) $\text{razu}(\mathcal{S})$ *is algebraically closed*
(94-3) $\text{razu}(\mathcal{S})$ *is additive:* $\varphi, \psi \in \text{razu}(\mathcal{S}) \Longrightarrow \varphi + \psi \in \text{razu}(\mathcal{S})$
(94-4) $\text{razu}(\mathcal{S})$ *is a linear subspace of* E_0
(94-5) $\text{razu}(\mathcal{S})$ *is a Razumikhin chain.*

Proof. The equivalence between all conditions excepting the first one is a direct consequence of

$\text{razu}(\mathcal{S})$ is a Razumikhin chain, iff
$\text{razu}(\mathcal{S})$ is *algebraically closed*: $\varphi, \xi \in \text{razu}(\mathcal{S}) \Longrightarrow \varphi - \xi \in \text{razu}(\mathcal{S})$.

It remains to prove that one of these – additivity, for example – is equivalent with transitivity of $\mathcal{S}$ (or, equivalently: the transitivity of $(\perp)$). If $\text{razu}(\mathcal{S})$ is additive, then $(\perp)$ is transitive; for, if $\varphi \perp \xi$ and $\xi \perp \lambda$, then

$$\varphi - \lambda = \varphi - \xi + \xi - \lambda \in \text{razu}(\mathcal{S}) + \text{razu}(\mathcal{S}) \subseteq \text{razu}(\mathcal{S}).$$

Conversely, assume that $(\perp)$ is transitive; and let φ, ξ be two elements in $\text{razu}(\mathcal{S})$. By definition, $K(0) \perp \varphi$, $\varphi \perp (\varphi + \xi)$. So, by transitivity, $K(0) \perp (\varphi + \xi)$; whence (again by definition), $\varphi + \xi \in \text{Raz}(\mathcal{S})$; and the conclusion follows. $\square$

The following consequence of these developments is to be noted. Let again $\mathcal{S} : E_0 \to E$ be a continuous linear operator. Remember that, a basic particular case when the couple $(\mathcal{S}, \mathcal{K})$ is an isometric system is

the triple $(\mathcal{S}, E, K)$ is strongly admissible (see above).

This, in view of K being (d, D)-isometric, reduces to

(SK-inv) $(\mathcal{S}, K)$ is invertible: $\mathcal{S}(Ku) = u$, $\forall u \in E$.

Suppose in the following that this condition holds (as well as the transitive one). A first consequence of the written condition is

$\mathcal{S}(\mathcal{K}) = \{\mathcal{S}(Ku); u \in E\} = E$; so that, $\mathcal{K}$ is *full.*

Further, a second consequence of the same writes

$(\forall u \in E)$: $||\mathcal{S}(Ku)|| = ||u|| = ||Ku||_0$;

which, in turn gives

$\mathcal{K} = \{Ku; u \in E\} \subseteq \text{razu}(iso)(\mathcal{S}) \subseteq \text{razu}(\mathcal{S})$;

so that, combining with the above,

(S-full) $\mathcal{S}(\text{razu}(\mathcal{S})) = E$; referred to as: $\mathcal{S}$ is *full.*

Summing up, $\mathcal{K}$ and $\text{razu}(\mathcal{S})$ are full Razumikhin chains of E_0 with

$\mathcal{K} \subseteq \text{razu}(\mathcal{S})$; hence, $\mathcal{K} = \text{razu}(\mathcal{S})$

if we take some auxiliary facts into account.

As a conclusion of these, we have that, whenever $\mathcal{S} : E_0 \to E$ fulfills

(n-t-i) $\mathcal{S}$ is continuous, linear, transitive, and $(\mathcal{S}, K)$ is invertible,

any existence/uniqueness result over the full Razumikhin chain $\text{razu}(\mathcal{S})$ is identical – in a trivial way – with the corresponding one over the full Razumikhin chain $\mathcal{K}$. This, in the particular case of $\mathcal{S} = \mathcal{S}_c$ (see above) means that all recent PPF dependent fixed point results in Agarwal et al. [8], Cho et al. [53], Cirić et al. [60], Dhage [73], [74], Harjani et al. [109], Hussain et al. [116], Kaewcharoen [136], Khan and Jhade [143], Kutbi and Sintunavarat [157], Paknazar et al. [183], are in fact reducible to corresponding fixed point theorems – especially, under the lines in Samet et al. [209] – over the ambient Banach space. Some other aspects may be found in Drici et al. [80]; see also Farajzadeh and Kaewcharoen [89].

It remains now to discuss the case of the invertible condition upon $(\mathcal{S}, K)$ being not fulfilled; clearly, the methodological basis to be considered is Isometric BLR theorem (the linear case). The most natural substitute of the underlying condition is that of $\mathcal{S} : E_0 \to E$ fulfilling

(nex-f-tr) $\mathcal{S}$ is nonexpansive, full and transitive.

By the nonexpansive condition, we have $\text{razu}(\mathcal{S}) = \text{razu}(iso)(\mathcal{S})$. In addition, by the transitive condition, $(\perp)$ is an equivalence in E_0; and, for each $\varphi \in E_0$, its equivalence class is $\widetilde{\varphi} = \varphi + \text{razu}(\mathcal{S}) = \varphi + \text{razu}(iso)(\mathcal{S})$.

Proposition 4.56. *Under the posed conditions,*

(95-1) each equivalence class is a full Razumikhin chain of E_0
(95-2) each full Razumikhin chain of E_0 *is an equivalence class.*

Proof. i): Clearly, $\varphi + \text{razu}(\mathcal{S})$ is a Razumikhin chain, in view of

$$(\varphi + \text{razu}(\mathcal{S})) - (\varphi + \text{razu}(\mathcal{S})) = \text{razu}(\mathcal{S}) - \text{razu}(\mathcal{S}) = \text{razu}(\mathcal{S}).$$

In addition, by the full property,

$$\mathcal{S}(\varphi + \text{razu}(\mathcal{S})) = \mathcal{S}(\varphi) + \mathcal{S}(\text{razu}(\mathcal{S})) = \mathcal{S}(\varphi) + E = E;$$

and we are done.

ii): Let L_0 be a full Razumikhin chain of E_0. By the chain property, L_0 is then included in an equivalence class: $L_0 \subseteq \varphi + \text{razu}(\mathcal{S})$, for some $\varphi \in E_0$. On the other hand, from the preceding stage, $\varphi + \text{razu}(\mathcal{S})$ is a full Razumikhin chain: $\mathcal{S}(\varphi + \text{razu}(\mathcal{S}) = E$. Combining with a preceding result, yields $L_0 = \varphi + \text{razu}(\mathcal{S})$; and the assertion follows. □

Summing up, each full Razumikhin chain L_0 appearing in Isometric BLR theorem is a translate of the Razumikhin class $\text{razu}(\mathcal{S})$. Further technical aspects may be found in Hammad and De la Sen [108]; see also Turinici [259, Sec. 4.3].

4.5 Meir-Keeler Maps in Order Pseudometric Spaces

Abstract. Some fixed point results are given for a class of Meir-Keeler contractive maps acting on order pseudometric spaces. Technical connections with the related statements due to Roldán and Shahzad [Fixed Point Th. Appl., 2014, 2014:218] are also being discussed.

Subject-Class 2020. 47H10 (Primary), 54H25 (Secondary).

Keywords. Order pseudometric space, Picard operator, fixed point, convergent and Cauchy sequence, Meir-Keler contractive property, Boyd-Wong and Matkowski admissible function, normal pair.

4.5.1 *Dependent Choice principles*

See Ch. 1.

4.5.2 *Conv-Cauchy structures*

See Ch. 1.

4.5.3 *Geometric-asymptotic relations*

See Ch. 1.

4.5.4 *Banach contractions and relatives*

See Ch. 1.

4.5.5 *Order pseudometric spaces*

Let X be a nonempty set. Call the subset Y of X, *almost singleton* (in short: *asingleton*) provided $[y_1, y_2 \in Y$ implies $y_1 = y_2]$; and *singleton* if, in addition, Y is nonempty; note that in this case $Y = \{y\}$, for some $y \in X$.

(A) In the following, the concept of *(general) conv-Cauchy structure* will be introduced. Let $\mathcal{S}(X)$ stand for the class of all sequences (x_n) in X. By a (sequential) *convergence structure* on X we mean any part $\mathcal{C}$ of $\mathcal{S}(X) \times X$ with the property (cf. Kasahara [141]):

(conv-1) $\mathcal{C}$ is *hereditary*:
$((x_n); x) \in \mathcal{C} \Longrightarrow ((y_n); x) \in \mathcal{C}$, for each subsequence (y_n) of (x_n).

Then, $((x_n); x) \in \mathcal{C}$ writes $x_n \xrightarrow{\mathcal{C}} x$; and reads: (x_n), $\mathcal{C}$-converges to x; also referred to as: x is the $\mathcal{C}$-*limit* of (x_n). The set of all such x is denoted

$\mathcal{C} - \lim_n(x_n)$; or, $\lim_n(x_n)$ (if $\mathcal{C}$ is understood);

when it is nonempty then (x_n) is called *$\mathcal{C}$-convergent.* The following (optional) supplementary conditions upon this structure are to be considered:

(conv-2) $\mathcal{C}$ is *reflexive*: for each $u \in X$,
the constant sequence $(x_n = u; n \geq 0)$ fulfills $x_n \xrightarrow{\mathcal{C}} u$
(conv-3) $\mathcal{C}$ is *separated*:
$\mathcal{C} - \lim_n(x_n)$ is an asingleton, for each sequence (x_n) in X.

In the last case, $x_n \xrightarrow{\mathcal{C}} z$ will be also written as $\lim_n(x_n) = z$; and if $z \in X$ is generic, then (x_n) is called *$\mathcal{C}$-strongly-convergent.*

Further, let us say that the subset $\mathcal{H} \subseteq \mathcal{S}(X)$ is a (sequential) *Cauchy structure* on X, provided (cf. Turinici [244])

(Cauchy-1) $\mathcal{H}$ is *hereditary*:
$(x_n) \in \mathcal{H}$ implies $(y_n) \in \mathcal{H}$, for each subsequence (y_n) of (x_n).

In this case, any sequence (x_n) in $\mathcal{H}$ will be referred to as a *$\mathcal{H}$-Cauchy* one; and, if no confusion can arise, we may denote $\mathcal{H} = \mathit{Cauchy}(X)$. The following (optional) condition upon this class is to be considered here:

(Cauchy-2) $\mathcal{H}$ is *reflexive*:
for each $u \in X$, the constant sequence $(x_n = u; n \geq 0)$ is in $\mathcal{H}$.

Finally, given the convergence structure $\mathcal{C}$ and the Cauchy structure $\mathcal{H}$ on X, we say that the couple $(\mathcal{C}, \mathcal{H})$ is a *conv-Cauchy structure* over X. The following properties of this object are to be discussed:

(conv-Cauchy-1) $(\mathcal{C}, \mathcal{H})$ is *regular*:
each $\mathcal{C}$-convergent sequence is $\mathcal{H}$-Cauchy
(conv-Cauchy-2) $(\mathcal{C}, \mathcal{H})$ is *complete*:
each $\mathcal{H}$-Cauchy sequence is $\mathcal{C}$-convergent
(conv-Cauchy-3) $(\mathcal{C}, \mathcal{H})$ is *separated*:
$\mathcal{C} - \lim_n(x_n)$ is an asingleton, for each $\mathcal{H}$-Cauchy sequence (x_n).

Some concrete examples will be given a bit further.

(B) Let X be a nonempty set. Further, let $(\leq)$ be a *quasi-order* (reflexive and transitive relation) over X; as well as a map $d : X \times X \to R$; referred to as a *pseudometric* on X; then, $(X, \leq, d)$ will be called a *quasi-ordered pseudometric space.* A basic couple of convergence structures over this triple may be described along the lines below.

I) Given the sequence (x_n) in X and the point $x \in X$, define

(x_n), $\mathcal{B}$-*converges* to x (written as: $x_n \xrightarrow{\mathcal{B}} x$), provided
$x_n \le x$, for all n.

The set of all such points x will be denoted $\mathcal{B} - \lim_n(x_n)$ or $\mathrm{ubd}(x_n)$; and referred to as the *upper bound* of $(x_n; n \ge 0)$; if it is nonempty, then (x_n) is called *bounded from above*. Note that, by this very definition

(ub-1) $\mathcal{B}$ is *hereditary*:
$x_n \xrightarrow{\mathcal{B}} x$ implies $y_n \xrightarrow{\mathcal{B}} x$, for each subsequence (y_n) of (x_n);

so that, $\mathcal{B}$ is indeed a convergence structure over X. In addition,

(ub-2) $\mathcal{B}$ is *reflexive*:
for each $u \in X$, the constant sequence $(x_n = u; n \ge 0)$ fulfills $x_n \xrightarrow{\mathcal{B}} u$.

Unfortunately, $\mathcal{B}$ is not in general separated.

II) Given the sequence (x_n) in X and the point $x \in X$, we say that (x_n), d-*converges* to x (written as: $x_n \xrightarrow{d} x$), provided

$$\forall \varepsilon > 0, \exists i = i(\varepsilon)\text{: } i \le n \Longrightarrow d(x_n, x) \le \varepsilon.$$

The set of all such points $x \in X$ will be denoted

$d - \lim_n(x_n)$; or simply, $\lim_n(x_n)$, if d is understood;

if it is nonempty, then (x_n) is called d-*convergent*. Clearly,

(conv-her) $(\xrightarrow{d})$ is *hereditary*:
$x_n \xrightarrow{d} x$ implies $y_n \xrightarrow{d} x$, for each subsequence (y_n) of (x_n);

so, $(\xrightarrow{d})$ is indeed a convergence over X. However, $(\xrightarrow{d})$ is neither reflexive nor separated, in general.

Further, call the sequence (x_n) in X, d-*Cauchy* when

$$\forall \varepsilon > 0, \exists j = j(\varepsilon)\text{: } j \le m \le n \Longrightarrow d(x_m, x_n) \le \varepsilon;$$

the class of all such sequences will be denoted as $Cauchy(d)$. Clearly,

(Cauchy-her) $Cauchy(d)$ is *hereditary*: (x_n) is d-Cauchy
implies (y_n) is d-Cauchy, for each subsequence (y_n) of (x_n);

so that, $Cauchy(d)$ is indeed a Cauchy structure on X. Unfortunately, $Cauchy(d)$ is not in general reflexive.

Now, under a previously introduced terminology, we say that the couple $((\xrightarrow{d}), Cauchy(d))$ is a *conv-Cauchy structure* over $(X, \leq, d)$. Note that, by the general setting (relative to d), compatible conditions like

(regu-1) $((\xrightarrow{d}), Cauchy(d))$ is regular
(d-convergent $\Longrightarrow$ d-Cauchy)
(regu-2) $((\xrightarrow{d}), Cauchy(d))$ is complete
(d-Cauchy $\Longrightarrow$ d-convergent)

are not in general valid.

III) Finally, as a direct combination of the preceding convergence objects, one may introduce the *product* convergence structure

$$x_n \xrightarrow{d\mathcal{B}} x, \text{ provided } x_n \xrightarrow{d} x \text{ and } x_n \xrightarrow{\mathcal{B}} x.$$

In fact, the hereditary property of $(\xrightarrow{d\mathcal{B}})$ is clear; so that, our terminology is adequate; but, in general, $(\xrightarrow{d\mathcal{B}})$ is neither reflexive nor separated.

(C) Passing to the specific part of our developments, we have to introduce a lot of compatible conditions relative to the couple $(d, \leq)$ above; the following ones are important for us:

(op-1) (d is *order-positive*): $x \leq y$ implies $d(x, y) \geq 0$;
hence, in particular, $(d(x, x) \geq 0, \forall x \in X)$
(op-2) (d is *order triangular*):
$d(x, z) \leq d(x, y) + d(y, z)$, whenever $x \leq y \leq z$
(op-3) (d is *strong triangular*):
$d(y, z) \leq d(x, y) + d(x, z)$, for all $x, y, z \in X$
(op-4) (d is *Altman triangular*):
$d(y, z) \leq d(x, y) + d(x, z)$, whenever $x \leq y$, $x \leq z$.

When (op-1) and (op-2) hold, d is called a $(\leq)$-*pseudometric* on X; and the triple $(X, \leq, d)$ is referred to as a $(\leq)$-*pseudometric space*. In this context, (op-3) and (op-4) (related to the developments in Altman [15]) may be viewed as extra conditions to be imposed upon such objects.

Let in the following $(X, \leq, d)$ be a $(\leq)$-pseudometric space. Concerning the introduced concepts, the following fact is useful in applications. Given $i \geq 0$, call the sequence $(x_n; n \geq 0)$, (d, i)-*asymptotic* provided

$$\forall \varepsilon > 0, \exists j = j(\varepsilon): \ j \leq m \Longrightarrow d(x_m, x_{m+i}) \leq \varepsilon.$$

Clearly, any d-Cauchy sequence $(x_n; n \geq 0)$ is (d, i)-asymptotic, for each $i \geq 0$. The reciprocal is not in general true. However, the following partial converse is holding. Call the sequence (x_n) in X, *d-stationary* provided

there exists $k \geq 0$, such that $k \leq m \Longrightarrow d(x_m, x_{m+1}) = 0$.

Proposition 4.57. *Suppose that the ascending sequence $(x_n; n \geq 0)$ is $(d, 0)$-asymptotic and d-stationary. Then, $(x_n; n \geq 0)$ is d-Cauchy.*

Proof. From the d-stationary property, there exists $k \geq 0$ such that

$$k \leq m \Longrightarrow d(x_m, x_{m+1}) = 0;$$

and this, by the triangular inequality, gives (as d is order-positive)

$d(x_m, x_n) = 0$, whenever $k \leq m < n$.

Let $\varepsilon > 0$ be arbitrary fixed. As $(x_n; n \geq 0)$ is $(d, 0)$-asymptotic, there must be some $h = h(\varepsilon) \geq k$, such that

$$h \leq m \Longrightarrow d(x_m, x_m) \leq \varepsilon.$$

Combining these facts, one derives an inclusion like

$(k \leq)\ h \leq m \leq n$ implies $d(x_m, x_n) \leq \varepsilon$;

and the proof is complete. □

Finally, let us introduce the relations over X

$u \sim v$ (resp.: $u \equiv v$) iff
$x_n \xrightarrow{d} u$, $x_n \xrightarrow{d} v$ (resp.: $x_n \xrightarrow{d\mathcal{B}} u$, $x_n \xrightarrow{d\mathcal{B}} v$),
for some ascending sequence $(x_n; n \geq 0)$ in X.

Clearly, $(\sim)$ is symmetric; but in general, it is neither reflexive nor transitive; a similar conclusion is to be derived for $(\equiv)$. Moreover, $(\equiv)$ is coarser that $(\sim)$, as it can be directly seen.

The following simple fact will be useful for us.

Proposition 4.58. *The generic inclusion holds (for each $u, v \in X$):*

d is strong triangular and $u \sim v$ implies $[d(u, v) \leq 0,\ d(v, u) \leq 0]$,
d is Altman triangular and $u \equiv v$ implies $[d(u, v) \leq 0,\ d(v, u) \leq 0]$.

Proof. Let $u, v \in X$ be such that $u \sim v$ (resp., $u \equiv v$). By definition, there exists some ascending sequence (x_n) in X, with

$$x_n \xrightarrow{d} u,\ x_n \xrightarrow{d} v \text{ (resp.: } x_n \xrightarrow{d\mathcal{B}} u,\ x_n \xrightarrow{d\mathcal{B}} v).$$

Combining with the strong (resp., Altman) triangular property gives

$$(\forall n)\colon d(u,v) \leq d(x_n,u) + d(x_n,v),\ d(v,u) \leq d(x_n,v) + d(x_n,u)$$
$$(\text{resp., } d(u,v) \leq d(x_n,u) + d(x_n,v),\ d(v,u) \leq d(x_n,v) + d(x_n,u))$$

Passing to limit as $n \to \infty$, one derives the needed conclusion. □

4.5.6 *Meir-Keeler contractions*

Let X be a nonempty set. Take a quasi-order $(\leq)$ over it; and let $d : X \times X \to R$ be a $(\leq)$-pseudometric on X; then, $(X, \leq, d)$ is called a $(\leq)$-*pseudometric space*. Let also $(\sim)$ and $(\equiv)$ stand for the relations (over X) we just introduced; remember that $(\equiv)$ is coarser than $(\sim)$. Call the subset Y of X,

(as-1) $(d, \leq; \sim)$-*almost-singleton*
(in short: $(d, \leq; \sim)$-*asingleton*), when
$y_1, y_2 \in Y$, $y_1 \leq y_2$, $y_1 \sim y_2$ imply $d(y_1, y_2) = 0$;
(as-2) $(d, \leq; \equiv)$-*almost-singleton*
(in short: $(d, \leq; \equiv)$-*asingleton*), when
$y_1, y_2 \in Y$, $y_1 \leq y_2$, $y_1 \equiv y_2$ imply $d(y_1, y_2) = 0$.

(A) Having these precise, take some selfmap $T \in \mathcal{F}(X)$. Denote

$$\text{Fix}(T; \sim) = \{z \in X; z \sim Tz\},\ \text{Fix}(T; \equiv) = \{z \in X; z \equiv Tz\};$$

each element of these will be referred to as a $(d, \leq; \sim)$-*fixed point* of T and $(d, \leq; \equiv)$-*fixed point* of T, respectively; note that, by the preceding relationships, one has $\text{Fix}(T; \sim) \supseteq \text{Fix}(T; \equiv)$.

In the following, sufficient conditions are given under which such elements exist. To do this, we start from the working hypotheses

(s-prog) T is $(\leq)$-*semi-progressive* $(X(T, \leq) := \{x \in X; x \leq Tx\} \neq \emptyset)$
(incr) T is $(\leq)$-*increasing* ($x \leq y$ implies $Tx \leq Ty$).

The basic directions under which the investigations be conducted are described by the list below, comparable with the one in Turinici [246]:

od-pic-0) We say that T is *fix-*$(d, \leq; \sim)$*-asingleton* when $\text{Fix}(T; \sim)$ is $(d, \leq; \sim)$-asingleton; and *fix-*$(d, \leq; \equiv)$*-asingleton* provided $\text{Fix}(T; \equiv)$ is $(d, \leq ; \equiv)$-asingleton

od-pic-1) We say that T is a *Picard operator* (modulo $(d, \leq)$) when, for each $x \in X(T, \leq)$, $(T^n x; n \geq 0)$ is d-Cauchy

od-pic-2) We say that T is a *strong Picard operator* (modulo $(d, \leq)$) if, for each $x \in X(T, \leq)$, $(T^n x; n \geq 0)$ is d-convergent with $\lim_n(T^n x) \subseteq \text{Fix}(T; \sim)$

od-pic-3) We say that T is a *Bellman Picard operator* (modulo $(d, \leq)$) when, for each $x \in X(T, \leq)$, $(T^n x; n \geq 0)$ is d-convergent, with (in addition) $\lim_n(T^n x) \subseteq \text{Fix}(T; \equiv)$ and $(T^n x \leq \lim_n(T^n x), \forall n)$.

Here, for $U, V \in \exp[X]$, we define:

$U \leq V$ iff $u \leq v$, for each $u \in U$, $v \in V$;

this relation over $\exp[X]$ is transitive but not in general reflexive.

The sufficient (regularity) conditions for such properties are being founded on *ascending orbital* concepts (in short: (a-o)-concepts). Namely, call the sequence $(z_n; n \geq 0)$ in X

$(\leq)$-*ascending*, if $z_i \leq z_j$ for $i \leq j$;
T-*orbital*, when $(z_n = T^n x; n \geq 0)$, for some $x \in X$;

the intersection of these notions is just the precise one.

od-reg-1) Call X, *(a-o,d)-complete* provided (for each (a-o)-sequence) d-Cauchy $\Longrightarrow$ d-convergent

od-reg-2) We say that T is *(a-o,d)-continuous*, if: $((z_n)$ is (a-o)-sequence and $z_n \xrightarrow{d} z)$ imply $Tz_n \xrightarrow{d} Tz$

od-reg-3) Call $(\leq)$, *(a-o,d)-selfclosed*, if: $((z_n)$ is (a-o)-sequence and $z_n \xrightarrow{d} z)$ imply $(z_n \leq z$, for all $n)$.

When the orbital properties are ignored, we are talking about *ascending concepts* (in short: *a-concepts*). In particular, when $(\leq) = X \times X$, the list of such notions is comparable with the one proposed by Rus [203, Ch. 2, Sec. 2.2]; because, in this case, $X(T, \leq) = X$.

(B) As a completion of these facts, we must now formulate the specific metrical contractive conditions upon our data. There are two basic conditions of this type.

I) The former of these is expressed as

(comp) T is $(d, \leq)$-*compatible*:
$x \leq y$ and $d(x, y) = 0$ imply $d(Tx, Ty) = 0$.

Note that, in the case of

d is reflexive-sufficient: $d(x, y) = 0 \Longleftrightarrow x = y$, this condition is trivial, in view of

$$(\forall x, y \in X)\colon d(x,y) = 0 \Longleftrightarrow x = y,\ d(Tx,Ty) = 0 \Longleftrightarrow Tx = Ty.$$

But, in our $(\leq)$-pseudometric setting, the posed condition is not trivial.

II) The latter of these is, essentially, a quasi-order variant of the developments in Meir and Keeler [168], subsequently refined by Matkowski [166] and Ćirić [59].

Let us say that T is *Meir-Keeler $(d,\leq)$-contractive*, if

(mk-1) $(x \leq y,\ d(x,y) > 0)$ imply $d(Tx,Ty) < d(x,y)$;
referred to as: T is $(d,\leq)$*-strictly-nonexpansive*
(mk-2) $\forall \varepsilon > 0, \exists \delta > 0$: $(x \leq y,\ \varepsilon < d(x,y) < \varepsilon+\delta) \Longrightarrow d(Tx,Ty) \leq \varepsilon$;
expressed as: T has the *Meir-Keeler property* (modulo $(d,\leq)$).

The following simple facts will be in effect for us.

Proposition 4.59. *Under the above conventions,*

(61-1) *If T is $(d,\leq)$-compatible and $(d,\leq)$-strictly-nonexpansive, then*

T is $(d,\leq)$-nonexpansive: $d(Tx,Ty) \leq d(x,y)$, $\forall x,y \in X$, $x \leq y$

(61-2) *If T is $(d,\leq)$-strictly-nonexpansive, and has the Meir-Keeler property (modulo $(d,\leq)$), then*

$\forall \varepsilon > 0, \exists \delta > 0$: $[x \leq y,\ 0 < d(x,y) < \varepsilon + \delta] \Longrightarrow d(Tx,Ty) \leq \varepsilon$;
expressed as: T has the extended Meir-Keeler property *(modulo $(d,\leq)$)*

(61-3) *If T is $(d,\leq)$-compatible, $(d,\leq)$-strictly-nonexpansive, and has the Meir-Keeler property (modulo $(d,\leq)$), then*

$\forall \varepsilon > 0, \exists \delta > 0$: $[x \leq y,\ d(x,y) < \varepsilon + \delta] \Longrightarrow d(Tx,Ty) \leq \varepsilon$;
expressed as: T has the complete Meir-Keeler property *(modulo $(d,\leq)$).*

A geometric version of these concepts may be given as below. Remember that given the relation $\Omega \in \exp(R^0_+ \times R^0_+)$ we introduced the property

(u-diag) Ω is *upper diagonal*: $(t,s) \in \Omega$ implies $t < s$;

the class of all these will be denoted as $\mathrm{udiag}(R^0_+)$. Then, at the level of $\mathrm{udiag}(R^0_+)$, define the geometric type concept

(g-mk) Ω has the *geometric Meir-Keeler property*:
$\forall \varepsilon > 0, \exists \delta > 0$: $[(t,s) \in \Omega,\ \varepsilon < s < \varepsilon + \delta] \Longrightarrow t \leq \varepsilon$;

as well as the asymptotic type one

(a-mk) Ω is *asymptotic Meir-Keeler*:
there are no strictly descending sequences (t_n) and (s_n) in R^0_+ and no elements ε in R^0_+, with $((t_n, s_n) \in \Omega, \forall n)$ and $(t_n \to \varepsilon+, s_n \to \varepsilon+)$.

Here, for each sequence (r_n) in R and each element $r \in R$, we denoted

$r_n \to r+$ (resp., $r_n \to r-$), when $r_n \to r$
and $[r_n > r$ (resp., $r_n < r$), $\forall n]$.

Note that, according to an auxiliary statement, this geometric concept is equivalent with its asymptotic counterpart.

Given the relation $\Omega \in \exp(R^0_+ \times R^0_+)$, let us say that the selfmap T is $(d, \leq; \Omega)$-*contractive*, provided

(Om-con) $(d(Tx, Ty), d(x, y)) \in \Omega$,
$\forall x, y \in X$, $x \leq y$, $d(x, y) > 0$, $d(Tx, Ty) > 0$.

Proposition 4.60. *Suppose that the selfmap T is $(d, \leq; \Omega)$-contractive where the relation $\Omega \in \exp(R^0_+ \times R^0_+)$ is upper diagonal and geometric Meir-Keeler. Then, T is Meir-Keeler $(d, \leq)$-contractive.*

Proof. i): Let $x, y \in X$ be such that

$x \leq y$ (hence, $Tx \leq Ty$) and $d(x, y) > 0$.

If $d(Tx, Ty) = 0$, then $d(Tx, Ty) < d(x, y)$. Suppose now that $d(Tx, Ty) > 0$. By definition, $(d(Tx, Ty), d(x, y)) \in \Omega$; and this, along with Ω = upper diagonal, yields $d(Tx, Ty) < d(x, y)$. Putting these together, one derives that T is $(d, \leq)$-strictly-nonexpansive.

ii): Let $\varepsilon > 0$ be arbitrary fixed; and $\delta > 0$ be the number assured by the geometric Meir-Keeler property for Ω. Then, let $x, y \in X$ be such that

$x \leq y$ (hence, $Tx \leq Ty$) and $\varepsilon < d(x, y) < \varepsilon + \delta$.

As before, if $d(Tx, Ty) = 0$, then $d(Tx, Ty) < \varepsilon$. Suppose now that $d(Tx, Ty) > 0$. By definition, $(d(Tx, Ty), d(x, y)) \in \Omega$; and this, along with the geometric Meir-Keeler property of Ω, gives $d(Tx, Ty) \leq \varepsilon$. Summing up, T has the Meir-Keeler property (modulo $(d, \leq)$).

As a consequence of these two stages, one thus derives that T is Meir-Keeler $(d, \leq)$-contractive; and the proof is complete. □

In the following, a kind of reciprocal for this result is formulated. Given the selfmap T (endowed with the general properties we already specified), let $\Omega = \Omega[d, \leq; T]$ stand for the associated relation over R_+^0:

$\Omega = \{(d(Tx, Ty), d(x, y)); x \leq y, d(x, y) > 0, d(Tx, Ty) > 0\}$;
or, in other words: $(t, s) \in \Omega$ iff
$t = d(Tx, Ty)$, $s = d(x, y)$, where $x \leq y$, $d(x, y) > 0$, $d(Tx, Ty) > 0$.

Proposition 4.61. *Under these conventions, we have*

(63-1) *If T is Meir-Keeler $(d, \leq)$-contractive, then the attached relation $\Omega := \Omega[d, \leq; T]$ over R_+^0 is upper diagonal and geometric Meir-Keeler*

(63-2) *T is Meir-Keeler $(d, \leq)$-contractive iff the attached relation $\Omega := \Omega[d, \leq; T]$ over R_+^0 is upper diagonal and geometric Meir-Keeler.*

Proof. i): Suppose that T is Meir-Keeler $(d, \leq)$-contractive; we have to establish that $\Omega = \Omega[d, \leq; T]$ is upper diagonal and geometric Meir-Keeler. There are two steps to be passed.

i-1): Let $(t, s) \in R_+^0 \times R_+^0$ be such that $(t, s) \in \Omega$; hence (by definition)

$t = d(Tx, Ty)$, $s = d(x, y)$,
where $x \leq y$, $d(x, y) > 0$, $d(Tx, Ty) > 0$.

From the $(d, \leq)$-strictly-nonexpansive property of T, we must have $d(Tx, Ty) < d(x, y)$; or, equivalently, $t < s$; which shows that the relation Ω is upper diagonal.

i-2): Let $\varepsilon > 0$ be arbitrary fixed; and $\delta > 0$ be associated by the Meir-Keeler property for T. Further, let $(t, s) \in R_+^0 \times R_+^0$ be such that

$(t, s) \in \Omega$ and $\varepsilon < s < \varepsilon + \delta$.

From the first part of this, we have

$t = d(Tx, Ty)$, $s = d(x, y)$,
where $x \leq y$, $d(x, y) > 0$, $d(Tx, Ty) > 0$;

so, combining with the second part of the same

$x \leq y$, and $\varepsilon < d(x, y) < \varepsilon + \delta$.

By the Meir-Keeler-property for T, we get

$d(Tx, Ty) \leq \varepsilon$; that is (under our notation): $t \leq \varepsilon$;

so that, Ω has the geometric Meir-Keeler property.

ii): Suppose that the associated relation $\Omega = \Omega[d, \leq; T]$ over R^0_+ is upper diagonal and geometric Meir-Keeler. By the definition of this object, T is $(d, \leq; \Omega)$-contractive. Combining with the preceding result, one derives that T is Meir-Keeler $(d, \leq)$-contractive; and the conclusion follows. □

As a consequence of this, the Meir-Keeler $(d, \leq)$–contractive property of the selfmap T is finally reducible to the upper diagonal and geometric Meir-Keeler properties for the associated relation $\Omega[d, \leq; T]$. Concerning this aspect, remember that various particular cases of such objects were treated in a previous place. These yield two basic examples of such contractions with a practical finality.

(I) Let $\mathcal{F}(re)(R^0_+, R)$ stand for the class of all $\chi \in \mathcal{F}(R^0_+, R)$ with

χ is *regressive*: $\chi(t) < t$, for all $t \in R^0_+$.

The following extra conditions imposed upon this class will be considered

(MK-a) χ is *Meir-Keeler admissible*:
$\forall \varepsilon > 0, \exists \delta > 0$, such that $(\varepsilon < s < \varepsilon + \delta)$ implies $\chi(s) \leq \varepsilon$.
(M-a) χ is *Matkowski admissible*:
for each (t_n) in R^0_+ with $(t_{n+1} \leq \chi(t_n), \forall n)$ we have $\lim_n t_n = 0$.

As we already stressed, these are equivalent in (ZF-AC+DC).

Now, given $\chi \in \mathcal{F}(R^0_+, R)$, we call T, *Matkowski* $(d, \leq; \chi)$*-contractive* if

(M-contr) $d(Tx, Ty) \leq \chi(d(x, y))$,
$\forall x, y \in X$, $x \leq y$, $d(x, y) > 0$, $d(Tx, Ty) > 0$.

(This convention is taken from the developments in Matkowski [164].)

Proposition 4.62. *Assume that T is Matkowski $(d, \leq; \chi)$-contractive, where $\chi \in \mathcal{F}(re)(R^0_+, R)$ fulfills*

χ is Meir-Keeler admissible (or, equivalently: Matkowski admissible).

Then, necessarily, T is Meir-Keeler $(d, \leq)$-contractive in (ZF-AC+DC).

Proof. By a previous result, the associated relation $\Omega := \Omega[\chi]$ over R^0_+

$(t, s \in R^0_+)$: $t\Omega s$ iff $t \leq \chi(s)$

is upper diagonal and geometric Meir-Keeler. This, along with

T is Matkowski $(d, \leq; \chi)$-contractive iff T is $(d, \leq; \Omega)$-contractive

gives the desired fact. □

Some important classes of such functions are given below.

I-1) For any $\chi \in \mathcal{F}(re)(R^0_+, R)$ and any $s \in R^0_+$, put

$\Lambda^+\chi(s) = \inf_{\varepsilon>0} H(s+)(\varepsilon)$;
where $H(s+)(\varepsilon) = \sup \chi(]s, s+\varepsilon[)$, $\varepsilon > 0$.

From the regressive property of χ, these limit quantities fulfill

$$-\infty \leq \Lambda^+\chi(s) \leq s, \text{ for all } s \in R^0_+.$$

Call $\chi \in \mathcal{F}(re)(R^0_+, R)$, *Boyd-Wong admissible* [35], if

(bw-adm) $\Lambda^+\chi(s) < s$, for all $s > 0$.

In particular, $\chi \in \mathcal{F}(re)(R^0_+, R)$ has such a property whenever

χ is right-usc on R^0_+: $\Lambda^+\chi(s) \leq \chi(s)$, $\forall s \in R^0_+$.

This, in turn, is fulfilled when χ is right-continuous on R^0_+; for, in such a case, $[\Lambda^+\chi(s) = \chi(s)$, for all $s > 0]$. Returning to the general case, note that, by a previous result,

(for each $\chi \in \mathcal{F}(re)(R^0_+, R)$):
χ is Boyd-Wong admissible implies χ is Meir-Keeler admissible.

I-2) Denote, for simplicity

$\mathcal{F}(re, in)(R^0_+, R) =$ the class of all increasing $\chi \in \mathcal{F}(re)(R^0_+, R)$.

Then, let us put, for each $\chi \in \mathcal{F}(re, in)(R^0_+, R)$ and each $s > 0$,

$\chi(s+0) = \lim_{t\to s+} \chi(t)$; (referred to as: the *right limit* of χ at s).

As χ is (in particular) increasing, this limit exists (as an element of R_+) and the representation formula is valid

$$\Lambda^+\chi(s) = \chi(s+0) = \inf\{\chi(t); t > s\}, \forall s \in R^0_+.$$

Note that, in this case,

(for each $\chi \in \mathcal{F}(re, in)(R^0_+, R)$):
χ is Boyd-Wong admissible iff $(\chi(s+0) < s, \forall s > 0)$
(referred to as: χ is *strongly regressive*).

Moreover, (cf. the developments in Matkowski [164])

(M-admi) (for each $\chi \in \mathcal{F}(re, in)(R^0_+, R)$):
χ is Matkowski admissible iff $\varphi^n(t) \to 0$ as $n \to \infty$, for all $t > 0$, whenever this iterative sequence exists.

[Here, for each $n \geq 0$, χ^n stands for the n-th iterate of χ.]

(II) Let (ψ, φ) be a pair of functions over $\mathcal{F}(R^0_+, R)$, endowed with

(nc-1) (ψ, φ) is a *normal couple*:
ψ is increasing and φ is *strictly positive* [$\varphi(t) > 0$, $\forall t > 0$].

The following extra condition involving this normal couple will be considered here for technical purposes

(nc-2) φ is asymptotic positive:
for each strictly descending sequence $(t_n; n \geq 0)$ in R^0_+ and each $\varepsilon > 0$ with $t_n \to \varepsilon+$, we have $\limsup_n(\varphi(t_n)) > 0$.

Given the couple ψ, φ of functions over $\mathcal{F}(R^0_+, R)$, let us say that T is *Rhoades $(d, \leq; \psi, \varphi)$-contractive*, provided (cf. Rhoades [197])

(R-contr) $\psi(d(Tx, Ty)) \leq \psi(d(x, y)) - \varphi(d(x, y))$,
$\forall x, y \in X$, $x \leq y$, $d(x, y) > 0$, $d(Tx, Ty) > 0$.

Proposition 4.63. *Suppose that T is Rhoades $(d, \leq; \psi, \varphi)$-contractive, for a couple (ψ, φ) of functions in $\mathcal{F}(R^0_+, R)$, with*

(ψ, φ) is normal with φ = asymptotic positive.

Then, T is Meir-Keeler $(d, \leq)$-contractive in the reduced axiomatic system (ZF-AC+DC).

Proof. By a previous result, the associated relation $\Omega := \Omega[\psi, \varphi]$ over R^0_+

$(t, s \in R^0_+)$: $t\Omega s$ iff $\psi(t) \leq \psi(s) - \varphi(s)$

is upper diagonal and geometric Meir-Keeler. This, and

T is Rhoades $(d, \leq; \psi, \varphi)$-contractive iff T is $(d, \leq; \Omega)$-contractive

give the desired fact. □

4.5.7 *Main result*

Let X be a nonempty set. Take a quasi-order $(\leq)$ over it; and let $d : X \times X \to R$ be a $(\leq)$-pseudometric on X; then, $(X, \leq, d)$ is called a $(\leq)$-*pseudometric space.* Further, let T be a selfmap of X; supposed to be $(\leq)$-semi-progressive and $(\leq)$-increasing. The basic directions and regularity conditions under which the problem of determining elements in $\text{Fix}(T; \sim)$ and/or $\text{Fix}(T; \equiv)$ is to be solved, were already precise; and the contractive framework was settled.

Our main result in this exposition is

Theorem 4.24. *Assume that T is $(d, \leq)$-compatible and $(d, \leq; \Omega)$-contractive, where the relation $\Omega \in \exp(R^0_+ \times R^0_+)$ fulfills*

Ω is upper diagonal and geometric Meir-Keeler.

In addition, let X be (a-o,d)-complete. Then,

(71-a) *T is fix-$(d, \leq; \sim)$-asingleton when (in addition to these), d is strong triangular*

(71-b) *T is fix-$(d, \leq; \equiv)$-asingleton when (in addition to these), d is Altman triangular*

(71-c) *T is a strong Picard operator (modulo $(d, \leq)$) when (in addition), T is (a-o,d)-continuous*

(71-d) *T is a Bellman Picard operator (modulo $(d, \leq)$) when (in addition), $(\leq)$ is (a-o,d)-selfclosed.*

Proof. There are several steps to be discussed.

Part 0. i): Suppose that d is strong triangular; and let $u, v \in \text{Fix}(T; \sim)$ be such that $u \leq v$ and $u \sim v$. From the former relation, $d(u, v) \geq 0$. Further, by the latter relation and a previous result, $d(u, v) \leq 0$. Putting these together yields $d(u, v) = 0$; and the assertion follows.

ii): Suppose that d is Altman triangular; and let $u, v \in \text{Fix}(T; \equiv)$ be such that $u \leq v$ and $u \equiv v$. From the former relation, $d(u, v) \geq 0$. Further, by the latter relation and a previous result, $d(u, v) \leq 0$. Putting these together yields $d(u, v) = 0$.

Passing to the remaining parts, take some $x_0 \in X(T, \leq)$; and put $(x_n = T^n x_0; n \geq 0)$; this is an ascending orbital sequence in X.

Part 1. For each $i \geq 0$, denote $(\rho(i)(n) := d(x_n, x_{n+i}); n \geq 0)$; clearly, this is a sequence in R_+. By the $(d, \leq)$-nonexpansive property,

$$\rho(i)(n+1) = d(Tx_n, Tx_{n+i}) \leq d(x_n, x_{n+i}) = \rho(i)(n), \ \forall n;$$

so that, $(\rho(i)(n); n \geq 0)$ is descending. The alternative

(alt-1) $\rho(i)(k) = 0$ for some $k \geq 0$

gives (by the positivity of our sequence)

$\rho(i)(n) = 0$, $\forall n \geq k$; whence, $\rho(i)(n) \to 0$ as $n \to \infty$.

Suppose that the opposite alternative is true:

(alt-2) $\rho(i)(n) > 0$, for all $n \geq 0$.

From the $(d, \leq)$-strict-nonexpansive property,

$$\rho(i)(n+1) = d(Tx_n, Tx_{n+i}) < d(x_n, x_{n+i}) = \rho(i)(n), \ \forall n;$$

so that, the sequence $(\rho(i)(n); n \geq 0)$ is strictly descending. As a consequence, $\rho(i) := \lim_n(\rho(i)(n))$ exists as an element of R_+. Assume by contradiction that $\rho(i) > 0$; and let $\delta > 0$ be the number given by the Meir-Keeler $(d, \leq)$-contractive condition upon T. By definition, there exists a rank $n(\delta)$ such that

$n \geq n(\delta)$ implies $\rho(i) < \rho(i)(n) = d(x_n, x_{n+i}) < \rho(i) + \delta$.

This, by the Meir-Keeler contractive condition, yields (for the same n)

$$(\rho(i) <)\rho(i)(n+1) = d(Tx_n, Tx_{n+i}) \leq \rho(i);$$

a contradiction. Hence, $\rho(i) = 0$; so that,

$\rho(i)(n) := d(x_n, x_{n+i}) \to 0$, as $n \to \infty$, for each $i \geq 0$;
meaning: $(x_n; n \geq 0)$ is (d, i)-asymptotic, for each $i \geq 0$.

Part 2. We now show that (x_n) is d-Cauchy. Let $\varepsilon > 0$ be arbitrary fixed; and $\delta > 0$ be associated by the Meir-Keeler $(d, \leq)$-contractive property; without loss, one may assume that $\delta < \varepsilon$. By the (d, i)-asymptotic property (for each $i \geq 0$), there exists a rank $n(\delta) \geq 0$, such that

(d-asy) $(d(x_n, x_{n+i}) < \delta/2, \ \forall n \geq n(\delta))$, for $i \in \{0, 1\}$.

We claim, via ordinary induction, that the relation below holds

(d-C;i) $[d(x_n, x_{n+i}) < \varepsilon + \delta/2, \ \forall n \geq n(\delta)]$,

for all $i \geq 0$; wherefrom, $(x_n; n \geq 0)$ is d-Cauchy. The case of $i \in \{0, 1\}$ is evident, by the preceding evaluation (d-asy). Assume that (d-C;i) holds for all $i \in \{0, ..., j\}$, where $j \geq 1$; we must establish that (d-C;j+1) holds too. So, let $n \geq n(\delta)$ be arbitrary fixed. From the inductive hypothesis, $d(x_n, x_{n+j}) < \varepsilon + \delta/2$. By the complete Meir-Keeler property, one gets

$$d(x_{n+1}, x_{n+j+1}) = d(Tx_n, Tx_{n+j}) \leq \varepsilon.$$

This, along with the order triangular inequality, gives

$$d(x_n, x_{n+j+1}) \leq d(x_n, x_{n+1}) + d(x_{n+1}, x_{n+j+1}) < \varepsilon + \delta/2;$$

and conclusion follows.

Part 3. As X is (a-o,d)-complete, $x_n \xrightarrow{d} z$, for some $z \in X$. There are two cases to be discussed.

Case 1. Suppose that T is (a-o,d)-continuous. Then, $y_n := Tx_n \xrightarrow{d} Tz$ as $n \to \infty$. On the other hand, $(y_n = x_{n+1}; n \geq 0)$ is an ascending subsequence of (x_n); whence $y_n \xrightarrow{d} z$ as $n \to \infty$. Putting these together yields $z \sim Tz$; that is: $z \in \mathrm{Fix}(T; \sim)$; which tells us that T is a strong Picard operator (modulo $(d, \leq)$).

Case 2. Suppose that $(\leq)$ is (a-o,d)-selfclosed. By the convergence property, we derive

$$\text{(p-1)} \quad (x_n \leq z, \forall n); \text{ whence } x_n \xrightarrow{d\mathcal{B}} z.$$

As $(y_n = x_{n+1}; n \geq 0)$ is a subsequence of (x_n), we must have $y_n \xrightarrow{d\mathcal{B}} z$. On the other hand, as T is $(\leq)$-increasing,

$$\text{(p-2)} \quad (y_n = x_{n+1} \leq Tz, \forall n); \text{ that is: } y_n \xrightarrow{\mathcal{B}} Tz.$$

Moreover, by the $(d, \leq)$-nonexpansive property,

$$(0 \leq d(y_n, Tz) \leq d(x_n, z), \forall n); \text{ so, } y_n \xrightarrow{d} Tz \text{ (via } x_n \xrightarrow{d} z);$$

wherefrom (by a preceding relation), $y_n \xrightarrow{d\mathcal{B}} Tz$. This, along with $y_n \xrightarrow{d\mathcal{B}} z$ gives $z \equiv Tz$; that is: $z \in \mathrm{Fix}(T; \equiv)$); which tells us that T is a Bellman Picard operator (modulo $(d, \leq)$). The proof is thereby complete. □

In particular, when $d(.,.)$ is a metric of X, we have

$$\mathrm{Fix}(T; \sim) = \mathrm{Fix}(T; \equiv) = \mathrm{Fix}(T)(:= \{z \in X; z = Tz\});$$

and then, Theorem 4.24 is a partial extension of the related statement in Turinici [246]. Moreover, when $(\leq) = X \times X$, the main result above extends in a direct way the standard contribution in this area, due to Meir and Keeler [168].

Finally, note that alternate extensions of these facts are possible, when the contractive condition is taken as in Leader [161]. In this case, the extra requirements to be admitted follow the ones imposed by the *partial metric spaces* context due to Matthews [167]; we do not give details.

4.5.8 *Further aspects*

Let in the following $(X, \leq, d)$ be a $(\leq)$-pseudometric space; and T be a selfmap of X; supposed to be $(\leq)$-semi-progressive and $(\leq)$-increasing. As already precise, the geometric (or, equivalently: Meir-Keeler) condition appearing in Theorem 4.24 admits two basic versions; and this, in turn, yields two particular cases of our main result. For both practical and theoretical reasons, it would be useful to highlight them; further aspects involving the obtained facts are also discussed.

Case 1. As a first consequence of Theorem 4.24, we get

Theorem 4.25. *Assume that the mapping $T : X \to X$ is $(d, \leq)$-compatible and Matkowski $(d, \leq; \chi)$-contractive, for a certain function $\chi \in \mathcal{F}(re)(R^0_+, R)$, with*

χ is Meir-Keeler admissible (or, equivalently: Matkowski admissible).

In addition, let X be (a-o,d)-complete. Then,

(81-a) *T is fix-$(d, \leq; \sim)$-asingleton when (in addition) d has the strong triangular property*

(81-b) *T is fix-$(d, \leq; \equiv)$-asingleton when (in addition) d has the Altman triangular property*

(81-c) *T is strong Picard (modulo $(d, \leq)$) if (in addition to this), T is supposed to be (a-o,d)-continuous*

(81-d) *T is a Bellman Picard operator (modulo $(d, \leq)$) when (in addition to this), $(\leq)$ is (a-o,d)-selfclosed.*

In particular, let $d(.,.)$ be a metric on X, in the usual sense. Then, the following sub-cases are of interest:

S-case 1) Suppose that $(\leq) = X \times X$ (the trivial quasi-order of X). Then, Theorem 4.25 includes the Boyd-Wong result [35] when $\chi \in \mathcal{F}(re)(R^0_+, R)$ is Boyd-Wong admissible; and, respectively, Matkowski's result [164] when $\chi \in \mathcal{F}(re)(R^0_+, R)$ is increasing and Matkowski admissible.

S-case 2) Suppose that $(\leq)$ is an order on X. Then, Theorem 4.25 includes the results in Turinici [236]. But, as precise, these results include the related ones in Ran and Reurings [192] or Nieto and Rodriguez-Lopez [179]; hence, so does Theorem 4.25.

Case 2. As a second consequence of Theorem 4.24, we have

Theorem 4.26. *Assume that the selfmap T is $(d, \le)$-compatible and Rhoades $(d, \le; \psi, \varphi)$-contractive, for a certain couple (ψ, φ) of functions over $\mathcal{F}(R^0_+, R)$, with*

(ψ, φ) is normal and φ = asymptotic positive.

In addition, let X be (a-o,d)-complete. Then,

(82-a) *T is fix-$(d, \le; \sim)$-asingleton when (in addition) d has the strong triangular property*

(82-b) *T is fix-$(d, \le; \equiv)$-asingleton if (in addition) d has the Altman triangular property*

(82-c) *T is strong Picard (modulo $(d, \le)$) if (in addition), T is supposed to be (a-o,d)-continuous*

(82-d) *T is a Bellman Picard operator (modulo $(d, \le)$) when (in addition to this), $(\le)$ is (a-o,d)-selfclosed.*

In particular, when $(\le) = X \times X$ (the *trivial quasi-order* on X), Theorem 4.26 extends the main result in Dutta and Choudhury [84]; see also Lakzian and Samet [159]. On the other hand, when $(\le)$ is a (partial) order on X, the obtained result is comparable with the one in Roldán and Shahzad [199]. In fact, an effective generalization of the quoted statement is possible; we shall discuss these facts elsewhere.

Bibliography

[1] M. Abbas, B. Ali and S. Romaguera, *Fixed and periodic points of generalized contractions in metric spaces,* Fixed Point Th. Appl., 2013, 2013:243.

[2] M. Abbas, D. Ilić and T. Nazir, *Iterative approximation of fixed points of generalized weak Presić type k-step iterative methods for a class of operators,* Filomat, 29 (2015), 713–724.

[3] S. Abian and A. B. Brown, *A theorem on partially ordered sets, with applications to fixed point theorems,* Canadian J. Math., 13 (1961), 78–82.

[4] M. Abtahi, *Fixed point theorems for Meir-Keeler type contractions in metric spaces,* Fixed Point Th., 17 (2016), 225–236.

[5] M. Abtahi, Z. Kadelburg and S. Radenović, *Fixed point theorems of Cirić-Matkowski-type contractions in ν-generalized metric spaces,* RACSAM, 111 (2017), 57–64.

[6] R. P. Agarwal, M. A. El-Gebeily and D. O'Regan, *Generalized contractions in partially ordered metric spaces,* Appl. Anal., 87 (2008), 109–116.

[7] R. P. Agarwal and M. Khamsi, *Extension of Caristi's fixed point theorem to vector valued metric spaces,* Nonlin. Anal., 74 (2011), 141–145.

[8] R. P. Agarwal, P. Kumam and W. Sintunavarat, *PPF dependent fixed point theorems for an α_c-admissible non-self mapping in the Razumikhin class,* Fixed Point Theory Appl., 2013, 2013:280.

[9] R. P. Agarwal, M. Meehan and D. O'Regan, *Fixed Point Theory and Applications,* Cambridge Univ. Press, Cambridge, 2004.

[10] A. Alam and M. Imdad, *Nonlinear contractions in metric spaces under locally T-transitive binary relations,* Arxiv, 1512-00348-v4, 2 Mar 2016.

[11] B. Alamri, T. Suzuki and L. A. Khan, *Caristi's fixed point theorem and Subrahmanyam's fixed point theorem in ν-generalized metric spaces,* J. Function Spaces, Volume 2015, Article ID 709391.

[12] P. S. Alexandrov, *An Introduction to Set Theory and Topology* (Russian), Nauka, Moscow, 1977.

[13] H. H. Alsulami, E. Karapinar, F. Khojasteh and A.-F. Roldán, *A proposal to the study of contractions in quasi-metric spaces,* Discrete Dyn. Nat. Soc., Volume 2014, Article ID 269286.

[14] M. Altman, *An integral test for series and generalized contractions,* Amer.

Math. Monthly, 82 (1975), 827–829.

[15] M. Altman, *A generalization of the Brezis-Browder principle on ordered sets,* Nonlinear Anal., 6 (1982), 157–165.

[16] A. Amini-Harandi, *Fixed and coupled fixed points of a new type set-valued contractive mappings in complete metric spaces,* Fixed Point Th. Appl., 2012, 2012:215.

[17] V. G. Angelov, *Fixed Points in Uniform Spaces and Applications,* Cluj University Press, Cluj-Napoca, 2009.

[18] H. Argoubi, B. Samet and C. Vetro, *Nonlinear contractions involving simulation functions in a metric space with a partial order,* J. Nonlin. Sci. Appl., 8 (2015), 1082–1094.

[19] D. Ariza-Ruiz and A. Jimenez-Melado, *A continuation method for weakly contractive mappings under the interior condition,* Fixed Point Th. Appl., Volume 2009, Article ID 809315.

[20] D. Ariza-Ruiz and A. Jimenez-Melado, *A continuation method for weakly Kannan maps,* Fixed Point Th. Appl., Volume 2010, Article ID 321594.

[21] A. Azam and M. Arshad, *Kannan fixed point theorem on generalized metric spaces,* J. Nonlinear Sci. Appl., 1 (2008), 45–48.

[22] S. Banach, *Sur les opérations dans les ensembles abstraits et leur application aux équations intégrales,* Fund. Math., 3 (1922), 133–181.

[23] R. Batra and S. Vashistha, *Fixed point theorem for F_w-contractions in complete metric spaces,* J. Nonlin. Anal. Appl., Volume 2013, Article ID jnaa-00211.

[24] M. Becheanu, A. Dincă, I. D. Ion, C. Niţă, I. Purdea, N. Radu, M. Ştefănescu and C. Vraciu, *Algebra for Teacher Training* (Romanian), Ed. Didactică Ped., Bucureşti, 1983.

[25] P. Bernays, *A system of axiomatic set theory: Part III. Infinity and enumerability analysis,* J. Symbolic Logic, 7 (1942), 65–89.

[26] P. Bernays, *Axiomatic set theory,* North-Holland Publ. Co., Amsterdam, 1958.

[27] S. R. Bernfeld, V. Lakshmikantham and Y. M. Reddy, *Fixed point theorems of operators with PPF dependence in Banach spaces,* Applicable Anal., 6 (1977), 271–280.

[28] M. Berzig, *Coincidence and common fixed point results on metric spaces endowed with an arbitrary binary relation and applications,* J. Fixed Point Th. Appl., 12 (2013), 221–238.

[29] M. Berzig, E. Karapinar and A. Roldán, *Discussion on generalized-$(\alpha\psi, \beta\varphi)$-contractive mappings via generalized altering distance function and related fixed point theorems,* Abstr. Appl. Anal., Volume 2014, Article ID 259768.

[30] C. Bessaga, *On the converse of the Banach fixed point principle,* Colloq. Math., 7 (1959), 41–43.

[31] F. Bojor, *Fixed points of Kannan mappings in metric spaces endowed with a graph,* An. Şt. Univ. Ovidius Constanţa (Mat.), 20 (2012), 31–40.

[32] N. Bourbaki, *Sur le théorème de Zorn,* Archiv Math., 2 (1949/1950), 434–437.

[33] N. Bourbaki, *General Topology (Chapters 1–4)*, Springer, Berlin, 1995.

[34] N. Bourbaki, *General Topology (Chapters 5–10)*, Springer, Berlin, 1989.

[35] D. W. Boyd and J. S. W. Wong, *On nonlinear contractions*, Proc. Amer. Math. Soc., 20 (1969), 458–464.

[36] A. Branciari, *A fixed point theorem of Banach-Caccioppoli type on a class of generalized metric spaces*, Publ. Math. Debrecen, 57 (2000), 31–37.

[37] G. Brauer, *Sets of convergence for series defined by iteration*, Canad. Math. Bull., 9 (1966), 83–87.

[38] H. Brezis and F. E. Browder, *A general principle on ordered sets in nonlinear functional analysis*, Advances Math., 21 (1976), 355–364.

[39] A. Brøndsted, *Fixed points and partial orders*, Proc. Amer. Math. Soc., 60 (1976), 365–366.

[40] F. E. Browder, *On the convergence of successive approximations for nonlinear functional equations*, Indag. Math., 30 (1968), 27–35.

[41] F. E. Browder, *Remarks on fixed point theorems of contractive type*, Nonlin. Anal., 3 (1979), 657–661.

[42] F. E. Browder, *Erratum to the paper "Remarks on fixed point theorems of contractive type"*, Nonlin. Anal., 5 (1981), 111.

[43] N. Brunner, *Topologische Maximalprinzipien*, Zeitschr. Math. Logik Grundl. Math., 33 (1987), 135–139.

[44] I. Cabrera, J. Harjani and K. Sadarangani, *A fixed point theorem for contractions of rational type in partially ordered metric spaces*, Ann. Univ. Ferrara, 59 (2013), 251–258.

[45] J. Caristi and W. A. Kirk, *Geometric fixed point theory and inwardness conditions*, The Geometry of Metric and Linear Spaces (Michigan State Univ., 1974), pp. 74–83, Lecture Notes Math. vol. 490, Springer, Berlin, 1975.

[46] S. Chandok, *Some common fixed point results for rational type contraction mappings in partially ordered metric spaces*, Math. Bohemica, 138 (2013), 407–413.

[47] S. Chandok, B. S. Choudhury and N. Metiya, *Fixed point results in ordered metric spaces for rational type expressions with auxiliary functions*, J. Egyptian Math. Soc., 23 (2015), 95–101.

[48] S. S. Chang, Y. J. Cho, B. S. Lee, J. S. Jung and S. M. Kang, *Coincidence point theorems and minimization theorems in fuzzy metric spaces*, Fuzzy Sets Syst., 88 (1997), 119–127.

[49] Y.-Z. Chen, *A Prešić type contractive condition and its applications*, Nonlin. Anal., 71 (2009), 2012–2017.

[50] C. M. Chen and W. Y. Sun, *Periodic points and fixed points for the weaker $(\phi - \varphi)$-contractive mappings in complete generalized metric spaces*, J. Appl. Math., Volume 2012, Article ID 856974.

[51] M. Cherichi and B. Samet, *Fixed point theorems on ordered gauge spaces with applications to integral equations*, Fixed Point Th. Appl., 2012, 2012:13.

[52] A. Chiş and R. Precup, *Continuation theory for general contractions in gauge spaces*, Fixed Point Th. Appl., 2004, 2004:3.

[53] Y. J. Cho, S. M. Kang and P. Salimi, *Some PPF dependent fixed point theorems for generalized $\alpha - F$-contractions in Banach spaces and applications,* Mathematics, 2018, 6, 267.

[54] Y. J. Cho, P. P. Murthy and G. Jungck, *A common fixed point theorem of Meir and Keeler type,* Int. J. Math. Mathematical Sci., 16 (1993), 669–674.

[55] B. S. Choudhury and A. Kundu, *A Kannan-like contraction in partially ordered spaces,* Demonstr. Math., 46 (2013), 327–334.

[56] B. S. Choudhury, P. Konar, B. E. Rhoades and N. Metiya, *Fixed point theorems for generalized weakly contractive mapping,* Nonlin. Anal., 74 (2011), 2116–2126.

[57] S. Chouhan, A. Tiwari and M. Solanki, *Some fixed point result in metric spaces for rational expression,* Math. Theory Modelling, 4 (2014), 77–84.

[58] P. Chuasuk and A. Kaewcharoen, *Fixed point results for generalized rational type contractions with PPF dependence,* Int. J. Pure Appl. Math., 96 (2014), 197–212.

[59] L. B. Ciric, *A new fixed-point theorem for contractive mappings,* Publ. Inst. Math., 30(44) (1981), 25–27.

[60] L. B. Cirić, S. M. Alsulami, P. Salimi and P. Vetro, *PPF dependent fixed point results for triangular α_c-admissible mappings,* The Sci. World J., Volume 2014, Article ID 673647.

[61] L. Cirić, N. Cakić, M. Rajović and J. S. Ume, *Monotone generalized nonlinear contractions in partially ordered metric spaces,* Fixed Point Th. Appl., Volume 2008, Art. ID 131294.

[62] L. B. Cirić and S. B. Presić, *On Presić type generalization of the Banach contraction mapping principle,* Acta Math. Comenianae, 76 (2007), 143–147.

[63] P. J. Cohen, *Set Theory and the Continuum Hypothesis,* Benjamin, New York, 1966.

[64] P. Collaco and J. C. E Silva, *A complete comparison of 25 contractive definitions,* Nonlin. Anal., 30 (1997), 441–476.

[65] V. Conserva and S. Rizzo, *Maximal elements in a class of order complete metric subspaces,* Math. Japonica, 37 (1992), 515–518.

[66] O. Costinescu, *Elements of General Topology* (Romanian), Ed. Tehnică, Bucureşti, 1970.

[67] R. Cristescu, *Topological Vector Spaces,* Noordhoff Intl. Publishers, Leyden (The Netherlands), 1977.

[68] S. Dancs, M. Hegedus and P. Medvegyev, *A general ordering and fixed-point principle in complete metric space,* Acta Sci. Math. (Szeged), 46 (1983), 381–388.

[69] P. Das and L. K. Dey, *Fixed point of contractive mappings in generalized metric spaces,* Math. Slovaca, 59 (2009), 499–504.

[70] B. K. Dass and S. Gupta, *An extension of Banach contraction principle through rational expressions,* Indian J. Pure Appl. Math., 6 (1975), 1455–1458.

[71] K. Deimling, *Nonlinear Functional Analysis,* Springer, Berlin, 1985.

[72] D. Dey and M. Saha, *On fixed point theorems in generalized metric space,*

Preprint, 2013.

[73] B. C. Dhage, *Fixed point theorems with PPF dependence and functional differential equations,* Fixed Point Th., 13 (2012), 439–452.

[74] B. Dhage, *On some common fixed point theorems with PPF dependence in Banach spaces,* J. Nonlin. Sci. Appl., 5 (2012), 220–232.

[75] C. Di Bari and C. Vetro, *Common fixed point theorems for weakly compatible maps satisfying a general contractive condition,* Int. J. Math. Mathematical Sci., Volume 2008, Article ID 891375.

[76] C. Di Bari and C. Vetro, *Common fixed points in generalized metric spaces,* Appl. Math. Comput., 218 (2012), 7322–7325.

[77] C. Di Bari and P. Vetro, *φ-pairs and common fixed points in cone metric spaces,* Rend. Circ. Mat. Palermo, 57 (2008), 279–285.

[78] J. Dodu and M. Morillon, *The Hahn-Banach property and the Axiom of Choice,* Math. Logic Quarterly, 45 (1999), 299–314.

[79] J. Drewniak and V. Drobot, *On the slow convergence of iterations of a function,* Opuscula Math., 6 (1990), 41–58.

[80] Z. Drici, F. A. McRae and J. Vasundara Devi, *Fixed point theorem in partially ordered metric spaces for operators with PPF dependence,* Nonlin. Anal., 67 (2007), 641–647.

[81] W.-S. Du, *On coincidence point and fixed point theorems for nonlinear multivalued maps,* Topology Appl., 159 (2012), 49–56.

[82] W.-S. Du and F. Khojasteh, *New results and generalizations for approximate fixed point property and their applications,* Abstr. Appl. Anal., Volume 2014, Article ID 581267.

[83] N. V. Dung and V. I. Hang, *A fixed point theorem for generalized F-contraction on complete metric space,* Vietnam J. Math., 43 (2015), 743–755.

[84] P. N. Dutta and B. S. Choudhury, *A generalisation of contraction principle in metric spaces,* Fixed Point Th. Appl., 2008, 2008:8, Article ID 406368.

[85] I. Ekeland, *On the variational principle,* J. Math. Anal. Appl., 47 (1974), 324–353.

[86] I. Ekeland, *Nonconvex minimization problems,* Bull. Amer. Math. Soc. (New Series), 1 (1979), 443–474.

[87] R. Engelking, *General Topology,* Heldermann Verlag, Berlin, 1989.

[88] J. X. Fang, *The variational principle and fixed point theorems in certain topological spaces,* J. Math. Anal. Appl., 202 (1996), 398–412.

[89] A. Farajzadeh and A. Kaewcharoen, *On fixed point theorems for mappings with PPF dependence,* J. Ineq. Appl., 2014, 2014:372.

[90] A. Fora, A. Bellour and A. Al-Bsoul, *Some results in fixed point theory concerning generalized metric spaces,* Mat. Vesnik, 61 (2009), 203–208.

[91] M. Frigon, *Fixed point results in Fréchet and gauge spaces and applications to differential and integral equations,* Lect. Notes. Nonlin. Anal., 16 (2017), 9–91.

[92] A. Fulga, *On (ψ, φ)-rational contractions,* Symmetry, 2020, 12, 723.

[93] A. Fulga and A. Proca, *A new generalization of Wardowski fixed point theorem in complete metric spaces,* Adv. Th. Nonlin. Anal. Appl., 1 (2017),

57–63.

[94] L. Găvruţa, P. Găvruţa and F. Khojasteh, *Two classes of Meir-Keeler contractions,* Arxiv, 1405-5034-v1, 20 May 2014.

[95] R. George and M. S. Khan, *On Presić type extension of Banach contraction principle,* Int. J. Math. Anal., 5 (2011), 1019–1024.

[96] M. A. Geraghty, *On contractive mappings,* Proc. Amer. Math. Soc., 40 (1973), 604–608.

[97] N. Gheorghiu, *Contraction theorem in uniform spaces* (Romanian), Stud. Cerc. Mat., 19 (1967), 119–122.

[98] N. Gheorghiu, *Fixed point theorems in uniform spaces,* An. Şt. Univ. A. I. Cuza Iaşi (Sect. I-a, Mat.), 28 (1982), 17–18.

[99] N. Gheorghiu and E. Rotaru, *A fixed point theorem in uniform spaces,* An. Şt. Univ. A. I. Cuza Iaşi (Sect. I-a, Mat), 18 (1972), 311–314.

[100] N. Gheorghiu et M. Turinici, *Equations intégrales dans les espaces localement convexes,* Rev. Roum. Math. Pures Appl., 23 (1978), 33–40.

[101] K. Goedel, *The consistency of the continuum hypothesis,* Princeton Univ. Press, Princeton, 1940.

[102] A. Goepfert, H. Riahi, C. Tammer and C. Zălinescu, *Variational Methods in Partially Ordered Spaces,* Canad. Math. Soc. Books Math., vol. 17, Springer, New York, 2003.

[103] A. Goepfert, C. Tammer and C. Zălinescu, *On the vectorial Ekeland's variational principle and minimal points in product spaces,* Nonlin. Anal., 39 (2000), 909–922.

[104] A. Granas and C. D. Horvath, *On the order-theoretic Cantor theorem,* Taiwanese J. Math., 4 (2000), 203–213.

[105] A. Gupta, *Fixed points of a new type of contractive mappings in G-metric spaces,* Int. J. Adv. Math., 1 (2014), 56–61.

[106] P. R. Halmos, *Naive Set Theory,* Van Nostrand Reinhold Co., New York, 1960.

[107] A. Hamel, *Variational Principles on Metric and Uniform Spaces,* (Habilitation Thesis), Martin-Luther University, Halle-Wittenberg (Germany), 2005.

[108] H. A. Hammad and M. De la Sen, *PPF dependent fixed point results for new multi-valued generalized F-contraction in the Razumikhin class with an application,* Mathematics, 2019, 7, 52.

[109] J. Harjani, B. Lopez and K. Sadarangani, *A fixed point theorem for mappings satisfying a contractive condition of rational type on a partially ordered metric space,* Abstr. Appl. Anal., Volume 2010, Article ID 190701.

[110] M. Hasse and L. Michler, *Theorie der Kategorien,* VEB Deutscher Verlag Wissenschaften, Berlin, 1966.

[111] S. Heikkila, *Existence results for operator equations in partially ordered sets and applications,* Nonlin. Anal., 59 (2004), 1033–1044.

[112] T. L. Hicks and B. E. Rhoades, *Fixed point theory in symmetric spaces with applications to probabilistic spaces,* Nonlin. Anal. (A), 36 (1999), 331–344.

[113] P. Hitzler, *Generalized metrics and topology in logic programming semantics,* PhD Thesis, Natl. Univ. Ireland, Univ. College Cork, 2001.

[114] S. Hong, *Fixed points of discontinuous multivalued increasing operators in Banach spaces with applications,* J. Math. Anal. Appl., 282 (2003), 151–162.

[115] S. Hong, *Solvability of nonlinear impulsive Volterra integral inclusions and functional differential inclusions,* J. Math. Anal. Appl., 295 (2004), 331–340.

[116] N. Hussain, S. Khaleghizadeh, P. Salimi and F. Akbar, *New fixed point results with PPF dependence in Banach spaces endowed with a graph,* Abstr. Appl. Anal., Volume 2013, Article ID 827205.

[117] D. H. Hyers, G. Isac and T. M. Rassias, *Topics in Nonlinear Analysis and Applications,* World Sci. Publ., Singapore, 1997.

[118] M. Imdad, R. Gubran, M. Arif and D. Gopal, *An observation on α-type F-contractions and some ordered-theoretic fixed point results,* Math. Sci., 11 (2017), 247–255.

[119] G. Isac, *The Ekeland's principle and Pareto ε-efficiency,* Multi–Objective Programming and Goal Programming (M. Tamiz Ed.), pp. 148–163, [L. Notes Econ. Math. Systems vol. 432], Springer, Berlin, 1996.

[120] J. Jachymski, *Common fixed point theorems for some families of mappings,* Indian J. Pure Appl. Math., 25 (1994), 925–937.

[121] J. Jachymski, *Equivalent conditions and the Meir-Keeler type theorems,* J. Math. Anal. Appl., 194 (1995), 293–303.

[122] J. Jachymski, *Equivalence of some contractivity properties over metrical structures,* Proc. Amer. Math. Soc., 125 (1997), 2327–2335.

[123] J. Jachymski, *A short proof of the converse to the contraction mapping principle and some related results,* Topol. Meth. Nonlin. Anal., 15 (2000), 179–186.

[124] J. R. Jachymski, *Converses to the fixed point theorems of Zermelo and Caristi,* Nonlin. Anal., 52 (2003), 1455–1463.

[125] J. Jachymski, *The contraction principle for mappings on a metric space with a graph,* Proc. Amer. Math. Soc., 136 (2008), 1359–1373.

[126] J. Jachymski, *Equivalent conditions for generalized contractions on (ordered) metric spaces,* Nonlin. Anal., 74 (2011), 768–774.

[127] M. Jleli and B. Samet, *The Kannan's fixed point theorem in a cone rectangular metric space,* J. Nonlin. Sci. Appl., 2 (2009), 161–167.

[128] M. Jleli and B. Samet, *Remarks on G-metric spaces and fixed point theorems,* Fixed Point Th. Appl., 2012, 2012:210.

[129] M. Jleli and B. Samet, *A new generalization of the Banach contraction principle,* J. Ineq. Appl., 2014, 2014:38.

[130] C. F. K. Jung, *On generalized complete metric spaces,* Bull. Amer. Math. Soc., 75 (1969), 113–116.

[131] G. F. Jungck, *Commuting mappings and fixed points,* Amer. Math. Monthly, 83 (1976), 261–263.

[132] G. F. Jungck, *Compatible mappings and common fixed points,* Int. J. Math. Mathematical Sci., 9 (1986), 771–778.

[133] O. Kada, T. Suzuki and W. Takahashi, *Nonconvex minimization theorems and fixed point theorems in complete metric spaces,* Math. Japonica, 44

(1996), 381–391.

[134] Z. Kadelburg, P. Kumam, S. Radenović and W. Sintunavarat, *Common coupled fixed point theorems for Geraghty type contraction mappings using monotone property,* Fixed Point Th. Appl., 2015, 2015:27.

[135] Z. Kadelburg and S. Radenović, *On generalized metric spaces: A survey,* TWMS J. Pure Appl. Math., 5 (2014), 3–13.

[136] A. Kaewcharoen, *PPF dependent common fixed point theorems for mappings in Banach spaces,* J. Inequal. Appl., 2013, 2013:287.

[137] B. G. Kang and S. Park, *On generalized ordering principles in nonlinear analysis,* Nonlinear Anal., 14 (1990), 159–165.

[138] R. Kannan, *Some results on fixed points,* Bull. Calcutta Math. Soc., 60 (1968), 71–76.

[139] E. Karapinar and M. Berzig, *Fixed point results for* $(\alpha\psi, \beta\varphi)$*-contractive mappings for a generalized altering distance,* Fixed Point Th. Appl., 2013, 2013:205.

[140] E. Karapinar, A. Fulga and R. P. Agarwal, *A survey: F-contractions with related fixed point results,* J. Fixed Point Th. Appl., (2020), 20–69.

[141] S. Kasahara, *On some generalizations of the Banach contraction theorem,* Publ. Res. Inst. Math. Sci. Kyoto Univ., 12 (1976), 427–437.

[142] M. S. Khan, M. Berzig and B. Samet, *Some convergence results for iterative sequences of Presić type and applications,* Advances Diff. Eqs., 2012, 2012:38.

[143] M. S. Khan and P. K. Jhade, *On a fixed point theorem with PPF dependence in the Razumikhin class,* Gazi Univ. J. Science, 28 (2015), 211–219.

[144] M. S. Khan, M. Swaleh and S. Sessa, *Fixed point theorems by altering distances between the points,* Bull. Aust. Math. Soc., 30 (1984), 1–9.

[145] P. Q. Khanh, *On Caristi-Kirk's theorem and Ekeland's variational principle for Pareto extrema,* Bull. Polish Acad. Sci. (Math.), 37 (1989), 33–39.

[146] F. Khojasteh, S. Shukla and S. Radenović, *Formulization of many contractions via the simulation functions,* Arxiv, 1109-3021-v2, 13 Aug 2013.

[147] L. Kikina, K. Kikina and K. Gjino, *A new fixed point theorem on generalized quasimetric spaces,* ISRN Math. Analysis, Volume 2012, Article ID 457846.

[148] J. Kincses and V. Totik, *Theorems and counterexamples on contractive mappings,* Math. Balkanica, 4 (1999), 69–99.

[149] W. A. Kirk and M. Saliga, *The Brezis-Browder order principle and extensions of Caristi's theorem,* Nonlin. Anal., 47 (2001), 2765–2778.

[150] W. A. Kirk and N. Shahzad, *Generalized metrics and Caristi's theorem,* Fixed Point Th. Appl., Volume 2013, 2013:129.

[151] W. A. Kirk and N. Shahzad, *Correction: Generalized metrics and Caristi's theorem,* Fixed Point Th. Appl., Volume 2014, 2014:177.

[152] M. A. Krasnoselskii and V. Ya. Stetsenko, *Towards a theory of equations with concave operators* (Russian), Sib. Math. Zh., 10 (1969), 565–572.

[153] P. Kumam, F. Rouzkard, M. Imdad and D. Gopal, *Fixed point theorems on ordered metric spaces through a rational contraction,* Abstr. Appl. Anal., Volume 2013, Article ID 206515.

[154] S. Kumaresan, *Topology of metric spaces,* Alpha Science International Ltd.,

Harrow (U.K.), 2005.

[155] K. Kuratowski, *Topology,* Academic Press, New York, 1966.

[156] K. Kuratowski and A. Mostowski, *Set Theory,* PWN (Polish Scientific Publishers), Warsaw, 1968.

[157] M. A. Kutbi and W. Sintunavarat, *On sufficient conditions for the existence of past-present-future dependent fixed point in the Razumikhin class and application,* Abstr. Appl. Anal., Volume 2014, Article ID 342687.

[158] V. Lakshmikantham and L. Cirić, *Coupled fixed point theorems for nonlinear contractions in partially ordered metric spaces,* Nonlin. Anal., 70 (2009), 4341–4349.

[159] H. Lakzian and B. Samet, *Fixed points of (ψ, φ)-weakly contractive mappings in generalized metric spaces,* Appl. Math. Letters, 25 (2012), 902–906.

[160] A. Latif, *Generalized Caristi's fixed point theorems,* Fixed Point Th. Appl., Volume 2009, Article ID: 170140.

[161] S. Leader, *Fixed points for general contractions in metric spaces,* Math. Japonica, 24 (1979), 17–24.

[162] T.-C. Lim, *On characterizations of Meir-Keeler contractive maps,* Nonlin. Anal., 46 (2001), 113–120.

[163] W. A. J. Luxemburg, *On the convergence of successive approximations in the theory of ordinary differential equations (II),* Indagationes Math., 20 (1958), 540–546.

[164] J. Matkowski, *Integrable solutions of functional equations,* Dissertationes Math., Vol. 127, Polish Sci. Publ., Warsaw, 1975.

[165] J. Matkowski, *Fixed point theorems for mappings with a contractive iterate at a point,* Proc. Amer. Math. Soc., 62 (1977), 344–348.

[166] J. Matkowski, *Fixed point theorems for contractive mappings in metric spaces,* Časopis Pěst. Mat., 105 (1980), 341–344.

[167] S. G. Matthews, *Partial metric topology (Proc. 8th Summer Conf. Gen. Top. Appl.),* Ann. New York Acad. Sci., 728 (1994), 183–197.

[168] A. Meir and E. Keeler, *A theorem on contraction mappings,* J. Math. Anal. Appl., 28 (1969), 326–329.

[169] H. Moench, *Boundary-value problems for nonlinear ordinary differential equations of second order in Banach spaces,* Nonlin. Anal., 4 (1980), 985–999.

[170] B. Mohammadi, S. Rezapour and N. Shahzad, *Some results on fixed points of $\alpha - \psi$-Cirić generalized multifunctions,* Fixed Point Th. Appl., 2013, 2013:124.

[171] G. H. Moore, *Zermelo's Axiom of Choice: its Origin, Development and Influence,* Springer, New York, 1982.

[172] S. Moradi and A. Farajzadeh, *On the fixed point of $(\psi - \varphi)$-weak and generalized $(\psi - \varphi)$-weak contraction mappings,* Appl. Math. Letters, 25 (2002), 1257–1262.

[173] Y. Moskhovakis, *Notes on Set Theory,* Springer, New York, 2006.

[174] Z. Mustafa, H. Aydi and E. Karapinar, *Mixed g-monotone property and quadruple fixed point theorems in partially ordered metric spaces,* Fixed Point Th. Appl., 2012, 2012:71.

[175] S. B. Nadler Jr., *Multi-valued contraction mappings,* Pacific J. Math., 30 (1969), 475–488.

[176] I. P. Natanson, *Theory of Functions of a Real Variable, Volume II (Chs 10–17),* Frederick Ungar Publishing Co., New York, 1960.

[177] I. P. Natanson, *Theory of Functions of a Real Variable, Volume I (Chs 1–9),* Frederick Ungar Publishing Co., New York, 1964.

[178] A. B. Nemeth, *A nonconvex vector minimization problem,* Nonlin. Anal., 10 (1986), 669–678.

[179] J. J. Nieto and R. Rodriguez-Lopez, *Contractive mapping theorems in partially ordered sets and applications to ordinary differential equations,* Order, 22 (2005), 223–239.

[180] D. O'Regan, *Fixed point theorems of Moench type for weakly sequentially upper semicontinuous maps,* Bull. Austral. Math. Soc., 61 (2000), 439–449.

[181] M. Păcurar, *Approximating common fixed points of Presić-Kannan type operators by a multi-step iterative method,* An. Şt. Univ. Ovidius Constanţa (Mat.), 17 (2009), 153–168.

[182] M. Păcurar, *Fixed points of almost Presić operators by a k-step iterative method,* An. Şt. Univ. A. I. Cuza Iaşi (Mat.), 57 (2011), 199–210.

[183] M. Paknazar, M. A. Kutbi, M. Demma and P. Salimi, *PPF dependent fixed point in modified Razumikhin class with applications,* J. Nonlin. Sci. Appl., 9 (2016), 270–286.

[184] S. Park, *On general contractive conditions,* J. Korean Math. Soc., 17 (1980), 131–140.

[185] H. K. Pathak, *A fixed point theorem on Banach space and its applications,* Commun. Fac. Sci. Univ. Ankara (Series A), 42 (1993), 61–67.

[186] M. Piticari, *Successive approximations method and fixed point principle* (Romanian), Diss. Thesis, A. I. Cuza Univ., Iaşi, 1982.

[187] T. Precupanu, *Linear Topological Spaces and Fundamentals of Convex Analysis* (Romanian), Ed. Acad. Române, Bucureşti, 1992.

[188] S. B. Presić, *Sur une classe d'inéquations aux différences finies et sur la convergence de certaines suites,* Publ. Inst. Math. (Beograd) (N.S.), 5(19) (1965), 75–78.

[189] P. D. Proinov, *Fixed point theorems for generalized contractive mappings in metric spaces,* J. Fixed Point Theory Appl., (2020), 22:21.

[190] J.-H. Qiu, *A pre-order principle and set-valued Ekeland variational principle,* J. Math. Anal. Appl., 419 (2014), 904–937.

[191] S. Radenović, Z. Kadelburg, D. Jandrlić and A. Jandrlić, *Some results on weakly contractive maps,* Bull. Iranian Math. Soc., 38 (2012), 625–645.

[192] A. C. M. Ran and M. C. Reurings, *A fixed point theorem in partially ordered sets and some applications to matrix equations,* Proc. Amer. Math. Soc., 132 (2004), 1435–1443.

[193] K. P. R. Rao, M. M. Ali and B. Fisher, *Some Presić type generalizations of the Banach contraction principle,* Math. Moravica, 15 (2011), 41–47.

[194] S. Reich, *Fixed points of contractive functions,* Boll. Un. Mat. Ital., 5 (1972), 26–42.

[195] B. E. Rhoades, *A comparison of various definitions of contractive mappings,*

Trans. Amer. Math. Soc., 226 (1977), 257–290.

[196] B. E. Rhoades, *Common fixed points of compatible set-valued mappings,* Publ. Math. Debrecen, 48 (1996), 237–240.

[197] B. E. Rhoades, *Some theorems on weakly contractive maps,* Nonlin. Anal., 47 (2001), 2683–2693.

[198] A.-F. Roldán, E. Karapinar, C. Roldán and J. Martinez-Moreno, *Coincidence point theorems on metric spaces via simulation method,* Fixed Point Th. Appl., 2015, 2015:98.

[199] A.-F. Roldán and N. Shahzad, *Some fixed/coincidence point theorems under (ψ, φ)-contractivity conditions without an underlying metric structure,* Fixed Point Th. Appl., 2014, 2014:218.

[200] P. Rozoveanu, *Ekeland's variational principle for vector valued functions,* Math. Reports [St. Cerc. Mat.], 2(52) (2000), 351–366.

[201] H. Rubin and J. E. Rubin, *Equivalents of the Axiom of Choice II,* North Holland, Amsterdam, 1985.

[202] I. A. Rus, *An iterative method for the solution of the equation $x = f(x, ..., x)$,* Mathematica (Rev. Anal. Num. Th. Approx.), 10 (1981), 95–100.

[203] I. A. Rus, *Generalized Contractions and Applications,* Cluj University Press, Cluj-Napoca, 2001.

[204] P. Salimi, A. Latif and N. Hussain, *Modified $\alpha - \psi$-contractive mappings with applications,* Fixed Point Th. Appl., 2013, 2013:151.

[205] A. S. Saluja, R. A. Rashwan, D. Magarde and P. K. Jhade, *Some result in ordered metric spaces for rational type expressions,* Facta Univ. Niš (Ser. Math. Inform.), 31 (2016), 125–138.

[206] B. Samet, *Coupled fixed point theorems for a generalized Meir-Keeler contraction in partially ordered metric spaces,* Nonlinear Anal., 72 (2010), 4508–4517.

[207] B. Samet, *Discussion on "A fixed point theorem of Banach-Caccioppoli type on a class of generalized metric spaces" by A. Branciari,* Publ. Math. Debrecen, 76 (2010), 493–494.

[208] B. Samet and M. Turinici, *Fixed point theorems on a metric space endowed with an arbitrary binary relation and applications,* Commun. Math. Anal., 13 (2012), 82–97.

[209] B. Samet, C. Vetro and F. Vetro, *Remarks on G-metric spaces,* Int. J. Analysis, Volume 2012, Article ID 917158.

[210] B. Samet, C. Vetro and H. Yazidi, *A fixed point theorem for a Meir-Keeler type contraction through rational expression,* J. Nonlin. Sci. Appl., 6 (2013), 162–169.

[211] E. Schechter, *Handbook of Analysis and its Foundation,* Academic Press, New York, 1997.

[212] S. Sessa, B. E. Rhoades and M. S. Khan, *On common fixed points of compatible mappings in metric and Banach spaces,* Int. J. Math. Mathematical Sci., 11 (1988), 375–392.

[213] M. Sgroi and C. Vetro, *Multi-valued F-contractions and the solution of certain functional and integral equations,* Filomat, 27 (2013), 1259–1268.

[214] N. Shahzad and S. Shukla, *Set-valued G-Presić operators on metric spaces endowed with a graph and fixed point theorems,* Fixed Point Th. Appl., 2015, 2015:24.

[215] S. Shukla, S. Radenović and S. Pantelić, *Some fixed point theorems for Presić-Hardy-Rogers type contractions in metric spaces,* J. Math., Volume 2013, Article ID 295093.

[216] W. Sierpinski, *Cardinal and Ordinal Numbers,* PWN (Polish Scientific Publishers), Warsaw, 1965.

[217] T. Som, *Some fixed point theorems of metric and Banach spaces,* Indian J. Pure Appl. Math., 16 (1985), 575–585.

[218] T. Suzuki, *Generalized metric spaces do not have the compatible topology,* Abstr. Appl. Anal., Volume 2014, Article ID 458098.

[219] T. Suzuki, *Discussion of several contractions by Jachymski's approach,* Fixed Point Th. Appl., 2016, 2016:91.

[220] T. Suzuki and C. Vetro, *Three existence theorems for weak contractions of Matkowski type,* Int. J. Math. Stat., 6 (2010), 110–120.

[221] A. Tarski, *Axiomatic and algebraic aspects of two theorems on sums of cardinals,* Fund. Math., 35 (1948), 79–104.

[222] A. Tarski, *A lattice-theoretical fixpoint theorem and its applications,* Pacific J. Math., 5 (1955), 285–309.

[223] M. R. Tasković, *Some results in the fixed point theory,* Publ. Inst. Math. (N.S.), 20(34) (1976), 231–242.

[224] M. R. Tasković, *On a question of priority regarding a fixed point theorem in a Cartesian product of metric spaces,* Math. Moravica, 15 (2011), 69–71.

[225] D. Tataru, *Viscosity solutions of Hamilton-Jacobi equations with unbounded nonlinear terms,* J. Math. Anal. Appl., 163 (1992), 345–392.

[226] V. Timofte, *New tests for positive iteration series,* Real Analysis Exchange, 30 (2004/2005), 799–812.

[227] M. Turinici, *Fixed points of implicit contraction mappings,* An. Şt. Univ. A. I. Cuza Iaşi (S I-a, Mat), 22 (1976), 177–180.

[228] M. Turinici, *Nonlinear contractions and applications to Volterra functional equations,* An. Şt. Univ. A. I. Cuza Iaşi (S I-a, Mat), 23 (1977), 43–50.

[229] M. Turinici, *Abstract monotone mappings and applications to functional differential equations,* Rend. Accad. Naz. Lincei (Serie 8), 66 (1979), 189–193.

[230] M. Turinici, *Maximal elements in a class of order complete metric spaces,* Math. Japonica, 25 (1980), 511–517.

[231] M. Turinici, *Maximality principles and mean value theorems,* An. Acad. Brasil. Cienc., 53 (1981), 653–655.

[232] M. Turinici, *Multivalued contractions and applications to functional differential equations,* Acta Math. Acad. Sci. Hungaricae, 37 (1981), 147–151.

[233] M. Turinici, *A generalization of Brezis-Browder's ordering principle,* An. Şt. Univ. A. I. Cuza Iaşi (S I-a: Mat), 28 (1982), 11–16.

[234] M. Turinici, *A generalization of Altman's ordering principle,* Proc. Amer. Math. Soc., 90 (1984), 128–132.

[235] M. Turinici, *Fixed points for monotone iteratively local contractions,*

Demonstratio Math., 19 (1986), 171–180.

[236] M. Turinici, *Abstract comparison principles and multivariable Gronwall-Bellman inequalities,* J. Math. Anal. Appl., 117 (1986), 100–127.

[237] M. Turinici, *Pseudometric extensions of the Brezis-Browder ordering principle,* Math. Nachrichten, 130 (1987), 91–103.

[238] M. Turinici, *Metric variants of the Brezis-Browder ordering principle,* Demonstr. Math., 22 (1989), 213–228.

[239] M. Turinici, *A monotone version of the variational Ekeland's principle,* An. Şt. Univ. A. I. Cuza Iaşi (S. I-a: Mat.) 36 (1990), 329–352.

[240] M. Turinici, *Fixed point results on abstract ordered sets,* Le Matematiche, 49 (1994), 25–34.

[241] M. Turinici, *Pseudometric versions of the Caristi-Kirk fixed point theorem,* Fixed Point Theory (Cluj-Napoca), 5 (2004), 147–161.

[242] M. Turinici, *Function variational principles and coercivity,* J. Math. Anal. Appl., 304 (2005), 236–248.

[243] M. Turinici, *Brezis-Browder principles in general separable sets,* Libertas Math., 26 (2006), 31–47.

[244] M. Turinici, *Function pseudometric VP and applications,* Bul. Inst. Polit. Iaşi (Sect.: Mat., Mec. Teor., Fiz.), 53(57) (2007), 393–411.

[245] M. Turinici, *Functional versions of the Caristi-Kirk theorem,* Revista Union Mat. Arg., 50 (2009), 87–97.

[246] M. Turinici, *Ran-Reurings theorems in ordered metric spaces,* J. Indian Math. Soc., 78 (2011), 207–214.

[247] M. Turinici, *Functional contractions in local Branciari metric spaces,* Romai J., 8 (No. 2) (2012), 189–199.

[248] M. Turinici, *Wardowski implicit contractions in metric spaces,* Arxiv, 1211-3164-v2, 15 Sep 2013.

[249] M. Turinici, *Functional contractive maps in triangular symmetric spaces,* J. Math., Volume 2013, Article ID 637521.

[250] M. Turinici, *PPF dependent fixed points in A-closed Razumikhin classes,* Romai J., 10 (No. 1) (2014), 205–220.

[251] M. Turinici, *Contraction Maps in Ordered Metrical Structures,* Mathematics Without Boundaries [Surveys Interdisc. Research] (P. M. Pardalos and T. M. Rassias, Eds.), pp. 533–575, Springer Science+Business Media, New York, 2014.

[252] M. Turinici, *Sequential maximality principles,* Mathematics Without Boundaries [Surveys Pure Math.] (T. M. Rassias and P. M. Pardalos, Eds.), pp. 515–548, Springer, New York, 2014.

[253] M. Turinici, *Variational principles in gauge spaces,* Optimization in Science and Engineering (T. M. Rassias, C. A. Floudas and S. Butenko, Eds.), pp. 503–542, Springer, New York, 2014.

[254] M. Turinici, *Implicit contractive maps in ordered metric spaces,* Topics in Mathematical Analysis and Applications (T. M. Rassias and L. Toth, Eds.), pp. 715–746, Springer International Publishing, Switzerland, 2014.

[255] M. Turinici, *Contractive maps in locally transitive relational metric spaces,* The Sci. World J., Volume 2014, Article ID 169358.

[256] M. Turinici, *Modern Directions in Metrical Fixed Point Theory,* Pim Editorial House, Iaşi, 2016.

[257] M. Turinici, *Contraction maps in pseudometric structures,* Essays in Mathematics and its Applications (T. M. Rassias and P. M. Pardalos, Eds.), pp. 513–562, Springer Intl. Publ., Switzerland, 2016.

[258] M. Turinici, *Selected Topics in Metrical Fixed Point Theory (Revised Ed.),* Pim Editorial House, Iaşi, 2017.

[259] M. Turinici, *Reports in Metrical Fixed Point Theory,* Pim Editorial House, Iaşi, 2020.

[260] I. Vaisman, *The Fundamentals of Mathematics* (Romanian), Ed. Didactică Ped., Bucureşti, 1968.

[261] J. Vujaković, S. Mitrović, M. Pavlović and S. Radenović, *On recent results concerning F-contraction in generalized metric spaces,* Mathematics, 2020, 8, 767.

[262] S. S. Wagstaff, Jr, *Sequences not containing an infinite arithmetic progression,* Proc. Amer. Math. Soc., 36 (1972), 395–397.

[263] W. Walter, *Remarks on a paper of F. Browder about contractions,* Nonlin. Anal., 5 (1981), 21–25.

[264] D. Wardowski, *Fixed points of a new type of contractive mappings in complete metric spaces,* Fixed Point Th. Appl., 2012, 2012:94.

[265] D. Wardowski, *Solving existence problems via F-contractions,* Proc. Amer. Math. Soc., 146 (2018), 1585–1598.

[266] D. Wardowski and N. Van Dung, *Fixed points of F-weak contractions on complete metric spaces,* Dem. Math., 47 (2014), 146–155.

[267] E. S. Wolk, *On the principle of dependent choices and some forms of Zorn's lemma,* Canad. Math. Bull., 26 (1983), 365–367.

[268] J. S. W. Wong, *Generalizations of the converse of the contraction mapping principle,* Canad. J. Math., 18 (1966), 1095–1104.

[269] Z. Xue, *Fixed point theorems for generalized weakly contractive mappings,* Bull. Austral. Math. Soc., 93 (2016), 321–329.

[270] C.-C. Yeh, *Some fixed point theorems in L-spaces,* Indian J. Pure Appl. Math., 9 (1978), 993–995.

[271] L. Zhao and J. Sun, *Caristi's fixed point theorem in vector lattices,* Preprint, 2012.

[272] C. K. Zhong, *A generalization of Ekeland's variational principle and application to the study of the relation between the weak P.S. condition and coercivity,* Nonlin. Anal., 29 (1997), 1421–1431.

[273] J. Zhu, J. J. Cho and S. M. Kang, *Equivalent contractive conditions in symmetric spaces,* Comput. Math. Appl., 50 (2005), 1621–1628.

[274] J. Zhu, X. Fan and S. Zhang, *Fixed points of increasing operators and solutions of nonlinear impulsive integro-differential equations in Banach space,* Nonlin. Anal., 42 (2000), 599–611.

[275] J. Zhu and S. J. Li, *Generalization of ordering principles and applications,* J. Optim. Th. Appl., 132 (2007), 493–507.

Index

addition (sum), 21, 311
additive relation, 103
additive set, 488
admissible chain, 87, 190
admissible cone metrical space, 298
admissible conv-Cauchy structure, 281
admissible couple, 400
admissible quadruple, 467
admissible relation, 192
Alexandrov theorem, 315
algebraically closed set, 488
almost Bellman Picard operator, 418
almost non right oscillating system, 208
almost nright set, 195
almost Picard map, 341
almost regular structure, 278, 282
almost semi-diametral map, 369, 385, 419
almost semimetric, 32
almost-selfclosed relation, 194, 418
altering rational contraction, 252
Altman triangular pseudometric, 494
amorphous relation, 8
anti-selfclosed order, 325
antisymmetric relation, 8
aprogression, 85
ascending orbital full sequence, 226, 246
ascending orbital sequence, 189
ascending sequence, 368, 392, 418, 497
ascending uadmissible regular map, 193
asingleton, 3, 31, 167, 188, 225, 280, 364, 391, 399, 417, 445, 491, 496
asymptotic Bari-Vetro couple, 123
asymptotic bilateral separable, 106, 142, 421
asymptotic dominating couple, 118, 127
asymptotic dominating function, 121, 144
asymptotic expansive function, 118, 144
asymptotic left separable, 106, 142, 421
asymptotic manageable couple, 136
asymptotic Meir-Keeler contractive map, 446
asymptotic Meir-Keeler relation, 106, 142, 169, 421, 499
asymptotic negative function, 131
asymptotic plane normal couple, 125
asymptotic positive function, 115, 125, 144, 179, 252, 373, 425, 503
asymptotic regressive function, 123
asymptotic regular cone, 294
asymptotic regular set, 318
asymptotic regular structure, 353
asymptotic sequence, 35, 36, 95, 233,

293, 318, 353, 395, 494
asymptotic set, 195
asymptotic simulation couple, 131
asymptotic subunitary function, 136
Axiom of Choice, 7, 27
Axiom of Regularity, 6
Axiom of Russell, 5
Axiom of Separation, 4

B-ascending sequence, 193
balanced basis, 291
Banach fixed point theorem, 184
Banach lattice, 305
Banach space, 480
basis, 150
Bellman cfp-cp-Picard couple, 401
Bellman cp-Picard couple, 401
Bellman Picard operator, 226, 246, 368, 392, 497
bilateral superior/inferior limit, 203
bilateral-closed order, 309
bilateral-selfclosed order, 325
boundary, 156
bounded above (below), 24
bounded left oscillating couple, 115
boundedly monotone function, 70
Boyd-Wong admissible function, 61, 76, 176, 228, 348, 448, 502
Boyd-Wong contractive map, 175
Boyd-Wong limiting Cauchy Criterion, 43
Boyd-Wong local Cauchy Criterion, 43
Boyd-Wong non-Cauchy Criterion, 38
Boyd-Wong relation, 112
Boyd-Wong relational non-Cauchy Criterion, 98
Boyd-Wong system, 42, 102
Boyd-Wong-Matkowski fixed point theorem, 176
Brøndsted order, 361
Burali-Forti paradox, 312

cardinal, 314
Cauchy net, 318, 323
Cauchy regular cone, 294
Cauchy regular structure, 282, 353
Cauchy sequence, 34, 95, 233, 293, 353
Cauchy sequentially inductive structure, 282
Cauchy structure, 322, 338, 367, 492
Cauchy-continuous map, 226
Cauchy-selfclosed quasi-order, 226
Cauchy-separated symmetric space, 225
chain, 232, 315
Cirić-Presić contractive map, 262
closed set, 148, 295, 355
closure (adherence), 148, 156, 293, 333, 338, 346, 356
cluster, 156
cofinal subset, 37, 98
coincidence point, 399, 461
common fixed point, 399
compact countable iterations, 308
compact set, 157, 326
compact space, 157
compact values map, 308
compatible couple, 400
compatible map, 375, 428, 497
complement function, 78, 348
complementary coercive function, 78, 348
complementary normal function, 49
complete cone metrical space, 298
complete conv-Cauchy structure, 281, 492
complete function, 74
complete Meir-Keeler property, 248, 498
complete pseudometric space, 284
complete relation, 138
complete space, 194, 226, 246, 368, 392, 401, 418, 445, 497
composition rule, 20
concatenation order, 313
cone metric space, 297
cone metrical Cauchy structure, 298
cone metrical conv-Cauchy structure, 298

cone metrical convergence structure, 297
conical modulus function, 303
conjugate map, 364
constant class, 471, 480
continuous function, 152, 205
continuous linear operator, 482
continuous map, 194, 246, 368, 392, 401, 418, 445, 497
continuous operator, 460, 472, 480
contractive couple, 402
contractive map, 253
contractive operator, 394, 461
conv-bd-regular structure, 347
conv-Cauchy structure, 233, 367, 492
convergence structure, 31, 322, 338, 355, 365, 491
convergent net, 318, 323, 333
convergent sequence, 31, 33, 50, 232, 280, 292, 492
convex cone, 291
cotopological space, 148
countable set, 312, 316
couple (ordered pair), 3
cp-asingleton couple, 400
cp-involutive map, 401
cp-Picard couple, 400
cp-self-progressive map, 401
cp-singleton couple, 400

Denumerable Axiom of Choice, 29
Dependent Choice principle, 27, 28
descending asymptotic set, 195
diagonal asymptotic expansive function, 126
diagonal asymptotic negative function, 131
diagonal asymptotic plane normal couple, 125
diagonal asymptotic positive function, 125
diagonal asymptotic simulation couple, 131
Diagonal Dependent Choice principle, 29
diagonal fixed point, 259
diagonal map, 259
diametral map, 171, 451
discrete metric space, 362
divergent sequence, 31, 33
duality formula, 156

effectively countable set, 312
Ekeland variational principle, 286, 359
element, 2
empty set, 3
equivalence, 9
equivalence class, 404, 460, 490
equivalent conv-Cauchy structures, 47
expansive linear operator, 482
extended convergence, 50
extended Meir-Keeler property, 498

factor space, 404
Fang space, 307, 322
Finite Dependent Choice property, 27
finite set, 147, 312
finitely-asymptotic regular couple, 118, 126
first element, 9, 25
first kind ordinal, 313
first variable continuous quasi-metric, 366
first variable contractive map, 347
first variable decreasing relation, 103, 139
first variable lsc quasi-metric, 366
first variable sup-closed relation, 103
fix positive map, 374, 426
fix semi-diametral map, 374, 388, 426
fix singular map, 375, 427
fix-asingleton map, 167, 183, 188, 226, 245, 260, 339, 368, 392, 418, 496
fix-positive map, 388
fix-singleton map, 167, 183, 188, 226, 245, 260, 339, 368, 392, 418, 496
fix-transversal map, 348
fixd-asingleton map, 260
fixd-singleton map, 260
Fréchet norm, 292
full operator, 489

full Razumikhin chain, 489
full sequence, 160, 167, 368, 395, 418, 445
full set, 295, 459
fully orbital element, 430
function, 3
function asymptotic sequence, 277
function strongly-contractive map, 428
function type point asymptotic triple, 209
function type right asymptotic triple, 209
functional contraction, 202

gauge space, 164, 338
generalized metric, 90
generalized pseudometric, 279
geometric bilateral separable, 104, 140, 371, 421
geometric contractive map, 169, 371, 422, 499
geometric left separable, 104, 140, 371, 421
geometric Meir-Keeler property, 104, 140, 169, 371, 421, 499
globally orbital transitive map, 430
globally scattered map, 430
globally sub-aprogression map, 430
Grothendieck universe, 311

Hausdorff separated space, 156, 333
homogeneous function, 264
homotopic map, 347

increasing couple, 400
increasing function, 267, 503
increasing map, 226, 245, 308, 367, 391, 418, 429, 496
inductive, 9
inf-lattice order, 361
inf-proper function, 359
inferior (superior) limit, 50
infinitely-asymptotic regular couple, 118, 127
int-normal function, 71
int-subnormal function, 71
interior, 156, 338
intersection of empty set, 4
invertible couple, 467, 471, 482
irreflexive relation, 8
isometric map, 474
isometric Razumikhin chain, 473
isometric system, 473
iterative fix-asingleton map, 183
iterative fix-singleton map, 183
iterative sequence, 461
iterative strong Picard map, 183

Jleli-Samet relation, 122

Kuratowski closure, 148

last element, 25
lattice normed space, 303
left admissible couple, 229
left regulated function, 205
Leibniz equality rule, 2
lexicographic order, 313
lim-right-normal function, 69
lim-right-subnormal function, 69
limit-bounded oscillating couple, 115, 252
Lipschitz property, 362
local Branciari metric space, 232
locally bounded function, 252
locally int-normal function, 67
locally int-subnormal function, 67
lsc function, 359

manageable function, 138
Matkowski admissible function, 49, 57, 75, 143, 177, 228, 268, 341, 347, 393, 448, 501
Matkowski admissible relation, 103, 140
Matkowski contractive map, 373, 501
Matkowski regular map, 247, 252
maximal element, 347, 353, 361
Meir-Keeler admissible function, 57, 75, 143, 176, 227, 341, 347, 373, 393, 424, 448, 501

Meir-Keeler contractive map, 168, 227, 247, 370, 421, 446, 498
Meir-Keeler fixed point theorem, 171
Meir-Keeler geometric fixed point theorem, 174
Meir-Keeler property, 168, 227, 247, 370, 421, 446, 498
metric, 32, 157
metric space, 259
metrizable space, 292
metrizable uniform space, 324
multiplication (product), 22

naive Cantor set theory, 13, 310
narrow relatively compact set, 161
natural number, 13, 311
neighborhood, 149, 291
neighborhoods subsystem, 150
neighborhoods system, 149
nnormal set, 196
non fix positive map, 374
non fix separated map, 427
non point oscillating couple, 208
non point oscillating system, 209
non right oscillating couple, 208
non right oscillating system, 208
non-degenerated proper subset, 291
non-diagonally admissible function, 57, 75
non-separable formula, 14
non-telescopic sequence, 193, 249, 378, 395, 452
nonexpansive couple, 467
nonexpansive operator, 458, 498
nonmaximal element, 347
normal couple, 115, 144, 179, 229, 252, 373, 400, 425, 449, 503
normal family, 147
normal function, 46
normally countable chain, 316
normed lattice, 303
npoint set, 196
nright set, 195

open set, 148
orbital (iterative) sequence, 167, 188, 339, 368, 418, 445, 497
orbital admissible map, 247
orbital strict nonexpansive map, 247
orbitally bounded map, 451
orbitally regular asymptotic map, 370, 386, 420
orbitally singular asymptotic map, 171, 370, 386, 420, 452
orbitally strongly regular asymptotic map, 370, 386, 420
order asingleton, 245
order function maximal element, 278
order segment, 295
order singleton, 245
order triangular pseudometric, 494
order-complete space, 339
order-continuous map, 339
order-positive pseudometric, 494
order-pseudometric space, 504
ordered metric space, 347
ordinal, 312
ordinal net, 313
original Meir-Keeler contractive map, 168, 446

partial order, 9, 311
partially ordered almost uniform space, 318
partially ordered Fang space, 307, 325, 329
partially ordered topological space, 333
Peano system, 17, 311
Picard admissible map, 247
Picard contractive operator, 247
Picard operator, 167, 226, 245, 260, 339, 368, 392, 418, 497
Picard sequence, 194
plane normal couple, 125
plane Rhoades relation, 125
point sequence, 195
positive map, 171, 369, 384, 419, 426, 451
positive section, 111
PPF dependent fixed point, 461

Presić contractive map, 261
Presić operator, 260
product convergence structure, 494
proper relation, 27, 400
pseudo-metrizable almost uniformity, 320
pseudo-metrizable uniform space, 324
pseudometric, 32, 157, 307, 322, 492
pseudometric Cauchy structure, 281
pseudometric conv-Cauchy structure, 281
pseudometric convergence structure, 280
pseudometric maximal element, 281

quasi admissible function, 265
quasi-metric, 32
quasi-metric space, 364
quasi-order, 9, 226, 277, 492
quasi-order convergence, 356
quasi-order maximal element, 284
quasi-order pseudometric fixed point, 496
quasi-order pseudometric space, 494
quasi-ordered local Branciari metric space, 236
quasi-ordered metric space, 245, 391, 399
quasi-ordered pseudometric space, 281, 353, 492
quasi-ordered quasi-metric space, 367
quasi-ordered real topological vector space, 291

Razumikhin chain, 459
Razumikhin iso-chain, 459
Razumikhin isometric relation, 458
Razumikhin standard relation, 458
real topological vector space, 291
reduced Zermelo-Fraenkel system, 28
reflexive pseudometric, 32, 157, 280
reflexive relation, 8
reflexive sufficient function, 144
regressive function, 48, 57, 143, 267, 424, 448, 501
regular chain, 232
regular cone, 294
regular conv-Cauchy structure, 492
regular operation, 206
regular structure, 354
regulated function, 205
relation, 3, 8, 103
relational metric space, 188, 417, 429
relational semimetric space, 94
relatively aprogression, 86
relatively compact net, 326
relatively compact set, 161, 326
relatively sub-aprogression, 86
Rhoades admissible function, 58
Rhoades contractive map, 179, 373, 503
Rhoades relation, 115
Rhoades-Dutta-Choudhury fixed point theorem, 180
right accessible structure, 347
right and point oscillation, 203
right compatible couple, 207
right complementary inverse, 78
right continuous complementary coercive function, 352
right inferior/superior limit, 175, 203
right limit, 65, 175
right locally strict positive function, 179
right normal function, 113
right regulated function, 205
right sequence, 195
right sequentially positive function, 229, 449
right-closed order, 308
right-usc function, 228

scattered chain, 90, 429
second kind (limit) ordinal, 313
second variable contractive map, 347
second variable increasing relation, 103, 139
second variable inf-closed relation, 103
Selected Dependent Choice principle, 29
selfclosed order, 325

selfclosed quasi-order, 246, 284, 368, 392, 401, 497
semi Bellman Picard sequence, 194
semi Picard operator, 245
semi Picard sequence, 193
semi regular operation, 206
semi-closure, 165
semi-diameter, 369, 419, 437
semi-diametral map, 369, 385, 419
semi-homogeneous function, 263
semi-normal family, 147
semi-progressive couple, 400
semi-progressive map, 226, 245, 308, 367, 391, 418, 429, 496
semi-regressive function, 46
semi-Wardowski function, 180
semimetric, 32, 47, 157, 337
semimetric space, 285, 341
separable structure, 317
separated cone metrical space, 297
separated conv-Cauchy structure, 492
separated convergence, 31, 280
separated map, 227
separated pseudometric, 33
separated quasi-metric, 365
sequential Cauchy structure, 32
sequential conv-Cauchy structure, 32, 164, 338, 346
sequentially Boyd-Wong admissible function, 62, 77
sequentially compact set, 159
sequentially continuous function, 161
sequentially inductive structure, 278
sequentially Rhoades admissible function, 62
sequentially zero-compatible function, 47
set, 2
set contraction, 202
set-contradictory formula, 14
similar couple, 400
simulation couple, 130
simulation function, 134
singleton, 3, 31, 167, 188, 225, 280, 364, 391, 399, 417, 445, 491
solid cone, 303
spectrum, 96, 189, 430
sphere (open, closed), 157, 362
standard Razumikhin chain, 465, 473
standard Razumikhin relation, 465
standard system, 466
stationary sequence, 495
strict order, 9
strict subsequence, 54
strictly contractive map, 498
strictly nonexpansive map, 168, 227, 247, 370, 421, 446
strictly positive function, 179, 425, 503
strong cfp-cp-Picard couple, 400
strong cp-Picard couple, 400
strong Picard operator, 167, 183, 226, 246, 260, 368, 392, 418, 497
strong Presić operator, 260
strong triangular pseudometric, 494
strongly Boyd-Wong admissible function, 61, 76
strongly Cauchy sequence, 282
strongly directed family, 151
strongly fix singular map, 375, 427
strongly Matkowski admissible function, 57, 75, 143, 268
strongly Matkowski admissible relation, 103, 140
strongly orbitally singular asymptotic map, 452
strongly Picard operator, 339
strongly Picard sequence, 194
strongly reduced Zermelo-Fraenkel system, 7, 26
strongly regressive function, 78, 393
strongly regular operation, 206
strongly separable structure, 317
strongly-compatible map, 375, 427
strongly-convergent sequence, 492
sub-aprogression, 85, 91, 429
subadditive function, 46
subbasis, 152
subnet, 314
subordinated couple, 482
subsequence of a net, 327
subunitary couple, 263

successor, 6, 85
sufficient almost uniformity, 320
sufficient pseudometric, 32, 157, 280
super-additive function, 113
super-admissible chain, 88
supremum topology, 163, 337, 346
symmetric pseudometric, 32, 157, 280
symmetric relation, 8

telescopic bounded map, 171, 369, 384, 419
telescopic sequence, 193, 194
telescopic subunitary couple, 263
topological space, 148, 333
total asymptotic sequence, 35, 233
total family, 150, 152
totally ordered set, 9, 25
trace of relation, 96
transitive operator, 488
transitive relation, 8, 460
translation invariant, 96
triangular pseudometric, 32, 157, 280
trivial relation, 8

uadmissible ascending orbital sequence, 189
uadmissible ascending regular map, 189, 200
uadmissible relation, 96
uadmissible sequence, 96, 189
unbounded chain, 85, 429
unbounded convergence, 50
uniform continuity modulus, 162
uniform space, 323
uniformly admissible relation, 93
uniformly continuous function, 162
unilateral Razumikhin isometric set, 464, 480
unilateral Razumikhin set, 464, 480
upper bound convergence, 493
upper countable chain, 316
upper diagonal relation, 103, 139, 169, 370, 421, 498

Wardowski function, 181
weakly asymptotic regular set, 318
weakly regular structure, 354
weakly sequentially inductive structure, 282
weakly sequentially regular structure, 354
well ordered set, 9, 25, 312

Zermelo-Fraenkel axiomatic system, 1, 311
zero first relation, 139
zero second relation, 139
zero vertical function, 180
zero-compatible function, 47
zero-contractive function, 68
zero-full set, 295
Zorn order, 361

www.ingramcontent.com/pod-product-compliance
Lightning Source LLC
LaVergne TN
LVHW010340220926
848484LV00004BB/942